C*-Algebras and Finite-Dimensional Approximations

GRADUATE STUDIES IN MATHEMATICS 88

C*-Algebras and Finite-Dimensional Approximations

Nathanial P. Brown
Narutaka Ozawa

AMERICAN MATHEMATICAL SOCIETY
Providence, Rhode Island

2000 *Mathematics Subject Classification.* Primary 46L05, 46L06, 46L07, 46L10, 46L35, 46L55, 05C25, 22D25, 43A07.

For additional information and updates on this book, visit
www.ams.org/bookpages/gsm-88

Library of Congress Cataloging-in-Publication Data

Brown, Nathanial P. (Nathanial Patrick), 1972–
C*-algebras and finite-dimensional approximations / Nathanial P. Brown, Narutaka Ozawa.
p. cm. — (Graduate studies in mathematics, ISSN 1065-7339 ; v. 88)
Includes bibliographical references and index.
ISBN-13: 978-0-8218-4381-9 (alk. paper)
ISBN-10: 0-8218-4381-8 (alk. paper)
1. C*-algebras. 2. Finite groups. I. Ozawa, Narutaka, 1974– II. Title.

QA326 .B758 2008
512′.556—dc22 2007060566

Softcover ISBN: 978-1-4704-7977-0

♾ The paper used in this book is acid-free and falls within the guidelines
established to ensure permanence and durability.
Visit the AMS home page at `http://www.ams.org/`

12 11 10 9 8 7 6 5 4 3   30 29 28 27 26 25

Contents

Preface

This is a book about C*-algebras, various types of approximation, and a few of the surprising applications that have been recently discovered. In short, we will study *approximation theory in the context of operator algebras.*

Approximation is ubiquitous in mathematics; when the object of interest cannot be studied directly, we approximate by tractable relatives and pass to a limit. In our context this is particularly important because C*-algebras are (almost always) infinite dimensional and we can say precious little without the help of approximation theory. Moreover, most concrete examples enjoy some sort of finite-dimensional localization; hence it is very important to exploit these features to the fullest. Indeed, over the years approximation theory has been at the heart of many of the deepest, most important results: Murray and von Neumann's uniqueness theorem for the hyperfinite II_1-factor and Connes's remarkable extension to the injective realm; Haagerup's discovery that reduced free group C*-algebras have the metric approximation property; Higson and Kasparov's resolution of the Baum-Connes conjecture for Haagerup's groups; Popa's work on subfactors and Cartan subalgebras; Voiculescu's whole free entropy industry, which is defined via approximation; Elliott's classification program, which collapses without approximate intertwining arguments; and one can't forget the influential work of Choi, Effros, and Kirchberg on nuclear and exact C*-algebras.

Approximation is everywhere; it is powerful, important, the backbone of countless breakthroughs. We intend to celebrate it. This subject is a functional analyst's delight, a beautiful mixture of hard and soft analysis, pure joy for the technically inclined. Our wheat may be other texts' chaff, but we see no reason to hide our infatuation with the grace and power which is approximation theory. We don't mean to suggest that mastering technicalities is the point of operator algebras – it isn't. We simply hope to elevate them from a necessary ally to a revered friend. Also, one shouldn't think these pages are a one-stop shopping place for all aspects of approximation theory – they aren't. The main focus is nuclearity and exactness, with several related concepts and a few applications thrown in for good measure.

From the outset of this project, we were torn between writing user-friendly notes which students would appreciate – many papers in this subject are notoriously difficult to read – or sticking to an expert-oriented, research-monograph level of exposition. In the end, we decided to split the difference. Part 1 of these notes is written with the beginner in mind, someone who just finished a first course in operator algebras (C*- and W*-algebras). We wanted the basic theory to be accessible to students working on their own; hence Chapters 2 - 10 have a lot of detail and proceed at a rather slow pace.[1] Chapters 11 - 17 and all of the appendices are written at a higher level, something closer to that found in the literature.

Here is a synopsis of the contents.

Part 1: Basic Theory

The primary objective here is an almost-comprehensive treatment of nuclearity and exactness.[2] Playing the revisionist-historian role, we define these classes in terms of finite-dimensional approximation properties and later demonstrate the tensor product characterizations. We also study several related ideas which contribute to, and benefit from, nuclearity and exactness.

The first chapter is just a collection of results that we need for later purposes. We often utilize the interplay between C*-algebras and von Neumann algebras; hence this chapter reviews a number of "basic" facts on both sides. (Some are not so basic and others are so classical that many students never learn them.)

Chapter 2 contains definitions, simple exercises designed to get the reader warmed up, and a few basic examples (AF algebras, C*-algebras of amenable groups, type I algebras).

---

[1]Except for a few sections in Chapters 4 and 5, where much more is demanded of the reader. This was necessary to keep the book to a reasonable length.

[2]The most egregious omission is probably Kirchberg's $\mathcal{O}_2$-embedding theorem for separable exact C*-algebras. We felt there were not enough general (i.e., outside of the classification program) applications to warrant including the difficult proof. The paper **[107]** is readily available and has a self-contained, well-written proof. Rørdam's book **[168]** has a nearly complete proof and a forthcoming book of Kirchberg and Wassermann will certainly contain all the details. Another significant omission is a discussion of general locally compact groups; we stick to the discrete case. The ideas are adequately exposed in this setting and we don't think beginners benefit from more generality.

In Chapter 3 we give a long introduction to the theory of C*-tensor products. Most of the chapter is devoted to definitions and a thorough discussion of the subtleties which make C*-tensor products both interesting and hazardous. However, the last two sections contain important theorems, taking us back to the original definitions of nuclearity and exactness.

In the next two chapters we show that many natural examples of C*-algebras admit some sort of finite-dimensional approximants. In Chapter 4 we discuss a number of general constructions which one finds in the literature (crossed products by amenable actions, free products, etc.). Chapter 5 is an introduction to exact discrete groups and some related topics which are relevant to noncommutative geometry. Both of these chapters contain redundancies in the sense that we start with special cases and gradually tack on generality. The Bourbakians may protest, but we feel this approach is pedagogically superior.

Someone who works through Chapters 2 - 5 will have a pretty good feel for most aspects of nuclearity and exactness. There is, however, one important permanence property which requires much more work: Both nuclearity and exactness pass to quotients. In some sense, the next four chapters are required to prove these fundamental facts. This doesn't mean we've taken the most direct route, however. On the contrary, we take our sweet time and present a number of related approximation properties which are of independent interest and play crucial roles in the quotient results.

Chapter 6 contains the basics of amenable tracial states. These "invariant means" on C*-algebras can be characterized in terms of approximation or tensor products. They also yield a simple proof of the deep fact that every finite injective von Neumann algebra is semidiscrete.

In Chapter 7 we study quasidiagonal C*-algebras. They are also defined via approximation, but the flavor is quite different from nuclearity or exactness. Most of the basic theory is presented, including Voiculescu's homotopy invariance theorem, though much of it isn't necessary for applications to exactness. (For this we only need Dadarlat's approximation theorem for exact quasidiagonal C*-algebras; see Section 7.5.)

This leads naturally to Chapter 8: AF Embeddability. For applications, the most important fact is that every exact C*-algebra is a subquotient of an AF C*-algebra. We give the proof in the beginning of the chapter so those only interested in exactness can quickly proceed forward. For others, we have included the homotopy invariance theorem for AF embeddability and a short survey of related results.

In Chapter 9 we put all the pieces together, completing the basic-theory portion of the book. The main result gives two more tensor product characterizations of exactness, from which corollaries flow: Exact C*-algebras are locally reflexive (another important finite-dimensional approximation property), nuclearity and exactness pass to quotients, and a few others.

Finally, we conclude Part 1 with a chapter summarizing permanence properties. This is just for ease of reference, in case one forgets whether or not extensions of exact C*-algebras are exact.

Part 2: Special Topics

The next four chapters are a disjoint collection of related concepts. They are logically independent and meant to spark the reader's interest – much more could be written about any one of them.

Chapter 11 is primarily about simple quasidiagonal C*-algebras. Motivated by Elliott's classification program, we spend time discussing the generalized inductive limit approach (of Blackadar and Kirchberg) to nuclear quasidiagonal C*-algebras. We also prove a theorem of Popa, showing that quasidiagonality is often detectable internally. Finally, we present Connes's amazing uniqueness theorem for the injective II_1-factor, exploiting Popa's techniques.

Chapter 12 introduces some properties of discrete groups that have been extremely important over the years. First, we discuss Kazhdan's property (T), prove that $\mathrm{SL}(3,\mathbb{Z})$ has this property, and demonstrate Connes's result that II_1-factors with property (T) have few outer automorphisms. Next, we define Haagerup's approximation property – the antithesis of property (T) – and prove that a group which acts properly on a tree (e.g., a free group) enjoys this property. The latter sections of this chapter discuss related approximation properties and their interrelations.

Chapter 13 – on Lance's weak expectation property and the local lifting property for C*-algebras – gives a streamlined approach to some of Kirchberg's influential work around these ideas. We also reproduce Junge and Pisier's theorem on the tensor product of $\mathbb{B}(\ell^2)$ with itself.

Part 2 concludes with Chapter 14: Weakly Exact von Neumann Algebras. This concept was first suggested by Kirchberg; the theorems and proofs are similar to C*-results found in Part 1 of the book. It is not yet clear if this theory will bear fruit like its C*-predecessor, but it seemed like a natural topic to include.

Part 3: Applications

The last three chapters, comprising Part 3, are devoted to applications. We hope to convince you that approximation properties are useful; seemingly unrelated problems will crack wide open when pried with the right technical tool.

Chapter 15 contains solidity and prime factorization results for certain group von Neumann algebras. The solidity results generalize one of the celebrated achievements of free probability theory, while the prime factorization results are natural analogues of some spectacular recent work in dynamical systems. Both depend in a crucial way on some of the C*-ideas and tensor product techniques contained in Part 1.

Chapter 16 resolves a problem in single operator theory which, at present, appears to require exact quasidiagonal C*-algebras. We need the fact that exactness implies local reflexivity – one of the deepest, most difficult theorems in C*-algebras – and it is hard to imagine an operator-theoretic proof which could circumvent this fact.

The final chapter is based on some work of Simon Wassermann. He observed that property (T) groups together with quasidiagonal ideas lead to natural examples for which the Brown-Douglas-Fillmore semigroup is not a group. Approximation properties, or the lack thereof, are at the heart of the argument.

So, that's what you'll find in this book. To the student: We hope these notes are reasonably accessible and a helpful introduction to an area of active research. To the veteran: We hope this will be a useful reference for C*-approximation theory. As mentioned earlier, Kirchberg and Wassermann are working on an exact C*-algebra text and there will certainly be overlap between these notes and theirs. However, the emphasis and selection of topics will likely differ; with any luck, the union of our books will satisfy the needs of most. We should also mention that a web page correcting this book's inevitable errors can be found at www.ams.org/bookpages/gsm-88.

This project would have been impossible without the support of our academic institutions and the encouragement of many friends and colleagues. Thank you. The first author gratefully acknowledges NSF support; the second author thanks JSPS, NSF, and the Sloan Foundation. Finally, we are deeply indebted to Reiji Tomatsu for his meticulous note-taking during the second author's lectures on some of the material covered in this book; Tyrone Crisp, Steve Hair, and Naokazu Mizuta for their fastidious proofreading; and Kenley Jung and Takeshi Katsura for some helpful questions, remarks, and suggestions.

Nate and Taka, 2007

Chapter 1

Fundamental Facts

We stated in the preface that these notes should be accessible to anyone with a "first course" in operator algebras under their belt. An excellent first course would consist of the material contained in [**127**], for example, and we assume familiarity with that book. However, we'll need numerous other facts that may or may not have made it into *your* first course; the purpose of this chapter is to summarize the requisite results.

It goes without saying that advanced students and seasoned researchers should skip this chapter, referring back if necessary. Indeed, the only things required before starting Chapter 2 are basic properties of completely positive maps and Arveson's Extension Theorem (Sections 1.5 and 1.6). We advise the novice to nail down this material and then to jump ahead to Chapter 2 – mathematics books need not be read linearly.[1]

1.1. Notation

We use $\mathcal{H}$, $\mathcal{K}$ and $\mathcal{L}$ to denote generic complex Hilbert spaces. The n-dimensional Hilbert space is usually denoted ℓ^2_n, while ℓ^2 is the separable, infinite-dimensional Hilbert space. Here $\mathbb{B}(\mathcal{H})$ is the algebra of bounded linear operators on $\mathcal{H}$, $\mathbb{K}(\mathcal{H})$ denotes the compacts and $Q(\mathcal{H}) = \mathbb{B}(\mathcal{H})/\mathbb{K}(\mathcal{H})$ is the Calkin algebra. (An abstract copy of the compacts will be denoted by $\mathbb{K}$.) Also, Tr is the canonical (typically unbounded, densely defined) trace on $\mathbb{B}(\mathcal{H})$. We let $\mathcal{S}_1$ (resp. $\mathcal{S}_2$) denote the trace class (resp. Hilbert-Schmidt) operators, with canonical norm $\|T\|_1 = \mathrm{Tr}(|T|)$ (resp. $\|T\|_2 = \sqrt{\mathrm{Tr}(|T|^2)}$).

[1] If you feel obliged to go carefully through this chapter, be advised that tensor products are required in a few places. Thus, first reading the beginning of Chapter 3 might be necessary. How's that for nonlinear?

Here $\mathbb{M}_n(\mathbb{C})$ is the $n \times n$ complex matrices and tr is its unique tracial state. A collection of $n \times n$ matrices $\{e_{i,j}\}_{1\le i,j\le n}$ is called a *system of matrix units* if $e_{i,j}e_{s,t} = \delta_{j,s}e_{i,t}$. It is often convenient to index our matrices over a finite set F, in which case we write $\mathbb{M}_F(\mathbb{C})$ $(= \mathbb{B}(\ell^2(F)))$ and let $\{e_{p,q}\}_{p,q\in F}$ denote the canonical matrix units.

We reserve A, B, C and D for C*-algebras while M and N will typically denote von Neumann algebras. We let A_{sa} be the self-adjoint elements, A_1 the closed unit ball and A_+ the positive elements in A. The symbols E and F will denote operator systems (or operator spaces). We usually use I and J for ideals in C*-algebras (e.g. $I \triangleleft A$), though they are occasionally index sets too. All ideals are assumed closed and two-sided.

The set of states on A – positive linear functionals of norm one – will be denoted $S(A)$. If $\varphi \in S(A)$, we let $L^2(A,\varphi)$ be the GNS (Gelfand-Naimark-Segal) Hilbert space and $\pi_\varphi \colon A \to \mathbb{B}(L^2(A,\varphi))$ be the GNS representation. For an element $a \in A$, we let $\hat{a} \in L^2(A,\varphi)$ denote its natural image.

1.2. C*-algebras

Quasicentral approximate units. Quasicentral approximate units are an indispensable tool. See [**53**, Theorem I.9.16] for a proof of the following fact.

Theorem 1.2.1. *Let $I \triangleleft A$ be an ideal. Then I has an approximate unit $\{e_i\} \subset I$ such that $\|e_i a - a e_i\| \to 0$, as $i \to \infty$, for all $a \in A$. In fact, if $\{f_k\} \subset I$ is any approximate unit for I, then a quasicentral approximate unit can always be extracted from its convex hull.*

We don't need it too many times, but it is worth mentioning that quasicentral approximate units allow a particular type of approximate decomposition.

Proposition 1.2.2. *Let A be unital and $\{e_i\} \subset I \triangleleft A$ be a quasicentral approximate unit. For every pair $a, b \in A$ such that $a - b \in I$ we have*

$$\left\|a - \left((1-e_i)^{\frac{1}{2}} b (1-e_i)^{\frac{1}{2}} + e_i^{\frac{1}{2}} a e_i^{\frac{1}{2}}\right)\right\| \to 0.$$

Proof. First notice that for every $x \in A$ and polynomial p we have $\|p(e_i)x - xp(e_i)\| \to 0$ (by some standard estimates). Since continuous functions can be approximated by polynomials, it follows that $\|e_i^{\frac{1}{2}} x - x e_i^{\frac{1}{2}}\| \to 0$ and $\|(1-e_i)^{\frac{1}{2}} x - x(1-e_i)^{\frac{1}{2}}\| \to 0$.

Next, observe that

$$\|(1-e_i)^{\frac{1}{2}} a (1-e_i)^{\frac{1}{2}} - (1-e_i)^{\frac{1}{2}} b (1-e_i)^{\frac{1}{2}}\| = \|(1-e_i)^{\frac{1}{2}}(a-b)(1-e_i)^{\frac{1}{2}}\| \to 0$$

since $\lim_i \|(1-e_i)x\|$ is equal to the norm of $x+I \in A/I$. Putting these observations together, we obtain asymptotic approximations

$$(1-e_i)^{\frac{1}{2}}b(1-e_i)^{\frac{1}{2}}+e_i^{\frac{1}{2}}ae_i^{\frac{1}{2}} \approx (1-e_i)^{\frac{1}{2}}a(1-e_i)^{\frac{1}{2}}+e_i^{\frac{1}{2}}ae_i^{\frac{1}{2}} \approx a(1-e_i)+ae_i = a$$

and the proof is complete. □

Uniqueness of GNS representations. Hopefully you already know the uniqueness statement for GNS representations, but here is a technical variation (with exactly the same proof).

Proposition 1.2.3. *Let $\varphi \in S(A)$ be a state on A, $A_0 \subset A$ be a norm dense $*$-subalgebra and $\rho\colon A_0 \to \mathbb{B}(\mathcal{H})$ be a $*$-homomorphism with the property that there exists a unit vector $v \in \mathcal{H}$ such that $A_0 v$ is dense in $\mathcal{H}$ and $\varphi(x) = \langle \rho(x)v, v\rangle$ for all $x \in A_0$. Then ρ extends to a representation of A (which is unitarily equivalent to π_φ).*

Proof. One defines a linear map $U\colon \hat{A}_0 \to A_0 v$ by declaring $U\hat{a} = \rho(a)v$. Check that this is well-defined and isometric from a dense subspace of $L^2(A,\varphi)$ to a dense subspace of $\mathcal{H}$; thus it extends uniquely to a unitary. The extension of ρ is obtained by conjugating π_φ by this unitary. □

1.3. Von Neumann algebras

Though these notes are primarily concerned with C*-algebras, we will need von Neumann algebras from time to time. The C*-purists should be forewarned that we intend to delve into W*-theory whenever possible (even when it isn't absolutely necessary).

Structure of von Neumann algebras. The basic decomposition theory of von Neumann algebras will be important. We won't give any proper definitions, but thanks to well-known theorems our approach is legal (i.e., our definitions are equivalent to the "real" definitions; see [**183**, Definition V.1.17]).

We let $\prod_j B_j$ denote the ℓ^∞-direct sum of C*-algebras $\{B_j\}_{j\in J}$, i.e., the set of tuples $(b_j)_{j\in J}$ such that $b_j \in B_j$ and $\sup_j \|b_j\| < \infty$.

Definition 1.3.1. The von Neumann algebra M is *type* I if it is isomorphic to

$$\prod_{j\in J} \mathcal{A}_j \bar{\otimes} \mathbb{B}(\mathcal{H}_j)$$

for some set J of cardinal numbers, where each $\mathcal{A}_j$ is an abelian von Neumann algebra and $\mathcal{H}_j$ is a Hilbert space of dimension j.[2]

[2]If you haven't seen it, the definition of von Neumann tensor products is given in Remark 3.3.5.

Definition 1.3.2. The von Neumann algebra M is *type* II_1 if it has no summand of type I and there exists a separating family of normal tracial states (i.e., for every $0 < x \in M$ there exists a normal tracial state τ on M such that $\tau(x) > 0$).

Roughly speaking, the next type is just an increasing union of II_1 corners.

Definition 1.3.3. The von Neumann algebra M is *type* II_∞ if M has no summand of type I or II_1 but there exists an increasing net of projections $\{p_i\}_{i\in I} \subset M$, converging strongly to 1_M, such that p_iMp_i is of type II_1 for every $i \in I$.

Finally, a von Neumann algebra is said to be of *type* III if it has no summand of any of the types defined above. The following decomposition theorem is fundamental ([**183**, Theorem V.1.19]).

Theorem 1.3.4. *Every von Neumann algebra M has a unique decomposition*

$$M \cong M_{\mathrm{I}} \oplus M_{\mathrm{II}_1} \oplus M_{\mathrm{II}_\infty} \oplus M_{\mathrm{III}}$$

as a direct sum of algebras of type I, II_1, II_∞ *and* III *(some of these summands may be* 0*).*

Preduals and Sakai's Theorem. Recall that $\mathbb{B}(\mathcal{H})$ is canonically isomorphic to the dual Banach space of the trace class operators $\mathcal{S}_1 \subset \mathbb{B}(\mathcal{H})$. Hence every von Neumann algebra $M \subset \mathbb{B}(\mathcal{H})$ is also a dual Banach space (namely, the dual of the quotient of $\mathcal{S}_1$ by the pre-annihilator of M). A fundamental result of Sakai (see [**183**, Corollary III.3.9]) implies that the induced weak-$*$ topology is canonical, i.e., independent of the normal representation $M \subset \mathbb{B}(\mathcal{H})$. (Recall that a map $\varphi\colon M \to N$ of von Neumann algebras is *normal* if $\varphi(\sup x_i) = \sup \varphi(x_i)$ for all norm bounded, monotone increasing nets of self-adjoint elements $\{x_i\} \subset M_{sa}$.)

Theorem 1.3.5. *For a von Neumann algebra M, let M_* be the Banach space of normal linear functionals on M. Then M is (isometrically) isomorphic to the dual of M_*. Moreover, M_* is the unique predual in the sense that if X is a Banach space with the property that M is isometrically isomorphic to X^*, then X is isometrically isomorphic to M_*.*

Definition 1.3.6. The canonical weak-$*$ topology on M (coming from M_*) is called the *ultraweak topology*.

Point-ultraweak limits. Let M be a von Neumann algebra and M_* be its predual. For a Banach space X, let $\mathbb{B}(X, M)$ be the bounded linear maps from X to M. It turns out that $\mathbb{B}(X, M)$ also has a predual. Let $\mathbb{B}(X, M)_* \subset \mathbb{B}(X, M)^*$ be the closed linear span of the linear functionals

$x \otimes \xi \in \mathbb{B}(X, M)^*$, where $x \in X$, $\xi \in M_*$ and $x \otimes \xi$ is defined by $x \otimes \xi(T) = \xi(T(x))$. Then $\mathbb{B}(X, M)$ is isometrically isomorphic to the dual of $\mathbb{B}(X, M)_*$, whence it receives a weak-$*$ topology. On bounded sets, this topology agrees with the *point-ultraweak (aka point-σ-weak) topology.* That is, for a bounded net convergence works as follows:

$$T_\lambda \to T \iff \xi(T_\lambda(x)) \to \xi(T(x)), \forall x \in X, \forall \xi \in M_*.$$

Thus the unit ball of $\mathbb{B}(X, M)$ is compact, by Alaoglu's Theorem, in the point-ultraweak topology. Hence we obtain the following theorem (cf. Theorem A.8).

Theorem 1.3.7. *Let X be a Banach space, M be a von Neumann algebra and $T_\lambda\colon X \to M$ be a bounded net of linear maps. Then $\{T_\lambda\}_{\lambda\in\Lambda}$ has a cluster point in the point-ultraweak topology.*

Representation theory. In contrast to the C*-case, representation theory of von Neumann algebras is almost trivial: one can cut by a projection in the commutant and that's about it. Of course, the precise statement is slightly more complicated (see [**183**, Theorem IV.5.5]).

Theorem 1.3.8. *Let $M \subset \mathbb{B}(\mathcal{H})$ be a von Neumann algebra and $\pi\colon M \to \mathbb{B}(\mathcal{K})$ be a normal representation. There exist a Hilbert space $\tilde{\mathcal{K}}$ and a projection $P_\pi \in \mathbb{B}(\mathcal{H} \otimes \tilde{\mathcal{K}})$ such that P_π commutes with $M \otimes 1 \subset \mathbb{B}(\mathcal{H} \otimes \tilde{\mathcal{K}})$ and π is unitarily equivalent to the representation $M \to P_\pi \mathbb{B}(\mathcal{H} \otimes \tilde{\mathcal{K}}) P_\pi$, $m \mapsto P_\pi(m \otimes 1)$.*

1.4. Double duals

The Banach space double dual of a C*-algebra A is a wild beast; it should be approached with humility, even trepidity. Whatever it takes, though, one must become acquainted with A^{**} as it's an extremely useful universe in which to work. See [**183**, Section III.2] for more.

The enveloping von Neumann algebra. Recall that the *universal representation* of a C*-algebra A is

$$\pi_u = \bigoplus_{\varphi\in S(A)} \pi_\varphi \colon A \to \mathbb{B}\Big(\bigoplus_{\varphi\in S(A)} L^2(A, \varphi) \Big) = \mathbb{B}(\mathcal{H}_u).$$

By definition, the *enveloping von Neumann algebra of A* is the double commutant $\pi_u(A)''$. Thanks to the next result, we need not distinguish between the double dual and the enveloping von Neumann algebra; we'll use A^{**} to denote either throughout this book.

Theorem 1.4.1. *The enveloping von Neumann algebra of A is isometrically isomorphic to the double dual A^{**}. Hence the ultraweak topology on $\pi_u(A)''$ ($= A^{**}$) restricts to the weak topology on A (by Sakai's Theorem).*

Here is an often used consequence: If $a_i, a \in A$ and $a_i \to a$ in the ultraweak topology, then a belongs to the *norm* closure of the convex hull of the a_i's (thanks to the Hahn-Banach Theorem).

Central covers. Since every representation can be decomposed as a direct sum of cyclic representations (i.e., GNS representations), it is easily seen that A^{**} enjoys the following *universal property*: for each nondegenerate representation $\pi\colon A \to \mathbb{B}(\mathcal{H})$ there exists a unique normal extension $\tilde{\pi}\colon A^{**} \to \mathbb{B}(\mathcal{H})$ such that $\tilde{\pi}|_A = \pi$ and $\tilde{\pi}(A^{**}) = \pi(A)''$. The kernel of $\tilde{\pi}$ is weakly closed (by normality); hence it's a von Neumann algebra. As such, it has a unit e_π which is a central projection in A^{**}.

Definition 1.4.2. Let $\pi\colon A \to \mathbb{B}(\mathcal{H})$ be a nondegenerate representation. The *central cover of* π, denoted $c(\pi)$, is defined to be $e_\pi^\perp = 1_{A^{**}} - e_\pi$.

The following isomorphisms are immediate from the definition:

$$c(\pi)A^{**} = c(\pi)A^{**}c(\pi) \cong \tilde{\pi}(A^{**}) = \pi(A)''.$$

We only need them on a few occasions, but here are some necessary facts.

Proposition 1.4.3. *If π_1 and π_2 are* irreducible *representations, then the following are equivalent:*

(1) $c(\pi_1)c(\pi_2) \neq 0$;
(2) $c(\pi_1) = c(\pi_2)$;
(3) *π_1 and π_2 are unitarily equivalent.*

Proposition 1.4.4. *For two representations $\pi\colon A \to \mathbb{B}(\mathcal{H})$ and $\rho\colon A \to \mathbb{B}(\mathcal{K})$, the following are equivalent:*

(1) $c(\pi)c(\rho) = 0$;
(2) $(\pi \oplus \rho)(A)'' = \pi(A)'' \oplus \rho(A)''$.

Two representations $\pi\colon A \to \mathbb{B}(\mathcal{H})$ and $\rho\colon A \to \mathbb{B}(\mathcal{K})$ are said to be *quasi-equivalent* if there exists an isomorphism $\theta\colon \pi(A)'' \to \rho(A)''$ such that $\theta(\pi(a)) = \rho(a)$ for all $a \in A$. Of course, unitarily equivalent representations are quasi-equivalent in this sense, but the converse is false. (Any representation of $\pi(A)''$ which is not unitarily equivalent to the original – for example, one could modify the commutant – will yield a quasi-equivalent representation of A which is not unitarily equivalent to π.) Here are two simple facts.

Proposition 1.4.5. *The representations π and ρ are quasi-equivalent if and only if $c(\pi) = c(\rho)$. Since every central projection in A^{**} defines a representation of A, there is a one-to-one correspondence between central projections $p \in \mathcal{Z}(A^{**})$ and quasi-equivalence classes of representations.*

Proposition 1.4.6. *The representation π is quasi-equivalent to a subrepresentation of ρ if and only if $c(\pi) \leq c(\rho)$.*

Lusin's Theorem, excision and Glimm's Lemma. The (difficult) proof of the following noncommutative extension of Lusin's Theorem can be found in [**183**] (II.4.15) or [**142**] (2.7.3).

Theorem 1.4.7 (Lusin's Theorem). *Let $A \subset \mathbb{B}(\mathcal{H})$ be a nondegenerate C*-algebra with $A'' = M$. For every finite set of vectors $\mathfrak{F} \subset \mathcal{H}$, $\varepsilon > 0$, projection $p_0 \in M$ and self-adjoint $y \in M$, there exist a self-adjoint $x \in A$ and a projection $p \in M$ such that $p \leq p_0$, $\|p(h) - p_0(h)\| < \varepsilon$ for all $h \in \mathfrak{F}$, $\|x\| \leq \min\{2\|yp_0\|, \|y\|\} + \varepsilon$ and $xp = yp$.*

We will need a slight sharpening of Kadison's Transitivity Theorem.

Corollary 1.4.8 (Strong transitivity). *Let $A^{**} \subset \mathbb{B}(\mathcal{H}_u)$ be the universal representation and $\pi \colon A \to \mathbb{B}(\mathcal{H})$ be an irreducible representation with normal extension $\tilde{\pi}$ to A^{**}. For each self-adjoint $a \in A^{**}$ and finite-rank projection $Q \in \mathbb{B}(\mathcal{H})$, we can find a self-adjoint net $(c_i)_{i\in I} \subset A$ such that $c_i \to a$ in the strong operator topology, $\|c_i\| \leq \|a\| + 1$ and $\tilde{\pi}(a)Q = \pi(c_i)Q$, for all $i \in I$. If $0 \leq a \leq Q$ (in the decomposition $A^{**} = \mathbb{B}(\mathcal{H}) \oplus (1 - c(\pi))A^{**}$), then the c_i's can be taken positive.*

Proof. Let $\mathfrak{F} \subset \mathcal{H}_{\mathcal{U}}$ be any finite set of vectors and $\tilde{Q} \in \mathbb{B}(\mathcal{H})$ be any finite-rank projection dominating Q. Applying Lusin's Theorem to a, $\tilde{Q} \oplus (1 - c(\pi))$ and any $\varepsilon > 0$, we can find $c \in A$ and a projection $P \in A^{**}$ such that

(1) $P \leq \tilde{Q} \oplus (1 - c(\pi))$;

(2) $\|P(v) - \tilde{Q} \oplus (1 - c(\pi))(v)\| < \varepsilon$, for all $v \in \mathfrak{F}$;

(3) $aP = cP$;

(4) $\|c\| \leq \|a\| + 1$.

Writing $P = c(\pi)P \oplus (1 - c(\pi))P$, we claim that it is no loss of generality to assume $Q \leq c(\pi)P = \tilde{Q}$; it is easily seen that this implies the lemma. The fact that $P \leq \tilde{Q} \oplus (1 - c(\pi))$ implies $c(\pi)P \leq \tilde{Q}$. On the other hand, $\mathfrak{F}$ can be any finite set of vectors – hence we could throw in a basis for the range of $\tilde{Q}$. But then for small ε we would have $\|c(\pi)P - \tilde{Q}\| < 1$, which implies $c(\pi)P = \tilde{Q}$ as desired.

Now, suppose that $0 \leq a \leq Q$. Applying the first part of the proof to $a^{1/2}$, we find a self-adjoint net (b_i) such that $b_i \to a^{1/2}$, $\|b_i\| \leq (\|a\| + 1)^{1/2}$

and $\tilde{\pi}(a^{1/2})Q = \pi(b_i)Q$, for all $i \in I$. Since a and Q commute, one has

$$\pi(b_i^2)Q = \pi(b_i)\big(\tilde{\pi}(a^{1/2})Q\big) = \big(\pi(b_i)Q\big)\tilde{\pi}(a^{1/2}) = \tilde{\pi}(a)Q.$$

Hence $c_i = b_i^2$ satisfy the required conditions. □

Lemma 1.4.9. *Let A be a C*-algebra, φ be a pure state and $L = \{a \in A : \varphi(a^*a) = 0\}$ be the associated left ideal. Then we have $\ker\varphi = L + L^*$.*

Proof. Let $x \in \ker\varphi$ be nonzero. Let $(\pi, \mathcal{H}, \xi)$ be the GNS triplet and $\mathcal{K}$ be the 2-dimensional subspace spanned by ξ and $\pi(x)\xi$. Since $\xi \perp \pi(x)\xi$, Kadison's Transitivity Theorem provides us with a positive element $b \in A$ such that $\pi(b)\xi = \xi$ and $\pi(b)(\pi(x)\xi) = 0$. It follows that $bx \in L$ and $(1-b)x \in L^*$. □

Here is half of the Akemann-Anderson-Pedersen Excision Theorem (see [**1**] for the other half).

Theorem 1.4.10 (Excision). *Let A be a C*-algebra and φ be a pure state. There exists a net $(e_i) \subset A$ such that $0 \leq e_i \leq 1$, $\varphi(e_i) = 1$ and $\lim_i \|e_i a e_i - \varphi(a)e_i^2\| = 0$ for every $a \in A$.*

Proof. We first assume that A is unital. Let L be the left ideal associated to φ and (c_i) be a right approximate unit for L (i.e., (c_i) is an approximate unit for the hereditary subalgebra $L \cap L^*$ and for every $a \in L$ we have $\|a - ac_i\| \to 0$). Let $e_i = 1 - c_i$. Since $a - \varphi(a) \in \ker\varphi = L + L^*$, we have $\lim \|e_i(a - \varphi(a))e_i\| = 0$ and $\varphi(e_i) = 1$.

Now suppose that A is nonunital and take e_i as above for $\tilde{A}$. Let (b_j) be a quasicentral approximate unit for A such that $\varphi(b_j) = 1$. (Existence follows from Kadison's Transitivity Theorem.) Then, $b_j e_i b_j$ does the job. □

Here is a nonstandard proof of a fundamental fact.

Lemma 1.4.11 (Glimm's Lemma). *Let $A \subset \mathbb{B}(\mathcal{H})$ be a separable C*-algebra containing no nonzero compact operators on $\mathcal{H}$. If φ is a state on A, then there exist orthonormal vectors (ξ_n) such that $\omega_{\xi_n}(a) \to \varphi(a)$ for all $a \in A$, where $\omega_{\xi_n}(T) = \langle T\xi_n, \xi_n\rangle$.*

Proof. [3] Let $\mathfrak{F} \subset A$ be a finite subset of norm-one elements, $\varepsilon > 0$, $\mathcal{K}_0 \subset \mathcal{H}$ be finite dimensional and $P_{\mathcal{K}_0}$ be the orthogonal projection onto $\mathcal{K}_0$. It suffices to show the existence of a unit vector $\xi \in \mathcal{K}_0^\perp$ such that $|\omega_\xi(a) - \varphi(a)| < 6\varepsilon$ for all $a \in \mathfrak{F}$. By the Krein-Milman Theorem, there exists a convex combination $\psi = \sum_{k=1}^n \lambda_k \psi_k$ of pure states ψ_k such that $\varphi \approx_{\mathfrak{F},\varepsilon} \psi$. By excision, for each k, there exists a norm-one positive element $e_k \in A$

[3] We thank Akitaka Kishimoto for showing us this short proof.

such that $\|e_k(a - \psi_k(a))e_k\| < \varepsilon$ for every $a \in \mathfrak{F}$. Let $P_{\mathcal{K}_0}^{\perp} = 1 - P_{\mathcal{K}_0}$ and notice that $P_{\mathcal{K}_0}^{\perp} e_1 P_{\mathcal{K}_0}^{\perp} - e_1$ is a compact operator; hence $\|P_{\mathcal{K}_0}^{\perp} e_1 P_{\mathcal{K}_0}^{\perp}\| = 1$.

Let $\zeta_1 \in \mathcal{K}_0^{\perp}$ be a unit vector such that $\|e_1\zeta_1 - \zeta_1\| < \varepsilon$. Now, let $\mathcal{K}_1$ be the finite-dimensional subspace spanned by $\mathcal{K}_0$ and $\{a\zeta_1\}_{a\in\mathfrak{F}} \cup \{a^*\zeta_1\}_{a\in\mathfrak{F}}$. Reasoning as above, there exists a unit vector $\zeta_2 \in \mathcal{K}_1^{\perp}$ such that $\|e_2\zeta_2 - \zeta_2\| < \varepsilon$. Repeating this n times, we get vectors $\zeta_1, \ldots, \zeta_n$ and set $\xi = \sum \lambda_k^{1/2}\zeta_k$. Then, for every $a \in \mathfrak{F}$, we have

$$\begin{aligned}\omega_\xi(a) &= \sum \lambda_k \omega_{\zeta_k}(a) \\ &\approx_{2\varepsilon} \sum \lambda_k \omega_{e_k\zeta_k}(a) \\ &\approx_{\varepsilon} \sum \lambda_k \psi_k(a)\|e_k\zeta_k\|^2 \\ &\approx_{2\varepsilon} \psi(a) \approx_{\varepsilon} \varphi(a).\end{aligned}$$

□

1.5. Completely positive maps

Completely positive maps (and their cousins, completely bounded maps) are the heart and soul of C*-approximation theory. For a positively complete treatment of these morphisms see [**141**].

Definitions, examples and Stinespring's Theorem.

Definition 1.5.1. An *operator system* E is a closed self-adjoint subspace of a unital C*-algebra A such that $1_A \in E$. The $n \times n$ matrices over E, $\mathbb{M}_n(E)$, inherit an order structure from $\mathbb{M}_n(A)$: an element in $\mathbb{M}_n(E)$ is positive if and only if it is positive in $\mathbb{M}_n(A)$. Note that the existence of a unit guarantees that E is spanned by positive elements.

A map φ from an operator system E to a (not necessarily unital) C*-algebra B is said to be *completely positive* if $\varphi_n \colon \mathbb{M}_n(E) \to \mathbb{M}_n(B)$, defined by

$$\varphi_n([a_{i,j}]) = [\varphi(a_{i,j})],$$

is positive (i.e., maps positive matrices to positive matrices) for every n. We denote by $\mathrm{CP}(E, B)$ the set of completely positive maps from an operator system E into B.

Following well-established precedent, we use c.p. to abbreviate "completely positive," u.c.p. for "unital completely positive" and c.c.p. for "contractive completely positive."

Example 1.5.2. A $*$-homomorphism π between C*-algebras is c.p. since the inflations π_n are also $*$-homomorphisms (hence preserve positivity). More generally, a map φ of the form $\varphi(a) = V^*\pi(a)V$ for some $*$-homomorphism

π and an operator V is c.p. (One should verify this. Don't forget that $a \geq 0 \Leftrightarrow a = x^*x$.) A positive linear functional f on an operator system E is c.p. Indeed, for $\xi = (\xi_1, \dots, \xi_n) \in \ell_n^2$ and $a = [a_{i,j}] \geq 0$ in $\mathbb{M}_n(E)$ we have

$$\langle f_n(a)\xi, \xi\rangle = f(\sum_{i,j=1}^{n} \bar{\xi}_i\xi_j a_{i,j}) = f(\begin{bmatrix}\bar{\xi}_1 & \cdots & \bar{\xi}_n\end{bmatrix} \begin{bmatrix} a_{11} & \cdots & a_{1n} \\ \vdots & \ddots & \vdots \\ a_{n1} & \cdots & a_{nn}\end{bmatrix} \begin{bmatrix}\xi_1 \\ \vdots \\ \xi_n\end{bmatrix}) \geq 0.$$

The transpose map on $\mathbb{M}_n(\mathbb{C})$ is positive but not c.p., since its norm increases after inflation (cf. Theorem 1.5.3 and Proposition 3.5.1).

Directly generalizing the GNS construction, we have Stinespring's Dilation Theorem for c.p. maps. The details of the proof can be found in many places; however, we need the explicit construction and hence we reproduce the main ingredients.

Theorem 1.5.3 (Stinespring). *Let A be a unital C^*-algebra and $\varphi\colon A \to \mathbb{B}(\mathcal{H})$ be a c.p. map. Then, there exist a Hilbert space $\widehat{\mathcal{H}}$, a $*$-representation $\pi\colon A \to \mathbb{B}(\widehat{\mathcal{H}})$ and an operator $V\colon \mathcal{H} \to \widehat{\mathcal{H}}$ such that*

$$\varphi(a) = V^*\pi(a)V$$

for every $a \in A$. In particular, $\|\varphi\| = \|V^*V\| = \|\varphi(1)\|$ (which, applied to φ_n, implies $\|\varphi_n\| = \|\varphi(1)\|$ as well).

Proof. Define a sesquilinear form $\langle \cdot\, , \cdot \rangle$ on $A \odot \mathcal{H}$ (this is the algebraic tensor product – see Chapter 3) by

$$\langle \sum_j b_j \otimes \eta_j, \sum_i a_i \otimes \xi_i\rangle = \sum_{i,j} \langle \varphi(a_i^* b_j)\eta_j, \xi_i\rangle_{\mathcal{H}}.$$

This form turns out to be positive semidefinite and so one mods out by the zero subspace and completes to get a Hilbert space $\widehat{\mathcal{H}}$ (just as in the usual GNS construction). We denote by $(\sum_i a_i \otimes \xi_i)^\wedge$ the element in $\widehat{\mathcal{H}}$ corresponding to $\sum_i a_i \otimes \xi_i \in A \odot \mathcal{H}$. Let $V\colon \mathcal{H} \to \widehat{\mathcal{H}}$ be the contraction defined by

$$V(\xi) = (1_A \otimes \xi)^\wedge.$$

For $a \in A$, we define a linear operator $\pi(a)$ on $(A \odot \mathcal{H})^\wedge \subset \widehat{\mathcal{H}}$ by

$$\pi(a)\Big((\sum_i b_i \otimes \xi_i)^\wedge\Big) = (\sum_i ab_i \otimes \xi_i)^\wedge.$$

As expected, π is a $*$-representation such that $\varphi(a) = V^*\pi(a)V$ for every $a \in A$. □

Remark 1.5.4 (Nonunital Stinespring). Stinespring's Dilation Theorem holds for non-unital C^*-algebras too. This follows from Proposition 2.2.1 in the next chapter, for example.

Remarks 1.5.5. We call the triplet $(\pi, \widehat{\mathcal{H}}, V)$ in Theorem 1.5.3 a *Stinespring dilation* of φ. When φ is unital, $V^*V = \varphi(1) = 1$ and hence V is an isometry in this case. The projection $VV^* \in \mathbb{B}(\widehat{\mathcal{H}})$ is called the *Stinespring projection.* In general there could be many different Stinespring dilations, but we may always assume that a dilation $(\pi, \widehat{\mathcal{H}}, V)$ is minimal in the sense that $\pi(A)V\mathcal{H}$ is dense in $\widehat{\mathcal{H}}$ (which holds for the construction used in the proof above). Under this minimality condition, a Stinespring dilation is unique up to unitary equivalence.

When we come to c.p. maps and *maximal* tensor products, the following result will be crucial (see Theorem 3.5.3): If $(\pi, \widehat{\mathcal{H}}, V)$ is a minimal Stinespring dilation of $\varphi\colon A \to \mathbb{B}(\mathcal{H})$, then the commutant $\varphi(A)' \subset \mathbb{B}(\mathcal{H})$ also lifts to $\mathbb{B}(\widehat{\mathcal{H}})$.

Proposition 1.5.6. *Let $(\pi, \widehat{\mathcal{H}}, V)$ be the minimal Stinespring dilation of a c.c.p. map $\varphi\colon A \to \mathbb{B}(\mathcal{H})$. Then, there exists a $*$-homomorphism*

$$\rho\colon \varphi(A)' \to \pi(A)' \subset \mathbb{B}(\widehat{\mathcal{H}})$$

such that

$$\varphi(a)x = V^*\pi(a)\rho(x)V$$

for every $a \in A$ and $x \in \varphi(A)'$.

Proof. For $x \in \varphi(A)'$, we define a linear operator $\rho(x)$ on the span of $\pi(A)V\mathcal{H}$ by

$$\rho(x)\Big(\sum_i \pi(a_i)V\xi_i\Big) = \sum_i \pi(a_i)Vx\xi_i.$$

Once we prove that $\rho(x)$ is well-defined and bounded for every $x \in \varphi(A)'$, it is not too hard to check that ρ gives rise to a $*$-representation of $\varphi(A)'$ on $\widehat{\mathcal{H}}$ such that $\rho(\varphi(A)') \subset \pi(A)'$ and $\varphi(a)x = V^*\pi(a)\rho(x)V$ for every $a \in A$ and $x \in \varphi(A)'$.

So, let $x \in \varphi(A)'$ and $\sum_i \pi(a_i)V\xi_i \in \pi(A)V\mathcal{H}$ be given. If we set $\xi = [\xi_1, \ldots, \xi_n]^T \in \mathcal{H}^n$ and let $\operatorname{diag}(x)$ denote the $n \times n$ matrix with x's down the diagonal and zeroes elsewhere, then we have

$$\begin{aligned}
\|\rho(x)\sum_i \pi(a_i)V\xi_i\|^2_{\widehat{\mathcal{H}}} &= \sum_{i,j} \langle x^*\varphi(a_i^*a_j)x\xi_j, \xi_i\rangle_{\mathcal{H}} \\
&= \langle \operatorname{diag}(x)^*\varphi_n([a_i^*a_j]) \operatorname{diag}(x)\xi, \xi\rangle_{\mathcal{H}^n} \\
&\le \|x\|^2\langle \varphi_n([a_i^*a_j])\xi, \xi\rangle_{\mathcal{H}^n} \\
&= \|x\|^2\|\sum_i \pi(a_i)V\xi_i\|^2_{\widehat{\mathcal{H}}},
\end{aligned}$$

where we use the fact that $\operatorname{diag}(x)$ and $\varphi_n([a_i^*a_j]) \in \mathbb{M}_n(\varphi(A))$ commute in the third line above. Therefore, we have $\|\rho(x)\| \le \|x\|$ as desired. □

Multiplicative domains.

Proposition 1.5.7. *Let A and B be* C*-*algebras and* $\varphi\colon A \to B$ *be a c.c.p. map.*

(1) *(Schwarz Inequality) The inequality* $\varphi(a)^*\varphi(a) \le \varphi(a^*a)$ *holds for every* $a \in A$.

(2) *(Bimodule Property) Given* $a \in A$, *if* $\varphi(a^*a) = \varphi(a)^*\varphi(a)$ *and* $\varphi(aa^*) = \varphi(a)\varphi(a)^*$, *then* $\varphi(ba) = \varphi(b)\varphi(a)$ *and* $\varphi(ab) = \varphi(a)\varphi(b)$ *for every* $b \in A$.

(3) *The subspace*

$$A_\varphi = \{a \in A : \varphi(a^*a) = \varphi(a)^*\varphi(a) \text{ and } \varphi(aa^*) = \varphi(a)\varphi(a)^*\}$$

is a C*-*subalgebra of* A.

Proof. Let $B \subset \mathbb{B}(\mathcal{H})$ be a faithful $*$-representation and $(\pi, \widehat{\mathcal{H}}, V)$ be a Stinespring dilation of $\varphi\colon A \to B \subset \mathbb{B}(\mathcal{H})$. Then, for every $a \in A$, we have

$$\varphi(a^*a) - \varphi(a)^*\varphi(a) = V^*\pi(a)^*(1_{\widehat{\mathcal{H}}} - VV^*)\pi(a)V \ge 0$$

since V is a contraction. This proves (1). Moreover, $\varphi(a^*a) - \varphi(a)^*\varphi(a) = 0$ is equivalent to $(1_{\widehat{\mathcal{H}}} - VV^*)^{1/2}\pi(a)V = 0$, which in turn implies

$$\varphi(ba) - \varphi(b)\varphi(a) = V^*\pi(b)(1_{\widehat{\mathcal{H}}} - VV^*)\pi(a)V = 0$$

for every $b \in A$. By symmetry, this proves (2). Assertion (3) follows from (2). □

Definition 1.5.8. Let $\varphi\colon A \to B$ be a c.c.p. map. The C*-subalgebra A_φ in Proposition 1.5.7 is called the *multiplicative domain* of φ.

Evidently A_φ is the largest subalgebra of A on which φ restricts to a $*$-homomorphism. Note also that if $\|a\| \le 1$ and $\varphi(a)$ is a unitary element, then a is in the multiplicative domain of φ (by the Schwarz inequality).

Conditional expectations. Conditional expectations are important examples of c.c.p. maps. Here's the definition:

Definition 1.5.9. Let $B \subset A$ be C*-algebras. A *projection from A onto B* is a linear map $E\colon A \to B$ such that $E(b) = b$ for every $b \in B$. A *conditional expectation from A onto B* is a c.c.p. projection E from A onto B such that $E(bxb') = bE(x)b'$ for every $x \in A$ and $b, b' \in B$ (i.e., E is a B-bimodule map).

Theorem 1.5.10 (Tomiyama). *Let $B \subset A$ be* C*-*algebras and E be a projection from A onto B. Then, the following are equivalent:*

(1) *E is a conditional expectation;*

(2) *E is c.c.p.;*

(3) *E is contractive.*

Proof. We only have to prove that the last condition implies the first, so assume E is contractive. Passing to double duals, we may assume that A and B are von Neumann algebras. We first prove that E is a B-bimodule map. Since von Neumann algebras are the (norm) closed linear span of their projections, it suffices to check the module property on projections. Let $p \in B$ be a projection and let $p^\perp = 1_A - p$. Since $pE(p^\perp x) = E(pE(p^\perp x))$ for every $x \in A$, we have that for any $t \in \mathbb{R}$,

$$\begin{aligned}(1+t)^2\|pE(p^\perp x)\|^2 &= \|pE(p^\perp x + tpE(p^\perp x))\|^2 \\ &\leq \|p^\perp x + tpE(p^\perp x)\|^2 \\ &\leq \|p^\perp x\|^2 + t^2\|pE(p^\perp x)\|^2.\end{aligned}$$

It follows that $\|pE(p^\perp x)\|^2 + 2t\|pE(p^\perp x)\|^2 \leq \|p^\perp x\|^2$ for all $t \in \mathbb{R}$; hence, $pE(p^\perp x) = 0$. In particular, $E(1_B^\perp x) = 1_B E(1_B^\perp x) = 0$. The same reasoning shows $(1_B - p)E(px) = 0$. It follows that

$$E(px) = pE(px) = pE(x - p^\perp x) = pE(x)$$

for every projection $p \in B$ and $x \in A$. Switching to the other side, one shows $E(xp) = E(x)p$ as well – hence E is a B-bimodule map.

Since E is unital – indeed, $bE(1_A) = E(b) = b$ for any $b \in B$ – and contractive, E is necessarily positive (this follows from the corresponding fact for functionals). To prove that E is c.p., let a positive element $[x_{i,j}] \in \mathbb{M}_n(A)$ be given. Let $\pi\colon B \to \mathbb{B}(\mathcal{H})$ be any $*$-representation with a cyclic vector ξ. Then, for any $b_1, \ldots, b_n \in B$, we have

$$\sum_{i,j}\langle \pi(E(x_{i,j}))\pi(b_j)\xi, \pi(b_i)\xi\rangle = \langle \pi(E(\sum_{i,j} b_i^* x_{i,j} b_j))\xi, \xi\rangle \geq 0$$

since $\sum_{i,j} b_i^* x_{i,j} b_j \geq 0$ in A. It follows that $[\pi(E(x_{i,j}))]_{i,j} \geq 0$ in $\mathbb{M}_n(\pi(B))$. Since π is an arbitrary cyclic representation, we conclude $[E(x_{i,j})]_{i,j} \geq 0$ in $\mathbb{M}_n(B)$. □

The following basic fact is extremely useful.

Lemma 1.5.11. *Let M be a von Neumann algebra with a faithful normal tracial state τ and let $1_M \in N \subset M$ be a von Neumann subalgebra. Then, there exists a unique trace-preserving, normal conditional expectation E from M onto N.*

Proof. The restriction of τ to N will also be denoted by τ. Let $a, y \in M$ be arbitrary and $a = u|a|$ be the polar decomposition of a. We claim that

$$|\tau(ya)| = |\tau(yu|a|)| \leq \tau(yu|a|u^*y^*)^{1/2}\tau(|a|)^{1/2} \leq \|y\|\tau(|a|).$$

Indeed, the first inequality is due to Cauchy-Schwarz, while the second follows from the general fact that $\tau(x|a|x^*) = \tau(|a|^{1/2}x^*x|a|^{1/2}) \leq \|x^*x\|\tau(|a|)$, for any $x \in M$.

For each $a \in N$, define $\tau_a \in N_*$ by $\tau_a(y) = \tau(ya)$. The inequality above implies that $\|\tau_a\| = \tau(|a|)$. Also note that $\{\tau_a : a \in N\}$ is a norm-dense linear subspace in N_*. (If it were not dense, we could find $0 \neq n \in N$ such that $\tau_a(n) = 0$ for all $a \in N$, which is impossible since τ is faithful.)

Now we construct the map $E\colon M \to N$: For each $x \in M$, define $E(x) \in N = (N_*)^*$ to be the unique linear functional such that $E(x)(\tau_a) = \tau(xa)$, for all $a \in N$. (Recall that $|\tau(xa)| \leq \|x\|\tau(|a|) = \|x\|\|\tau_a\|$; hence $\|E(x)\| \leq \|x\|$.) Note that

$$\tau(E(x)a) = \tau_a(E(x)) = E(x)(\tau_a) = \tau(xa),$$

for all $a \in N$. A routine exercise shows that E is a trace-preserving normal projection from M onto N; since E is also contractive, it must be a conditional expectation.

To prove uniqueness, assume E' is another trace-preserving conditional expectation. Then for every $x \in M$ and $a \in N$, we have

$$\tau(E'(x)a) = \tau(E'(xa)) = \tau(xa) = \tau(E(xa)) = \tau(E(x)a),$$

and hence $E' = E$. □

With the same hypotheses as the last lemma, let $L^2(M, \tau)$ be the GNS Hilbert space for (M, τ) and $L^2(N, \tau)$ be the Hilbert subspace corresponding to N. Then, the trace-preserving conditional expectation E extends to the orthogonal projection e_N from $L^2(M, \tau)$ onto $L^2(N, \tau)$ in such a way that $e_N x e_N = E(x)e_N$ for every $x \in M$. In fact, one can give an alternate proof of the lemma by observing that $e_N x e_N \in N$, for every $x \in M$, as an element in $\mathbb{B}(L^2(N, \tau))$. More precisely, one must know that the commutant of the right N action on $L^2(N, \tau)$ coincides with N (see Section 6.1) and then check that $e_N x e_N$ commutes with the right N action.

The case of matrices. When either the domain or range is a matrix algebra, there are useful one-to-one correspondences which we will often invoke. Proofs are included for completeness, but the important part is the explicit maps defining the correspondences.

Proposition 1.5.12. *Let A be a* C*-*algebra and* $\{e_{i,j}\}$ *be matrix units of* $\mathbb{M}_n(\mathbb{C})$. *A map* $\varphi\colon \mathbb{M}_n(\mathbb{C}) \to A$ *is c.p. if and only if* $[\varphi(e_{i,j})]$ *is positive in* $\mathbb{M}_n(A)$. *In other words,*

$$\mathrm{CP}(\mathbb{M}_n(\mathbb{C}), A) \ni \varphi \longmapsto [\varphi(e_{i,j})] \in \mathbb{M}_n(A)_+$$

is a bijective correspondence.

Proof. Since $[e_{i,j}] \in \mathbb{M}_n(\mathbb{M}_n(\mathbb{C}))$ is positive (it's a multiple of a rank-one projection), the "only if" part is trivial. To prove the "if" part, assume $a = [\varphi(e_{i,j})] \geq 0$ in $\mathbb{M}_n(A)$ and let $a^{1/2} = [b_{i,j}]$. It follows that

$$\varphi(e_{i,j}) = \sum_{k=1}^{n} b_{k,i}^* b_{k,j}.$$

Let $A \subset \mathbb{B}(\mathcal{H})$ be a faithful representation and define $V\colon \mathcal{H} \to \ell_n^2 \otimes \ell_n^2 \otimes \mathcal{H}$ by

$$V\xi = \sum_{j,k=1}^{n} \zeta_j \otimes \zeta_k \otimes b_{k,j}\xi,$$

where $\{\zeta_j\}_{j=1}^n$ is the standard orthonormal basis for ℓ_n^2. Then, for $T = [t_{i,j}] \in \mathbb{M}_n(\mathbb{C})$, we have

$$\begin{aligned}
\langle V^*(T \otimes 1 \otimes 1)V\eta, \xi\rangle &= \langle (T \otimes 1 \otimes 1)V\eta, V\xi\rangle \\
&= \sum_{i,j,k,l=1}^{n} \langle T\zeta_j, \zeta_i\rangle \langle \zeta_k, \zeta_l\rangle \langle b_{k,j}\eta, b_{l,i}\xi\rangle \\
&= \sum_{i,j=1}^{n} t_{i,j} \langle \sum_{k=1}^{n} b_{k,i}^* b_{k,j}\eta, \xi\rangle \\
&= \langle \varphi(\sum_{i,j} t_{i,j} e_{i,j})\eta, \xi\rangle \\
&= \langle \varphi(T)\eta, \xi\rangle
\end{aligned}$$

for every $\xi, \eta \in \mathcal{H}$. Therefore, $\varphi(T) = V^*(T \otimes 1 \otimes 1)V$ for every $T \in \mathbb{M}_n(\mathbb{C})$ and φ is c.p. □

Example 1.5.13. Let $a_1, \ldots, a_n \in A$ be given and define a linear map $\varphi\colon \mathbb{M}_n(\mathbb{C}) \to A$ by $\varphi(e_{i,j}) = a_i a_j^*$. The previous result easily implies that φ is completely positive. Indeed,

$$\begin{aligned}
[\varphi(e_{i,j})] &= \begin{bmatrix} a_1a_1^* & a_1a_2^* & \cdots & a_1a_n^* \\ a_2a_1^* & a_2a_2^* & \cdots & a_2a_n^* \\ \vdots & \vdots & \ddots & \vdots \\ a_na_1^* & a_na_2^* & \cdots & a_na_n^* \end{bmatrix} \\
&= \begin{bmatrix} a_1 & 0 & \cdots & 0 \\ a_2 & 0 & \cdots & 0 \\ \vdots & \vdots & \ddots & \vdots \\ a_n & 0 & \cdots & 0 \end{bmatrix} \begin{bmatrix} a_1 & 0 & \cdots & 0 \\ a_2 & 0 & \cdots & 0 \\ \vdots & \vdots & \ddots & \vdots \\ a_n & 0 & \cdots & 0 \end{bmatrix}^* \\
&\geq 0.
\end{aligned}$$

There is a similar characterization of c.p. maps from a C*-algebra A into $\mathbb{M}_n(\mathbb{C})$. For a linear map $\varphi\colon A \to \mathbb{M}_n(\mathbb{C})$, we define a linear functional $\widehat{\varphi}$ on $\mathbb{M}_n(A)$ by

$$\widehat{\varphi}([a_{i,j}]) = \sum_{i,j=1}^{n} \varphi(a_{i,j})_{i,j}.$$

The notation $\varphi(a_{i,j})_{i,j}$ means the $(i,j)^{\text{th}}$ entry of the matrix $\varphi(a_{i,j})$. Yes, the formula is slightly complicated, but it is explicit and very important.

Proposition 1.5.14. *Let A be a unital C*-algebra. A map $\varphi\colon A \to \mathbb{M}_n(\mathbb{C})$ is c.p. if and only if $\widehat{\varphi}$ is positive on $\mathbb{M}_n(A)$. In other words,*

$$\mathrm{CP}(A, \mathbb{M}_n(\mathbb{C})) \ni \varphi \longmapsto \widehat{\varphi} \in \mathbb{M}_n(A)^*_+$$

is a bijective correspondence.

Proof. Let $\{\zeta_i\}_{i=1}^n$ be the standard orthonormal basis for ℓ^2_n and let $\zeta = [\zeta_1, \ldots, \zeta_n]^T \in (\ell^2_n)^n$. Since

$$\widehat{\varphi}([a_{i,j}]) = \langle \varphi_n([a_{i,j}])\zeta, \zeta \rangle$$

for $[a_{i,j}] \in \mathbb{M}_n(A)$, positivity of φ_n implies that of $\widehat{\varphi}$. This proves the "only if" part. To prove the "if" part, assume $\widehat{\varphi}$ is positive and let $(\pi, \mathcal{H}, \xi)$ be the GNS triplet of $\widehat{\varphi}$. Let $\{e_{i,j}\}$ be the standard matrix units for $\mathbb{M}_n(\mathbb{C})$, which we also view as elements in $\mathbb{M}_n(A)$. Then, for the operator $V\colon \ell^2_n \to \mathcal{H}$ defined by $V\zeta_j = \pi(e_{1,j})\xi$, it is not hard to check

$$\varphi(a) = V^*\pi(\begin{bmatrix} a & & \\ & \ddots & \\ & & a \end{bmatrix})V.$$

It follows that φ is c.p. □

Lemma 1.5.15. *Let $E \subset A$ be an operator subsystem and $\psi\colon E \to \mathbb{C}$ be a positive linear functional. Then $\|\psi\| = \psi(1)$. Hence, any norm-preserving extension of ψ to A is also positive.*

Proof. Fix $x \in E$ such that $\|x\| \le 1$ and $|\psi(x)|$ is close to $\|\psi\|$. Multiplying by a complex scalar of norm one, we may assume that $0 < \psi(x) \in \mathbb{R}$. Since positive maps are automatically self-adjoint, we have

$$\psi(x) = \frac{1}{2}\psi(x + x^*)$$

and thus we may assume x is self-adjoint. However, in this case we have the operator inequality $x \le \|x\|1$ and hence

$$\psi(x) \le \psi(1)\|x\|.$$

Since a functional satisfying the equation $\psi(1) = \|\psi\|$ is necessarily positive, we are done. □

Corollary 1.5.16. *Let $E \subset A$ be an operator subsystem and $\varphi\colon E \to \mathbb{M}_n(\mathbb{C})$ be a c.p. map. Then φ extends to a c.p. map $A \to \mathbb{M}_n(\mathbb{C})$.*

Proof. Given φ, we can define a linear functional $\widehat{\varphi}$ on $\mathbb{M}_n(E)$ (as above) and it is positive (same proof as above). By the previous lemma, we can extend to a positive functional on all of $\mathbb{M}_n(A)$ and then apply the one-to-one correspondence in reverse to get our desired extension. □

1.6. Arveson's Extension Theorem

The following theorem, due to Arveson, is absolutely fundamental and probably gets invoked more than any other result in these notes.

Theorem 1.6.1. *Let A be a unital C*-algebra and $E \subset A$ be an operator subsystem. Then, every c.c.p. map $\varphi\colon E \to \mathbb{B}(\mathcal{H})$ extends to a c.c.p. map $\bar{\varphi}\colon A \to \mathbb{B}(\mathcal{H})$.*

Proof. Let $P_i \in \mathbb{B}(\mathcal{H})$ be an increasing net of finite-rank projections which converge to the identity in the strong operator topology. For each i, we regard the c.c.p. map $\varphi_i\colon E \to P_i\mathbb{B}(\mathcal{H})P_i$, $\varphi_i(e) = P_i\varphi(e)P_i$ as taking values in a matrix algebra. Thus, by Corollary 1.5.16, we may assume that each φ_i is actually defined on all of A. Now we regard φ_i as taking values in $\mathbb{B}(\mathcal{H})$ and apply compactness of the unit ball of $\mathbb{B}(A, \mathbb{B}(\mathcal{H}))$ in the point-ultraweak topology (Theorem 1.3.7) to find a cluster point $\Phi\colon A \to \mathbb{B}(\mathcal{H})$. It is readily verified that Φ is c.p. and extends φ. □

Remark 1.6.2 (Injectivity and Arveson's Theorem)**.** Arveson's Extension Theorem is equivalent to the statement that $\mathbb{B}(\mathcal{H})$ is injective in the category of operator systems with c.c.p. maps as morphisms. This is even true in the category of operator spaces with completely bounded maps as morphisms.

It follows from Arveson's Theorem that a von Neumann algebra $M \subset \mathbb{B}(\mathcal{H})$ is *injective* if and only if there is a conditional expectation from $\mathbb{B}(\mathcal{H})$ onto M. It also follows that injectivity is independent of the choice of faithful representation $M \subset \mathbb{B}(\mathcal{H})$.

Corollary 1.6.3. *Let $E \subset \mathbb{B}(\mathcal{H})$ be an ultraweakly closed operator system and let $\varphi\colon E \to \mathbb{M}_n(\mathbb{C})$ be a c.c.p. map. There exists a net of ultraweakly continuous c.c.p. maps $\varphi_\lambda\colon E \to \mathbb{M}_n(\mathbb{C})$ which converges to φ in the point-norm topology (i.e., $\|\varphi_\lambda(x) - \varphi(x)\| \to 0$ for all $x \in E$).*

Proof. By Arveson's Extension Theorem, we may assume that $E = \mathbb{B}(\mathcal{H})$. Since φ is c.p., the corresponding functional $\widehat{\varphi} \in \mathbb{M}_n(\mathbb{B}(\mathcal{H}))^*$ is positive. Hence, there exists a net $\widehat{\varphi}_\lambda$ of positive normal linear functionals which converges pointwise to $\widehat{\varphi}$. Then, the corresponding c.p. maps $\varphi_\lambda\colon \mathbb{B}(\mathcal{H}) \to \mathbb{M}_n(\mathbb{C})$ are normal and converge to φ in the point-norm topology (which is easily seen from explicit form of the correspondence). Unfortunately the φ_λ's need not be contractive, but they are "almost contractive" (since $\varphi_\lambda(1) \to \varphi(1)$ in norm); hence we can fiddle with their norms a bit to correct this deficiency. □

1.7. Voiculescu's Theorem

Voiculescu's Theorem is analogous to the Hahn-Banach Theorem in two ways: It gets used all of the time; and it really refers to a collection of related results and corollaries.[4] Here, we collect all the forms we need, though we only prove those which haven't yet appeared in a book.

Finite-dimensional case. Exploiting the duality between c.p. maps $A \to \mathbb{M}_n(\mathbb{C})$ and states on $\mathbb{M}_n(A)$, it is not too hard to deduce the next result from Glimm's lemma (Lemma 1.4.11).

Proposition 1.7.1. *Let $\mathcal{H}$ be separable, $1 \in A \subset \mathbb{B}(\mathcal{H})$ be a separable* C**-algebra and $\varphi\colon A \to \mathbb{M}_n(\mathbb{C})$ be a u.c.p. map such that $\varphi|_{A\cap\mathbb{K}(\mathcal{H})} = 0$. Then there exist isometries $V_k\colon \ell^2_n \to \mathcal{H}$ with the following properties:*

(1) *the ranges of the V_k's are pairwise orthogonal;*
(2) $\|\varphi(a) - V_k^* a V_k\| \to 0$ *for every $a \in A$.*

General case.

Definition 1.7.2. Two maps $\pi\colon A \to \mathbb{B}(\mathcal{H})$ and $\sigma\colon A \to \mathbb{B}(\mathcal{K})$ are called *approximately unitarily equivalent* if there is a sequence of unitary operators $U_n\colon \mathcal{H} \to \mathcal{K}$ such that

$$\|\sigma(a) - U_n\pi(a)U_n^*\| \to 0$$

for all $a \in A$. If it also happens that $\sigma(a) - U_n\pi(a)U_n^*$ is a compact operator, for all $a \in A$ and $n \in \mathbb{N}$, then we say that π and σ are *approximately unitarily equivalent relative to the compacts.*

Note that approximate unitary equivalence relative to the compacts is a *much* stronger notion as it implies that after passing to the Calkin algebra, the representations π and σ are actually unitarily equivalent. See [**11**] or [**53**, Corollary II.5.5] for a proof of the next result.

Theorem 1.7.3 (Voiculescu's Theorem)**.** *Let $\mathcal{H}$ and $\mathcal{K}$ be separable Hilbert spaces and $A \subset \mathbb{B}(\mathcal{H})$ be a separable* C**-algebra such that $1_\mathcal{H} \in A$. Let $\iota\colon A \hookrightarrow \mathbb{B}(\mathcal{H})$ denote the canonical inclusion and let $\rho\colon A \to \mathbb{B}(\mathcal{K})$ be any unital representation such that $\rho|_{A\cap\mathbb{K}(\mathcal{H})} = 0$. Then ι and $\iota \oplus \rho$ are approximately unitarily equivalent relative to the compacts.*

[4] Thirdly, some authors assume familiarity with all possible formulations and don't bother explaining which version is being invoked.

Definition 1.7.4. A representation $\pi\colon A \to \mathbb{B}(\mathcal{H})$ is called *essential* if $\pi(A)$ contains no nonzero compact operators.

Essential representations are easy to construct: if $\pi\colon A \to \mathbb{B}(\mathcal{H})$ is any representation, then its infinite inflation (i.e., the direct sum of infinitely many copies of π) will be essential.

Corollary 1.7.5. *Let $\pi_i\colon A \to \mathbb{B}(\mathcal{H}_i)$, $i = 1, 2$, be faithful essential representations. If A is unital and both π_1, π_2 are unital, then they are approximately unitarily equivalent relative to the compacts. If A is nonunital, then π_1 and π_2 are always approximately unitarily equivalent relative to the compacts.*

In particular, the previous corollary implies that if A is simple and unital, then it has precisely one unital representation, up to approximate unitary equivalence relative to the compacts, since all representations will be faithful and essential.

Technical variations. We'll need some technical variations of Voiculescu's Theorem, but they require a bit more terminology.

If $\pi\colon \mathbb{B}(\mathcal{H}) \to Q(\mathcal{H})$ is the canonical mapping onto the Calkin algebra, A is a unital C*-algebra and $\varphi\colon A \to \mathbb{B}(\mathcal{H})$ is a unital completely positive map, then we say that φ is a *representation modulo the compacts* if $\pi \circ \varphi\colon A \to Q(\mathcal{H})$ is a *-homomorphism. If $\pi \circ \varphi$ is injective, then we say that φ is a *faithful representation modulo the compacts.* In this situation we define constants $\eta_\varphi(a)$ by

$$\eta_\varphi(a) = 2\max\{\|\varphi(a^*a) - \varphi(a^*)\varphi(a)\|^{1/2}, \|\varphi(aa^*) - \varphi(a)\varphi(a^*)\|^{1/2}\}$$

for every $a \in A$.

Theorem 1.7.6. *Let A be a unital separable C*-algebra and $\varphi\colon A \to \mathbb{B}(\mathcal{H})$ be a faithful representation modulo the compacts on a separable space $\mathcal{H}$. If $\sigma\colon A \to \mathbb{B}(\mathcal{K})$ is any faithful unital essential representation on a separable space $\mathcal{K}$, then there exist unitaries $U_n\colon \mathcal{H} \to \mathcal{K}$ such that*

$$\limsup_{n\to\infty} \|\sigma(a) - U_n\varphi(a)U_n^*\| \le \eta_\varphi(a)$$

for every $a \in A$.

Proof. It suffices to show the existence of a representation σ satisfying the conclusion of the theorem, since all such representations are approximately unitarily equivalent.

Let $\rho\colon A \to \mathbb{B}(\mathcal{L})$ be the Stinespring dilation of φ, $V\colon \mathcal{H} \to \mathcal{L}$ be the associated isometry, $P = VV^* \in \mathbb{B}(\mathcal{L})$ be the Stinespring projection and

$P^\perp = 1 - P$. It follows from the identity

$$(P^\perp \rho(a)P)^*(P^\perp \rho(a)P) = V\big(\varphi(a^*a) - \varphi(a^*)\varphi(a)\big)V^*$$

that

$$\|P^\perp \rho(a)P\| = \|\varphi(a^*a) - \varphi(a^*)\varphi(a)\|^{1/2},$$

for all $a \in A$.

Now write $\mathcal{L} = P\mathcal{L} \oplus P^\perp\mathcal{L}$ and decompose the representation ρ accordingly. That is, consider the matrix decomposition

$$\rho(a) = \begin{bmatrix} \rho(a)_{11} & \rho(a)_{12} \\ \rho(a)_{21} & \rho(a)_{22} \end{bmatrix},$$

where $\rho(a)_{21} = P^\perp \rho(a)P$ and $\rho(a)_{12} = \rho(a^*)^*_{21}$. Thanks to orthogonal domains and ranges, the norm of the matrix

$$\begin{bmatrix} 0 & \rho(a)_{12} \\ \rho(a)_{21} & 0 \end{bmatrix}$$

is equal to $\frac{1}{2}\eta_\phi(a)$.

Now comes the trick. We consider the Hilbert space $P^\perp\mathcal{L} \oplus P\mathcal{L}$ and the representation $\rho'\colon A \to \mathbb{B}(P^\perp\mathcal{L} \oplus P\mathcal{L})$ given in matrix form as

$$\rho'(a) = \begin{bmatrix} \rho(a)_{22} & \rho(a)_{21} \\ \rho(a)_{12} & \rho(a)_{11} \end{bmatrix}.$$

Using the obvious identification of the Hilbert spaces

$$P\mathcal{L} \oplus \Big(\bigoplus_{\mathbb{N}} P^\perp\mathcal{L} \oplus P\mathcal{L}\Big) \quad \text{and} \quad \bigoplus_{\mathbb{N}} \mathcal{L} = \bigoplus_{\mathbb{N}} (P\mathcal{L} \oplus P^\perp\mathcal{L}),$$

a standard calculation shows that

$$\|\rho(a)_{11} \oplus \rho'^{\infty}(a) - \rho^\infty(a)\| \le \eta_\varphi(a)$$

for all $a \in A$, where $\rho'^\infty = \bigoplus_{\mathbb{N}} \rho'$ and $\rho^\infty = \bigoplus_{\mathbb{N}} \rho$. Note also that $\rho(a)_{11} = V\varphi(a)V^*$.

Let $C = \varphi(A) + \mathbb{K}(\mathcal{H})$ and observe that C is actually a separable, unital C*-subalgebra of $\mathbb{B}(\mathcal{H})$ with $\pi(C) \cong A$ (again, $\pi\colon \mathbb{B}(\mathcal{H}) \to Q(\mathcal{H})$ is the quotient map). Note that $\iota \oplus \rho'^\infty \circ \pi$ is approximately unitarily equivalent relative to the compacts to ι, where $\iota\colon C \hookrightarrow \mathbb{B}(\mathcal{H})$ is the inclusion. Let $W_n\colon \mathcal{H} \to \mathcal{H} \oplus (\bigoplus_{\mathbb{N}}(P^\perp\mathcal{L} \oplus P\mathcal{L}))$ be unitaries such that

$$\|\varphi(a) \oplus \rho'^\infty(a) - W_n\varphi(a)W_n^*\| \to 0$$

for all $a \in A$.

Let

$$\tilde{V}\colon \mathcal{H} \oplus \Big(\bigoplus_{\mathbb{N}} (P^\perp\mathcal{L} \oplus P\mathcal{L})\Big) \to \bigoplus_{\mathbb{N}} \mathcal{L}$$

be the unitary $V \oplus 1$. Note that $\tilde{V}(\varphi(a) \oplus \rho'^{\infty}(a))\tilde{V}^* = V\varphi(a)V^* \oplus \rho'^{\infty}(a) = \rho(a)_{11} \oplus \rho'^{\infty}(a)$. We now complete the proof by defining $\mathcal{K} = \bigoplus_{\mathbb{N}} \mathcal{L}$, $\sigma = \rho^{\infty} = \bigoplus_{\mathbb{N}} \rho$, and $U_n = \tilde{V}W_n \colon \mathcal{H} \to \bigoplus_{\mathbb{N}} \mathcal{L} = \mathcal{K}$. □

Corollary 1.7.7. *Let $\varphi\colon A \to \mathbb{M}_n(\mathbb{C}) \subset \mathbb{B}(\mathcal{K})$ be a u.c.p. map where $\mathbb{M}_n(\mathbb{C}) \subset \mathbb{B}(\mathcal{K})$ is a unital inclusion and $\mathcal{K}$ is infinite dimensional. Let $\pi\colon A \to \mathbb{B}(\mathcal{H})$ be a faithful unital essential representation. Then there exists a sequence of unitaries $U_n\colon \mathcal{H} \to \mathcal{H} \oplus \mathcal{K}$ such that*

$$\limsup_{n\to\infty} \|(\pi(a) \oplus \varphi(a)) - U_n\pi(a)U_n^*\| \le \eta_{\varphi}(a)$$

for every $a \in A$.

Proof. Note that if $\mathcal{K}$ had finite dimension, then this result would follow from the previous result, since $\pi\oplus\varphi$ would be a faithful homomorphism modulo the compacts. However, we have assumed $\mathcal{K}$ to be infinite dimensional; hence there is something to prove.

Let $\tilde{\varphi}\colon A \to \mathbb{M}_n(\mathbb{C}) = \mathbb{B}(\ell^2_n)$ be the map φ but now regarded as taking values in $\mathbb{B}(\ell^2_n)$ (instead of $\mathbb{B}(\mathcal{K})$). As noted above, we can find unitaries $V_n : \mathcal{H} \to \mathcal{H} \oplus \ell^2_n$ such that

$$\limsup_{n\to\infty} \|(\pi(a) \oplus \tilde{\varphi}(a)) - V_n\pi(a)V_n^*\| \le \eta_{\varphi}(a)$$

for every $a \in A$.

Since $\mathbb{M}_n(\mathbb{C}) \subset \mathbb{B}(\mathcal{K})$ is a unital inclusion, we can find an isomorphism $\mathbb{B}(\mathcal{K}) \cong \mathbb{B}(\ell^2_n \otimes \mathcal{L})$ that maps $\mathbb{M}_n(\mathbb{C}) \subset \mathbb{B}(\mathcal{K})$ to $\mathbb{M}_n(\mathbb{C}) \otimes 1_{\mathcal{L}}$. Under this isomorphism we may identify φ with $\tilde{\varphi}\otimes 1_{\mathcal{L}}$ and hence the unitaries $V_n \otimes 1_{\mathcal{L}}$ will conjugate $\pi\otimes 1_{\mathcal{L}}$ to $(\pi\oplus\tilde{\varphi})\otimes 1_{\mathcal{L}}$ which we further identify with $\pi\otimes 1_{\mathcal{L}}\oplus\varphi$. The proof is finished once we observe that π is approximately unitarily equivalent to $\pi \otimes 1_{\mathcal{L}}$ and $\pi \otimes 1_{\mathcal{L}} \oplus \varphi$ is approximately unitarily equivalent to $\pi \oplus \varphi$. □

There is one more version of Voiculescu's Theorem that we'll need, but not until the end of the book. See [**53**, II.5.3] for a proof.

Theorem 1.7.8. *Let $A \subset \mathbb{B}(\mathcal{H})$ be a separable C*-algebra and $\varphi\colon A \to \mathbb{B}(\mathcal{K})$ be a c.c.p. map such that $\varphi(x) = 0$ for all $x \in A \cap \mathbb{K}(\mathcal{H})$. Then there exist isometries $V_k\colon \mathcal{K} \to \mathcal{H}$ such that $\varphi(a) - V_k^*aV_k \in \mathbb{K}(\mathcal{K})$ and $\lim \|\varphi(a) - V_k^*aV_k\| = 0$, for all $a \in A$.*

Part 1

Basic Theory

Chapter 2

Nuclear and Exact C*-Algebras: Definitions, Basic Facts and Examples

Nuclearity and exactness have dominated the C*-scene for quite a while. We will define these classes in terms of the nuclearity of certain maps; historically they were defined via tensor products, as we'll see in Sections 3.8 and 3.9.

For the most part, this chapter consists of easy propositions and examples. We have attempted to lay out, as simply as possible, the main themes, subtleties and techniques. Sections 2.1 and 2.3 contain numerous exercises which newcomers are highly encouraged to work through. Most are quite easy, but we aren't trying to insult your intelligence. These exercises tend to get used without explanation in the literature (and this book), so we thought it might be helpful to isolate them from the get-go.

2.1. Nuclear maps

The following definition is the cornerstone of nuclearity and exactness.

Definition 2.1.1. A map $\theta\colon A \to B$ is called *nuclear* if there exist c.c.p. maps $\varphi_n\colon A \to \mathbb{M}_{k(n)}(\mathbb{C})$ and $\psi_n\colon \mathbb{M}_{k(n)}(\mathbb{C}) \to B$ such that $\psi_n \circ \varphi_n \to \theta$ in the point-norm topology:

$$\|\psi_n \circ \varphi_n(a) - \theta(a)\| \to 0,$$

for all $a \in A$.

Note that nuclear maps are automatically c.c.p. If A, B and θ are unital, we'll soon see that the φ_n's and ψ_n's can be replaced with u.c.p. maps (Proposition 2.2.6).

Though they won't get used as much as the C*-version, we will need nuclear maps in the von Neumann algebra context too.

Definition 2.1.2. If A is a C*-algebra and N is a von Neumann algebra, a map $\theta\colon A \to N$ is called *weakly nuclear* if there exist c.c.p. maps $\varphi_n\colon A \to \mathbb{M}_{k(n)}(\mathbb{C})$ and $\psi_n\colon \mathbb{M}_{k(n)}(\mathbb{C}) \to N$ such that $\psi_n \circ \varphi_n \to \theta$ in the point-ultraweak topology:

$$\eta(\psi_n \circ \varphi_n(a)) \to \eta(\theta(a)),$$

for all $a \in A$ and all normal functionals $\eta \in N_*$.

As in the C*-context, when θ is a weakly nuclear unital map, we can replace c.c.p. approximations by u.c.p. maps; we can also arrange normality (whether or not θ is normal). See Propositions 2.2.7 and 2.2.8 in the next section.

Remark 2.1.3. It follows from Sakai's predual uniqueness theorem that when checking point-ultraweak convergence of bounded nets, it always suffices to check convergence on certain vector functionals. That is, if $N \subset \mathbb{B}(\mathcal{H})$ is any faithful normal representation and $\Omega \subset \mathcal{H}$ is any set of vectors whose linear span is dense in $\mathcal{H}$, then $\psi_n \circ \varphi_n \to \theta$ in the point-ultraweak topology if and only if

$$\langle \psi_n \circ \varphi_n(a)v, w\rangle \to \langle \theta(a)v, w\rangle$$

for all $a \in A$ and $v, w \in \Omega$. If the set Ω is a linear subspace, then the polarization identity implies that one only need check the positive vector functionals arising from Ω.

A fundamental subtlety which makes nuclear maps irritating and/or interesting is the *dependence on the range*. More precisely, it often happens that a map $\theta\colon A \to B$ is not nuclear, but after embedding B into some larger algebra C, it becomes nuclear. In fact, this is the difference between exact and nuclear C*-algebras.

We will have to wait for C*-examples, but this phenomenon is readily seen in the von Neumann algebra context. Indeed, there are many concrete examples of von Neumann algebras $M \subset \mathbb{B}(\mathcal{H})$ for which the identity map $\mathrm{id}_M\colon M \to M$ is *not* weakly nuclear (cf. Theorem 2.6.8); however, the natural inclusion $M \hookrightarrow \mathbb{B}(\mathcal{H})$ is *always* weakly nuclear.

Proposition 2.1.4. *Let $M \subset \mathbb{B}(\mathcal{H})$ be a von Neumann algebra. The natural inclusion map $M \hookrightarrow \mathbb{B}(\mathcal{H})$ is always weakly nuclear.*

Proof. Let $\{P_i\}_{i\in I}$ be a net of finite-rank projections which increases to the identity (i.e., $i \leq j \Rightarrow P_i \leq P_j$ and $\|P_i v - v\| \to 0$ for all $v \in \mathcal{H}$). If P_i has rank $k(i)$, then we define c.c.p. maps $\varphi_i \colon M \to \mathbb{M}_{k(i)}(\mathbb{C}) \cong P_i\mathbb{B}(\mathcal{H})P_i$ by compression (i.e., $\varphi_i(T) = P_iTP_i$) and we let $\psi_i \colon \mathbb{M}_{k(i)}(\mathbb{C}) \to \mathbb{B}(\mathcal{H})$ be the natural inclusion maps. Since the predual of $\mathbb{B}(\mathcal{H})$ is the trace class operators, a routine exercise shows that these maps converge to the identity (on all of $\mathbb{B}(\mathcal{H})$, in fact) in the point-ultraweak topology and hence $M \hookrightarrow \mathbb{B}(\mathcal{H})$ is weakly nuclear. □

Here are some simple, but very useful, observations to get you warmed up. As mentioned above, we will use these exercises – often without reference – throughout this book.

Exercises

Exercise 2.1.1 (Finite sets and ε's). Show that $\theta\colon A \to B$ is nuclear if and only if for each finite set $\mathfrak{F} \subset A$ and $\varepsilon > 0$ there exist $n \in \mathbb{N}$ and c.c.p. maps $\varphi\colon A \to \mathbb{M}_n(\mathbb{C})$, $\psi\colon \mathbb{M}_n(\mathbb{C}) \to B$ such that $\|\theta(a) - \psi\circ\varphi(a)\| < \varepsilon$ for all $a \in \mathfrak{F}$.[1]

Exercise 2.1.2 (Matrices versus finite-dimensional algebras). Prove that $\theta\colon A \to B$ is nuclear if and only if there exist c.c.p. maps $\varphi_n\colon A \to C_n$ and $\psi_n\colon C_n \to B$, where the C_n's are finite-dimensional C*-algebras, such that $\psi_n \circ \varphi_n \to \theta$ in the point-norm topology. (Hint: Find integers $k(n)$ such that there is a unital embedding $C_n \subset \mathbb{M}_{k(n)}(\mathbb{C})$ and construct a conditional expectation $\mathbb{M}_{k(n)}(\mathbb{C}) \to C_n$.)

Exercise 2.1.3 (Restriction to subalgebras). If $\theta\colon A \to B$ is nuclear and $C \subset A$ is a C*-subalgebra, then $\theta|_C\colon C \to B$ is also nuclear. (It need not be true that $\theta|_C\colon C \to \theta(C)$ is nuclear. But don't try to construct examples until we prove that subalgebras of nuclear C*-algebras need not be nuclear.)

Exercise 2.1.4 (Compositions). If c.c.p. maps $\theta\colon A \to B$ and $\sigma\colon B \to C$ are given and either θ or σ is nuclear, then so is the composition $\sigma \circ \theta$.

Exercise 2.1.5 (Special case of compositions). If A is a C*-algebra such that the identity map on A is nuclear, then any other c.c.p. map $\theta\colon A \to B$ is also nuclear.

Exercise 2.1.6 (Special case of compositions). Assume $A \subset \mathbb{B}(\mathcal{H})$ is a concretely represented C*-algebra such that the inclusion map $A \hookrightarrow \mathbb{B}(\mathcal{H})$ is nuclear. Show that if $\theta\colon A \to \mathbb{B}(\mathcal{K})$ is any c.c.p. map, then θ is nuclear. (Hint: You will need Arveson's Extension Theorem.)

---

[1] You may have noticed that we haven't specified in Definition 2.1.1 whether one should work with sequences or nets. Of course, one should use nets in general and show that sequences suffice in the separable setting. However, the main point of this exercise is that nuclearity is really a *local* property and hence there is little difference between the separable and nonseparable worlds. This exercise also tests whether or not you know how to construct a net.

Exercise 2.1.7 (Dependence on range). If $\theta\colon A \to B$ is nuclear and $C \subset B$ is a C*-subalgebra with the properties that (a) $\theta(A) \subset C$ and (b) there exists a conditional expectation $\Phi\colon B \to C$, then $\theta\colon A \to C$ is also nuclear.

Exercise 2.1.8 (Dependence on range). If $\theta\colon A \to B$ is nuclear and $C \subset B$ is a C*-subalgebra with the properties that (a) $\theta(A) \subset C$ and (b) there exist a sequence of c.c.p. maps $\Phi_n\colon B \to C$ such that $\Phi_n|_C \to \mathrm{id}_C$ in the point-norm topology, then $\theta\colon A \to C$ is also nuclear.

Definition 2.1.5. If A is unital, an extension $0 \to J \to A \xrightarrow{\pi} A/J \to 0$ is called *locally split* if for each finite-dimensional operator system $E \subset A/J$ there exists a u.c.p. map $\sigma\colon E \to A$ such that $\pi \circ \sigma = \mathrm{id}_E$.

Exercise 2.1.9 (Quotients). Let $\theta\colon A \to B$ be a unital nuclear map such that $\theta|_J = 0$ for some ideal $J \lhd A$. First show that θ descends to a u.c.p. map $\dot{\theta}\colon A/J \to B$. Next, show that if $0 \to J \to A \xrightarrow{\pi} A/J \to 0$ is locally split, then $\dot{\theta}$ is nuclear. (Hint: You will need Arveson's Extension Theorem again.)

2.2. Nonunital technicalities

Many arguments are more transparent in the presence of units; hence we will reduce to this case whenever possible. The purpose of this section is to collect a number of technical facts needed for this reduction. The results are important and anyone wishing to work in this field should read the proofs at some point in his or her life, but we don't recommend spending too much time here.

The first issue is the unitization of a c.c.p. map.

Proposition 2.2.1. *Assume A is nonunital, B is unital and $\varphi\colon A \to B$ is a c.c.p. map. Then φ extends to a u.c.p. map $\tilde{\varphi}\colon \tilde{A} \to B$ by the formula*

$$\tilde{\varphi}(a + \lambda 1_{\tilde{A}}) = \varphi(a) + \lambda 1_B,$$

where $\tilde{A}$ denotes the unitization of A.

Proof. "All" we have to prove is that $\tilde{\varphi}$ is also completely positive. To do this, we first consider the double adjoint map $\varphi^{**}\colon A^{**} \to B^{**}$. Identifying double duals with enveloping von Neumann algebras, one checks that φ^{**} maps positive operators to positive operators. Since $\mathbb{M}_n(C^{**}) \cong (\mathbb{M}_n(C))^{**}$ for any C*-algebra C, it follows that φ^{**} is also completely positive.[2] Identifying $\tilde{A}$ with $A + \mathbb{C}1_{A^{**}} \subset A^{**}$, we have thus extended φ to a c.p. map $\tilde{A} \to B^{**}$. Of course, if we get lucky and $\varphi^{**}(1_{A^{**}})$ happens to be the unit

[2] There are a number of things to check here, due to the identifications involved, but it all works.

of B, then we are done. Though that won't be the case in general, we do have $0 \leq \varphi^{**}(1_{A^{**}}) \leq \|\varphi\| 1_{B^{**}} \leq 1_B$.

Now we are in a position to check that $\tilde{\varphi}$ is completely positive. Let $0 \leq [a_{ij} + \lambda_{ij} 1_{A^{**}}] \in \mathbb{M}_n(\tilde{A})$ be given. Then $\tilde{\varphi}_n([a_{ij} + \lambda_{ij} 1_{A^{**}}]) = [\varphi(a_{ij}) + \lambda_{ij} 1_B]$ and it suffices to show that $\tilde{\varphi}_n([a_{ij} + \lambda_{ij} 1_{A^{**}}]) \geq (\varphi^{**})_n([a_{ij} + \lambda_{ij} 1_{A^{**}}])$, since we already observed that φ^{**} is completely positive. But,

$$\begin{aligned}\tilde{\varphi}_n([a_{ij} + \lambda_{ij} 1_{A^{**}}]) - (\varphi^{**})_n([a_{ij} + \lambda_{ij} 1_{A^{**}}]) &= [\lambda_{ij}(1_B - \varphi^{**}(1_{A^{**}}))] \\ &= [\lambda_{ij} 1_B]\operatorname{diag}(1_B - \varphi^{**}(1_{A^{**}})),\end{aligned}$$

where $\operatorname{diag}(1_B - \varphi^{**}(1_{A^{**}})) \in \mathbb{M}_n(B^{**})$ is the diagonal matrix with constant entries $1_B - \varphi^{**}(1_{A^{**}})$ down the diagonal. Since $\mathbb{M}_n(\mathbb{C})$ is a quotient of $\mathbb{M}_n(\tilde{A})$ and $0 \leq [a_{ij} + \lambda_{ij} 1_{A^{**}}] \in \mathbb{M}_n(\tilde{A})$, it follows that $[\lambda_{ij} 1_B] \geq 0$. Note also that this scalar matrix commutes with $\operatorname{diag}(1_B - \varphi^{**}(1_{A^{**}})) \geq 0$. Since the product of commuting positive operators is still positive, it follows that $[\lambda_{ij} 1_B]\operatorname{diag}(1_B - \varphi^{**}(1_{A^{**}})) \geq 0$ as desired. □

Remark 2.2.2. The norm of $\tilde{\varphi}$ may be larger than that of φ (if and only if $\|\varphi\| < 1$), but the fact that it is unital usually outweighs this deficiency. Though we won't need it, note that the proof above still works if one requires that $1_{\tilde{A}} \mapsto \|\varphi\| 1_B$, and this produces a map with the same norm as that of φ.

We will also need the following extension result.

Lemma 2.2.3. *If A is unital, B is unital and $\varphi\colon A \to B$ is c.c.p., then one can extend φ to a u.c.p. map $\tilde{\varphi}\colon A \oplus \mathbb{C} \to B$ by*

$$\tilde{\varphi}(a \oplus \lambda) = \varphi(a) + \lambda(1_B - \varphi(1_A)).$$

Proof. First note that $0 \leq 1_B - \varphi(1_A)$ since φ is positive and contractive. Hence $\lambda \mapsto \lambda(1_B - \varphi(1_A))$ defines a c.p. map $\mathbb{C} \to B$. It is a general (and easily verified) fact that the sum of two c.p. maps is again c.p.; thus the proof is complete. □

Proposition 2.2.4. *Assume $\theta\colon A \to B$ is nuclear. If A is nonunital and B is unital, then the u.c.p. extension given by Proposition 2.2.1 is also nuclear. If both A and B are nonunital, then the unique unital extension $\tilde{\theta}\colon \tilde{A} \to \tilde{B}$ is also nuclear.*

Proof. Let $\varphi_n\colon A \to \mathbb{M}_{k(n)}(\mathbb{C})$ and $\psi_n\colon \mathbb{M}_{k(n)}(\mathbb{C}) \to B$ be c.c.p. maps converging in the point-norm topology to θ. In the case that A is nonunital but B has a unit we first extend the maps φ_n to u.c.p. maps

$$\tilde{\varphi}_n\colon \tilde{A} \to \mathbb{M}_{k(n)}(\mathbb{C}) \oplus \mathbb{C}$$

by regarding $\mathbb{M}_{k(n)}(\mathbb{C})$ as a subalgebra of $\mathbb{M}_{k(n)}(\mathbb{C}) \oplus \mathbb{C}$ and applying Proposition 2.2.1. We then use the previous lemma to extend the ψ_n's to u.c.p.

maps $\tilde{\psi}_n \colon \mathbb{M}_{k(n)}(\mathbb{C}) \oplus \mathbb{C} \to B$ and one checks that $\{\tilde{\psi}_n \circ \tilde{\varphi}_n\}$ converges in the point-norm topology to $\tilde{\theta}$.[3]

In the case that both A and B are nonunital we get u.c.p. maps $\tilde{\varphi}_n \colon \tilde{A} \to \mathbb{M}_{k(n)}(\mathbb{C}) \oplus \mathbb{C}$ just as in the previous paragraph. Regarding B as a subalgebra of $\tilde{B}$ and the ψ_n's as taking values in $\tilde{B}$, we again invoke the previous lemma to get u.c.p. maps $\tilde{\psi}_n \colon \mathbb{M}_{k(n)}(\mathbb{C}) \oplus \mathbb{C} \to \tilde{B}$. Another standard computation completes the proof. □

The previous result will usually allow us to assume that our algebras are unital. In most cases, we can further reduce to the case of unital maps.

Lemma 2.2.5. *If $\tilde{\varphi} \colon A \to \mathbb{M}_n(\mathbb{C})$ is c.p. and A is unital, then there exists a u.c.p. map $\varphi \colon A \to \mathbb{M}_n(\mathbb{C})$ such that for all $a \in A$ we have*

$$\tilde{\varphi}(a) = \tilde{\varphi}(1_A)^{\frac{1}{2}} \varphi(a) \tilde{\varphi}(1_A)^{\frac{1}{2}}.$$

Proof. In the case that $\tilde{\varphi}(1_A)$ is an invertible matrix this result is trivial as one simply defines

$$\varphi(a) = \tilde{\varphi}(1_A)^{-\frac{1}{2}} \tilde{\varphi}(a) \tilde{\varphi}(1_A)^{-\frac{1}{2}}$$

for all $a \in A$. The general case is more technical but similarly simple.

If P denotes the projection onto the kernel of $\tilde{\varphi}(1_A)$ and $P^\perp = 1 - P$ is the orthogonal complement, then

$$\tilde{\varphi}(a) = P^\perp \tilde{\varphi}(a) = \tilde{\varphi}(a) P^\perp$$

for all $a \in A$. Evidently it suffices to see this in the case that $0 \leq a \leq 1_A$ and then it is a consequence of the fact that $0 \leq \tilde{\varphi}(a) \leq \tilde{\varphi}(1_A)$ (since this implies the kernel of $\tilde{\varphi}(1_A)$ is contained in the kernel of $\tilde{\varphi}(a)$).

Applying the trick from the first part of the proof, we can find a u.c.p. map $\varphi_1 \colon A \to P^\perp \mathbb{M}_n(\mathbb{C}) P^\perp$ as in the statement of the lemma. To complete the proof, we just take any state $\eta \colon A \to \mathbb{C}$ and define a u.c.p. map $\varphi \colon A \to \mathbb{M}_n(\mathbb{C})$ by $\varphi(a) = \varphi_1(a) \oplus \eta(a) P$. □

Proposition 2.2.6. *If $\theta \colon A \to B$ is a unital nuclear map, then there exist u.c.p. maps $\varphi_n \colon A \to \mathbb{M}_{k(n)}(\mathbb{C})$ and $\psi_n \colon \mathbb{M}_{k(n)}(\mathbb{C}) \to B$ such that $\psi_n \circ \varphi_n \to \theta$ in the point-norm topology.*

Proof. Let $\tilde{\varphi}_n \colon A \to \mathbb{M}_{k(n)}(\mathbb{C})$ and $\tilde{\psi}_n \colon \mathbb{M}_{k(n)}(\mathbb{C}) \to B$ be c.c.p. maps whose compositions converge in the point-norm topology to θ. By the previous lemma we can find u.c.p. maps $\varphi_n \colon A \to \mathbb{M}_{k(n)}(\mathbb{C})$ such that

$$\tilde{\varphi}_n(a) = \tilde{\varphi}_n(1_A)^{\frac{1}{2}} \varphi_n(a) \tilde{\varphi}_n(1_A)^{\frac{1}{2}}$$

for all $a \in A$. These will be the first replacements we need.

[3]Of course, one now applies Exercise 2.1.2 to complete the proof.

The right maps to replace the ψ_n's are slightly technical to describe but there is nothing deep about the remainder of the proof. First we note that since θ is a unital map,

$$\|1_B - \tilde{\psi}_n \circ \tilde{\varphi}_n(1_A)\| \to 0.$$

Hence for all sufficiently large n, $\tilde{\psi}_n \circ \tilde{\varphi}_n(1_A)$ is a positive invertible element and some standard functional calculus shows that

$$\|1_B - \tilde{\psi}_n(\tilde{\varphi}_n(1_A))^{-\frac{1}{2}}\| \to 0.$$

To ease notation, we let

$$b_n = \tilde{\psi}_n(\tilde{\varphi}_n(1_A))^{-\frac{1}{2}} \in B$$

and

$$Y_n = \tilde{\varphi}_n(1_A)^{\frac{1}{2}} \in \mathbb{M}_{k(n)}(\mathbb{C}).$$

The u.c.p. maps $\psi_n \colon \mathbb{M}_{k(n)}(\mathbb{C}) \to B$ we are after are given by

$$\psi_n(T) = b_n \tilde{\psi}_n(Y_n T Y_n) b_n,$$

for all $T \in \mathbb{M}_{k(n)}(\mathbb{C})$. Since

$$\psi_n \circ \varphi_n(a) = b_n \tilde{\psi}_n(\tilde{\varphi}_n(a)) b_n$$

and $\|1_B - b_n\| \to 0$, it follows that $\psi_n \circ \varphi_n \to \theta$ in the point-norm topology. □

Proposition 2.2.7. *If A is a C*-algebra, N is a von Neumann algebra and we are given a unital weakly nuclear map $\theta\colon A \to N$, then there exist u.c.p. maps $\varphi_n\colon A \to \mathbb{M}_{k(n)}(\mathbb{C})$ and $\psi_n\colon \mathbb{M}_{k(n)}(\mathbb{C}) \to N$ such that $\psi_n \circ \varphi_n \to \theta$ in the point-ultraweak topology.*

Proof. Let $\tilde{\varphi}_n\colon A \to \mathbb{M}_{k(n)}(\mathbb{C})$ and $\tilde{\psi}_n\colon \mathbb{M}_{k(n)}(\mathbb{C}) \to N$ be c.c.p. maps whose compositions converge to θ in the point-ultraweak topology and, as before, let $\varphi_n\colon A \to \mathbb{M}_{k(n)}(\mathbb{C})$ be u.c.p. maps such that

$$\tilde{\varphi}_n(a) = \tilde{\varphi}_n(1_A)^{\frac{1}{2}} \varphi_n(a) \tilde{\varphi}_n(1_A)^{\frac{1}{2}}$$

for all $a \in A$.

Since $\tilde{\psi}_n(\tilde{\varphi}_n(1)) \le \tilde{\psi}_n(1) \le 1_N$ and $\tilde{\psi}_n(\tilde{\varphi}_n(1)) \to 1_N$ in the ultraweak topology, it follows that

$$b_n = 1_N - \tilde{\psi}_n(\tilde{\varphi}_n(1))$$

is a sequence of positive operators tending ultraweakly to zero. Let ρ_n be any sequence of states on $\mathbb{M}_{k(n)}(\mathbb{C})$ and define linear maps $\psi_n\colon \mathbb{M}_{k(n)}(\mathbb{C}) \to N$ by

$$\psi_n(T) = \rho_n(T) b_n + \tilde{\psi}_n(\tilde{\varphi}_n(1)^{\frac{1}{2}} T \tilde{\varphi}_n(1)^{\frac{1}{2}}).$$

Since the sum of positive operators is positive, it is easy to check that the ψ_n's are u.c.p. Since $b_n \to 0$, it follows that $\psi_n \circ \varphi_n \to \theta$ in the point-ultraweak topology. □

Proposition 2.2.8. *Let M and N be von Neumann algebras and $\theta\colon M \to N$ be a unital weakly nuclear map. Then, there exist normal u.c.p. maps $\varphi_n\colon M \to \mathbb{M}_{k(n)}(\mathbb{C})$ and $\psi_n\colon \mathbb{M}_{k(n)}(\mathbb{C}) \to N$ such that $\psi_n \circ \varphi_n \to \theta$ in the point-ultraweak topology.*

Proof. Fix a finite set $\mathfrak{F} \subset M$, a finite set of normal functionals $\chi \subset N_*$ and $\varepsilon > 0$. By the last proposition, we can find u.c.p. maps $\tilde{\varphi}\colon M \to \mathbb{M}_k(\mathbb{C})$ and $\psi\colon \mathbb{M}_k(\mathbb{C}) \to N$ such that

$$|\eta(\theta(m)) - \eta(\psi \circ \tilde{\varphi}(m))| < \varepsilon$$

for all $m \in \mathfrak{F}$ and $\eta \in \chi$. Of course ψ is automatically normal so we only have to replace $\tilde{\varphi}$ with a normal u.c.p. map.

By Corollary 1.6.3 we can find a net of normal u.c.p. maps $\varphi_\lambda\colon M \to \mathbb{M}_k(\mathbb{C})$ which converge to $\tilde{\varphi}$ in the point-norm topology. This completes the proof. □

2.3. Nuclear and exact C*-algebras

Definition 2.3.1. A C*-algebra A is *nuclear* if the identity map $\mathrm{id}_A\colon A \to A$ is nuclear.

Sometimes nuclear C*-algebras are called *amenable*, which comes from a cohomological characterization ([**42**], [**76**]); sometimes they are said to have the *completely positive approximation property* (CPAP), which refers to the existence of diagrams

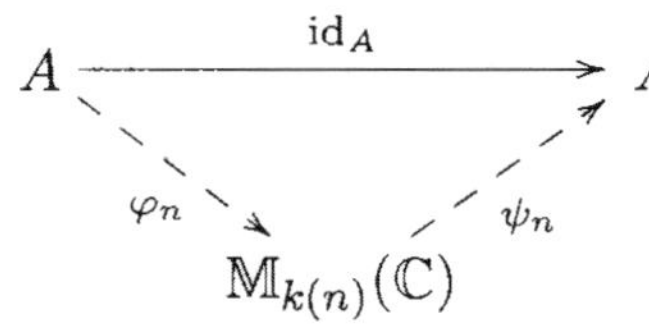

which asymptotically commute pointwise. Though never convenient when writing, it is often very helpful to draw diagrams as above. For example, Arveson's Extension Theorem gets used all the time and sketching diagrams helps one understand which maps can be extended, where the extensions are defined and where they take their values. (For example, Exercise 2.1.9 is completely transparent if one draws the right diagram.)

Definition 2.3.2. A C*-algebra A is *exact* if there exists a faithful representation $\pi\colon A \to \mathbb{B}(\mathcal{H})$ such that π is nuclear.

Diagrammatically, a C*-algebra is exact if there exist a faithful representation $\pi\colon A \to \mathbb{B}(\mathcal{H})$ and diagrams

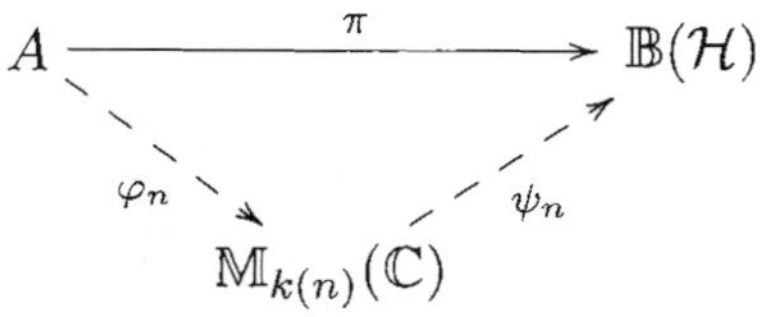

which asymptotically commute pointwise. Exact C*-algebras are sometimes called *nuclearly embeddable.* Note that a C*-algebra is exact if and only if there exists a nuclear embedding into some C*-algebra (since we could faithfully represent that C*-algebra to get a faithful nuclear representation).

Returning to the issue of dependence on the range, let $\pi\colon A \to \mathbb{B}(\mathcal{H})$ be a faithful representation; then, A is nuclear if and only if π is nuclear when regarded as taking values in $\pi(A)$, while A is exact if and only if π is nuclear when regarded as taking values in $\mathbb{B}(\mathcal{H})$. In particular, nuclearity implies exactness (the converse is false).

It turns out that the von Neumann algebra analogue of exactness is slightly tricky to formulate. If one gives the obvious adaptation of Definition 2.3.2, then Proposition 2.1.4 would imply that every von Neumann algebra enjoys this property (and hence this isn't the right W*-notion). We will study the proper analogue in Chapter 14. However, nuclearity is easily adapted to von Neumann algebras as follows.

Definition 2.3.3. A von Neumann algebra M is called *semidiscrete* if the identity map $\mathrm{id}_M\colon M \to M$ is weakly nuclear.

A fundamental theorem of Connes states that for separable factors the notion of semidiscreteness is equivalent to numerous other conditions including injectivity and hyperfiniteness. We will discuss this deep and important result later; it turns out to have important consequences in the theories of nuclear and exact C*-algebras (see Theorem 9.3.3). We have a more modest goal at the moment, namely, the observation that semidiscreteness of the double dual implies nuclearity of the algebra. The proof requires two simple, but very useful, lemmas. The first is a general functional analytic fact.

Lemma 2.3.4. *Let A be a Banach space, $\mathbb{B}(A)$ be the space of all bounded linear maps from A to A and $\mathcal{C} \subset \mathbb{B}(A)$ be any* convex *set. Then the point-weak and point-norm closures of $\mathcal{C}$ coincide.*

Proof. Let $T \in \mathbb{B}(A)$ be a map such that there exists a net $\{T_i\}_{i\in I} \subset \mathcal{C}$ with the property that for every $a \in A$ and functional $\eta \in A^*$,

$$\eta(T_i(a)) \to \eta(T(a)).^4$$

[4] In other words, assume T belongs to the point-weak closure of $\mathcal{C}$.

We must show that for each finite set $\mathfrak{F} = \{a_1, \ldots, a_k\} \subset A$ and $\varepsilon > 0$ there exists $S \in \mathcal{C}$ such that $\|S(a_j) - T(a_j)\| < \varepsilon$ for $1 \leq j \leq k$. This is a standard application of the Hahn-Banach Theorem since the net of maps

$$T_i \oplus \cdots \oplus T_i \ (k\text{-fold direct sum})$$

converges point-weakly to

$$T \oplus \cdots \oplus T$$

in the Banach space $\mathbb{B}(A \oplus \cdots \oplus A)$ (take your favorite ℓ^p-norm on $A \oplus \cdots \oplus A$). Thus the element

$$T(a_1) \oplus \cdots \oplus T(a_k)$$

belongs to the weak closure of

$$\{T_i(a_1) \oplus \cdots \oplus T_i(a_k)\}_{i \in I}$$

and hence, by the Hahn-Banach Theorem, also to the norm closure of the convex hull. This implies that we can find a single convex combination of the T_i's which is simultaneously close (in norm) to T on all of $\mathfrak{F}$. □

The second lemma is awkward to state if we don't introduce some terminology.

Definition 2.3.5. We will say a c.p. map $\theta \colon A \to A$ is *factorable* if there exist c.p. maps $\varphi \colon A \to \mathbb{M}_n(\mathbb{C})$ and $\psi \colon \mathbb{M}_n(\mathbb{C}) \to A$ such that $\theta = \psi \circ \varphi$. We call $(\varphi, \psi, \mathbb{M}_n(\mathbb{C}))$ a *factorization* of θ.

Note the absence of restrictions on the norms of φ and ψ (which will be handy later).

Lemma 2.3.6. *For any* C*-*algebra A the set of factorable maps $A \to A$ is convex.*[5]

Proof. Let $\theta_1, \theta_2 \colon A \to A$ be factorable maps and $0 < \lambda < 1$ be given. If factorizations $(\varphi_i, \psi_i, \mathbb{M}_{n(i)}(\mathbb{C}))$ of θ_i are given, then we claim that $\lambda\theta_1 + (1-\lambda)\theta_2$ factorizes through the finite-dimensional C*-algebra

$$\mathbb{M}_{n(1)}(\mathbb{C}) \oplus \mathbb{M}_{n(2)}(\mathbb{C}).$$

Indeed, factorizations are given by

$$\varphi_1 \oplus \varphi_2 \colon A \to \mathbb{M}_{n(1)}(\mathbb{C}) \oplus \mathbb{M}_{n(2)}(\mathbb{C})$$

and $\mathbb{M}_{n(1)}(\mathbb{C}) \oplus \mathbb{M}_{n(2)}(\mathbb{C}) \to A$,

$$T \oplus S \mapsto \lambda\psi_1(T) + (1-\lambda)\psi_2(S).$$

Since the set of positive operators is a cone, this last map is c.p. Now one completes the proof as in Exercise 2.1.2. □

[5] One could also consider factorable maps with values in a different C*-algebra, but not much changes.

Remark 2.3.7. The proof shows that the set of factorable maps which have *contractive* factorizations (i.e., $\theta = \psi \circ \varphi$ where both φ and ψ are contractive) is also convex.

Proposition 2.3.8. *If A^{**} is semidiscrete, then A is nuclear.*[6]

Proof. Assume first that A is unital and fix a finite set $\mathfrak{F} \subset A$, a finite set $\chi \subset A^* = (A^{**})_*$ and $\varepsilon > 0$. By the previous lemmas, it suffices to show that id_A belongs to the point-weak closure of the contractive factorable maps – i.e., there exist c.c.p. maps $\varphi\colon A \to \mathbb{M}_n(\mathbb{C})$ and $\psi\colon \mathbb{M}_n(\mathbb{C}) \to A$ such that

$$|\eta(\psi \circ \varphi(a)) - \eta(a)| < \epsilon$$

for all $a \in \mathfrak{F}$ and $\eta \in \chi$.

Since A^{**} is semidiscrete, we can apply Proposition 2.2.7 to find u.c.p. maps $\varphi\colon A^{**} \to \mathbb{M}_n(\mathbb{C})$ and $\psi'\colon \mathbb{M}_n(\mathbb{C}) \to A^{**}$ such that $|\eta(\psi' \circ \varphi(a)) - \eta(a)| < \epsilon$ for all $a \in \mathfrak{F}$ and $\eta \in \chi$. Of course, if ψ' happens to take values in A, then we are done – so this is what we'll arrange.

Using the duality between c.p. maps $\mathbb{M}_n(\mathbb{C}) \to A^{**}$ and positive elements in $\mathbb{M}_n(A^{**})$ (Proposition 1.5.12) and the fact that positive elements in $\mathbb{M}_n(A)$ are ultraweakly dense in the positive part of $\mathbb{M}_n(A^{**})$, it is not hard to find a net of c.p. maps $\psi_\lambda\colon \mathbb{M}_n(\mathbb{C}) \to A$ such that $\psi_\lambda \to \psi'$ in the point-ultraweak topology. Unfortunately the ψ_λ's need not be contractive. However, since φ and ψ' are unital maps, we do have that $\{\psi_\lambda(1_{\mathbb{M}_n(\mathbb{C})})\} \subset A$ is a net converging weakly to 1_A. Hence, going far enough out in the net and taking an appropriate convex combination, we can find a c.p. map $\psi''\colon \mathbb{M}_n(\mathbb{C}) \to A$ such that for all $a \in \mathfrak{F}$ and $\eta \in \chi$,

$$|\eta(\psi'' \circ \varphi(a)) - \eta(a)| < \varepsilon$$

and $\|\psi''(1_{\mathbb{M}_n(\mathbb{C})}) - 1_A\| < \varepsilon$. To get a contractive map, we now replace ψ'' by

$$\psi(T) = \frac{1}{\|\psi''(1_{\mathbb{M}_n(\mathbb{C})})\|}\psi''(T)$$

and one easily verifies that $\psi \circ \varphi$ is close to the identity on $\mathfrak{F}$ for the prescribed finite set of functionals.

In the nonunital case one first shows that a C*-algebra is nuclear if and only if its unitization is nuclear (cf. Exercise 2.3.4). Then one observes that $(\tilde{A})^{**} = A^{**} \oplus \mathbb{C}$ (in general, if $J \lhd B$ is an ideal, then $B^{**} \cong J^{**} \oplus (B/J)^{**}$), which easily implies that $(\tilde{A})^{**}$ is semidiscrete whenever A^{**} is semidiscrete. □

Here are a few more useful observations.

[6]The converse also holds, but it requires the equivalence of injectivity and semidiscreteness; see Theorem 9.3.3.

Exercises

Exercise 2.3.1 (Finite sets and ε's). Give the proper "local" formulations of nuclearity and exactness. What additional ingredient needs to be localized for semidiscreteness?

Exercise 2.3.2 (General subalgebras). It is easily seen that exactness passes to subalgebras, but the same is not true for nuclearity. Where does the proof break down?

Exercise 2.3.3 (Subalgebras with conditional expectations). If A is nuclear and $B \subset A$ is a C*-subalgebra such that there exists a conditional expectation $\Phi \colon A \to B$, then B is also nuclear. Formulate a similar result for semidiscreteness. In particular, prove that an arbitrary von Neumann subalgebra of a separable semidiscrete algebra of type II_1 is again semidiscrete. (Hint: See Lemma 1.5.11 for the II_1 case.)

Exercise 2.3.4 (Hereditary subalgebras and nuclearity). Prove that a hereditary subalgebra of a nuclear C*-algebra is again nuclear. In particular, nuclearity passes to ideals. (Hint: If e_n is an approximate unit of a hereditary subalgebra $B \subset A$, then the c.c.p. maps $\Phi_n \colon A \to B$ defined by $\Phi_n(a) = e_n a e_n$ have the property that $\|\Phi_n(b) - b\| \to 0$ for all $b \in B$.)

Exercise 2.3.5 (Unitizations). Show that a nonunital C*-algebra A is nuclear (resp. exact) if and only if the unitization $\tilde{A}$ is nuclear (resp. exact).

Exercise 2.3.6 (Direct sums). Prove that a finite direct sum $A_1 \oplus \cdots \oplus A_k$ is nuclear (resp. exact) if and only if each A_i is nuclear (resp. exact). It turns out that the ℓ^∞-direct sum of nuclear C*-algebras need not be exact. In fact,

$$\prod_{n \in \mathbb{N}} \mathbb{M}_n(\mathbb{C}) = \{(x_n) : x_n \in \mathbb{M}_n(\mathbb{C}), \sup_n \|x_n\| < \infty\}$$

is not exact.[7] What happens with the c_0-direct sum (i.e., sequences tending to zero in norm)?

Exercise 2.3.7 (Locally nuclear). Prove that a C*-algebra which is "locally nuclear" is nuclear. That is, if for each finite set $\mathfrak{F} \subset A$ and $\varepsilon > 0$ one can find a nuclear subalgebra $B \subset A$ such that B almost contains $\mathfrak{F}$, within ε in norm, then A is nuclear. In particular, the class of nuclear C*-algebras is closed under taking inductive limits with injective connecting maps.[8] (Hint: You'll need Arveson's Extension Theorem.)

---

[7]This follows from the fact that there exist nonexact residually finite-dimensional C*-algebras, such as the full group C*-algebra of a free group (see Theorem 7.4.1).

[8]Actually, injectivity of the connecting maps is not necessary, but then you need to know that nuclearity passes to quotients – a deep, hard fact. Exact C*-algebras are also closed under arbitrary inductive limits.

Exercise 2.3.8 (Separable versus nonseparable). Show that a C*-algebra is exact if and only if all of its separable subalgebras are exact. The nuclear case is trickier but uses a technique which can be useful in other contexts: A is nuclear if and only if each separable subalgebra $B \subset A$ is contained in a separable nuclear subalgebra $C \subset A$. (Hint for the "only if" part of the nuclear case: If $B \subset A$ is a separable subalgebra and a finite set $\mathfrak{F} \subset B$ and $\varepsilon > 0$ are given, then we can find c.c.p. maps $\varphi\colon A \to \mathbb{M}_n(\mathbb{C})$ and $\psi\colon \mathbb{M}_n(\mathbb{C}) \to A$ such that $\|x - \psi \circ \varphi(x)\| < \varepsilon$ for all $x \in \mathfrak{F}$. Letting B_1 be the C*-algebra generated by B and $\psi \circ \varphi(A)$, we have that B_1 satisfies a local form of nuclearity on the set $\mathfrak{F}$ and *it's still separable.* Hence we can take a larger finite set from B_1, a smaller $\varepsilon > 0$ and repeat the procedure. If done carefully, the norm closure of an increasing sequence of subalgebras constructed this way will be nuclear.)

One of the great advantages (and disadvantages) of the class of exact C*-algebras is that they are defined via an *external* approximation property (i.e., the approximating maps take values outside the algebra). This is a disadvantage when one wants to study their fine structure. However, external approximation has the advantage of being easier to verify in some situations. We already saw one example of this in Exercise 2.3.2; Exercise 2.3.10 below gives another example. We need a very useful preliminary fact.

Exercise 2.3.9 (Independence of representation). Assume that $A \subset \mathbb{B}(\mathcal{H})$ is a concretely represented exact C*-algebra. Show that there exist c.c.p. maps $\varphi_n\colon A \to \mathbb{M}_{k(n)}(\mathbb{C})$ and $\psi_n\colon \mathbb{M}_{k(n)}(\mathbb{C}) \to \mathbb{B}(\mathcal{H})$ such that $\|a - \psi_n \circ \varphi_n(a)\| \to 0$ for all $a \in A$. (Hint: Arveson's Extension Theorem.)

Exercise 2.3.10 (Externally locally exact). Assume $A \subset \mathbb{B}(\mathcal{H})$ is a concretely represented C*-algebra which is "externally locally exact" in the sense that for each finite set $\mathfrak{F} \subset A$ and $\varepsilon > 0$ there exists an exact C*-algebra $B \subset \mathbb{B}(\mathcal{H})$ which almost contains $\mathfrak{F}$, within ε in norm. Then A is also exact.

The exercise above fails badly for nuclear C*-algebras since subalgebras of nuclear C*-algebras need not be nuclear (see Remark 4.4.4 and Corollary 9.4.5).

Exercise 2.3.11. Assume $\varphi_n\colon A \to A_n$ and $\psi_n\colon A_n \to A$ are c.c.p. maps such that $\psi_n \circ \varphi_n \to \mathrm{id}_A$ in the point-norm topology. Prove that if each A_n is nuclear, then so is A.

Exercise 2.3.12. Assume $A \subset \mathbb{B}(\mathcal{H})$ and there exist c.c.p. maps $\varphi_n\colon A \to A_n$ and $\psi_n\colon A_n \to \mathbb{B}(\mathcal{H})$ such that $\psi_n \circ \varphi_n \to \mathrm{id}_A$ in the point-norm topology. Use Arveson's Extension Theorem to show that A is exact if each of the A_n's is exact.

Exercise 2.3.13. The proof of Proposition 2.3.8 goes through verbatim if one only assumes that the natural inclusion $A \hookrightarrow A^{**}$ is weakly nuclear. Do you agree?

Exercise 2.3.14 (Lance's WEP). A C*-algebra A is said to have the *weak expectation property* (WEP) if there exists a u.c.p. map $\Phi\colon \mathbb{B}(\mathcal{H}_u) \to A^{**}$ such that $\Phi(a) = a$ for all $a \in A$, where $A \subset A^{**} \subset \mathbb{B}(\mathcal{H}_u)$ is the universal representation of A. Prove that A is nuclear if and only if A is exact and has the WEP. (Hint: Use Arveson's Extension Theorem and point-ultraweak limits – Theorem 1.3.7 – to prove that nuclearity implies the WEP.)

Prove also that A has the WEP if and only if for any faithful representation $A \subset \mathbb{B}(\mathcal{H})$, there exists a u.c.p. map $\Phi\colon \mathbb{B}(\mathcal{H}) \to A''$ such that $\Phi(a) = a$, for all $a \in A$.

Exercise 2.3.15 (Semidiscrete implies injective). Prove that every semidiscrete von Neumann algebra is injective.

2.4. First examples

It turns out that most algebras built out of abelian and finite-dimensional C*-algebras will be nuclear. In K-theoretic terms, this already provides a huge class of examples. On the other hand, von Neumann algebras provide nice examples of nonexact (hence nonnuclear) algebras, so we discuss this as well.

Exercise 2.1.2 obviously implies that every finite-dimensional C*-algebra is nuclear. Since inductive limits of nuclear algebras are nuclear (Exercise 2.3.7), we have the following fact.

Proposition 2.4.1. *Approximately finite-dimensional (AF) algebras are nuclear.*

Proof. By definition, AF algebras are inductive limits of finite-dimensional C*-algebras. □

A slightly less trivial, though equally fundamental, example is that of abelian algebras.

Proposition 2.4.2. *Every abelian* C*-*algebra is nuclear.*

Proof. It suffices to prove this in the unital case – i.e., we may assume that $A = C(X)$ for some compact Hausdorff space X. Despite what topologists say, partitions of unity were invented specifically to prove this proposition.

If a finite set $\mathfrak{F} \subset A$ and $\varepsilon > 0$ are given, then we can find a finite open cover $\{U_1, \ldots, U_n\}$ of X with the property that for each $f \in \mathfrak{F}$ and $1 \leq i \leq n$

we have

$$|f(x) - f(y)| < \epsilon$$

for any pair of points $x, y \in U_i$. Let $y_i \in U_i$ be arbitrarily chosen and $\{\sigma_1, \ldots, \sigma_n\}$ be a partition of unity subordinate to the cover $\{U_1, \ldots, U_n\}$. Define $\varphi\colon A \to \mathbb{C}^n$ by $\varphi(f) = (f(y_1), \ldots, f(y_n))$. Evidently φ is a unital $*$-homomorphism, hence a u.c.p. map. We then define $\psi\colon \mathbb{C}^n \to A$ by

$$(d_1, \ldots, d_n) \mapsto \sum_{i=1}^{n} d_i \sigma_i.$$

It is not hard to see that the map ψ is positive and it is a general fact that if either the range or domain of a positive map is an abelian C*-algebra, then the map is automatically c.p. ([**141**, Theorem 3.9 and Theorem 3.11]).[9] By Exercise 2.1.2, we are left to estimate $\|f - \psi \circ \phi(f)\|$:

$$\|f - \psi \circ \phi(f)\| = \|(\sum_{i=1}^{n} \sigma_i)f - \sum_{i=1}^{n} f(y_i)\sigma_i\| = \|\sum_{i=1}^{n}(f - f(y_i)1)\sigma_i\| \leq \epsilon$$

for every $f \in \mathfrak{F}$. □

Remark 2.4.3. Nuclearity is sometimes regarded as the noncommutative analogue of having a partition of unity. Though the analogy is not perfect, it has substance and the proof above explains this point of view.

Corollary 2.4.4. *For every locally compact Hausdorff space X and natural number $n \in \mathbb{N}$, $\mathbb{M}_n(C_0(X))$ is nuclear.*

Proof. More generally, it is easily seen that if A is nuclear, then so is $\mathbb{M}_n(A)$. □

Corollary 2.4.5. *Every approximately homogeneous (AH) algebra is nuclear.*

Proof. By definition, an AH algebra is an inductive limit, with injective connecting maps, of algebras A_k, where each A_k is a finite direct sum of algebras of the form

$$P\mathbb{M}_n(C_0(X))P,$$

[9] In our setting the proof is straightforward. We identify $\mathbb{M}_k(\mathbb{C}^n)$ with $\mathbb{M}_k(\mathbb{C}) \oplus \cdots \oplus \mathbb{M}_k(\mathbb{C})$ (direct sum n times) and it is clear that

$$\psi_k(T_1 \oplus \cdots \oplus T_k) = \sum_{l=1}^{n} T_l \otimes \sigma_l,$$

where $T \otimes \sigma \in \mathbb{M}_k(C(X)) \cong C(X, \mathbb{M}_k(\mathbb{C}))$ is the matrix-valued function $x \mapsto \sigma(x)T$. Evidently this implies that ψ is completely positive.

and where $P \in \mathbb{M}_n(C_0(X))$ is a projection. Hence various exercises imply this corollary.[10] □

Remark 2.4.6. It follows from the corollary above and some work of Elliott and Gong that for any countable dimension group G_0 and any other countable abelian group G_1 we can find an AH algebra A such that $\mathrm{K}_0(A) \cong G_0$ and $\mathrm{K}_1(A) \cong G_1$ (cf. [**65**]). In fact their result is much sharper, but our point is that we have shown the existence of a large class of nuclear C*-algebras.

As might be expected, it turns out our C*-notions impose tremendous restrictions on von Neumann algebras. Since we haven't yet demonstrated the existence of nonexact C*-algebras, a tiny bit of faith will be required of the reader.

Lemma 2.4.7. *Let $k(n)$ be a sequence of integers tending to infinity. Then the von Neumann algebra*

$$M = \prod_{n\in\mathbb{N}} \mathbb{M}_{k(n)}(\mathbb{C}) = \{(x_n) : x_n \in \mathbb{M}_{k(n)}(\mathbb{C}), \sup_n \|x_n\| < \infty\}$$

is not exact.

Proof. Eventually we'll prove that there exists a separable nonexact C*-algebra A which has a sequence of representations $\pi_n \colon A \to \mathbb{M}_{j(n)}(\mathbb{C})$ such that $\bigoplus_n \pi_n$ is faithful; see Corollary 3.7.12 and Theorem 7.4.1. With this in hand, the remainder of the proof is simple. Indeed, the fact that $k(n) \to \infty$ ensures that there is enough room to embed A into M by putting a separating family of finite-dimensional representations into the tail of M (i.e., for a given representation $\pi \colon A \to \mathbb{M}_j(\mathbb{C})$ we simply take a nonunital embedding $\mathbb{M}_j(\mathbb{C}) \subset \mathbb{M}_{k(n)}(\mathbb{C})$, for sufficiently large n). Hence M contains a nonexact subalgebra and thus is not exact (since exactness passes to subalgebras). □

Recall that a projection $p \in M$ is called *abelian* if pMp is an abelian algebra. It is a fact that no von Neumann algebra of type II or III contains an abelian projection (this is required in the proof of the type decomposition explained in Section 1.3).

Lemma 2.4.8. *Assume a von Neumann algebra M has no abelian projections. Then for every $n \in \mathbb{N}$ there is an embedding $\mathbb{M}_n(\mathbb{C}) \hookrightarrow M$.*

Proof. It suffices to show the existence of pairwise orthogonal projections $p_1, \ldots, p_n \in M$ with the property that there exist partial isometries $v_1, \ldots, v_n$ such that $v_i^* v_i = p_1$ and $v_i v_i^* = p_i$ for $i = 1, \ldots, n$ (i.e., $p_i \sim p_j$ for all i, j); defining $e_{i,j} = v_i v_j^*$, we get a system of matrix units and hence $C^*(\{e_{i,j} : 1 \le i, j \le n\}) \cong \mathbb{M}_n(\mathbb{C})$.

---

[10] In case one hasn't studied these things before, $P\mathbb{M}_n(C_0(X))P$ is a hereditary subalgebra of $\mathbb{M}_n(C_0(X))$.

Actually, it suffices to show that we can always find two orthogonal projections p_1, p_2 which are Murray-von Neumann equivalent. Suppose we could prove this and p_1, p_2 were such projections. Since M has no abelian projections, the same is true of $p_2 M p_2$ and hence we would be able to find equivalent orthogonal projections $q_1, q_2 \leq p_2$. Transporting the q_i's back underneath p_1, we would thus have four equivalent orthogonal projections. Applying the procedure again underneath q_2 would give us eight such projections, and so on.

So, how can we construct two orthogonal equivalent projections? Simply take a noncentral projection $p \in M$ and find some $m \in M$ such that $pm(1-p) \neq 0$. The partial isometry in the polar decomposition of this operator will have orthogonal support and range projections. □

Proposition 2.4.9. *Let M be a von Neumann algebra. Then the following are equivalent:*

(1) *M is a finite direct sum of finite homogeneous von Neumann algebras (that is,*

$$M \cong \mathbb{M}_{k(1)}(\mathcal{A}_1) \oplus \cdots \oplus \mathbb{M}_{k(j)}(\mathcal{A}_j)$$

where each $\mathcal{A}_i$ is an abelian von Neumann algebra);

(2) *M is a nuclear* C*-*algebra;*

(3) *M is an exact* C*-*algebra.*

Proof. Thanks to Corollary 2.4.4 and the fact that every nuclear C*-algebra is exact, we only have to prove the implication (3) ⇒ (1).

Assume first that M is not type I – i.e., has a summand N without abelian projections. Since N is infinite dimensional, we can find a pairwise orthogonal sequence of projections $p_1, p_2, \ldots$ in N. None of the corners $p_i N p_i$ contains an abelian projection (since N doesn't) and hence the previous lemma provides an embedding $\mathbb{M}_i(\mathbb{C}) \hookrightarrow p_i N p_i$. Evidently this yields an embedding

$$\prod_n \mathbb{M}_n(\mathbb{C}) \hookrightarrow \prod_n p_n N p_n \hookrightarrow N \subset M.$$

Since exactness passes to subalgebras, and we've seen that $\prod_n \mathbb{M}_n(\mathbb{C})$ is not exact, it follows that M can't be exact.

Hence we've reduced to the type I case: $M = \prod_{i \in I} \mathcal{A}_i \bar{\otimes} \mathbb{B}(\mathcal{H}_i)$. Since $\mathbb{B}(\mathcal{H})$, for an infinite-dimensional Hilbert space $\mathcal{H}$, also contains a copy of $\prod_n \mathbb{M}_n(\mathbb{C})$, it follows that each cardinal number i must be a natural number. Similar reasoning shows that only a finite number of the remaining $\mathcal{A}_i$'s can be nonzero, so the proof is complete. □

2.5. C*-algebras associated to discrete groups

This section contains a bare-bones introduction to an important class of C*-algebras. There is nothing about nuclearity or exactness until the next section, to which you should proceed if already familiar with the basics of group C*-algebras.

For a discrete group Γ we let $\lambda\colon \Gamma \to \mathbb{B}(\ell^2(\Gamma))$ denote the *left regular representation*: $\lambda_s(\delta_t) = \delta_{st}$ for all $s, t \in \Gamma$, where $\{\delta_t : t \in \Gamma\} \subset \ell^2(\Gamma)$ is the canonical orthonormal basis. There is also a *right regular representation* $\rho\colon \Gamma \to \mathbb{B}(\ell^2(\Gamma))$, defined by $\rho_s(\delta_t) = \delta_{ts^{-1}}$. Note that λ and ρ are unitarily equivalent; the intertwining unitary is defined by $U\delta_t = \delta_{t^{-1}}$.

We denote the *group ring* of Γ by $\mathbb{C}[\Gamma]$. By definition, it is the set of formal sums

$$\sum_{s\in\Gamma} a_s s,$$

where only finitely many of the scalar coefficients $a_s \in \mathbb{C}$ are nonzero, and multiplication is defined by

$$(\sum_{s\in\Gamma} a_s s)(\sum_{t\in\Gamma} a_t t) = \sum_{s,t\in\Gamma} a_s a_t st.$$

The group ring $\mathbb{C}[\Gamma]$ acquires an involution by declaring $(\sum_{s\in\Gamma} a_s s)^* = \sum_{s\in\Gamma} \overline{a_s} s^{-1}$. Note that the left regular representation can be extended to an injective $*$-homomorphism $\mathbb{C}[\Gamma] \to \mathbb{B}(\ell^2(\Gamma))$, which we also denote by λ. Evidently, there is a one-to-one correspondence between unitary representations of Γ and $*$-representations of $\mathbb{C}[\Gamma]$.

Both amenable and exact groups are defined in terms of their canonical actions on $\ell^\infty(\Gamma)$. For $f \in \ell^\infty(\Gamma)$ and $s \in \Gamma$ we let $s.f \in \ell^\infty(\Gamma)$ be the function $s.f(t) = f(s^{-1}t)$; simple calculations show that $f \mapsto s.f$ defines a group action of Γ on $\ell^\infty(\Gamma)$. An important fact is that this action is *spatially implemented by the left regular representation.* That is, if we regard $\ell^\infty(\Gamma) \subset \mathbb{B}(\ell^2(\Gamma))$ as multiplication operators (i.e., $f\delta_t = f(t)\delta_t$), then a calculation shows

$$\lambda_s f \lambda_s^* = s.f$$

for all $f \in \ell^\infty(\Gamma)$ and $s \in \Gamma$.

The *reduced* C*-algebra of Γ, denoted $C^*_\lambda(\Gamma)$,[11] is the completion of $\mathbb{C}[\Gamma]$ with respect to the norm

$$\|x\|_r = \|\lambda(x)\|_{\mathbb{B}(\ell^2(\Gamma))}.$$

Though isomorphic to $C^*_\lambda(\Gamma)$, it is sometimes useful to consider $C^*_\rho(\Gamma)$, which is just the closure of $\mathbb{C}[\Gamma]$ in the right regular representation.

[11] You will also see $C^*_r(\Gamma)$ in the literature.

Example 2.5.1. If $\Gamma = \mathbb{Z}$, then $C^*_\lambda(\Gamma) = C(\mathbb{T})$, the continuous functions on the circle. Indeed, the Fourier transform identifies $\ell^2(\mathbb{Z})$ with $L^2(\mathbb{T}, \text{Lebesgue})$ and one checks that this unitary takes $C^*_\lambda(\mathbb{Z})$ to (continuous) multiplication operators. More generally, for every abelian group Γ, Pontryagin duality gives an identification of $C^*_\lambda(\Gamma)$ with $C(\hat{\Gamma})$, the continuous functions on the dual group.

The *full* (or *universal*) *group* C*-algebra of Γ, denoted $C^*(\Gamma)$, is the completion of $\mathbb{C}[\Gamma]$ with respect to the norm

$$\|x\|_u = \sup_\pi \|\pi(x)\|,$$

where the supremum is taken over all (cyclic) $*$-representations $\pi\colon \mathbb{C}[\Gamma] \to \mathbb{B}(\mathcal{H})$. (Note that, since unitaries have norm one, the supremum is finite.) Evidently $C^*(\Gamma)$ enjoys the following *universal property.*

Proposition 2.5.2. *Let $u\colon \Gamma \to \mathbb{B}(\mathcal{H})$ be any unitary representation of Γ. Then, there is a unique $*$-homomorphism $\pi_u\colon C^*(\Gamma) \to \mathbb{B}(\mathcal{H})$ such that $\pi_u(s) = u_s$ for all $s \in \Gamma$.*

In particular, notice that $C^*(\Gamma)$ always has a character – i.e., a one-dimensional representation – coming from the trivial representation $\Gamma \to \mathbb{C} = \mathbb{B}(\mathbb{C})$. (This also defines a tracial state on $C^*(\Gamma)$ such that $s \mapsto 1$ for all $s \in \Gamma$.) The reduced group C*-algebra $C^*_\lambda(\Gamma) \subset \mathbb{B}(\ell^2(\Gamma))$ always has a faithful trace (which is more important than the trivial trace on $C^*(\Gamma)$ described above).

Proposition 2.5.3. *The vector state $x \mapsto \langle x\delta_e, \delta_e\rangle$ defines a faithful tracial state on $C^*_\lambda(\Gamma)$.*

Proof. A simple calculation shows this state to be tracial.

Clearly ρ_s commutes with every operator in $C^*_\lambda(\Gamma)$ (since this is easily seen on the generators $\lambda_g \in C^*_\lambda(\Gamma)$). It follows that δ_e is a *separating* vector, meaning that $x\delta_e = y\delta_e$ if and only if $x = y$ (for all $x, y \in C^*_\lambda(\Gamma)$). Indeed, if $x\delta_e = y\delta_e$, then

$$x\delta_s = \rho_s^* x\delta_e = \rho_s^* y\delta_e = y\delta_s,$$

for all $s \in \Gamma$. Since such vectors span $\ell^2(\Gamma)$, it follows that $x = y$. With this observation, faithfulness is simple: If $0 \le x \in C^*_\lambda(\Gamma)$ and $0 = \langle x\delta_e, \delta_e\rangle$, then $0 = \|x^{1/2}\delta_e\|$ and this implies $x^{1/2} = 0$. Thus $x = 0$ too. □

The *group von Neumann algebra of* Γ is defined to be

$$L(\Gamma) = C^*_\lambda(\Gamma)'' \subset \mathbb{B}(\ell^2(\Gamma)).$$

Though we won't give the proof until Chapter 6, a fundamental theorem of Murray and von Neumann states that $L(\Gamma)$ is the commutant of the right regular representation $\rho\colon \Gamma \to \mathbb{B}(\ell^2(\Gamma))$ – i.e., $L(\Gamma) = \rho(\mathbb{C}[\Gamma])'$ and $L(\Gamma)' = \rho(\mathbb{C}[\Gamma])''$ (see Theorem 6.1.4).

Another way of describing $L(\Gamma)$ is as the set of $T \in \mathbb{B}(\ell^2(\Gamma))$ such that *T is constant down the diagonals* – meaning that for every $s, t, x, y \in \Gamma$ such that $ts^{-1} = yx^{-1}$, we have $\langle T\delta_s, \delta_t\rangle = \langle T\delta_x, \delta_y\rangle$.[12] A simple calculation shows that every unitary $\lambda_s \in \mathbb{B}(\ell^2(\Gamma))$ is constant down all diagonals; hence any finite linear combination has this property; thus anything in the weak closure $C^*_\lambda(\Gamma)'' = L(\Gamma)$ does too. The converse, that every such operator is a weak limit of something in $C^*_\lambda(\Gamma)$, uses the bicommutant theorem. Indeed, assume $T \in \mathbb{B}(\ell^2(\Gamma))$ and assume there exist scalars $\{\alpha_s\}_{s\in\Gamma} \subset \mathbb{C}$ such that $\langle T\delta_g, \delta_h\rangle = \alpha_{hg^{-1}}$ for all $g, h \in \Gamma$. A simple calculation shows $\langle T\rho_s\delta_g, \delta_h\rangle = \langle \rho_s T\delta_g, \delta_h\rangle$, for all $s \in \Gamma$, and hence $T \in \rho(\mathbb{C}[\Gamma])' = L(\Gamma)$.

Here is a useful consequence of these remarks.

Proposition 2.5.4. *Assume Γ is countably infinite and $\pi\colon C^*_\lambda(\Gamma) \to \mathbb{B}(\mathcal{H})$ is a representation on a separable Hilbert space $\mathcal{H}$. Then $\iota \oplus \pi$ is approximately unitarily equivalent to ι, where $\iota\colon C^*_\lambda(\Gamma) \to \mathbb{B}(\ell^2(\Gamma))$ is the canonical inclusion. In particular, every state on $C^*_\lambda(\Gamma)$ can be approximated by vector states coming from $\ell^2(\Gamma)$.*

Proof. To invoke Voiculescu's Theorem (version 1.7.3), we only need to know that $C^*_\lambda(\Gamma) \cap \mathbb{K}(\ell^2(\Gamma)) = \{0\}$. In fact, this is even true for $L(\Gamma)$ since no nonzero compact operator can be constant down the diagonals. (The details are left to the reader.) □

Another important representation-theoretic fact is Fell's absorbtion principle. Roughly, it states that the left regular representation absorbs all other representations tensorially.[13]

Theorem 2.5.5 (Fell's absorption principle). *Let π be a unitary representation of Γ on $\mathcal{H}$. Then, $\lambda \otimes \pi$ is unitarily equivalent to $\lambda \otimes 1_{\mathcal{H}}$.*

The proof amounts to writing down the proper unitary and doing some calculations. We'll do the first part: Define a unitary U on $\ell^2(\Gamma) \otimes \mathcal{H}$ by

$$\sum_t \delta_t \otimes \xi_t \mapsto \sum_t \delta_t \otimes \pi(t)\xi_t.$$

In certain situations it is important to know about *positive definite functions* on Γ.

---

[12] "What? Why is that 'constant down the diagonals'?" you may wonder. Well, if $\Gamma = \mathbb{Z}$ and you write down the matrix of such an operator (with respect to the canonical basis), you'll see that it really is constant down the diagonals.

[13] Fell's principle and the proof of Theorem 2.5.11 depend on the basics of tensor products (found in the first three sections of Chapter 3).

Definition 2.5.6. A function $\varphi\colon \Gamma \to \mathbb{C}$ is said to be *positive definite* if the matrix

$$[\varphi(s^{-1}t)]_{s,t\in F} \in \mathbb{M}_F(\mathbb{C})$$

is positive for every finite set $F \subset \Gamma$.

Fix a positive definite function φ and let $C_c(\Gamma)$ be the finitely supported functions on Γ. Define a sesquilinear form $C_c(\Gamma) \times C_c(\Gamma) \to \mathbb{C}$ by

$$\langle f, g\rangle_\varphi = \sum_{s,t\in\Gamma} \varphi(s^{-1}t)f(t)\overline{g(s)}.$$

This form is positive semidefinite. Indeed, if $f \in C_c(\Gamma)$ has support F, then

$$\langle f, f\rangle_\varphi = \sum_{s,t\in\Gamma} \varphi(s^{-1}t)f(t)\overline{f(s)} = \big\langle [\varphi(s^{-1}t)]_{s,t\in F}(f), (f)\big\rangle,$$

where the inner product on the right is the standard one on $\ell^2(F)$. Since φ is positive definite, $\langle f, f\rangle_\varphi \geq 0$ as asserted. Hence we can mod out by the zero elements and complete to get a Hilbert space $\ell^2_\varphi(\Gamma)$. For $f \in C_c(\Gamma)$ we let $\hat{f} \in \ell^2_\varphi(\Gamma)$ denote its natural image. Here's a GNS construction for the present context.

Definition 2.5.7. If φ is a positive definite function on Γ, then $\lambda^\varphi\colon \Gamma \to \mathbb{B}(\ell^2_\varphi(\Gamma))$ is the unitary representation given by $\lambda^\varphi_s(\hat{f}) = \widehat{s.f}$, where $s.f(t) = f(s^{-1}t)$, for all $t \in \Gamma$.[14]

Note that

$$\langle \lambda^\varphi_s \hat{\delta}_e, \hat{\delta}_e\rangle_\varphi = \langle \hat{\delta}_s, \hat{\delta}_e\rangle_\varphi = \varphi(s),$$

for all $s \in \Gamma$, and hence we recover φ from the vector functional $\langle \cdot\, \hat{\delta}_e, \hat{\delta}_e\rangle$.

Perhaps the construction of $\ell^2_\varphi(\Gamma)$ seems familiar? It should. Suppose φ is a positive linear functional on $C^*(\Gamma)$. Then $s \mapsto \varphi(s)$ is a positive definite function on Γ: for $s_1, \ldots, s_n \in \Gamma$ we have

$$[\varphi(s_i^{-1}s_j)]_{i,j} = (\mathrm{id}_n \otimes \varphi)\left(\begin{bmatrix} s_1 & s_2 & \cdots & s_n \\ 0 & 0 & \cdots & 0 \\ \vdots & \vdots & \ddots & \vdots \\ 0 & 0 & \cdots & 0 \end{bmatrix}^* \begin{bmatrix} s_1 & s_2 & \cdots & s_n \\ 0 & 0 & \cdots & 0 \\ \vdots & \vdots & \ddots & \vdots \\ 0 & 0 & \cdots & 0 \end{bmatrix} \right),$$

which is positive since φ is a c.p. map. It is a simple exercise to show that the GNS space of $C^*(\Gamma)$ with respect to φ is nothing but $\ell^2_\varphi(\Gamma)$.

Why then have we introduced positive definite functions? It is often necessary to work only with Γ, instead of some bloated C*-algebra. For example, here is a useful application.

---

[14] It isn't hard, just tedious, to check that this is really a unitary representation.

Proposition 2.5.8. *Let $\Lambda \subset \Gamma$ be a subgroup. There is a canonical inclusion*

$$C^*(\Lambda) \subset C^*(\Gamma).$$

Proof. By universality, there is a canonical $*$-homomorphism $\pi\colon C^*(\Lambda) \to C^*(\Gamma)$. Our task is to show it's injective.

We may assume Λ is countable. Then $C^*(\Lambda)$ has a faithful state φ which we think of as a positive definite function on Λ. Now extend φ to all of Γ by defining $\varphi(s) = 0$, for all $s \notin \Lambda$. Our proof will be complete once we see why this extension is positive definite on Γ (since the GNS representation of $C^*(\Lambda)$ with respect to φ will be a subrepresentation of π composed with the GNS representation of $C^*(\Gamma)$ with respect to the extension of φ to Γ).

The key observation is that $[\varphi(s^{-1}t)]_{s,t\in\mathfrak{F}}$ is block diagonal with respect to the left coset decomposition – i.e., if $s, t \in \Gamma$ and belong to different cosets ($\Leftrightarrow s^{-1}t \notin \Lambda$), then $\varphi(s^{-1}t) = 0$ – for every finite set $\mathfrak{F} \subset \Gamma$. Since a block diagonal matrix is positive if and only if every block is positive, we may assume that $F \subset g\Lambda$ for some fixed $g \in \Gamma$. Checking positive definiteness is now trivial. □

An analogous result holds in the reduced case too. The key point is that right cosets give a direct sum decomposition

$$\ell^2(\Gamma) \cong \bigoplus \ell^2(\Lambda g)$$

and hence the left regular representation of Γ, when restricted to Λ, is a multiple of the left regular representation of Λ (multiplicity equals the number of cosets). This implies

Proposition 2.5.9. *If $\Lambda \subset \Gamma$ is a subgroup, then $C^*_\lambda(\Lambda) \subset C^*_\lambda(\Gamma)$ canonically.*

We will need one more important fact about positive definite functions: they naturally give rise to completely positive maps at the C*-level. First a bit more notation.

Definition 2.5.10. Let $\varphi\colon \Gamma \to \mathbb{C}$ be a function. We define a corresponding linear functional $\omega_\varphi\colon \mathbb{C}[\Gamma] \to \mathbb{C}$ by

$$\omega_\varphi(\sum_{t\in\Gamma} \alpha_t t) = \sum_{t\in\Gamma} \varphi(t)\alpha_t$$

and *multiplier* $m_\varphi\colon \mathbb{C}[\Gamma] \to \mathbb{C}[\Gamma]$ by

$$m_\varphi(\sum_{t\in\Gamma} \alpha_t t) = \sum_{t\in\Gamma} \varphi(t)\alpha_t t.$$

Theorem 2.5.11. *Let $\varphi\colon \Gamma \to \mathbb{C}$ be a function with $\varphi(e) = 1$. The following are equivalent:*

(1) *the function φ is positive definite;*

(2) *there exists a unitary representation λ_φ of Γ on a Hilbert space $\mathcal{H}_\varphi$ and a unit vector ξ_φ such that*

$$\varphi(s) = \langle \lambda_\varphi(s)\xi_\varphi, \xi_\varphi \rangle;$$

(3) *the functional ω_φ extends to a state on $C^*(\Gamma)$;*

(4) *the multiplier m_φ extends to a u.c.p. map on either $C^*(\Gamma)$ or $C^*_\lambda(\Gamma)$, or extends to a normal u.c.p. map on $L(\Gamma)$.*

Proof. (1) $\Rightarrow$ (2): This follows from Definition 2.5.7.

(2) $\Rightarrow$ (3): Trivial.

(3) $\Rightarrow$ (4): First we handle the von Neumann algebra case. We can identify ω_φ with a vector state in the universal representation $C^*(\Gamma) \subset \mathbb{B}(\mathcal{H})$ (i.e., the direct sum of all GNS representations). By Fell's absorption principle, there is a unitary operator which conjugates $C^*_\lambda(\Gamma) \otimes 1$ onto the "diagonal" subalgebra of $C^*_\lambda(\Gamma) \otimes C^*(\Gamma)$; that is, the mapping $\sigma\colon C^*_\lambda(\Gamma) \to C^*_\lambda(\Gamma) \otimes C^*(\Gamma)$ defined by

$$\sum_t \alpha_t \lambda_t \mapsto \sum_t \alpha_t(\lambda_t \otimes t)$$

is a $*$-homomorphism and it extends to a normal $*$-homomorphism (also denoted σ) from $L(\Gamma)$ into $L(\Gamma) \bar{\otimes} \mathbb{B}(\mathcal{H})$ (since Fell's principle is spatially implemented). Notice that m_φ coincides with the continuous u.c.p. map

$$(\mathrm{id}_{L(\Gamma)} \otimes \omega_\varphi) \circ \sigma\colon L(\Gamma) \to L(\Gamma),$$

and this completes the von Neumann case (which evidently implies the reduced C*-algebra case as well).

For $C^*(\Gamma)$ we consider the diagonal map $C^*(\Gamma) \to C^*(\Gamma) \otimes C^*(\Gamma)$, $s \mapsto s \otimes s$, and repeat the argument above.

(4) $\Rightarrow$ (1): If m_φ is u.c.p., then for any finite sequence $s_1, \ldots, s_n \in \Gamma$,

$$[\varphi(s_i^{-1}s_j)]_{ij} = \mathrm{diag}(s_1, \ldots, s_n)[m_\varphi(s_i^{-1}s_j)]_{ij}\,\mathrm{diag}(s_1^{-1}, \ldots, s_n^{-1})$$

is positive since $[s_i^{-1}s_j]_{ij} \in \mathbb{M}_n(\mathbb{C}(\Gamma))$ is positive. □

The proof of the following corollary amounts to checking whether or not a particular function is positive definite, which we leave to you.

Corollary 2.5.12. *If $\Lambda \subset \Gamma$ is an inclusion of groups and $C^*_\lambda(\Lambda) \subset C^*_\lambda(\Gamma)$ the resulting inclusion of C*-algebras, then there is a conditional expectation $E^\Gamma_\Lambda\colon C^*_\lambda(\Gamma) \to C^*_\lambda(\Lambda)$ obtained by "throwing away" all group elements outside*

Λ. That is, $E^{\Gamma}_{\Lambda}(\lambda(s)) = \chi_{\Lambda}(s)\lambda(s)$ for $s \in \Gamma$, where χ_{Λ} is the characteristic function of Λ. The same result holds for universal C**-algebras.*

2.6. Amenable groups

Amenable groups admit approximately $10^{10^{10}}$ different characterizations; our goal in this section is to present a few that can be proved without too much effort.[15]

Definition 2.6.1. A group Γ is *amenable* if there exists a state μ on $\ell^{\infty}(\Gamma)$ which is invariant under the left translation action: for all $s \in \Gamma$ and $f \in \ell^{\infty}(\Gamma)$, $\mu(s.f) = \mu(f)$.

Such a state μ is called an *invariant mean.*

Definition 2.6.2. For a discrete group Γ, we let $\mathrm{Prob}(\Gamma)$ be the space of all probability measures on Γ:

$$\mathrm{Prob}(\Gamma) = \{\mu \in \ell^1(\Gamma) : \mu \geq 0 \text{ and } \sum_{t\in\Gamma} \mu(t) = 1\}.$$

Note that the left translation action of Γ on $\ell^{\infty}(\Gamma)$ leaves the subspace $\mathrm{Prob}(\Gamma)$ invariant; hence we can also use $\mu \mapsto s.\mu$ to denote the canonical action of Γ on $\mathrm{Prob}(\Gamma)$.

Definition 2.6.3. We say Γ has an *approximate invariant mean* if for any finite subset $E \subset \Gamma$ and $\varepsilon > 0$, there exists $\mu \in \mathrm{Prob}(\Gamma)$ such that

$$\max_{s\in E} \|s.\mu - \mu\|_1 < \varepsilon.$$

Recall that the *symmetric difference* of two sets E and F, denoted $E\triangle F$, is $E \cup F \setminus E \cap F$.

Definition 2.6.4. We say Γ satisfies the *Følner condition* if for any finite subset $E \subset \Gamma$ and $\varepsilon > 0$, there exists a finite subset $F \subset \Gamma$ such that

$$\max_{s\in E} \frac{|sF \triangle F|}{|F|} < \varepsilon,$$

where $sF = \{st : t \in F\}$.[16] A sequence of finite sets $F_n \subset \Gamma$ such that

$$\frac{|sF_n \triangle F_n|}{|F_n|} \to 0$$

for every $s \in \Gamma$ is called a *Følner sequence.*

---

[15]We'll see several more in later chapters.

[16]Since $sF \triangle F = [sF \setminus (sF \cap F)] \cup [F \setminus (sF \cap F)]$, it follows that $\frac{|sF\triangle F|}{|F|} = 2 - 2\frac{|F\cap sF|}{|F|}$. Hence the Følner condition is equivalent to requiring $\max_{s\in E} \frac{|sF\cap F|}{|F|} > 1 - \varepsilon/2$, which is often how it gets used in our context.

Note that this implies the existence of an approximate invariant mean given by normalized characteristic functions. Indeed, if χ_F is the characteristic function over F, then $\frac{1}{|F|}\chi_F \in \text{Prob}(\Gamma)$ and a computation confirms that

$$\|s.(\frac{1}{|F|}\chi_F) - \frac{1}{|F|}\chi_F\|_1 = \frac{|sF \triangle F|}{|F|}.$$

It turn out that all the definitions above give rise to the same class of groups. Before the proof, however, a few examples might be nice.

Example 2.6.5 (Elementary amenable groups). It is not hard to see that finite groups are amenable (take the state which maps $\chi_{\{s\}}$ to $1/|\Gamma|$, for each group element). So are abelian groups, as the Markov-Kakutani fixed point theorem easily implies. (There is an alternate proof below.) It is also true that the class of amenable groups is closed under taking subgroups, extensions, quotients and inductive limits. (These all make excellent exercises.) Hence anything built out of finite or abelian groups, using the four operations above, is also amenable; by definition, these are the *elementary amenable groups*. In particular, all solvable (hence all nilpotent) groups are amenable.

Example 2.6.6 (Groups with subexponential growth). A group Γ is said to have *subexponential growth* if $\limsup |E^n|^{1/n} = 1$ for every finite subset $E \subset \Gamma$. ($E^n = \{g_1 g_2 \cdots g_n : g_i \in E\}$.) It is clear that if a particular finite set E satisfies the above condition, then every subset $F \subset E^n$ will too. Hence if Γ is generated by a finite subset $E \subset \Gamma$ as a semigroup, then it suffices to check the growth condition only for E.

Such groups are amenable. To see this, we construct an increasing sequence $E_0 \subset E_1 \subset E_2 \subset \cdots$ of finite subsets of G, whose union equals Γ, such that $E_n^{-1} = E_n$, $E_m E_n \subset E_{m+n}$, and $\liminf |E_n|^{1/n} = 1$. (Start with any finite set, keep throwing in group elements, and then take higher powers as in the definition of subexponential growth.) It turns out that some subsequence of $\{E_n\}$ must be a Følner sequence. Indeed, for any $g \in E_k$, we have $gE_{n-k} \subset E_n$, and thus $|gE_n \cap E_n| \geq |gE_{n-k}| = |E_{n-k}|$. The proof of the ratio test, from elementary calculus, contains the following general fact:

$$\liminf_{n\to\infty} \frac{a_n}{a_{n-k}} \leq \liminf_{n\to\infty} a_n^{k/n},$$

for $a_n \geq 0$ and any fixed $k \in \mathbb{N}$. Applying the reciprocal of this inequality, we have

$$\limsup_{n\to\infty} \frac{|gE_n \cap E_n|}{|E_n|} \geq \limsup_{n\to\infty} \frac{|E_{n-k}|}{|E_n|} \geq \limsup_{n\to\infty} \frac{1}{|E_n|^{k/n}} = 1.$$

It is a fun combinatorial exercise to show all abelian groups have subexponential growth.

Here is the simplest example of something nonamenable.

Example 2.6.7 (Nonabelian free groups). The free group $\mathbb{F}_2$ of rank two is not amenable. Let $a, b \in \mathbb{F}_2$ be the free generators and set

$$A^+ = \{\text{all reduced words starting with } a\} \subset \mathbb{F}_2.$$

Similarly, let A^- be the reduced words beginning with a^{-1} and likewise define B^+ and B^-. Then, for $C = \{1, b, b^2, \ldots\} \subset \mathbb{F}_2$, we have

$$\begin{aligned}\mathbb{F}_2 &= A^+ \sqcup A^- \sqcup (B^+ \setminus C) \sqcup (B^- \cup C)\\ &= A^+ \sqcup aA^-\\ &= b^{-1}(B^+ \setminus C) \sqcup (B^- \cup C).\end{aligned}$$

This kind of decomposition is said to be *paradoxical*.[17] Note that the existence of an invariant mean μ on $\ell^\infty(\Gamma)$ would lead to a contradiction:

$$\begin{aligned}1 = \mu(1) &= \mu(\chi_{A^+}) + \mu(\chi_{A^-}) + \mu(\chi_{B^+\setminus C}) + \mu(\chi_{B^-\cup C})\\ &= \mu(\chi_{A^+}) + \mu(a.\chi_{A^-}) + \mu(b^{-1}.\chi_{B^+\setminus C}) + \mu(\chi_{B^-\cup C})\\ &= \mu(\chi_{A^+} + a.\chi_{A^-} + b^{-1}.\chi_{B^+\setminus C} + \chi_{B^-\cup C})\\ &= 2\mu(1) = 2.\end{aligned}$$

Since amenability passes to subgroups, it follows that all nonabelian free groups (on any number of generators) are nonamenable.

Here is a small sample of the known characterizations of amenable groups.

Theorem 2.6.8. *Let Γ be a discrete group. The following are equivalent:*

(1) *Γ is amenable;*

(2) *Γ has an approximate invariant mean;*

(3) *Γ satisfies the Følner condition;*

(4) *the trivial representation τ_0 is weakly contained in the regular representation λ (i.e., there exist unit vectors $\xi_i \in \ell^2(\Gamma)$ such that $\|\lambda_s(\xi_i) - \xi_i\| \to 0$ for all $s \in \Gamma$);*

(5) *there exists a net (φ_i) of finitely supported positive definite functions on Γ such that $\varphi_i \to 1$ pointwise;*

(6) *$C^*(\Gamma) = C^*_\lambda(\Gamma)$;*

(7) *$C^*_\lambda(\Gamma)$ has a character (i.e., one-dimensional representation);*

[17] This paradoxical decomposition leads to the famous Banach-Tarski paradox. See Eric Weisstein's website *Mathworld* (mathworld.wolfram.com) for more.

(8) *for any finite subset* $E \subset \Gamma$, *we have*

$$\|\frac{1}{|E|}\sum_{s\in E}\lambda_s\| = 1;$$

(9) $C^*_\lambda(\Gamma)$ *is nuclear;*[18]

(10) $L(\Gamma)$ *is semidiscrete.*

Proof. (1) $\Rightarrow$ (2): Take an invariant mean μ on $\ell^\infty(\Gamma)$. Being the predual of $\ell^\infty(\Gamma)$, $\ell^1(\Gamma)$ is dense in $\ell^\infty(\Gamma)^*$ and thus we can find a net (μ_i) in $\mathrm{Prob}(\Gamma)$ which converges to μ in the $\sigma(\ell^\infty(\Gamma)^*, \ell^\infty(\Gamma))$-topology. Note that for each $s \in \Gamma$, the net $(s.\mu_i - \mu_i)$ converges to zero weakly in $\ell^1(\Gamma)$ (not just weak* in $\ell^\infty(\Gamma)^*$). Hence, for any finite subset $E \subset \Gamma$, the weak closure of the convex subset $\bigoplus_{s\in E}\{s.\mu - \mu : \mu \in \mathrm{Prob}(\Gamma)\}$ contains zero. Since the weak and norm closures coincide, by the Hahn-Banach Theorem, assertion (2) follows.

(2) $\Rightarrow$ (3): Let a finite subset $E \subset \Gamma$ and $\varepsilon > 0$ be given. Choose $\mu \in \mathrm{Prob}(\Gamma)$ such that

$$\sum_{s\in E}\|s.\mu - \mu\|_1 < \varepsilon.$$

Given a positive function $f \in \ell^1(\Gamma)$ and $r \geq 0$, we define a set $F(f,r) = \{t \in \Gamma : f(t) > r\}$ and let $\chi_{F(f,r)}$ be the characteristic function of this set. For a pair of positive functions $f, h \in \ell^1(\Gamma)$ and $t \in \Gamma$, observe that $|\chi_{F(f,r)}(t) - \chi_{F(h,r)}(t)| = 1$ if and only if r lies between the numbers $f(t)$ and $h(t)$. If both f and h are bounded above by 1, it follows that

$$|f(t) - h(t)| = \int_0^1 |\chi_{F(f,r)}(t) - \chi_{F(h,r)}(t)|dr.$$

Applying this observation to μ and $s.\mu$, we get

$$\begin{aligned}
\|s.\mu - \mu\|_1 &= \sum_{t\in\Gamma}|s.\mu(t) - \mu(t)| \\
&= \sum_{t\in\Gamma}\int_0^1 |\chi_{F(s.\mu,r)}(t) - \chi_{F(\mu,r)}(t)|dr \\
&= \int_0^1 \Big(\sum_{t\in\Gamma}|\chi_{sF(\mu,r)}(t) - \chi_{F(\mu,r)}(t)|\Big)dr \\
&= \int_0^1 |sF(\mu,r) \bigtriangleup F(\mu,r)|dr.
\end{aligned}$$

[18]This is also equivalent to knowing $C^*(\Gamma)$ is nuclear. Thanks to condition (6), one direction is immediate, but we don't yet have the tools for the other. Once we introduce The Trick (Section 3.6), it will be trivial by considering the right regular representation.

Hence we have

$$\varepsilon \int_0^1 |F(\mu, r)|dr = \varepsilon > \sum_{s\in E} \|s\mu - \mu\|_1 = \int_0^1 \sum_{s\in E} |sF(\mu, r) \triangle F(\mu, r)|dr.$$

Thus for some r we must have

$$\sum_{s\in E} |sF(\mu, r) \triangle F(\mu, r)| < \varepsilon |F(\mu, r)|,$$

which shows that $F(\mu, r)$ is almost invariant under translation by the elements in E.

(3) $\Rightarrow$ (4): Let (F_i) be a Følner sequence and $\xi_i = |F_i|^{-1/2}\chi_{F_i}$ be the normalized characteristic functions of the F_i's (viewed as unit vectors in $\ell^2(\Gamma)$). The same calculation used in the ℓ^1 context (see the paragraph after Definition 2.6.4) shows that $\|\lambda_s(\xi_i) - \xi_i\|_{\ell^2(\Gamma)} \to 0$ for every $s \in \Gamma$.

(4) $\Rightarrow$ (5): Consider the vector states $x \mapsto \langle x\xi_i, \xi_i\rangle$. As noted in the previous section, these restrict to positive definite functions on Γ and obviously tend to 1 pointwise. To make them finitely supported, one simply forces each ξ_i to be a finitely supported ℓ^2 function.

(5) $\Rightarrow$ (6): Take a net (φ_i) as in condition (5). By Theorem 2.5.11, the multipliers m_{φ_i} (resp. $\tilde{m}_{\varphi_i}$) are u.c.p. on $C^*(\Gamma)$ (resp. $C^*_\lambda(\Gamma)$). We note that $\lambda \circ m_{\varphi_i} = \tilde{m}_{\varphi_i} \circ \lambda$ on $C^*(\Gamma)$ since the two maps are continuous and coincide on the dense subspace $\mathbb{C}[\Gamma]$. Observe that $m_{\varphi_i}(x) \to x$ for every $x \in C^*(\Gamma)$ since this is true for $x \in \mathbb{C}[\Gamma]$. Now suppose $x \in C^*(\Gamma)$ and $\lambda(x) = 0$. Then, we have

$$\lambda(m_{\varphi_i}(x)) = \tilde{m}_{\varphi_i}(\lambda(x)) = 0$$

for every i. But since φ_i is finitely supported, we have $m_{\varphi_i}(x) \in \mathbb{C}[\Gamma]$, and hence $\lambda(m_{\varphi_i}(x)) = 0$ implies $m_{\varphi_i}(x) = 0$. Therefore, $x = \lim_i m_{\varphi_i}(x) = 0$ and the $*$-homomorphism $\lambda\colon C^*(\Gamma) \to C^*_\lambda(\Gamma)$ is injective.

(6) $\Rightarrow$ (7): The trivial representation $\Gamma \to \mathbb{C}$ extends to $C^*(\Gamma) = C^*_\lambda(\Gamma)$.

(7) $\Rightarrow$ (1): Let $\tau\colon C^*_\lambda(\Gamma) \to \mathbb{C}$ be any $*$-homomorphism, but regard it as a state. Extending to $\mathbb{B}(\ell^2(\Gamma))$, we may assume that τ is also defined on $\ell^\infty(\Gamma) \subset \mathbb{B}(\ell^2(\Gamma))$. Since the left translation action is spatially implemented,

$$\tau(s.f) = \tau(\lambda_s f \lambda_s^*) = \tau(\lambda_s)\tau(f)\overline{\tau(\lambda_s)} = \tau(f)$$

for all $s \in \Gamma$ and $f \in \ell^\infty(\Gamma)$ (the unitaries λ_s belong to the multiplicative domain of τ). Hence, τ is an invariant mean as desired.

At this point we have shown the first seven conditions to be equivalent.

(4) $\Leftrightarrow$ (8): The $\Rightarrow$ direction is easy. For the converse, it suffices to show that if E is a finite *symmetric* set (meaning $E = E^{-1}$) satisfying condition (8), then E generates an amenable group. In this situation, the norm-one

operator $S = \frac{1}{|E|}\sum_{s\in E}\lambda_s$ is self-adjoint. Thus, for any $\varepsilon > 0$, we can find a unit vector $\xi \in \ell^2(\Gamma)$ such that $|\langle S\xi, \xi\rangle| > 1-\epsilon$. Letting $|\xi|$ be the pointwise absolute value of ξ, a straightforward calculation confirms

$$1-\varepsilon < |\langle S\xi,\xi\rangle| \leq \langle S|\xi|, |\xi|\rangle = \frac{1}{|E|}\sum_{s\in E}\langle \lambda_s|\xi|, |\xi|\rangle.$$

Since the cardinality of E is fixed, by taking ε sufficiently small, we deduce that all the numbers $\langle \lambda_s|\xi|, |\xi|\rangle$ must be close to 1; hence the norms $\|\lambda_s|\xi| - |\xi|\|$ are small, for all $s \in E$.

(1) $\Rightarrow$ (9): Let $F_k \subset \Gamma$ be a sequence of Følner sets. For each k let $P_k \in \mathbb{B}(\ell^2(\Gamma))$ be the orthogonal projection onto the finite-dimensional subspace spanned by $\{\delta_g : g \in F_k\}$. Identify $P_k\mathbb{B}(\ell^2(\Gamma))P_k$ with the matrix algebra $\mathbb{M}_{F_k}(\mathbb{C})$ and let $\{e_{p,q}\}_{p,q\in F_k}$ be the canonical matrix units of $\mathbb{M}_{F_k}(\mathbb{C})$. One can check that for each $s \in \Gamma$ we have $e_{p,p}\lambda_s e_{q,q} = 0$ unless $sq = p$, and $e_{p,p}\lambda_s e_{q,q} = e_{p,q}$ if $sq = p$. Since $P_k = \sum_{p\in F_k} e_{p,p}$, we have

$$P_k\lambda_s P_k = \sum_{p,q\in F_k} e_{p,p}\lambda_s e_{q,q} = \sum_{p\in F_k\cap sF_k} e_{p,s^{-1}p}.$$

Let $\varphi_k\colon C^*_\lambda(\Gamma) \to \mathbb{M}_{F_k}(\mathbb{C})$ be the u.c.p. map defined by $x \mapsto P_k x P_k$. Now define a map $\psi_k\colon \mathbb{M}_{F_k}(\mathbb{C}) \to C^*_\lambda(\Gamma)$ by sending

$$e_{p,q} \mapsto \frac{1}{|F_k|}\lambda_p\lambda_{q^{-1}}.$$

Evidently this map is unital; it is also completely positive, being (a scalar multiple) of the form described in Example 1.5.13.

The φ_k's and ψ_k's do the trick. Since the linear span of $\{\lambda_s : s \in \Gamma\}$ is norm dense in $C^*_\lambda(\Gamma)$, it suffices to check that $\|\lambda_s - \psi_k\circ\varphi_k(\lambda_s)\| \to 0$ for all $s \in \Gamma$. This follows from the definition of Følner sets together with the following computation:

$$\psi_k\circ\varphi_k(\lambda_s) = \psi_k\Big(\sum_{p\in F_k\cap sF_k} e_{p,s^{-1}p}\Big) = \sum_{p\in F_k\cap sF_k}\frac{1}{|F_k|}\lambda_s = \frac{|F_k\cap sF_k|}{|F_k|}\lambda_s.$$

Hence the reduced group C*-algebra is nuclear.

(1) $\Rightarrow$ (10): The maps constructed above also prove semidiscreteness of $L(\Gamma)$. According to Remark 2.1.3, it suffices to show that for every $x \in L(\Gamma)$ and $g, h \in \Gamma$,

$$\langle \psi_k\circ\varphi_k(x)\delta_g, \delta_h\rangle \to \langle x\delta_g, \delta_h\rangle.$$

If $x \in L(\Gamma)$ is given, then we can find unique scalars $\{\alpha_s\}_{s\in\Gamma}$ such that $\langle x\delta_g, \delta_h\rangle = \alpha_s$, whenever $hg^{-1} = s$. A computation shows

$$\psi_k\circ\varphi_k(x) = \sum_{s\in\Gamma}\alpha_s\frac{|F_k\cap sF_k|}{|F_k|}\lambda_s.$$

It follows that for each fixed pair $g, h \in \Gamma$,

$$\langle \psi_k \circ \varphi_k(x)\delta_g, \delta_h\rangle = \langle \sum_{s\in\Gamma} \alpha_s \frac{|F_k \cap sF_k|}{|F_k|} \lambda_s \delta_g, \delta_h\rangle = \alpha_{hg^{-1}} \frac{|F_k \cap hg^{-1}F_k|}{|F_k|}$$

converges to $\langle x\delta_g, \delta_h\rangle = \alpha_{hg^{-1}}$ as $k \to \infty$.

(9) $\Rightarrow$ (1): Let $\varphi_n \colon C^*_\lambda(\Gamma) \to \mathbb{M}_{k(n)}(\mathbb{C})$ and $\psi_n \colon \mathbb{M}_{k(n)}(\mathbb{C}) \to C^*_\lambda(\Gamma)$ be u.c.p. maps converging to $\mathrm{id}_{C^*_\lambda(\Gamma)}$ in the point-norm topology. By Arveson's Extension Theorem we may assume that the φ_n's are actually defined on all of $\mathbb{B}(\ell^2(\Gamma))$. In other words, letting $\Phi_n = \psi_n \circ \varphi_n$, we have u.c.p. maps $\Phi_n \colon \mathbb{B}(\ell^2(\Gamma)) \to C^*_\lambda(\Gamma)$ such that $\Phi_n(x) \to x$ for all $x \in C^*_\lambda(\Gamma)$. Taking a point-ultraweak limit point of $\{\Phi_n\}$ (see Theorem 1.3.7), we get a u.c.p. map $\Phi \colon \mathbb{B}(\ell^2(\Gamma)) \to L(\Gamma)$ which restricts to the identity on $C^*_\lambda(\Gamma)$. This is all we need to show amenability of Γ.

Let τ be the canonical vector trace on $L(\Gamma)$ and consider the state

$$\eta = \tau \circ \Phi$$

on $\mathbb{B}(\ell^2(\Gamma))$. Restricting to $\ell^\infty(\Gamma) \subset \mathbb{B}(\ell^2(\Gamma))$, we get an invariant mean. Indeed, for any $T \in \mathbb{B}(\ell^2(\Gamma))$ and $s \in \Gamma$ we have

$$\eta(\lambda_s T \lambda_s^*) = \tau\big(\lambda_s \Phi(T) \lambda_s^*\big) = \tau(\Phi(T)) = \eta(T),$$

where the first equality uses the fact that Φ restricts to the identity on $C^*_\lambda(\Gamma)$ (hence $C^*_\lambda(\Gamma)$ falls in the multiplicative domain of Φ) and the second uses the fact that τ is a trace. Thus if $T \in \ell^\infty(\Gamma)$, we have $\eta(s.T) = \eta(\lambda_s T \lambda_s^*) = \eta(T)$, since left translation is spatially implemented.

(10) $\Rightarrow$ (1): Semidiscreteness allows one to construct a u.c.p. map $\Phi \colon \mathbb{B}(\ell^2(\Gamma)) \to L(\Gamma)$ which restricts to the identity on $L(\Gamma)$ (i.e., semidiscreteness implies injectivity – Exercise 2.3.15). This is more than enough to imply amenability of Γ, as we saw above. □

Remark 2.6.9. This theorem not only shows that amenable groups give rise to a very natural class of nuclear C*-algebras, but it also gives our first examples of nonnuclear C*-algebras (since there are plenty of nonamenable groups).

Remark 2.6.10. A similar theorem holds in the locally compact case, but not everything generalizes. For example, nuclearity or semidiscreteness need not imply amenability in general; Connes proved in [**41**] that if G_0 denotes the connected component of G and if G/G_0 is amenable, then $C^*_\lambda(G)$ is nuclear and $L(G)$ is semidiscrete. In particular, all connected Lie groups have nuclear reduced C*-algebras (though they need not be amenable).

2.7. Type I C*-algebras

We now consider the class of C*-algebras characterized by having nice representation theory (in Glimm's sense) and which typically arise in (nondiscrete) group representation theory. Our objective is to show that every C*-algebra of type I is nuclear. In fact, nuclearity can be used to characterize these algebras[19] but this will have to wait as it requires, among other things, the nontrivial fact that nuclearity passes to quotients.

We begin with some von Neumann algebraic preliminaries. Recall that every von Neumann algebra of type I is of the form

$$\prod_{i\in I} \mathcal{A}_i \bar{\otimes} \mathbb{B}(\mathcal{H}_i)$$

where I is a set of cardinal numbers, each $\mathcal{A}_i$ is an abelian von Neumann algebra and $\mathcal{H}_i$ is a Hilbert space of dimension i (see Definition 1.3.1). Our first task is to show that all of these algebras are semidiscrete.

Lemma 2.7.1. *Let $M \subset \mathbb{B}(\mathcal{H})$ be a von Neumann algebra and assume there exists a net of projections $p_\lambda \in M$ such that $p_\lambda \to 1_{\mathcal{H}}$ in the strong operator topology and each of the corners $p_\lambda M p_\lambda$ is semidiscrete. Then M is also semidiscrete.*

Proof. Given finite sets $\mathfrak{F} \subset M$, $\Omega \subset \mathcal{H}$ and $\varepsilon > 0$, we must show that there exist c.c.p. maps $\varphi\colon M \to \mathbb{M}_n(\mathbb{C})$ and $\psi\colon \mathbb{M}_n(\mathbb{C}) \to M$ such that

$$|\langle \psi \circ \varphi(m)\xi, \eta\rangle - \langle m\xi, \eta\rangle| < \varepsilon$$

for all $m \in \mathfrak{F}$ and $\xi, \eta \in \Omega$ (see Remark 2.1.3). To do this, we first take some projection $p \in M$ such that pMp is semidiscrete and $\|p\xi - \xi\| < \varepsilon$ for all $\xi \in \Omega$. We then find c.c.p. maps $\tilde{\varphi}\colon pMp \to \mathbb{M}_n(\mathbb{C})$ and $\psi\colon \mathbb{M}_n(\mathbb{C}) \to M$ such that

$$|\langle \psi \circ \tilde{\varphi}(pmp)\xi, \eta\rangle - \langle pmp\xi, \eta\rangle| < \varepsilon$$

for all $m \in \mathfrak{F}$ and $\xi, \eta \in \Omega$. A standard bit of estimating shows that the c.c.p. map $\varphi\colon M \to \mathbb{M}_n(\mathbb{C})$ given by $\varphi(m) = \tilde{\varphi}(pmp)$ together with ψ satisfies the required inequality (with a slightly larger ε). □

Proposition 2.7.2. *Every von Neumann algebra of type* I *is semidiscrete.*

Proof. First let us handle the special case $\mathcal{A} \bar{\otimes} \mathbb{B}(\mathcal{H})$, where $\mathcal{A}$ is abelian. If $\mathcal{H}$ happens to be finite dimensional, then $\mathcal{A} \bar{\otimes} \mathbb{B}(\mathcal{H}) \cong \mathbb{M}_n(\mathcal{A})$ is a nuclear C*-algebra, by Corollary 2.4.4, and this evidently implies semidiscreteness. For infinite-dimensional $\mathcal{H}$ we first represent $\mathcal{A} \bar{\otimes} \mathbb{B}(\mathcal{H}) \subset \mathbb{B}(\mathcal{K} \otimes \mathcal{H})$, where $\mathcal{A} \subset \mathbb{B}(\mathcal{K})$ is any faithful normal representation. For a set $\Omega \subset \mathcal{H}$ we let $q_\Omega \in \mathbb{B}(\mathcal{H})$ be the projection onto the span of Ω. Then we note that the net

[19] A is type I if and only if every subalgebra of A is nuclear (Corollary 9.4.5).

of projections $p_\Omega = 1 \otimes q_\Omega$, where the index set is the set of all finite subsets $\Omega \subset \mathcal{H}$ partially ordered by inclusion, converges to the identity operator in the strong operator topology. On the other hand, each of the corners

$$p_\Omega \mathcal{A} \bar{\otimes} \mathbb{B}(\mathcal{H}) p_\Omega$$

is a nuclear C*-algebra (being isomorphic to $\mathbb{M}_n(\mathcal{A})$, where n is the dimension of the span of Ω) and hence the lemma above implies that $\mathcal{A} \bar{\otimes} \mathbb{B}(\mathcal{H})$ is semidiscrete, as desired.

The proposition then follows from the structure theorem for general type I von Neumann algebras, together with the fact that a direct product of semidiscrete von Neumann algebras is again semidiscrete. (This latter fact is a nice exercise, using the lemma above.) □

Not wishing to dwell on classical material, we define our way to the result we're after.

Definition 2.7.3. A C*-algebra is *type* I if its double dual is a type I von Neumann algebra.

The standard warning that type I von Neumann algebras are *not* type I as C*-algebras is probably in order. Indeed, we are about to show that type I C*-algebras are always nuclear and we saw in Proposition 2.4.9 that von Neumann algebras rarely enjoy this property.

Proposition 2.7.4. *Type* I *C*-algebras are nuclear.*

Proof. If A is type I, then A^{**} is a type I von Neumann algebra. By Proposition 2.7.2, A^{**} is semidiscrete and hence Proposition 2.3.8 implies that A is nuclear. □

Remark 2.7.5. For those who think of type I C*-algebras in terms of composition series, we note that this structure doesn't yield a simple proof of nuclearity. Indeed, one would still need to know that extensions of nuclear algebras are nuclear. It is possible to give a C*-proof of this fact, but it is no simpler or more enlightening than the proof given above (in our not-completely-objective opinion).

Having defined our way out of a discussion of type I C*-algebras, it is only proper that giving examples becomes difficult.[20]

Definition 2.7.6. A is called *subhomogeneous* if there exists $n \in \mathbb{N}$ such that every irreducible representation of A is on a Hilbert space of dimension less than or equal to n.

[20] For a comprehensive treatment of type I C*-algebras see [**142**].

Proposition 2.7.7. *If A is subhomogeneous, then A is type* I *(hence nuclear).*

Proof. For each natural number j let I_j be the set of pure states with j-dimensional GNS representations and let

$$\pi_j \colon A \to \prod_{i \in I_j} \mathbb{M}_j(\mathbb{C})$$

be the direct sum of all the corresponding GNS maps. There is a natural isomorphism[21]

$$\prod_{i \in I_j} \mathbb{M}_j(\mathbb{C}) \cong \mathbb{M}_j(\ell^\infty(I_j))$$

and hence we get an embedding

$$A \overset{\pi_1 \oplus \cdots \oplus \pi_k}{\hookrightarrow} \ell^\infty(I_1) \oplus \mathbb{M}_2(\ell^\infty(I_2)) \oplus \cdots \oplus \mathbb{M}_k(\ell^\infty(I_k))$$

for some sufficiently large k, since the direct sum of all irreducible representations is always faithful. Hence the double dual of A is isomorphic to a subalgebra of

$$\Big(\ell^\infty(I_1) \oplus \cdots \oplus \mathbb{M}_k(\ell^\infty(I_k))\Big)^{**} \cong \ell^\infty(I_1)^{**} \oplus \cdots \oplus \mathbb{M}_k(\ell^\infty(I_k)^{**}).$$

Since exactness passes to subalgebras, Proposition 2.4.9 tells us that the double dual of A is also a von Neumann algebra of type I. □

Exercises

Exercise 2.7.1 (Semidiscreteness and direct sums)**.** Show that if M_i, $i \in I$, is a collection of semidiscrete von Neumann algebras, then the direct sum

$$\prod_{i \in I} M_i = \{(x_i) : \sup_{i \in I} \|x_i\| < \infty\}$$

is also semidiscrete.

Exercise 2.7.2 (Semidiscreteness and inductive limits)**.** It is extremely easy to show that increasing unions of nuclear or exact C*-algebras are again nuclear or, respectively, exact (Exercise 2.3.7). Try to prove that if M has an increasing sequence of semidiscrete subalgebras $M_1 \subset M_2 \subset \cdots$ whose union is ultraweakly dense in M, then M is also semidiscrete.[22]

[21] This well-known fact makes a nice exercise. Or, refer to Lemma 3.9.4.

[22] If you succeed, without quoting difficult theorems, then you should publish the proof! It turns out that M will be semidiscrete in this case but the point of the exercise is that the obvious adaptation of the analogous C*-result fails, and it is instructive to see where the argument breaks down.

2.8. References

Proposition 2.3.8 is due to Effros-Lance; they are largely responsible for the systematic study of tensor products and approximation properties [**60**]. Amenable groups arose from von Neumann's work on the Banach-Tarski paradox. Theorem 2.6.8, the only really substantial result in this chapter, is the work of many hands. See [**139**] for more, including history and references, on this important class of groups.

Chapter 3

Tensor Products

This chapter is devoted to tensor products in the C*-context. Those with a low tolerance for operator algebras often find it unfortunate that the theory is subtle, tricky and requires great care. We prefer to embrace this fact. Indeed, if tensor products behaved too nicely, then operator algebras would be far less interesting. *That* would be unfortunate.

Unlike the previous chapter, this one contains several major theorems. Of course, there are also lots of preliminaries, remarks, exercises and discussions of various land mines that one should avoid. Sections 3.1, 3.2 and 3.3 lay out the basics of tensor products, starting in the algebraic realm and moving to the analytic world. For the experts these three sections just establish notation. In Section 3.4 comes the first important theorem: the spatial norm is the smallest possible C*-norm on the tensor product of two C*-algebras. Section 3.5 contains some important continuity results for maps on tensor products; we use these facts over and over. In Sections 3.6 and 3.7 we investigate two subtleties which make C*-tensor theory quite different from the algebraic variety. Though these sections are mostly about examples and counterexamples, there is one important fact that deserves mention: The Trick described in Proposition 3.6.5 is extremely useful and will be used many times. Perhaps the two most important theorems come in Sections 3.8 and 3.9 where we give tensor product characterizations of the classes of nuclear and, respectively, exact C*-algebras.

3.1. Algebraic tensor products

For those who haven't studied algebraic tensor products, we now describe how they are constructed and some of their basic properties. There are three facts which help us rationalize the absence of formal proof in this section: All of the results are simple, some hints and tricks are explained and we are confident that the reader can fill the gaps with minimal effort. Of course, one could consult an algebra book for the proofs, but this would be a waste of time – it is just as easy to give the proofs as to look them up.

If X and Y are two vector spaces, then the algebraic tensor product of X and Y is a new vector space which is both a useful tool and an interesting object in its own right.[1] Though not common in the algebraic literature, we will use $X \odot Y$ to denote the algebraic tensor product.

The construction of $X \odot Y$ goes as follows. Regard the Cartesian product $X \times Y$ as a discrete space and consider the vector space of compactly (i.e., finitely) supported functions $C_c(X \times Y)$. Let $\chi_{(x,y)} \in C_c(X \times Y)$ be the characteristic function over the point $(x, y) \in X \times Y$ and note that the collection of all such characteristic functions is a basis for $C_c(X \times Y)$. Now define a linear subspace $K \subset C_c(X \times Y)$ as the subspace spanned by elements of the following four types:

(1) $\chi_{(x_1+x_2,y)} - \chi_{(x_1,y)} - \chi_{(x_2,y)}$,
(2) $\chi_{(x,y_1+y_2)} - \chi_{(x,y_1)} - \chi_{(x,y_2)}$,
(3) $\lambda\chi_{(x,y)} - \chi_{(\lambda x,y)}$ and
(4) $\lambda\chi_{(x,y)} - \chi_{(x,\lambda y)}$.

Definition 3.1.1. Given vector spaces X and Y, their algebraic tensor product is the quotient vector space

$$X \odot Y = C_c(X \times Y)/K.$$

The image of an element $\chi_{(x,y)} \in C_c(X \times Y)$ under the canonical quotient map $C_c(X \times Y) \to X \odot Y$ is called an *elementary tensor* and is denoted $x \otimes y$.

Note that $X \odot Y$ is spanned by the elementary tensors (but they are not a basis).

There are really only two things that one needs to know about tensor products in order to handle most issues that arise. The first is the *tensor calculus* and follows easily from the definition of the space K.

Proposition 3.1.2 (Tensor calculus). *The following identities hold for all vectors and scalars:*

(1) $(x_1 + x_2) \otimes y = x_1 \otimes y + x_2 \otimes y$ *and* $x \otimes (y_1 + y_2) = x \otimes y_1 + x \otimes y_2$.
(2) $\lambda(x \otimes y) = (\lambda x) \otimes y = x \otimes (\lambda y)$.

[1] All of this can be done in much greater generality – for modules over rings – but for our purposes the reader may assume that everything happens over $\mathbb{C}$.

Note that the vector space structures on $X \odot Y$ and $X \times Y$ are completely different. For example, in $X \times Y$ we have $(x_1, y_1) + (x_2, y_2) = (x_1 + x_2, y_1 + y_2)$ while there is no way to "simplify" $x_1 \otimes y_1 + x_2 \otimes y_2$ (in general) and

$$(x_1 + x_2) \otimes (y_1 + y_2) = x_1 \otimes y_1 + x_1 \otimes y_2 + x_2 \otimes y_1 + x_2 \otimes y_2.$$

In many proofs it will suffice to work only with elementary tensors, but this is because they form a spanning set for $X \odot Y$ – one must not forget that $X \odot Y$ contains a lot more than just the elementary tensors.

The other crucial fact about tensor products is their *universal property*; they are designed to turn bilinear maps $X \times Y \to Z$ into linear maps $X \odot Y \to Z$. Moreover, $X \odot Y$ is the unique vector space, up to isomorphism, with this property. Before making this precise, first note that the natural mapping

$$X \times Y \to X \odot Y, \ (x, y) \mapsto x \otimes y$$

is not linear – it is bilinear (since this map factors through $C_c(X \times Y)$ and the subspace K is specifically designed for bilinearity).[2]

Proposition 3.1.3 (Universality). *For any vector space Z and any bilinear map $\sigma\colon X \times Y \to Z$, there exists a unique linear map $\dot{\sigma}\colon X \odot Y \to Z$ such that*

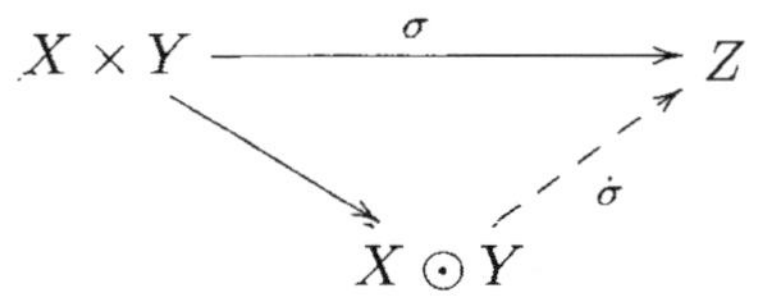

commutes (i.e., $\dot{\sigma}(x \otimes y) = \sigma((x, y))$ for all $x \in X$, $y \in Y$).

Though important, we prefer to leave it to the reader to formulate and prove uniqueness of $X \odot Y$. As usual, the rough version states that any other vector space enjoying the same universal property as $X \odot Y$ must be isomorphic to $X \odot Y$.

The following mappings on tensor products will be indispensable.

Proposition 3.1.4 (Tensor product maps). *If $\varphi\colon W \to Y$ and $\psi\colon X \to Z$ are linear maps, then there is a unique linear map*

$$\varphi \odot \psi\colon W \odot X \to Y \odot Z$$

such that $\varphi \odot \psi(w \otimes x) = \varphi(w) \otimes \psi(x)$ for all $w \in W$, $x \in X$.

Proposition 3.1.5 (Product maps). *If C is an algebra and $\varphi\colon X \to C$, $\psi\colon Y \to C$ are linear maps, then there is a unique linear map*

$$\varphi \times \psi\colon X \odot Y \to C$$

such that $\varphi \times \psi(x \otimes y) = \varphi(x)\psi(y)$ for all $x \in X$, $y \in Y$.

[2] Note also that the map $X \times Y \to X \odot Y$ is (usually) not onto. For example, you can't hit all of the vectors of the form $x_1 \otimes y_1 + x_2 \otimes y_2$ (unless X or Y is one dimensional).

Both of these propositions are simple applications of universality. For example, if $\varphi\colon X \to C$, $\psi\colon Y \to C$ are linear maps, then

$$X \times Y \to C, \ (x, y) \mapsto \varphi(x)\psi(y)$$

is obviously a bilinear map.

An important special case which deserves highlighting is that of functionals.

Corollary 3.1.6 (Tensor product functionals). *If $\varphi\colon X \to \mathbb{C}$, $\psi\colon Y \to \mathbb{C}$ are linear functionals, then there is a unique linear functional*

$$\varphi \odot \psi\colon X \odot Y \to \mathbb{C}$$

such that $\varphi \odot \psi(x \otimes y) = \varphi(x)\psi(y)$ for all $x \in X$, $y \in Y$.

We have mixed notation in the corollary above both to cause confusion and to see if you're paying attention. Luckily it is justified by the fact that the map

$$\mathbb{C} \odot \mathbb{C} \to \mathbb{C}, \ \alpha \otimes \beta \mapsto \alpha\beta$$

is an isomorphism; hence the functional case follows either from Proposition 3.1.4 or Proposition 3.1.5.

Though we only need it once, the following corollary should be mentioned now. It follows from the previous result because if $\varphi\colon X \to \mathbb{C}$ is a conjugate linear functional (i.e., scalars come out with complex conjugates), then the map $x \mapsto \overline{\varphi(x)}$ is linear.

Corollary 3.1.7 (Conjugate linear functionals). *If $\varphi\colon X \to \mathbb{C}$ and $\psi\colon Y \to \mathbb{C}$ are conjugate linear functionals, there is a conjugate linear tensor product functional $\varphi \odot \psi\colon X \odot Y \to \mathbb{C}$ such that $\varphi \odot \psi(x \otimes y) = \varphi(x)\psi(y)$ for all $x \in X$, $y \in Y$.*

Unfortunately, it is not always easy to decide when a set of elementary tensors is linearly independent. Here is a useful sufficient condition, however.

Proposition 3.1.8 (Linear independence). *If $\{x_1, \dots, x_n\} \subset X$ are linearly independent, $\{y_1, \dots, y_n\} \subset Y$ are arbitrary and*

$$0 = \sum_{i=1}^{n} x_i \otimes y_i \in X \odot Y,$$

then $y_1 = y_2 = \cdots = y_n = 0$.

The proof of this becomes short and sweet if we use a little tensor product functional trick. Namely, let $\{\varphi_1, \dots, \varphi_n\} \subset X^*$ be a dual set of functionals (i.e., $\varphi_j(x_i) = \delta_{i,j}$) and let $\psi \in Y^*$ be arbitrary. Then for each $1 \leq j \leq n$ we have

$$0 = \varphi_j \odot \psi\Big(\sum_{i=1}^{n} x_i \otimes y_i\Big) = \psi(y_j).$$

Since this holds for all ψ, it follows that each y_j is the zero vector.

Corollary 3.1.9 (Bases). *If $\{x_i\}_{i\in I} \subset X$ and $\{y_j\}_{j\in J} \subset Y$ are bases, then*

$$\{x_i \otimes y_j\}_{(i,j)\in I\times J}$$

is a basis of $X \odot Y$. In particular, $\dim(X \odot Y) = \dim(X)\dim(Y)$.

Since the set of elementary tensors spans $X \odot Y$, it is easy to see that $\{x_i \otimes y_j\}_{(i,j)\in I\times J}$ also spans. That they are linearly independent follows from Proposition 3.1.8 together with an exercise in "collecting like terms".[3]

Though bases are important, it turns out that the following fact is often more convenient. The proof uses the previous two results.

Corollary 3.1.10 (Unique representation). *If $\{x_i\}_{i\in I} \subset X$ is a basis, then for each vector $v \in X \odot Y$ there is a unique set $\{y_i\}_{i\in I} \subset Y$ such that*

$$v = \sum_{i\in I} x_i \otimes y_i.$$ [4]

We now consider inclusions and exact sequences of algebraic tensor products. The results are easy, but the corresponding statements in the C*-setting are false.

Proposition 3.1.11 (Inclusions). *If $W \subset Y$ and $X \subset Z$ are subspaces, then there is a natural inclusion*

$$W \odot X \subset Y \odot Z.$$

Understanding the ambiguity of the previous proposition is 90 percent of the proof. The "natural inclusion" is just the mapping

$$\iota_W \odot \iota_X \colon W \odot X \to Y \odot Z$$

induced by the identity inclusions $\iota_W \colon W \hookrightarrow Y$ and $\iota_X \colon X \hookrightarrow Z$. Hence we are asserting that $\iota_W \odot \iota_X$ is injective – which is a special case of something more general.

Proposition 3.1.12 (Injectivity of tensor product maps). *If $\varphi \colon W \to Y$ and $\psi \colon X \to Z$ are injective linear maps, then*

$$\varphi \odot \psi \colon W \odot X \to Y \odot Z$$

is also injective.

---

[3] $\sum_{i,j} \lambda_{i,j}(x_i \otimes y_j) = \sum_i x_i \otimes (\sum_j \lambda_{i,j} y_j)$.

[4] Of course, only finitely many y_i's are nonzero.

To see that $\varphi \odot \psi$ is injective is a simple exercise using what we know about bases and linear independence.

Proposition 3.1.13 (Exact sequences). *If $0 \to X \xrightarrow{\iota} Y \xrightarrow{\pi} Z \to 0$ is a short exact sequence and W is a vector space, then the sequence*

$$0 \to X \odot W \xrightarrow{\iota \odot \mathrm{id}_W} Y \odot W \xrightarrow{\pi \odot \mathrm{id}_W} Z \odot W \to 0$$

is also exact. In particular, if $X \subset Y$ is a subspace and W is arbitrary, we have a canonical isomorphism

$$\frac{Y \odot W}{X \odot W} \cong (Y/X) \odot W.$$[5]

Proof. Proposition 3.1.12 implies $X \odot W \xrightarrow{\iota \odot \mathrm{id}_W} Y \odot W$ is injective while it is trivial that $Y \odot W \xrightarrow{\pi \odot \mathrm{id}_W} Z \odot W$ is surjective. Hence one only needs to check exactness in the middle. Since $X \odot W \subset \ker(\pi \odot \mathrm{id}_W)$, the linear map $\pi \odot \mathrm{id}_W$ induces a surjection $\tilde{\pi}\colon (Y \odot W)/(X \odot W) \to Z \odot W$. To prove injectivity of $\tilde{\pi}$, we construct a left inverse. Define $\sigma\colon Z \times W \to (Y \odot W)/(X \odot W)$ by

$$\sigma((z, w)) = y \otimes w + (X \odot W),$$

where $y \in Y$ is any element such that $\pi(y) = z$. It is clear that $\sigma((z, w))$ is independent of the choice of the lift y and that σ is bilinear. Hence by universality, there is a linear map $\dot{\sigma}\colon Z \odot W \to (Y \odot W)/(X \odot W)$ such that $\dot{\sigma}(z \otimes w) = \sigma((z, w))$ for every $z \in Z$ and $w \in W$. Evidently $\dot{\sigma} \circ \tilde{\pi} = \mathrm{id}_{(Y \odot W)/(X \odot W)}$, since this is clear for elementary tensors, so we're done.

□

Though the results above have been formulated for general vector spaces, we are primarily interested in C*-algebras. Let's put an involution on the tensor product.

Proposition 3.1.14 (Involution). *If A and B are C*-algebras, then $A \odot B$ carries a unique involution such that $(a \otimes b)^* = a^* \otimes b^*$ for all elementary tensors.*

Of course, the involution on all of $A \odot B$ is defined by

$$\sum_i a_i \otimes b_i \mapsto \sum_i a_i^* \otimes b_i^*$$

and to prove that this is well-defined, it suffices to show that if $\sum_i a_i \otimes b_i = 0$, then $\sum_i a_i^* \otimes b_i^* = 0$ as well. Expanding the a_i's out in terms of a basis for A and playing around with the tensor calculus and linear independence will show this to be true.

[5] A C*-analogue of this isomorphism can fail due to the existence of nonexact C*-algebras.

We complete the $*$-algebra structure on $A \odot B$ by defining multiplication. As with the involution we know how it must go; the proof amounts to showing it is well-defined.

Proposition 3.1.15 (Multiplication). *The tensor product $A \odot B$ has a multiplication defined by*

$$\Big(\sum_i a_i \otimes b_i\Big)\Big(\sum_j c_j \otimes d_j\Big) = \sum_{i,j} a_i c_j \otimes b_i d_j.$$

To prove that this multiplication works, we first consider $L(A \odot B)$, the vector space of all linear maps $A \odot B \to A \odot B$. If $M_a\colon A \to A$ is left multiplication by $a \in A$ and $M_b\colon B \to B$ is left multiplication by $b \in B$, then, thanks to Proposition 3.1.4, for every pair $(a, b) \in A \times B$ we have the tensor product map $M_a \otimes M_b \in L(A \odot B)$. It is routine to check that

$$A \times B \to L(A \odot B), \ (a, b) \mapsto M_a \otimes M_b$$

is a bilinear map. By universality there is a linear map $M\colon A \odot B \to L(A \odot B)$ such that $M(a \otimes b) = M_a \otimes M_b$. Finally one checks that the bilinear map

$$A \odot B \times A \odot B \to A \odot B, \ (x, y) \mapsto M(x)y$$

defines our multiplication.

We close this section with two simple consequences of the existence of product and tensor product maps (Propositions 3.1.4 and 3.1.5). The proofs are straightforward calculations.

Proposition 3.1.16 (Tensor product morphisms). *Given $*$-homomorphisms $\varphi\colon A \to C$ and $\psi\colon B \to D$, the tensor product map $\varphi \odot \psi\colon A \odot B \to C \odot D$ is also a $*$-homomorphism.*

Proposition 3.1.17 (Product morphisms). *Given two $*$-homomorphisms $\pi_A\colon A \to C$ and $\pi_B\colon B \to C$* with commuting ranges *(i.e., $[\pi_A(a), \pi_B(b)] = 0$ for all $a \in A$, $b \in B$), the product map $\pi_A \times \pi_B\colon A \odot B \to C$ is also a $*$-homomorphism.*

Exercises

Exercise 3.1.1. Observe that if C and D are both unital, then Proposition 3.1.16 is a special case of Proposition 3.1.17. How about in the nonunital case?

Exercise 3.1.2. Justify the following identifications (which, by the way, get used all of the time): $A \cong A \odot \mathbb{C} \cong A \odot \mathbb{C}1_B \subset A \odot B$.

Exercise 3.1.3. Prove that if A is a C*-algebra, then for any n and any choice of matrix units $\{e_{i,j}\}_{i,j=1}^n \subset \mathbb{M}_n(\mathbb{C})$ there is a $*$-algebra isomorphism

$\mathbb{M}_n(\mathbb{C}) \odot A \cong \mathbb{M}_n(A)$ given by

$$\sum_{i,j} e_{i,j} \otimes a_{i,j} \mapsto [a_{i,j}].$$

Exercise 3.1.4. For an arbitrary nonunital algebra C let $\tilde{C}$ denote the unitization. Assume A and B are both nonunital and discuss the relations between $A \odot B$, $\widetilde{(A \odot B)}$, $\tilde{A} \odot B$ and $\tilde{A} \odot \tilde{B}$. For example, which ones can be identified with subalgebras of the others? How about ideals? What are the corresponding quotients?

Exercise 3.1.5. Let A and B be C*-algebras. Prove that $A^* \odot B^*$ separates points of $A^{**} \odot B^{**}$ – i.e., show that for every $0 \neq x \in A^{**} \odot B^{**}$ there exist linear functionals $\varphi_1, \ldots, \varphi_n$ on A and $\psi_1, \ldots, \psi_n$ on B such that

$$\Big(\sum_{i=1}^{n} \varphi_i \odot \psi_i\Big)(x) \neq 0.$$

In fact, the elementary tensor product functionals $\{\varphi \odot \psi : \varphi \in A^*, \psi \in B^*\}$ separate points.

Though it is technical and not particularly interesting, we will need the following exercise later on (hence the long hint).

Exercise 3.1.6. Assume A is nonunital, $\pi_A\colon A \to \mathbb{B}(\mathcal{H})$ and $\pi_B\colon B \to \mathbb{B}(\mathcal{H})$ are $*$-homomorphisms with commuting ranges and the product homomorphism $\pi_A \times \pi_B\colon A \odot B \to \mathbb{B}(\mathcal{H})$ is injective. Prove that the product homomorphism $\tilde{\pi}_A \times \pi_B\colon \tilde{A} \odot B \to \mathbb{B}(\mathcal{H})$ is also injective. (Hint: If $x \in \tilde{A} \odot B$ is an element such that $\tilde{\pi}_A \times \pi_B(x) = 0$, then – since $A \odot B$ is an ideal in $\tilde{A} \odot B$ – for every $y \in A \odot B$ we have that $xy = 0$. Now write

$$x = \sum_{i=1}^{n} a_i \otimes b_i$$

where the b_i's are linearly independent. Let $\{e_k\} \subset A$ and $\{f_k\} \subset B$ be approximate units and we get that

$$0 = x(e_k \otimes f_k) = \sum_{i=1}^{n} (a_i e_k) \otimes (b_i f_k)$$

for all k. But for large k, $\{b_1 f_k, \ldots, b_n f_k\}$ must also be a linearly independent set.)

3.2. Analytic preliminaries

Our first goal is to make the tensor product of two Hilbert spaces $\mathcal{H} \odot \mathcal{K}$ into a Hilbert space. While $\mathcal{H} \odot \mathcal{K}$ does have a natural positive definite sesquilinear form, it is not complete (unless either $\mathcal{H}$ or $\mathcal{K}$ is finite dimensional) in the corresponding norm. Of course, this isn't a real problem.

Proposition 3.2.1 (Tensor product of Hilbert spaces). *If $\mathcal{H}$ and $\mathcal{K}$ are Hilbert spaces, then $\mathcal{H} \odot \mathcal{K}$ is a pre-Hilbert space with respect to the inner product*

$$\langle \sum_i h_i \otimes k_i, \sum_j h'_j \otimes k'_j \rangle = \sum_{i,j} \langle h_i, h'_j \rangle \langle k_i, k'_j \rangle.$$

The Hilbert space completion will be denoted $\mathcal{H} \otimes \mathcal{K}$.

Proof. First we must show that there is a sesquilinear map

$$\langle \cdot, \cdot \rangle \colon \mathcal{H} \odot \mathcal{K} \times \mathcal{H} \odot \mathcal{K} \to \mathbb{C}$$

satisfying the formula given above. The proof is similar to the proof of Proposition 3.1.15 so we only sketch the main points. First, let $(\mathcal{H} \odot \mathcal{K})^{*,c}$ denote the vector space of conjugate linear functionals on $\mathcal{H} \odot \mathcal{K}$. By Corollary 3.1.7, for each pair $(h, k) \in \mathcal{H} \times \mathcal{K}$ there is an element $f_{(h,k)} \in (\mathcal{H} \odot \mathcal{K})^{*,c}$ such that

$$f_{(h,k)} \Big(\sum_i v_i \otimes w_i \Big) = \sum_i \langle h, v_i \rangle \langle k, w_i \rangle$$

for all $\sum_i v_i \otimes w_i \in \mathcal{H} \odot \mathcal{K}$. One checks that the mapping $\mathcal{H} \times \mathcal{K} \to (\mathcal{H} \odot \mathcal{K})^{*,c}$, $(h, k) \mapsto f_{(h,k)}$, is bilinear and thus induces a linear map $M \colon \mathcal{H} \odot \mathcal{K} \to (\mathcal{H} \odot \mathcal{K})^{*,c}$. We then define the desired sesquilinear form by

$$\langle v, w \rangle = M(v)w$$

for all $v, w \in \mathcal{H} \odot \mathcal{K}$.

That we have a positive definite form on $\mathcal{H} \odot \mathcal{K}$ follows from the calculation

$$\langle \sum_i e_i \otimes k_i, \sum_i e_i \otimes k_i \rangle = \sum_{i,j} \langle e_i, e_j \rangle \langle k_i, k_j \rangle = \sum_i \|k_i\|^2,$$

whenever $\{e_i\} \subset \mathcal{H}$ is an orthonormal set of vectors. □

The following fact is a nice exercise. It does not follow from Corollaries 3.1.9 and 3.1.10, but the proof is similar.

Proposition 3.2.2 (Orthonormal bases and vector representations). *Suppose that $\{v_i\}_{i \in I} \subset \mathcal{H}$ and $\{w_j\}_{j \in J} \subset \mathcal{K}$ are orthonormal bases. Then*

$$\{v_i \otimes w_j\}_{(i,j) \in I \times J} \subset \mathcal{H} \otimes \mathcal{K}$$

is an orthonormal basis. Moreover, for each $x \in \mathcal{H} \otimes \mathcal{K}$ there is a unique set of vectors $\{k_i\} \subset \mathcal{K}$ such that

$$x = \sum_i v_i \otimes k_i.$$[6]

If $S \in \mathbb{B}(\mathcal{H})$ and $T \in \mathbb{B}(\mathcal{K})$, then we can consider the algebraic tensor product mapping $S \odot T \colon \mathcal{H} \odot \mathcal{K} \to \mathcal{H} \odot \mathcal{K}$. Naturally, this map is bounded and has the expected norm.

Proposition 3.2.3 (Tensor product operators). *If $S \in \mathbb{B}(\mathcal{H})$ and $T \in \mathbb{B}(\mathcal{K})$, then there is a unique linear operator $S \otimes T \in \mathbb{B}(\mathcal{H} \otimes \mathcal{K})$ such that*

$$S \otimes T(v \otimes w) = Sv \otimes Tw$$

for all $v \in \mathcal{H}$, $w \in \mathcal{K}$. Moreover, $\|S \otimes T\| = \|S\|\|T\|$.

Proof. Let us first consider the case that $S = 1_{\mathcal{H}}$. For any vector $x \in \mathcal{H} \odot \mathcal{K}$ we can find an orthonormal set $\{e_i\} \subset \mathcal{H}$ and unique vectors $\{k_i\} \subset \mathcal{K}$ such that $x = \sum_i e_i \otimes k_i$.[7] Now we compute

$$\begin{aligned} \|1_{\mathcal{H}} \odot T(x)\|^2 &= \|\sum_i e_i \otimes T(k_i)\|^2 \\ &= \sum_i \|T(k_i)\|^2 \\ &\leq \|T\|^2 \|x\|^2. \end{aligned}$$

It follows that $1_{\mathcal{H}} \odot T$ has a unique extension to a bounded linear operator on $\mathcal{H} \otimes \mathcal{K}$, which we denote by $1_{\mathcal{H}} \otimes T$, and its norm is bounded by $\|T\|$. Evidently the same holds for $S \odot 1_{\mathcal{K}}$. Hence we may define $S \otimes T$ to be $(S \otimes 1_{\mathcal{K}})(1_{\mathcal{H}} \otimes T)$ and we have the inequality

$$\|S \otimes T\| \leq \|S\|\|T\|.$$

One should also check that

$$S \otimes T|_{\mathcal{H} \odot \mathcal{K}} = S \odot T,$$

but this is easy.

To show the opposite inequality $\|S \otimes T\| \geq \|S\|\|T\|$, first note that the norm on $\mathcal{H} \otimes \mathcal{K}$ is a *cross norm*. In other words, for all $h \in \mathcal{H}$, $k \in \mathcal{K}$ we have $\|h \otimes k\| = \|h\|\|k\|$. This is obvious but has the nice consequence that

---

[6] Of course, this is no longer a finite sum and convergence is in the norm topology.

[7] Is this sum finite or infinite? Are we appealing to Proposition 3.2.2 or Corollary 3.1.10? Actually we can neither appeal to Proposition 3.2.2, since $\mathcal{H} \odot \mathcal{K}$ is not complete, nor to Corollary 3.1.10, since an orthonormal basis is not an algebraic basis. What are we to do? How about starting with an orthonormal set which spans a prescribed finite-dimensional subspace, then extending it to an algebraic basis and invoking Corollary 3.1.10. Or, if you prefer, appeal to Corollary 3.1.10 first, and then apply the Gram-Schmidt procedure to arrange orthogonality. Either way, we must pay attention.

if we take unit vectors $h_n \in \mathcal{H}$ and $k_n \in \mathcal{K}$ such that $\|S\| = \lim \|Sh_n\|$ and $\|T\| = \lim \|Tk_n\|$, then $h_n \otimes k_n \in \mathcal{H} \odot \mathcal{K}$ are still unit vectors and

$$\|(S \otimes T)h_n \otimes k_n\| = \|(Sh_n) \otimes (Tk_n)\| = \|Sh_n\|\|Tk_n\| \to \|S\|\|T\|.$$

This completes the proof. □

It is a simple, though slightly tedious, exercise to show that the tensor product operators above satisfy the same tensor calculus as elementary tensors in an algebraic tensor product. Moreover, the adjoint and multiplication are compatible as well. It would be tempting to think that $\mathbb{B}(\mathcal{H}) \odot \mathbb{B}(\mathcal{K})$ and $\mathbb{B}(\mathcal{H} \otimes \mathcal{K})$ are isomorphic algebras but sadly this isn't quite correct (though "up to weak closure" it is). However, we do have identifications $\mathbb{B}(\mathcal{H}) \cong \mathbb{B}(\mathcal{H}) \otimes \mathbb{C}1_{\mathcal{K}}$ and $\mathbb{B}(\mathcal{K}) \cong \mathbb{C}1_{\mathcal{H}} \otimes B(\mathcal{K})$ and, since $\mathbb{B}(\mathcal{H}) \otimes \mathbb{C}1_{\mathcal{K}}, \mathbb{C}1_{\mathcal{H}} \otimes B(\mathcal{K}) \subset \mathbb{B}(\mathcal{H} \otimes \mathcal{K})$ are commuting subalgebras, Proposition 3.1.17 provides us with a natural $*$-homomorphism

$$\mathbb{B}(\mathcal{H}) \odot \mathbb{B}(\mathcal{K}) \to \mathbb{B}(\mathcal{H} \otimes \mathcal{K}), \ \sum_i S_i \otimes T_i \mapsto \sum_i S_i \otimes T_i.$$

The following corollary is immediate from these remarks.

Corollary 3.2.4 (Tensor product morphisms). *Given two $*$-representations $\pi_A \colon A \to \mathbb{B}(\mathcal{H})$ and $\pi_B \colon B \to \mathbb{B}(\mathcal{K})$, there is an induced $*$-representation*

$$\pi_A \odot \pi_B \colon A \odot B \to \mathbb{B}(\mathcal{H} \otimes \mathcal{K})$$

such that $\pi_A \odot \pi_B(a \otimes b) = \pi_A(a) \otimes \pi_B(b)$ for all $a \in A$ and $b \in B$.[8]

We've noted that if $\pi_A \colon A \to C$ and $\pi_B \colon B \to C$ are $*$-homomorphisms with commuting ranges, then the product map $\pi_A \times \pi_B \colon A \odot B \to C$ is also a $*$-homomorphism. It is an important fact that in the case $C = \mathbb{B}(\mathcal{H})$ every $*$-homomorphism $A \odot B \to \mathbb{B}(\mathcal{H})$ arises this way. It turns out that the proof is completely trivial in the unital case and rather unpleasant in the absence of units. Hence, if one only cares about unital C*-algebras, then spend the three seconds required to work out Exercise 3.2.4 and skip the rest of this section.

Lemma 3.2.5. *Let $\pi \colon A \odot B \to \mathbb{B}(\mathcal{H})$ be a $*$-homomorphism. Then for each fixed $a \in A$, the linear map*

$$B \to \mathbb{B}(\mathcal{H}), \ b \mapsto \pi(a \otimes b)$$

is bounded.

[8]Note that $\otimes$ has different meanings on the left and right hand sides of $\pi_A \odot \pi_B(a \otimes b) = \pi_A(a) \otimes \pi_B(b)$. Of course, if you didn't notice this abuse of notation in the paragraph preceding this corollary, then it probably won't matter here either.

Proof. Since positive elements span, we may assume $a \geq 0$. In this case, for each positive $b \in B$ we have that $\pi(a \otimes b)$ is also a positive operator in $\mathbb{B}(\mathcal{H})$ since

$$\pi(a \otimes b) = \pi(a^{1/2} \otimes b^{1/2})\pi(a^{1/2} \otimes b^{1/2}) \geq 0.$$

By the Closed Graph Theorem it suffices to show that if $b_n \to 0$ and $\pi(a \otimes b_n) \to T \in B(H)$ (in norm), then $T = 0$. Since every bounded linear functional is a linear combination of positive functionals, it will follow that $T = 0$ if we can show that $\varphi(T) = 0$ for every positive functional φ.

So let φ be an arbitrary positive functional on $\mathbb{B}(\mathcal{H})$. Since $0 \leq b \Rightarrow 0 \leq \pi(a \otimes b)$, it follows that $b \mapsto \varphi \circ \pi(a \otimes b)$ defines a positive linear functional on B and, as such, is necessarily bounded. Hence

$$\varphi(T) = \lim \varphi(\pi(a \otimes b_n)) = \lim \varphi \circ \pi(a \otimes b_n) = 0$$

as desired. □

Theorem 3.2.6 (Restrictions). *Let* $\pi\colon A \odot B \to \mathbb{B}(\mathcal{H})$ *be a nondegenerate* $*$-*representation. Then there exist nondegenerate* $*$-*representations* $\pi_A\colon A \to \mathbb{B}(\mathcal{H})$ *and* $\pi_B\colon B \to \mathbb{B}(\mathcal{H})$, *with commuting ranges, such that*

$$\pi = \pi_A \times \pi_B.$$

Proof. The question is not how to define the restrictions – there is only one possibility – but rather, why the definition works. Since π is nondegenerate, the vectors $\pi(x)v$, $x \in A \odot B$ and $v \in \mathcal{H}$, are dense in $\mathcal{H}$. Hence we are forced to define

$$\pi_A(a)\big(\pi(x)v\big) = \pi\Big(\sum_i aa_i \otimes b_i\Big)v$$

for $a \in A$, $x = \sum_i a_i \otimes b_i \in A \odot B$ and $v \in \mathcal{H}$. We must show that this is well-defined and bounded (hence it extends to a bounded operator on all of $\mathcal{H}$).

To prove this, we let $\{e_n\}$ be an approximate unit for B and consider the (well-defined, bounded) operators $\pi(a \otimes e_n)$. For $x = \sum_i a_i \otimes b_i \in A \odot B$ and $v \in \mathcal{H}$ we have

$$\begin{aligned}\|\pi(a \otimes e_n)(\pi(x)v) - \pi\Big(\sum_i aa_i \otimes b_i\Big)v\| &= \Big\|\pi\Big(\sum_i aa_i \otimes (e_n b_i - b_i)\Big)v\Big\| \\ &\leq \sum_i M_i \|e_n b_i - b_i\| \to 0,\end{aligned}$$

where the M_i's are (a finite set of) constants depending on the elements aa_i and coming from Lemma 3.2.5. Since there is a uniform bound on the norms $\|\pi(a \otimes e_n)\|$ (Lemma 3.2.5 again), it follows that our definition of $\pi_A(a)$ is both well-defined and extends to a bounded operator on all $\mathcal{H}$.

Of course π_B is defined similarly and it is routine to check that we get a pair of $*$-homomorphisms with commuting ranges in this way. It is also evident that $\pi_A \times \pi_B = \pi$.

The last thing to check is that both π_A and π_B are nondegenerate. But if $\{f_j\} \subset A$ is an approximate unit for A, we have that

$$\pi_A(f_j)\big(\pi(x)v\big) = \pi(\sum_i f_j a_i \otimes b_i)v \stackrel{j\to\infty}{\longrightarrow} \pi(x)v,$$

for all $v \in \mathcal{H}$ and $x \in A \odot B$, by an argument analogous to that above (apply Lemma 3.2.5 to the finite set of b_i's). Thus $\pi_A(f_j) \to 1_{\mathcal{H}}$ in the strong operator topology – i.e., π_A is nondegenerate and the same holds for π_B. □

As usual, the restriction to nondegenerate representations is only for convenience. For a general representation $\pi\colon A \odot B \to \mathbb{B}(\mathcal{H})$ we can still define restrictions by first cutting to the closure of $\{\pi(x)v : x \in A \odot B, v \in \mathcal{H}\}$, applying the previous theorem and then extending by zero on the orthogonal vectors.

Exercises

Exercise 3.2.1. Let $\{v_j\}_{j\in J} \subset \mathcal{H}$ and $\{w_i\}_{i\in I} \subset \mathcal{K}$ be orthonormal bases. Show that

$$\mathcal{H} \otimes \mathcal{K} \cong \bigoplus_{i\in I} \mathcal{H} \cong \bigoplus_{j\in J} \mathcal{K}.$$

Using this isomorphism make the following statement rigorous: $\mathbb{B}(\mathcal{H} \otimes \mathcal{K})$ is just the $I \times I$ matrices with entries in $\mathbb{B}(\mathcal{H})$ which define bounded operators on $\bigoplus_i \mathcal{H}$. Under this identification what matrix does the operator $T \otimes 1$ correspond to? How about $1 \otimes S$?

Exercise 3.2.2. Show that

$$\left(\bigoplus_{i\in I} \mathcal{H}_i\right) \otimes \left(\bigoplus_{j\in J} \mathcal{K}_j\right) \cong \bigoplus_{i\in I, j\in J} \mathcal{H}_i \otimes \mathcal{K}_j.$$

Exercise 3.2.3. Wouldn't it be more appropriate to use the notation $\pi_A \times \pi_B$ in Corollary 3.2.4? Is this question rhetorical? If you are confused, see Lemma 3.3.9 in the next section.

Exercise 3.2.4. Show that if $\pi\colon A \odot B \to C$ is a $*$-homomorphism and both A and B are unital, then there exist $*$-homomorphisms $\pi_A\colon A \to C$ and $\pi_B\colon B \to C$, with commuting ranges, such that $\pi = \pi_A \times \pi_B$.

Exercise 3.2.5. Show that Theorem 3.2.6 still holds when $\mathbb{B}(\mathcal{H})$ is replaced by an arbitrary von Neumann algebra. The result fails, in general, if $\mathbb{B}(\mathcal{H})$

is replaced by a C*-algebra. Indeed, let A be a *unital* C*-algebra and let

$$SA = C_0(\mathbb{R}, A) = \{f : \mathbb{R} \to A : f \text{ is continuous and } \lim_{t\to\pm\infty} \|f(t)\| = 0\}$$

be the suspension of A. Show that there is a $*$-homomorphism $C_0(\mathbb{R}) \odot A \to SA$ defined on elementary tensors by $f \otimes a \mapsto f(\cdot)a$. This map cannot be restricted to A since SA has no projections and A is unital.

Exercise 3.2.6. Are the restrictions given by Theorem 3.2.6 unique? Let $\pi_1 \colon A \to \mathbb{B}(\mathcal{H})$ and $\pi_2 \colon B \to \mathbb{B}(\mathcal{H})$ be (possibly degenerate) representations with commuting ranges and let

$$\pi := \pi_1 \times \pi_2 \colon A \odot B \to \mathbb{B}(\mathcal{H}).$$

Find necessary and sufficient conditions which ensure that $\pi_A = \pi_1$ and $\pi_B = \pi_2$.

3.3. The spatial and maximal C*-norms

When A and B are C*-algebras, it can happen that numerous different norms make $A \odot B$ into a pre-C*-algebra. In other words, $A \odot B$ may carry more than one C*-norm.

Definition 3.3.1. A C*-*norm* $\|\cdot\|_\alpha$ on $A \odot B$ is a norm such that $\|xy\|_\alpha \le \|x\|_\alpha\|y\|_\alpha$, $\|x^*\|_\alpha = \|x\|_\alpha$ and $\|x^*x\|_\alpha = \|x\|_\alpha^2$ for all $x, y \in A \odot B$. We will let $A \otimes_\alpha B$ denote the completion of $A \odot B$ with respect to $\|\cdot\|_\alpha$.

The following example is both of fundamental importance and also illustrates the fact that even "trivial" examples in this subject can have subtleties which require care.

Proposition 3.3.2. *For each* C*-*algebra* A *there is a* C*-*norm on the algebraic tensor product* $\mathbb{M}_n(\mathbb{C}) \odot A$ *and it is unique.*

Proof. We assume the reader knows how to make $\mathbb{M}_n(A)$ into a C*-algebra and hence the existence of a C*-norm follows from the existence of an algebraic $*$-isomorphism (Exercise 3.1.3)

$$\mathbb{M}_n(\mathbb{C}) \odot A \cong \mathbb{M}_n(A).$$

Uniqueness is then a consequence of the fact that C*-algebras have unique norms since $\mathbb{M}_n(\mathbb{C}) \odot A$ is a C*-algebra with respect to the norm it gets from $\mathbb{M}_n(A)$.[9] □

[9]Depending on what the phrase "C*-algebras have unique norms" means to you, there may or may not be a subtlety here. If this statement only means, "Whenever an algebra B is a C*-algebra with respect to two norms $\|\cdot\|$ and $\|\cdot\|'$, then those norms agree," then the proof of uniqueness has a gap. Luckily, the more general statement, "If $(B, \|\cdot\|)$ is a C*-algebra and $(B, \|\cdot\|')$ is a pre-C*-algebra (i.e., not necessarily complete), then $\|\cdot\| = \|\cdot\|'$," is true and this is what we are using above.

It requires a little work, but it's a fact that C*-norms on algebraic tensor products always exist. Here are the two most natural candidates.

Definition 3.3.3. (Maximal norm) Given A and B, we define the *maximal* C*-*norm* on $A \odot B$ to be

$$\|x\|_{\max} = \sup\{\|\pi(x)\| : \pi \colon A \odot B \to \mathbb{B}(\mathcal{H}) \text{ a (cyclic) } *\text{-homomorphism}\}$$

for $x \in A \odot B$. We let $A \otimes_{\max} B$ denote the completion of $A \odot B$ with respect to $\|\cdot\|_{\max}$.

Definition 3.3.4. (Spatial norm) Let $\pi \colon A \to \mathbb{B}(\mathcal{H})$ and $\sigma \colon B \to \mathbb{B}(\mathcal{K})$ be *faithful* representations. Then the *spatial* (or *minimal*) C*-norm on $A \odot B$ is

$$\|\sum a_i \otimes b_i\|_{\min} = \|\sum \pi(a_i) \otimes \sigma(b_i)\|_{\mathbb{B}(\mathcal{H} \otimes \mathcal{K})}.$$

The completion of $A \odot B$ with respect to $\|\cdot\|_{\min}$ is denoted $A \otimes B$.[10]

Remark 3.3.5 (Von Neumann algebra tensor products). If $M \subset \mathbb{B}(\mathcal{H})$ and $N \subset \mathbb{B}(\mathcal{K})$ are von Neumann algebras, then there are a number of C*-norms that one can put on $M \odot N$. However, the norm completions won't be von Neumann algebras and researchers have virtually forgotten about the subject of C*-tensor products of von Neumann algebras. On the other hand, the *von Neumann algebraic* tensor product is still very important and is denoted by $M \bar{\otimes} N$. By definition, this is the von Neumann algebra generated by $M \otimes \mathbb{C}1_{\mathcal{K}} \subset \mathbb{B}(\mathcal{H} \otimes \mathcal{K})$ and $\mathbb{C}1_{\mathcal{H}} \otimes N \subset \mathbb{B}(\mathcal{H} \otimes \mathcal{K})$ – i.e., the weak closure of $M \otimes N \subset \mathbb{B}(\mathcal{H} \otimes \mathcal{K})$.

Remark 3.3.6 (Operator space tensor products). For completeness we also mention that one defines the spatial tensor product norm on operator systems (or spaces) in exactly the same way. Given X and Y, we take embeddings $X \subset \mathbb{B}(\mathcal{H})$ and $Y \subset \mathbb{B}(\mathcal{K})$ which induce the given operator space structures and then define $X \otimes Y$ to be the norm closure of the span of $\{x \otimes y \in \mathbb{B}(\mathcal{H} \otimes \mathcal{K}) : x \in X, y \in Y\}$. As we'll soon see for C*-algebras, $X \otimes Y$ is independent of the embeddings (so long as they induce the proper operator space structures, of course).

There are numerous technical points which one should worry about. The first is whether or not $\|\cdot\|_{\max}$ is even finite. This is the case thanks to the existence of restrictions (Theorem 3.2.6). Indeed, if $\pi : A \odot B \to \mathbb{B}(\mathcal{H})$ is a $*$-representation with restrictions π_A and π_B, then

$$\|\pi(a \otimes b)\| \leq \|\pi_A(a)\| \|\pi_B(b)\| \leq \|a\| \|b\|$$

for all elementary tensors. This implies that $\|x\|_{\max} < \infty$ for all $x \in A \odot B$.

[10] You will also see $A \otimes_{\min} B$ in the literature.

The remainder of this section is devoted to resolving the following technical issues.

(1) Are $\|\cdot\|_{\max}$ and $\|\cdot\|_{\min}$ norms (as opposed to seminorms)?[11]

(2) Is $\|\cdot\|_{\min}$ independent of the choice of faithful representations?

(3) Can one usually reduce the nonunital case to the unital case?

All three questions have affirmative answers, though none are completely obvious.

Let us first tackle the norm vs. seminorm question. The following *universal property of* $\|\cdot\|_{\max}$ implies that it suffices to show $\|\cdot\|_{\min}$ is a norm.

Proposition 3.3.7 (Universality). *If $\pi\colon A \odot B \to C$ is a $*$-homomorphism, then there exists a unique $*$-homomorphism $A \otimes_{\max} B \to C$ which extends π. In particular, any pair of $*$-homomorphisms with commuting ranges $\pi_A\colon A \to C$ and $\pi_B\colon B \to C$ induces a unique $*$-homomorphism*

$$\pi_A \times \pi_B\colon A \otimes_{\max} B \to C.$$

Proof. Faithfully representing C on some Hilbert space, this fact follows from the definition of $\|\cdot\|_{\max}$. □

Corollary 3.3.8. *The norm $\|\cdot\|_{\max}$ is the largest possible* C*-*norm on* $A \odot B$.

Proof. If $\|\cdot\|_\alpha$ is any other C*-norm on $A \odot B$, then, by universality, there is a (surjective) $*$-homomorphism $A \otimes_{\max} B \to A \otimes_\alpha B$. Hence, $\|x\|_\alpha \le \|x\|_{\max}$ for every $x \in A \odot B$. □

In particular, $\|\cdot\|_{\max}$ dominates $\|\cdot\|_{\min}$ and thus, if $\|x\|_{\min} = 0 \Rightarrow x = 0$, then it will follow that both $\|\cdot\|_{\max}$ and $\|\cdot\|_{\min}$ are honest norms.

Lemma 3.3.9. *The product $*$-homomorphism $\mathbb{B}(\mathcal{H}) \odot \mathbb{B}(\mathcal{K}) \to \mathbb{B}(\mathcal{H} \otimes \mathcal{K})$, induced by the commuting $*$-representations $\mathbb{B}(\mathcal{H}) \cong \mathbb{B}(\mathcal{H}) \otimes \mathbb{C}1_{\mathcal{K}} \subset \mathbb{B}(\mathcal{H}\otimes\mathcal{K})$ and $\mathbb{B}(\mathcal{K}) \cong \mathbb{C}1_{\mathcal{H}} \otimes \mathbb{B}(\mathcal{K}) \subset \mathbb{B}(\mathcal{H} \otimes \mathcal{K})$, is injective.*

Proof. We must show that if a finite sum of tensor product operators $\sum_i S_i \otimes T_i \in \mathbb{B}(\mathcal{H} \otimes \mathcal{K})$ is zero, then the corresponding sum of elementary tensors $\sum_i S_i \otimes T_i \in \mathbb{B}(\mathcal{H}) \odot \mathbb{B}(\mathcal{K})$ is also zero. We may assume that the operators $\{S_i\} \subset \mathbb{B}(\mathcal{H})$ are linearly independent.

If $0 = \sum_i S_i \otimes T_i \in \mathbb{B}(\mathcal{H} \otimes \mathcal{K})$, then for all vectors $v, w \in \mathcal{H}$ and $\xi, \eta \in \mathcal{K}$ we have

$$\langle (\sum_i S_i \otimes T_i) v \otimes \xi, w \otimes \eta \rangle = 0.$$

[11]Since both of these (semi)norms are defined via $*$-representations and honest C*-norms, an affirmative answer to this question will imply that both $\|\cdot\|_{\max}$ and $\|\cdot\|_{\min}$ are C*-norms.

Rearranging terms, we get

$$\begin{aligned}\langle(\sum_i S_i\otimes T_i)v\otimes\xi, w\otimes\eta\rangle &= \sum_i\langle S_i\otimes T_i(v\otimes\xi), w\otimes\eta\rangle\\ &= \sum_i\langle S_i v, w\rangle\langle T_i\xi,\eta\rangle\\ &= \langle(\sum_i\langle T_i\xi,\eta\rangle S_i)v, w\rangle.\end{aligned}$$

Since this holds for all $v, w \in \mathcal{H}$, it follows that the operator $\sum_i\langle T_i\xi,\eta\rangle S_i \in \mathbb{B}(\mathcal{H})$ is zero and hence, by linear independence, that each of the coefficients $\langle T_i\xi,\eta\rangle$ is zero. Since this holds for all $\xi,\eta\in\mathcal{K}$, it follows that $0 = T_i \in \mathbb{B}(\mathcal{K})$ for all i, and the proof is complete. □

Corollary 3.3.10. *For each $x \in A\odot B$, if $\|x\|_{\min} = 0$, then $x = 0$.*

Proof. If $\pi\colon A \to \mathbb{B}(\mathcal{H})$ and $\sigma\colon B \to \mathbb{B}(\mathcal{K})$ are faithful representations, then, by Proposition 3.1.12, the tensor product map

$$\pi\odot\sigma\colon A\odot B \to \mathbb{B}(\mathcal{H})\odot\mathbb{B}(\mathcal{K})$$

is also injective. Together with the previous lemma this implies the result. □

We now resolve the second technical question.

Proposition 3.3.11. *The spatial tensor product norm is independent of the choices of* faithful *representations $\pi\colon A\to\mathbb{B}(\mathcal{H})$ and $\sigma\colon B\to\mathbb{B}(\mathcal{K})$.*

Proof. For the moment we will let $\|\cdot\|_{\min}^{(\pi,\sigma)}$ denote the minimal norm with respect to π and σ. Evidently it suffices to prove that if $\sigma'\colon B\to\mathbb{B}(\mathcal{K}')$ is another faithful representation, then $\|\cdot\|_{\min}^{(\pi,\sigma)} = \|\cdot\|_{\min}^{(\pi,\sigma')}$.

For notational reasons it is slightly more convenient to give the proof in the separable setting. It is a simple exercise to net-ify the argument and deduce the general case. Let $P_1 \le P_2 \le \cdots$ be finite-rank projections in $\mathbb{B}(\mathcal{H})$ such that P_n has rank n and $\|P_n(h) - h\| \to 0$ for all $h\in\mathcal{H}$. Then it is not hard to show that for every $X\in\mathbb{B}(\mathcal{H}\otimes\mathcal{K})$ we have

$$\|X\| = \sup_n\{\|(P_n\otimes 1_{\mathcal{K}})X(P_n\otimes 1_{\mathcal{K}})\|\}.$$

Thus, if $\sum a_i\otimes b_i\in A\odot B$ is arbitrary, we have

$$\|\sum a_i\otimes b_i\|_{\min}^{(\pi,\sigma)} = \sup_n\{\|\sum(P_n\pi(a_i)P_n)\otimes\sigma(b_i)\|\}$$

and

$$\|\sum a_i\otimes b_i\|_{\min}^{(\pi,\sigma')} = \sup_n\{\|\sum(P_n\pi(a_i)P_n)\otimes\sigma'(b_i)\|\}.$$

But since $P_n\mathbb{B}(\mathcal{H})P_n$ is naturally isomorphic to $\mathbb{M}_n(\mathbb{C})$, we have

$$\|\sum(P_n\pi(a_i)P_n)\otimes\sigma(b_i)\| = \|\sum(P_n\pi(a_i)P_n)\otimes\sigma'(b_i)\|$$

for each n, since there is a unique C*-norm on $\mathbb{M}_n(\mathbb{C})\odot B$ (Proposition 3.3.2). □

Finally we present a result which allows many nonunital questions to be handled (relatively) painlessly. For a nonunital C*-algebra A we will let $\tilde{A}$ denote the unitization.

Corollary 3.3.12. *If A is nonunital, then any* C*-*norm $\|\cdot\|_\alpha$ on $A\odot B$ can be extended to a* C*-*norm on $\tilde{A}\odot B$. Hence, when both A and B are nonunital, any* C*-*norm $\|\cdot\|_\alpha$ on $A\odot B$ can be extended to a* C*-*norm on $\tilde{A}\odot\tilde{B}$.*

Proof. Let $\|\cdot\|_\alpha$ be a C*-norm on $A\odot B$ and $\pi\colon A\otimes_\alpha B\to\mathbb{B}(\mathcal{H})$ be a faithful representation of the completion of $A\odot B$ with respect to $\|\cdot\|_\alpha$. Let π_A and π_B be the restrictions given by Theorem 3.2.6. Since π is faithful, it follows that π_A is faithful and hence may be extended to a faithful $*$-homomorphism $\tilde{\pi}_A\colon\tilde{A}\to\mathbb{B}(\mathcal{H})$. Since the ranges of $\tilde{\pi}_A$ and π_B still commute, we can consider the product map $\tilde{\pi}_A\times\pi_B\colon\tilde{A}\odot B\to\mathbb{B}(\mathcal{H})$. If we knew that $\tilde{\pi}_A\times\pi_B$ were injective, then we would be done since the norm on $\mathbb{B}(\mathcal{H})$ would restrict to a C*-norm on $\tilde{\pi}_A\times\pi_B(\tilde{A}\odot B)\cong\tilde{A}\odot B$ which would agree with $\|\cdot\|_\alpha$ on $A\odot B$. If you didn't already do Exercise 3.1.6, now is the time. □

Exercises

Exercise 3.3.1. Show that both $\|\cdot\|_{\min}$ and $\|\cdot\|_{\max}$ are *commutative* tensor product norms – i.e., there are canonical isomorphisms $A\otimes B\cong B\otimes A$ and $A\otimes_{\max}B\cong B\otimes_{\max}A$.

Exercise 3.3.2. Show that both $\|\cdot\|_{\min}$ and $\|\cdot\|_{\max}$ are *associative* – i.e., there are canonical isomorphisms $(A\otimes B)\otimes C\cong A\otimes(B\otimes C)$ and $(A\otimes_{\max}B)\otimes_{\max}C\cong A\otimes_{\max}(B\otimes_{\max}C)$. How would you define the maximal or minimal tensor product of n algebras?

Exercise 3.3.3. Give an example of a $*$-representation $\pi\colon A\odot B\to\mathbb{B}(\mathcal{H})$ such that both π_A and π_B are injective but π is not. (Hint: Think finite dimensional and abelian.)

Exercise 3.3.4. Prove that if $\pi\colon A\to\mathbb{B}(\mathcal{H})$ and $\sigma\colon B\to\mathbb{B}(\mathcal{K})$ are arbitrary (not necessarily faithful) representations, then there exists a unique extending $*$-homomorphism $\pi\otimes\sigma\colon A\otimes B\to\mathbb{B}(\mathcal{H}\otimes\mathcal{K})$ such that $\pi\otimes\sigma(a\otimes b)=\pi(a)\otimes\sigma(b)$. (Hint: Dilate π and σ to faithful representations and then cut back down; see Exercise 3.2.2.)

Exercise 3.3.5. If $\pi\colon A \to C$ and $\sigma\colon B \to D$ are $*$-homomorphisms, prove that there is a unique $*$-homomorphism $\pi \otimes \sigma\colon A \otimes B \to C \otimes D$ such that $\pi \otimes \sigma(a \otimes b) = \pi(a) \otimes \sigma(b)$.

Exercise 3.3.6. Prove that $\mathbb{B}(\ell^2) \odot \mathbb{B}(\ell^2) \subset \mathbb{B}(\ell^2 \otimes \ell^2)$ (see Lemma 3.3.9) is not dense in norm.

3.4. Takesaki's Theorem

This section is devoted to the fact that $\|\cdot\|_{\min}$ is really the smallest possible C*-norm on $A \odot B$. The main ingredient in the proof is Proposition 3.4.7 – for any pair of functionals $\varphi \in A^*$, $\psi \in B^*$ and any C*-norm $\|\cdot\|_\alpha$ on $A \odot B$, the natural linear functional $\varphi \odot \psi$ is *automatically* continuous with respect to $\|\cdot\|_\alpha$. There are two routes to this result. We first give Takesaki's original argument – a lovely display of soft analysis – then, at the end of the section, we sketch an alternate proof based on excision.

The first lemma may already be familiar. If not, it is a fact worth remembering.

Lemma 3.4.1. *Let $S(A)$ denote the state space of a C*-algebra A. Assume that $\mathcal{S} \subset S(A)$ is a set of states with the property that for each* self-adjoint *$a \in A$ we have*

$$\|a\| = \sup_{\varphi \in \mathcal{S}}\{|\varphi(a)|\}.$$

Then the weak-$*$ closed convex hull of $\mathcal{S}$ contains $S(A)$ (hence is equal to $S(A)$ when A is unital).

Proof. Let S denote the closed convex hull of $\mathcal{S}$ and assume there is a state ψ which doesn't belong to S. By the Hahn-Banach Theorem we can find $a \in A$ and a real number t such that $\mathrm{Re}(\varphi(a)) < t < \mathrm{Re}(\psi(a))$ for every $\varphi \in S$. (Recall that A is the dual of A^* with respect to the weak-$*$ topology.) Replacing a with its real part, we may assume a is self-adjoint. Note that if a happened to be positive, then we would have our contradiction since $\psi(a) \leq \|a\|$. So the trick is to replace a with a positive element. In the unital case one considers $a + \|a\|1_A \geq 0$ and notes that

$$0 \leq \varphi(a + \|a\|1_A) = \varphi(a) + \|a\| < t + \|a\| < \psi(a + \|a\|1_A).$$

Hence

$$\|a+\|a\|1_A\| = \sup_{\varphi \in \mathcal{S}}\{|\varphi(a+\|a\|1_A)|\} \leq t+\|a\| < \psi(a+\|a\|1_A) \leq \|a+\|a\|1_A\|,$$

which gives the contradiction. In the nonunital case one passes to unitizations and applies the argument above. □

The next two results will allow us to deduce that certain states on tensor products can always be decomposed into a product of states.

Lemma 3.4.2. *If $A \subset C$, $B = A' \cap C$ is the relative commutant of A and if ξ is a state on C such that $\xi|_A$ is a pure state on A, then $\xi(ab) = \xi(a)\xi(b)$ for all $a \in A$, $b \in B$.*

Proof. By a standard reduction we may assume that $b \in B$ is a positive element of norm one and we will show $\xi(ab) = \xi(a)\xi(b)$ for all $a \in A$.

Case 1: Assume $\xi(b) = 0$. Then for all $a \in A$ we have that

$$|\xi(ab)|^2 = |\xi((ab^{1/2})(b^{1/2}))|^2 \leq \xi(aba^*)\xi(b),$$

by the Cauchy-Schwarz inequality. Hence $\xi(ab) = 0 = \xi(a)\xi(b)$.

Case 2: Assume $\xi(b) = 1$. Then (passing to unitizations, if necessary) $\xi(1-b) = 0$, so by Case 1 we have $0 = \xi(a(1-b)) = \xi(a) - \xi(ab)$. Thus $\xi(ab) = \xi(a) = \xi(a)\xi(b)$.

Case 3: Assume $0 < \xi(b) < 1$. Then we have

$$\xi(a) = \xi(b)\left(\frac{1}{\xi(b)}\xi(ab)\right) + (1-\xi(b))\left(\frac{1}{1-\xi(b)}\xi(a(1-b))\right).$$

This equation shows that the restriction of ξ to A is a convex combination of states on A. (Since b commutes with A, the functionals $a \mapsto \frac{1}{\xi(b)}\xi(ab)$ and $a \mapsto \frac{1}{1-\xi(b)}\xi(a(1-b))$ are states.) But we assumed this restriction to be pure and so it follows that $\xi(a) = \frac{1}{\xi(b)}\xi(ab)$ for all $a \in A$, so the proof is complete. □

Let $\|\cdot\|_\alpha$ be a C*-norm on $A \odot B$, $A \otimes_\alpha B$ be the completion and ξ be a state on $A \otimes_\alpha B$. We define the *restrictions* $\xi|_A$ and $\xi|_B$ as follows. Let $(\pi_\xi, \mathcal{H}_\xi, v_\xi)$ be the GNS triplet and $\pi_{\xi,A}$ and $\pi_{\xi,B}$ be the restriction homomorphisms given by Theorem 3.2.6. Then define $\xi|_A(a) = \langle \pi_{\xi,A}(a)v_\xi, v_\xi\rangle$ and $\xi|_B(b) = \langle \pi_{\xi,B}(b)v_\xi, v_\xi\rangle$.

Corollary 3.4.3. *Let $\|\cdot\|_\alpha$ be a C*-norm on $A \odot B$ and ξ be a state on $A \otimes_\alpha B$.*

(1) *If the restriction $\xi|_A$ is pure, then $\xi|_{A\odot B} = \xi|_A \odot \xi|_B$.*
(2) *If A is abelian and ξ is a pure state on $A \otimes_\alpha B$, then both $\xi|_A$ and $\xi|_B$ are pure and $\xi|_{A\odot B} = \xi|_A \odot \xi|_B$.*

Proof. Assume that $\xi|_A$ is pure. Applying the previous lemma to the vector state $T \mapsto \langle Tv_\xi, v_\xi\rangle$ on $\mathbb{B}(\mathcal{H}_\xi)$, it follows that

$$\begin{aligned}\xi(a \otimes b) &= \langle \pi_{\xi,A}(a)\pi_{\xi,B}(b)v_\xi, v_\xi\rangle \\ &= \langle \pi_{\xi,A}(a)v_\xi, v_\xi\rangle\langle \pi_{\xi,B}(b)v_\xi, v_\xi\rangle\end{aligned}$$

$$= \xi|_A(a)\xi|_B(b).$$

Hence $\xi(x) = \xi|_A \odot \xi|_B(x)$ for all $x \in A \odot B$.

To prove the second assertion, we assume A is abelian and ξ is a pure state on $A \otimes_\alpha B$. Thus the weak closure of $\pi_\xi(A \otimes_\alpha B)$ is all of $\mathbb{B}(\mathcal{H}_\xi)$. But $\pi_\xi(A \otimes_\alpha B)$ is contained in the C*-algebra generated by $\pi_{\xi,A}(A)$ and $\pi_{\xi,B}(B)$ and since A is abelian and $\pi_{\xi,A}$ and $\pi_{\xi,B}$ have commuting ranges, we see that $\pi_{\xi,A}(A)$ lives in the center of $\mathbb{B}(\mathcal{H}_\xi)$ – i.e., $\pi_{\xi,A}(A) = \mathbb{C}1$. It follows that $\xi|_A$ is a character (hence pure) and the weak closure of $\pi_{\xi,B}(B)$ is all of $\mathbb{B}(\mathcal{H}_\xi)$. Hence $\xi|_B$ is also pure and we are done. □

Lemma 3.4.4. *Let $\varphi \in S(A)$ and $\psi \in S(B)$. Then $\varphi \odot \psi : A \odot B \to \mathbb{C}$ is* algebraically positive *– i.e., $\varphi \odot \psi(x^*x) \geq 0$ for all $x \in A \odot B$.*

Proof. Let $(\pi_\varphi, \mathcal{H}_\varphi, v_\varphi)$ and $(\pi_\psi, \mathcal{H}_\psi, v_\psi)$ be the GNS representations. By Exercise 3.3.4 we have a representation $\pi_\varphi \otimes \pi_\psi \colon A \otimes B \to \mathbb{B}(\mathcal{H}_\varphi \otimes \mathcal{H}_\psi)$. A calculation shows that $\varphi \odot \psi(x)$ is equal to $\langle \pi_\varphi \otimes \pi_\psi(x) v_\varphi \otimes v_\psi, v_\varphi \otimes v_\psi \rangle$ for all $x \in A \odot B$, which implies the result. □

We now begin to show that product states are always continuous on $A \otimes_\alpha B$. The process starts by considering the abelian case and building up (via Lemma 3.4.1 and Corollary 3.4.3) to the general case. The target result is Proposition 3.4.7, but if reading the next two lemmas in order makes little sense, try jumping ahead to Proposition 3.4.7 and reading backwards.

Lemma 3.4.5. *Assume that both A and B are unital and abelian. Then for every C*-norm $\|\cdot\|_\alpha$ on $A \odot B$ and pair of* pure *states $\varphi \in S(A)$, $\psi \in S(B)$ the linear functional $\varphi \odot \psi : A \odot B \to \mathbb{C}$ extends to a state on $A \otimes_\alpha B$.*

Proof. Let $P(A)$ (resp. $P(B)$) denote the pure states on A (resp. B). Let

$$\mathcal{U} = \{(\varphi, \psi) \in P(A) \times P(B) : |\varphi \odot \psi(x)| \leq \|x\|_\alpha, \forall x \in A \odot B\}.$$

Then the lemma asserts that $\mathcal{U} = P(A) \times P(B)$.[12]

So assume that $\mathcal{U}$ is a proper subset of $P(A) \times P(B)$. A standard calculation shows that $\mathcal{U}$ is a closed subset of $P(A) \times P(B)$ (in the product of the weak-$*$ topologies) and hence we can find open sets $U_A \subset P(A)$ and $U_B \subset P(B)$ such that $(U_A \times U_B) \cap \mathcal{U} = \emptyset$. Let $0 \leq f \in A$ and $0 \leq g \in B$ be norm-one elements whose supports (via the Gelfand transform) live in U_A and U_B, respectively. Then $\varphi \odot \psi(f \otimes g) = 0$ for every $(\varphi, \psi) \in \mathcal{U}$.

This will lead to a contradiction since we can construct a pair $(\varphi, \psi) \in \mathcal{U}$ such that $\varphi \odot \psi(f \otimes g) > 0$. Indeed, let $\xi \in P(A \otimes_\alpha B)$ be a pure state such

[12]If $(\varphi, \psi) \in \mathcal{U}$, then $\varphi \odot \psi$ extends to a contractive linear functional on $A \otimes_\alpha B$. One then either appeals to algebraic positivity of $\varphi \odot \psi$ to deduce that the extension is a state or to the fact that norm-one functionals mapping the identity to 1 are necessarily positive.

that $\xi(f \otimes g) > 0$ and let $\xi|_A$ and $\xi|_B$ be the restrictions. Corollary 3.4.3 gives the desired conclusion since ξ is an extension of $\xi|_A \odot \xi|_B$. □

Lemma 3.4.6. *Let B be unital, A be unital and abelian and φ be a pure state on A. Then for every* C*-*norm $\|\cdot\|_\alpha$ on $A \odot B$ and every state $\psi \in S(B)$ the linear functional $\varphi \odot \psi : A \odot B \to \mathbb{C}$ extends to a state on $A \otimes_\alpha B$.*

Proof. Let $\mathcal{S}_\varphi = \{\psi \in S(B) : |\varphi \odot \psi(x)| \leq \|x\|_\alpha, \forall x \in A \odot B\}$. Note that $\mathcal{S}_\varphi$ is convex and weak-$*$ closed in $S(B)$. This lemma asserts that $\mathcal{S}_\varphi = S(B)$ and to show this, we invoke Lemma 3.4.1. So let $h \in B$ be any self-adjoint element and let $C = C^*(h, 1_B)$ be the unital C*-algebra generated by h. Let $A \otimes_\alpha C$ be the completion of $A \odot C$ with respect to the restriction of $\|\cdot\|_\alpha$. Let $\psi \in P(C)$ be a pure state such that $|\psi(h)| = \|h\|$ and let $\varphi \otimes \psi : A \otimes_\alpha C \to \mathbb{C}$ be the extension given by the last lemma. Note that $\varphi \otimes \psi$ is a pure state and hence extends to a state ξ on $A \otimes_\alpha B$. Since $\xi|_A = \varphi$ is pure, Corollary 3.4.3 tells us that ξ is an extension of $\varphi \odot \xi|_B$. Hence $\xi|_B \in \mathcal{S}_\varphi$. Moreover, $\|h\| = |\xi|_B(h)|$ since $\xi|_B(h) = \psi(h)$. Thus Lemma 3.4.1 implies that $\mathcal{S}_\varphi = S(B)$. □

We finally come to the main ingredient in the proof of Takesaki's result.

Proposition 3.4.7. *Let A and B be unital* C*-*algebras with states $\varphi \in S(A)$ and $\psi \in S(B)$. For any* C*-*norm $\|\cdot\|_\alpha$ on $A \odot B$, $\varphi \odot \psi$ extends to a state on $A \otimes_\alpha B$.*

Proof. We apply essentially the same trick as in the last lemma – but we must do it twice and with a bit of care. So first fix a *pure* state $\varphi \in S(A)$ and again let $\mathcal{S}_\varphi = \{\psi \in S(B) : |\varphi \odot \psi(x)| \leq \|x\|_\alpha, \forall x \in A \odot B\}$. The first step is to show that for every self-adjoint $h \in B$,

$$\|h\| = \sup_{\psi \in \mathcal{S}_\varphi} |\psi(h)|.$$

So choose a pure state ψ on $C = C^*(h, 1_B)$ such that $|\psi(h)| = \|h\|$. By the previous lemma we may extend $\varphi \odot \psi$ to a state on $A \otimes_\alpha C \subset A \otimes_\alpha B$. We further extend to a state ξ on $A \otimes_\alpha B$. Since $\xi|_A = \varphi$ is pure, Corollary 3.4.3 implies ξ is an extension of $\varphi \odot \xi|_B$ and as in the last lemma we conclude that $\mathcal{S}_\varphi = S(B)$.

Now we reverse the roles a bit. Fix a state $\psi \in S(B)$ and let $\mathcal{S}_\psi = \{\varphi \in S(A) : |\varphi \odot \psi(x)| \leq \|x\|_\alpha, \forall x \in A \odot B\}$. The previous paragraph shows $\mathcal{S}_\psi$ contains all the pure states of A. Thus, the Krein-Milman Theorem implies $\mathcal{S}_\psi = S(A)$, as desired. □

Theorem 3.4.8 (Takesaki). *For arbitrary* C*-*algebras A and B, $\|\cdot\|_{min}$ is the smallest* C*-*norm on $A \odot B$.*

Proof. Assume first that both A and B are unital and separable. Then we can find *faithful* states $\varphi \in S(A)$ and $\psi \in S(B)$. Let $\|\cdot\|_\alpha$ be any C*-norm on $A \odot B$ and by the previous lemma we may extend $\varphi \odot \psi$ to a state $\varphi \otimes_\alpha \psi$ on $A \otimes_\alpha B$. By uniqueness of GNS representations it follows that the representations $\pi_{\varphi\otimes_\alpha\psi}|_{A\odot B}$ and $\pi_\varphi \odot \pi_\psi$ are unitarily equivalent. Since we chose faithful states, both π_φ and π_ψ are faithful representations. Proposition 3.3.11 then implies that the norm closure of $\pi_\varphi \odot \pi_\psi(A \odot B)$ is isomorphic to $A \otimes B$. Hence $A \otimes B$ is also isomorphic to $\pi_{\varphi\otimes_\alpha\psi}(A \otimes_\alpha B)$ which, being a quotient of $A \otimes_\alpha B$, shows that $\|\cdot\|_{\min} \leq \|\cdot\|_\alpha$.

The nonunital case can be deduced from the unital one with the help of Corollary 3.3.12. In the nonseparable setting we first fix an arbitrary element

$$x = \sum_{i=1}^{n} a_i \otimes b_i \in A \odot B \subset A \otimes_\alpha B.$$

Letting $A_1 = C^*(a_1, \ldots, a_n)$ and $B_1 = C^*(b_1, \ldots, b_n)$, we have a natural inclusion $A_1 \otimes B_1 \subset A \otimes B$. (Just think about what this should mean and it will become an obvious consequence of Proposition 3.3.11 – otherwise see Proposition 3.6.1.) Thus the separable case implies that $\|x\|_{\min} \leq \|x\|_\alpha$ and we are done. □

The following corollary gets used, both explicitly and implicitly, all of the time. For example, in the literature it is often written that $A \odot B$ has a unique C*-norm if and only if $A \otimes_{\max} B = A \otimes B$.

Corollary 3.4.9. *For any A and B and any C*-norm $\|\cdot\|_\alpha$ on $A \odot B$ we have natural surjective $*$-homomorphisms*

$$A \otimes_{\max} B \to A \otimes_\alpha B \to A \otimes B.$$

It is a fact that every C*-norm on $A \odot B$ is a *cross norm* (i.e., $\|a \otimes b\|_\alpha = \|a\|\|b\|$ for all elementary tensors $a \otimes b \in A \odot B$). One can deduce this from Takesaki's Theorem (since Proposition 3.2.3 implies that $\otimes$ is a cross norm and $\otimes_{\max}$ is subcross). However, we give an independent proof.

Lemma 3.4.10. *If $\|\cdot\|_\alpha$ is a C*-norm on $A \odot B$, then it is a cross norm.*

Proof. Let $A \otimes_\alpha B \subset \mathbb{B}(\mathcal{H})$. By the C*-equation, it suffices to show $\|a \otimes b\| = 1$ for positive norm-one elements $a \in A$ and $b \in B$. Let $\varepsilon > 0$ be given and let $e = \chi_{[1-\varepsilon,1]}(a)$ and $f = \chi_{[1-\varepsilon,1]}(b)$ be spectral projections. The projections e and f commute (since they belong to the weak closures of the ranges of the commuting restriction maps) and we claim that $ef \neq 0$. To see this, note that functional calculus provides nonzero elements $a_0 \in C^*(a)$ and $b_0 \in C^*(b)$ such that $a_0 \leq e$ and $b_0 \leq f$; hence $0 \neq a_0 \otimes b_0 \leq ef$.

Thus, for a unit vector $\xi \in ef\mathcal{H}$, we have $\|(a \otimes b)(\xi) - \xi\| \leq 2\varepsilon$. Since ε was arbitrary, $\|a \otimes b\| = 1$. □

Another proof of Proposition 3.4.7 . We must show $\varphi \otimes \psi$ is continuous on $A \otimes_\alpha B$. By convexity arguments, we may assume that φ and ψ are pure. Find some excising nets (e_i) and (f_i) for φ and ψ, respectively (see Theorem 1.4.10). It follows that for $x = \sum_{k=1}^n a_k \otimes b_k \in A \odot B$, we have

$$\|x\|_\alpha \geq \lim_i \|(e_i \otimes f_i)x(e_i \otimes f_i)\| = \lim_i \|(\varphi \otimes \psi)(x)(e_i \otimes f_i)^2\| = |(\varphi \otimes \psi)(x)|.$$

□

Exercises

Exercise 3.4.1. Let $\pi\colon A \otimes B \to C$ be a $*$-homomorphism which is injective when restricted to $A \odot B$. Show that π must be injective on all of $A \otimes B$. Is this still true if one replaces $\|\cdot\|_{\mathrm{min}}$ by $\|\cdot\|_{\mathrm{max}}$?

Exercise 3.4.2. Prove that $A \otimes_\alpha B$ is simple if and only if $\|\cdot\|_\alpha = \|\cdot\|_{\mathrm{min}}$ and both A and B are simple. (Hint: For the "if" direction it suffices to show that every irreducible representation of $A \otimes B$ is faithful. But if $\pi\colon A \otimes B \to \mathbb{B}(\mathcal{H})$ is irreducible, then both $\pi_A(A)''$ and $\pi_B(B)''$ must be factors. One then uses the previous exercise and a theorem of Murray and von Neumann which asserts that if $M \subset \mathbb{B}(\mathcal{H})$ is a factor, then the product map

$$M \odot M' \to \mathbb{B}(\mathcal{H})$$

is injective (see [**183**, Proposition IV.4.20]).

3.5. Continuity of tensor product maps

Given how delicate the proof of Takesaki's Theorem is, it's no surprise that continuity of maps on tensor products requires some care. It turns out that nothing funny happens so long as one sticks to c.p. maps, but this is the largest class of maps which always behave well. To get a feel for what can go wrong, let's consider a finite-dimensional example.

Proposition 3.5.1. *Let $\varphi\colon \mathbb{M}_n(\mathbb{C}) \to \mathbb{M}_n(\mathbb{C})$ be the usual transpose map on the $n \times n$ matrices. Then φ is a unital, positive isometry but the norm of*

$$\varphi \otimes \mathrm{id}_{\mathbb{M}_n(\mathbb{C})}\colon \mathbb{M}_n(\mathbb{C}) \otimes \mathbb{M}_n(\mathbb{C}) \to \mathbb{M}_n(\mathbb{C}) \otimes \mathbb{M}_n(\mathbb{C})$$

is greater than or equal to n.[13]

[13] Actually, it is equal to n, but we won't need this fact.

Proof. It's well known, and easily verified, that the transpose map is a positive isometry. Let $\{e_{i,j}\}_{1\leq i,j\leq n}$ be a system of matrix units for $\mathbb{M}_n(\mathbb{C})$ and consider

$$S := \sum_{i,j=1}^{n} e_{i,j} \otimes e_{j,i}.$$

Evidently S is a permutation matrix – hence unitary – and has norm one. (If $\{\delta_k\}$ is an orthonormal basis, then $S(\delta_k \otimes \delta_l) = \delta_l \otimes \delta_k$.) On the other hand

$$\varphi \otimes \mathrm{id}_{\mathbb{M}_n(\mathbb{C})}(S) = \sum_{i,j=1}^{n} e_{j,i} \otimes e_{j,i}$$

and a straightforward computation shows that this matrix is equal to nP where P is the one-dimensional projection onto the span of the vector

$$v = \sum_{k=1}^{n} \delta_k \otimes \delta_k.$$

□

Unlike the case of states, this shows that the tensor product of norm-one maps need not have norm one – even on the 2×2 matrices when one map is the identity and the other is a positive unital isometry! The next result follows easily from the previous one.

Proposition 3.5.2. *Let $\varphi\colon A \to A$ be a positive, unital isometry. It can happen that $\varphi \odot \mathrm{id}_A\colon A \odot A \to A \odot A$ is unbounded. For example, let φ be the transpose map on the unitization of the compact operators.*

Not wanting to dwell on the problems that occur for more general maps, let's treat the case of c.p. maps and move on. We will need the following result approximately $\aleph_0$ times (maybe more).

Theorem 3.5.3 (Continuity of tensor product maps)**.** *Let $\varphi\colon A \to C$ and $\psi\colon B \to D$ be c.p. maps. Then the algebraic tensor product map*

$$\varphi \odot \psi\colon A \odot B \to C \odot D$$

extends to a c.p. (hence continuous) map on both the minimal and maximal tensor products. Moreover, letting $\varphi \otimes_{\max} \psi\colon A \otimes_{\max} B \to C \otimes_{\max} D$ and $\varphi \otimes \psi\colon A \otimes B \to C \otimes D$ denote the extensions, we have

$$\|\varphi \otimes_{\max} \psi\| = \|\varphi \otimes \psi\| = \|\varphi\|\|\psi\|.$$

Proof. We first handle the spatial tensor product case. Assume $C \subset \mathbb{B}(\mathcal{H})$ and $D \subset \mathbb{B}(\mathcal{K})$. Let $\pi_A\colon A \to \mathbb{B}(\tilde{\mathcal{H}})$, $\pi_B\colon B \to \mathbb{B}(\tilde{\mathcal{K}})$ be the Stinespring dilations of φ and ψ, respectively, and $V_A\colon \mathcal{H} \to \tilde{\mathcal{H}}$, $V_B\colon \mathcal{K} \to \tilde{\mathcal{K}}$ the associated bounded linear operators. By Exercise 3.3.4 there is a natural

$*$-homomorphism $\pi_A \otimes \pi_B \colon A \otimes B \to \mathbb{B}(\tilde{\mathcal{H}} \otimes \tilde{\mathcal{K}})$. Hence we may define $\varphi \otimes \psi \colon A \otimes B \to C \otimes D$ by the formula

$$\varphi \otimes \psi(x) = (V_A \otimes V_B)^* \pi_A \otimes \pi_B(x)(V_A \otimes V_B).$$

Note that on elementary tensors we have

$$\varphi \otimes \psi(a \otimes b) = (V_A^* \pi_A(a) V_A) \otimes (V_B^* \pi_B(b) V_B) = \varphi(a) \otimes \psi(b).$$

Hence $\varphi \otimes \psi$ really is a c.p. extension of $\varphi \odot \psi$ which takes values in $C \otimes D \subset \mathbb{B}(\mathcal{H} \otimes \mathcal{K})$. Finally note that the completely bounded norm (cf. Appendix B) satisfies

$$\|\varphi \otimes \psi\|_{\mathrm{cb}} \leq \|V_A \otimes V_B\|^2 = \|V_A\|^2 \|V_B\|^2 = \|\varphi\| \|\psi\|.$$

The other inequality is easy and will be left to the reader (consider elementary tensors and Lemma 3.4.10).

For the maximal tensor product, let's first tackle the case that $B = D$ and $\psi = \mathrm{id}_B$. Fix a faithful representation $C \otimes_{\max} B \subset \mathbb{B}(\mathcal{H})$. By the existence of restrictions, we may assume that $C \subset \mathbb{B}(\mathcal{H})$ and $B \subset \mathbb{B}(\mathcal{H})$ commute (and generate $C \otimes_{\max} B$) thus allowing us to regard φ as a c.p. map into $\mathbb{B}(\mathcal{H})$ with $B \subset \varphi(A)'$. Applying Stinespring to φ – also lifting B with the commutant $\varphi(A)'$ (Proposition 1.5.6) – we get a $*$-representation of $A \otimes_{\max} B$ (by universality) which we can cut to recover the original map $\varphi \odot \mathrm{id}_B \colon A \odot B \to C \otimes_{\max} B \subset \mathbb{B}(\mathcal{H})$ (just as in the spatial tensor product case above).

Since an arbitrary map $\varphi \odot \psi \colon A \odot B \to C \odot D$ can be decomposed as $(\varphi \odot \mathrm{id}_D) \circ (\mathrm{id}_A \odot \psi)$, the proof is complete. □

Remark 3.5.4. It is not hard to extend the previous result to *minimal* tensor products and c.b. maps. The corresponding result for maximal tensor products is not true ([**88**]).

The previous theorem extends to operator spaces (see Remark B.3; note that any operator space sits inside $\mathbb{B}(\mathcal{H})$).

Corollary 3.5.5. *If $\varphi \colon W \to Y$ and $\psi \colon X \to Z$ are c.b. maps (resp. u.c.p. maps) of operator spaces (resp. operator systems), then the tensor product map $\varphi \odot \psi$ extends uniquely to a c.b. map (resp. u.c.p. map)*

$$\varphi \otimes \psi \colon W \otimes X \to Y \otimes Z.$$

Moreover, $\|\varphi \otimes \psi\|_{\mathrm{cb}} = \|\varphi\|_{\mathrm{cb}} \|\psi\|_{\mathrm{cb}}$.

Since C*-norms are always cross norms (Lemma 3.4.10) and we can always extend c.p. maps to the minimal and maximal tensor products, the next result is a triviality (which we will use frequently and without reference).

Corollary 3.5.6. *Assume $\theta\colon A \to C$ and $\sigma\colon B \to D$ are c.c.p. maps and $\theta_n\colon A \to C$ are c.c.p. maps converging to θ in the point-norm topology. Then*

$$\theta_n \otimes_{\max} \sigma \to \theta \otimes_{\max} \sigma$$

and

$$\theta_n \otimes \sigma \to \theta \otimes \sigma$$

in the point-norm topology as well.

Exercise

Exercise 3.5.1. Let $\varphi\colon A \to \mathbb{B}(\mathcal{H})$ and $\psi\colon B \to \mathbb{B}(\mathcal{H})$ be c.p. maps with commuting ranges. Show that there exists a c.p. map

$$\varphi \times_{\max} \psi\colon A \otimes_{\max} B \to \mathbb{B}(\mathcal{H})$$

such that $\varphi \times_{\max} \psi(a \otimes b) = \varphi(a)\psi(b)$ for all $a \in A$ and $b \in B$.

Even for representations, the previous exercise fails when one tries to replace the maximal norm with the minimal norm. See Exercise 3.6.3 for a natural counterexample coming from discrete groups.

3.6. Inclusions and The Trick

In this section we discuss one of the important subtleties of C*-tensor products. We also introduce one of the great tensor product tricks, a technique so important that it should not be regarded as a trick, but rather The Trick.

The issue at hand is whether or not inclusions of C*-algebras give inclusions of tensor products. So long as one stays at the algebraic (i.e., pre-C*-algebra) level, nothing funny happens (Proposition 3.1.11). This simple fact implies that *spatial* tensor products are also well behaved in this regard.

Proposition 3.6.1. *If $A \subset B$ and C are C*-algebras, then there is a natural inclusion*

$$A \otimes C \subset B \otimes C.$$

Proof. Perhaps we should first point out what this proposition is really asserting. Since we have a natural algebraic inclusion

$$A \odot C \subset B \odot C,$$

one can ask what sort of norm we would get on $A \odot C$ if we took the spatial norm on $B \odot C$ and restricted it. This proposition asserts that we just get the spatial norm on $A \odot C$.

Having understood the meaning of the result, the proof is now an immediate consequence of Proposition 3.3.11. □

Applying this fact again on the right hand side implies that a pair of inclusions $A \subset B$ and $C \subset D$ induces a natural inclusion $A \otimes C \subset B \otimes D$.

For maximal tensor products the question then becomes: If $A \subset B$ and C are given, do we have a natural inclusion $A \otimes_{\max} C \subset B \otimes_{\max} C$? In general this turns out to be false and may seem a little puzzling at first. However, when reformulated at the algebraic level, it becomes clear what can go wrong. Indeed, what we are really asking is whether or not the maximal norm on $B \odot C$ restricts to the maximal norm on $A \odot C \subset B \odot C$. But the maximal norm is defined via a supremum over representations and since every representation of $B \odot C$ gives a representation of the smaller algebra $A \odot C$, it is clear that the supremum only over representations of $B \odot C$ will always be less than or equal to the supremum over all representations of $A \odot C$.

Having seen what the problem could be, it's not too hard to formulate a condition which ensures that inclusions behave nicely for maximal tensor products.

Proposition 3.6.2. *Let $A \subset B$ be an inclusion of C^*-algebras and assume that for every nondegenerate $*$-homomorphism $\pi \colon A \to \mathbb{B}(\mathcal{H})$ there exists a c.c.p. map $\varphi \colon B \to \pi(A)''$ such that $\varphi(a) = \pi(a)$ for all $a \in A$. Then for every C^*-algebra C there is a natural inclusion*

$$A \otimes_{\max} C \subset B \otimes_{\max} C.$$

Proof. By universality, we have a canonical $*$-homomorphism $A \otimes_{\max} C \to B \otimes_{\max} C$. Our goal is to show that if $x \in A \otimes_{\max} C$ is in the kernel of this map, then $x = 0$.

Let $\pi \colon A \otimes_{\max} C \to \mathbb{B}(\mathcal{H})$ be a faithful representation and $\pi_A \colon A \to \mathbb{B}(\mathcal{H})$, $\pi_C \colon C \to \mathbb{B}(\mathcal{H})$ be the restrictions given by Theorem 3.2.6. Note that $\pi_C(C) \subset \pi_A(A)'$ and hence the commuting inclusions $\pi_A(A)'' \hookrightarrow \mathbb{B}(\mathcal{H})$, $\pi_C(C) \hookrightarrow \mathbb{B}(\mathcal{H})$ induce, by universality, a product $*$-homomorphism

$$\pi_A(A)'' \otimes_{\max} \pi_C(C) \longrightarrow \mathbb{B}(\mathcal{H}).$$

Extend π_A to a c.c.p. map $\varphi \colon B \to \pi(A)''$ such that $\varphi(a) = \pi(a)$ for all $a \in A$. By Theorem 3.5.3 we have the following commutative diagram:

$$\begin{array}{ccc} B \otimes_{\max} C & \xrightarrow{\varphi \otimes_{\max} \pi_C} & \pi_A(A)'' \otimes_{\max} \pi_C(C) \\ \uparrow & & \downarrow \\ A \otimes_{\max} C & \xrightarrow{\quad\pi\quad} & \mathbb{B}(\mathcal{H}). \end{array}$$

The fact that π is faithful implies that the map on the left is also injective. □

We will soon introduce The Trick and provide the converse to the previous result, but first we consider two nice corollaries.

Corollary 3.6.3. *If $A \subset B$, A is nuclear and C is arbitrary, then we have a natural inclusion*

$$A \otimes_{\max} C \subset B \otimes_{\max} C.$$

Proof. Let $\pi\colon A \to \mathbb{B}(\mathcal{H})$ be a representation and $\varphi_n\colon A \to \mathbb{M}_{k(n)}(\mathbb{C})$, $\psi_n\colon \mathbb{M}_{k(n)}(\mathbb{C}) \to \pi(A)$ be c.c.p. maps converging to π in the point-norm topology.[14] By Arveson's Extension Theorem we may assume that each of the φ_n's is actually defined on all of B. Letting $\Phi\colon B \to \pi_A(A)''$ be any point-ultraweak cluster point of the maps $\psi_n \circ \varphi_n\colon B \to A \subset \pi_A(A)''$, we get the c.c.p. extension of π required to invoke Proposition 3.6.2. □

Corollary 3.6.4. *If $A \subset B$ is a hereditary subalgebra, then for every C we have a natural inclusion*

$$A \otimes_{\max} C \subset B \otimes_{\max} C.$$

Proof. If $\{e_n\} \subset A$ is an approximate unit, then the c.c.p. maps $\varphi_n\colon B \to A$, $\varphi_n(b) = e_n b e_n$ have the property that $\varphi_n(a) \to a$ for all $a \in A$. With this observation, the proof is similar to the previous corollary, so we leave the details to the reader. □

Proposition 3.6.5 (The Trick). *Let $A \subset B$ and C be C*-algebras, $\|\cdot\|_\alpha$ be a C*-norm on $B \odot C$ and $\|\cdot\|_\beta$ be the C*-norm on $A \odot C$ obtained by restricting $\|\cdot\|_\alpha$ to $A \odot C \subset B \odot C$. If $\pi_A\colon A \to \mathbb{B}(\mathcal{H})$, $\pi_C\colon C \to \mathbb{B}(\mathcal{H})$ are representations with commuting ranges and if the product *-homomorphism*

$$\pi_A \times \pi_C\colon A \odot C \to \mathbb{B}(\mathcal{H})$$

is continuous with respect to $\|\cdot\|_\beta$, then there exists a c.c.p. map $\varphi\colon B \to \pi_C(C)'$ which extends π_A.

Proof. Assume first that A, B and C are all unital and, moreover, that $1_A = 1_B$. Let

$$\pi_A \times_\beta \pi_C\colon A \otimes_\beta C \to \mathbb{B}(\mathcal{H})$$

be the extension of the product map to $A \otimes_\beta C$. Since $A \otimes_\beta C \subset B \otimes_\alpha C$, we apply Arveson's Extension Theorem to get a u.c.p. extension $\Phi\colon B \otimes_\alpha C \to \mathbb{B}(\mathcal{H})$. The desired map is just $\varphi(b) = \Phi(b \otimes 1_C)$.

To see that φ takes values in $\pi_C(C)'$ is a simple multiplicative domain argument. Indeed, $\mathbb{C}1_B \otimes C$ lives in the multiplicative domain of Φ since $\Phi|_{\mathbb{C}1_B \otimes C} = \pi_C$ is a *-homomorphism. Since $B \otimes \mathbb{C}1_C$ commutes with $\mathbb{C}1_B \otimes C$

[14] It's crucial that $\pi\colon A \to \pi(A)$ be nuclear; the result need not hold if $\pi(A)$ is replaced by $\mathbb{B}(\mathcal{H})$.

and u.c.p. maps are bimodule maps over their multiplicative domains, a simple calculation completes the proof.

The nonunital case is a bit more irritating but can be deduced from the unital case as follows. For a C*-algebra D, let $\tilde{D}$ be the unitization if D is nonunital and $\tilde{D} = D$ if D is already unital. For an arbitrary inclusion $A \subset B$ and auxiliary algebra C we may extend any C*-norm $\|\cdot\|_\alpha$ on $B \odot C$ to unitizations (Corollary 3.3.12) and get an inclusion $B \otimes_\alpha C \subset \tilde{B} \otimes_\alpha \tilde{C}$. Let $A_1 = A + \mathbb{C}1_{\tilde{B}}$ (which may or may not be the same as $\tilde{A}$) and note that $A_1 \odot C \subset \tilde{B} \otimes_\alpha \tilde{C}$. Hence $\|\cdot\|_\beta$ extends to a norm which yields an inclusion $A \otimes_\beta C \subset A_1 \otimes_\beta \tilde{C}$. The key observation is that $A \otimes_\beta C$ is an ideal in $A_1 \otimes_\beta \tilde{C}$ and hence any representation of $A \otimes_\beta C$ extends to a representation of $A_1 \otimes_\beta \tilde{C}$. Given this fact, it is easy to deduce the general case from the unital one proved above. □

At first glance, the utility of The Trick is far from obvious, but please be patient as the mileage one can get out of this simple observation is remarkable. Let us briefly explain what the point is and then we will give an application.

Given an inclusion $A \subset B$ and a representation $\pi\colon A \to \mathbb{B}(\mathcal{H})$, Arveson's Extension Theorem always allows one to extend π to a c.c.p. map $\varphi\colon B \to \mathbb{B}(\mathcal{H})$. When The Trick is applicable, it gives one the ability to better control the range of φ and this is how it gets used. As our first example we provide the converse of Proposition 3.6.2, promised earlier. An inclusion satisfying one of the following equivalent conditions is called *relatively weakly injective.*

Proposition 3.6.6. *Let $A \subset B$ be an inclusion. Then the following are equivalent:*

(1) *there exists a c.c.p. map $\varphi\colon B \to A^{**}$ such that $\varphi(a) = a$ for all $a \in A$;*

(2) *for every $*$-homomorphism $\pi\colon A \to \mathbb{B}(\mathcal{H})$ there exists a c.c.p. map $\varphi\colon B \to \pi(A)''$ such that $\varphi(a) = \pi(a)$ for all $a \in A$;*

(3) *for every C*-algebra C there is a natural inclusion*

$$A \otimes_{\max} C \subset B \otimes_{\max} C.$$

Proof. Since every representation of A extends to a normal representation of A^{**}, the equivalence of the first two statements is easy.

Assume condition (3) and let $\pi\colon A \to \mathbb{B}(\mathcal{H})$ be a representation. Let $C = \pi(A)'$ and, by universality, we can apply The Trick to the product map induced by the commuting representations $\pi\colon A \to \mathbb{B}(\mathcal{H})$ and $\pi(A)' \hookrightarrow \mathbb{B}(\mathcal{H})$. That's it. □

Definition 3.6.7. A C*-algebra $A \subset \mathbb{B}(\mathcal{H})$ is said to have Lance's *weak expectation property* (WEP) if there exists a u.c.p. map $\Phi\colon \mathbb{B}(\mathcal{H}) \to A^{**}$ such that $\Phi(a) = a$ for all $a \in A$.

A simple application of Arveson's Extension Theorem shows that the WEP is independent of the choice of faithful representation.

Corollary 3.6.8. *A C*-algebra A has the WEP if and only if for every inclusion $A \subset B$ and arbitrary C we have a natural inclusion $A \otimes_{\max} C \subset B \otimes_{\max} C$.*

Proof. Assume first that $A \subset \mathbb{B}(\mathcal{H})$ has the WEP and $A \subset B$. The inclusion $A \hookrightarrow \mathbb{B}(\mathcal{H})$ extends to a c.c.p. map $\Psi\colon B \to \mathbb{B}(\mathcal{H})$ by Arveson's Extension Theorem. Composing with Φ gives a map $B \to A^{**}$ which restricts to the identity on A and then Proposition 3.6.6 applies. The converse uses The Trick just as in the previous proposition. This time take $B = \mathbb{B}(\mathcal{H}_{\mathcal{U}})$, the universal representation of A, and $C = (A^{**})'$. □

We opened this section by claiming that inclusions of tensor products can be tricky and then proceeded to give several instances where they behave well. Here is an example where inclusions misbehave.

Proposition 3.6.9. *Let Γ be a discrete group. Then the following are equivalent:*

(1) *Γ is amenable;*

(2) *$C^*_\lambda(\Gamma)$ has the WEP;*

(3) *the natural inclusion $\iota\colon C^*_\lambda(\Gamma) \hookrightarrow \mathbb{B}(\ell^2(\Gamma))$ induces an injective tensor product map*

$$\iota \otimes_{\max} \mathrm{id}\colon C^*_\lambda(\Gamma) \otimes_{\max} C^*_\lambda(\Gamma) \hookrightarrow \mathbb{B}(\ell^2(\Gamma)) \otimes_{\max} C^*_\lambda(\Gamma).$$

In particular, the natural map

$$\iota \otimes_{\max} \mathrm{id}\colon C^*_\lambda(\Gamma) \otimes_{\max} C^*_\lambda(\Gamma) \to \mathbb{B}(\ell^2(\Gamma)) \otimes_{\max} C^*_\lambda(\Gamma)$$

has a nontrivial kernel for every nonamenable group.

Proof. (1) $\Rightarrow$ (2) follows from Theorem 2.6.8 and Exercise 2.3.14, while (2) $\Rightarrow$ (3) is immediate from Corollary 3.6.8.

For the final implication we use The Trick to produce a u.c.p. map $\Phi\colon \mathbb{B}(\ell^2(\Gamma)) \to L(\Gamma)$ such that $\Phi(x) = x$ for all $x \in C^*_\lambda(\Gamma)$. We already saw in the proof of Theorem 2.6.8 that this is enough to imply amenability of Γ.

So let $B = \mathbb{B}(\ell^2(\Gamma))$, $C = C^*_\lambda(\Gamma)$ and recall that the commutant of the right regular representation is $L(\Gamma)$. In other words, if $C^*_\lambda(\Gamma) \otimes_{\max} C^*_\lambda(\Gamma) \to \mathbb{B}(\ell^2(\Gamma))$ is the product of the left and right regular representations, then the extension $\varphi\colon \mathbb{B}(\ell^2(\Gamma)) \to \mathbb{B}(\ell^2(\Gamma))$ given by The Trick takes values in $L(\Gamma)$. □

It turns out that if $A \subset \mathbb{B}(\mathcal{H})$ is an exact C*-algebra, then the inclusion question we have been considering can fail badly. The proof requires the easy direction of a difficult theorem.

Lemma 3.6.10. [15] *Assume that $\theta\colon A \to B$ is a nuclear map. Then for every C*-algebra C the map $\theta \otimes_{\max} \mathrm{id}_C\colon A \otimes_{\max} C \to B \otimes_{\max} C$ factors through $A \otimes C$. That is, there exists a c.c.p. map $\Psi\colon A \otimes C \to B \otimes_{\max} C$ such that the diagram*

$$\begin{array}{ccc} A \otimes_{\max} C & \xrightarrow{\theta\otimes_{\max}\mathrm{id}_C} & B \otimes_{\max} C \\ & \searrow & \nearrow_{\Psi} \\ & A \otimes C & \end{array}$$

commutes, where $A \otimes_{\max} C \to A \otimes C$ is the canonical quotient map.

Proof. Let $\varphi_n\colon A \to \mathbb{M}_{k(n)}(\mathbb{C})$ and $\psi_n\colon \mathbb{M}_{k(n)}(\mathbb{C}) \to B$ be c.c.p. maps converging to θ in the point-norm topology. Due to the fact that there is a unique C*-norm on $\mathbb{M}_{k(n)}(\mathbb{C}) \odot C$, we get an approximately commuting diagram

$$\begin{array}{ccc} A \otimes_{\max} C & \xrightarrow{\theta\otimes_{\max}\mathrm{id}_C} & B \otimes_{\max} C \\ \downarrow & \searrow^{\varphi_n\otimes_{\max}\mathrm{id}_C} & \uparrow {\scriptstyle \psi_n\otimes_{\max}\mathrm{id}_C} \\ A \otimes C & \xrightarrow[\varphi_n\otimes\mathrm{id}_C]{} & \mathbb{M}_{k(n)}(\mathbb{C}) \otimes C. \end{array}$$

Hence we can define a sequence of c.c.p. maps $\Psi_n\colon A \otimes C \to B \otimes_{\max} C$ by

$$\Psi_n = (\psi_n \otimes_{\max} \mathrm{id}_C) \circ (\varphi_n \otimes \mathrm{id}_C).$$

It follows that the algebraic tensor product map $\theta \odot \mathrm{id}_C\colon A \odot C \to B \odot C$ is contractive from the spatial norm on $A \odot C$ to the maximal norm on $B \odot C$ (since $\Psi_n(x) \to \theta \odot \mathrm{id}_C(x)$ for all $x \in A \odot C$) and hence it extends to a contractive linear map $\Psi\colon A \otimes C \to B \otimes_{\max} C$. Finally, one checks that Ψ is the point-norm limit of the Ψ_n's, hence is completely positive. □

Proposition 3.6.11. *If $A \subset \mathbb{B}(\mathcal{H})$ is an exact C*-algebra and C is arbitrary, then the restriction of $\|\cdot\|_{\max}$ on $\mathbb{B}(\mathcal{H}) \odot C$ is always the spatial norm on $A \odot C$.*

Proof. When A is exact, any inclusion $A \subset \mathbb{B}(\mathcal{H})$ is nuclear and hence there is a c.c.p. map $\Psi\colon A \otimes C \to \mathbb{B}(\mathcal{H}) \otimes_{\max} C$ such that $\Psi(a \otimes c) = a \otimes c \in \mathbb{B}(\mathcal{H}) \odot C$

[15] The converse of this lemma is shown in Corollary 3.8.8.

for all $a \in A, c \in C$. Evidently Ψ is a $*$-homomorphism and it is injective since this is the case when restricted to $A \odot C$ (Exercise 3.4.1). □

The reader who is really paying attention may have noticed a glaring contradiction at this point. On the one hand, we already observed that inclusions behave nicely when the subalgebra is nuclear (Corollary 3.6.3). However, we also claimed that the previous proposition shows how badly inclusions can behave for exact C*-algebras. But nuclear C*-algebras are also exact, so how can this be?

Proposition 3.6.12. [16] *If A is nuclear, then for every* C**-algebra C there is a unique* C**-norm on $A \odot C$. In other words, the canonical quotient mapping*

$$A \otimes_{\max} C \to A \otimes C$$

is injective.

Proof. Apply Lemma 3.6.10 to $\theta = \mathrm{id}_A\colon A \to A$. □

Exercises

Exercise 3.6.1. Why is Corollary 3.6.3 an immediate consequence of Proposition 3.6.12?

Exercise 3.6.2. Assume A has the WEP and let $\pi\colon A \to \mathbb{B}(\mathcal{H})$ be a nondegenerate *faithful*[17] representation. Prove the existence of a u.c.p. map $\Phi\colon \mathbb{B}(\mathcal{H}) \to \pi(A)''$ such that $\Phi(\pi(a)) = \pi(a)$ for all $a \in A$. (There are two simple proofs: one using the definition of the WEP and Arveson's Extension Theorem and another based on The Trick.)

Exercise 3.6.3. Let Γ be a discrete group and let

$$\lambda \times \rho\colon C^*_\lambda(\Gamma) \odot C^*_\rho(\Gamma) \to \mathbb{B}(\ell^2(\Gamma))$$

be the product of the left and right regular representations. Prove that Γ is amenable if and only if $\lambda \times \rho$ is continuous with respect to the minimal tensor product norm.

Exercise 3.6.4. Show that if A is nuclear and $\pi\colon A \to \mathbb{B}(\mathcal{H})$ is any nondegenerate representation, then $\pi(A)'$ is injective. (Hint: If it takes you more than two lines, then you have missed the point! The Trick is your friend.)

Exercise 3.6.5. Let X be a locally compact Hausdorff space and $C_0(X)$ be the continuous functions vanishing at ∞. For a C*-algebra A we let

$$C_0(X, A) = \{f\colon X \to A : f \text{ is continuous and } f(\infty) = 0\}.$$

---

[16] This is also the easy direction of a hard result; see Theorem 3.8.7.

[17] This exercise can fail for noninjective representations. Giving counterexamples isn't trivial, though.

Show that there is a natural isomorphism

$$C_0(X) \otimes_{\max} A \cong C_0(X) \otimes A \cong C_0(X, A)$$

such that $h \otimes a$ maps to the function $x \mapsto h(x)a$. (Hint: A partition of unity argument will show density.)

3.7. Exact sequences

In this section we exhibit another peculiar feature of C*-tensor products, one related to exact sequences. Namely, we consider the analogues of Proposition 3.1.13 with our two C*-norms $\|\cdot\|_{\min}$ and $\|\cdot\|_{\max}$. In contrast to the previous section, the spatial norm is now the problem child.

Proposition 3.7.1. *If* $0 \to J \to A \to (A/J) \to 0$ *is an exact sequence, then for every* B*, the natural sequence*

$$0 \to J \otimes_{\max} B \to A \otimes_{\max} B \to (A/J) \otimes_{\max} B \to 0$$

is also exact.

Proof. Modulo one subtlety, the proof is fairly simple. First note that injectivity of $J \otimes_{\max} B \to A \otimes_{\max} B$ follows from Corollary 3.6.4, while $A \otimes_{\max} B \to (A/J) \otimes_{\max} B$ is surjective thanks to the fact that $*$-homomorphisms of C*-algebras always have closed range. Moreover, it is clear that $J \otimes_{\max} B$ is contained in the kernel of the map $A \otimes_{\max} B \to (A/J) \otimes_{\max} B$; hence the only question is whether or not $J \otimes_{\max} B$ is the whole kernel.

To see that this is the case, we first observe that there is a C*-norm $\|\cdot\|_\alpha$ on $(A/J) \odot B$ such that

$$\frac{A \otimes_{\max} B}{J \otimes_{\max} B} \cong (A/J) \otimes_\alpha B,$$

since exactness of the sequence

$$0 \to J \odot B \to A \odot B \to (A/J) \odot B \to 0$$

guarantees that we have a dense embedding

$$(A/J) \odot B \cong \frac{A \odot B}{J \odot B} \hookrightarrow \frac{A \otimes_{\max} B}{J \otimes_{\max} B}.^{18}$$

On the other hand, we also have a quotient (hence contractive) mapping

$$(A/J) \otimes_\alpha B \cong \frac{A \otimes_{\max} B}{J \otimes_{\max} B} \to (A/J) \otimes_{\max} B.$$

By maximality of $\|\cdot\|_{\max}$, it follows that $\|\cdot\|_\alpha = \|\cdot\|_{\max}$ and thus $J \otimes_{\max} B$ is precisely the kernel of $A \otimes_{\max} B \to (A/J) \otimes_{\max} B$.

[18] The norm $\|\cdot\|_\alpha$ is just the restriction of the quotient norm to this embedded copy of $(A/J) \odot B$.

Did you catch the subtlety? Since $J \otimes_{\max} B$ is much bigger than $J \odot B$, is it not possible that $(J \otimes_{\max} B) \cap (A \odot B)$ contains more than just $J \odot B$? This would imply that the map

$$\frac{A \odot B}{J \odot B} \to \frac{A \otimes_{\max} B}{J \otimes_{\max} B}$$

is not injective and we would have a fatal gap in the proof. Luckily the proof is correct as the fact that $(A/J) \odot B$ sits faithfully inside $(A/J) \otimes_{\max} B$ (which is a quotient of $\frac{A \otimes_{\max} B}{J \otimes_{\max} B}$) ensures that the map

$$\frac{A \odot B}{J \odot B} \to \frac{A \otimes_{\max} B}{J \otimes_{\max} B}$$

is injective. □

The reader who understood the previous proof should have no trouble demonstrating the following fact. In the literature this result is used frequently and without reference – we will do the same.

Proposition 3.7.2. *Given $J \lhd A$ and B, there is a* C*-*norm* $\|\cdot\|_\alpha$ *on* $(A/J) \odot B$ *such that*

$$\frac{A \otimes B}{J \otimes B} \cong (A/J) \otimes_\alpha B.$$

Moreover, the sequence

$$0 \to J \otimes B \to A \otimes B \to (A/J) \otimes B \to 0$$

is exact if and only if $\|\cdot\|_\alpha$ *is the spatial norm.*

It is only slightly harder to give an example of a nonexact tensor product sequence than it is to give an example where inclusions of maximal tensor products fail. We will do this at the end of the section, but let's first consider a number of cases where things go well. The following is a special case of the previous result.

Corollary 3.7.3. *If there exists a unique* C*-*norm on* $(A/J) \odot B$, *then the sequence*

$$0 \to J \otimes B \to A \otimes B \to (A/J) \otimes B \to 0$$

is exact.

The next fact follows from Corollary 3.7.3 and Proposition 3.6.12.

Corollary 3.7.4. *If either* A/J *or* B *is nuclear, then*

$$0 \to J \otimes B \to A \otimes B \to (A/J) \otimes B \to 0$$

is exact.

In the previous section we saw that for certain inclusions $A \subset B$, we have inclusions $A \otimes_{\max} C \subset B \otimes_{\max} C$ for every C. Our next result is in this same spirit, but we should first recall some terminology.

Definition 3.7.5. If A is unital, an extension $0 \to J \to A \xrightarrow{\pi} A/J \to 0$ is called *locally split* if for each finite-dimensional operator system $E \subset A/J$ there exists a u.c.p. map $\sigma\colon E \to A$ such that $\pi \circ \sigma = \mathrm{id}_E$.[19]

Proposition 3.7.6. [20] *Assume that* $0 \to J \to A \xrightarrow{\pi} A/J \to 0$ *is locally split. Then for every* B *the sequence*

$$0 \to J \otimes B \to A \otimes B \to (A/J) \otimes B \to 0$$

is exact.

Proof. We know that

$$\frac{A \otimes B}{J \otimes B} \cong (A/J) \otimes_\alpha B$$

for some C*-norm and our job is to show that $\|\cdot\|_\alpha$ is the spatial norm. By Takesaki's Theorem (Theorem 3.4.8) it suffices to show that if $y \in (A/J) \odot B$, then

$$\|y\|_\alpha \le \|y\|_{\min}.$$

Since $y \in (A/J) \odot B$ is a finite sum of elementary tensors, we can find a finite-dimensional operator system $X \subset A/J$ such that $y \in X \odot B \subset (A/J) \odot B$. By assumption, we can find a u.c.p. map $\theta\colon X \to A$ which lifts X and this induces a u.c.p. (hence contractive) map

$$\theta \otimes \mathrm{id}_B\colon X \otimes B \to A \otimes B.$$

The remainder of the proof is contained in the following diagram:

$$\begin{array}{ccccc} (A/J) \otimes B & & & & \\ \text{isometric} \uparrow \text{inclusion} & & & & \\ X \otimes B & \xrightarrow{\theta \otimes \mathrm{id}_B} & A \otimes B & \longrightarrow & \frac{A \otimes B}{J \otimes B} \cong (A/J) \otimes_\alpha B. \end{array}$$

Indeed, both maps on the bottom are contractions and since θ was a lifting, it follows that the element $y \in X \odot B \subset X \otimes B$ gets mapped to the element $y \in X \odot B \subset (A/J) \otimes_\alpha B$. Thus $\|y\|_\alpha \le \|y\|_{\min}$ as desired. □

Now let us reverse the roles and ask if some property of B will ensure that $0 \to J \otimes B \to A \otimes B \to (A/J) \otimes B \to 0$ is exact for all choices of A and ideals $J \lhd A$. We'll need a lemma.

[19] The nonunital definition replaces operator systems with operator spaces and u.c.p. maps with c.c. maps.

[20] This is the easy half of the Effros-Haagerup lifting theorem; see Theorem C.4.

Lemma 3.7.7. *If $J \lhd A$ is an ideal and $B \subset C$ is a C*-subalgebra, then*

$$(J \otimes C) \cap (A \otimes B) = J \otimes B.$$

Proof. The inclusion $\supset$ is immediate, so assume that $x \in (J\otimes C) \cap (A \otimes B)$. Let $\{e_n\} \subset J$ be an approximate unit and let us assume for the moment that C is unital. Then $\{e_n \otimes 1_C\} \subset J \otimes C$ is an approximate unit (use Lemma 3.4.10) and hence the elements $(e_n \otimes 1_C)x$ belong to $J \otimes B$ (since $x \in A \otimes B$) and converge in norm to x (since $x \in J \otimes C$). But $J \otimes B$ is norm closed and so we are finished in this case.

If C is not unital, then we can unitize and apply the argument above. □

Proposition 3.7.8. *If B is exact, then*

$$0 \to J \otimes B \to A \otimes B \to (A/J) \otimes B \to 0$$

is exact for all A and $J \lhd A$.[21]

Proof. If $J \lhd A$ is arbitrary and an element $x \in A \otimes B$ is in the kernel of $A \otimes B \to (A/J) \otimes B$, then we must convince ourselves that $x \in J \otimes B$.

Let $\pi \colon B \hookrightarrow \mathbb{B}(\mathcal{H})$ be a faithful representation and find some c.c.p. maps $\varphi_n \colon B \to \mathbb{M}_{k(n)}(\mathbb{C})$, $\psi_n \colon \mathbb{M}_{k(n)}(\mathbb{C}) \to \mathbb{B}(\mathcal{H})$ such that $\|\pi(b) - \psi_n \circ \varphi_n(b)\| \to 0$ for all $b \in B$. Consider the following commutative diagram:

$$\begin{array}{ccccccccc}
0 & \longrightarrow & J \otimes B & \longrightarrow & A \otimes B & \longrightarrow & (A/J) \otimes B & \longrightarrow & 0 \\
 & & \Big\downarrow{\scriptstyle \mathrm{id}\otimes\varphi_n} & & \Big\downarrow{\scriptstyle \mathrm{id}\otimes\varphi_n} & & \Big\downarrow{\scriptstyle \mathrm{id}\otimes\varphi_n} & & \\
0 & \longrightarrow & J \otimes \mathbb{M}_{k(n)}(\mathbb{C}) & \longrightarrow & A \otimes \mathbb{M}_{k(n)}(\mathbb{C}) & \longrightarrow & (A/J) \otimes \mathbb{M}_{k(n)}(\mathbb{C}) & \longrightarrow & 0 \\
 & & \Big\downarrow{\scriptstyle \mathrm{id}\otimes\psi_n} & & \Big\downarrow{\scriptstyle \mathrm{id}\otimes\psi_n} & & \Big\downarrow{\scriptstyle \mathrm{id}\otimes\psi_n} & & \\
0 & \longrightarrow & J \otimes \mathbb{B}(\mathcal{H}) & \longrightarrow & A \otimes \mathbb{B}(\mathcal{H}) & \longrightarrow & (A/J) \otimes \mathbb{B}(\mathcal{H}) & \longrightarrow & 0.
\end{array}$$

Since the middle row is exact, a bit of diagram chasing shows that

$$\mathrm{id} \otimes (\psi_n \circ \varphi_n)(x) \in J \otimes B(H)$$

for all n. Since $\|\mathrm{id} \otimes \pi(x) - \mathrm{id} \otimes (\psi_n \circ \varphi_n)(x)\| \to 0$, this shows that

$$\mathrm{id} \otimes \pi(x) \in (J \otimes \mathbb{B}(\mathcal{H})) \cap (A \otimes \pi(B)).$$

By the previous lemma, this completes the proof since

$$\mathrm{id} \otimes \pi \colon J \otimes B \to J \otimes \pi(B) = (J \otimes \mathbb{B}(\mathcal{H})) \cap (A \otimes \pi(B))$$

is an isomorphism. □

[21] This actually characterizes exact C*-algebras and explains the terminology (see Theorem 3.9.1).

Now let's give an example where a tensor product sequence fails to be exact.[22] Though elementary, the proof is somewhat delicate and requires a few preliminary facts.

A discrete group Γ is called *residually finite* if there exist subgroups $\Gamma \supset \Gamma_1 \supset \Gamma_2 \supset \cdots$ such that each Γ_i is a finite-index, normal subgroup of Γ and $\bigcap_n \Gamma_n = \{e\}$ (the neutral element of Γ). The key fact is that if Γ is residually finite, then the *minimal* tensor product of the *full* group C*-algebras $C^*(\Gamma) \otimes C^*(\Gamma)$ has a very special state coming from the left and right regular representations.

If Γ is any discrete group, then we can consider the product map

$$\lambda \times \rho \colon C^*(\Gamma) \odot C^*(\Gamma) \to \mathbb{B}(\ell^2(\Gamma))$$

induced by the (commuting) left and right regular representations. If we take some elements $x = \sum_{s\in\Gamma} \alpha_s s$, $y = \sum_{t\in\Gamma} \beta_t t \in C^*(\Gamma)$ (only finitely many coefficients nonzero), then a straightforward computation shows that

$$\langle \lambda \times \rho(x \otimes y)\delta_e, \delta_e \rangle = \sum_{s\in\Gamma} \alpha_s \beta_s.$$

Note that if Γ happened to be a finite group, then $C^*(\Gamma) \odot C^*(\Gamma) = C^*(\Gamma) \otimes C^*(\Gamma)$ and hence we would have a state on the minimal tensor product satisfying the formula above.

Lemma 3.7.9. *If Γ is residually finite, then there exists a state μ on $C^*(\Gamma) \otimes C^*(\Gamma)$ such that for finite sums*

$$x = \sum_{s\in\Gamma} \alpha_s s, \ y = \sum_{t\in\Gamma} \beta_t t \in C^*(\Gamma)$$

we have

$$\mu(x \otimes y) = \sum_{s\in\Gamma} \alpha_s \beta_s.$$

Proof. Since each Γ/Γ_n is a finite group, we have a sequence of states μ_n on $C^*(\Gamma/\Gamma_n) \otimes C^*(\Gamma/\Gamma_n)$ satisfying the formula above (for Γ/Γ_n). However we also have quotient mappings $\pi_n \colon C^*(\Gamma) \to C^*(\Gamma/\Gamma_n)$ and hence tensor product $*$-homomorphisms

$$\pi_n \otimes \pi_n \colon C^*(\Gamma) \otimes C^*(\Gamma) \to C^*(\Gamma/\Gamma_n) \otimes C^*(\Gamma/\Gamma_n).$$

Since the intersection of the Γ_n's is just the neutral element, a straightforward computation shows that any cluster point of the states $\mu_n \circ (\pi_n \otimes \pi_n)$ must satisfy the right formula. (Hence there is a unique cluster point of the $\mu_n \circ (\pi_n \otimes \pi_n)$'s.) □

[22] In Chapter 6, when we come to Kirchberg's factorization property, we'll revisit these ideas.

Proposition 3.7.10. *If Γ is residually finite, then the product map*

$$\lambda \times \rho \colon C^*(\Gamma) \odot C^*(\Gamma) \to \mathbb{B}(\ell^2(\Gamma))$$

induced by the left and right regular representations is continuous with respect to the spatial tensor product norm. (Compare with Exercise 3.6.3.)

Proof. Let $\pi \colon C^*(\Gamma) \otimes C^*(\Gamma) \to \mathbb{B}(\mathcal{H})$ be the GNS representation with respect to the state constructed in the previous lemma. Uniqueness of GNS representations implies that the (algebraic) representations

$$\pi|_{C^*(\Gamma)\odot C^*(\Gamma)} \colon C^*(\Gamma) \odot C^*(\Gamma) \to \mathbb{B}(\mathcal{H})$$

and

$$\lambda \times \rho \colon C^*(\Gamma) \odot C^*(\Gamma) \to \mathbb{B}(\ell^2(\Gamma))$$

are unitarily equivalent since $\delta_e \in \ell^2(\Gamma)$ is a cyclic vector for the algebra $\lambda \times \rho(C^*(\Gamma) \odot C^*(\Gamma))$ whose corresponding vector functional agrees with μ. This implies the C*-algebra generated by $\lambda \times \rho(C^*(\Gamma) \odot C^*(\Gamma))$ is a quotient of $C^*(\Gamma) \otimes C^*(\Gamma)$, so the proof is complete. □

Proposition 3.7.11. *Let Γ be a residually finite discrete group. Then the following are equivalent:*

(1) *Γ is amenable;*

(2) *$C^*(\Gamma)$ is exact;*

(3) *the sequence*

$$0 \to J \otimes C^*(\Gamma) \to C^*(\Gamma) \otimes C^*(\Gamma) \to C^*_\lambda(\Gamma) \otimes C^*(\Gamma) \to 0$$

is exact, where J is the kernel of the quotient map $C^*(\Gamma) \to C^*_\lambda(\Gamma)$.

Proof. (1) $\Rightarrow$ (2) follows from Theorem 2.6.8 since full and reduced C*-algebras of amenable groups always agree.

(2) $\Rightarrow$ (3) is a consequence of Proposition 3.7.8.

(3) $\Rightarrow$ (1): If

$$\frac{C^*(\Gamma) \otimes C^*(\Gamma)}{J \otimes C^*(\Gamma)} \cong C^*_\lambda(\Gamma) \otimes C^*(\Gamma),$$

then Proposition 3.7.10 implies that we have a $*$-homomorphism $\pi \colon C^*_\lambda(\Gamma) \otimes C^*(\Gamma) \to \mathbb{B}(\ell^2(\Gamma))$ such that $\pi(x \otimes y) = x\rho(y)$ for all $x \in C^*_\lambda(\Gamma)$ and $y \in C^*(\Gamma)$. Indeed, we know that the product map $\lambda \times \rho$ extends to a $*$-homomorphism which we will denote by

$$\lambda \times_{\min} \rho \colon C^*(\Gamma) \otimes C^*(\Gamma) \to \mathbb{B}(\ell^2(\Gamma)).$$

However, it is evident that $J \otimes C^*(\Gamma)$ belongs to the kernel of $\lambda \times_{\min} \rho$ and hence this $*$-homomorphism factors through

$$\frac{C^*(\Gamma) \otimes C^*(\Gamma)}{J \otimes C^*(\Gamma)} \cong C^*_\lambda(\Gamma) \otimes C^*(\Gamma),$$

so we get our representation π.

To finish off the proof, use The Trick to get a u.c.p. map $\Phi\colon \mathbb{B}(\ell^2(\Gamma)) \to L(\Gamma)$ such that $\Phi(x) = x$ for all $x \in C^*_\lambda(\Gamma)$. This is very similar to the proof of Proposition 3.6.9 (with maximal norms replaced by minimal ones and the right regular representation regarded as originating from $C^*(\Gamma)$), so we leave the details to you. □

As promised in the last chapter, we finally prove the existence of nonexact C*-algebras. This follows immediately from the previous result together with Proposition 3.7.8.

Corollary 3.7.12. *If Γ is a nonamenable residually finite group (e.g., $\mathbb{F}_n$ or $\mathrm{SL}(n, \mathbb{Z})$), then $C^*(\Gamma)$ is not exact since the sequence*

$$0 \to J \otimes C^*(\Gamma) \to C^*(\Gamma) \otimes C^*(\Gamma) \to C^*_\lambda(\Gamma) \otimes C^*(\Gamma) \to 0$$

is not exact.

It follows from this corollary that $\mathbb{B}(\ell^2)$ is not exact either (since exactness passes to subalgebras).

Exercises

Exercise 3.7.1. Use Lemma 3.7.7 to show that if the sequence

$$0 \to J \otimes C \to A \otimes C \to A/J \otimes C \to 0$$

is exact and $B \subset C$ is a C*-subalgebra, then

$$0 \to J \otimes B \to A \otimes B \to A/J \otimes B \to 0$$

is also exact.

Exercise 3.7.2 (Uniqueness of the extension in Corollary 3.3.12). Let α and β be C*-norms on $\tilde{A} \odot B$, where $\tilde{A}$ is the unitization of A. Prove that if $\alpha = \beta$ on $A \odot B$, then $\alpha = \beta$ on $\tilde{A} \odot B$. In particular if $A \odot B$ has a unique C*-norm, then so does $\tilde{A} \odot B$. Hint: Consider $\gamma = \max\{\alpha, \beta\}$ and use the 5 Lemma[23] on the following diagram:

$$\begin{array}{ccccccccc} 0 & \longrightarrow & A \otimes_\gamma B & \longrightarrow & \tilde{A} \otimes_\gamma B & \longrightarrow & \mathbb{C}1 \otimes B & \longrightarrow & 0 \\ & & \downarrow & & \downarrow & & \downarrow & & \\ 0 & \longrightarrow & A \otimes_\alpha B & \longrightarrow & \tilde{A} \otimes_\alpha B & \longrightarrow & \mathbb{C}1 \otimes B & \longrightarrow & 0. \end{array}$$

[23]See *Mathworld* (mathworld.wolfram.com/FiveLemma.html) if you're not familiar with the 5 Lemma.

Exercise 3.7.3. Let A be nonunital and assume that

$$0 \to J \otimes A \to B \otimes A \to B/J \otimes A \to 0$$

is exact. Show that the same sequence with A replaced by its unitization is also exact.

3.8. Nuclearity and tensor products

In this section we come back to the historical origins of nuclearity, recasting everything in terms of tensor products. The main theorems are not trivial and require a few technical preliminaries. The first three requisite results can be found in most basic operator algebra texts and may be known to the reader. However, experience suggests that they are unfamiliar to many; hence we'll state them as separate propositions. In order to keep sight of the forest, we recommend skipping the proofs of these trees at first; after the main results are absorbed, one can return and fill in the gaps.

Lemma 3.8.1. *Let $\mathbb{B}(\mathbb{B}(\mathcal{H}))$ be the set of bounded linear operators on $\mathbb{B}(\mathcal{H})$ and $\mathcal{C} \subset \mathbb{B}(\mathbb{B}(\mathcal{H}))$ be any convex set. Then the point-weak operator topology and point-strong operator topology closures of $\mathcal{C}$ coincide.*

Proof. Since a convex subset of $\mathbb{B}(\mathcal{H})$ has the same closure in the weak or strong operator topology, the proof of this lemma is very similar to Lemma 2.3.4. □

Proposition 3.8.2. *Let A be a unital C*-algebra, M be a von Neumann algebra and $\varphi\colon A \to M$ be a u.c.p. map. If φ belongs to the point-ultraweak closure of the factorable maps (Definition 2.3.5), then φ is weakly nuclear.*

Proof. The only issue is how to replace a net of maps which have norms wandering off to infinity with contractions. So let $\varphi_n\colon A \to \mathbb{M}_{k(n)}(\mathbb{C})$ and $\psi_n\colon \mathbb{M}_{k(n)}(\mathbb{C}) \to M$ be c.p. maps whose compositions converge in the point-strong operator topology to φ. (Since ultraweak convergence implies weak convergence, we have used the previous lemma to get a better topology.) As we have seen many times, Lemma 2.2.5 allows us to assume that φ_n is unital for every n.

Since both φ and the φ_n's are unital, it follows that $\psi_n(1) \to 1$ in the strong operator topology (though the norms of $\psi_n(1)$ may be large). Now fix $\delta > 0$ and let $P_n \in M$ be the spectral projection of $\psi_n(1)$ corresponding to the interval $[0, 1+\delta]$. Since $\|P_n\| = 1$, $P_n(\psi_n \circ \varphi_n(x) - \varphi(x)) \to 0$ in the strong operator topology, for every $x \in A$. Since $|\psi_n(1) - 1| \geq \delta P_n^{\perp}$, we have that $P_n \to 1$ in the strong operator topology. Hence, $P_n(\psi_n \circ \varphi_n(x)) \to \varphi(x)$ in the strong operator topology as well.

Define new c.p. maps by

$$\psi'_n(T) = P_n\psi_n(T)P_n$$

and note that $\|\psi'_n\| \le 1+\delta$. Since $\delta > 0$ was arbitrary, it suffices to show that $\psi'_n \circ \varphi_n \to \varphi$ in the point-strong operator topology.

Let $x \in A$ be any positive contraction. Then, since $P_n^\perp(\psi_n \circ \varphi_n(x))P_n^\perp \le P_n^\perp \psi_n(1)$, we have that $(\psi_n \circ \varphi_n(x))^{1/2}P_n^\perp \to 0$ in the strong operator topology. Moreover, $\|P_n(\psi_n \circ \varphi_n(x))^{1/2}\|^2 \le \|P_n\psi_n(1)P_n\| \le 1+\delta$. It follows that $P_n(\psi_n \circ \varphi_n(x))P_n^\perp \to 0$ in the strong operator topology. Therefore,

$$\psi'_n \circ \varphi_n(x) = P_n(\psi_n \circ \varphi_n(x)) - P_n(\psi_n \circ \varphi_n(x))P_n^\perp \to \varphi(x)$$

in the strong operator topology. Since positive elements span A, we are done. □

Proposition 3.8.3 (Radon-Nikodym). *Assume $f \in S(A)$ is a state on a C*-algebra A and let $\pi_f\colon A \to \mathbb{B}(L^2(A,f))$ be the GNS representation. If ξ is any positive linear functional on A such that $\xi \le f$ (i.e., $\xi(a) \le f(a)$ for all $0 \le a \in A$), then there exists a unique operator $y_\xi \in \pi_f(A)'$ such that $0 \le y_\xi \le 1$ and*

$$\xi(a) = \langle \pi_f(a)y_\xi\hat{1}, \hat{1}\rangle,$$

for all $a \in A$, where $\hat{1}$ is the canonical cyclic vector in $L^2(A,f)$ (though A is not assumed unital).

Proof. Here is a sketch of the proof. First define a positive definite sesquilinear form on $L^2(A,f)$ by

$$\langle \hat{a}, \hat{b}\rangle_\xi = \xi(b^*a).$$

One must check that this is well-defined (use Cauchy-Schwarz), bounded on a dense subspace and hence can be extended to all of $L^2(A,f)$. Then we can find a unique operator $y_\xi \in \mathbb{B}(L^2(A,f))$ such that

$$\langle \hat{a}, \hat{b}\rangle_\xi = \langle y_\xi\hat{a}, \hat{b}\rangle$$

and one must check that this operator has the desired properties. □

We need one more lemma before coming to the first major theorem. The statement is technical but amounts to saying that a certain property is independent of the representation of the von Neumann algebra involved.

Lemma 3.8.4. *Let A be a C*-algebra, $M \subset \mathbb{B}(\mathcal{H})$ be a von Neumann algebra and $\varphi\colon A \to M$ be a c.p. map. Assume that the product map*

$$\varphi \times \iota_{M'}\colon A \odot M' \to \mathbb{B}(\mathcal{H}),\ \varphi\Big(\sum_i a_i \otimes m'_i\Big) = \sum_i \varphi(a_i)m'_i,$$

is continuous with respect to the spatial tensor product norm and let $\pi\colon M \to \mathbb{B}(\mathcal{K})$ *be any normal representation. Then the product map*

$$(\pi \circ \varphi) \times \iota_{\pi(M)'}\colon A \odot \pi(M)' \to \mathbb{B}(\mathcal{K})$$

is also min-*continuous.*[24]

Proof. Any normal representation of M can be identified with the cut-down by a projection in the commutant of the representation $M \otimes 1_{\mathcal{K}} \subset \mathbb{B}(\mathcal{H} \otimes \mathcal{K})$. Hence it suffices to show that the product map with the commutant in this particular representation is min-continuous.

Since $(M \otimes 1_{\mathcal{K}})' \cap \mathbb{B}(\mathcal{H} \otimes \mathcal{K}) = M' \bar{\otimes} \mathbb{B}(\mathcal{K})$ – just think of $\mathbb{B}(\mathcal{H} \otimes \mathcal{K})$ as matrices with entries in $\mathbb{B}(\mathcal{H})$ – we thus have to show that

$$(\varphi \otimes 1_{\mathbb{B}(\mathcal{K})}) \times \iota_{M' \bar{\otimes} \mathbb{B}(\mathcal{K})}\colon A \odot (M' \bar{\otimes} \mathbb{B}(\mathcal{K})) \to \mathbb{B}(\mathcal{H} \otimes \mathcal{K})$$

is min-continuous. But, except for the horrific notation required, this is easy since $(\varphi \otimes 1_{\mathbb{B}(\mathcal{K})}) \times \iota_{M' \bar{\otimes} \mathbb{B}(\mathcal{K})}$ is a point-strong limit of min-continuous maps (with uniformly bounded norms). More precisely, if $P \in \mathbb{B}(\mathcal{K})$ is a *finite-rank* projection, then the map

$$(\varphi \otimes 1_{\mathbb{B}(P\mathcal{K})}) \times \iota_{M' \otimes \mathbb{B}(P\mathcal{K})}\colon A \odot (M' \otimes \mathbb{B}(P\mathcal{K})) \to \mathbb{B}(\mathcal{H} \otimes P\mathcal{K})$$

is min-continuous and its norm is bounded by $\|\varphi \times \iota_{M'}\|$ because it can be identified with

$$(\varphi \times \iota_{M'}) \otimes \mathrm{id}_{\mathbb{B}(P\mathcal{K})}\colon (A \odot M') \odot \mathbb{B}(P\mathcal{K}) \to \mathbb{B}(\mathcal{H} \otimes P\mathcal{K})$$

and Exercise 3.5.1 then comes into play. Finally, taking a net $\{P_\lambda\}$ of finite-rank projections which converge to $1_{\mathcal{K}}$ in the strong operator topology and fixing

$$x = \sum a_i \otimes T_i \in A \odot (M' \bar{\otimes} \mathbb{B}(\mathcal{K})),$$

it is easy to check that

$$(\varphi \otimes 1_{\mathbb{B}(P_\lambda\mathcal{K})}) \times \iota_{M' \otimes \mathbb{B}(P_\lambda\mathcal{K})}\big((1_{\mathcal{H}} \otimes P_\lambda)x(1_{\mathcal{H}} \otimes P_\lambda)\big) \to (\varphi \otimes 1_{\mathbb{B}(\mathcal{K})}) \times \iota_{M' \bar{\otimes} \mathbb{B}(\mathcal{K})}(x)$$

in the strong operator topology. This completes the proof. □

We are now ready for an important theorem of Kirchberg. Though the proof is not so long, it is delicate and deliciously technical. Bon appétit!

Theorem 3.8.5. *Let* $\varphi\colon A \to M \subset \mathbb{B}(\mathcal{H})$ *be a u.c.p. map from a unital* C*-*algebra* A *to a von Neumann algebra* M. *Then* φ *is weakly nuclear if and only if the product map* $\varphi \times \iota_{M'}\colon A \odot M' \to \mathbb{B}(\mathcal{H})$ *is continuous with respect to the spatial tensor product norm.*

---

[24] That is, "continuous with respect to the minimal norm."

Proof. ($\Rightarrow$) This is very similar to the proof of Lemma 3.6.10. If $\varphi_n\colon A \to \mathbb{M}_{k(n)}(\mathbb{C})$ and $\psi_n\colon \mathbb{M}_{k(n)}(\mathbb{C}) \to M$ are c.c.p. maps whose compositions converge point-ultraweakly to φ, then we can define a sequence of c.c.p. maps $\Phi_n\colon A \otimes M' \to \mathbb{B}(\mathcal{H})$ by $\Phi_n = (\psi_n \times_{\max} \iota_{M'}) \circ (\varphi_n \otimes \iota_{M'})$. In other words, we take the composition

$$A \otimes M' \overset{\varphi_n \otimes \iota_{M'}}{\longrightarrow} \mathbb{M}_{k(n)}(\mathbb{C}) \otimes M' \overset{\psi_n \times_{\max} \iota_{M'}}{\longrightarrow} \mathbb{B}(\mathcal{H}),$$

where the maximal product map on the right exists thanks to Exercise 3.5.1 (and uniqueness of the C*-norm on $\mathbb{M}_{k(n)}(\mathbb{C}) \odot M'$). Letting $\Phi\colon A \otimes M' \to \mathbb{B}(\mathcal{H})$ be a point-ultraweak cluster point of the Φ_n's, we get a c.c.p. map which evidently extends the product map $\varphi \times \iota_{M'}\colon A \odot M' \to \mathbb{B}(\mathcal{H})$; this direction of the proof is complete.

($\Leftarrow$) Now assume that $\varphi \times \iota_{M'}\colon A \odot M' \to \mathbb{B}(\mathcal{H})$ is min-continuous. According to Proposition 3.8.2 we only have to show that φ belongs to the point-ultraweak closure of the factorable maps. Thus we fix a finite set $\mathfrak{F} \subset A$, a finite set of normal states $\mathcal{S} \subset M_*$ and $\varepsilon > 0$. Let $f \in M_*$ be the average of all the states in $\mathcal{S}$ and note that f dominates a multiple of each $\xi \in \mathcal{S}$. Thus, by the Radon-Nikodym-type result above, we can find elements $y_\xi \in \pi(M)'$ such that

$$\xi(m) = \langle \pi(m) y_\xi \hat{1}_M, \hat{1}_M \rangle,$$

for all $m \in M$ and all $\xi \in \mathcal{S}$, where $\pi\colon M \to \mathbb{B}(L^2(M, f))$ is the GNS representation.

Since Lemma 3.8.4 tells us that the product map

$$(\pi \circ \varphi) \times \iota_{\pi(M)'}\colon A \odot \pi(M)' \to \mathbb{B}(L^2(M, f))$$

is also min-continuous, we can define a state μ on $A \otimes \pi(M)'$ by the formula

$$\mu(a \otimes y) = \langle \pi(\varphi(a)) y \hat{1}_M, \hat{1}_M \rangle.$$

Note that $\xi(\varphi(a)) = \mu(a \otimes y_\xi)$ for all $a \in A$ and $\xi \in \mathcal{S}$.

Let $A \otimes \pi(M)' \subset \mathbb{B}(\mathcal{K} \otimes L^2(M, f))$ be a representation containing no nonzero compacts (i.e., inflate any faithful representation of A to ensure $A \subset \mathbb{B}(\mathcal{K})$ contains no compacts). By Glimm's lemma (Lemma 1.4.11) we can approximate the functional μ by vector states on $\mathbb{B}(\mathcal{K} \otimes L^2(M, f))$. In particular, we can find an orthonormal set $\{v_1, \ldots, v_n\} \subset \mathcal{K}$ and $\{b_1, \ldots, b_n\} \subset M$ such that

$$|\mu(a \otimes y_\xi) - \langle a \otimes y_\xi (\sum_{i=1}^n v_i \otimes \hat{b}_i), \sum_{i=1}^n v_i \otimes \hat{b}_i \rangle| < \epsilon$$

for all $a \in \mathfrak{F}$ and all $\xi \in \mathcal{S}$.

Now let $P \in \mathbb{B}(\mathcal{K})$ be the orthogonal projection onto the span of the vectors $\{v_1, \dots, v_n\} \subset \mathcal{K}$ and define a u.c.p. map $\rho\colon A \to P\mathbb{B}(\mathcal{K})P$ by compression. The matrix of $\rho(a)$ with respect to $\{v_1, \dots, v_n\}$ is

$$[\langle av_j, v_i\rangle]_{i,j}.$$

Define a map $\psi\colon P\mathbb{B}(\mathcal{K})P \to M$ by declaring

$$\psi(e_{i,j}) = b_i^* b_j,$$

where $\{e_{i,j}\}$ is a set of matrix units relative to $\{v_1, \dots, v_n\}$. According to Example 1.5.13, ψ is a c.p. map (though it may have large norm).

For $a \in A$ and $\xi \in \mathcal{S}$ we compute

$$\xi(\psi \circ \rho(a)) = \xi\Big(\sum_{i,j=1}^{n} \langle av_j, v_i\rangle b_i^* b_j\Big) = \sum_{i,j=1}^{n} \langle av_j, v_i\rangle \xi(b_i^* b_j).$$

On the other hand, for $a \in \mathfrak{F}$ and $\xi \in \mathcal{S}$ we have

$$\begin{aligned}
\xi(\varphi(a)) &= \mu(a \otimes y_\xi) \\
&\overset{\epsilon}{\approx} \langle a \otimes y_\xi\big(\sum_{i=1}^{n} v_i \otimes \hat{b}_i\big), \sum_{i=1}^{n} v_i \otimes \hat{b}_i\rangle \\
&= \sum_{i,j=1}^{n} \langle av_j, v_i\rangle \langle y_\xi \hat{b}_j, \hat{b}_i\rangle \\
&= \sum_{i,j=1}^{n} \langle av_j, v_i\rangle \langle \pi(b_i^* b_j) y_\xi \hat{1}, \hat{1}\rangle \\
&= \sum_{i,j=1}^{n} \langle av_j, v_i\rangle \xi(b_i^* b_j) \\
&= \xi(\psi \circ \rho(a)).
\end{aligned}$$

Thus $\psi \circ \rho$ is a factorable map which is close to φ on our prescribed finite sets and this is all we needed to show. □

For dessert, we have a few important applications to choose from. The first, which is an immediate consequence of the bicommutant theorem and the previous result, shows that semidiscreteness of the commutant does not depend on the representation.

Corollary 3.8.6. *Let $M \subset \mathbb{B}(\mathcal{H})$ be a von Neumann algebra. Then M is semidiscrete if and only if M' is semidiscrete.*

From the C*-point of view, the next application is probably the most important.

Theorem 3.8.7 (Choi and Effros, Kirchberg). *For a* C*-*algebra A, the following statements are equivalent:*

(1) *A is nuclear;*

(2) *for every* C*-*algebra B, there is a unique* C*-*norm on $A \odot B$.*[25]

Proof. We already observed (1) $\Rightarrow$ (2) – see Proposition 3.6.12; thus we assume there is always a unique C*-norm on $A \odot B$. By Exercise 3.7.2 we may assume A is unital. Applying Theorem 3.8.5 with $\varphi = \iota\colon A \to A^{**}$, we see that the inclusion of A into its double dual is a weakly nuclear map. This implies A is nuclear (Proposition 2.3.8 and Exercise 2.3.13). □

Finally, we provide the converse to Lemma 3.6.10, as promised earlier.

Corollary 3.8.8. *Let $\varphi\colon A \to B$ be a u.c.p. map. Then φ is nuclear if and only if for every* C*-*algebra C, the maximal tensor product map*

$$\varphi \otimes_{\max} \mathrm{id}_C\colon A \otimes_{\max} C \to B \otimes_{\max} C$$

factors through $A \otimes C$.

Proof. Since we only have to show the "if" direction, let $B \subset B^{**} \subset \mathbb{B}(\mathcal{H}_u)$ be the universal representation and assume

$$\varphi \otimes_{\max} \mathrm{id}_{(B^{**})'}\colon A \otimes_{\max} (B^{**})' \to B \otimes_{\max} (B^{**})'$$

factors through $A \otimes (B^{**})'$. Composing this map with the product map $B \otimes_{\max} (B^{**})' \to \mathbb{B}(\mathcal{H}_u)$, we see that the product map

$$\varphi \times \iota_{(B^{**})'}\colon A \odot (B^{**})' \to \mathbb{B}(\mathcal{H}_u)$$

is min-continuous and hence φ is weakly nuclear as a map from A to B^{**}. As with the previous result, this is enough to imply nuclearity of φ.[26] □

The C*-purists may find our proof of Theorem 3.8.7 somewhat annoying since we relied so heavily on W*-machinery. There is an "alternate" proof which bypasses the W*-preliminaries and goes directly to Theorem 3.8.7. However, it is virtually identical to the proof of Theorem 3.8.5, thus not much easier.

Here is a sketch: Suppose A has the property that $A \odot C$ has a unique C*-norm for every C*-algebra C. Invoking Lemma 2.3.4, it suffices to show that id_A is in the point-weak closure of the factorable maps. Hence we fix a finite set $\mathfrak{F} \subset A$, finite set of states $\mathcal{S} \subset S(A)$ and $\varepsilon > 0$. Now we let f be the average of the states in $\mathcal{S}$, $\pi\colon A \to \mathbb{B}(L^2(A, f))$ be the GNS representation

[25] Originally this condition was the definition of nuclearity.

[26] Since φ takes values in B and is weakly nuclear as a map into B^{**}, this should now be routine.

and mimic exactly what we did in the proof of Theorem 3.8.5. It is a good exercise to check the remaining details.

Exercise

Exercise 3.8.1. Prove that extensions of nuclear C*-algebras are nuclear. That is, if $J \triangleleft A$ is a nuclear ideal and A/J is also nuclear, then so is A. (It is possible, but not so easy, to prove this directly from the definition of nuclearity. However, a tensor product argument, invoking the 5 Lemma, is very easy.)

3.9. Exactness and tensor products

Our only goal in this section is to prove the following deep and difficult theorem.

Theorem 3.9.1 (Kirchberg). *A* C*-*algebra A is exact if and only if for each* C*-*algebra B and ideal $J \triangleleft B$ the sequence*

$$0 \to J \otimes A \to B \otimes A \to (B/J) \otimes A \to 0$$

is exact – i.e., if and only if we always have a canonical isomorphism

$$\frac{B \otimes A}{J \otimes A} \cong (B/J) \otimes A.$$

We already observed the "only if" direction in Proposition 3.7.8; hence we are left to prove the converse. Throughout this section we say that A is $\otimes$-*exact*[27] if for each C*-algebra B and ideal $J \triangleleft B$ the sequence

$$0 \to J \otimes A \to B \otimes A \to (B/J) \otimes A \to 0$$

is exact; our goal is to show that every $\otimes$-exact C*-algebra is exact.

The proof is quite technical and requires meticulous care as there are a number of identifications that one must keep track of. Before diving into the details, let us give the main idea so you can keep the big picture in mind.

Suppose that $A \subset \mathbb{B}(\mathcal{H})$ is given, $P_n \in \mathbb{B}(\mathcal{H})$ are finite-rank projections converging to the identity (strong operator topology) and $E \subset A$ is a finite-dimensional operator system. We can always define u.c.p. maps $\varphi_n \colon A \to \mathbb{M}_{s(n)}(\mathbb{C}) \cong P_n\mathbb{B}(\mathcal{H})P_n$ by $\varphi_n(a) = P_n a P_n$ and, by finite-dimensionality, the restrictions $\varphi_n|_E \colon E \to \mathbb{M}_{s(n)}(\mathbb{C})$ will be injective linear maps for all sufficiently large n. Hence we can consider the inverse maps $\varphi_n^{-1}|_{\varphi_n(E)} : \varphi_n(E) \to E$. *Now assume a miracle occurs* and we know that

$$\|\varphi_n^{-1}|_{\varphi_n(E)}\|_{\mathrm{cb}} \to 1.$$

[27]This is just convenient terminology used primarily in this section.

Then a perturbation argument (Lemma 3.9.7 plus Arveson's Extension Theorem) would imply that we can find u.c.p. maps $\psi_n : \mathbb{M}_{s(n)}(\mathbb{C}) \to \mathbb{B}(\mathcal{H})$ such that

$$\|\psi_n(\varphi_n(x)) - \varphi_n^{-1}|_{\varphi_n(E)}(\varphi_n(x))\| = \|\psi_n(\varphi_n(x)) - x\| \to 0$$

for all $x \in E$. If this held for all finite-dimensional operator systems $E \subset A$, then we could evidently conclude that A is exact.

The miraculous fact (Proposition 3.9.6) is that if A is $\otimes$-exact, then the completely bounded norms of $\varphi_n^{-1}|_{\varphi_n(E)}$ really do tend to 1 for every finite-dimensional operator system $E \subset A$. The bulk of the proof of Theorem 3.9.1 goes into showing this.

Let's first show that $\otimes$-exactness is a local property. That is, a C*-algebra A is $\otimes$-exact if and only if all of its finite-dimensional operator subsystems are $\otimes$-exact. By definition, an operator system E is $\otimes$-exact if we have an isometric identification

$$\frac{E \otimes B}{E \otimes J} \cong E \otimes (B/J)$$

for all C*-algebras B and ideals $J \triangleleft B$. Note that there is always a contractive map

$$\frac{E \otimes B}{E \otimes J} \to E \otimes (B/J),$$

since the kernel of the contraction $E \otimes B \to E \otimes (B/J)$ contains $E \otimes J$.

Lemma 3.9.2. *If $E \subset A$ is an operator system and $J \triangleleft B$ is an ideal, then there is an isometric inclusion*

$$\frac{E \otimes B}{E \otimes J} \subset \frac{A \otimes B}{A \otimes J}.$$

Proof. We must show that if $x \in E \otimes B$, then its norm down in $\frac{A\otimes B}{A\otimes J}$ is equal to

$$\inf\{\|x + y\| : y \in E \otimes J\}.$$

This, however, is easily seen since the norm in $\frac{A\otimes B}{A\otimes J}$ is equal to

$$\lim \|x(1 \otimes (1 - e_i))\|,$$

where $\{e_i\}$ is an approximate unit for J. □

Proposition 3.9.3. *A C**-algebra A is $\otimes$-exact if and only if all its finite-dimensional operator subsystems are $\otimes$-exact.*

Proof. The "if" direction is not too hard, since the union of the subspaces $\frac{E\otimes B}{E\otimes J}$, where $E \subset A$ is a finite-dimensional operator subsystem, is dense in $\frac{A\otimes B}{A\otimes J}$.

The converse follows from the previous lemma since the top row of the commutative diagram

$$\begin{array}{ccc} \dfrac{A \otimes B}{A \otimes J} & \longrightarrow & A \otimes (B/J) \\ \uparrow & & \uparrow \\ \dfrac{E \otimes B}{E \otimes J} & \longrightarrow & E \otimes (B/J) \end{array}$$

is an isometric isomorphism whenever A is $\otimes$-exact. □

It follows that we may always assume separability.

The next result is quasi-obvious, there are just a bunch of identifications to check. For the record, $\bigoplus_n A_n$ denotes the c_0-direct sum, i.e., the set of sequences (a_n) such that $\lim_n \|a_n\| = 0$.

Lemma 3.9.4. *If $E \subset A$ is a finite-dimensional operator system and B_n are unital* C*-*algebras, then there is a u.c.p. isometric isomorphism*

$$E \otimes (\prod_n B_n) \to \prod_n (E \otimes B_n),$$

defined on elementary tensors by

$$e \otimes (b_n)_n \mapsto (e \otimes b_n)_n.$$

This map also gives an identification of $E \otimes (\bigoplus_n B_n)$ and $\bigoplus_n (E \otimes B_n)$.

Proof. Fix some concrete representations $B_n \subset \mathbb{B}(\mathcal{H}_n)$ and $A \subset \mathbb{B}(\mathcal{K})$. The natural "diagonal" embedding

$$\prod_n B_n \subset \mathbb{B}(\bigoplus_n \mathcal{H}_n)$$

induces an inclusion

$$A \otimes (\prod_n B_n) \subset \mathbb{B}\Big(\mathcal{K} \otimes (\bigoplus_n \mathcal{H}_n)\Big).$$

Similarly, we have

$$\prod_n A \otimes B_n \subset \prod_n \mathbb{B}(\mathcal{K} \otimes \mathcal{H}_n) \subset \mathbb{B}\Big(\bigoplus_n (\mathcal{K} \otimes \mathcal{H}_n)\Big).$$

The canonical isomorphism $\mathcal{K} \otimes (\bigoplus_n \mathcal{H}_n) \to \bigoplus_n (\mathcal{K} \otimes \mathcal{H}_n)$ gives an identification

$$\mathbb{B}\Big(\mathcal{K} \otimes (\bigoplus_n \mathcal{H}_n)\Big) \cong \mathbb{B}\Big(\bigoplus_n (\mathcal{K} \otimes \mathcal{H}_n)\Big).$$

What one must check is that this isomorphism takes

$$a \otimes (b_n)_n \in A \otimes (\prod_n B_n) \subset \mathbb{B}\Big(\mathcal{K} \otimes (\bigoplus_n \mathcal{H}_n)\Big)$$

to the element

$$(a \otimes b_n)_n \in \prod_n A \otimes B_n \subset \mathbb{B}\Big(\bigoplus_n (\mathcal{K} \otimes \mathcal{H}_n)\Big).$$

Thus we have constructed a $*$-homomorphism $A \otimes (\prod_n B_n) \to \prod_n A \otimes B_n$, but it won't be surjective if A is infinite dimensional.

However, everything is fine for E. Indeed, if we fix an algebraic basis $\{x_1, \ldots, x_m\} \subset E$, then an arbitrary sequence $(d_n)_n \in \prod_n E \otimes B_n$ has a unique representation as

$$(d_n)_n = (\sum_{j=1}^{m} x_j \otimes b_{j,n})_n = \sum_{j=1}^{m} (x_j \otimes b_{j,n})_n,$$

where we used the fact that $(b_{j,n})$ is a bounded sequence, for every j. But each of the sequences $(x_j \otimes b_{j,n})_n$ comes from $E \otimes (\prod_n B_n)$, so the proof is complete. □

With the previous identification in hand, the following is immediate from the definition of $\otimes$-exactness.

Lemma 3.9.5. *If E is a finite-dimensional operator system with algebraic basis $\{x_1, \ldots, x_m\}$ and B_n are unital C*-algebras, then there is a natural contractive linear mapping*

$$\frac{\prod_n (E \otimes B_n)}{\bigoplus_n (E \otimes B_n)} \to E \otimes \left(\frac{\prod_n B_n}{\bigoplus_n B_n}\right)$$

such that the image of an element $(\sum_{j=1}^{m} x_j \otimes b_{j,n})_n \in \prod_n (E \otimes B_n)$ under the mapping

$$\prod_n (E \otimes B_n) \to \frac{\prod_n (E \otimes B_n)}{\bigoplus_n (E \otimes B_n)} \to E \otimes \left(\frac{\prod_n B_n}{\bigoplus_n B_n}\right)$$

is

$$\sum_{j=1}^{m} x_j \otimes \left((b_{j,n})_n + \bigoplus_n B_n\right) \in E \otimes \left(\frac{\prod_n B_n}{\bigoplus_n B_n}\right).$$

If E is $\otimes$-exact, then this map is an isometric isomorphism.

In the proposition below, both A and the Hilbert space $\mathcal{H}$ are separable.

Proposition 3.9.6. *Let $A \subset \mathbb{B}(\mathcal{H})$ be a unital $\otimes$-exact* C**-algebra, $\{P_n\}$ be any increasing sequence of finite-rank projections converging strongly to $1_{\mathcal{H}}$ and $E \subset A$ be any finite-dimensional operator system. If $\varphi_n \colon E \to \mathbb{M}_{s(n)}(\mathbb{C}) \cong P_n\mathbb{B}(\mathcal{H})P_n$ are defined by $\varphi_n(x) = P_n x P_n$, then we have*

$$\|\varphi_n^{-1}|_{\varphi_n(E)}\|_{\mathrm{cb}} \to 1.$$ [28]

Proof. Just to ease notation, let us assume that the rank of P_n is n so that $P_n\mathbb{B}(\mathcal{H})P_n \cong \mathbb{M}_n(\mathbb{C})$. Modulo notational complications, the same proof works in general.

Proceeding by contradiction, we'll show that if $\lim_{n\to\infty} \|\varphi_n^{-1}|_{\varphi_n(E)}\|_{\mathrm{cb}} = \beta > 1$, then it is possible to construct an element

$$X \in \frac{\prod_n (E \otimes \mathbb{M}_{k(n)}(\mathbb{C}))}{\bigoplus_n (E \otimes \mathbb{M}_{k(n)}(\mathbb{C}))},$$

for a suitable choice of $k(n)$'s, such that $\|X\| = 1$ but under the mapping

$$\frac{\prod_n (E \otimes \mathbb{M}_{k(n)}(\mathbb{C}))}{\bigoplus_n (E \otimes \mathbb{M}_{k(n)}(\mathbb{C}))} \to E \otimes \left(\frac{\prod_n \mathbb{M}_{k(n)}(\mathbb{C})}{\bigoplus_n \mathbb{M}_{k(n)}(\mathbb{C})}\right)$$

X goes to an element of norm $\leq \beta^{-1} < 1$. Since A is $\otimes$-exact, Lemma 3.9.5 will give our contradiction.

Thus we assume that $\lim_{n\to\infty} \|\varphi_n^{-1}|_{\varphi_n(E)}\|_{\mathrm{cb}} = \beta > 1$. Since the maps $\varphi_n^{-1}|_{\varphi_n(E)}$ are expanding some elements (after tensoring with large enough matrices), it follows that the original maps φ_n have to be shrinking some elements (after tensoring with large enough matrices). In other words, we can find a sequence of integers $k(n)$ and an element $(X_n)_n \in \prod_n (E \otimes \mathbb{M}_{k(n)}(\mathbb{C}))$ such that $\|X_n\| = 1$ for all n and

$$\lim_{n\to\infty} \|\varphi_n \otimes \mathrm{id}_{k(n)}(X_n)\| = \beta^{-1} < 1.$$

Denote by

$$X \in \frac{\prod_n (E \otimes \mathbb{M}_{k(n)}(\mathbb{C}))}{\bigoplus_n (E \otimes \mathbb{M}_{k(n)}(\mathbb{C}))}$$

the canonical image of $(X_n)_n \in \prod_n (E \otimes \mathbb{M}_{k(n)}(\mathbb{C}))$.

Now apply the contractive map (Lemma 3.9.5)

$$\frac{\prod_n (E \otimes \mathbb{M}_{k(n)}(\mathbb{C}))}{\bigoplus_n (E \otimes \mathbb{M}_{k(n)}(\mathbb{C}))} \to E \otimes \left(\frac{\prod_n \mathbb{M}_{k(n)}(\mathbb{C})}{\bigoplus_n \mathbb{M}_{k(n)}(\mathbb{C})}\right)$$

---

[28] Since $P_n \leq P_{n+1}$, we have $\varphi_n = \varphi_n \circ \varphi_{n+1} = \varphi_{n+1} \circ \varphi_n$. Hence if φ_{n+1} compresses the norm of some operator, then φ_n can only shrink it further; that is, $1 \leq \|\varphi_{n+1}^{-1}|_{\varphi_{n+1}(E)}\|_{\mathrm{cb}} \leq \|\varphi_n^{-1}|_{\varphi_n(E)}\|_{\mathrm{cb}}$ for all n. In particular, this shows $\lim \|\varphi_n^{-1}|_{\varphi_n(E)}\|_{\mathrm{cb}}$ exists.

and let $\tilde{X}$ be the image of X. By definition of the spatial norm, we have an inclusion

$$E \otimes \left(\frac{\prod_n \mathbb{M}_{k(n)}(\mathbb{C})}{\bigoplus_n \mathbb{M}_{k(n)}(\mathbb{C})} \right) \subset \mathbb{B}(\mathcal{H} \otimes \mathcal{K})$$

for some (nonseparable) Hilbert space $\mathcal{K}$ and hence it is easy to see that

$$\|\tilde{X}\| = \sup_s \|\varphi_s \otimes \mathrm{id}(\tilde{X})\|,$$

where the norm on the right is computed in

$$\mathbb{M}_s(\mathbb{C}) \otimes \left(\frac{\prod_n \mathbb{M}_{k(n)}(\mathbb{C})}{\bigoplus_n \mathbb{M}_{k(n)}(\mathbb{C})} \right).$$

However, $\mathbb{M}_s(\mathbb{C})$ is $\otimes$-exact and so we can invert the isomorphism given by Lemma 3.9.5 to get an isomorphism

$$\mathbb{M}_s(\mathbb{C}) \otimes \left(\frac{\prod_n \mathbb{M}_{k(n)}(\mathbb{C})}{\bigoplus_n \mathbb{M}_{k(n)}(\mathbb{C})} \right) \to \frac{\prod_n (\mathbb{M}_s(\mathbb{C}) \otimes \mathbb{M}_{k(n)}(\mathbb{C}))}{\bigoplus_n (\mathbb{M}_s(\mathbb{C}) \otimes \mathbb{M}_{k(n)}(\mathbb{C}))}.$$

Now one must check that the element

$$(\varphi_s \otimes \mathrm{id}_{k(n)}(X_n))_n \in \prod_n (\mathbb{M}_s(\mathbb{C}) \otimes \mathbb{M}_{k(n)}(\mathbb{C}))$$

is a lift of the image of $\tilde{X}$ under the mapping

$$\begin{aligned} E \otimes \left(\frac{\prod_n \mathbb{M}_{k(n)}(\mathbb{C})}{\bigoplus_n \mathbb{M}_{k(n)}(\mathbb{C})} \right) &\to \mathbb{M}_s(\mathbb{C}) \otimes \left(\frac{\prod_n \mathbb{M}_{k(n)}(\mathbb{C})}{\bigoplus_n \mathbb{M}_{k(n)}(\mathbb{C})} \right) \\ &\to \frac{\prod_n (\mathbb{M}_s(\mathbb{C}) \otimes \mathbb{M}_{k(n)}(\mathbb{C}))}{\bigoplus_n (\mathbb{M}_s(\mathbb{C}) \otimes \mathbb{M}_{k(n)}(\mathbb{C}))}. \end{aligned}$$

It follows that

$$\begin{aligned} \|\tilde{X}\| &= \sup_s \|\varphi_s \otimes \mathrm{id}(\tilde{X})\| \\ &\leq \sup_s \left\| \left(\varphi_s \otimes \mathrm{id}_{k(n)}(X_n)\right)_n + \bigoplus_n (\mathbb{M}_s(\mathbb{C}) \otimes \mathbb{M}_{k(n)}(\mathbb{C})) \right\| \\ &= \sup_s \left(\limsup_{n \to \infty} \|\varphi_s \otimes \mathrm{id}_{k(n)}(X_n)\| \right), \end{aligned}$$

since the norm of $(c_n)_n + \bigoplus C_n \in (\prod C_n)/(\bigoplus C_n)$ is equal to $\limsup \|c_n\|$. But, $\varphi_s = \varphi_s \circ \varphi_n$ for all $n > s$ and hence

$$\sup_s \left(\limsup_{n \to \infty} \|\varphi_s \otimes \mathrm{id}_{k(n)}(X_n)\| \right) \leq \limsup_{n \to \infty} \|\varphi_n \otimes \mathrm{id}_{k(n)}(X_n)\|.$$

Putting this all together, we get

$$\|X\| > \beta^{-1} = \limsup_{n \to \infty} \|\varphi_n \otimes \mathrm{id}_{k(n)}(X_n)\| \geq \|\tilde{X}\|$$

and this contradicts $\otimes$-exactness of A. □

The final step is an immediate consequence of a basic c.b.-perturbation fact. Indeed, the following lemma is just a special case of Corollary B.11 since each of the maps $\varphi_n^{-1}|_{\varphi_n(E)}$ is already self-adjoint and the dimension of $\varphi_n(E)$ is fixed.

Lemma 3.9.7. *With assumptions and notation as in Proposition 3.9.6, it is possible to find u.c.p. maps* $\psi_n\colon \varphi_n(E) \to A$ *such that* $\|\psi_n - \varphi_n^{-1}|_{\varphi_n(E)}\| \to 0$.

The proof of Theorem 3.9.1 in the case that A is unital and separable is now basically complete. The only thing left to observe is that the u.c.p. maps $\psi_n\colon \varphi_n(E) \to A$ can be extended to u.c.p. maps $\mathbb{M}_{s(n)}(\mathbb{C}) \to \mathbb{B}(\mathcal{H})$; hence the argument sketched in the beginning of this section actually works. The general case is not too hard to deduce from the separable unital case, so we consider our work here finished.

Exercises

Exercise 3.9.1. If A is nonunital and $\otimes$-exact, then its unitization is also $\otimes$-exact (Exercise 3.7.3).

Exercise 3.9.2. Since a nonunital C*-algebra is exact if and only if every unital, separable subalgebra of its unitization is exact, use the previous exercise and Exercise 3.7.1 to prove the general case of Theorem 3.9.1.

Exercise 3.9.3. Prove that if $\mathbb{K}$ denotes the compact operators, then

$$\mathbb{K} \otimes \big(\prod_{\mathbb{N}} \mathbb{C}\big) \to \prod_{\mathbb{N}} (\mathbb{K} \otimes \mathbb{C})$$

is not surjective. In other words, finite-dimensionality is necessary in Lemma 3.9.4.

If you understand the proof of Theorem 3.9.1, the next three exercises should be easy.

Exercise 3.9.4. Let $A \subset \mathbb{B}(\mathcal{H})$ be exact and $E \subset A$ be a finite-dimensional operator system. Show that for every $\varepsilon > 0$ there exists a u.c.p. map $\varphi\colon A \to \mathbb{M}_n(\mathbb{C})$ and a c.b. map $\psi\colon \mathbb{M}_n(\mathbb{C}) \to \mathbb{B}(\mathcal{H})$ such that $\|\psi\|_{\mathrm{cb}} \le 1 + \varepsilon$ and $\psi \circ \varphi(x) = x$ for all $x \in E$. In other words, if one allows c.b. maps, then exactness is not an *approximate* factorization property – it is local factorization on the nose!

Exercise 3.9.5. Let $A \subset \mathbb{B}(\mathcal{H})$ be exact and $E \subset A$ be a finite-dimensional operator system. Show that if $\{v_i\} \subset \mathcal{H}$ is an orthonormal basis and P_n is the orthogonal projection onto the span of $\{v_1, \ldots, v_n\}$, then for all large n there exist u.c.p. maps $\psi_n\colon P_n\mathbb{B}(\mathcal{H})P_n \to \mathbb{B}(\mathcal{H})$ such that $\psi_n(P_n x P_n) \in A$ and $\|x - \psi_n(P_n x P_n)\| \le \varepsilon\|x\|$ for all $x \in E$. Why doesn't this imply nuclearity? Show that if (and only if) A is nuclear, one can force the entire range of each ψ_n into A.

Exercise 3.9.6. Prove that A is exact if and only if the sequence

$$0 \to \Big(\bigoplus_n \mathbb{M}_n(\mathbb{C})\Big) \otimes A \to \Big(\prod_n \mathbb{M}_n(\mathbb{C})\Big) \otimes A \to \left(\frac{\prod_n \mathbb{M}_n(\mathbb{C})}{\bigoplus_n \mathbb{M}_n(\mathbb{C})}\right) \otimes A \to 0$$

is exact.

Exercise 3.9.7. Prove that A is exact if and only if the sequence

$$0 \to \mathbb{K}(\mathcal{H}) \otimes A \to \mathbb{B}(\mathcal{H}) \otimes A \to \frac{\mathbb{B}(\mathcal{H})}{\mathbb{K}(\mathcal{H})} \otimes A \to 0$$

is exact for some infinite-dimensional Hilbert space $\mathcal{H}$.

Exercise 3.9.8. Assume $0 \to J \to A \to A/J \to 0$ is a locally split (Definition 3.7.5) short exact sequence. Prove that if J and A/J are exact, then so is A.[29] (Hint: Revisit Proposition 3.7.6 and then do a big diagram chase (aka the 3×3 Lemma) through

$$\begin{array}{ccccccccc}
 & & 0 & & 0 & & 0 & & \\
 & & \downarrow & & \downarrow & & \downarrow & & \\
0 & \longrightarrow & I \otimes J & \longrightarrow & B \otimes J & \longrightarrow & B/I \otimes J & \longrightarrow & 0 \\
 & & \downarrow & & \downarrow & & \downarrow & & \\
0 & \longrightarrow & I \otimes A & \longrightarrow & B \otimes A & \longrightarrow & B/I \otimes A & \longrightarrow & 0 \\
 & & \downarrow & & \downarrow & & \downarrow & & \\
0 & \longrightarrow & I \otimes A/J & \longrightarrow & B \otimes A/J & \longrightarrow & B/I \otimes A/J & \longrightarrow & 0 \\
 & & \downarrow & & \downarrow & & \downarrow & & \\
 & & 0 & & 0 & & 0 & &
\end{array}$$

Exercise 3.9.9. If one assumes a nontrivial tensor product fact, a very short proof of Kirchberg's Theorem can be given. Assume that there is a unique C*-norm on $C^*(\mathbb{F}_\infty) \odot \mathbb{B}(\mathcal{H})$ (this is a fact – see Section 13.2). Deduce Theorem 3.9.1. (Hint: If $A \subset \mathbb{B}(\mathcal{H})$ is $\otimes$-exact, then try to prove that $A \otimes_{\max} B \to \mathbb{B}(\mathcal{H}) \otimes_{\max} B$ factors through the spatial tensor product, for every B.)

3.10. References

Takesaki's Theorem was proved in [**182**]; the result is quite surprising, as the injective Banach space norm need not be the smallest, in general. Continuity of c.p. maps on maximal tensor products relies on [**9**], where Arveson introduced c.c. maps and proved his fundamental extension theorem. The Trick is

[29] Kirchberg has constructed counterexamples when the sequence is not locally split (see Remark 13.4.2).

adapted from [**113**], where Lance introduced the weak expectation property – inspired by Tomiyama's extensive work on conditional expectations. Tensor products and exact sequences were studied by several authors, notably S. Wassermann and Kirchberg. The equivalence between tensor product conditions and approximation properties, Theorems 3.8.7 and 3.9.1, come from [**40**], [**100**] and, respectively, [**106**]; our proof of Theorem 3.9.1 is due to Pisier [**149**].

Chapter 4

Constructions

There are numerous ways of creating new C*-algebras out of old ones. Our goal in this chapter is *not* to study the constructions per se, but rather to see how nuclearity and exactness behave. As such, the reader unfamiliar with crossed products, free products or Cuntz-Pimsner algebras will likely find our treatment lacking in several areas.

In recent years, there have been attempts to unify as many constructions as possible into a single point of view – i.e., to find a general procedure which contains several well-known constructions as special cases. Elegant and efficient as this generality may be, we take a different approach, meandering from the particular to the very general; this leads to redundancy, but we feel it benefits the novice. This also allows us to expose some techniques that may be useful in other contexts. For example, we construct explicit approximating maps on crossed products by $\mathbb{Z}$ while using a different approach – fixed point subalgebras of compact group actions – to prove nuclearity of graph algebras.

4.1. Crossed products

One of the most important constructions in operator algebra theory arises from noncommutative dynamical systems: the crossed product of a C*-algebra by a group action. Indeed, this idea goes back to von Neumann (in the context of measurable transformations of a measure space) and has since given rise to some of the most important modern examples of C*-algebras such as irrational rotation algebras, (stabilized) Cuntz algebras, (stabilized) graph algebras and others.

It turns out that nuclearity and exactness are reasonably well behaved under this construction and, in most cases, one can construct explicit finite-dimensional approximations on the crossed product. As such, we'll give a detailed description of these important examples, trying to stay as concrete as possible.

Definition 4.1.1. Let Γ be a discrete group and A be a C*-algebra. An *action* of Γ on A is a group homomorphism α from Γ into the group of $*$-automorphisms on A. A C*-algebra equipped with a Γ-action is called a *Γ-C*-algebra.*[1]

Our goal is to construct a single C*-algebra which encodes the action of Γ on A. In group theory, this procedure is well known and is called the *semidirect product*. We will adapt this idea and create an algebra $A \rtimes_\alpha \Gamma$ with the property that there is a copy of Γ inside the unitary group of $A \rtimes_\alpha \Gamma$ (at least when A is unital) and there is a natural inclusion $A \subset A \rtimes_\alpha \Gamma$ such that (a) $A \rtimes_\alpha \Gamma$ is generated by A and Γ and (b) $\alpha_g(a) = gag^*$ for all $a \in A$ and $g \in \Gamma$ (i.e., the action of Γ becomes inner).

For a Γ-C*-algebra A, we denote by $C_c(\Gamma, A)$ the linear space of finitely supported functions on Γ with values in A. A typical element S in $C_c(\Gamma, A)$ is written as a finite sum $S = \sum_{s\in\Gamma} a_s s$. We equip $C_c(\Gamma, A)$ with an α-twisted convolution product and $*$-operation as follows: for $S = \sum_{s\in\Gamma} a_s s, T = \sum_{t\in\Gamma} b_t t \in C_c(\Gamma, A)$ we declare

$$S *_\alpha T = \sum_{s,t\in\Gamma} a_s \alpha_s(b_t) st \text{ and } S^* = \sum_{s\in\Gamma} \alpha_{s^{-1}}(a_s^*) s^{-1}.$$

The twisted convolution is a generalization of the classical convolution of two $\ell^2(\mathbb{Z})$ functions, but the algebraic explanation of these formulas is perhaps more enlightening. Indeed, we are trying to turn $C_c(\Gamma, A)$ into a $*$-algebra where the action becomes inner and hence the definition above comes from the formal calculation

$$(\sum_{s\in\Gamma} a_s s)(\sum_{t\in\Gamma} b_t t) = \sum_{s,t\in\Gamma} a_s (s b_t s^*) st = \sum_{s,t\in\Gamma} a_s \alpha_s(b_t) st.$$

However you care to think about it, $C_c(\Gamma, A)$ is the smallest $*$-algebra which encodes the action of Γ on A. Note that when $A = \mathbb{C}$ and the action α is trivial, we simply recover the group ring $\mathbb{C}[\Gamma]$. Now the question is, "How shall we complete $C_c(\Gamma, A)$?" Just as for group C*-algebras, there are two natural choices, a universal and a reduced completion.

A *covariant representation* $(u, \pi, \mathcal{H})$ of the Γ-C*-algebra A consists of a unitary representation $(u, \mathcal{H})$ of Γ and a $*$-representation $(\pi, \mathcal{H})$ of A such

[1] Of course, there could be many different actions of Γ on A, giving rise to different Γ-C*-algebra structures.

that $u_s\pi(a)u_s^* = \pi(\alpha_s(a))$ for every $s \in \Gamma$ and $a \in A$. It is not hard to see that every covariant representation gives rise to a $*$-representation of $C_c(\Gamma, A)$ and, conversely, every (nondegenerate) $*$-representation of $C_c(\Gamma, A)$ arises this way. For a covariant representation $(u, \pi, \mathcal{H})$, we denote by $u \times \pi$ the associated $*$-representation of $C_c(\Gamma, A)$.

Definition 4.1.2. The *full crossed product* (sometimes called the "universal" crossed product) of a C*-dynamical system (A, α, Γ), denoted $A \rtimes_\alpha \Gamma$, is the completion of $C_c(\Gamma, A)$ with respect to the norm

$$\|x\|_u = \sup \|\pi(x)\|,$$

where the supremum is over all (cyclic) $*$-homomorphisms $\pi\colon C_c(\Gamma, A) \to \mathbb{B}(\mathcal{H})$.

Though it isn't completely obvious, we will soon see that there are lots of representations $C_c(\Gamma, A) \to \mathbb{B}(\mathcal{H})$. (In particular, $\|\cdot\|_u$ really is a norm, as opposed to seminorm, on $C_c(\Gamma, A)$ and hence we have a natural inclusion $C_c(\Gamma, A) \subset A \rtimes_\alpha \Gamma$.) Evidently our definition implies the following *universal property*.

Proposition 4.1.3 (Universal property). *For every covariant representation $(u, \pi, \mathcal{H})$ of a Γ-C*-algebra A, there is a $*$-homomorphism $\sigma\colon A \rtimes_\alpha \Gamma \to \mathbb{B}(\mathcal{H})$ such that*

$$\sigma(\sum_{s\in\Gamma} a_s s) = \sum_{s\in\Gamma} \pi(a_s)u_s,$$

for all $\sum_{s\in\Gamma} a_s s \in C_c(\Gamma, A)$.

To define the reduced crossed product, we begin with a *faithful* representation $A \subset \mathbb{B}(\mathcal{H})$. Define a new representation of A on $\mathcal{H} \otimes \ell^2(\Gamma)$ by

$$\pi(a)(v \otimes \delta_g) = (\alpha_{g^{-1}}(a)(v)) \otimes \delta_g,$$

where $\{\delta_g\}_{g\in G}$ is the canonical orthonormal basis of $\ell^2(\Gamma)$. Under the identification $\mathcal{H} \otimes \ell^2(\Gamma) \cong \bigoplus_{g\in\Gamma} \mathcal{H}$ we have simply taken the direct sum representation

$$\pi(a) = \bigoplus_{g\in\Gamma} \alpha_g^{-1}(a) \in \mathbb{B}(\bigoplus_{g\in\Gamma} \mathcal{H}).$$

The point of doing this is that now the left regular representation of Γ spatially implements the action α: for all elementary tensors we have

$$\begin{aligned}
(1 \otimes \lambda_s)\pi(a)(1 \otimes \lambda_s^*)(v \otimes \delta_g) &= (1 \otimes \lambda_s)\pi(a)(v \otimes \delta_{s^{-1}g}) \\
&= (1 \otimes \lambda_s)((\alpha_{g^{-1}s}(a)(v)) \otimes \delta_{s^{-1}g}) \\
&= (\alpha_{g^{-1}s}(a)(v)) \otimes \delta_g \\
&= (\alpha_{g^{-1}}(\alpha_s(a))(v)) \otimes \delta_g \\
&= \pi(\alpha_s(a))(v \otimes \delta_g).
\end{aligned}$$

Hence we get an induced covariant representation $(1 \otimes \lambda) \times \pi$, called a *regular representation.*[2]

Definition 4.1.4. The *reduced crossed product* of a C*-dynamical system (A, Γ, α), denoted $A \rtimes_{\alpha,r} \Gamma$, is defined to be the norm closure of the image of a regular representation $C_c(\Gamma, A) \to \mathbb{B}(\mathcal{H} \otimes \ell^2(\Gamma))$.

For notational simplicity, we will usually forget about the representation π and the fact that we had to inflate the left regular representation of Γ – i.e., we often (abuse notation slightly and) denote a typical element $x \in C_c(\Gamma, A) \subset A \rtimes_{\alpha,r} \Gamma$ as a finite sum $x = \sum_{s \in \Gamma} a_s \lambda_s$.

Though the following proposition should come as no surprise, the proof contains some important calculations.

Proposition 4.1.5. *The reduced crossed product $A \rtimes_{\alpha,r} \Gamma$ does not depend on the choice of the faithful representation $A \subset \mathbb{B}(\mathcal{H})$.*

Proof. The proof boils down to the fact that there is a unique C*-norm on $\mathbb{M}_n(A)$, just as in the proof of Proposition 3.3.11. For a finite set $F \subset \Gamma$, let $P \in \mathbb{B}(\ell^2(\Gamma))$ be the finite-rank projection onto the span of $\{\delta_g : g \in F\}$. Rather than compute the norm of $x \in \mathbb{B}(\mathcal{H} \otimes \ell^2(\Gamma))$, we will cut by the (infinite-rank) projections $1 \otimes P$ and show that the norm of the compression is independent of the representation $A \subset \mathbb{B}(\mathcal{H})$ – taking a limit over finite sets in Γ, we conclude the same for x.

Let $\{e_{p,q}\}_{p,q \in F}$ be the canonical matrix units of $P\mathbb{B}(\ell^2(\Gamma))P \cong \mathbb{M}_F(\mathbb{C})$ and fix some arbitrary elements $a \in A$ and $s \in \Gamma$. Let $\pi \colon A \to \mathbb{B}(\mathcal{H} \otimes \ell^2(\Gamma))$ be a regular representation. Our first claim is that

$$(1 \otimes P)\pi(a) = (1 \otimes P)\pi(a)(1 \otimes P) = \sum_{q \in F} \alpha_q^{-1}(a) \otimes e_{q,q}.$$

This is clear if one thinks of $\pi(a)$ as a diagonal matrix in $\mathbb{B}(\bigoplus_{g \in \Gamma} \mathcal{H})$; in the tensor product picture we have

$$\pi(a) = \sum_{q \in \Gamma} \alpha_q^{-1}(a) \otimes e_{q,q},$$

where convergence is in the strong operator topology.

Thus we see that

$$\begin{aligned}(1 \otimes P)\pi(a)(1 \otimes \lambda_s)(1 \otimes P) &= \Big(\sum_{q \in F} \alpha_q^{-1}(a) \otimes e_{q,q}\Big)(1 \otimes P\lambda_s P) \\ &= \Big(\sum_{q \in F} \alpha_q^{-1}(a) \otimes e_{q,q}\Big)\Big(\sum_{p \in F \cap sF} 1 \otimes e_{p,s^{-1}p}\Big)\end{aligned}$$

[2]Regular representations are easily seen to be injective on $C_c(\Gamma, A)$; hence the universal norm really is a norm.

$$= \sum_{p \in F \cap sF} \alpha_p^{-1}(a) \otimes e_{p,s^{-1}p} \in A \otimes \mathbb{M}_F(\mathbb{C}).$$

Now if $x = \sum a_s\lambda_s \in C_c(\Gamma, A) \subset \mathbb{B}(\mathcal{H} \otimes \ell^2(\Gamma))$, then we have

$$(1 \otimes P)x(1 \otimes P) = \sum_{s \in \Gamma} \sum_{p \in F \cap sF} \alpha_p^{-1}(a_s) \otimes e_{p,s^{-1}p} \in A \otimes \mathbb{M}_F(\mathbb{C})$$

and thus the norm of $(1 \otimes P)x(1 \otimes P)$ does not depend on the embedding $A \subset \mathbb{B}(\mathcal{H})$. □

The following description of positive elements is sometimes handy.

Corollary 4.1.6. *An element $x = \sum_{s\in\Gamma} a_s\lambda_s \in C_c(\Gamma, A)$ is positive in $A \rtimes_{\alpha,r} \Gamma$ if and only if for any finite sequence $s_1, \ldots, s_n \in \Gamma$, the operator matrix $[\alpha_{s_i}^{-1}(a_{s_i s_j^{-1}})]_{i,j} \in \mathbb{M}_n(A)$ is positive.*

Proof. Since an operator is positive if and only if its compression by any projection is positive, the result follows from a calculation above:

$$(1 \otimes P)x(1 \otimes P) = \sum_{s \in \Gamma} \sum_{p \in F \cap sF} \alpha_p^{-1}(a_s) \otimes e_{p,s^{-1}p} \in A \otimes \mathbb{M}_F(\mathbb{C}).$$

Indeed, if $F = \{s_1, \ldots, s_n\}$, then we can identify this double sum with the operator matrix in the statement of the corollary. (Let $p = s_i$ and $s_j = s^{-1}p$.) □

Here is a C*-dynamical version of Fell's absorption principle, with identical proof.

Proposition 4.1.7 (Fell's absorbtion principle II). *If $(u, \mathrm{id}_A, \mathcal{H})$ is a covariant representation (i.e., $A \subset \mathbb{B}(\mathcal{H})$ and the action α is spatially implemented in this representation), then the covariant representation*

$$(u \otimes \lambda, \mathrm{id}_A \otimes 1, \mathcal{H} \otimes \ell^2(\Gamma))$$

is unitarily equivalent to a regular representation. In particular, we have a natural $*$-isomorphism

$$C^*\big((u \otimes \lambda)(\Gamma), A \otimes 1\big) \cong A \rtimes_{\alpha,r} \Gamma.$$

Proof. Let $(u, \mathrm{id}_A, \mathcal{H})$ be a covariant representation and define a unitary U on $\mathcal{H} \otimes \ell^2(\Gamma)$ by $U(\xi \otimes \delta_t) = u_t\xi \otimes \delta_t$. One checks that

$$U(1 \otimes \lambda_s)U^* = (u_s \otimes \lambda_s) \quad \text{and} \quad U(\sum_t \alpha_t^{-1}(a) \otimes e_{t,t})U^* = a \otimes 1$$

for every $s \in \Gamma$ and $a \in A$. □

We close this section with the existence of conditional expectations. First, a lemma.

Lemma 4.1.8. *Let ψ be a faithful state on B. Then* $\mathrm{id}_A \otimes \psi\colon A \otimes B \to A$ *is faithful.*

Proof. Observe that $\{f \otimes g : f \in A^*,\ g \in B^*\} \subset (A \otimes B)^*$ separates the points of $A \otimes B$. Indeed, $A \otimes B \subset \mathbb{B}(\mathcal{H} \otimes \mathcal{K})$ and vector states arising from elementary tensors $h \otimes k \in \mathcal{H} \otimes \mathcal{K}$ separate all of $\mathbb{B}(\mathcal{H} \otimes \mathcal{K})$.

So, if $x \in (A \otimes B)_+$ is nonzero, we can find a state φ on A such that $(\varphi \otimes \mathrm{id}_B)(x) \in B$ is nonzero (and positive). Since ψ is faithful, we have

$$0 < \psi((\varphi \otimes \mathrm{id}_B)(x)) = \varphi((\mathrm{id}_A \otimes \psi)(x)),$$

which implies $(\mathrm{id}_A \otimes \psi)(x)$ is nonzero. □

Proposition 4.1.9. *The map $E\colon C_c(\Gamma, A) \to A$, $E(\sum_s a_s \lambda_s) = a_e$, extends to a faithful conditional expectation from $A \rtimes_{\alpha,r} \Gamma$ onto A. In particular,*

$$\max_{s\in\Gamma} \|a_s\|_A \leq \|\sum_{s\in\Gamma} a_s\lambda_s\|_{A\rtimes_{\alpha,r}\Gamma}.$$

Proof. Let $(u, \mathrm{id}_A, \mathcal{H})$ be a covariant representation. By Fell's absorption principle, we may view $A \rtimes_{\alpha,r} \Gamma$ as the C*-algebra generated by $A \otimes 1$ and $(u \otimes \lambda)(\Gamma)$ – in particular, it is a subalgebra of $\mathbb{B}(\mathcal{H}) \otimes C^*_r(\Gamma)$. The key observation is that in this representation our map E is nothing but the restriction of $\mathrm{id}_{\mathbb{B}(\mathcal{H})} \otimes \tau$, where τ is the canonical faithful tracial state on $C^*_r(\Gamma)$ (which is clear since $\tau(\lambda_s) = 0$, whenever $s \neq e$). Thus the previous lemma implies that E is faithful.

Finally, note that $a_s = E(z\lambda_s^*)$ for $z = \sum_s a_s\lambda_s$. This implies the asserted inequality, so the proof is complete. □

Remark 4.1.10. Note that $E\colon A \rtimes_{\alpha,r} \Gamma \to A$ is Γ-*equivariant*: $E(\lambda_s z \lambda_s^{-1}) = \alpha_s(E(z))$ for every $s \in \Gamma$ and $z \in A \rtimes_{\alpha,r} \Gamma$.

More generally, if α and β are actions of Γ on sets X and, respectively, Y, we will say a map $\Phi\colon X \to Y$ is Γ-*equivariant* if $\Phi \circ \alpha_g = \beta_g \circ \Phi$ for all $g \in \Gamma$.

Exercises

Exercise 4.1.1. Let $\tau\colon \Gamma \to \mathrm{Aut}(\mathbb{C})$ be the trivial action. Since $\mathbb{C} = \mathbb{B}(\mathbb{C})$, use Proposition 4.1.5 to show that $\mathbb{C} \rtimes_{\tau,r} \Gamma \cong C^*_r(\Gamma)$. While you are at it, observe that $\mathbb{C} \rtimes_\tau \Gamma \cong C^*(\Gamma)$.

Exercise 4.1.2. Prove that if $\tau\colon \Gamma \to \mathrm{Aut}(A)$ is the trivial action (i.e., $\tau_g = \mathrm{id}_A$ for all $g \in \Gamma$), then

$$A \rtimes_{\tau,r} \Gamma \cong A \otimes C^*_r(\Gamma).$$

What is the corresponding result for universal crossed products? (Hint: Think of universal properties.)

When proving that a particular map into a crossed product is *completely* positive, it often suffices to prove positivity. We will use the following exercise (taking B to be a matrix algebra) for this reduction.

Exercise 4.1.3. If $\alpha\colon \Gamma \to \mathrm{Aut}(A)$ is an action and $\tau\otimes\alpha\colon \Gamma \to \mathrm{Aut}(B\otimes A)$ is defined by $(\tau\otimes\alpha)_g = \mathrm{id}_B \otimes \alpha_g$, then

$$(B\otimes A)\rtimes_{\tau\otimes\alpha,r}\Gamma \cong B\otimes(A\rtimes_{\alpha,r}\Gamma).$$

What is the corresponding result for universal crossed products?

Exercise 4.1.4. Let (A,α) and (B,β) be Γ-C*-algebras and $\varphi\colon A\to B$ be a c.c.p. map which is Γ-equivariant. Prove that the map $\tilde{\varphi}\colon C_c(\Gamma,A)\to C_c(\Gamma,B)$, defined by $\tilde{\varphi}(\sum_s a_s s) = \sum_s \varphi(a_s)s$, extends to a c.c.p. map from $A\rtimes_{\alpha,r}\Gamma$ into $B\rtimes_{\beta,r}\Gamma$ (resp. from $A\rtimes_\alpha\Gamma$ into $B\rtimes_\beta\Gamma$). (Hint: The reduced case is easy. For the full case, you'll need a Γ-equivariant Stinespring Dilation Theorem.)

4.2. Integer actions

We now specialize to the case of $\mathbb{Z}$ actions, where things are easier to digest. Our goal is to construct explicit approximating maps on the crossed product. This approach is a little boorish, but it has been very important for other purposes (e.g. noncommutative entropy theory or calculating Haagerup invariants).

Suppose $A\subset\mathbb{B}(\mathcal{H})$ and $\alpha\in\mathrm{Aut}(A)$ is an automorphism. We also use α to denote the induced action of $\mathbb{Z}$ given by $n\mapsto\alpha^n$. Let $[k,n] = \{k,k+1,k+2,\dots,n\}$ be the interval of integers from k to n. The finite subsets $F_n=[0,n]$ will play an important role as they happen to be a *Følner sequence*: for each $k\in\mathbb{Z}$,

$$\frac{|(k+F_n)\cap F_n|}{|F_n|} = \frac{|[k,n+k]\cap[0,n]|}{n+1} = \frac{n-k+1}{n+1}\to 1,$$

as $n\to\infty$, where $|\cdot|$ denotes cardinality. It turns out that we can easily construct approximating maps by cutting to Følner sets and then mapping back to the crossed product. We will need a few simple lemmas.

Lemma 4.2.1. *Let A be a C*-algebra and let $n\in\mathbb{N}$. Every positive element in $\mathbb{M}_n(A)$ is a sum of n elements of the form $[a_i^*a_j]_{i,j=1}^n$.*

Proof. Take an arbitrary positive element $x\in\mathbb{M}_n(A)$ and decompose it as a product $x=[b_{ij}]^*[b_{ij}]$. Now one writes

$$[b_{ij}] = A_1 + A_2 + \cdots + A_n$$

and

$$[b_{ij}]^* = A_1^* + A_2^* + \cdots + A_n^*,$$

where

$$A_1 = \begin{bmatrix} b_{11} & b_{12} & \cdots & b_{1n} \\ 0 & 0 & \cdots & 0 \\ \vdots & \vdots & \ddots & \vdots \\ 0 & 0 & \cdots & 0 \end{bmatrix}, A_2 = \begin{bmatrix} 0 & 0 & \cdots & 0 \\ b_{21} & b_{22} & \cdots & b_{2n} \\ \vdots & \vdots & \ddots & \vdots \\ 0 & 0 & \cdots & 0 \end{bmatrix},$$

and so on. Since $A_j^* A_i = 0$ whenever $i \neq j$, a straightforward calculation completes the proof. □

Lemma 4.2.2. *If A is a Γ-C*-algebra and $F \subset \Gamma$ is a finite set, then for each set $\{a_p\}_{p\in F} \subset A$, the element*

$$\sum_{p,q\in F} \alpha_p(a_p^* a_q)\lambda_{pq^{-1}} \in C_c(\Gamma, A)$$

is positive as an element in $A \rtimes_\alpha \Gamma$ (or $A \rtimes_{\alpha,r} \Gamma$).

Proof. The element in question is equal to $(\sum_{p\in F} a_p\lambda_{p^{-1}})^*(\sum_{p\in F} a_p\lambda_{p^{-1}})$. □

Here are the approximating maps we're after.

Lemma 4.2.3. *If A is a Γ-C*-algebra and $F \subset \Gamma$ is a finite set, then there exist c.c.p. maps $\varphi\colon A \rtimes_{\alpha,r} \Gamma \to A \otimes \mathbb{M}_F(\mathbb{C})$ and $\psi\colon A \otimes \mathbb{M}_F(\mathbb{C}) \to C_c(\Gamma, A) \subset A \rtimes_{\alpha,r} \Gamma$ such that for all $a \in A$ and $s \in \Gamma$ we have*

$$\psi \circ \varphi(a\lambda_s) = \frac{|F \cap sF|}{|F|} a\lambda_s.$$

Proof. In the proof of Proposition 4.1.5 we saw that there is a c.c.p. map $\varphi\colon A \rtimes_{\alpha,r} \Gamma \to A \otimes \mathbb{M}_F(\mathbb{C})$ such that

$$\varphi(a\lambda_s) = \sum_{p\in F\cap sF} \alpha_p^{-1}(a) \otimes e_{p,s^{-1}p}.$$

It suffices to prove that $\psi\colon A \otimes \mathbb{M}_F(\mathbb{C}) \to C_c(\Gamma, A) \subset A \rtimes_{\alpha,r} \Gamma$ defined by

$$\psi(a \otimes e_{p,q}) = \frac{1}{|F|}\alpha_p(a)\lambda_{pq^{-1}}$$

is a c.c.p. map, as a simple calculation confirms the asserted formula.

In fact, it suffices to prove that ψ is positive since Exercise 4.1.3 provides a natural commutative diagram

$$\begin{array}{ccc} \mathbb{M}_n(\mathbb{C}) \otimes (A \otimes \mathbb{M}_F(\mathbb{C})) & \xrightarrow{\cong} & (\mathbb{M}_n(\mathbb{C}) \otimes A) \otimes \mathbb{M}_F(\mathbb{C}) \\ \downarrow & & \downarrow \\ \mathbb{M}_n(\mathbb{C}) \otimes (A \rtimes_{\alpha,r} \Gamma) & \xrightarrow{\cong} & (\mathbb{M}_n(\mathbb{C}) \otimes A) \rtimes_{\tau\otimes\alpha,r} \Gamma. \end{array}$$

By Lemma 4.2.1, we only need to check that for every set $\{a_p\}_{p\in F} \subset A$, $\psi(\sum a_p^* a_q \otimes e_{p,q}) \geq 0$. But

$$\psi(\sum_{p,q\in F} a_p^* a_q \otimes e_{p,q}) = \sum_{p,q\in F} \frac{1}{|F|}\alpha_p(a_p^* a_q)\lambda_{pq^{-1}},$$

so the previous lemma completes the proof. □

Theorem 4.2.4. *For an automorphism* $\alpha \in \mathrm{Aut}(A)$*, the following statements are true:*

(1) $A \rtimes_\alpha \mathbb{Z} = A \rtimes_{\alpha,r} \mathbb{Z}$*;*

(2) A *is nuclear if and only if* $A \rtimes_\alpha \mathbb{Z}$ *is nuclear;*

(3) A *is exact if and only if* $A \rtimes_\alpha \mathbb{Z}$ *is exact.*

Proof. Proof of (1): It suffices to show that there exist c.c.p. maps

$$\Psi_n \colon A \rtimes_{\alpha,r} \mathbb{Z} \to A \rtimes_\alpha \mathbb{Z}$$

such that $\|x - \Psi_n \circ \pi(x)\|_u \to 0$ for all $x \in C_c(\mathbb{Z}, A) \subset A \rtimes_\alpha \mathbb{Z}$, where

$$\pi \colon A \rtimes_\alpha \mathbb{Z} \to A \rtimes_{\alpha,r} \mathbb{Z}$$

is the canonical quotient map (coming from universality).

The key observation is that the proof of Lemma 4.2.3 is algebraic. In other words, if $F_n = [0, n] \subset \mathbb{Z}$ is a Følner sequence and φ_n, ψ_n are the corresponding maps constructed in Lemma 4.2.3, then we can define c.c.p. maps $\Psi_n \colon A \rtimes_{\alpha,r} \mathbb{Z} \to A \rtimes_\alpha \mathbb{Z}$ by $\Psi_n = \psi_n \circ \varphi_n$, but simply regarding the ψ_n's as taking values in the universal crossed product (as opposed to the reduced one, since the range of ψ_n is contained in $C_c(\mathbb{Z}, A)$). The formula in Lemma 4.2.3 still holds, and hence for $x = \sum_{k\in\mathbb{Z}} a_k k \in C_c(\mathbb{Z}, A)$ we have

$$\|x - \Psi_n(\pi(x))\|_{A\rtimes_\alpha\mathbb{Z}} = \|\sum_{k\in\mathbb{Z}}(1 - \frac{|F_n \cap (k + F_n)|}{|F_n|})a_k k\|_{A\rtimes_\alpha\mathbb{Z}} \to 0$$

since only finitely many a_k's are nonzero.

Proofs of (2) and (3): Both of the "if" directions are trivial since there is a conditional expectation $A \rtimes_\alpha \mathbb{Z} \to A$.

For the other direction one should first review Exercises 2.3.11 and 2.3.12. Indeed, another way of stating Lemma 4.2.3 is that there exist c.c.p. maps $\varphi_n \colon A \rtimes_\alpha \mathbb{Z} \to A \otimes \mathbb{M}_{k(n)}(\mathbb{C})$ and $\psi_n \colon A \otimes \mathbb{M}_{k(n)}(\mathbb{C}) \to A \rtimes_\alpha \mathbb{Z}$ such that $\psi_n \circ \varphi_n \to \mathrm{id}$ in the point-norm topology. Since $A \otimes \mathbb{M}_{k(n)}(\mathbb{C})$ is nuclear (resp. exact) whenever A is nuclear (resp. exact), we are done. □

Recall that for a real number $\theta > 0$, the rotation algebra A_θ is defined to be $C(\mathbb{T}) \rtimes_{\alpha_\theta} \mathbb{Z}$ where α_θ is the automorphism induced by a rotation of the circle through an angle of $2\pi\theta$ radians.

Corollary 4.2.5. *The rotation algebras A_θ are nuclear.*

At this point, the reader may feel lied to – we said we would stick to $\mathbb{Z}$-actions in order to stay concrete and then proceeded to keep $\mathbb{Z}$ out of the picture until the very end. Here is a corollary of our deceit (the proof is identical to the proof of Theorem 4.2.4).

Theorem 4.2.6. *For any amenable group Γ and action $\alpha\colon \Gamma \to \mathrm{Aut}(A)$, the following statements hold:*

(1) $A \rtimes_\alpha \Gamma = A \rtimes_{\alpha,r} \Gamma$;

(2) *A is nuclear if only if $A \rtimes_\alpha \Gamma$ is nuclear;*

(3) *A is exact if and only if $A \rtimes_\alpha \Gamma$ is exact.*

4.3. Amenable actions

We now step up the generality ladder and consider crossed products by amenable actions – i.e., the group involved need not be amenable, but we require it to act nicely. When defined "appropriately" (not the definition usually found in the literature, but an equivalent one that makes our present work easier) and done abstractly, finding approximating maps on a crossed product by an amenable action is only slightly harder than the case of $\mathbb{Z}$ actions.

Given a Γ-C*-algebra A, we put a third norm on the (α-twisted) convolution algebra $C_c(\Gamma, A)$: for finitely supported functions $S, T\colon \Gamma \to A$ we define

$$\langle S, T\rangle = \sum S(g)^* T(g) \in A$$

and

$$\|S\|_2 = \|\langle S, S\rangle\|^{1/2}.$$

The informed reader will notice that we have made a Hilbert C*-module – more on that subject in Section 4.6. The Cauchy-Schwarz inequality holds in this context: $\|\langle S, T\rangle\|_A \leq \|S\|_2 \|T\|_2$, for all $S, T \in C_c(\Gamma, A)$.[3]

Definition 4.3.1. An action $\alpha\colon \Gamma \to \mathrm{Aut}(A)$ on a unital C*-algebra A is *amenable* if there exist finitely supported functions $T_i\colon \Gamma \to A$ with the following properties:

(1) $0 \leq T_i(g) \in \mathcal{Z}(A)$ (the center of A) for all $i \in \mathbb{N}$ and $g \in \Gamma$;

(2) $\langle T_i, T_i\rangle = \sum_{g\in\Gamma} T_i(g)^2 = 1_A$;

[3] This is a general fact about Hilbert modules, but here we only need the case that A is abelian. If $A = C(X)$, the asserted inequality follows from the usual Cauchy-Schwarz inequality, applied pointwise in X.

(3) $\|s *_\alpha T_i - T_i\|_2 \to 0$ for all $s \in \Gamma$, where $s \in C_c(\Gamma, A)$ is the function which sends $s \mapsto 1_A$ and all other group elements to zero.[4]

The functions T_i will replace the Følner sets we used in the previous section.

Lemma 4.3.2. *Let A be a Γ-C*-algebra and $T\colon \Gamma \to A$ be a finitely supported function such that $0 \le T(g) \in \mathcal{Z}(A)$ for all $g \in \Gamma$ and $\sum_g T(g)^2 = 1_A$. Then,*

(1) $T *_\alpha T^*(s) = \sum_{p \in F \cap sF} T(p)\alpha_s(T(s^{-1}p))$, *where F is the support of T, and*

(2) $\|1_A - T *_\alpha T^*(s)\| \le \|T - s *_\alpha T\|_2$, *for all $s \in \Gamma$.*

Proof. Statement (1) is a trivial calculation, using the fact that $T(g)^* = T(g)$ for all $g \in \Gamma$.

To prove the second statement, we first note that $s *_\alpha T(p) = \alpha_s(T(s^{-1}p))$ for all $p \in \Gamma$. Now we compute

$$\begin{aligned} 1_A - T *_\alpha T^*(s) &= \sum_{p\in\Gamma} T(p)^2 - \sum_{p\in\Gamma} T(p)\alpha_s(T(s^{-1}p)) \\ &= \sum_{p\in\Gamma} T(p)\Big(T(p) - \alpha_s(T(s^{-1}p))\Big) \\ &= \langle T, T - s *_\alpha T\rangle. \end{aligned}$$

Hence the desired inequality follows from the Cauchy-Schwarz inequality, since $\|T\|_2 = 1$. □

Here is the analogue of Lemma 4.2.3 for crossed products by amenable actions.

Lemma 4.3.3. *Let A be a unital Γ-C*-algebra and $T\colon \Gamma \to A$ be a finitely supported function with support F, such that $0 \le T(g) \in \mathcal{Z}(A)$ for all $g \in \Gamma$ and $\sum_g T(g)^2 = 1_A$. Then, there exist u.c.p. maps $\varphi\colon A \rtimes_{\alpha,r} \Gamma \to A \otimes \mathbb{M}_F(\mathbb{C})$ and $\psi\colon A \otimes \mathbb{M}_F(\mathbb{C}) \to A \rtimes_{\alpha,r} \Gamma$ such that for all $s \in \Gamma$ and $a \in A$,*

$$\psi \circ \varphi(a\lambda_s) = \big(T *_\alpha T^*(s)\big)a\lambda_s.$$

Proof. We already have a u.c.p. compression map $\varphi\colon A \rtimes_{\alpha,r} \Gamma \to A \otimes \mathbb{M}_F(\mathbb{C})$ such that

$$\varphi(a\lambda_s) = \sum_{p\in F\cap sF} \alpha_p^{-1}(a) \otimes e_{p,s^{-1}p} \in A \otimes \mathbb{M}_F(\mathbb{C}).$$

[4]This definition comes from the characterization of (classical) amenability in terms of weak containment of the trivial representation in the left regular representation.

Define

$$X = \sum_{p \in F} \alpha_p^{-1}(T(p)) \otimes e_{p,p}$$

and note that $X = X^*$. Hence compression by X is a c.p. map and a computation confirms that

$$X\varphi(a\lambda_s)X = \sum_{p \in F \cap sF} \alpha_p^{-1}(T(p)a)\alpha_{s^{-1}p}^{-1}(T(s^{-1}p)) \otimes e_{p,s^{-1}p}.$$

We know the map $A \otimes M_F(\mathbb{C}) \to A \rtimes_{\alpha,r} \Gamma$ defined by

$$b \otimes e_{x,y} \mapsto \alpha_x(b)\lambda_{xy^{-1}}$$

is u.c.p., so we get another u.c.p. map $\psi\colon A \otimes \mathbb{M}_F(\mathbb{C}) \to A \rtimes_{\alpha,r} \Gamma$ by composing it with compression by X:

$$\psi\colon A \otimes \mathbb{M}_F(\mathbb{C}) \overset{X \cdot X}{\to} A \otimes \mathbb{M}_F(\mathbb{C}) \overset{b\otimes e_{x,y} \mapsto \alpha_x(b)\lambda_{xy^{-1}}}{\longrightarrow} A \rtimes_{\alpha,r} \Gamma.$$

Finally, since $T(g) \in \mathcal{Z}(A)$ for all $g \in \Gamma$, we have

$$\begin{aligned}\psi \circ \varphi(a\lambda_s) &= \sum_{p \in F \cap sF} \alpha_p\Big(\alpha_p^{-1}(T(p)a)\alpha_{s^{-1}p}^{-1}(T(s^{-1}p))\Big)\lambda_s \\ &= \Big(\sum_{p \in F \cap sF} T(p)\alpha_s(T(s^{-1}p))\Big)a\lambda_s \\ &= \big(T *_\alpha T^*(s)\big)a\lambda_s.\end{aligned}$$

□

The proof of the next theorem is but a tiny perturbation of that given for Theorem 4.2.4. We leave the details to the reader.

Theorem 4.3.4. *For any amenable action of Γ on A, the following statements hold:*

(1) $A \rtimes_\alpha \Gamma = A \rtimes_{\alpha,r} \Gamma$;

(2) *A is nuclear if only if $A \rtimes_\alpha \Gamma$ is nuclear;*

(3) *A is exact if and only if $A \rtimes_\alpha \Gamma$ is exact.*

We call a compact[5] space X a Γ-*space* if it is equipped with an action of Γ (by homeomorphisms). Let $x \mapsto s.x$ denote the action of $s \in \Gamma$ on $x \in X$. To help distinguish, we let $\alpha_s\colon C(X) \to C(X)$ denote the induced automorphism of $C(X)$ (i.e., $\alpha_s(f)(x) = f(s^{-1}.x)$). The notion of an amenable action comes from classical (i.e., abelian) dynamical systems. As already mentioned, our definition at the C*-level is not very common in the literature. Here is a more popular version.

[5]As usual, compactness includes the Hausdorff axiom.

Definition 4.3.5. An action of Γ on a compact space X is called (topologically) *amenable* (or, equivalently, X is an *amenable* Γ*-space*) if there exists a net of continuous maps $m_i\colon X \to \mathrm{Prob}(\Gamma)$, such that for each $s \in \Gamma$,

$$\lim_{i\to\infty}\Big(\sup_{x\in X}\|s.m_i^x - m_i^{s.x}\|_1\Big) = 0,$$

where $s.m_i^x(g) = m_i^x(s^{-1}g)$.[6]

Remark 4.3.6. Let $\mathrm{Prob}(X)$ be the set of all regular Borel probability measures on X. In Proposition 5.2.1 we will show for a countable group Γ that amenability can be reformulated as: For any finite subset $E \subset \Gamma$, $\varepsilon > 0$ and any $m \in \mathrm{Prob}(X)$, there exists a Borel map $\mu\colon X \to \mathrm{Prob}(\Gamma)$ (i.e., the function $X \to \mathbb{R}$, $x \mapsto \mu^x(t)$, is Borel for every $t \in \Gamma$) such that

$$\max_{s\in E}\int_X \|s.\mu^x - \mu^{s.x}\|_1\, dm(x) < \varepsilon.$$

Lemma 4.3.7. *An action $\alpha\colon \Gamma \to \mathrm{Homeo}(X)$ is amenable if and only if the induced action on $C(X)$ is amenable in the sense of Definition 4.3.1.*

Proof. The proofs of both directions are similar. First assume the action is amenable in the sense of Definition 4.3.5. Let $m_i\colon X \to \mathrm{Prob}(\Gamma)$ be a sequence of continuous maps such that for each $s \in \Gamma$,

$$\lim_{i\to\infty}\big(\sup_{x\in X}\|s.m_i^x - m_i^{s.x}\|_1\big) = 0.$$

Define $S_i\colon \Gamma \to C(X)$ by

$$S_i(g)(x) = m_i^x(g).$$

Then for each $x \in X$ we have

$$\sum_g S_i(g)(x) = \sum_g m_i^x(g) = 1,$$

since m_i^x is a probability measure. Defining $\tilde{T}_i(g) = \sqrt{S_i(g)}$, it follows that for each i,

$$\langle \tilde{T}_i, \tilde{T}_i\rangle = \sum_g \tilde{T}_i(g)^2 = 1_{C(X)}.$$[7]

Of course, the $\tilde{T}_i$'s are not finitely supported (we will fix that later) but note that for each $x \in X$,

$$(s *_\alpha \tilde{T}_i)(g)(x) = \alpha_s\big(\tilde{T}_i(s^{-1}g)\big)(x) = \tilde{T}_i(s^{-1}g)(s^{-1}.x) = \sqrt{s.m_i^{s^{-1}.x}(g)}.$$

[6]By definition, $\mathrm{Prob}(\Gamma)$ is the set of probability measures on Γ – which we identify with the set of positive, norm-one elements in $\ell^1(\Gamma)$. Continuity means with respect to the restriction of the weak-$*$ topology on $\ell^1(\Gamma)$. In other words, $m\colon X \to \mathrm{Prob}(\Gamma)$ is continuous if and only if for each convergent net $x_i \to x \in X$ we have $m^{x_i}(g) \to m^x(g)$ for all $g \in \Gamma$.

[7]Note that Dini's Theorem implies this sum converges uniformly, since everything is positive.

Using the inequality $(a-b)^2 \leq |a^2 - b^2|$ for all positive numbers a, b, we then get

$$\begin{aligned}
\|s *_\alpha \tilde{T}_i - \tilde{T}_i\|_2^2 &= \sup_{x\in X} \Big(\sum_{g\in\Gamma} |\sqrt{s.m_i^{s^{-1}.x}(g)} - \sqrt{m_i^x(g)}|^2 \Big) \\
&\leq \sup_{x\in X} \Big(\sum_{g\in\Gamma} |s.m_i^{s^{-1}.x}(g) - m_i^x(g)| \Big) \\
&\overset{x=s.y}{=} \sup_{y\in X} \Big(\sum_{g\in\Gamma} |s.m_i^{y}(g) - m_i^{s.y}(g)| \Big) \\
&= \sup_{y\in X} \|s.m_i^y - m_i^{s.y}\|_1 \to 0.
\end{aligned}$$

Hence the $\tilde{T}_i$'s have the right properties, except for finite support. Fixing this problem is easy once we prove the following claim.

Claim. If $T\colon \Gamma \to C(X)$ is a positive function such that $\langle T, T\rangle = 1_{C(X)}$, then there exists a sequence of *finitely supported* positive functions $T_n\colon \Gamma \to C(X)$ such that $\langle T_n, T_n\rangle = 1_{C(X)}$ for all n and

$$\|s *_\alpha T_n - T_n\|_2 \to \|s *_\alpha T - T\|_2,$$

for all $s \in \Gamma$.

To prove this claim, we let $F_n \subset F_{n+1}$ be a sequence of finite subsets of Γ such that $\bigcup F_n = \Gamma$. Since

$$\sum_{g\in\Gamma} T(g)^2 = 1_{C(X)}$$

and convergence is uniform, it follows that for all sufficiently large n,

$$\sum_{g\in F_n} T(g)^2 > 0,$$

meaning bounded uniformly away from 0. Hence we can define T_n by declaring

$$T_n(g) = \sqrt{\frac{1}{\sum_{g\in F_n} T(g)^2}}\, T(g),$$

for all $g \in F_n$ and $T_n(g) = 0$ if $g \notin F_n$. Tedious and unenlightening calculations (left to the diligent few) show that these functions do the trick.

To prove the opposite direction of Lemma 4.3.7, one basically reverses the procedure above. That is, define $m_i^x(g) = T_i(g)^2(x)$ and calculate away. It should be noted that the Cauchy-Schwarz inequality gets used in the following way:

$$\sum |a_i^2 - b_i^2| = \sum |a_i - b_i|(a_i + b_i) \leq \|(a_i) - (b_i)\|_2 \|(a_i) + (b_i)\|_2.$$

□

The proof of this lemma shows that if X is an amenable Γ-space, then we can assume each map $m_i \colon X \to \mathrm{Prob}(\Gamma)$ has the property that there exists a finite set $F_i \subset \Gamma$ with $\mathrm{supp}(m_i^x) \subset F_i$, for every $x \in X$. Here is a direct proof of this fact.

Lemma 4.3.8. *Let $m \colon X \to \mathrm{Prob}(\Gamma)$ be a continuous map. Then, for any $\varepsilon > 0$, there exist $\tilde{m} \colon X \to \mathrm{Prob}(\Gamma)$ and a finite subset $F \subset \Gamma$ such that $\mathrm{supp}\, \tilde{m}^x \subset F$ and $\|m^x - \tilde{m}^x\|_1 < \varepsilon$ for all $x \in X$.*

Proof. For every finite subset $F \subset \Gamma$, let $U(F) = \{x \in X : \|m^x \chi_F\|_1 > 1 - \varepsilon/2\} \subset X$, where χ_F is the characteristic function of F. It is easily seen that $\{U(F)\}_F$ is an open cover of X which is upward directed. Since X is compact, there exists F such that $X = U(F)$. It follows that $\tilde{m}^x = m^x \chi_F + \|m^x \chi_{\Gamma \setminus F}\|_1 \delta_e$ has the desired property. □

Exercises

Exercise 4.3.1. Assume $\Gamma \times \Gamma$ acts on $C(X)$ and there exist two $\Gamma \times \Gamma$-invariant subalgebras $A, B \subset C(X)$ such that (a) $\Gamma \times \{e\}|_A$ is amenable while $\{e\} \times \Gamma|_A$ is trivial and (b) $\Gamma \times \{e\}|_B$ is trivial while $\{e\} \times \Gamma|_B$ is amenable. Prove that the action of $\Gamma \times \Gamma$ on $C(X)$ is amenable. (In addition to helping cement Definition 4.3.5 in your mind, this exercise will be needed later; see Corollary 5.3.19.)

Exercise 4.3.2. Find an elementary proof, based on Fell's absorbtion principle, of the fact that $A \rtimes \Gamma = A \rtimes_r \Gamma$, whenever the action is amenable. (Hint: Fix an embedding $A \rtimes \Gamma \subset \mathbb{B}(\mathcal{H})$ and use the definition of amenable action to construct some isometries from $\mathcal{H}$ to $\mathcal{H} \otimes \ell^2(\Gamma)$.)

4.4. $X \rtimes \Gamma$-algebras

The previous section says precious little about when an action is amenable. The point was that *if* it is, then crossed products are well behaved. It turns out that determining amenability is inextricably intertwined with nuclearity (and all tangled up with exactness too). This section is devoted to exposing this connection, in a slightly more general context.

Instead of considering the $*$-algebra of finitely supported functions from Γ to $C(X)$, it is often convenient to think of compactly supported functions $X \times \Gamma \to \mathbb{C}$. That is, since any compact subset of $X \times \Gamma$ is contained in $X \times F$, for some finite subset $F \subset \Gamma$, we can identify $C_c(\Gamma, C(X))$ with $C_c(X \times \Gamma)$ – an element $\sum_{s \in \Gamma} f_s s \in C_c(\Gamma, C(X))$ corresponds to $f \in C_c(X \times \Gamma)$, where $f(x, s) = f_s(x)$. One checks that our α-twisted convolution and adjoint on $C_c(\Gamma, C(X))$, when transferred to $C_c(X \times \Gamma)$, look like

$$(g *_\alpha f)(x, s) = \sum_{t \in \Gamma} g(x, t) f(t^{-1}.x, t^{-1}s) \quad \text{and} \quad f^*(x, s) = \overline{f(s^{-1}.x, s^{-1})}.$$

Definition 4.4.1. We say a function $h\colon X \times \Gamma \to \mathbb{C}$ is of *positive type* if for any finite sequence $s_1, \ldots, s_n \in \Gamma$ and $x \in X$, the matrix $[h(s_i.x, s_i s_j^{-1})]_{i,j} \in \mathbb{M}_n(\mathbb{C})$ is positive.

Since an element $T \in \mathbb{M}_n(C(X)) = C(X, \mathbb{M}_n(\mathbb{C}))$ is positive if and only if the matrix $T(x)$ is positive for all points $x \in X$, Corollary 4.1.6 implies that a function $h \in C_c(X \times \Gamma)$ is of positive type if and only if it is positive when regarded as an element in $C_c(\Gamma, C(X)) \subset C(X) \rtimes_{\alpha,r} \Gamma$.

Definition 4.4.2. Let X be a compact topological Γ-space. A Γ-C*-algebra A is called a *$(X \rtimes \Gamma)$-C*-algebra* if $C(X)$ is embedded in the center of A in such a way that $\alpha_s(C(X)) = C(X)$ and $\alpha_s(f)(x) = f(s^{-1}.x)$ for every $s \in \Gamma$, $f \in C(X)$ and $x \in X$. In other words, there is an equivariant embedding of $C(X)$ into the center of A.

Theorem 4.4.3. *Let X be a compact topological Γ-space. The following are equivalent:*

(1) *the action of Γ on X is amenable;*

(2) *for any finite subset $F \subset \Gamma$ and $\varepsilon > 0$, there exists a positive type function $h \in C_c(X \times \Gamma)$ such that*
$$\max_{s \in F} \sup_{x \in X} |h(x,s) - 1| < \varepsilon;$$

(3) *for any $(X \rtimes \Gamma)$-C*-algebra A, we have $A \rtimes_\alpha \Gamma = A \rtimes_{\alpha,r} \Gamma$;*

(4) *the C*-algebra $C(X) \rtimes_{\alpha,r} \Gamma$ is nuclear.*

Proof. (2) $\Rightarrow$ (1): Let $h \in C_c(X \times \Gamma)$ be a positive-type function as in condition (2). Since h is positive in the C*-algebra $C(X) \rtimes_{\alpha,r} \Gamma$, there exists $g \in C_c(X \times \Gamma)$ such that $\|g^* *_\alpha g - h\| < \varepsilon$, where $\|\cdot\|$ is the C*-norm on $C(X) \rtimes_{\alpha,r} \Gamma$. Let $E\colon C(X) \rtimes_{\alpha,r} \Gamma \to C(X)$ be the standard conditional expectation. Since $E(g^* *_\alpha g) \approx E(h) \approx 1$ in $C(X)$, replacing g with $g *_\alpha E(g^* *_\alpha g)^{-1/2}$ (and changing $\varepsilon > 0$), we may assume that $E(g^* *_\alpha g) = 1$. We define $T\colon \Gamma \to C(X)$ by $T(t)(x) = |g(t^{-1}.x, t^{-1})|$. Then,
$$\begin{aligned}\langle T, s *_\alpha T\rangle(x) &= \Big(\sum_{t\in\Gamma} T(t)\alpha_s(T(s^{-1}t))\Big)(x)\\ &= \sum_{t\in\Gamma} |g(t^{-1}.x, t^{-1})|\,|g(t^{-1}.x, t^{-1}s)|\\ &\geq \Big|\sum_{t\in\Gamma} g^*(x,t)g(t^{-1}.x, t^{-1}s)\Big|\\ &= |(g^* *_\alpha g)(x,s)|,\end{aligned}$$

where equality holds if $s = e$. In particular, $\langle T, T\rangle = E(g^* *_\alpha g) = 1$. Note also that $\langle s *_\alpha T, s *_\alpha T\rangle = 1$ and thus

$$\|s *_\alpha T - T\|_2^2 = \|2 - \langle T, s *_\alpha T\rangle - \langle s *_\alpha T, T\rangle\|,$$

which will be close to zero whenever $|(g^* *_\alpha g)(x,s)|$ is close to one (uniformly in x). But this happens whenever the positive-type function h is uniformly close to one, so the proof is complete.

(1) $\Rightarrow$ (3): This follows immediately from Lemma 4.3.7 and Theorem 4.3.4.

(3) $\Rightarrow$ (4): Let B be any unital C*-algebra, but regard it as a Γ-C*-algebra with the trivial Γ-action τ. Then, $C(X) \otimes B$ is a $(X \rtimes \Gamma)$-C*-algebra and

$$\begin{aligned}(C(X) \rtimes_\alpha \Gamma) \otimes_{\max} B &= (C(X) \otimes_{\max} B) \rtimes_{\alpha\otimes\tau} \Gamma \\ &= (C(X) \otimes B) \rtimes_{\alpha\otimes\tau, r} \Gamma \\ &= (C(X) \rtimes_\alpha \Gamma) \otimes B,\end{aligned}$$

where the first equality follows from universal considerations, the second uses nuclearity of $C(X)$ and our assumption that reduced and universal crossed products are isomorphic, while the last line uses Exercise 4.1.3 and our assumption one more time. Therefore $C(X) \rtimes_\alpha \Gamma$ is nuclear.

(4) $\Rightarrow$ (2): Let a finite subset $F \subset \Gamma$ and $\varepsilon > 0$ be given. Since the crossed product $C(X) \rtimes_{\alpha,r} \Gamma$ is nuclear, there exist u.c.p. maps

$$\varphi\colon C(X) \rtimes_{\alpha,r} \Gamma \to \mathbb{M}_n(\mathbb{C}), \quad \psi\colon \mathbb{M}_n(\mathbb{C}) \to C(X) \rtimes_{\alpha,r} \Gamma \ \text{ and } \ \theta = \psi \circ \varphi$$

such that $\|\theta(\lambda_s) - \lambda_s\| < \varepsilon$ for $s \in F$. By construction we have an inclusion $C(X) \rtimes_{\alpha,r} \Gamma \subset \mathbb{B}(\mathcal{H} \otimes \ell^2(\Gamma))$; hence we may assume that φ is the compression onto $\mathbb{B}(\mathcal{H}_0 \otimes \ell^2(\tilde{F}))$ for some finite subset $\tilde{F} \subset \Gamma$ and some finite-dimensional subspace $\mathcal{H}_0 \subset \mathcal{H}$ (Exercise 3.9.5). It follows that $\theta(\lambda_s) = 0$ if $s \notin \tilde{F}\tilde{F}^{-1}$ (i.e., if $s\tilde{F} \cap \tilde{F} = \emptyset$). Denote by $E\colon C(X) \rtimes_{\alpha,r} \Gamma \to C(X)$ the canonical conditional expectation and define $h \in C_c(X \times \Gamma)$ by $h(x,s) = h_s(x)$, where $h_s = E(\theta(\lambda_s)\lambda_s^{-1}) \in C(X)$. For every $s \in F$, we have

$$\|1 - h_s\|_{C(X)} = \|E((\lambda_s - \theta(\lambda_s))\lambda_s^{-1})\| \le \|\lambda_s - \theta(\lambda_s)\| < \varepsilon.$$

Moreover, h is of positive type. Indeed, for any $s_1, \ldots, s_n \in \Gamma$, we have

$$\begin{aligned}[\alpha_{s_i^{-1}}(h_{s_i s_j^{-1}})]_{i,j} &= \Big[\alpha_{s_i^{-1}}\Big(E(\theta(\lambda_{s_i s_j^{-1}})\lambda_{s_j s_i^{-1}})\Big)\Big]_{i,j} \\ &= \Big[E\big(\lambda_{s_i}^* \theta(\lambda_{s_i}\lambda_{s_j}^*)\lambda_{s_j}\big)\Big]_{i,j} \\ &= E\Big(\operatorname{diag}(\lambda_{s_1}, \ldots, \lambda_{s_n})^* \theta\Big([\lambda_{s_i}\lambda_{s_j}^*]_{i,j}\Big) \operatorname{diag}(\lambda_{s_1}, \ldots, \lambda_{s_n})\Big)\end{aligned}$$

since E is Γ-equivariant. But this latter matrix is positive since E and θ are c.p. □

Remark 4.4.4. This theorem provides natural examples of subalgebras of nuclear C*-algebras which are not nuclear. Indeed, if Γ is any nonamenable group which admits an amenable action on some compact space X, then $C^*_\lambda(\Gamma) \subset C(X) \rtimes \Gamma$ is a nonnuclear subalgebra (see Theorem 2.6.8). Such groups abound, as we will see in the next chapter.

We close this section with a few non-C*-characterizations of amenable actions on compact spaces.

Proposition 4.4.5. *Let X be a compact topological Γ-space. The following are equivalent:*

(1) *the action is amenable;*

(2) *for any finite subset $E \subset \Gamma$ and $\varepsilon > 0$, there exists a continuous map $\xi\colon X \to \ell^2(\Gamma)$ such that $\|\xi_x\|_2 = 1$ for all $x \in X$ and*

$$\max_{s\in E} \sup_{x\in X} \|s.\xi_x - \xi_{s.x}\|_2 < \varepsilon;$$

(3) *for any finite subset $E \subset \Gamma$ and $\varepsilon > 0$, there exist a finite subset $F \subset \Gamma$ and a family of nonnegative continuous functions $(f_t)_{t\in F}$ on X such that $\sum_{t\in F} f_t^2 = 1$ and*

$$\max_{s\in E} \sup_{x\in X} |1 - \sum_{t\in F\cap s^{-1}F} f_t(s^{-1}.x) f_{st}(x)| < \varepsilon.$$

Proof. The assertion (1) $\Leftrightarrow$ (2) follows from the fact that $\mu \mapsto \mu^{1/2}$ is a uniform homeomorphism from $\mathrm{Prob}(\Gamma)$ into $\ell^2(\Gamma)$ (see the estimates used in the proof of Lemma 4.3.7). For (2) $\Leftrightarrow$ (3), we define $f_t(x) = \xi_x(t)$ (and *vice versa*). Then for every $x \in X$, $t \mapsto f_t(x)$ defines a unit vector in $\ell^2(\Gamma)$ and a calculation shows that

$$\max_{s\in E} \sup_{x\in X} |1 - \sum_{t\in F\cap s^{-1}F} f_t(s^{-1}.x) f_{st}(x)| = \max_{s\in E} \sup_{x\in X} |1 - \langle s^{-1}.\xi_x, \xi_{s^{-1}.x}\rangle|,$$

where the set F comes from Lemma 4.3.8. Since unit vectors in Hilbert space are close in norm if and only if their inner product is almost one, the proof is complete. □

Exercises

Exercise 4.4.1. Prove that any action of an amenable group on a compact space is amenable. Prove that the trivial action of Γ on a one-point set is amenable if and only if Γ is amenable.

For the next three exercises, assume that Γ acts amenably on X.

Exercise 4.4.2. If Γ_0 is a subgroup of Γ, then Γ_0 also acts amenably on X.

Exercise 4.4.3. Prove that if Y is a compact Γ-space with a Γ-equivariant continuous map $f\colon Y \to X$, then the action of Γ on Y is also amenable.

Exercise 4.4.4. Show that if X admits a Γ-invariant probability measure, then Γ itself is amenable. (Hint for C*-enthusiasts: The nuclear C*-algebra $C(X) \rtimes \Gamma$ has a tracial state in this case.)

4.5. Compact group actions and graph C*-algebras

In this section we will give a simple proof of the fact that the C*-algebras arising from directed graphs are always nuclear. Our goal here is not to study these important examples properly; we want to introduce a new method of proving nuclearity in a specific context (setting the stage for the next section where the same technique will be used for Cuntz-Pimsner algebras). We will try to reach nuclearity as quickly as possible and hence quote a few results without proof. The interested reader can consult [**163**] for the things we omit.

So far we have only considered actions of discrete groups on C*-algebras, but there are plenty of important examples of nondiscrete group actions. The main idea we wish to advertise is that when a C*-algebra admits an action of a *compact* group, the question of nuclearity (or exactness) is reduced to understanding the fixed point algebra.

If K is a locally compact group, then an *action of K on A* is a homomorphism $\alpha\colon K \to \mathrm{Aut}(A)$ which is continuous in the point-norm topology (i.e., $g \mapsto \alpha_g(a)$ is a continuous map, from the given topology on K to the norm topology on A, for every $a \in A$). The *fixed point subalgebra A^α* is defined to be the set of $a \in A$ such that $\alpha_g(a) = a$ for all $g \in K$. If K happens to be a *compact* group, then there is always a conditional expectation $E_\alpha\colon A \to A^\alpha$ given by

$$E_\alpha(a) = \int_K \alpha_z(a)\, dz,$$

where the integration is with respect to Haar measure on K. We note that E_α is faithful. Indeed, let $a \in A_+ \setminus \{0\}$ and choose a state φ on A with $\varphi(a) > 0$. Then, the function $K \ni z \mapsto \varphi(\alpha_z(a))$ is nonzero and nonnegative. Thus, $\varphi(E_\alpha(a)) = \int_K \varphi(\alpha_z(a))\, dz > 0$, proving $E_\alpha(a) \neq 0$.

The following fact is well known, having been exploited by Cuntz and many others.

Proposition 4.5.1. *Let A and B be C*-algebras, α and β be actions of a compact group K on A and B, respectively, and $\pi\colon A \to B$ be an equivariant $*$-homomorphism. Then, π is injective if and only if it's injective on the fixed point algebra A^α.*

Proof. If $a \in A_+$ and $\pi(a) = 0$, then $\pi(E_\alpha(a)) = E_\beta(\pi(a)) = 0$. □

Theorem 4.5.2. *Let A be a C^*-algebra and α be an action of a compact group K on A. Then, A is nuclear (resp. exact) if and only if A^α is nuclear (resp. exact).*

Proof. Since A^α is the range of a conditional expectation, the "only if" parts are easy.

Let's show that nuclearity of A^α implies the same for A. For an arbitrary B, we can tensor with the trivial action $\alpha \otimes \mathrm{id}_B$ to get actions of K on both $A \otimes_{\max} B$ and $A \otimes B$. Since A^α is the range of a conditional expectation, there is a canonical inclusion $A^\alpha \otimes_{\max} B \subset A \otimes_{\max} B$ (Proposition 3.6.2). The key observation is that

$$(A \otimes_{\max} B)^{\alpha \otimes \mathrm{id}_B} = A^\alpha \otimes_{\max} B,$$

since the conditional expectation evidently maps elementary tensors into $A^\alpha \otimes_{\max} B$ (and linearity and continuity forces everything else into the same algebra). The same argument works for minimal tensor products and hence we have the commutative diagram

$$\begin{array}{ccc} A \otimes_{\max} B & \longrightarrow & A \otimes B \\ \uparrow & & \uparrow \\ A^\alpha \otimes_{\max} B & \xrightarrow{\cong} & A^\alpha \otimes B, \end{array}$$

where the bottom row is an isomorphism of fixed point algebras. Thus the previous proposition implies injectivity of the top row, so we see that A must be nuclear.

The case that A^α is exact uses a similar argument. For arbitrary B and ideal $J \lhd B$ we consider the commutative diagram

$$\begin{array}{ccc} \frac{A \otimes B}{A \otimes J} & \longrightarrow & A \otimes (B/J) \\ \uparrow & & \uparrow \\ \frac{A^\alpha \otimes B}{A^\alpha \otimes J} & \xrightarrow{\cong} & A^\alpha \otimes (B/J). \end{array}$$

One must again show that the bottom row consists of the fixed point subalgebras of the appropriate actions – recall that there is an embedding $A \odot (B/J) \subset \frac{A \otimes B}{A \otimes J}$ and this dense subalgebra evidently gets pushed into $A^\alpha \odot (B/J) \subset \frac{A^\alpha \otimes B}{A^\alpha \otimes J}$ under the conditional expectation onto the fixed point subalgebra – and conclude the proof as above. □

We now apply this theorem to the class of C^*-algebras arising from directed graphs. By definition, a directed graph $\mathfrak{G} = (V, E, s, r)$ consists of a set V of vertices, a set E of edges and two maps $s, r \colon E \to V$, called the source and range maps ($s(e)$ is the vertex at which an edge e begins, and $r(e)$ is the vertex at which it ends). An operator algebraist likes to think of the vertices as being (pairwise orthogonal) projections on a Hilbert space and each directed edge corresponding to a partial isometry going from the source projection to something underneath the range projection.[8] It is not hard to show that such projections and partial isometries can always be constructed (on a suitable direct sum of Hilbert spaces) and hence we can appeal to universal nonsense to construct a largest C*-algebra generated by such elements. More precisely, we have the following result (see [**163**, Proposition 1.21]).

Theorem 4.5.3. *Let $\mathfrak{G} = (V, E, s, r)$ be a row finite[9] graph. Then there exists a C*-algebra $C^*(\mathfrak{G})$ with the following properties:*

(1) *for each vertex $v \in V$ there is a projection $p_v \in C^*(\mathfrak{G})$ and the p_v's are pairwise orthogonal;*

(2) *for each edge $e \in E$ there is a partial isometry $s_e \in C^*(\mathfrak{G})$ such that $s_e^* s_e = p_{s(e)}$;*

(3) *for each $v \in V$, if $r^{-1}(v) \neq \emptyset$, then $p_v = \sum_{e \in r^{-1}(v)} s_e s_e^*$ – i.e., p_v is the (finite) sum of the range projections of the partial isometries coming from edges going into v;*

(4) *$C^*(\mathfrak{G}) = C^*(\{p_v : v \in V\}, \{s_e : e \in E\})$ and is universal in the sense that for any collection of projections $\{P_v : v \in V\}$ and partial isometries $\{S_e : e \in E\}$ satisfying the three conditions above, there is a $*$-homomorphism*

$$C^*(\mathfrak{G}) \to C^*(\{P_v : v \in V\}, \{S_e : e \in E\})$$

such that $p_v \mapsto P_v$, for all $v \in V$, and $s_e \mapsto S_e$ for all $e \in E$.

The C*-algebra $C^*(\mathfrak{G})$ is called the *graph* C*-*algebra* of the graph $\mathfrak{G}$. It follows easily from universality that $C^*(\mathfrak{G})$ admits a canonical *gauge action* $\mathbb{T} \to \mathrm{Aut}(C^*(\mathfrak{G}))$. That is, if $z \in \mathbb{C}$ has modulus one, then the partial isometries $\{zs_e : e \in E\}$ satisfy the same relations as $\{s_e : e \in E\}$; hence there is a $*$-isomorphism $\gamma_z : C^*(\mathfrak{G}) \to C^*(\mathfrak{G})$ such that $p_v \mapsto p_v$ and $s_e \mapsto zs_e$. One checks that we have point-norm continuity, hence a group action of $\mathbb{T}$ on $C^*(\mathfrak{G})$. According to Theorem 4.5.2, we would know that $C^*(\mathfrak{G})$ is nuclear if we could show the fixed point subalgebra of the gauge action to be nuclear. If you have studied Cuntz algebras already, then this should come as no surprise. (Actually, Cuntz algebras are nice examples of graph

[8] Some papers have the partial isometries go the other way, but we prefer to travel in the direction of the edge.

[9] Row finite means that every vertex has at most finitely many edges coming into it – i.e., $r^{-1}(v)$ is a finite subset of E, for all $v \in V$. It is possible that a vertex has infinitely many edges going out of it, however.

algebras. Can you find the finite graph which yields $\mathcal{O}_2$? Hint: It has a single vertex!)

Though the details require plenty of calculations, here is a description of the fixed point subalgebra of the gauge action on $C^*(\mathfrak{G})$. Consider the norm closure of the set

$$\text{span}\{(s_{e_1}s_{e_2}\cdots s_{e_n})(s_{f_1}s_{f_2}\cdots s_{f_n})^* : e_i, f_j \in E, 1 \leq i, j \leq n \in \mathbb{N}\}.$$

Thanks to the relations in the definition of $C^*(\mathfrak{G})$, it can be shown that this is actually a C*-subalgebra (similar to the case of Cuntz algebras); it is clearly contained in the fixed point subalgebra of the gauge action; and, using the conditional expectation, it is not too hard to show that this is the entire fixed point subalgebra (since the conditional expectation maps words into this set). One then must prove that we are actually looking at an AF algebra which, though slightly more technical, is also similar to the case of Cuntz algebras. See Corollary 3.3, and the discussion after it, in [**163**] for more details.

If you are willing to believe this, then we have the following corollary.

Corollary 4.5.4. *For any row finite graph $\mathfrak{G}$, there is a canonical gauge action $\mathbb{T} \to \text{Aut}(C^*(\mathfrak{G}))$ whose fixed point subalgebra is AF. It follows that $C^*(\mathfrak{G})$ is nuclear.*

In [**163**, Remark 4.3], an alternate proof for nuclearity is given. It turns out that $\mathbb{K} \otimes C^*(\mathfrak{G}) \cong A \rtimes_\alpha \mathbb{Z}$ for an AF algebra A (in fact, $A = C^*(\mathfrak{G}) \rtimes_\gamma \mathbb{T}$). This proof has the added benefit of showing that graph algebras always satisfy the Universal Coefficient Theorem of Rosenberg and Schochet ([**170**]).

4.6. Cuntz-Pimsner algebras

We now wish to study nuclearity and exactness in the context of a very general construction due to Mihai Pimsner. Most of this section is spent describing the construction and proving requisite facts. The main theorem for us – both nuclearity and exactness are preserved by this construction: Theorem 4.6.25 – comes at the very end. Unfortunately, tons of technical results are needed to get there. In order to keep this section less than 1,000 pages, we will be a bit terse. For those primarily interested in groups and their actions, there is little harm in jumping to Chapter 5 and referring back as necessary.

Preliminaries on Hilbert C*-modules. Pimsner's construction requires an understanding of the basic theory of Hilbert modules. We state the facts we need; see [**114**] for proofs.

Let A be a C*-algebra and $\mathcal{H}$ be a linear space which is a right A-module. An *A-valued inner product* on $\mathcal{H}$ is a map

$$\mathcal{H} \times \mathcal{H} \ni (\xi, \eta) \mapsto \langle \xi, \eta \rangle \in A$$

satisfying

(1) $\langle \cdot, \cdot \rangle$ is linear in the *second* variable,

(2) $\langle \xi, \eta a \rangle = \langle \xi, \eta \rangle a$ for every $\xi, \eta \in \mathcal{H}$ and $a \in A$,

(3) $\langle \xi, \eta \rangle^* = \langle \eta, \xi \rangle$ for every $\xi, \eta \in \mathcal{H}$,

(4) $\langle \xi, \xi \rangle \geq 0$; and $\langle \xi, \xi \rangle = 0$ implies $\xi = 0$.

An *A-valued semi-inner product* is a map which satisfies all the above conditions except for the second part of (4). Let $\mathcal{H}$ be a right A-module with an A-valued inner product. Mimicking Hilbert space, we define $\|\xi\| = \|\langle \xi, \xi \rangle\|^{1/2}$. Note that the Cauchy-Schwarz inequality extends to this context. Indeed, the operator inequality

$$\langle \xi, \eta \rangle^* \langle \xi, \eta \rangle \geq \langle \xi, \eta \rangle^* \frac{\langle \xi, \xi \rangle}{\|\xi\|^2} \langle \xi, \eta \rangle,$$

plus a calculation, shows that

$$0 \leq \langle \|\xi\|^{-1}\xi\langle \xi, \eta \rangle - \|\xi\|\eta, \|\xi\|^{-1}\xi\langle \xi, \eta \rangle - \|\xi\|\eta \rangle \leq \|\xi\|^2 \langle \eta, \eta \rangle - \langle \xi, \eta \rangle^* \langle \xi, \eta \rangle.$$

Hence, $\|\langle \xi, \eta \rangle\| \leq \|\xi\|\|\eta\|$ for all $\xi, \eta \in \mathcal{H}$. It follows that $\|\xi\| = \|\langle \xi, \xi \rangle\|^{1/2}$ defines a norm on $\mathcal{H}$ which satisfies $\|\xi a\| \leq \|\xi\|\|a\|$ and $\|\langle \xi, \eta \rangle\| \leq \|\xi\|\|\eta\|$. We say that $\mathcal{H}$ is a Hilbert A-module if it is complete with respect to this norm. Every right A-module with an A-valued semi-inner product can be promoted to a bona fide Hilbert A-module by separation and completion.

For Hilbert A-modules $\mathcal{H}$ and $\mathcal{K}$, we denote by $\mathbb{B}(\mathcal{H}, \mathcal{K})$ the set of adjointable operators from $\mathcal{H}$ to $\mathcal{K}$. We simply write $\mathbb{B}(\mathcal{H})$ for the C*-algebra $\mathbb{B}(\mathcal{H}, \mathcal{H})$. For every $\xi, \eta \in \mathcal{H}$, we define the "rank-one" operator $\theta_{\xi,\eta} \colon \mathcal{H} \to \mathcal{H}$ by $\theta_{\xi,\eta}(\zeta) = \xi\langle \eta, \zeta \rangle$. Then $\theta_{\xi,\eta} \in \mathbb{B}(\mathcal{H})$ and we have the following calculus: $\theta_{\xi_1+\xi_2,\eta_1+\eta_2} = \theta_{\xi_1,\eta_1} + \theta_{\xi_1,\eta_2} + \theta_{\xi_2,\eta_1} + \theta_{\xi_2,\eta_2}$, $\theta_{\xi,\eta}^* = \theta_{\eta,\xi}$ and $x\theta_{\xi,\eta} = \theta_{x\xi,\eta}$ for every $\xi, \eta \in \mathcal{H}$ and $x \in \mathbb{B}(\mathcal{H})$. We define the C*-algebra $\mathbb{K}(\mathcal{H})$ of "compact operators" as the closed linear span of $\{\theta_{\xi,\eta} : \xi, \eta \in \mathcal{H}\}$. Note that $\mathbb{K}(\mathcal{H})$ is a closed two-sided ideal of $\mathbb{B}(\mathcal{H})$.

The simplest Hilbert A-module is A itself with the A-valued inner product $\langle a, b \rangle = a^*b$. We often write $\hat{a}$ for $a \in A$ when it is viewed as an element in the Hilbert A-module A. We note that $\mathbb{K}(A) \cong A$ and $\mathbb{B}(A) \cong \mathcal{M}(A)$, where $\mathcal{M}(A)$ is the multiplier algebra of A. If $\mathcal{H}_i$, $i \in I$, are a set of Hilbert A-modules, then the algebraic direct sum $\bigoplus^{\mathrm{alg}} \mathcal{H}_i$ has an A-valued inner product: $\langle (\xi_i)_{i \in I}, (\eta_i)_{i \in I} \rangle = \sum_{i \in I} \langle \xi_i, \eta_i \rangle$. The completion of $\bigoplus^{\mathrm{alg}} \mathcal{H}_i$ is denoted by $\bigoplus \mathcal{H}_i$. We also write $\mathcal{H}^{\oplus n}$ for the n-fold direct sum $\mathcal{H} \oplus \cdots \oplus \mathcal{H}$

and note that $\mathcal{H}^{\oplus n}$ is canonically isomorphic to $\ell_n^2 \otimes \mathcal{H}$ equipped with the A-valued inner product

$$\langle \xi_1 \otimes \xi_2, \eta_1 \otimes \eta_2 \rangle = \langle \xi_1, \eta_1 \rangle \langle \xi_2, \eta_2 \rangle,$$

where the inner product on ℓ_n^2 is linear in the second variable. The same thing holds if we replace ℓ_n^2 with ℓ^2.

Let $1 \in D \subset A$ be unital C*-algebras with a conditional expectation E from A onto D. Then, A is naturally a right D-module and $\langle a, b \rangle = E(a^*b)$ is a D-valued semi-inner product. We denote by $L^2(A, E)$ the Hilbert D-module obtained from A by separation and completion and by $\hat{a} \in L^2(A, E)$ the vector corresponding to $a \in A$. Left multiplication by elements in A defines a $*$-representation $\pi_E \colon A \to \mathbb{B}(L^2(A, E))$; $\pi_E(a)\hat{b} = \widehat{ab}$. For $\xi_E = \hat{1} \in L^2(A, E)$, we have

$$E(a) = \langle \xi_E, \pi_E(a)\xi_E \rangle$$

for every $a \in A$. Hence it is quite natural to call $(\pi_E, L^2(A, E), \xi_E)$ the GNS representation for (A, E). We say the conditional expectation E is *nondegenerate* if π_E is faithful (or, equivalently, $a = 0$ if and only if $E(xay) = 0$ for all $x, y \in A$). Since the conditional expectation $E \colon A \to D$ extends to an orthogonal projection E from $L^2(A, E)$ onto $\xi_E D$, we have $L^2(A, E) \cong \xi_E D \oplus L^2(A, E)^o$, where $L^2(A, E)^o = \ker E \subset L^2(A, E)$. We note that $A \cap \ker E$ is dense in $L^2(A, E)^o$. Indeed, if $x_n \in A$ and $\widehat{x_n} \to \eta \in L^2(A, E)^o$, then $(x_n - E(x_n))^\wedge \to \eta$. Since $L^2(A, E)^o$ is invariant under $\pi_E(D)$, we may restrict $\pi_E|_D$ to get a $*$-representation $\pi_E^o \colon D \to \mathbb{B}(L^2(A, E)^o)$. In other words, $L^2(A, E)^o$ is a C*-correspondence over D.

By definition, an A-B *C*-correspondence* is a Hilbert B-module $\mathcal{H}$ together with a faithful $*$-representation $\pi_\mathcal{H} \colon A \to \mathbb{B}(\mathcal{H})$; it is called a C*-correspondence over A when $B = A$. (NB: Sometimes these are called C*-bimodules in the literature. Also, some authors do not require faithfulness in the definition.) The $*$-representation $\pi_\mathcal{H}$ is referred to as the *left action* of A on $\mathcal{H}$. We often omit $\pi_\mathcal{H}$ when there is no confusion. We say that the C*-correspondence $\mathcal{H}$ is *full* if the ideal $\{\langle \xi, \eta \rangle : \xi, \eta \in \mathcal{H}\}$ is dense in B. It is *nondegenerate* if $\pi_\mathcal{H}(A)\mathcal{H}$ is dense in $\mathcal{H}$ (in fact it equals $\mathcal{H}$, by Cohen's factorization theorem – Theorem 4.6.4). If $\{\mathcal{H}_i\}_{i \in I}$ is a collection of A-B C*-correspondences, then the direct sum $\bigoplus \mathcal{H}_i$ is naturally an A-B C*-correspondence with left action given by $\pi_{\bigoplus \mathcal{H}_i} = \bigoplus \pi_{\mathcal{H}_i}$. The simplest C*-correspondence is the *identity correspondence* which is the Hilbert A-module A with the left action given by multiplication from the left.

Now we introduce the interior tensor product of C*-correspondences. See Chapter 4 of [**114**] for proofs of the following claims. Let $\mathcal{H}, \mathcal{K}$ be, respectively, A-B and B-C C*-correspondences. Then, the algebraic tensor product $\mathcal{H} \odot \mathcal{K}$ is naturally a right C-module, together with the C-valued semi-inner product defined by

$$\langle \xi' \otimes \eta', \xi \otimes \eta \rangle = \langle \eta', \pi_{\mathcal{K}}(\langle \xi', \xi \rangle)\eta \rangle.$$

We denote by $\mathcal{H} \otimes_B \mathcal{K}$ the Hilbert C-module obtained from $\mathcal{H} \odot \mathcal{K}$ by separation and completion. The vector in $\mathcal{H} \otimes_B \mathcal{K}$ corresponding to $\xi \otimes \eta \in \mathcal{H} \odot \mathcal{K}$ is still denoted by $\xi \otimes \eta$. Note that the kernel of $\mathcal{H} \odot \mathcal{K} \to \mathcal{H} \otimes_B \mathcal{K}$ is

$$\operatorname{span}\{(\xi b) \otimes \eta - \xi \otimes (\pi_{\mathcal{K}}(b)\eta) : \xi \in \mathcal{H},\ \eta \in \mathcal{K} \text{ and } b \in B\}.$$

There is a natural faithful $*$-representation $\mathbb{B}(\mathcal{H}) \to \mathbb{B}(\mathcal{H} \otimes_B \mathcal{K})$ given by $x \mapsto x \otimes 1$, where $(x \otimes 1)(\xi \otimes \eta) = (x\xi) \otimes \eta$ for $\xi \in \mathcal{H}$, $\eta \in \mathcal{K}$ and $x \in \mathbb{B}(\mathcal{H})$. In particular, $\mathcal{H} \otimes_B \mathcal{K}$ is an A-C C*-correspondence. There is also a natural $*$-representation $\pi_{\mathcal{K}}(B)' \cap \mathbb{B}(\mathcal{K}) \to \mathbb{B}(\mathcal{H} \otimes_B \mathcal{K})$, $y \mapsto 1 \otimes y$, where $(1 \otimes y)(\xi \otimes \eta) = \xi \otimes (y\eta)$. Indeed, we only need to check that $1 \otimes y$ is bounded:

$$\|(1 \otimes y)\sum_{i=1}^{n} \xi_i \otimes \eta_i\| = \|\langle Y\tilde{\eta}, XY\tilde{\eta}\rangle\|^{1/2},$$

where $\tilde{\eta} = [\eta_1 \cdots \eta_n]^T \in \mathcal{K}^{\oplus n}$, $X = [\pi_{\mathcal{K}}(\langle \xi_i, \xi_j \rangle)]_{i,j} \in \mathbb{B}(\mathcal{K}^{\oplus n})$ and $Y = \operatorname{diag}(y, \cdots, y) \in \mathbb{B}(\mathcal{K}^{\oplus n})$. But since $X \geq 0$ commutes with Y, we have

$$\|\langle Y\tilde{\eta}, XY\tilde{\eta}\rangle\| \leq \|Y\|^2 \|X^{1/2}\tilde{\eta}\|^2 = \|y\|^2 \|\sum_{i=1}^{n} \xi_i \otimes \eta_i\|^2.$$

The interior tensor product $\otimes_B$ has several nice properties, including associativity and a distribution law with respect to direct sums. For each $\xi \in \mathcal{H}$, we define a map $T_\xi \colon \mathcal{K} \to \mathcal{H} \otimes_B \mathcal{K}$ by $T_\xi(\eta) = \xi \otimes \eta$. Then, $T_\xi \in \mathbb{B}(\mathcal{K}, \mathcal{H} \otimes_B \mathcal{K})$ with $T_\xi^*(\zeta \otimes \eta) = \pi_{\mathcal{K}}(\langle \xi, \zeta \rangle)\eta$. In particular,

$$T_\xi^* T_{\xi'} = \pi_{\mathcal{K}}(\langle \xi, \xi' \rangle) \text{ and } T_{\xi'} T_\xi^* = \theta_{\xi', \xi} \otimes 1.$$

We note that for the identity correspondence A over A, there are natural identifications

$$\mathcal{H} \otimes_A A \cong \mathcal{H} \text{ and } A \otimes_A \mathcal{H} \cong \pi_{\mathcal{H}}(A)\mathcal{H}.$$

Lemma 4.6.1. *Let $\mathcal{H}$ be a Hilbert A-module and let $\xi_1, \ldots, \xi_n, \eta_1, \ldots, \eta_n \in \mathcal{H}$. Then we have*

$$\|\sum_{i=1}^{n} \theta_{\xi_i, \eta_i}\|_{\mathbb{K}(\mathcal{H})} = \|[\langle \xi_i, \xi_j \rangle]_{ij}^{1/2} [\langle \eta_i, \eta_j \rangle]_{ij}^{1/2}\|_{\mathbb{M}_n(A)},$$

where $[\langle \xi_i, \xi_j \rangle]_{ij}$ and $[\langle \eta_i, \eta_j \rangle]_{ij}$ are positive elements in $\mathbb{M}_n(A)$.

Proof. For every $\xi \in \mathcal{H}$, we denote by $T_\xi \in \mathbb{B}(A, \mathcal{H})$ the operator given by $T_\xi(\hat{a}) = \xi a$. Then $T_\xi^* T_\eta = \langle \xi, \eta \rangle$ and $\theta_{\xi,\eta} = T_\xi T_\eta^*$. It follows that for every $\xi_1, \ldots, \xi_n, \eta_1, \ldots, \eta_n \in \mathcal{H}$,

$$\sum_{i=1}^{n} \theta_{\xi_i, \eta_i} = \sum_{i=1}^{n} T_{\xi_i} T_{\eta_i}^* = T_{\tilde{\xi}} T_{\tilde{\eta}}^*,$$

where $T_{\tilde{\xi}} = [T_{\xi_1} \cdots T_{\xi_n}] \in \mathbb{B}(A^{\oplus n}, \mathcal{H})$ and likewise for $T_{\tilde{\eta}}$. Note that for operators x and y,

$$\|xy^*\| = \|yx^*xy^*\|^{1/2} = \|(x^*x)^{1/2}y^*y(x^*x)^{1/2}\|^{1/2} = \|(x^*x)^{1/2}(y^*y)^{1/2}\|.$$

Thus,

$$\|\sum_{i=1}^{n} \theta_{\xi_i,\eta_i}\| = \|(T_{\tilde{\xi}}^*T_{\tilde{\xi}})^{1/2}(T_{\tilde{\eta}}^*T_{\tilde{\eta}})^{1/2}\| = \|[\langle\xi_i,\xi_j\rangle]_{ij}^{1/2}[\langle\eta_i,\eta_j\rangle]_{ij}^{1/2}\|.$$

This completes the proof. □

Remark 4.6.2. Note the following special case: $\|\theta_{\xi,\xi}\| = \|\xi\|^2$, for all $\xi \in \mathcal{H}$. It follows that if (e_i) is an approximate unit for $\mathbb{K}(\mathcal{H})$, then

$$\|\xi - e_i\xi\|^2 = \|\theta_{\xi - e_i\xi, \xi - e_i\xi}\| = \|\theta_{\xi,\xi} - \theta_{\xi,\xi}e_i - e_i\theta_{\xi,\xi} + e_i\theta_{\xi,\xi}e_i\| \to 0,$$

for every $\xi \in \mathcal{H}$.

Proposition 4.6.3. *Let $\mathcal{H}$ be a Hilbert A-module and B be a C*-algebra. Let $\pi\colon A \to B$ be a $*$-homomorphism and $\tau\colon \mathcal{H} \to B$ be a linear map such that $\tau(\xi)^*\tau(\eta) = \pi(\langle\xi,\eta\rangle)$ for every $\xi,\eta \in \mathcal{H}$. Then, the $*$-homomorphism $\sigma_\tau\colon \mathbb{K}(\mathcal{H}) \to B$, defined by*

$$\sum_{i=1}^{n} \theta_{\xi_i,\eta_i} \mapsto \sum_{i=1}^{n} \tau(\xi_i)\tau(\eta_i)^*,$$

is continuous and satisfies $\sigma_\tau(x)\tau(\xi) = \tau(x\xi)$ for every $x \in \mathbb{K}(\mathcal{H})$ and $\xi \in \mathcal{H}$. Moreover, σ_τ is injective whenever π is injective.

Proof. It is not hard to see that $\tau(\xi a) = \tau(\xi)\pi(a)$ for every $\xi \in \mathcal{H}$ and $a \in A$ and that σ_τ is a well-defined $*$-homomorphism on the $*$-subalgebra of "finite-rank operators." Let $\xi_1,\ldots,\xi_n,\eta_1,\ldots,\eta_n \in \mathcal{H}$ be given. By Lemma 4.6.1, we have

$$\|\sum_{i=1}^{n} \theta_{\xi_i,\eta_i}\| = \|[\langle\xi_i,\xi_j\rangle]_{ij}^{1/2}[\langle\eta_i,\eta_j\rangle]_{ij}^{1/2}\|.$$

On the other hand, the proof of Lemma 4.6.1 shows

$$\begin{aligned}\|\sum_{i=1}^{n} \tau(\xi_i)\tau(\eta_i)^*\| &= \|[\tau(\xi_i)^*\tau(\xi_j)]_{ij}^{1/2}[\tau(\eta_i)^*\tau(\eta_j)]_{ij}^{1/2}\| \\ &= \|\pi\big([\langle\xi_i,\xi_j\rangle]_{ij}^{1/2}[\langle\eta_i,\eta_j\rangle]_{ij}^{1/2}\big)\| \le \|\sum_{i=1}^{n} \theta_{\xi_i,\eta_i}\|.\end{aligned}$$

This shows that σ_τ is contractive and that σ_τ is isometric whenever π is. □

Let A be a C*-algebra. A (left) Banach A-module is a Banach space X which is an A-module satisfying $\|ax\| \le \|a\|\|x\|$ for every $a \in A$ and $x \in X$. The following is a special case of the Cohen factorization theorem for Banach modules.

Theorem 4.6.4. *Let A be a C*-algebra and X be a Banach A-module. Then, the subset*

$$AX = \{ax : a \in A,\ x \in X\} \subset X$$

is a closed A-submodule of X.

Proof. (Pedersen) Letting Y be the closed linear span of AX, we'll show $Y = AX$. It is easily verified that for every $y \in Y$ and approximate unit (e_i) of A, we have $e_i y \to y$. This implies Y can be viewed as a module over the unitization $\tilde{A}$. Indeed, for every $a \in A$, $\lambda \in \mathbb{C}$ and $y \in Y$, we have

$$\|(a + \lambda 1)y\| = \lim \|(a + \lambda 1)e_i y\| \leq \lim \|(a + \lambda 1)e_i\| \|y\| = \|a + \lambda 1\| \|y\|.$$

Let $z \in Y$ and $\varepsilon > 0$ be given. Set $a_0 = 1_{\tilde{A}}$, $y_0 = z$ and define inductively

$$a_n = a_{n-1} - 2^{-n}(1 - e_n) \in \tilde{A}, \quad y_n = a_n^{-1} z \in Y.$$

Then, $a_n = 2^{-n} + \sum_{k=1}^{n} 2^{-k} e_k \geq 2^{-n}$ and hence $\|2^{-n} a_n^{-1}\| \leq 1$. Moreover,

$$y_n - y_{n-1} = a_n^{-1}(a_{n-1} - a_n) y_{n-1} = 2^{-n} a_n^{-1}(1 - e_n) y_{n-1}.$$

Choosing the e_n's carefully, we can assume that $\|y_n - y_{n-1}\| < 2^{-n}\varepsilon$ for all n. Let $a = \lim a_n = \sum_{k=1}^{\infty} 2^{-k} e_k \in A$ and $y = \lim y_n \in Y$. (Note that $0 \leq a \leq 1$, $\|y - z\| < \varepsilon$.) Then $z = \lim a_n y_n = ay$, as desired. □

Let $\mathcal{H}$ be a C*-correspondence over B and $J \lhd B$ be an ideal. We set

$$\mathcal{H}J = \{\zeta b : \zeta \in \mathcal{H},\ b \in J\}.$$

By Cohen's factorization theorem, $\mathcal{H}J$ is a closed right B-submodule of $\mathcal{H}$. Hence, $\mathcal{H}J$ is again a C*-correspondence over B. Moreover, $\zeta \in \mathcal{H}J$ if and only if $\langle \zeta, \zeta \rangle \in J$ (one direction is trivial and the other uses approximate units).

Toeplitz and Cuntz-Pimsner algebras. Now we move to the main topic of this section. Let $\mathcal{H}$ be a C*-correspondence over A. For notational convenience, we view A as a subalgebra of $\mathbb{B}(\mathcal{H})$ and stop writing $\pi_{\mathcal{H}}$. We set $\mathcal{H}^{\otimes 0} = A$ and $\mathcal{H}^{\otimes n} = \mathcal{H} \otimes_A \cdots \otimes_A \mathcal{H}$, the n-fold tensor product. The *full Fock space* over $\mathcal{H}$ is the C*-correspondence $\mathcal{F}(\mathcal{H})$ over A, defined by

$$\mathcal{F}(\mathcal{H}) = \bigoplus_{n \geq 0} \mathcal{H}^{\otimes n}.$$

Abusing notation, we also view $A \subset \mathbb{B}(\mathcal{F}(\mathcal{H}))$ omitting $\pi_{\mathcal{F}(\mathcal{H})}$. Via the identification $\mathcal{H} \otimes_A \mathcal{F}(\mathcal{H}) \cong \bigoplus_{n \geq 1} \mathcal{H}^{\otimes n} \subset \mathcal{F}(\mathcal{H})$, for each $\xi \in \mathcal{H}$ we define $T_\xi \in \mathbb{B}(\mathcal{F}(\mathcal{H}))$:

$$T_\xi(\hat{a}) = \xi a \quad \text{and} \quad T_\xi(\xi_1 \otimes \cdots \otimes \xi_n) = \xi \otimes \xi_1 \otimes \cdots \otimes \xi_n.$$

These operators T_ξ are called *creation operators* and satisfy the relations

$$T_{\alpha\xi + \eta} = \alpha T_\xi + T_\eta, \quad T_{a\xi b} = a T_\xi b, \quad \text{and} \quad T_\xi^* T_\eta = \langle \xi, \eta \rangle$$

for every $\alpha \in \mathbb{C}$, $\xi, \eta \in \mathcal{H}$ and $a, b \in A$. Let $\sigma_{\mathcal{F}} \colon \mathbb{B}(\mathcal{H}) \to \mathbb{B}(\mathcal{F}(\mathcal{H}))$ be the $*$-representation given by $\sigma_{\mathcal{F}}(x) = 0_{\mathcal{H}^{\otimes 0}} \oplus (x \otimes 1_{\mathcal{F}(\mathcal{H})})$; i.e.,

$$\sigma_{\mathcal{F}}(x)\hat{a} = 0 \quad \text{and} \quad \sigma_{\mathcal{F}}(x)(\xi_1 \otimes \cdots \otimes \xi_n) = (x\xi_1) \otimes \xi_2 \otimes \cdots \otimes \xi_n.$$

Note that $\sigma_{\mathcal{F}}|_{\mathbb{K}(\mathcal{H})}$ is nothing but the representation σ_T defined in Proposition 4.6.3. Let $P_0 \in \mathbb{B}(\mathcal{F}(\mathcal{H}))$ be the orthogonal projection onto $\mathcal{H}^{\otimes 0}$. Evidently P_0 commutes with every $a \in A$ and

$$a = P_0 a P_0 + \sigma_{\mathcal{F}}(a) \in \mathbb{B}(\mathcal{F}(\mathcal{H})).$$

For $\hat{a} \in \mathcal{H}^{\otimes 0}$, we set $T_{\hat{a}} = a \in \mathbb{B}(\mathcal{F}(\mathcal{H}))$; for $\mu = \xi_1 \otimes \cdots \otimes \xi_m \in \mathcal{H}^{\otimes m}$, we set $T_\mu = T_{\xi_1} \cdots T_{\xi_m} \in \mathbb{B}(\mathcal{F}(\mathcal{H}))$. Observe that this notation is compatible with

$$T_\mu \colon \mathcal{F}(\mathcal{H}) \to \mathcal{H}^{\otimes m} \otimes_A \mathcal{F}(\mathcal{H}) \cong \bigoplus_{n \geq m} \mathcal{H}^{\otimes n} \subset \mathcal{F}(\mathcal{H}).$$

Then, for every $\mu \in \mathcal{H}^{\otimes m}$ and $\nu \in \mathcal{H}^{\otimes n}$ $(m, n \geq 0)$, we have

$$\theta_{\mu,\nu} = T_\mu P_0 T_\nu^* \in \mathbb{K}(\mathcal{F}(\mathcal{H})).$$

Definition 4.6.5. Let $\mathcal{H}$ be a C*-correspondence over A. The (augmented) Toeplitz-Pimsner algebra $\mathcal{T}(\mathcal{H})$ is the C*-subalgebra of $\mathbb{B}(\mathcal{F}(\mathcal{H}))$ generated by A and $\{T_\xi : \xi \in \mathcal{H}\}$.[10]

We record the main identities that hold in $\mathcal{T}(\mathcal{H})$.

Theorem 4.6.6. *Let $\mathcal{T}(\mathcal{H}) = C^*(A \cup \{T_\xi : \xi \in \mathcal{H}\})$ be the Toeplitz-Pimsner algebra of a C*-correspondence $\mathcal{H}$ over A.*

(1) *For every $\alpha \in \mathbb{C}$, $\xi, \eta \in \mathcal{H}$ and $a, b \in A$, we have*

$$T_{\alpha\xi+\eta} = \alpha T_\xi + T_\eta, \; T_{a\xi b} = a T_\xi b \quad \text{and} \quad T_\xi^* T_\eta = \langle \xi, \eta \rangle.$$

(2) *We have $\mathcal{T}(\mathcal{H}) = \overline{\operatorname{span}}\{T_\mu T_\nu^* : \mu \in \mathcal{H}^{\otimes m}, \nu \in \mathcal{H}^{\otimes n},\ m, n \geq 0\}$ and there exists a nondegenerate conditional expectation $E_{\mathcal{H}}$ from $\mathcal{T}(\mathcal{H})$ onto A such that $E_{\mathcal{H}}(T_\mu T_\nu^*) = 0$ for every $\mu \in \mathcal{H}^{\otimes m}$ and $\nu \in \mathcal{H}^{\otimes n}$ with $(m, n) \neq 0$.*

(3) *There is an action γ (called the* gauge action*) of $\mathbb{T}$ on $\mathcal{T}(\mathcal{H})$ such that*

$$\gamma_z(a) = a \quad \text{and} \quad \gamma_z(T_\xi) = z T_\xi$$

for every $z \in \mathbb{T} = \{z \in \mathbb{C} : |z| = 1\}$, $a \in A$ and $\xi \in \mathcal{H}$.

Proof. The conditional expectation $E_{\mathcal{H}} \colon \mathcal{T}(\mathcal{H}) \to A$ in assertion (2) is given by compression to $\mathcal{H}^{\otimes 0}$:

$$E_{\mathcal{H}}(x) = P_0 x P_0 \in \mathbb{B}(\mathcal{H}^{\otimes 0}) \cong A \subset \mathcal{T}(\mathcal{H}).$$

---

[10] This definition is not exactly the same as Pimsner's in the nonfull case – i.e., when $\{\langle \xi, \eta \rangle : \xi, \eta \in \mathcal{H}\}$ does not generate A – hence the (augmented) terminology. However, most examples are full, and then our definition agrees with Pimsner's.

Since $\mathcal{F}(\mathcal{H}) = \overline{\operatorname{span}}\, \mathcal{T}(\mathcal{H})\mathcal{H}^{\otimes 0}$, it coincides with the GNS Hilbert A-module for $(\mathcal{T}(\mathcal{H}), E_{\mathcal{H}})$ by uniqueness of GNS representation. This shows nondegeneracy of $E_{\mathcal{H}}$. For assertion (3), observe that each γ_z is implemented by a unitary operator $U_z \in \mathbb{B}(\mathcal{F}(\mathcal{H}))$ defined by $U_z = \bigoplus_{n\geq 0} z^n$ on $\mathcal{F}(\mathcal{H}) = \bigoplus_{n\geq 0} \mathcal{H}^{\otimes n}$. (Although $z \mapsto U_z$ is not norm-continuous, $z \mapsto \gamma_z = \operatorname{Ad} U_z$ is continuous.) □

Since the Toeplitz-Pimsner algebra $\mathcal{T}(\mathcal{H})$ is too large for many purposes, we define the Cuntz-Pimsner algebra $\mathcal{O}(\mathcal{H})$ to be a natural quotient of $\mathcal{T}(\mathcal{H})$. Let $I_{\mathcal{H}} = A \cap \mathbb{K}(\mathcal{H})$ in $\mathbb{B}(\mathcal{H})$. Since $I_{\mathcal{H}}$ is an ideal in A and $\mathcal{F}(\mathcal{H})I_{\mathcal{H}}$ is a $\mathbb{B}(\mathcal{F}(\mathcal{H}))$ invariant C*-sub-correspondence of $\mathcal{F}(\mathcal{H})$, the C*-algebra

$$\mathbb{K}(\mathcal{F}(\mathcal{H})I_{\mathcal{H}}) = \overline{\operatorname{span}}\{\theta_{\xi,\eta} : \xi, \eta \in \mathcal{F}(\mathcal{H})I_{\mathcal{H}}\}$$

is an ideal of $\mathbb{B}(\mathcal{F}(\mathcal{H}))$.

Lemma 4.6.7. *We have an inclusion $\mathbb{K}(\mathcal{F}(\mathcal{H})I_{\mathcal{H}}) \subset \mathcal{T}(\mathcal{H})$. Moreover, $\mathbb{K}(\mathcal{F}(\mathcal{H})I_{\mathcal{H}})$ is globally invariant under the gauge action.*

Proof. Let $\mu \in \mathcal{H}^{\otimes n}$, $\nu \in \mathcal{H}^{\otimes m}$ $(m, n \geq 0)$ and $x \in I_{\mathcal{H}}$. Then, we have

$$\theta_{\mu x,\nu} = T_\mu P_0 x P_0 T_\nu^* = T_\mu(x - \sigma_{\mathcal{F}}(x))T_\nu^*.$$

Since $\sigma_{\mathcal{F}}(\mathbb{K}(\mathcal{H})) = \overline{\operatorname{span}}\{T_\xi T_\eta^* : \xi, \eta \in \mathcal{H}\} \subset \mathcal{T}(\mathcal{H})$, we have $\theta_{\mu x,\nu} \in \mathcal{T}(\mathcal{H})$. Since $\mathcal{F}(\mathcal{H})I_{\mathcal{H}}$ is invariant under the unitary operators U_z defined in the proof of Theorem 4.6.6, the second assertion follows. □

Denote by $Q_I \colon \mathbb{B}(\mathcal{F}(\mathcal{H})) \to \mathbb{B}(\mathcal{F}(\mathcal{H}))/\mathbb{K}(\mathcal{F}(\mathcal{H})I_{\mathcal{H}})$ the quotient map and write $S_\xi = Q_I(T_\xi)$. Since $\mathbb{K}(\mathcal{F}(\mathcal{H})I_{\mathcal{H}})$ is an ideal of $\mathbb{B}(\mathcal{F}(\mathcal{H}))$, it is an ideal of $\mathcal{T}(\mathcal{H})$ as well. Note that Q_I is injective on A since A acts diagonally on $\mathcal{F}(\mathcal{H}) = \bigoplus \mathcal{H}^{\otimes n}$.

Definition 4.6.8. Let $\mathcal{H}$ be a C*-correspondence over A. The (augmented) Cuntz-Pimsner algebra $\mathcal{O}(\mathcal{H})$ of $\mathcal{H}$ is $Q_I(\mathcal{T}(\mathcal{H}))$, the C*-algebra generated by A and $\{S_\xi : \xi \in \mathcal{H}\}$.

We record the main identities that hold in $\mathcal{O}(\mathcal{H})$.

Theorem 4.6.9. *Let $\mathcal{O}(\mathcal{H}) = C^*(A \cup \{S_\xi : \xi \in \mathcal{H}\})$ be the Cuntz-Pimsner algebra of a C*-correspondence $\mathcal{H}$ over A.*

(1) *For every $\alpha \in \mathbb{C}$, $\xi, \eta \in \mathcal{H}$ and $a, b \in A$, we have*

$$S_{\alpha\xi+\eta} = \alpha S_\xi + S_\eta,\ S_{a\xi b} = aS_\xi b \ \text{ and } \ S_\xi^* S_\eta = \langle \xi, \eta\rangle.$$

(2) *If $a \in I_{\mathcal{H}}$, then we have*

$$a = \sigma_S(a),$$

where $\sigma_S \colon \mathbb{K}(\mathcal{H}) \to \mathcal{O}(\mathcal{H})$ is the $*$-homomorphism given in Proposition 4.6.3.

(3) *There is an action* γ *(called the* gauge action*) of* $\mathbb{T}$ *on* $\mathcal{O}(\mathcal{H})$ *such that*

$$\gamma_z(a) = a \quad \textit{and} \quad \gamma_z(S_\xi) = zS_\xi$$

for every $z \in \mathbb{T}$, $a \in A$ *and* $\xi \in \mathcal{H}$.

Identity (2) follows from the fact that $a - \sigma_{\mathcal{F}}(a) = \theta_{\xi,\xi} \in \mathbb{K}(\mathcal{F}(\mathcal{H})I_{\mathcal{H}})$, where $a \in I_{\mathcal{H}}$ with $a \geq 0$ and $\xi = \widehat{a^{1/2}} \in \mathcal{H}^{\otimes 0}I_{\mathcal{H}}$. The other facts descend from the Toeplitz-Pimsner algebra.

Example 4.6.10 (The Toeplitz algebra and Cuntz algebras). We view ℓ^2_n as a C*-correspondence over $\mathbb{C}$. Then, $\mathcal{T}(\ell^2_n)$ is generated by isometries $T_1, \ldots, T_n$ (we write $T_i = T_{\delta_i}$ for simplicity) with orthogonal ranges. One checks that $e = 1 - \sum_{i=1}^n T_iT_i^*$ is the rank-one projection onto $\mathbb{C} \subset \mathcal{F}(\ell^2_n)$ and the corresponding vector state $\omega = E_{\ell^2_n}$ is called the *vacuum state*. The ideal $I_{\mathcal{H}}$ coincides with $\mathbb{C}$ and $\mathcal{O}_n = \mathcal{O}(\ell^2_n) = \mathcal{T}(\ell^2_n)/\mathbb{K}(\mathcal{F}(\mathcal{H}))$ is generated by isometries $S_1, \ldots, S_n$ such that $\sum_{i=1}^n S_iS_i^* = 1$. The C*-algebra $\mathcal{T}(\mathbb{C})$ is the classical Toeplitz algebra – the C*-algebra generated by the unilateral shift on $\ell^2(\mathbb{N})$ – and $\mathcal{O}_n$ are the celebrated Cuntz algebras.

Example 4.6.11 (Bimodules from u.c.p. maps). Let φ be a u.c.p. map on a unital C*-algebra A such that $\ker\varphi$ does not contain a nonzero ideal. The algebraic tensor product $A \odot A$ is naturally an A-A bimodule, with left and right actions given by $a \cdot (x \otimes y) \cdot b = ax \otimes yb$. We equip $A \odot A$ with an A-valued semi-inner product

$$\langle a_1 \otimes b_1, a_2 \otimes b_2 \rangle = b_1^* \varphi(a_1^* a_2) b_2.$$

(It is routine to check that the product is positive semidefinite.) Let $\mathcal{H}_A^\varphi$ be the C*-correspondence obtained from $A \odot A$ by separation and completion. As usual, we denote by $\widehat{a \otimes b}$ the vector in $\mathcal{H}_A^\varphi$ that is represented by $a \otimes b \in A \odot A$. Let $\xi = \widehat{1 \otimes 1}$. Since $A\xi A$ has dense linear span in $\mathcal{H}_A^\varphi$, the Toeplitz-Pimsner algebra $\mathcal{T}(\mathcal{H}_A^\varphi)$ is generated by A and an isometry $T = T_\xi$ which satisfies the relation $T^*aT = \varphi(a)$ for every $a \in A$.

Example 4.6.12 (Crossed products by $\mathbb{Z}$). Now suppose that φ is a *-automorphism. Then, $\widehat{a \otimes b} = \widehat{1 \otimes \varphi(a)b}$ for every $a, b \in A$ and $\theta_{\xi,\xi} = 1$ in $\mathbb{B}(\mathcal{H}_A^\varphi)$. Hence, $U = S_\xi$ is a unitary element in the Cuntz-Pimsner algebra $\mathcal{O}(\mathcal{H}_A^\varphi)$ which satisfies $U^*aU = \varphi(a)$ for $a \in A$. It is not difficult to see that $\mathcal{O}(\mathcal{H}_A^\varphi) \cong A \rtimes_\varphi \mathbb{Z}$ (especially after we prove universality).

Example 4.6.13 (Correspondences from graphs). Let $\mathfrak{G} = (V, E, s, r)$ be a directed graph without a *source*[11] – i.e., there is no vertex v such that

[11] To treat a graph with sources, one needs to work with correspondences whose left action is not faithful.

$r^{-1}(v) = \emptyset$. Then the set $c_{00}(E)$ of all finitely supported functions on E is naturally a $c_0(V)$-bimodule, with the left and right action given by

$$(f \cdot \xi \cdot g)(e) = f(r(e))\xi(e)g(s(e))$$

for $\xi \in c_{00}(E)$ and $f, g \in c_0(V)$. We equip $c_{00}(E)$ with a $c_0(V)$-valued inner product by

$$\langle \xi, \eta \rangle(v) = \sum_{e \in s^{-1}(v)} \overline{\xi(e)}\eta(e).$$

(If $v \in V$ is a *sink*, i.e., $s^{-1}(v) = \emptyset$, we set $\langle \xi, \eta \rangle(v) = 0$.) It is routine to check that $c_{00}(E)$ gives rise to a C*-correspondence $\mathcal{H}_{\mathfrak{G}}$ over $c_0(V)$ by separation and completion. The C*-correspondence $\mathcal{H}_{\mathfrak{G}}$ is full if and only if the graph does not have a sink. We now assume that the graph is row finite. Then,

$$\delta_v = \sum_{e \in r^{-1}(v)} \theta_{\delta_e, \delta_e} \in \mathbb{K}(\mathcal{H}_{\mathfrak{G}})$$

for every $v \in V$. The Cuntz-Pimsner algebra $\mathcal{O}(\mathcal{H}_{\mathfrak{G}})$ is generated by projections $\{\delta_v : v \in V\}$ and partial isometries $\{S_{\delta_e} : e \in E\}$ such that

$$S_{\delta_e}^* S_{\delta_e} = \delta_{s(e)} \quad \text{and} \quad \sum_{e \in r^{-1}(v)} S_{\delta_e} S_{\delta_e}^* = \delta_v.$$

Therefore, by universality, there exists a $*$-homomorphism from the graph C*-algebra $C^*(\mathfrak{G})$ onto $\mathcal{O}(\mathcal{H}_{\mathfrak{G}})$ which maps the p_v's to the δ_v's and the s_e's to the S_{δ_e}'s. It follows from gauge-invariant uniqueness (Theorem 4.6.20 below) that this $*$-homomorphism is actually a $*$-isomorphism.

Universality of Pimsner algebras.

Definition 4.6.14. Let $\mathcal{H}$ be a C*-correspondence over A. A *representation* of $\mathcal{H}$ on a C*-algebra B is a pair (π, τ), where $\pi \colon A \to B$ is a $*$-homomorphism and $\tau \colon \mathcal{H} \to B$ is a linear map such that

$$\tau(a\xi b) = \pi(a)\tau(\xi)\pi(b) \quad \text{and} \quad \tau(\xi)^*\tau(\eta) = \pi(\langle \xi, \eta \rangle).$$

We denote by $C^*(\pi, \tau)$ the C*-subalgebra of B generated by $\pi(A)$ and $\tau(\mathcal{H})$. A representation $(\tilde{\pi}, \tilde{\tau})$ of $\mathcal{H}$ is *universal* if for any other representation (π, τ) of $\mathcal{H}$, there is a (continuous) $*$-homomorphism from $C^*(\tilde{\pi}, \tilde{\tau})$ to $C^*(\pi, \tau)$ sending $\tilde{\pi}(a)$ to $\pi(a)$ and $\tilde{\tau}(\xi)$ to $\tau(\xi)$.

We observe that if (π, τ) is a representation of $\mathcal{H}$, then

$$\|\tau(\xi)\| = \|\pi(\langle \xi, \xi \rangle)\|^{1/2} \leq \|\xi\|$$

for every $\xi \in \mathcal{H}$. In particular, τ is isometric if π is. Thus, considering a suitable direct sum, one can show that a universal representation $(\tilde{\pi}, \tilde{\tau})$ of $\mathcal{H}$ always exists; we will see that the canonical representation $(\pi_{\mathcal{F}(\mathcal{H})}, T)$ of $\mathcal{H}$ on $\mathcal{T}(\mathcal{H})$ is universal.

Let (π, τ) be a representation of a C*-correspondence $\mathcal{H}$ over A. Then, for every $n \geq 1$, we define $\tau_n : \mathcal{H}^{\otimes n} \to C^*(\pi, \tau)$ by

$$\tau_n(\xi_1 \otimes \cdots \otimes \xi_n) = \tau(\xi_1) \cdots \tau(\xi_n)$$

on elementary tensors. It is not hard to see that $\tau_n(\mu)^* \tau_n(\nu) = \pi(\langle \mu, \nu \rangle)$ for every $\mu, \nu \in \mathcal{H}^{\otimes n}$ and the pair (π, τ_n) is a well-defined representation of $\mathcal{H}^{\otimes n}$ on $C^*(\pi, \tau)$. We define C*-subalgebras $B_n \subset C^*(\pi, \tau)$ by $B_0 = \pi(A)$ and

$$B_n = \overline{\operatorname{span}}\{\tau_n(\mu)\tau_n(\nu)^* : \mu, \nu \in \mathcal{H}^{\otimes n}\} = \sigma_n(\mathbb{K}(\mathcal{H}^{\otimes n}))$$

for $n \geq 1$, where σ_n is the $*$-homomorphism defined in Proposition 4.6.3. Recall that σ_n is injective if π is. Now define an ideal $I_{(\pi,\tau)}$ of A by

$$I_{(\pi,\tau)} = \pi^{-1}(\pi(A) \cap B_1) \subset A$$

and an increasing sequence of subspaces $B_{\leq n}$ of $C^*(\pi, \tau)$ by

$$B_{\leq n} = B_0 + B_1 + \cdots + B_n \subset C^*(\pi, \tau).$$

Lemma 4.6.15. *Suppose that π is faithful. Then, $I_{(\pi,\tau)} \subset I_{\mathcal{H}}$. Moreover, we have $\pi(a) = \sigma_\tau(a)$ for every $a \in I_{(\pi,\tau)}$.*

Proof. Let $a \in I_{(\pi,\tau)}$ and choose $x \in \mathbb{K}(\mathcal{H})$ such that $\pi(a) = \sigma_\tau(x)$. Then, for any $\xi \in \mathcal{H}$ we have

$$\tau(a\xi) = \pi(a)\tau(\xi) = \sigma_\tau(x)\tau(\xi) = \tau(x\xi).$$

Since τ is injective, $a\xi = x\xi$ for all $\xi \in \mathcal{H}$. This implies $a = x \in A \cap \mathbb{K}(\mathcal{H}) = I_{\mathcal{H}}$. □

Lemma 4.6.16. *For every $n \geq 1$, the subspace $B_{\leq n}$ is a C*-subalgebra of $C^*(\pi, \tau)$ and B_n is an ideal of $B_{\leq n}$. Moreover,*

$$B_{\leq n} \cap B_{n+1} = B_n \cap B_{n+1} = \sigma_n(\mathbb{K}(\mathcal{H}^{\otimes n} I_{(\pi,\tau)})).$$

Proof. Recall the general fact that if C and J are C*-subalgebras of D such that $CJC \subset J$, then $C + J$ is a C*-subalgebra of D. (Indeed, it is clear that $\overline{C + J}$ is a C*-algebra containing J as an ideal. Since the image of C in $\overline{C + J}/J$ is a dense C*-subalgebra, it must coincide with $\overline{C + J}/J$, or equivalently $\overline{C + J} = C + J$.) Hence, by induction, it suffices to show that $B_{\leq n-1} B_n B_{\leq n-1} \subset B_n$ for every $n \geq 1$, which is obvious from the definition.

Let $\mu, \nu \in \mathcal{H}^{\otimes n}$ and $a \in I_{(\pi,\tau)}$. Then, $\sigma_n(\theta_{\mu a, \nu}) \in B_n$, but also

$$\sigma_n(\theta_{\mu a, \nu}) = \tau_n(\mu)\pi(a)\tau_n(\nu)^* \in B_{n+1}$$

since $\pi(a) \in B_1$. This proves $\sigma_n(\mathbb{K}(\mathcal{H}^{\otimes n} I_{(\pi,\tau)})) \subset B_n \cap B_{n+1}$. Conversely, let $x \in B_{\leq n} \cap B_{n+1}$ be given and $\mu, \nu \in \mathcal{H}^{\otimes n}$ be arbitrary. Then, $x \in B_{\leq n}$ implies that $\tau_n(\mu)^* x \tau_n(\nu) \in \pi(A)$, while $x \in B_{n+1}$ implies that $\tau_n(\mu)^* x \tau_n(\nu) \in$

B_1. It follows that $\tau_n(\mu)^* x \tau_n(\nu) \in \pi(I_{(\pi,\tau)})$. Hence, for any "finite-rank operator" $e = \sum_i \theta_{\nu_i,\mu_i}$ on $\mathcal{H}^{\otimes n}$, we have

$$\sigma_n(e) x \sigma_n(e) = \sum_{i,j} \tau_n(\nu_i)\big(\tau_n(\mu_i)^* x \tau_n(\nu_j)\big)\tau_n(\mu_j)^* \in \sigma_n(\mathbb{K}(\mathcal{H}^{\otimes n} I_{(\pi,\tau)}))$$

since $\tau_n(\mu_i)^* x \tau_n(\nu_j) \in \pi(I_{(\pi,\tau)})$ for every i, j. Now, let e_λ be an approximate unit of $\mathbb{K}(\mathcal{H}^{\otimes n})$ consisting of "finite-rank operators." Then, for any $\mu \in \mathcal{H}^{\otimes n}$, we have

$$\lim_\lambda \sigma_n(e_\lambda)\tau_n(\mu) = \lim_\lambda \tau_n(e_\lambda \mu) = \tau_n(\mu),$$

since τ_n is contractive and $\lim e_\lambda \mu = \mu$ (Remark 4.6.2). It follows that $\lim_\lambda \sigma_n(e_\lambda)\tau_m(\mu) = \tau_m(\mu)$ for every $m \geq n$ and $\mu \in \mathcal{H}^{\otimes m}$. Since $x \in B_{n+1}$, we have

$$x = \lim_\lambda \sigma_n(e_\lambda) x \sigma_n(e_\lambda) \in \sigma_n(\mathbb{K}(\mathcal{H}^{\otimes n} I_{(\pi,\tau)})).$$

This proves that $B_{\leq n} \cap B_{n+1} \subset \sigma_n(\mathbb{K}(\mathcal{H}^{\otimes n} I_{(\pi,\tau)}))$. □

Note that $I_{(\pi_{\mathcal{F}(\mathcal{H})},T)} = \{0\}$ for the Fock space representation $(\pi_{\mathcal{F}(\mathcal{H})}, T)$ of $\mathcal{H}$ on $\mathcal{T}(\mathcal{H})$. Indeed there is a conditional expectation onto $\pi_{\mathcal{F}(\mathcal{H})}(A)$ that annihilates B_1 (see Theorem 4.6.6). It follows that $I_{(\tilde{\pi},\tilde{\tau})} = \{0\}$ for a universal representation $(\tilde{\pi}, \tilde{\tau})$ since $\pi_{\mathcal{F}(\mathcal{H})}$ is injective.

We say the representation (π, τ) *admits a gauge action* if there is an action β of $\mathbb{T} = \{z \in \mathbb{C} : |z| = 1\}$ on $C^*(\pi, \tau)$ such that

$$\beta_z(\pi(a)) = \pi(a) \quad\text{and}\quad \beta_z(\tau(\xi)) = z\tau(\xi)$$

for every $z \in \mathbb{T}$, $a \in A$ and $\xi \in \mathcal{H}$. Evidently the representation $(\pi_{\mathcal{F}(\mathcal{H})}, T)$ of $\mathcal{H}$ on $\mathcal{T}(\mathcal{H})$ admits a gauge action. A universal representation also admits a gauge action because of universality. We now describe the fixed point algebra $C^*(\pi, \tau)^\beta$ of a gauge action β.

Lemma 4.6.17. *Let (π, τ) be a representation that admits a gauge action β. Then, we have*

$$C^*(\pi, \tau)^\beta = \overline{\bigcup_{n\geq 0} B_{\leq n}}.$$

Proof. Recall that $C^*(\pi, \tau)^\beta$ is the range of the conditional expectation E_β defined by

$$E_\beta(x) = \int_{\mathbb{T}} \beta_z(x)\, dz.$$

For every $\mu \in \mathcal{H}^{\otimes m}$ and $\nu \in \mathcal{H}^{\otimes n}$, we have

$$E_\beta(\tau_m(\mu)\tau_n(\nu)^*) = \Big(\int_{\mathbb{T}} z^{m-n}\, dz\Big)\tau_m(\mu)\tau_n(\nu)^* = \delta_{m,n}\tau_m(\mu)\tau_n(\nu)^*.$$

Since $C^*(\pi, \tau) = \overline{\operatorname{span}}(A \cup \{\tau_m(\mu)\tau_n(\nu)^*\})$, the assertion follows. □

Here is the *gauge-invariant uniqueness theorem* for Toeplitz-Pimsner algebras.

Theorem 4.6.18. *Let $\mathcal{H}$ be a C*-correspondence over A and (π, τ) be a representation of $\mathcal{H}$ such that π is faithful. Assume that*

$$\pi(A) \cap \overline{\operatorname{span}}\{\tau(\xi)\tau(\eta)^* : \xi, \eta \in \mathcal{H}\} = \{0\}$$

and (π, τ) admits a gauge action. Then, the representation (π, τ) is universal. In particular, the representation $(\pi_{\mathcal{F}(\mathcal{H})}, T)$ of $\mathcal{H}$ on $\mathcal{T}(\mathcal{H})$ is universal.

Proof. Let $(\tilde{\pi}, \tilde{\tau})$ be a universal representation with gauge action $\tilde{\beta}$. By Proposition 4.5.1, it suffices to show that the canonical surjection

$$Q \colon C^*(\tilde{\pi}, \tilde{\tau}) \to C^*(\pi, \tau)$$

is injective on the fixed point algebra $C^*(\tilde{\pi}, \tilde{\tau})^{\tilde{\beta}}$. By Lemma 4.6.17, it suffices to show that Q is injective (isometric) on $\tilde{B}_{\leq n}$ for every n. We prove this by induction. First, Q is injective on $\tilde{B}_{\leq 0} \cong A$ by assumption. Now, let $n \geq 0$ and suppose that Q is injective on $\tilde{B}_{\leq n}$. Then, by Lemma 4.6.16 and the assumption that $I_{(\pi,\tau)} = \{0\}$, we have the commutative diagram

$$\begin{array}{ccccccccc}
0 & \longrightarrow & \tilde{B}_{n+1} & \longrightarrow & \tilde{B}_{\leq n+1} & \longrightarrow & \tilde{B}_{\leq n} & \longrightarrow & 0 \\
 & & \big\downarrow{\scriptstyle Q} & & \big\downarrow{\scriptstyle Q} & & \big\downarrow{\scriptstyle Q} & & \\
0 & \longrightarrow & B_{n+1} & \longrightarrow & B_{\leq n+1} & \longrightarrow & B_{\leq n} & \longrightarrow & 0
\end{array}$$

whose rows are split exact. Since $\tilde{B}_{n+1} \cong \mathbb{K}(\mathcal{H}^{\otimes(n+1)}) \cong B_{n+1}$, the left vertical arrow is injective. The right arrow is injective by our inductive hypothesis. Hence, by the 5 Lemma, the middle arrow is injective too. □

Definition 4.6.19. Let $\mathcal{H}$ be a C*-correspondence over A. A representation (π, τ) of $\mathcal{H}$ is *covariant* if $\pi(a) = \sigma_\tau(a)$ for every $a \in I_{\mathcal{H}} = A \cap \mathbb{K}(\mathcal{H})$, where σ_τ is the $*$-homomorphism defined in Proposition 4.6.3. A covariant representation $(\tilde{\pi}, \tilde{\tau})$ of $\mathcal{H}$ is *universal* if for any other covariant representation (π, τ) of $\mathcal{H}$, there is a (continuous) $*$-homomorphism from $C^*(\tilde{\pi}, \tilde{\tau})$ to $C^*(\pi, \tau)$ sending $\tilde{\pi}(a)$ to $\pi(a)$ and $\tilde{\tau}(\xi)$ to $\tau(\xi)$.

We note that the representation $(Q_I \circ \pi_{\mathcal{F}(\mathcal{H})}, S)$ of $\mathcal{H}$ on $\mathcal{O}(\mathcal{H})$ is covariant and admits a gauge action. Universal covariant representations also admit gauge actions.

Cuntz-Pimsner algebras also enjoy *gauge-invariant uniqueness.*

Theorem 4.6.20. *Let $\mathcal{H}$ be a C*-correspondence over A and (π, τ) be a covariant representation of $\mathcal{H}$ such that π is faithful. Suppose that (π, τ) admits a gauge action. Then, the covariant representation (π, τ) is universal.*

In particular, the covariant representation $(Q_I \circ \pi_{\mathcal{F}(\mathcal{H})}, S)$ *of* $\mathcal{H}$ *on* $\mathcal{O}(\mathcal{H})$ *is universal.*

Proof. The proof is similar to that of Theorem 4.6.18, except $I_{(\pi,\tau)} = I_{\mathcal{H}}$ by Lemma 4.6.15 and covariance. Indeed, with the same notation, it suffices to show that Q is injective on $\tilde{B}_{\leq n}$ for every n. As before, this requires induction.

First, Q is injective on $\tilde{B}_{\leq 0} \cong A$ by assumption. Now, let $n \geq 0$ and suppose that Q is injective on $\tilde{B}_{\leq n}$. Then, by Lemma 4.6.16, we have the commutative diagram

$$\begin{array}{ccccccccc} 0 & \longrightarrow & \tilde{B}_{n+1} & \longrightarrow & \tilde{B}_{\leq n+1} & \longrightarrow & \tilde{B}_{\leq n}/(\tilde{B}_{\leq n} \cap \tilde{B}_{n+1}) & \longrightarrow & 0 \\ & & \downarrow Q & & \downarrow Q & & \downarrow \dot{Q} & & \\ 0 & \longrightarrow & B_{n+1} & \longrightarrow & B_{\leq n+1} & \longrightarrow & B_{\leq n}/(B_{\leq n} \cap B_{n+1}) & \longrightarrow & 0 \end{array}$$

whose rows are exact. Since $\tilde{B}_{n+1} \cong \mathbb{K}(\mathcal{H}^{\otimes(n+1)}) \cong B_{n+1}$, the left vertical arrow is injective. The right arrow is injective since Q is injective on $\tilde{B}_{\leq n}$, by the induction hypothesis, and Q maps $\tilde{B}_{\leq n} \cap \tilde{B}_{n+1} \cong \mathbb{K}(\mathcal{H}^{\otimes n} I_{\mathcal{H}})$ onto $B_{\leq n} \cap B_{n+1}$. Hence, the middle arrow is also injective by the 5 Lemma. This completes the proof. □

We have several corollaries of the gauge-invariant uniqueness theorems.

Corollary 4.6.21. *Let* $\mathcal{H}$ *be a* C*-*correspondence over* A. *Let* $B \subset A$ *and* $\mathcal{K} \subset \mathcal{H}$ *be a closed subspace such that* $B\mathcal{K}B \subset \mathcal{K}$ *and* $\langle \xi, \eta \rangle \in B$ *for every* $\xi, \eta \in \mathcal{K}$. *Viewing* $\mathcal{K}$ *as a* C*-*correspondence over* B, *we have a natural inclusion* $\mathcal{T}(\mathcal{K}) \subset \mathcal{T}(\mathcal{H})$.

Corollary 4.6.22. *Let* $\mathcal{H}$ *be a* C*-*correspondence over* A. *Let* α *be a* $*$-*automorphism of* A *and* U *be an isometric surjection on* $\mathcal{H}$ *such that*

$$U(a\xi b) = \alpha(a)U(\xi)\alpha(b) \quad \text{and} \quad \langle U\xi, U\eta \rangle = \alpha(\langle \xi, \eta \rangle)$$

for every $\xi, \eta \in \mathcal{H}$ *and* $a, b \in A$. *Then, there exists a* $*$-*automorphism (called a Bogoljubov automorphism)* α_U *on* $\mathcal{T}(\mathcal{H})$ *such that*

$$\alpha_U(a) = \alpha(a) \quad \text{and} \quad \alpha_U(T_\xi) = T_{U(\xi)}$$

for every $a \in A$ *and* $\xi \in \mathcal{H}$. *The same thing holds for* $\mathcal{O}(\mathcal{H})$.

Proof. Let α and U be given as above. Then, the pair of maps $\alpha\colon A \ni a \mapsto \alpha(a) \in \mathcal{T}(\mathcal{H})$ and $T \circ U\colon \mathcal{H} \ni \xi \mapsto T_{U(\xi)} \in \mathcal{T}(\mathcal{H})$ is a representation of $\mathcal{H}$ and induces a $*$-homomorphism $\alpha_U\colon \mathcal{T}(\mathcal{H}) \to \mathcal{T}(\mathcal{H})$. The same construction for α^{-1} and U^{-1} gives rise to the inverse of α_U and hence α_U is an automorphism. To prove the assertion for $\mathcal{O}(\mathcal{H})$, we have to check the covariance property of the representation $(\alpha, S \circ U)$ of $\mathcal{H}$ into $\mathcal{O}(\mathcal{H})$. For this, we observe first that $\mathrm{Ad}_U\colon \mathbb{B}(\mathcal{H}) \ni x \mapsto UxU^{-1} \in \mathbb{B}(\mathcal{H})$ is a well-defined $*$-automorphism such that $\mathrm{Ad}_U|_A = \alpha$. Moreover, for $\xi, \eta \in \mathcal{H}$, we have

$$\begin{aligned}\mathrm{Ad}_U(\theta_{\xi,\eta})\zeta &= U(\xi\langle\eta, U^{-1}(\zeta)\rangle) = U(\xi)\alpha(\langle\eta, U^{-1}(\zeta)\rangle)\\ &= U(\xi)\langle U(\eta), \zeta\rangle = \theta_{U(\xi),U(\eta)}\zeta\end{aligned}$$

and hence $\mathrm{Ad}_U(\theta_{\xi,\eta}) = \theta_{U(\xi),U(\eta)}$. It follows that $\alpha(I_{\mathcal{H}}) = \mathrm{Ad}_U(I_{\mathcal{H}}) \subset I_{\mathcal{H}}$ and that $\sigma_{S\circ U} = \sigma_S \circ \mathrm{Ad}_U$ on $\mathbb{K}(\mathcal{H})$. Therefore, if $a \in I_{\mathcal{H}}$, we have

$$\sigma_{S\circ U}(a) = (\sigma_S \circ \mathrm{Ad}_U)(a) = \sigma_S(\alpha(a)) = \alpha(a),$$

where the last equality follows from the covariance of (id, S). This proves covariance of the representation $(\alpha, S\circ U)$ of $\mathcal{H}$ into $\mathcal{O}(\mathcal{H})$; the rest of the proof is similar to the Toeplitz-Pimsner case. □

The following technical result will be handy in the future.

Proposition 4.6.23. *Let A and B be unital* C*-*algebras with a common unital subalgebra D and nondegenerate conditional expectations $E_D^A\colon A \to D$ and $E_D^B\colon B \to D$. Let α be a $*$-automorphism on D. Define maps $\varphi = \alpha \circ E_D^A\colon A \to D \subset A$ and $\psi = \alpha \circ E_D^B\colon B \to D \subset B$. As in Example 4.6.11, we let $\mathcal{T}(\mathcal{H}_A^\varphi)$ be the universal* C*-*algebra generated by A and T, subject to the relation $T^*aT = \varphi(a)$; let $\mathcal{T}(\mathcal{H}_B^\psi)$ be the corresponding algebra for B. Assume there is a u.c.p. map $\theta\colon A \to B$ such that $\theta|_D = \mathrm{id}_D$ and $E_D^B \circ \theta = E_D^A$. Then, there is a u.c.p. map*

$$\Theta\colon \mathcal{T}(\mathcal{H}_A^\varphi) \to \mathcal{T}(\mathcal{H}_B^\psi)$$

such that

$$\Theta(a_0Ta_1\cdots Ta_kT^*\cdots T^*a_n) = \theta(a_0)T\theta(a_1)\cdots T\theta(a_k)T^*\cdots T^*\theta(a_n)$$

for every $a_0, \ldots, a_n \in A$.

Proof. Let $\mathcal{K} = L^2(B, E_D^B)$ be the C*-correspondence over D with left action given by $\pi_{\mathcal{K}}(d)\widehat{b} = \widehat{\alpha(d)b}$. Let $\mathcal{F}(\mathcal{K})^o = \bigoplus_{n\geq 1}\mathcal{K}^{\otimes n}$, where $\otimes$ is the interior tensor product over D. Since B naturally acts on $\mathcal{K}$ (though this action does not coincide with $\pi_{\mathcal{K}}$ on D), we have $B \subset \mathbb{B}(\mathcal{F}(\mathcal{K})^o)$ (acting on the first tensor component). Let $T_B \in \mathbb{B}(\mathcal{F}(\mathcal{K})^o)$ be the shift isometry: $T_B(\zeta_1 \otimes \cdots \otimes \zeta_n) = \widehat{1} \otimes \zeta_1 \otimes \cdots \otimes \zeta_n$. Then, we have

$$T_B^* bT_B = (\pi_{\mathcal{K}} \circ E_D^B)(b) = \psi(b)$$

for every $b \in B$. Since the family of unitary operators $U_z = \bigoplus_{n\geq 1} z^n$ implements the gauge action and the compression to $\mathcal{K}^{\otimes 1} \subset \mathcal{F}(\mathcal{K})^o$ separates B from the "compact operators" $\overline{\mathrm{span}}(BT_BBT_B^*B)$, we have $\mathcal{T}(\mathcal{H}_B^\psi) \cong C^*(B, T_B) \subset \mathbb{B}(\mathcal{F}(\mathcal{K})^o)$, by the gauge-invariant uniqueness theorem.

Let $\mathcal{H} = A \otimes_\theta L^2(B, E_D^B)$ be the Hilbert D-module obtained from $A \odot B$ equipped with the D-valued semi-inner product

$$\langle a_1 \otimes b_1, a_2 \otimes b_2 \rangle = E_D^B(b_1^* \theta(a_1^* a_2) b_2).$$

We will view $\mathcal{H}$ as a C*-correspondence over D with left action determined by $\pi_{\mathcal{H}}(d)\widehat{a \otimes b} = (\alpha(\widehat{d)a) \otimes b}$. Let $\mathcal{F}(\mathcal{H})^o = \bigoplus_{n \geq 1} \mathcal{H}^{\otimes n}$ and consider the shift isometry $T_A \in \mathbb{B}(\mathcal{F}(\mathcal{H})^o)$ given by $T_A(\zeta_1 \otimes \cdots \otimes \zeta_n) = \widehat{1 \otimes 1} \otimes \zeta_1 \otimes \cdots \otimes \zeta_n$. Since A naturally acts on $\mathcal{H}$, we have $A \subset \mathbb{B}(\mathcal{F}(\mathcal{H})^o)$. One checks that $T_A^* a T_A = (\pi_{\mathcal{H}} \circ E_D^B \circ \theta)(a) = \varphi(a)$ for every $a \in A$. As in the first paragraph we have $\mathcal{T}(\mathcal{H}_A^\varphi) \cong C^*(A, T_A) \subset \mathbb{B}(\mathcal{F}(\mathcal{H})^o)$. The map $B \ni b \mapsto 1 \otimes b \in A \odot B$ gives rise to an isometry $V \in \mathbb{B}(\mathcal{K}, \mathcal{H})$ with $V^*(\widehat{a \otimes b}) = \widehat{\theta(a)b}$. (Use the Schwarz-type inequality $[\theta(a_i)^* \theta(a_j)] = \theta([a_1^* \cdots a_n^*]^T)\theta([a_1 \cdots a_n]) \leq \theta([a_1^* \cdots a_n^*]^T [a_1 \cdots a_n]) = [\theta(a_i^* a_j)]$ to check that V^* is bounded.) Now one verifies that $\pi_{\mathcal{H}}(d)V = V\pi_{\mathcal{K}}(d)$ for every $d \in D$. Hence,

$$V_{\mathcal{F}} \colon \mathcal{F}(\mathcal{K})^o \ni \zeta_1 \otimes \cdots \otimes \zeta_n \mapsto (V\zeta_1) \otimes \cdots \otimes (V\zeta_n) \in \mathcal{F}(\mathcal{H})^o$$

is a well-defined adjointable isometry (like every other sentence in this proof, we leave it as an exercise to check this). We claim that $T_A^* a V_{\mathcal{F}} = V_{\mathcal{F}} T_B^* \theta(a)$ for every $a \in A$. Indeed,

$$\begin{aligned} T_A^* a V_{\mathcal{F}}(\widehat{b_1} \otimes \cdots \otimes \widehat{b_n}) &= T_A^*\big((\widehat{a \otimes b_1}) \otimes (\widehat{1 \otimes b_2}) \otimes \cdots \otimes (\widehat{1 \otimes b_n})\big) \\ &= \big(\pi_{\mathcal{H}}(E_D^B(\theta(a)b_1))(\widehat{1 \otimes b_2})\big) \otimes \cdots \otimes (\widehat{1 \otimes b_n}) \\ &= (1 \otimes \pi_{\mathcal{K}}(E_D^B(\theta(a)b_1))b_2)^\wedge \otimes \cdots \otimes (\widehat{1 \otimes b_n}) \\ &= V_{\mathcal{F}} T_B^* \theta(a)(\widehat{b_1} \otimes \cdots \otimes \widehat{b_n}). \end{aligned}$$

It follows that $V_{\mathcal{F}}^* a T_A = \theta(a) T_B V_{\mathcal{F}}^*$ as well. Similarly, one has $V_{\mathcal{F}}^* a V_{\mathcal{F}} = \theta(a)$ for $a \in A$. Therefore

$$\begin{aligned} &V_{\mathcal{F}}^*(a_0 T_A a_1 \cdots T_A a_k T_A^* \cdots T_A^* a_n) V_{\mathcal{F}} \\ &\qquad = \theta(a_0) T_B \theta(a_1) \cdots T_B \theta(a_k) T_B^* \cdots T_B^* \theta(a_n) \end{aligned}$$

and $\Theta(x) = V_{\mathcal{F}}^* x V_{\mathcal{F}}$ is the desired u.c.p. map. □

Let $\mathcal{H}$ be a C*-correspondence over A and B be a C*-algebra. The algebraic tensor product $\mathcal{H} \odot B$ is naturally equipped with an $A \otimes B$-valued semi-inner product:

$$\langle \xi \otimes \hat{a}, \eta \otimes \hat{b} \rangle = \langle \xi, \eta \rangle_{\mathcal{H}} \otimes a^* b \in A \otimes B.$$

By separation and completion, we obtain a Hilbert $A \otimes B$-module $\mathcal{H} \otimes B$. We observe that $\mathbb{B}(\mathcal{H}) \otimes B \subset \mathbb{B}(\mathcal{H} \otimes B)$ naturally, where $\mathbb{B}(\mathcal{H}) \odot B$ acts on $\mathcal{H} \otimes B$ by

$$(x \otimes a)(\xi \otimes \hat{b}) = x\xi \otimes \widehat{ab}.$$

(See Exercise 4.6.4.) Hence $\mathcal{H}\otimes B$ is a C*-correspondence over $A\otimes B$. This correspondence is called the *exterior tensor product* of $\mathcal{H}$ and B.

Lemma 4.6.24. *Let $\mathcal{H}$ be a C*-correspondence over A and B be a C*-algebra. There are natural isomorphisms $\mathcal{T}(\mathcal{H}\otimes B)\cong\mathcal{T}(\mathcal{H})\otimes B$ and $\mathcal{O}(\mathcal{H}\otimes B)\cong\mathcal{O}(\mathcal{H})\otimes B$.*

Proof. If we denote by $\tau\colon\mathcal{H}\hookrightarrow\mathcal{T}(\mathcal{H})$ the canonical inclusion, then $\tau\otimes\mathrm{id}_B$ is a representation of the C*-correspondence $\mathcal{H}\otimes B$ into $\mathcal{T}(\mathcal{H})\otimes B$ which admits a gauge action $\gamma_z\otimes\mathrm{id}_B$. Observe that $A\otimes B$ does not contain nonzero "compact operators," because the conditional expectation $E_{\mathcal{H}}\otimes\mathrm{id}_B\colon\mathcal{T}(\mathcal{H})\otimes B\to A\otimes B$ kills all such elements. Hence the first assertion follows from Theorem 4.6.18. The proof of $\mathcal{O}(\mathcal{H}\otimes B)\cong\mathcal{O}(\mathcal{H})\otimes B$ is similar. It suffices to show that the pair $A\otimes B\hookrightarrow\mathcal{O}(\mathcal{H})\otimes B$ and $S\otimes\mathrm{id}_B\colon\mathcal{H}\otimes B\hookrightarrow\mathcal{O}(\mathcal{H})\otimes B$ is covariant. Since $\theta_{\xi\otimes\hat{a},\eta\otimes\hat{b}}=\theta_{\xi,\eta}\otimes a^*b$, we have $\mathbb{K}(\mathcal{H}\otimes B)=\mathbb{K}(\mathcal{H})\otimes B$ and

$$I_{\mathcal{H}\otimes B}=(\mathbb{K}(\mathcal{H})\otimes B)\cap(A\otimes B)\subset\mathbb{B}(\mathcal{H})\otimes B.$$

Let $x\in I_{\mathcal{H}\otimes B}$ be given and ω be any state on B. We note that $(\mathrm{id}\otimes\omega)(x)\in\mathbb{K}(\mathcal{H})\cap A=I_{\mathcal{H}}$. Since $\sigma_{S\otimes\mathrm{id}_B}=\sigma_S\otimes\mathrm{id}_B$, we have

$$(\mathrm{id}\otimes\omega)\circ\sigma_{S\otimes\mathrm{id}_B}(x)=\sigma_S((\mathrm{id}\otimes\omega)(x))=(\mathrm{id}\otimes\omega)(x)$$

by covariance of $S\colon\mathcal{H}\hookrightarrow\mathcal{O}(\mathcal{H})$. Since ω was arbitrary, $\sigma_{S\otimes\mathrm{id}_B}(x)=x$. □

We've finally come to the point of this section!

Theorem 4.6.25. *Let $\mathcal{H}$ be a C*-correspondence over A. Then, the Toeplitz-Pimsner algebra $\mathcal{T}(\mathcal{H})$ is nuclear (resp. exact) if and only if A is.*

Proof. The "only if" direction is trivial since there is a conditional expectation $\mathcal{T}(\mathcal{H})\to A$. So, assume A is nuclear (resp. exact). Let $\pi\colon\mathcal{T}(\mathcal{H})\to D$ be a $*$-homomorphism; it turns out that π is nuclear if $\pi|_A$ is (and this implies both the nuclear and exact cases). Let B be any C*-algebra. Since the $*$-homomorphism $\tilde{\pi}=(\pi|_A)\otimes\mathrm{id}_B\colon A\otimes B\to D\otimes_{\max}B$ is continuous, the representation

$$\tilde{\tau}\colon\mathcal{H}\otimes B\ni\sum\xi_k\otimes\widehat{b_k}\mapsto\sum\pi(\tau(\xi_k))\otimes b_k\in D\otimes_{\max}B$$

is also continuous and well-defined. Hence, by universality of $\mathcal{T}(\mathcal{H}\otimes B)$, the representation $(\tilde{\pi},\tilde{\tau})$ induces a $*$-homomorphism $\mathcal{T}(\mathcal{H}\otimes B)$ into $D\otimes_{\max}B$. Combined with the previous lemma, this shows that $\pi\otimes\mathrm{id}$ is continuous from $\mathcal{T}(\mathcal{H})\otimes B$ into $D\otimes_{\max}B$. Thus Corollary 3.8.8 yields the desired conclusion. □

It follows that $\mathcal{O}(\mathcal{H})$ is nuclear (resp. exact) whenever A is nuclear (resp. exact), since nuclearity (and exactness) pass to quotients (Corollaries 9.4.3

and 9.4.4). Unfortunately there is presently no direct proof of this fact, however.

There is another proof of nuclearity/exactness of $\mathcal{T}(\mathcal{H})$, which we now sketch. By Theorem 4.5.2, it suffices to show that the fixed point algebra $\mathcal{T}(\mathcal{H})^\gamma$ of the gauge action γ is nuclear/exact. As in the proof of Theorem 4.6.18, this reduces to showing that every $B_{\leq n}$ is nuclear/exact. Since each $B_{\leq n}$ is a split extension of $B_{\leq n-1}$ by $B_n \cong \mathbb{K}(\mathcal{H}^{\otimes n})$, we are further reduced to the nuclearity/exactness of $\mathbb{K}(\mathcal{H}^{\otimes n})$. The exercises below finish off the proof.

Exercises

Exercise 4.6.1. A C*-subalgebra $A_0 \subset A$ is said to be *full* if A_0 generates A as an ideal – i.e., $\overline{\mathrm{span}}(AA_0A) = A$. Let A_0 be a full and hereditary C*-subalgebra of A and I be an ideal of A. Prove that $A_0 \cap I$ is a full, hereditary C*-subalgebra of I. (Generous hint: Suppose, by contradiction, that $A_0 \cap I$ is not full in I. Then, there is a nonzero nondegenerate $*$-representation π of I such that $\pi(A_0 \cap I) = \{0\}$. We may regard π as a $*$-representation of A. If we denote by $(e_i)_i$ an approximate unit of I, then we have $\pi(a_0^* a_0) = \lim \pi(a_0^* e_i a_0) = 0$ for every $a_0 \in A_0$, where the limit is taken in the strong operator topology. This contradicts the fullness of A_0. Pretty generous, eh?)

Exercise 4.6.2. Let A_0 be a full, hereditary C*-subalgebra of A. Prove that A is nuclear (resp. exact) if and only if A_0 is. (Another generous hint: Let $\pi\colon A \to D$ be a $*$-homomorphism. Prove that π is nuclear if $\pi|_{A_0}$ is. To do this, let B be any C*-algebra. Since $A_0 \subset A$ is hereditary, we have $A_0 \otimes_{\max} B \subset A \otimes_{\max} B$. Moreover, one checks that $A_0 \otimes_{\max} B$ is full and hereditary. Let $I = \ker(A \otimes_{\max} B \to A \otimes B)$. Then, by the previous exercise,

$$\ker(A_0 \otimes_{\max} B \to A_0 \otimes B) = I \cap (A_0 \otimes_{\max} B)$$

is full in I. Therefore, the $*$-homomorphism $\pi \otimes_{\max} \mathrm{id}_B\colon A \otimes_{\max} B \to D \otimes_{\max} B$ factors through $A \otimes B$, if one knows the same thing holds for $\pi|_{A_0}$.)

Exercise 4.6.3. Let $\mathcal{H}$ be a Hilbert A-module. Prove that $\mathbb{K}(\mathcal{H})$ is nuclear (resp. exact) if A is. (Why stop now? Here is a proof: Let $I = \overline{\mathrm{span}}\{\langle \xi, \eta \rangle : \xi, \eta \in \mathcal{H}\} \subset A$. Since I is an ideal, I is nuclear (resp. exact) if A is. Hence, we may assume that $I = A$, i.e., the Hilbert A-module $\mathcal{H}$ is full. It is not hard to see that both $\mathbb{K}(\mathcal{H})$ and A are full hereditary C*-subalgebras of $\mathbb{K}(\mathcal{H} \oplus A)$. Now apply the previous exercise.)

Exercise 4.6.4. Let $\mathcal{H} \otimes B$ be the exterior tensor product. Prove that the natural inclusion $\mathbb{B}(\mathcal{H}) \odot B \hookrightarrow \mathbb{B}(\mathcal{H} \otimes B)$ is min-continuous. (Hint: Realize $\mathcal{H}$ as a corner of $\mathbb{B}(\mathcal{H} \oplus A)$.)

Exercise 4.6.5. Let $\mathcal{H}$ be a C*-correspondence over A and assume that the quotient map from A onto $A/I_{\mathcal{H}}$ is locally liftable. Prove that the quotient map from $\mathcal{T}(\mathcal{H})$ onto $\mathcal{O}(\mathcal{H})$ is locally liftable. (Hint: Use Theorem C.4.)

4.7. Reduced amalgamated free products

Thanks to Voiculescu's groundbreaking noncommutative probability theory, the free product construction is all over the C*-literature these days. Our only goal (realized in the next section) is a theorem of Dykema which asserts that *reduced* free products preserve exactness. As with Pimsner's construction, a lot of technical preliminaries are required, and for the remainder of this chapter most sentences should be understood as nontrivial exercises.

Construction and freeness. Let A_i, $i \in I$, be unital C*-algebras, each of which contains a unital copy of some fixed C*-algebra D. Assume that for each i there exist nondegenerate conditional expectations E_i from A_i onto D. In this situation, one can construct *the reduced amalgamated free product* C*-algebra $*_D(A_i, E_i)$. The simplest, though highly nontrivial, case is when $D = \mathbb{C}$ and E_i are states.

With $1 \in D \subset A_i$ and $E_i \colon A_i \to D$ as above, we denote by $(\pi_i, \mathcal{H}_i, \xi_i)$ the GNS representations for (A_i, E_i). Note that each π_i is faithful, because of nondegeneracy. The Hilbert D-submodule $\xi_i D \subset \mathcal{H}_i$ is isomorphic to D and $\mathcal{H}_i \cong \xi_i D \oplus \mathcal{H}_i^o$. We recall that $A_i^o = A_i \cap \ker E_i$ satisfies $DA_i^oD \subset A_i^o$ and $\mathcal{H}_i^o = \overline{A_i^o\xi_i}$ is a C*-correspondence over D. We define the free product Hilbert D-module $(\mathcal{H}, \xi) = *(\mathcal{H}_i, \xi_i)$ by

$$\mathcal{H} = \xi D \oplus \bigoplus_{n\geq 1} \bigoplus_{i_1 \neq \cdots \neq i_n} \mathcal{H}_{i_1}^o \otimes_D \cdots \otimes_D \mathcal{H}_{i_n}^o.$$

Here, ξD is the trivial Hilbert D-module D with $\xi = \hat{1}$ and the notation "$i_1 \neq \cdots \neq i_n$" means that $i_1 \neq i_2$, $i_2 \neq i_3, \ldots, i_{n-1} \neq i_n$. For each $i \in I$, let

$$\mathcal{H}(i) = \xi D \oplus \bigoplus_{n\geq 1} \bigoplus_{\substack{i\neq i_1 \\ i_1 \neq \cdots \neq i_n}} \mathcal{H}_{i_1}^o \otimes_D \cdots \otimes_D \mathcal{H}_{i_n}^o$$

and define an isomorphism $U_i \in \mathbb{B}(\mathcal{H}_i \otimes_D \mathcal{H}(i), \mathcal{H})$ by the identifications

$$U_i \colon \begin{array}{c} \xi_i D \otimes_D \xi D \\ \mathcal{H}_i^o \otimes_D \xi D \\ \xi_i D \otimes_D (\mathcal{H}_{i_1}^o \otimes_D \cdots \otimes_D \mathcal{H}_{i_n}^o) \\ \mathcal{H}_i^o \otimes_D (\mathcal{H}_{i_1}^o \otimes_D \cdots \otimes_D \mathcal{H}_{i_n}^o) \end{array} \xrightarrow{\cong} \begin{array}{c} \xi D \\ \mathcal{H}_i^o \\ \mathcal{H}_{i_1}^o \otimes_D \cdots \otimes_D \mathcal{H}_{i_n}^o \\ \mathcal{H}_i^o \otimes_D \mathcal{H}_{i_1}^o \otimes_D \cdots \otimes_D \mathcal{H}_{i_n}^o. \end{array}$$

We define a $*$-representation $\lambda_i \colon A_i \to \mathbb{B}(\mathcal{H})$ by $\lambda_i(x) = U_i(\pi_i \otimes 1)(x)U_i^*$. Observe that $\lambda_i|_D$ is the canonical left action of D on the C*-correspondence $\mathcal{H}$ over D, and hence $\lambda_i|_D = \lambda_j|_D$ for every i and j.

Definition 4.7.1 (Reduced amalgamated free product)**.** The reduced amalgamated free product $(A, E) = \ast_D(A_i, E_i)$ is the C*-subalgebra A of $\mathbb{B}(\mathcal{H})$ generated by $\bigcup_{i\in I} \lambda_i(A_i)$, together with the conditional expectation from A onto D given by $E(x) = \langle \xi, x\xi \rangle$.

The construction $*_D$ is commutative and associative, if you can fathom the notation required to check this. Direct calculation shows that $E_i = E \circ \lambda_i$ on A_i. Also, λ_i is injective (because E_i is nondegenerate); hence we often omit λ_i and view A_i as a subalgebra of A. Here are the main properties of the reduced amalgamated free product:

Theorem 4.7.2. *Let $1 \in D \subset A_i$ be unital C*-algebras with nondegenerate conditional expectations E_i from A_i onto D, and let $(A, E) = \ast_D(A_i, E_i)$.*

1. *There is an inclusion $1 \in D \subset A$ and a nondegenerate conditional expectation $E\colon A \to D$.*
2. *There are inclusions $D \subset A_i \subset A$ which are compatible on D, and A is generated by $\bigcup A_i$ as a C*-algebra.*
3. *One has $E|_{A_i} = E_i$ for every i, and the C*-subalgebras A_i are free over D in (A, E). Namely,*
$$E(a_1 \cdots a_n) = 0$$
for every $a_j \in A^o_{i_j}$ with $i_1 \neq \cdots \neq i_n$.

Moreover, the above conditions uniquely characterize the reduced amalgamated free product (A, E).

To verify the third assertion, we need the following fact; sanity prevents us from TeXing the proof.

Lemma 4.7.3. *Let $a \in A^o_i$ and $\zeta_j \in \mathcal{H}^o_{i_j}$ with $i_1 \neq \cdots \neq i_n$. Then,*

$$a(\zeta_1 \otimes \cdots \otimes \zeta_n) = \begin{cases} (a\zeta_1 - \xi_i\langle \xi_i, a\zeta_1\rangle) \otimes \zeta_2 \otimes \cdots \otimes \zeta_n & \\ \qquad + \langle \xi_i, a\zeta_1\rangle \zeta_2 \otimes \cdots \otimes \zeta_n & \text{if } i_1 = i, \\ a\xi_i \otimes \zeta_1 \otimes \cdots \otimes \zeta_n & \text{if } i_1 \neq i. \end{cases}$$

(When $i_1 = i$ and $n = 1$, the term $\langle \xi_i, a\zeta_1\rangle\zeta_2$ should be understood as $\xi\langle \xi_i, a\zeta_1\rangle$.)

In passing, we note that there is a conditional expectation from A onto A_i which preserves E. Indeed, if $V_i \in \mathbb{B}(\mathcal{H}_i, \mathcal{H})$ is the natural isometry given by $V_i\xi_i = \xi$ and $V_i\zeta = \zeta$ for $\zeta \in \mathcal{H}^o_i$, then $x \mapsto V_i^* x V_i$ defines the desired conditional expectation from A onto A_i.

Proof of Theorem 4.7.2. We only prove the last assertion. Let (A', E') be another pair which satisfies conditions (1)–(3). We denote by $\mathcal{A}$ the algebraic amalgamated free product[12] of A_i's over D and let $\pi\colon \mathcal{A} \to A$ and $\pi'\colon \mathcal{A} \to A'$ be the canonical representations. Because of conditions (1) and (2) and uniqueness of the GNS representation, it suffices to show that $E' \circ \pi' = E \circ \pi$. Since $A_i = D + A_i^o$ and $DA_i^oD \subset A_i^o$, we have

$$\mathcal{A} = \operatorname{span}(D \cup \bigcup_{n\geq 1} \bigcup_{i_1 \neq \cdots \neq i_n} A_{i_1}^o \cdots A_{i_n}^o).$$

Hence, property (3) implies that $x - (E \circ \pi)(x) \in \operatorname{span}(\bigcup A_{i_1}^o \cdots A_{i_n}^o)$ for every $x \in \mathcal{A}$. It follows that $(E' \circ \pi')(x) = (E' \circ \pi')((E \circ \pi)(x)) = (E \circ \pi)(x)$, again by condition (3), and we are done. □

Example 4.7.4. An inclusion of discrete groups $\Lambda \leq \Gamma$ gives rise to an inclusion $C^*_\lambda(\Lambda) \subset C^*_\lambda(\Gamma)$, and there is a conditional expectation $E^\Gamma_\Lambda\colon C^*_\lambda(\Gamma) \to C^*_\lambda(\Lambda)$ (Corollary 2.5.12).

Let $\Gamma = *_\Lambda \Gamma_i$ be an amalgamated free product of discrete groups Γ_i over a common subgroup Λ (cf. Appendix E). Then, we have

$$(C^*_\lambda(\Gamma), E^\Gamma_\Lambda) = *_{C^*_\lambda(\Lambda)}(C^*_\lambda(\Gamma_i), E^{\Gamma_i}_\Lambda).$$

This is easily checked by verifying the conditions of Theorem 4.7.2 and invoking uniqueness. It is also useful to see the isomorphism at the Hilbert space/C*-module level. As an important special case, we have natural isomorphisms $C^*_\lambda(\mathbb{F}_n) \cong C^*(\mathbb{Z}) * \cdots * C^*(\mathbb{Z})$ – the n-fold free product, taken with respect to the canonical trace on $C^*(\mathbb{Z})$ – for all $n \in \mathbb{N} \cup \{\infty\}$.

Example 4.7.5 (Speicher). Let $\mathcal{H}_i$ be C*-correspondences over A. Then, we have

$$(\mathcal{T}(\bigoplus \mathcal{H}_i), E_{\bigoplus \mathcal{H}_i}) \cong *_A(\mathcal{T}(\mathcal{H}_i), E_{\mathcal{H}_i}).$$

Proof. We will show the $\mathcal{T}(\mathcal{H}_i)$'s are free over A in $(\mathcal{T}(\bigoplus \mathcal{H}_i), E_{\bigoplus \mathcal{H}_i})$ (which suffices, by Theorem 4.7.2). Since

$$\ker E_{\mathcal{H}_i} = \overline{\operatorname{span}}\{T_\mu T_\nu^* : \mu \in \mathcal{H}_i^{\otimes m},\ \nu \in \mathcal{H}_i^{\otimes n} \text{ with } (m,n) \neq (0,0)\},$$

this reduces to showing

$$E_{\bigoplus \mathcal{H}_i}(T_{\mu_1}T_{\nu_1}^* \cdots T_{\mu_k}T_{\nu_k}^*) = 0$$

for $\mu_j \in \mathcal{H}_{i_j}^{\otimes m_j}$ and $\nu_j \in \mathcal{H}_{i_j}^{\otimes n_j}$ with $(m_j, n_j) \neq (0,0)$ and $i_1 \neq \cdots \neq i_k$. But this easily follows from the facts that $E_{\bigoplus \mathcal{H}_i}(T_\xi) = 0$ and $T_\xi^* T_\eta = 0$, whenever $\xi \in \mathcal{H}_i$ and $\eta \in \mathcal{H}_j$ with $i \neq j$. □

[12]That is, the universal algebra generated by copies of A_i which agree on D. It has the universal property that if $\mathcal{B}$ is any algebra containing a unital copy of D and $A_i \to \mathcal{B}$ are D-preserving homomorphisms, then they extend uniquely to a homomorphism from $\mathcal{A}$ into $\mathcal{B}$. If you prefer, here's a categorical definition: An amalgamated free product over D is the coproduct in the category whose objects are algebras containing unital copies of D and whose morphisms are D-preserving homomorphisms.

Exercises

Exercise 4.7.1. Let $(A, E) = *_D(A_i, E_i)$, where $1 \in D \subset A_i$ are unital C*-algebras with faithful conditional expectations E_i from A_i onto D. Prove that E is faithful on A. (Hint: For the natural isometry $V_i \in \mathbb{B}(\mathcal{H}_i, \mathcal{H})$, we have $V_i^* A V_i = A_i$. Assume that $x \in A$ is given such that $x(\widehat{a}_1 \otimes \cdots \otimes \widehat{a}_n) \neq 0$, and prove that $x(\widehat{a}_1 \otimes \cdots \otimes \widehat{a}_{n-1}) \neq 0$.)

Exercise 4.7.2. Let $(A, E) = *_D(A_i, E_i)$, where $1 \in D \subset A_i$ are unital C*-algebras with nondegenerate conditional expectations E_i from A_i onto D. Suppose that there exists a tracial state τ_D on D such that every $\tau_D \circ E_i$ is tracial on A_i. Prove that $\tau_D \circ E$ is tracial on A. (Hint: Induction on the length of words.)

4.8. Maps on reduced amalgamated free products

Let $1 \in D \subset A_i$, $i = 1, 2$, be unital C*-algebras with nondegenerate conditional expectations E_i from A_i onto D. Consider the u.c.p. map

$$\varphi\colon A_1 \oplus A_2 \ni (a_1 \oplus a_2) \mapsto E_2(a_2) \oplus E_1(a_1) \in D \oplus D \subset A_1 \oplus A_2$$

on $A_1 \oplus A_2$ and the C*-correspondence $\mathcal{H}^{\varphi}_{A_1 \oplus A_2}$ over $A_1 \oplus A_2$ that is given in Example 4.6.11. We will describe a faithful representation of $\mathcal{T}(\mathcal{H}^{\varphi}_{A_1 \oplus A_2})$. Let $\mathcal{H}_i = L^2(A_i, E_i)$ and $\xi_i = \widehat{1_{A_i}} \in \mathcal{H}_i$. Consider the Hilbert D-module $\mathcal{K} = \mathcal{K}_1 \oplus \mathcal{K}_2$, where

$$\mathcal{K}_i = \bigoplus_{n \geq 1} \bigoplus_{i_1 = i,\ i_1 \neq \cdots \neq i_n} \mathcal{H}_{i_1} \otimes_D \cdots \otimes_D \mathcal{H}_{i_n}.$$

Let $T \in \mathbb{B}(\mathcal{K})$ be the isometry which exchanges $\mathcal{K}_1$ and $\mathcal{K}_2$, and acts like a creation operator:

$$T(\zeta_1 \otimes \cdots \otimes \zeta_n) = \xi_i \otimes \zeta_1 \otimes \cdots \otimes \zeta_n,$$

where $i = 1$ or 2 is suitably chosen. We have $T^* \zeta_1 = 0$ and

$$T^*(\zeta_1 \otimes \cdots \otimes \zeta_n) = \langle \xi_{i_1}, \zeta_1 \rangle \zeta_2 \otimes \cdots \otimes \zeta_n$$

for $n \geq 2$. We regard A_i as a C*-subalgebra of $\mathbb{B}(\mathcal{K}_i)$ acting on the first tensor component. Thus, $A_1 \oplus A_2 \subset \mathbb{B}(\mathcal{K})$. It is easily seen that

$$T^*(a_1 \oplus a_2) T = E_2(a_2) \oplus E_1(a_1)$$

for every $a_1 \oplus a_2 \in A_1 \oplus A_2$. Since the family of unitary operators $U_z = \bigoplus_{n \geq 1} z^n$ implements the gauge action and the compression to $\mathcal{H}_1 \oplus \mathcal{H}_2 \subset \mathcal{K}$ separates $A_1 \oplus A_2$ from the "compact operators" $\overline{\mathrm{span}}((A_1 \oplus A_2)T(A_1 \oplus A_2)T^*(A_1 \oplus A_2))$, the gauge-invariant uniqueness theorem implies:

Lemma 4.8.1. *The* C*-*algebra* $\mathcal{T}(\mathcal{H}^{\varphi}_{A_1 \oplus A_2})$ *is canonically* $*$-*isomorphic to the* C*-*subalgebra* $C^*(A_1 \oplus A_2, T)$ *of* $\mathbb{B}(\mathcal{K})$.

Let $p = 1 - T^2(T^*)^2 \in C^*(A_1 \oplus A_2, T)$ be the orthogonal projection onto

$$\bigoplus_{i \neq j} \Big(\mathcal{H}_i \oplus (\mathcal{H}_i \otimes_D \mathcal{H}_j) \oplus \big((\mathcal{H}_i \otimes_D \mathcal{H}_j \ominus \xi_i D \otimes \xi_j D) \otimes_D \mathcal{K}_i \big) \Big),$$

and recall that $\mathcal{H}_i = \xi_i D \oplus \mathcal{H}_i^o$. Let $u = p(T + T^*)p \in C^*(A_1 \oplus A_2, T)$ and observe that u is a self-adjoint partial isometry with $u^2 = p$ that interchanges $p\mathcal{K}_1$ and $p\mathcal{K}_2$:

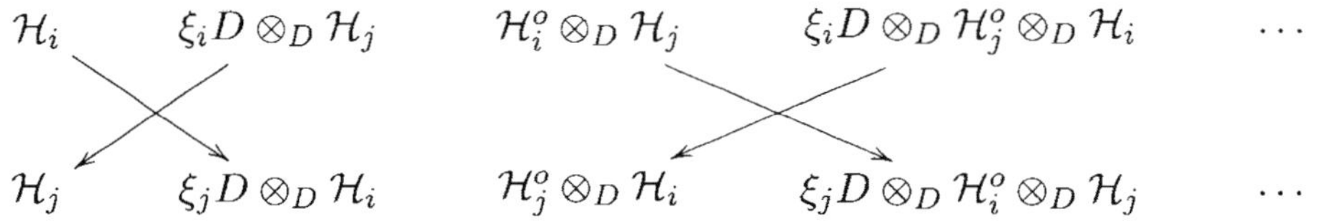

where the arrows are either T or T^*. We define u.c.p. maps

$$\psi_i \colon A_i \ni a \mapsto pap + uau \in pC^*(A_1 \oplus A_2, T)p$$

and note that $\psi_1(d) = \psi_2(d)$ for $d \in D$.

Theorem 4.8.2. *There exists a u.c.p. map*

$$\Psi \colon A_1 *_D A_2 \to pC^*(A_1 \oplus A_2, T)p$$

such that $\Psi(d) = \psi_1(d)$ *for* $d \in D$ *and*

$$\Psi(a_1 \cdots a_n) = \psi_{i_1}(a_1) \cdots \psi_{i_n}(a_n)$$

for $a_k \in A_{i_k}^o$ *with* $i_1 \neq \cdots \neq i_n$. *Moreover, there exists a $*$-homomorphism*

$$\pi \colon C^*(\Psi(A_1 *_D A_2)) \to A_1 *_D A_2$$

such that $\pi \circ \Psi = \mathrm{id}$. *In particular,* Ψ *is a complete order isomorphism.*

Proof. We decompose $p\mathcal{K}_i = \mathcal{K}_{i,1} \oplus \mathcal{K}_{i,2} \oplus \mathcal{K}_{i,3} \oplus \mathcal{K}_{i,4}$ as follows. The first component is

$$\mathcal{K}_{i,1} = \mathcal{H}_i \oplus \bigoplus \mathcal{H}_i \otimes_D \mathcal{H}_j^o \otimes_D \cdots \otimes_D \mathcal{H}_j^o \otimes_D \mathcal{H}_i;$$

in other words, all the tensors beginning and ending with vectors from $\mathcal{H}_i$ and having vectors from the $\mathcal{H}_k^o$'s in the middle. The second is

$$\mathcal{K}_{i,2} = (\mathcal{H}_i \otimes_D \mathcal{H}_j) \oplus \bigoplus \mathcal{H}_i \otimes_D \mathcal{H}_j^o \otimes_D \cdots \otimes_D \mathcal{H}_i^o \otimes_D \mathcal{H}_j;$$

this is very similar to $\mathcal{K}_{i,1}$, except we begin with $\mathcal{H}_i$ and end with $\mathcal{H}_j$. Recall that $\mathcal{H}_i = \xi_i D \oplus \mathcal{H}_i^o$, and note that no tensors in the first two components contain $\xi_i D$ or $\xi_j D$ in the middle – only on the ends. Our next two components take care of this. The third component is

$$\mathcal{K}_{i,3} = \bigoplus \mathcal{H}_i \otimes_D \mathcal{H}_j^o \otimes_D \cdots \otimes_D \mathcal{H}_j^o \ddot{\otimes}_D \xi_i D \otimes_D \mathcal{K}_j,$$

where the $\ddot{\otimes}$ is just a marker to indicate when the first $\xi_i D$ in the middle appears, and the direct sum is taken over the positions of $\ddot{\otimes}$. Finally,

$$\mathcal{K}_{i,4} = \mathcal{H}_i^o \ddot{\otimes}_D \xi_j D \otimes_D \mathcal{K}_i \oplus \bigoplus \mathcal{H}_i \otimes_D \mathcal{H}_j^o \otimes_D \cdots \otimes_D \mathcal{H}_i^o \ddot{\otimes}_D \xi_j D \otimes_D \mathcal{K}_i$$

consists of those tensor products such that $\xi_j D$ appears before $\xi_i D$ in the middle.

We observe that the (partial) isometry u interchanges $\mathcal{K}_{i,1}$ (resp. $\mathcal{K}_{i,3}$) and $\mathcal{K}_{j,2}$ (resp. $\mathcal{K}_{j,4}$) for $i \neq j$ and that each $\mathcal{K}_{i,l}$ is invariant under $\psi_k(a)$, for any i, k and l.

Via standard identifications $\xi_i D \otimes_D \mathcal{H}_j^o \cong \mathcal{H}_j^o$ and $\mathcal{H}_j^o \otimes_D \xi_i D \cong \mathcal{H}_j^o$, we have a natural isomorphism $V_{i,1} \colon \mathcal{K}_{i,1} \to \mathcal{H}$, where $(\mathcal{H}, \xi) = (\mathcal{H}_1, \xi_1) * (\mathcal{H}_2, \xi_2)$ is the free product Hilbert D-module. We let $A = A_1 *_D A_2$ and define a u.c.p. map $\Psi_{i,1} \colon A \to \mathbb{B}(\mathcal{K}_{i,1})$ by $\Psi_{i,1}(x) = V_{i,1}^* x V_{i,1}$. It is routine to check that

$$V_{i,1}^* \lambda_i(a_i) V_{i,1} = (a_i)|_{\mathcal{K}_{i,1}} = \psi_i(a_i)|_{\mathcal{K}_{i,1}},$$
$$V_{i,1}^* \lambda_j(a_j) V_{i,1} = (u a_j u)|_{\mathcal{K}_{i,1}} = \psi_j(a_j)|_{\mathcal{K}_{i,1}}$$

for every $a_i \in A_i$ and $a_j \in A_j$ $(i \neq j)$. Therefore, $\Psi_{i,1}$ is a $*$-isomorphism such that

$$\Psi_{i,1}(a_1 \cdots a_n) = \psi_{i_1}(a_1) \cdots \psi_{i_n}(a_n)|_{\mathcal{K}_{i,1}}$$

for any $a_k \in A_{i_k}$. The same holds for $\mathcal{K}_{i,2}$.

Let

$$\mathcal{H}_{(j)} = \bigoplus_{n \geq 1} \bigoplus_{\substack{i_1 \neq \cdots \neq i_n \\ i_n = j}} \mathcal{H}_{i_1}^o \otimes_D \cdots \otimes_D \mathcal{H}_{i_n}^o \subset \mathcal{H}.$$

Via the identification $\xi_i D \otimes_D \mathcal{H}_j^o \cong \mathcal{H}_j^o$, we have a natural isomorphism

$$V_{i,3} \colon \mathcal{K}_{i,3} \to \mathcal{H}_{(j)} \ddot{\otimes}_D \xi_i D \otimes_D \mathcal{K}_j \subset \mathcal{H} \ddot{\otimes}_D (\xi_i D \otimes \mathcal{K}_j).$$

Define $\Psi_{i,3} \colon A \to \mathbb{B}(\mathcal{K}_{i,3})$ by $\Psi_{i,3}(x) = V_{i,3}^*(x \ddot{\otimes} 1) V_{i,3}$ and we will prove that

$$(\diamond) \qquad \Psi_{i,3}(\lambda_{i_1}(a_1) \cdots \lambda_{i_n}(a_n)) = \psi_{i_1}(a_1) \cdots \psi_{i_n}(a_n)|_{\mathcal{K}_{i,3}}$$

for $a_k \in A_{i_k}^o$ with $i_1 \neq \cdots \neq i_n$. Let $a \in A_i^o$ and $b \in A_j^o$ be given $(i \neq j)$. With the help of Lemma 4.7.3, we first check, for $\zeta_k \in \mathcal{H}_{i_k}^o$ with $i = i_1 \neq i_2 \neq \cdots$, that

$$\begin{aligned}
&\Psi_{i,3}(\lambda_i(a))(\zeta_1 \otimes \zeta_2 \otimes \cdots \ddot{\otimes} \xi_i \otimes \cdots) \\
&= V_{i,3}^*\Big(\big((a\zeta_1 - \xi_i\langle \xi_i, a\zeta_1\rangle) \otimes \zeta_2 + \langle \xi_i, a\zeta_1\rangle \zeta_2\big) \otimes \cdots \ddot{\otimes} \xi_i \otimes \cdots\Big) \\
&= a\zeta_1 \otimes \zeta_2 \otimes \cdots \ddot{\otimes} \xi_i \otimes \cdots \\
&= p(a\zeta_1 \otimes \zeta_2 \otimes \cdots \ddot{\otimes} \xi_i \otimes \cdots) \\
&= \psi_i(a)(\zeta_1 \otimes \zeta_2 \otimes \cdots \ddot{\otimes} \xi_i \otimes \cdots)
\end{aligned}$$

and

$$\begin{aligned}\Psi_{i,3}(\lambda_j(b))(\zeta_1 \otimes \zeta_2 \otimes \cdots \ddot{\otimes}\, \xi_i \otimes \cdots) &= V_{i,3}^*\big((b\xi_j) \otimes \zeta_1 \otimes \zeta_2 \otimes \cdots \ddot{\otimes}\, \xi_i \otimes \cdots\big) \\ &= \xi_i \otimes (b\xi_j) \otimes \zeta_1 \otimes \zeta_2 \otimes \cdots \ddot{\otimes}\, \xi_i \otimes \cdots \\ &= \psi_j(b)(\zeta_1 \otimes \zeta_2 \otimes \cdots \ddot{\otimes}\, \xi_i \otimes \cdots).\end{aligned}$$

We next check, for $\zeta_k \in \mathcal{H}_{i_k}^o$ with $i \neq i_1 \neq \cdots \neq i_m$, that

$$\begin{aligned}\Psi_{i,3}(\lambda_i(a))(\xi_i \otimes \zeta_1 \otimes \cdots \otimes \zeta_m \ddot{\otimes}\, \xi_i \otimes \cdots)& \\ &= V_{i,3}^*(a\xi_i \otimes \zeta_1 \otimes \cdots \otimes \zeta_m \ddot{\otimes}\, \xi_i \otimes \cdots) \\ &= a\xi_i \otimes \zeta_1 \otimes \cdots \otimes \zeta_m \ddot{\otimes}\, \xi_i \otimes \cdots \\ &= \psi_i(a)(\xi_i \otimes \zeta_1 \otimes \cdots \otimes \zeta_m \ddot{\otimes}\, \xi_i \otimes \cdots)\end{aligned}$$

and if $m = 1$,

$$\begin{aligned}&\Psi_{i,3}(\lambda_j(b))(\xi_i \otimes \zeta_1 \otimes \cdots \otimes \zeta_m \ddot{\otimes}\, \xi_i \otimes \cdots) \\ &= V_{i,3}^*\Big(\big((b\zeta_1 - \xi_j\langle\xi_j, b\zeta_1\rangle) + \xi\langle\xi_j, b\zeta_1\rangle\big) \ddot{\otimes}\, \xi_i \otimes \cdots\Big) \\ &= \xi_i \otimes (b\zeta_1 - \xi_j\langle\xi_j, b\zeta_1\rangle) \ddot{\otimes}\, \xi_i \otimes \cdots \\ &= u\big((b\zeta_1) \otimes \cdots \otimes \zeta_m \ddot{\otimes}\, \xi_i \otimes \cdots\big) \\ &= \psi_j(b)(\xi_i \otimes \zeta_1 \otimes \cdots \otimes \zeta_m \ddot{\otimes}\, \xi_i \otimes \cdots),\end{aligned}$$

while if $m > 1$,

$$\begin{aligned}&\Psi_{i,3}(\lambda_j(b))(\xi_i \otimes \zeta_1 \otimes \cdots \otimes \zeta_m \ddot{\otimes}\, \xi_i \otimes \cdots) \\ &= V_{i,3}^*\Big(\big((b\zeta_1 - \xi_j\langle\xi_j, b\zeta_1\rangle) \otimes \zeta_2 + \langle\xi_j, b\zeta_1\rangle\zeta_2\big) \otimes \cdots \otimes \zeta_m \ddot{\otimes}\, \xi_i \otimes \cdots\Big) \\ &= \big(\xi_i \otimes (b\zeta_1 - \xi_j\langle\xi_j, b\zeta_1\rangle) \otimes \zeta_2 + \langle\xi_j, b\zeta_1\rangle\zeta_2\big) \otimes \cdots \otimes \zeta_m \ddot{\otimes}\, \xi_i \otimes \cdots \\ &= u\big((b\zeta_1) \otimes \cdots \otimes \zeta_m \ddot{\otimes}\, \xi_i \otimes \cdots\big) \\ &= \psi_j(b)(\xi_i \otimes \zeta_1 \otimes \cdots \otimes \zeta_m \ddot{\otimes}\, \xi_i \otimes \cdots).\end{aligned}$$

Hence, $\Psi_{i,3}(\lambda_i(a)) = \psi_i(a)$ for every $a \in A_i^o$ and $\Psi_{i,3}(\lambda_j(b)) = \psi_j(b)$ for every $b \in A_j^o$.

Claim. Let $\Phi_i \colon \mathbb{B}(\mathcal{H}) \to \mathbb{B}(\mathcal{H}_{(j)})$ be the compression. Then, Φ_i is "multiplicative on reduced words," i.e., for every $a_k \in A_{i_k}^o$ with $i_1 \neq \cdots \neq i_n$, one has

$$\Phi_i(\lambda_{i_1}(a_1) \cdots \lambda_{i_n}(a_n)) = \Phi_i(\lambda_{i_1}(a_1)) \cdots \Phi_i(\lambda_{i_n}(a_n)).$$

The proof is by induction. Let $a_k \in A_{i_k}^o$ with $i_1 \neq \cdots \neq i_{n+1}$ be given. If $i_{n+1} \neq j$, then

$$\lambda_{i_{n+1}}(a_{n+1})P_{\mathcal{H}_{(j)}} = P_{\mathcal{H}_{(j)}}\lambda_{i_{n+1}}(a_{n+1})P_{\mathcal{H}_{(j)}}$$

and we are done. Now, suppose that $i_{n+1} = j$. Then, we have

$$\lambda_{i_{n+1}}(a_{n+1})P_{\mathcal{H}_{(j)}} = (P_{\xi D} + P_{\mathcal{H}_{(j)}})\lambda_{i_{n+1}}(a_{n+1})P_{\mathcal{H}_{(j)}}.$$

But since $\lambda_{i_1}(a_1)\cdots\lambda_{i_n}(a_n)\xi D \subset \mathcal{H}_{i_1}^o \otimes_D \cdots \otimes_D \mathcal{H}_{i_n}^o$ and $i_n \neq j$, we have that

$$P_{\mathcal{H}_{(j)}}\lambda_{i_1}(a_1)\cdots\lambda_{i_n}(a_n)P_{\xi D} = 0.$$

Therefore, in both cases, we have

$$\Phi_i(\lambda_{i_1}(a_1)\cdots\lambda_{i_{n+1}}(a_{n+1})) = \Phi_i(\lambda_{i_1}(a_1)\cdots\lambda_{i_n}(a_n))\Phi_i(\lambda_{i_{n+1}}(a_{n+1})).$$

Thus induction will complete the proof of the claim.

Now ($\diamond$) follows from this claim and the preceding calculation. The same holds for $\mathcal{K}_{i,4}$.

Finally, we set $\Psi = \bigoplus_{i,l}\Psi_{i,l}\colon A \to \mathbb{B}(\mathcal{K})$. Then, Ψ is a u.c.p. map into $pC^*(A_1 \oplus A_2, T)p$ with the desired properties. We can define a $*$-homomorphism $\pi\colon C^*(\Psi(A)) \to A$ by $\pi(x) = V_{i,1}xV_{i,1}^* \in \mathbb{B}(\mathcal{H})$. □

Here are three important consequences. The first states that exactness is preserved by reduced free products.

Corollary 4.8.3. *Let $1 \in D \subset A_i$ $(i \in I)$ be unital* C**-algebras with nondegenerate conditional expectations E_i from A_i onto D. Then, the reduced amalgamated free product $(A, E) = \ast_D(A_i, E_i)$ is exact if every A_i is exact.*

Proof. Since the amalgamated free product construction is associative and exactness is closed under inductive limits, we may assume that $I = \{1, 2\}$. By Theorem 4.6.25, the C*-algebra $C^*(A_1 \oplus A_2, T) \cong \mathcal{T}(\mathcal{H}^{\varphi}_{A_1\oplus A_2})$ is exact if $A_1 \oplus A_2$ is. Since $A = A_1 *_D A_2$ is a quotient of $C^*(\Psi(A)) \subset pC^*(A_1 \oplus A_2, T)p$ with the u.c.p. splitting Ψ, exactness of A follows from that of $C^*(A_1 \oplus A_2, T)$. □

The following corollary is rather easy if the conditional expectations are faithful.

Corollary 4.8.4. *Let $1 \in D \subset A_i$ and conditional expectations E_i from A_i onto D be given. Let $1 \in D_B \subset B_i \subset A_i$ be such that $E_i(B_i) = D_B$. Assume E_i and $E_i^B = E_i|_{B_i}$ are nondegenerate. Then,*

$$(B_1, E_1^B) *_{D_B} (B_2, E_2^B) \cong C^*(B_1, B_2) \subset (A_1, E_1) *_D (A_2, E_2).$$

Proof. Let $A = (A_1, E_1) *_D (A_2, E_2)$ and $B = (B_1, E_1^B) *_{D_B} (B_2, E_2^B)$. Because of freeness, it is clear that the GNS representation ρ of $C^*(B_1, B_2)$ with respect to the conditional expectation E of A yields B. We have to show that ρ is faithful. If E is faithful, then this is trivial; otherwise, it is not so simple. This is why we need Theorem 4.8.2. We construct the inverse of ρ as follows. The C*-algebra B is completely order isomorphic to the C*-subalgebra of $pC^*(B_1 \oplus B_2, T)p$. But $pC^*(B_1 \oplus B_2, T)p$ is canonically $*$-isomorphic to the C*-subalgebra of $pC^*(A_1 \oplus A_2, T)p$ by Corollary 4.6.21.

Since A is a quotient of a C*-subalgebra of $pC^*(A_1 \oplus A_2, T)p$, the composition of these maps is the desired inverse of ρ. □

Finally, we deduce the existence of free products of c.p. maps.

Theorem 4.8.5. *Let $1 \in D \subset A_i$ and nondegenerate conditional expectations E_i^A from A_i onto D be given. Assume $1 \in D \subset B_i$ and assume there exist nondegenerate conditional expectations E_i^B from B_i onto D. Let $\theta_i \colon A_i \to B_i$ be u.c.p. maps such that $(\theta_i)|_D = \mathrm{id}_D$ and $E_i^B \circ \theta_i = E_i^A$. Then, there is a u.c.p. map*

$$\Theta \colon (A_1, E_1^A) *_D (A_2, E_2^A) \to (B_1, E_1^B) *_D (B_2, E_2^B)$$

such that $\Theta|_D = \mathrm{id}_D$ and

$$\Theta(a_1 \cdots a_n) = \theta_{i_1}(a_1) \cdots \theta_{i_n}(a_n)$$

for $a_j \in A_{i_j}^o$ with $i_1 \neq \cdots \neq i_n$.

Proof. By Theorem 4.8.2, we have unital complete order embeddings

$$(A_1, E_1^A) *_D (A_2, E_2^A) \subset p_A C^*(A_1 \oplus A_2, T_A) p_A,$$
$$(B_1, E_1^B) *_D (B_2, E_2^B) \subset p_B C^*(B_1 \oplus B_2, T_B) p_B.$$

We claim that the restriction of $\Theta \colon C^*(A_1 \oplus A_2, T_A) \to C^*(B_1 \oplus B_2, T_B)$ given by applying Proposition 4.6.23 to $\theta_1 \oplus \theta_2$ is the desired u.c.p. map. Since p_A and u_A are polynomials in T_A and T_A^*, for $a_j \in A_{i_j}^o$ with $i_1 \neq \cdots \neq i_n$, we have

$$\Psi_A(a_1 \cdots a_n) = f(T_A, T_A^*, a_1, \ldots, a_n),$$

where $f(T_A, T_A^*, a_1, \ldots, a_n)$ is a linear combination of monomials in which each $a_1, \ldots, a_n$ appears once in this order. Since $T_A^* a T_A = \alpha(E_D^A(a)) = 0$ for all $a \in A_{i_j}^o$ and $A_1 A_2 = \{0\}$, every monomial is of the form

$$T^{m_0} a_1 T^{m_1} \cdots T^{m_{k-1}} a_k (T^*)^{m_k} \cdots a_n (T^*)^{m_n}.$$

The same holds for $\Psi_B(\theta_{i_1}(a_1) \cdots \theta_{i_n}(a_n))$. Hence, we have

$$\Theta(\Psi_A(a_1 \cdots a_n)) = f(T_B, T_B^*, \theta_{i_1}(a_1), \ldots, \theta_{i_n}(a_n)) = \Psi_B(\theta_{i_1}(a_1) \cdots \theta_{i_n}(a_n))$$

by Proposition 4.6.23. □

We have seen that nuclearity typically is not preserved by free products (e.g., $C_\lambda^*(\mathbb{F}_2) = C(\mathbb{T}) * C(\mathbb{T})$). However, when dealing with pure states, things are different. We need a technical lemma of Kishimoto and Sakai.

Lemma 4.8.6. *Let A be a unital nuclear C*-algebra, $(\pi, \mathcal{H})$ be an irreducible representation, $\mathcal{K} \subset \mathcal{H}$ be a finite-dimensional subspace and $P_{\mathcal{K}} \in \mathbb{B}(\mathcal{H})$ be the orthogonal projection onto $\mathcal{K}$. Then, there exist nets of u.c.p. maps $\alpha_n \colon A \to \mathbb{M}_{k(n)}(\mathbb{C})$, $\beta_n \colon \mathbb{M}_{k(n)}(\mathbb{C}) \to A$ and isometries $V_n \colon \mathcal{K} \to \ell_{k(n)}^2$ such that the net $\beta_n \circ \alpha_n$ converges to the identity of A in the point-norm topology* *and such that* $V_n^*\alpha_n(a)V_n = P_{\mathcal{K}}\pi(a)P_{\mathcal{K}}$, $V_n\pi(\beta_n(x))V_n^* = V_nV_n^*xV_nV_n^*$ *for all* $a \in A$ *and* $x \in \mathbb{M}_{k(n)}(\mathbb{C})$.

Proof. Let Ω be the set of all (possibly noncontractive) c.p. maps θ on A of the form $\theta = \beta \circ \alpha$, where α is such that

$$\alpha(a) = P_{\mathcal{K}'}\pi(a)P_{\mathcal{K}'} \oplus \alpha'(a) \in \mathbb{B}(\mathcal{K}' \oplus \mathcal{L})$$

for some finite-dimensional Hilbert spaces $\mathcal{K}', \mathcal{L}$ with $\mathcal{K} \subset \mathcal{K}' \subset \mathcal{H}$ and some u.c.p. map $\alpha' \colon A \to \mathbb{B}(\mathcal{L})$ and $\beta \colon \mathbb{B}(\mathcal{K}' \oplus \mathcal{L}) \to A$ is such that

$$\pi(\beta(1))P_{\mathcal{K}} = P_{\mathcal{K}} \text{ and } P_{\mathcal{K}}\pi(\beta(x))P_{\mathcal{K}} = P_{\mathcal{K}}xP_{\mathcal{K}}$$

for all $x \in \mathbb{B}(\mathcal{K}' \oplus \mathcal{L})$. (We apologize for the abuse of notation.) Observe that Ω is a convex set. Indeed, if $\theta_i = \beta_i \circ \alpha_i$ $(i = 1, 2)$ with $\alpha_i \colon A \to \mathbb{B}(\mathcal{K}_i' \oplus \mathcal{L}_i)$ and $0 < t < 1$ are given, then $t\theta_1 + (1-t)\theta_2 = \beta \circ \alpha$, where

$$\alpha(a) = P_{\mathcal{K}'}\pi(a)P_{\mathcal{K}'} \oplus \big(\alpha_1'(a) \oplus \alpha_2'(a)\big) \in \mathbb{B}(\mathcal{K}' \oplus (\mathcal{L}_1 \oplus \mathcal{L}_2))$$

for the finite-dimensional Hilbert space $\mathcal{K}'$ spanned by $\mathcal{K}_1'$ and $\mathcal{K}_2'$ and

$$\beta(x) = t\beta_1(P_{\mathcal{K}_1' \oplus \mathcal{L}_i}\, x\, P_{\mathcal{K}_1' \oplus \mathcal{L}_i}) + (1-t)\beta_2(P_{\mathcal{K}_2' \oplus \mathcal{L}_2}\, x\, P_{\mathcal{K}_2' \oplus \mathcal{L}_2}).$$

We claim that it suffices to show the identity map id_A of A is in the point-weak closure of Ω. Indeed, if this were true, then, taking convex combinations, we could find a net $\theta_n = \beta_n \circ \alpha_n$ in Ω which converges to id_A in the point-norm topology (Lemma 2.3.4). In particular, $b_n = \theta_n(1) \approx 1$ for sufficiently large n. Hence, we can approximate (and replace) β_n with the u.c.p. map $\tilde{\beta}_n(\,\cdot\,) = b_n^{-1/2}\beta_n(\,\cdot\,)b_n^{-1/2}$. Since $b_n = \beta_n(1)$ is such that $\pi(b_n) = P_{\mathcal{K}} + P_{\mathcal{K}}^{\perp}\pi(b_n)P_{\mathcal{K}}^{\perp}$, the u.c.p. map $\tilde{\beta}_n$ still satisfies $P_{\mathcal{K}}\pi(\tilde{\beta}_n(x))P_{\mathcal{K}} = P_{\mathcal{K}}xP_{\mathcal{K}}$ for all x, showing the $\tilde{\beta}_n$'s to be as desired.

It is left to show that id_A is in the point-weak closure of Ω. Let $p \in A^{**}$ be the central projection supporting the irreducible representation $(\pi, \mathcal{H})$. It follows that $pA^{**} = \mathbb{B}(\mathcal{H})$ canonically. Since A is nuclear, there exist nets of u.c.p. maps $\alpha_n' \colon A \to \mathbb{B}(\mathcal{L}_n)$, $\beta_n' \colon \mathbb{B}(\mathcal{L}_n) \to (1-p)A^{**}$, with $\mathcal{L}_n$ finite-dimensional, such that the net $\beta_n' \circ \alpha_n'$ converges to the $*$-homomorphism $A \to (1-p)A^{**}, a \mapsto (1-p)a$, in the point-norm topology. Let $\mathcal{K}_n$ be an increasing net of finite-dimensional subspaces of $\mathcal{H}$ containing $\mathcal{K}$ and with dense union (all index sets may be assumed the same). We define c.p. maps $\alpha_n \colon A \to \mathbb{B}(\mathcal{K}_n \oplus \mathcal{L}_n)$ and $\beta_n \colon \mathbb{B}(\mathcal{K}_n \oplus \mathcal{L}_n) \to A^{**}$ by

$$\alpha_n(a) = P_{\mathcal{K}_n}\pi(a)P_{\mathcal{K}_n} \oplus \alpha_n'(a) \in \mathbb{B}(\mathcal{K}_n \oplus \mathcal{L}_n)$$

and

$$\beta_n(x) = P_{\mathcal{K}_n}xP_{\mathcal{K}_n} \oplus \beta_n'(P_{\mathcal{L}_n}xP_{\mathcal{L}_n}) \in \mathbb{B}(\mathcal{H}) \oplus (1-p)A^{**} = A^{**}.$$

It is clear that the net $\beta_n \circ \alpha_n$ converges to id_A in the point-ultraweak topology and that $\pi(\beta_n(x)) = P_{\mathcal{K}_n}xP_{\mathcal{K}_n}$ for all x. By the bijective correspondence described in Proposition 1.5.12, the c.p. map β_n corresponds to a positive element $a_n \in \mathbb{B}(\mathcal{K}_n \oplus \mathcal{L}_n) \otimes A^{**}$. We note that the c.p. map $\pi \circ \beta_n$ corresponds to

$$(\mathrm{id} \otimes \pi)(a_n) \in (P_{\mathcal{K}_n} \otimes P_{\mathcal{K}_n}) \left(\mathbb{B}(\mathcal{K}_n \oplus \mathcal{L}_n) \otimes \mathbb{B}(\mathcal{H})\right) (P_{\mathcal{K}_n} \otimes P_{\mathcal{K}_n}).$$

Since $\mathrm{id} \otimes \pi$ is an irreducible representation of $\mathbb{B}(\mathcal{K}_n \oplus \mathcal{L}_n) \otimes A$, it follows from strong transitivity (Corollary 1.4.8) that $a_n^{1/2} \in \mathbb{B}(\mathcal{K}_n \oplus \mathcal{L}_n) \otimes A^{**}$ is approximated, in the point-ultrastrong topology, by a net $(c_{n,i})_i$ of self-adjoint elements in $\mathbb{B}(\mathcal{K}_n \oplus \mathcal{L}_n) \otimes A$ such that $\|c_{n,i}\| \leq \|a_n^{1/2}\| + 1$ and

$$(\mathrm{id} \otimes \pi)(c_{n,i}) \, (1 \otimes P_{\mathcal{K}_n}) = (\mathrm{id} \otimes \pi)(a_n^{1/2}) \, (1 \otimes P_{\mathcal{K}_n}).$$

Then, the positive elements $a_{n,i} = c_{n,i}^2$ in $\mathbb{B}(\mathcal{K}_n \oplus \mathcal{L}_n) \otimes A$ converge to a_n in the point-ultrastrong topology and satisfy the equation

$$(\mathrm{id} \otimes \pi)(a_{n,i}) \, (1 \otimes P_{\mathcal{K}_n}) = (\mathrm{id} \otimes \pi)(a_n) \, (1 \otimes P_{\mathcal{K}_n}).$$

Now the c.p. maps $\beta_{n,i} \colon \mathbb{B}(\mathcal{K}_n \oplus \mathcal{L}_n) \to A$ corresponding to $a_{n,i}$ converge to β_n in the point-ultraweak topology and satisfy

$$\pi(\beta_{n,i}(x)) P_{\mathcal{K}} = \pi(\beta_n(x)) P_{\mathcal{K}} = P_{\mathcal{K}_n} x P_{\mathcal{K}}$$

for all $x \in \mathbb{B}(\mathcal{K}_n \oplus \mathcal{L}_n)$. This completes the proof. □

Theorem 4.8.7. *Let (A_i, φ_i) $(i = 1, 2)$ be unital* C**-algebras with states φ_i whose GNS representations are faithful. Let $(A, \varphi) = (A_1, \varphi_1) * (A_2, \varphi_2)$ be the reduced amalgamated free product* C**-algebra. Suppose that both A_i are nuclear and at least one of the φ_i's is pure. Then A is nuclear.*

Proof. We may assume that φ_1 is pure. By Lemma 4.8.6, there exist nets of u.c.p. maps $\alpha_n \colon A \to \mathbb{M}_{k(n)}(\mathbb{C})$, $\beta_n \colon \mathbb{M}_{k(n)}(\mathbb{C}) \to A$ and vector states ω_n on $\mathbb{M}_{k(n)}(\mathbb{C})$ such that $\beta_n \circ \alpha_n$ converges to id_A, $\omega_n \circ \alpha_n = \varphi_1$ and $\varphi_1 \circ \beta_n = \omega_n$. By Theorem 4.8.5, we may define the free product u.c.p. maps

$$\tilde{\alpha}_n = \alpha_n * \mathrm{id}_{A_2} \colon A \to (\mathbb{M}_{k(n)}(\mathbb{C}), \omega_n) * (A_2, \varphi_2)$$

and $\tilde{\beta}_n = \beta_n * \mathrm{id}_{A_2}$. Since $(\mathbb{M}_{k(n)}(\mathbb{C}), \omega_n) * (A_2, \varphi_2)$ is nuclear (see Exercise 4.8.2) and $\tilde{\beta}_n \circ \tilde{\alpha}_n = (\beta_n \circ \alpha_n) * \mathrm{id}_{A_2}$ converges point-norm to id_A, we are done. □

Exercises

Exercise 4.8.1. Let $1 \in D \subset A$ be C*-algebras with a nondegenerate conditional expectation E_D^A from A onto D. Let $\varphi = E_D^A \colon A \to D \subset A$ and consider $\mathcal{T}(\mathcal{H}_A^\varphi)$ as in Example 4.6.11. Prove that there is a natural isomorphism

$$(\mathcal{T}(\mathcal{H}_A^\varphi), E_D^A \circ E_{\mathcal{H}_A^\varphi}) \cong (A, E_D^A) *_D (\mathcal{T}(\mathbb{C}) \otimes D, \omega \otimes \mathrm{id}_D),$$

where $\mathcal{T}(\mathbb{C})$ is the Toeplitz algebra and ω is the vacuum state. (Really long hint, which some consider a proof: It is not hard to see that the nondegeneracy of E_D^A and $E_{\mathcal{H}_A^\varphi}$ implies that of $E_D^A \circ E_{\mathcal{H}_A^\varphi}$. Since $\mathcal{T}(\mathcal{H}_A^\varphi)$ is generated by A and $\mathcal{T}(\mathcal{H}_D^\varphi) \cong \mathcal{T}(\mathbb{C}) \otimes D$, it suffices to show that A and $\mathcal{T}(\mathcal{H}_D^\varphi)$ are free over D in $(\mathcal{T}(\mathcal{H}_A^\varphi), E_D^A \circ E_{\mathcal{H}_A^\varphi})$, by Theorem 4.7.2. Since $dT = T\varphi(d)$ for every $d \in D$, we have

$$\ker E_{\mathcal{H}_D^\varphi} = \overline{\mathrm{span}}\{dT^m(T^*)^n : d \in D,\ (m,n) \neq (0,0)\} \subset \mathcal{T}(\mathcal{H}_D^\varphi).$$

Thus, it suffices to show that

$$(E_D^A \circ E_{\mathcal{H}_A^\varphi})(a_0 T^{m_1}(T^*)^{n_1} a_1 \cdots T^{m_k}(T^*)^{n_k} a_k) = 0$$

for $k \geq 1$, $(m_j, n_j) \neq (0,0)$ and $a_j \in \ker E_D^A$ (except that a_0 and a_k are possibly 1). But this easily follows from the fact that $T^*aT = E_D^A(a) = 0$ for $a \in \ker E_D^A$ and Theorem 4.6.6.)

Exercise 4.8.2. Let $1 \in D \subset A$ be C*-algebras with a nondegenerate conditional expectation E_D^A from A onto D. Prove that

$$(A, E_D^A) *_D (\mathbb{M}_n(\mathbb{C}) \otimes D, \omega \otimes \mathrm{id}_D)$$

is nuclear if A is nuclear and ω is a pure state on $\mathbb{M}_n(\mathbb{C})$. (Hint: The previous exercise says free products with the Toeplitz algebra – using the vacuum state – preserve nuclearity, by Theorem 4.6.25. Thus, given $\mathbb{M}_n(\mathbb{C})$ and a vector state, it would suffice to find a really nice embedding into the Toeplitz algebra, since Theorem 4.8.5 would provide some c.p. maps to work with.)

4.9. References

Amenable actions were introduced by Zimmer in the measurable context (cf. [**200**]). They were transported into topological/C*-terms by Anantharaman-Delaroche [**3**], where Theorem 4.4.3 was proved. This naturally led to amenable groupoids; see [**6**] for a detailed study. Pimsner's influential class of algebras was introduced in [**146**]; our treatment is highly influenced by Katsura's work [**99**], where he dealt with C*-correspondences with noninjective left action. The gauge-invariant uniqueness theorem is due to Fowler, Muhly and Raeburn [**66**]. Theorem 4.6.25 is due to Dykema and Shlyakhtenko [**58**], where they gave a new proof of Dykema's theorem on exactness of amalgamated free products: Corollary 4.8.3 ([**57**]). The existence of free product maps was established, in the present generality, by Blanchard and Dykema [**20**]. For more on free products and free probability, see the monograph [**193**].

Chapter 5

Exact Groups and Related Topics

We begin this chapter by defining exact groups and proving that this concept is equivalent to acting amenably on some compact space. The next three sections deal with examples; that is, we show how to construct amenable actions in several interesting cases. The last two sections contain a discussion of two natural generalizations of groups: coarse metric spaces and groupoids. We don't give proper introductions to these important topics; we only focus on approximation issues.

5.1. Exact groups

Definition 5.1.1. A discrete group is *exact* if its reduced group C*-algebra is exact.[1]

Though the proof is beyond the scope of these notes (see [**73**]), the following result shows that most familiar groups are known to be exact. Recall that a *linear group* is any subgroup of the invertible matrices over some field.

Theorem 5.1.2 (Guentner, Higson and Weinberger). *Linear groups are exact.*

To begin our reformulations of exactness, we introduce some terminology. Let Γ be a discrete group and $E \subset \Gamma$ be a finite subset. The *tube of*

[1] This definition is not the original one – see the end of this section.

width E is the subset $\mathrm{Tube}(E)$ in $\Gamma \times \Gamma$ given by

$$\mathrm{Tube}(E) = \{(s,t) \in \Gamma \times \Gamma : st^{-1} \in E\}.^2$$

By the generic term *tube* we mean a tube of width E for some finite subset $E \subset \Gamma$. The *uniform algebra* (or the *uniform Roe algebra*) $C^*_u(\Gamma)$ of Γ is the C*-subalgebra of $\mathbb{B}(\ell^2(\Gamma))$ generated by $C^*_\lambda(\Gamma)$ and $\ell^\infty(\Gamma)$. Thinking of operators in $\mathbb{B}(\ell^2(\Gamma))$ as infinite matrices indexed by Γ, it is instructive to convince yourself of the following fact: $x = [x_{s,t}]_{s,t\in\Gamma} \in \mathbb{B}(\ell^2(\Gamma))$ belongs to the $*$-algebra generated by $\lambda(\mathbb{C}[\Gamma])$ and $\ell^\infty(\Gamma)$ if and only if x is supported in a tube (i.e., there exists a finite set $E \subset \Gamma$ such that $x_{s,t} = 0$ whenever $(s,t) \notin \mathrm{Tube}(E)$).[3]

It turns out that the uniform Roe algebra is an old friend incognito.

Proposition 5.1.3. *Let* $\alpha\colon \Gamma \to \mathrm{Aut}(\ell^\infty(\Gamma))$ *be the left translation action. Then*

$$C^*_u(\Gamma) \cong \ell^\infty(\Gamma) \rtimes_{\alpha,r} \Gamma.$$

Proof. We may apply the construction of $\ell^\infty(\Gamma) \rtimes_{\alpha,r} \Gamma$ to any faithful representation of $\ell^\infty(\Gamma)$, so we start with the canonical inclusion $\ell^\infty(\Gamma) \subset \mathbb{B}(\ell^2(\Gamma))$. (You may want to review Section 4.1 for the concrete construction of reduced crossed products.)

Define a unitary $U\colon \ell^2(\Gamma)\otimes\ell^2(\Gamma) \to \ell^2(\Gamma)\otimes\ell^2(\Gamma)$ by $U(\delta_x\otimes\delta_y) = \delta_x\otimes\delta_{yx}$. Now we compute

$$\begin{aligned} U\pi(f)(\delta_s \otimes \delta_t) &= U((\alpha_t^{-1}(f)\delta_s) \otimes \delta_t) \\ &= U((f(ts)\delta_s) \otimes \delta_t) \\ &= f(ts)\delta_s \otimes \delta_{ts} \\ &= \delta_s \otimes (f(ts)\delta_{ts}) \\ &= (1 \otimes f)(U(\delta_s \otimes \delta_t)). \end{aligned}$$

It follows that $U\pi(f)U^* = 1 \otimes f$ for all $f \in \ell^\infty(\Gamma)$. A similar calculation shows that U commutes with $1 \otimes \lambda_g$ for all $g \in \Gamma$ and hence

$$U(\ell^\infty(\Gamma) \rtimes_{\alpha,r} \Gamma)U^* = \mathbb{C}1 \otimes C^*(\ell^\infty(\Gamma), C^*_\lambda(\Gamma)) \cong C^*_u(\Gamma).$$

□

Definition 5.1.4. A bounded function $k\colon \Gamma \times \Gamma \to \mathbb{C}$ is called a *positive definite kernel* if the matrix $[k(s,t)]_{s,t\in\mathfrak{F}}$ is positive for any finite subset $\mathfrak{F} \subset \Gamma$.

[2]One should be careful about st^{-1} and $s^{-1}t$. We use here the *right* invariant tube so that $\lambda(s)$ is supported on a tube. However, when we deal with the Cayley graph later, we use the *left* invariant metric to make the left multiplication action isometric.

[3]If you aren't familiar with this point of view, it is good to start with $\ell^\infty(\Gamma)$; all of these operators are supported in $\mathrm{Tube}(\{e\})$. Next consider an element from the group ring $\lambda(\mathbb{C}[\Gamma])$. Such an operator is "constant down the diagonals," so which tube is it supported in?

Remark 5.1.5. Note that if k is a positive definite kernel *supported in a tube* (i.e., $k(s,t) = 0$ for all (s,t) outside some tube), then we can identify k with an element in $C_u^*(\Gamma)$, since $[k(s,t)]_{s,t\in\Gamma}$ obviously defines a bounded operator on $\ell^2(\Gamma)$. Since an operator is positive if and only if all compressions by finite-rank projections are positive matrices, we see that there is a one-to-one correspondence between positive definite kernels supported in tubes and positive operators in the $*$-algebra generated by $\lambda(\mathbb{C}[\Gamma])$ and $\ell^\infty(\Gamma)$.

The following is largely a translation of the results from Section 4.4. The only new addition is (1) $\Rightarrow$ (2).

Theorem 5.1.6. *Let Γ be a discrete group. The following are equivalent:*

(1) *Γ is exact;*

(2) *for any finite subset $E \subset \Gamma$ and $\varepsilon > 0$, there exists a positive definite kernel $k\colon \Gamma \times \Gamma \to \mathbb{C}$ whose support is contained in a tube and such that*
$$\sup\{|k(s,t) - 1| : (s,t) \in \operatorname{Tube}(E)\} < \varepsilon;$$

(3) *for any finite subset $E \subset \Gamma$ and $\varepsilon > 0$, there exist a finite subset $F \subset \Gamma$ and $\zeta\colon \Gamma \to \ell^2(\Gamma)$ such that $\|\zeta_t\| = 1$, $\operatorname{supp}\zeta_t \subset Ft$ for every $t \in \Gamma$ and*
$$\sup\{\|\zeta_s - \zeta_t\| : (s,t) \in \operatorname{Tube}(E)\} < \varepsilon;$$

(4) *for any finite subset $E \subset \Gamma$ and $\varepsilon > 0$, there exist a finite subset $F \subset \Gamma$ and $\mu\colon \Gamma \to \operatorname{Prob}(\Gamma)$ such that $\operatorname{supp}\mu_t \subset Ft$ for every $t \in \Gamma$ and*
$$\sup\{\|\mu_s - \mu_t\| : (s,t) \in \operatorname{Tube}(E)\} < \varepsilon;$$

(5) *$C_u^*(\Gamma)$ is nuclear.*

Proof. (1) $\Rightarrow$ (2): Let a finite subset $E \subset \Gamma$ and $\varepsilon > 0$ be given. It follows from Exercise 3.9.5 that we can find a finite subset $F \subset \Gamma$ with the following property: If $\varphi\colon \mathbb{B}(\ell^2(\Gamma)) \to \mathbb{B}(\ell^2(F))$ is compression by the projection onto $\ell^2(F) \subset \ell^2(\Gamma)$, then there is a u.c.p. map $\psi\colon \mathbb{B}(\ell^2(F)) \to \mathbb{B}(\ell^2(\Gamma))$ such that, setting $\theta = \psi \circ \varphi$, one has
$$\max_{s\in E} \|\theta(\lambda(s)) - \lambda(s)\| < \varepsilon.$$
We define a kernel $k\colon \Gamma \times \Gamma \to \mathbb{C}$ by
$$k(s,t) = \langle \theta(\lambda(st^{-1}))\delta_t, \delta_s\rangle.$$
This kernel is positive definite. Indeed, for all $s_1, \ldots, s_n \in \Gamma$ and $\alpha_1, \ldots, \alpha_n \in \mathbb{C}$, one has
$$\sum_{i,j} k(s_i, s_j)\bar{\alpha}_i\alpha_j = \sum_{i,j}\langle \theta(\lambda(s_is_j^{-1}))\alpha_j\delta_{s_j}, \alpha_i\delta_{s_i}\rangle$$

$$= \langle \theta(\begin{bmatrix} \lambda(s_i s_j^{-1}) \end{bmatrix}_{i,j}) \begin{bmatrix} \alpha_1\delta_{s_1} \\ \vdots \\ \alpha_n\delta_{s_n} \end{bmatrix}, \begin{bmatrix} \alpha_1\delta_{s_1} \\ \vdots \\ \alpha_n\delta_{s_n} \end{bmatrix} \rangle$$

$$\geq 0$$

since θ is c.p. and $[\lambda(s_i s_j^{-1})]_{i,j} \geq 0$ in $\mathbb{M}_n(C^*_\lambda(\Gamma))$ (Example 1.5.13). Since $\varphi(\lambda(s)) = 0$ when $sF \cap F = \emptyset$, the support of the kernel k is contained in $\mathrm{Tube}(FF^{-1})$. Moreover, if $(s,t) \in \mathrm{Tube}(E)$, then

$$k(s,t) = \langle \theta(\lambda(st^{-1}))\delta_t, \delta_s \rangle \approx_\varepsilon \langle \lambda(st^{-1})\delta_t, \delta_s \rangle = 1.$$

(2) $\Rightarrow$ (3): Let a finite subset $E \subset \Gamma$ and $\varepsilon > 0$ be given and take a kernel k satisfying condition (2). Denote by $a_k \in C^*_u(\Gamma)$ the positive operator corresponding to the kernel k (see Remark 5.1.5). Since $a_k^{1/2} \in C^*_u(\Gamma)$, there exists $b \in \mathbb{B}(\ell^2(\Gamma))$ with $\|a_k - b^*b\| < \varepsilon$ such that $\operatorname{supp} b \subset \mathrm{Tube}(F)$ for some finite subset $F \subset \Gamma$. Set $\eta_t = b\delta_t \in \ell^2(\Gamma)$. We note that $\|\eta_t\|^2 \approx_\varepsilon \langle a_k\delta_t, \delta_t\rangle = k(t,t)$. It follows that $\operatorname{supp} \eta_t \subset Ft$ for every $t \in \Gamma$ and

$$\langle \eta_t, \eta_s \rangle = \langle b^*b\delta_t, \delta_s \rangle \approx_\varepsilon \langle a_k\delta_t, \delta_s \rangle = k(s,t).$$

Hence, $\zeta_t = \eta_t / \|\eta_t\|$ satisfies condition (3) (with ε modified).

(3) $\Leftrightarrow$ (4): Observe that the map $\ell^2(\Gamma) \ni \zeta \mapsto |\zeta|^2 \in \ell^1(\Gamma)$ is uniformly continuous on the unit sphere.

(4) $\Rightarrow$ (5): Thanks to Proposition 5.1.3 and Theorem 4.4.3, it suffices to show the translation action of Γ on $\ell^\infty(\Gamma)$ is amenable. So, fix a finite symmetric set $E \subset \Gamma$ and $\varepsilon > 0$. Choose some $\mu\colon \Gamma \to \mathrm{Prob}(\Gamma)$ as in condition (4).

Define a function $T\colon \Gamma \to \ell^\infty(\Gamma)$ by $T(g)(x) = \sqrt{\mu_x(x^{-1}g)}$. As in the proof of Lemma 4.3.7, one checks that for every $s \in \Gamma$,

$$\|s *_\alpha T - T\|_2^2 = \sup_{x\in\Gamma} \Big(\sum_{g\in\Gamma} |\sqrt{\mu^{s^{-1}x}(x^{-1}g)} - \sqrt{\mu^x(x^{-1}g)}|^2 \Big)$$

$$\leq \sup_{x\in\Gamma} \|\mu^{s^{-1}x} - \mu^x\|_1.$$

However, $(s^{-1}x, x) \in \mathrm{Tube}(E)$ whenever $s^{-1} \in E$ and hence we see that $\|s *_\alpha T - T\|_2$ is small for all $s \in E$. The proof of Lemma 4.3.7 shows that we can replace T with a function of finite support, and this completes the proof.

(5) $\Rightarrow$ (1): Trivial. □

Here is a purely dynamical characterization of exactness.

Theorem 5.1.7. *For a discrete group Γ, the following are equivalent:*

(1) *Γ is exact;*

(2) *the left translation action of* Γ *on the Stone-Čech compactification* $\beta\Gamma$ *is amenable (i.e., for any finite subset* $E \subset \Gamma$ *and* $\varepsilon > 0$*, there exist a finite subset* $F \subset \Gamma$ *and a function* $m\colon \Gamma \to \mathrm{Prob}(F) \subset \mathrm{Prob}(\Gamma)$ *such that*

$$\sup\{\|s.m_t - m_{st}\| : s \in E,\ t \in \Gamma\} \leq \varepsilon);$$

(3) Γ *acts amenably on some compact topological space.*

Moreover, if Γ *is countable and exact, then* Γ *acts amenably on a compact metrizable space.*

Proof. First observe that amenability of the canonical action on $\beta\Gamma$ is equivalent to the parenthetical remark in condition (2), by Lemma 4.3.8 and the universal property of the Stone-Čech compactification. Now, it is easy to see that condition (2) is equivalent to condition (4) in Theorem 5.1.6 via the correspondence $\mu \leftrightarrow m$ given by $m_t(x) = \mu_t(x^{-1}t)$. Hence, we get (1) $\Rightarrow$ (2). (This also follows from the identification $C(\beta\Gamma) = \ell^\infty(\Gamma)$ of Γ-C*-algebras, Proposition 5.1.3 and Theorem 4.4.3.) The implication (2) $\Rightarrow$ (3) is immediate; (3) $\Rightarrow$ (1) follows from the inclusion $C^*_\lambda(\Gamma) \subset C(X) \rtimes_r \Gamma$ and Theorem 4.4.3.

The countability/metrizability assertion is standard fare once we recall that $C(Y)$ is separable if and only if Y is metrizable. Indeed, assume Γ acts on $C(X)$ amenably and let $T_n\colon \Gamma \to C(X)$ be a sequence of functions satisfying the definition of amenable action. Since Γ is countable, the C*-algebra A generated by the functions $\{T_n(g) : n \in \mathbb{N}, g \in \Gamma\}$ is separable. The algebra generated by A and all its translates by Γ is separable, abelian and Γ-invariant. Finally, the T_n's take values in this algebra, so Γ acts amenably on it. □

There are several different ways of proving exactness for free groups. For example, it follows from work of Choi that $C^*_\lambda(\mathbb{F}_2)$ embeds into a Cuntz algebra ([**36**]). Since subalgebras of nuclear C*-algebras are exact, this shows exactness of all free groups.[4] Here is a straightforward proof based on measures.

Proposition 5.1.8. *Free groups are exact.*

Proof. It suffices to show $\mathbb{F}_2$ is exact. For $t \in \mathbb{F}_2$ of length l with reduced form $t = t_1 \cdots t_l$, we denote $t(k) = t_1 \cdots t_k$ for $k \leq l$ and $t(k) = t$ for $k > l$.

[4] Perhaps you haven't seen the fact that all free groups contain one another as subgroups? It suffices to show $\mathbb{F}_2$ contains a copy of $\mathbb{F}_\infty$. Let $a, b \in \mathbb{F}_2$ be free generators and check that the group elements $b^n a b^{-n}$ are all free.

Fix $N \in \mathbb{N}$ and define $m\colon \mathbb{F}_2 \to \mathrm{Prob}(\mathbb{F}_2)$ by

$$m_t = \frac{1}{N} \sum_{k=0}^{N-1} \delta_{t(k)}.$$

Clearly, all m_t are supported on the finite set of elements in $\mathbb{F}_2$ with length $< N$. An instructive calculation left to the reader confirms that $\|s.m_t - m_{st}\| \leq 2|s|/N$ for every $s, t \in \mathbb{F}_2$. Letting $N \to \infty$ completes the proof. $\square$

The proof above is more transparent when viewed geometrically in the Cayley graph of $\mathbb{F}_2$. Moreover, a geometric point of view will be crucial in the next two sections, so let's develop our intuition by proving that free groups act amenably on their ideal boundaries.

Fix $r \in \mathbb{N}$ and let $\mathbb{F}_r = \langle g_1, \ldots, g_r \rangle$ be the rank-r free group (think of the $r = 2$ case for now). Then, its *ideal boundary* is the set

$$\partial \mathbb{F}_r = \{(a_i) \in \prod_{\mathbb{N}} \{g_1, g_1^{-1}, \ldots, g_r, g_r^{-1}\} : \forall i \in \mathbb{N}, a_{i+1} \neq a_i^{-1}\}.$$

The complement of $\partial\mathbb{F}_r$ (in $\prod_{\mathbb{N}}\{g_1, g_1^{-1}, \ldots, g_r, g_r^{-1}\}$) is easily seen to be open in the product topology; hence $\partial\mathbb{F}_r$ is compact. For geometric intuition, it is better to identify $\partial\mathbb{F}_r$ with the set of infinite paths in the Cayley graph of $\mathbb{F}_r$ which start at the neutral element. Indeed, given $x = (x_i) \in \partial\mathbb{F}_r$, we first think of x as the infinite word $x_1x_2x_3\cdots$ (note that this is in reduced form, since no cancellation occurs); then we identify this word with the path determined by the sequence of vertices $\{x_1, x_1x_2, x_1x_2x_3, \ldots\}$ in the Cayley graph of $\mathbb{F}_r$. Thinking of $\partial\mathbb{F}_r$ as infinite reduced words, it is easy to see that $\mathbb{F}_r$ acts continuously on $\partial\mathbb{F}_r$ by left multiplication (and rectifying possible cancellation).

For $x \in \partial\mathbb{F}_r$ with reduced word form $x = x_1x_2\cdots$, we set $x(0) = e$ and $x(k) = x_1 \cdots x_k$ for all $k > 0$. Fix $N \in \mathbb{N}$ and define a continuous map $\mu\colon \partial\mathbb{F}_r \to \mathrm{Prob}(\mathbb{F}_r)$ by

$$\mu^x = \frac{1}{N} \sum_{k=0}^{N-1} \delta_{x(k)}.$$

Looking at the Cayley graph in Figure 1, μ^x is just the normalized characteristic function of the first N steps along the infinite path determined by x. Here's an important observation/exercise: For each $s \in \mathbb{F}_r$ and $x \in \partial\mathbb{F}_r$, there exists a unique geodesic path starting at s and eventually merging with the path determined by $s.x \in \partial\mathbb{F}_r$ (see Figure 2); moreover, $s.\mu^x$ is just the normalized characteristic function of the first N steps along this geodesic.

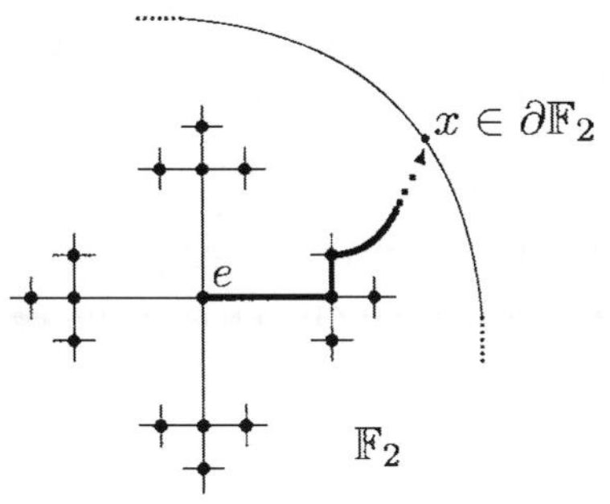

Figure 1. The Cayley graph of $\mathbb{F}_2$ and the boundary $\partial\mathbb{F}_2$

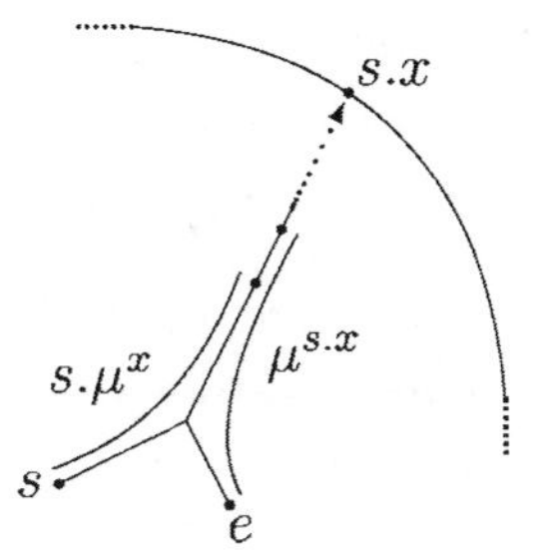

Figure 2. Amenability of $\mathbb{F}_2$ acting on $\partial\mathbb{F}_2$

With this geometric picture in mind, one checks that $\|s.\mu^x - \mu^{s.x}\| \leq 2|s|/N$ for all $s \in \mathbb{F}_r$ and $x \in \partial\mathbb{F}_r$. Letting $N \to \infty$, this shows that $\mathbb{F}_r$ acts amenably on its ideal boundary.

The original definition. Kirchberg and Wassermann introduced exactness for groups in [**108**]. Their definition involves short exact sequences of Γ-C*-algebras – that is, an algebra A, an action $\alpha\colon \Gamma \to \mathrm{Aut}(A)$ and an ideal $I \lhd A$ such that $\alpha_g(I) = I$, for every $g \in \Gamma$. In this situation, I and A/I inherit Γ-algebra structures denoted by $\alpha|_I$ and $\dot{\alpha}$, respectively, and we refer to $0 \to I \to A \to A/I \to 0$ as a *short exact sequence of* Γ*-algebras.*

Definition 5.1.9. We say Γ is *KW-exact* if for every short exact sequence of Γ-algebras, $0 \to I \to A \to A/I \to 0$, the sequence

$$0 \to I \rtimes_r \Gamma \to A \rtimes_r \Gamma \to (A/I) \rtimes_r \Gamma \to 0$$

is also exact.[5]

Just as for short exact sequences and *maximal* tensor products (see Proposition 3.7.1), it is not too hard to show that *full* crossed products always preserve exactness, due to the fact that

$$\frac{A \rtimes \Gamma}{I \rtimes \Gamma}$$

contains a Γ-equivariant copy of A/I. (Consider this an exercise.)

Theorem 5.1.10. *A discrete group* Γ *is exact if and only if it is KW-exact.*

[5]We can evidently identify $I \rtimes_r \Gamma$ with an ideal in $A \rtimes_r \Gamma$, by independence of representation. Moreover, there is a canonical surjective $*$-homomorphism $A \rtimes_r \Gamma \to (A/I) \rtimes_r \Gamma$ whose kernel contains $I \rtimes_r \Gamma$. See Exercise 4.1.4.

Proof. Assume first that Γ is KW-exact and let $0 \to I \to A \to A/I \to 0$ be an arbitrary short exact sequence. We must show that

$$0 \to I \otimes C^*_\lambda(\Gamma) \to A \otimes C^*_\lambda(\Gamma) \to (A/I) \otimes C^*_\lambda(\Gamma) \to 0$$

is also exact (Theorem 3.9.1). This, however, is trivial because $0 \to I \to A \to A/I \to 0$ can be thought of as a short exact sequence of Γ-algebras by equipping A with the trivial action (and recalling that $A \otimes C^*_\lambda(\Gamma) \cong A \rtimes_r \Gamma$ in this case – see Exercise 4.1.2).

For the converse, assume that Γ is exact and let $0 \to I \to A \to A/I \to 0$ be an arbitrary short exact sequence of Γ-algebras. To prove that the sequence of reduced crossed products is exact, we will exploit the fact that full crossed products preserve exactness.

By Fell's absorption principle (Proposition 4.1.7), there is an embedding $\pi_A \colon A \rtimes_r \Gamma \hookrightarrow (A \rtimes \Gamma) \otimes C^*_\lambda(\Gamma)$ such that

$$\pi_A(a\lambda_s) = as \otimes \lambda_s,$$

for all $a \in A$ and $s \in \Gamma$ (just apply Fell's principle to any faithful representation of the full crossed product). One easily checks that

$$\begin{array}{ccccccccc}
0 & \longrightarrow & I \rtimes \Gamma \otimes C^*_\lambda(\Gamma) & \longrightarrow & A \rtimes \Gamma \otimes C^*_\lambda(\Gamma) & \longrightarrow & (A/I) \rtimes \Gamma \otimes C^*_\lambda(\Gamma) & \longrightarrow & 0 \\
 & & {\scriptstyle \pi_I}\Big\uparrow & & {\scriptstyle \pi_A}\Big\uparrow & & {\scriptstyle \pi_{A/I}}\Big\uparrow & & \\
0 & \longrightarrow & I \rtimes_r \Gamma & \longrightarrow & A \rtimes_r \Gamma & \longrightarrow & A/I \rtimes_r \Gamma & \longrightarrow & 0
\end{array}$$

is a commutative diagram. Note that the top row is exact since we assumed Γ to be exact.

It turns out that there is also a c.c.p. map $\Phi_A \colon (A \rtimes \Gamma) \otimes C^*_\lambda(\Gamma) \to A \rtimes_r \Gamma$ such that $\Phi_A \circ \pi_A = \mathrm{id}_{A \rtimes_r \Gamma}$ and we have a commutative diagram

$$\begin{array}{ccccccccc}
0 & \longrightarrow & I \rtimes \Gamma \otimes C^*_\lambda(\Gamma) & \longrightarrow & A \rtimes \Gamma \otimes C^*_\lambda(\Gamma) & \longrightarrow & (A/I) \rtimes \Gamma \otimes C^*_\lambda(\Gamma) & \longrightarrow & 0 \\
 & & {\scriptstyle \Phi_I}\Big\downarrow & & {\scriptstyle \Phi_A}\Big\downarrow & & {\scriptstyle \Phi_{A/I}}\Big\downarrow & & \\
0 & \longrightarrow & I \rtimes_r \Gamma & \longrightarrow & A \rtimes_r \Gamma & \longrightarrow & A/I \rtimes_r \Gamma & \longrightarrow & 0.
\end{array}$$

We will construct Φ_A below, but first one should chase through the two diagrams above to see that the bottom row is also exact; hence the proof is complete once we know that Φ_A exists.

The construction of Φ_A, though elementary, is somewhat tedious. It is sufficient to find a c.c.p. map $(A \rtimes_r \Gamma) \otimes C^*_\lambda(\Gamma) \to A \rtimes_r \Gamma$ such that $a\lambda_s \otimes \lambda_g \mapsto 0$ if $s \neq g$ and $a\lambda_s \otimes \lambda_g \mapsto a\lambda_s$ if $s = g$. Indeed, if successful, then composing with the canonical quotient map $(A \rtimes \Gamma) \otimes C^*_\lambda(\Gamma) \to (A \rtimes_r \Gamma) \otimes C^*_\lambda(\Gamma)$ will give the desired map Φ_A.

Starting from a faithful representation $A \subset \mathbb{B}(\mathcal{H})$, we represent $A \rtimes_r \Gamma$ faithfully on $\mathcal{H} \otimes \ell^2(\Gamma)$ via the induced regular representation (Definition

4.1.4). Thus we have an embedding $(A\rtimes_r\Gamma)\otimes C^*_\lambda(\Gamma)\subset\mathbb{B}(\mathcal{H}\otimes\ell^2(\Gamma)\otimes\ell^2(\Gamma))$. Letting $V\colon\ell^2(\Gamma)\to\ell^2(\Gamma)\otimes\ell^2(\Gamma)$ be the isometry such that $V(\delta_t)=\delta_t\otimes\delta_t$ for all $t\in\Gamma$, an unenlightening (read: unpleasant) calculation shows that

$$(1_{\mathcal{H}}\otimes V^*)(a\lambda_s\otimes\lambda_g)(1_{\mathcal{H}}\otimes V)=a\lambda_s\in A\rtimes_r\Gamma\subset\mathbb{B}(\mathcal{H}\otimes\ell^2(\Gamma))$$

whenever $s=g$, and it equals zero otherwise. □

Much more can be found in [**108**] and [**109**]. Interestingly, it isn't yet known whether the "only if" direction of the theorem above extends to locally compact groups.

Exercises

Exercise 5.1.1. Prove the following permanence properties:

(1) subgroups of exact groups are exact;

(2) exactness is preserved by amalgamated free products;

(3) an increasing union of exact groups is exact. (Warning: Arbitrary inductive limits of exact groups apparently need not be exact. Neither of the authors understand the proof, but Gromov claims – and experts seem to agree – that one can construct a nonexact group as an inductive limit of hyperbolic groups.)

Exercise 5.1.2. It is also easy to show that extensions of exact groups are exact.

Just kidding. The first three permanence properties really are trivial consequences of what we know about C*-algebras. The extension problem is not.

Proposition 5.1.11. *Let Γ be a discrete group, Λ be a normal subgroup and $\bar\Gamma=\Gamma/\Lambda$. If X is a compact amenable $\bar\Gamma$-space and Y is a compact Γ-space which is amenable as a Λ-space, then $X\times Y$ (with the diagonal Γ-action) is an amenable Γ-space. In particular, an extension of exact groups is exact.*[6]

Proof. Let a finite subset $E\subset\Gamma$ and $\varepsilon>0$ be given. For $s\in\Gamma$, we denote by $\bar s$ the corresponding element in $\bar\Gamma$. We can find a continuous map $\xi\colon X\to\operatorname{Prob}(\bar\Gamma)$ such that $F=\bigcup_{x\in X}\operatorname{supp}\xi^x$ is finite and

$$\max_{s\in E}\sup_{x\in X}\|\bar s.\xi^x-\xi^{\bar s.x}\|<\varepsilon.$$

Choose a cross section $\sigma\colon\bar\Gamma\to\Gamma$ (i.e., $\overline{\sigma(p)}=p$ for all $p\in\bar\Gamma$). Since $\sigma(p)^{-1}s^{-1}\sigma(\bar sp)\in\Lambda$ for every $s\in\Gamma$ and $p\in\bar\Gamma$, we can find a continuous

[6]Let Γ be a group and $\Lambda\le\Gamma$ be a subgroup. If Λ is exact, then the left multiplication action of Λ on $\beta\Gamma$ is amenable. Indeed, there exists a Λ-equivariant continuous map from $\beta\Gamma$ onto $\beta\Lambda$.

map $\eta\colon Y \to \operatorname{Prob}(\Lambda) \subset \operatorname{Prob}(\Gamma)$ such that

$$\max_{s\in E}\max_{p\in F}\sup_{y\in Y} \|(\sigma(p)^{-1}s^{-1}\sigma(\bar{s}p)).\eta^y - \eta^{\sigma(p)^{-1}s^{-1}\sigma(\bar{s}p).y}\| < \varepsilon.$$

We define $\mu\colon X\times Y \to \operatorname{Prob}(\Gamma)$ by

$$\mu^{x,y} = \sum_{p\in F} \xi^x(p)\sigma(p).\eta^{\sigma(p)^{-1}y}.$$

Then, μ is continuous and one checks that

$$\begin{aligned}
\mu^{\bar{s}.x,s.y} &= \sum_{q\in F} \xi^{\bar{s}.x}(q)\sigma(q).\eta^{\sigma(q)^{-1}s.y}\\
&\approx_\varepsilon \sum_{q\in\bar{\Gamma}} \xi^x(\bar{s}^{-1}q)\sigma(q).\eta^{\sigma(q)^{-1}s.y}\\
&= \sum_{p\in F} \xi^x(p)s\sigma(p)\big(\sigma(p)^{-1}s^{-1}\sigma(\bar{s}p)\big).\eta^{\sigma(\bar{s}p)^{-1}s.y}\\
&\approx_\varepsilon \sum_{p\in F} \xi^x(p)s\sigma(p).\eta^{\sigma(p)^{-1}.y}\\
&= s.\mu^{x,y}
\end{aligned}$$

for all $s\in E$ and $(x,y)\in X\times Y$. □

5.2. Groups acting on trees

For any Γ-space K, we denote the stabilizer subgroup of $a\in K$ by $\Gamma^a = \{s\in\Gamma : s.a = a\}$. Our goal here is to show that a group acting on a tree is exact whenever all the vertex stabilizers are exact. This gives an alternate proof of the fact that amalgamated free products of exact groups are exact, since $\Gamma = \Gamma_1 *_\Lambda \Gamma_2$ acts on a tree in such a way that the vertex stabilizers are conjugates of Γ_1 or Γ_2. (See Appendix E.)

Our first result is inspired by, and generalizes, Proposition 5.1.11. The flexibility provided by Borel maps makes it very useful.

Proposition 5.2.1. *Let Γ be a countable group, X a compact Γ-space and K a countable Γ-space. Assume that the stabilizer subgroups Γ^a are exact, for all $a\in K$, and that there exists a net of Borel maps $\zeta_n\colon X\to \operatorname{Prob}(K)$ (i.e., the function $X\ni x\mapsto \zeta_n^x(a)\in\mathbb{R}$ is Borel for every $a\in K$) such that*

$$\lim_n \int_X \|s.\zeta_n^x - \zeta_n^{s.x}\|\,dm(x) = 0$$

for every $s\in\Gamma$ and every regular Borel probability measure m on X. Then Γ is exact.

Moreover, if X is amenable as a Γ^a-space for every $a\in K$, then X is an amenable Γ-space.

Proof. We first claim that for every $\varepsilon > 0$ and finite subset $E \subset \Gamma$, there exists a continuous map $\eta\colon X \to \operatorname{Prob}(K)$ such that

$$\max_{s\in E}\sup_{x\in X} \|s.\eta^x - \eta^{s.x}\| < \varepsilon.$$

Let $E \subset \Gamma$ be a fixed finite symmetric subset containing e. For every continuous map $\eta\colon X \to \operatorname{Prob}(K)$, we define $f_\eta \in C(X)$ by

$$f_\eta(x) = \sum_{s\in E} \|s.\eta^x - \eta^{s.x}\| = \sum_{s\in E}\sum_{a\in K} |\eta^x(s^{-1}.a) - \eta^{s.x}(a)|.$$

Observe that $f_{\sum_k \alpha_k\zeta_k} \leq \sum_k \alpha_k f_{\zeta_k}$ for every $\alpha_k \geq 0$ with $\sum_k \alpha_k = 1$. Hence, it suffices to show that 0 is in the norm-closed convex hull of $\{f_\eta : \eta\colon X \to \operatorname{Prob}(K) \text{ is continuous}\}$. By the Hahn-Banach separation theorem, it actually suffices to show 0 is in the weak closure of this set. That is, by the Riesz representation theorem, we must show that for every finite set of regular Borel probability measures $\mu_1, \ldots, \mu_n$ on X, there exists a continuous function $\eta\colon X \to \operatorname{Prob}(K)$ such that $\int f_\eta \, d\mu_i < \varepsilon$, for $i = 1, \ldots, n$.

Letting $m = \frac{1}{n}\sum \mu_i$, a little thought reveals that we really only have to find η such that $\int f_\eta \, dm < \varepsilon$ (for a smaller ε than that above). So, let $\varepsilon > 0$ be arbitrary. By assumption, there exists a Borel map $\zeta\colon X \to \operatorname{Prob}(K)$ such that

$$\sum_{s\in E}\int_X \|s.\zeta^x - \zeta^{s.x}\| \, dm(x) < \frac{\varepsilon}{9}.$$

Fubini's Theorem and the fact that $\zeta^{s.x}$ is a probability measure implies

$$1 = \int_X \Big(\sum_{a\in K} \zeta^{s.x}(a)\Big) \, dm(x) = \sum_{a\in K}\int_X \zeta^{s.x}(a) \, dm(x),$$

for every s. Hence we can find a finite subset $F \subset K$ such that

$$\sum_{s\in E}\int_X \sum_{a\in\Gamma\setminus F} \zeta^{s.x}(a) \, dm(x) < \frac{\varepsilon}{9}.$$

By Lusin's Theorem (applied to the measure $\sum_{s\in E} s.m$) we can approximate, for each $a \in F$, the Borel function $x \mapsto \zeta^x(a)$ by a continuous function $x \mapsto \eta^x(a)$ so that

$$\sum_{s\in E}\sum_{a\in F}\int_X |\eta^{s.x}(a) - \zeta^{s.x}(a)| \, dm(x) < \frac{\varepsilon}{9}.$$

Now fix $a_0 \in K \setminus F$ and define $\eta^x(a_0) = 1 - \sum_{a\in F} \eta^x(a)$, for every $x \in X$. For $b \notin F \cup \{a_0\}$ we define $\eta^x(b) = 0$ for all $x \in X$. We may assume that $\eta^x(a_0) \geq 0$ and regard η as a continuous map $\eta\colon X \to \operatorname{Prob}(K)$ such that $\operatorname{supp}\eta^x \subset F \cup \{a_0\}$ for all $x \in X$. It follows that

$$\sum_{s\in E}\int_X \|\eta^{s.x} - \zeta^{s.x}\| \, dm(x) < \frac{4\varepsilon}{9}.$$

This implies

$$\int_X f_\eta(x)\,dm(x) < \int_X \sum_{s\in E} \|s.\zeta^x - \zeta^{s.x}\|\,dm(x) + \frac{8\varepsilon}{9} < \varepsilon$$

and we obtain the claim.

Now, let a finite subset $E \subset \Gamma$ and $\varepsilon > 0$ be given. By our work above, there exists a continuous map η such that $\sup_{x\in X} \|s.\eta^x - \eta^{s.x}\| < \varepsilon$ for every $s \in E$. We may assume that there exists a finite subset $F \subset \Gamma$ such that $\operatorname{supp}\eta^x \subset F$ for all $x \in X$. Picking one point out of every orbit, we can find a Γ-fundamental domain $V \subset K$ – i.e., K decomposes into the disjoint union $\bigsqcup_{v\in V} \Gamma v$ – and let $v\colon K \to V$ be the corresponding projection (v takes every element in an orbit to its representative in V). Next we fix a cross section $\sigma\colon K \to \Gamma$ such that $a = \sigma(a).v(a)$ for every $a \in K$. Since the map v is constant along orbits, $\sigma(s.a)^{-1}s\sigma(a) \in \Gamma^{v(a)}$ for every $s \in \Gamma$ and $a \in K$. For each $v \in V$, we set

$$E^v = \{\sigma(s.a)^{-1}s\sigma(a) : a \in F \cap \Gamma v \text{ and } s \in E\} \subset \Gamma^v.$$

Let Y be a compact Γ-space which is amenable as a Γ^v-space for every $v \in V$. (Such Y always exists when each Γ^a is exact – take $Y = \beta\Gamma$). The proof of our proposition will be complete once we see that $X \times Y$ is an amenable Γ-space (with the diagonal action). Indeed, this will imply Γ is exact; moreover, if we can take $Y = X$, then the diagonal embedding $X \hookrightarrow X \times X$ is Γ-equivariant and continuous – hence amenability of X will follow from amenability of the diagonal action on $X \times X$. Since Y is Γ^v-amenable and E^v is finite, there exists a continuous map $\nu_v\colon Y \to \operatorname{Prob}(\Gamma)$ such that

$$\max_{s\in E^v} \sup_{y\in Y} \|s.\nu_v^y - \nu_v^{s.y}\| < \varepsilon.$$

Now, we define $\mu\colon X \times Y \to \operatorname{Prob}(\Gamma)$ by

$$\mu^{x,y} = \sum_{a\in F} \eta^x(a)\,\sigma(a).\nu_{v(a)}^{\sigma(a)^{-1}.y}.$$

The map μ is clearly continuous. Moreover, we have

$$\begin{aligned}
s.\mu^{x,y} &= \sum_{a\in K} \eta^x(a)\,s\sigma(a).\nu_{v(a)}^{\sigma(a)^{-1}.y} \\
&= \sum_{a\in F} \eta^x(a)\,\sigma(s.a).\big(\sigma(s.a)^{-1}s\sigma(a).\nu_{v(a)}^{\sigma(a)^{-1}.y}\big) \\
&\approx_\varepsilon \sum_{a\in K} \eta^x(a)\,\sigma(s.a).\nu_{v(s.a)}^{\sigma(s.a)^{-1}s.y} \\
&\approx_\varepsilon \sum_{a\in K} \eta^{s.x}(s.a)\,\sigma(s.a).\nu_{v(s.a)}^{\sigma(s.a)^{-1}s.y}
\end{aligned}$$

$$= \mu^{s.x,s.y}$$

for every $s \in E$ and $(x, y) \in X \times Y$. □

Remark 5.2.2. This result really does generalize the fact that extensions of exact groups are exact. Let Γ be a group and $\Lambda \triangleleft \Gamma$ be a normal subgroup such that Λ and Γ/Λ are exact. The hypotheses of Proposition 5.2.1 are satisfied with $X = \beta(\Gamma/\Lambda)$ and $K = \Gamma/\Lambda$.

The hard work is essentially over. We now recall lots of definitions and prove a few well-known facts about trees, compactifications and groups acting on these objects.

Let $\mathbf{T}$ be a tree, which we identify (as a metric space) with its vertex set (see Appendix E). A finite or infinite sequence $x(0), x(1), \ldots$ in $\mathbf{T}$ is called a *geodesic path* if $d(x(n), x(m)) = |n - m|$ for every n and m.[7] For convenience, if $(x(n))_{n=0}^{N}$ is a finite geodesic path, then we extend it to an infinite sequence by setting $x(m) = x(N)$ for every $m \geq N$; we still call this sequence a (finite) geodesic, even though it isn't, strictly speaking. Two geodesic paths x and x' are *equivalent* if they eventually flow together, i.e., if there exist $m_0, n_0 \in \mathbb{N}$ such that $x(m_0 + n) = x'(n_0 + n)$ for every $n \geq 0$. We can (and will) identify $\mathbf{T}$ with a subset of the equivalence classes of geodesics: every point x in $\mathbf{T}$ is identified with the equivalence class of geodesic paths ending at x. The *ideal boundary* $\partial\mathbf{T}$ of $\mathbf{T}$ is defined as the set of all equivalence classes of infinite geodesic paths. We define the *compactification of the tree* $\mathbf{T}$ to be $\bar{\mathbf{T}} = \mathbf{T} \sqcup \partial\mathbf{T}$ (a topology will be described shortly). If $(x(n))_n$ is a geodesic path with equivalence class $x \in \bar{\mathbf{T}}$, then we say the geodesic path $(x(n))_n$ *connects* $x(0)$ with x. For a bi-infinite geodesic path $(x(n))_{n=-\infty}^{\infty}$, we let $x(\infty) \in \partial\mathbf{T}$ (resp. $x(-\infty) \in \partial\mathbf{T}$) be the equivalence class of the geodesic path $(x(n))_{n\geq 0}$ (resp. $(x(-n))_{n\geq 0}$), and we say $(x(n))_n$ *connects* $x(-\infty)$ with $x(\infty)$.

Lemma 5.2.3 (See also Lemma E.2). *Let $x \in \mathbf{T}$ and $y \in \partial\mathbf{T}$. Then, there exists a unique geodesic connecting x with y.*

Proof. Pictorially, the proof is totally transparent. Here's the recipe in words: Let $(y(n))$ be a representative of y and let $(w(j))_{j=1}^{N}$ be a finite geodesic connecting x with $y(0)$ (which exists, since $\mathbf{T}$ is connected). Let $N_0 \leq N$ be the first integer such that there exists n_0 with $w(N_0) = y(n_0)$ – i.e., find the first point of intersection of the two geodesics. Define a new geodesic $(z(k))$ by $z(k) = w(k)$ for $1 \leq k \leq N_0$ and $z(k) = y(n_0 + (k - N_0))$ for $k > N_0$. Evidently $(z(k))$ is a geodesic connecting x with y.

Uniqueness of $(z(k))$ follows from the fact that $\mathbf{T}$ is a tree – any other geodesic connecting x with y would yield a loop in $\mathbf{T}$. □

---

[7]In a tree, a path is geodesic if and only if it never backtracks.

The lemma above is really a special case. Indeed, essentially the same proof yields the following fundamental fact (left to the reader): If $x, y \in \bar{\mathbf{T}}$, then there exists a geodesic connecting x with y (since $\mathbf{T}$ is connected), and it is unique (since $\mathbf{T}$ is a tree); this path will be denoted $[x, y]$.

If $[x, w_0]$ is a finite geodesic path and $[w_0, y]$ is any other geodesic, then we let $[x, w_0] \cup [w_0, y]$ denote the concatenation of these two paths (which is equal to $[x, y]$, of course). The following important lemma will be used repeatedly.

Lemma 5.2.4. *Given $x, y, z \in \bar{\mathbf{T}}$, $[x, y] \cap [y, z] \cap [z, x]$ is a singleton (i.e., there exists a unique point $w_0 \in \mathbf{T}$ such that $[x, y] = [x, w_0] \cup [w_0, y]$, $[x, z] = [x, w_0] \cup [w_0, z]$ and $[y, z] = [y, w_0] \cup [w_0, z]$).*

Proof. Again, the proof is trivial pictorially, so we only state the main idea. First note that $[x, y]$ and $[z, y]$ are equivalent geodesics. Letting w_0 be the first point of intersection of $[x, y]$ and $[z, y]$, the remainder of the proof is routine. □

We're now ready to topologize $\bar{\mathbf{T}}$. For $x \in \bar{\mathbf{T}}$ and a finite subset $F \subset \mathbf{T}$, we define

$$U(x; F) = \{x\} \cup \{y \in \bar{\mathbf{T}} : [x, y] \cap F = \emptyset\}.$$

One checks that $\{U(x; F)\}_{x,F}$ forms a basis for a topology (if $x \in U(x_1, F_1) \cap U(x_2, F_2)$, then Lemma 5.2.4 implies that $U(x, F_1 \cup F_2) \subset U(x_1, F_1) \cap U(x_2, F_2)$) and that the resulting topology is Hausdorff (given x, y, and any point $z_0 \neq x, y$ on the geodesic $[x, y]$, Lemma 5.2.4 implies $U(x, \{z_0\}) \cap U(y, \{z_0\}) = \emptyset$; and if x and y are adjacent, then $U(x, \{y\}) \cap U(y, \{x\}) = \emptyset$). This topology is very visual: cut finitely many edges in $\mathbf{T}$ and the connected components of $\bar{\mathbf{T}}$ are open (first verify this when only one edge is cut). Finally, it is worth checking that for a point $x \in \mathbf{T}$, the set $\{x\}$ is open if and only if x has finite degree.

Proposition 5.2.5. *The topological space $\bar{\mathbf{T}}$ is compact and any automorphism (i.e., isometric bijection) of the tree $\mathbf{T}$ extends to a homeomorphism of $\bar{\mathbf{T}}$.*

Proof. We must show that any net $(x_\alpha)_{\alpha \in A}$ in $\bar{\mathbf{T}}$ has an accumulation point (by Theorem A.8).

Fix a base point $o \in \mathbf{T}$ and identify every $x_\alpha \in \bar{\mathbf{T}}$ with the unique geodesic path connecting o to x_α. Let N be the largest integer (possibly 0 or ∞) such that there exist $x(0), \ldots, x(N)$ satisfying

$$\bigcap_{n=0}^{N} \{\alpha \in A : x_\alpha(n) = x(n)\} \in \mathcal{U}.$$

We observe that for each n there exists at most one $x(n)$ such that $\{\alpha : x_\alpha(n) = x(n)\} \in \mathcal{U}$ and such that $(x(n))_{n=0}^N$ is a geodesic path. Thus, if $N = \infty$, then the boundary point represented by $(x(n))_{n=0}^\infty$ is an accumulation point. On the other hand, if $N < \infty$, then $x(N)$ is an accumulation point. Thus $\bar{\mathbf{T}}$ is compact.

If s is an automorphism of $\mathbf{T}$, then it naturally acts on $\bar{\mathbf{T}} = \mathbf{T} \sqcup \partial\mathbf{T}$. Namely, for every $x \in \bar{\mathbf{T}}$, we define $s.x \in \bar{\mathbf{T}}$ to be the equivalence class of the geodesic path $(s.x(n))_n$, where $(x(n))_n$ is a representative of x. It is routine to check that this is a homeomorphism. □

Lemma 5.2.6. *Let* $\mathbf{T}$ *be a countable tree with fixed base point* o. *There exists a sequence of Borel maps*

$$\zeta_n \colon \bar{\mathbf{T}} \to \mathrm{Prob}(\mathbf{T})$$

such that

$$\sup_{x \in \bar{\mathbf{T}}} \|s.\zeta_n^x - \zeta_n^{s.x}\| \le \frac{2d(s.o, o)}{n}$$

for every automorphism s *on* $\mathbf{T}$.

Proof. As before, we identify every $x \in \bar{\mathbf{T}}$ with the unique geodesic path $(x(n))_n$ connecting o to x. (Recall our convention that $x(k) = x$ when $x \in \mathbf{T}$ and $k \ge \mathrm{dist}(x, o)$.) The maps ζ_n, defined by

$$\zeta_n^x = \frac{1}{n} \sum_{k=0}^{n-1} \delta_{x(k)} \in \mathrm{Prob}(\mathbf{T}),$$

satisfy the desired inequality. Indeed, $s.\zeta_n^x(p) > 0$ if and only if $s^{-1}.p$ is one of the first n points in the geodesic from o to x; equivalently, p is one of the first n points in the geodesic from $s.o$ to $s.x$. Similarly, $\zeta_n^{s.x}(q) > 0$ if and only if q is one of the first n points in the geodesic from o to $s.x$. Hence, for $n > d(s.o, o)$ we have cancellation in the difference $s.\zeta_n^x - \zeta_n^{s.x}$, because the geodesics from $s.o$ to $s.x$ and o to $s.x$ are equivalent.

Finally, it is easy to see that the ζ_n's are Borel since the set $\{x \in \bar{\mathbf{T}} : x(k) = z\}$ is clopen in $\bar{\mathbf{T}}$ for every $k \ge 0$ and $z \in \mathbf{T}$. □

Theorem 5.2.7. *Let* Γ *be a countable group and* $\mathbf{T}$ *be a countable tree on which* Γ *acts. If every vertex stabilizer* Γ^x *of* $x \in \mathbf{T}$ *is exact, then* Γ *is exact. In particular, an amalgamated free product of exact groups is exact.*

Proof. This follows from Proposition 5.2.1 and Lemma 5.2.6. □

We close this section with a result that won't be needed until much later in the book – but it makes sense to prove it now.

Lemma 5.2.8. *Let Γ be a group and* $\mathbf{T}$ *be a tree on which* Γ *acts. Let* (s_n) *be a net in* Γ *such that* $s_n \notin s\Lambda$ *eventually*[8] *for every* $s \in \Gamma$ *and every edge stabilizer* Λ. *If* $s_n.x \to z$ *for some* $x \in \mathbf{T}$ *and* $z \in \bar{\mathbf{T}}$, *then* $s_n.y \to z$ *for every* $y \in \mathbf{T}$.

Proof. We consider the open neighborhood of z given by a finite set F of edges in $\mathbf{T}$. It suffices to show that the geodesic paths $[s_n.x, s_n.y]$ between $s_n.x$ and $s_n.y$ do not cross F eventually. Since $[s_n.x, s_n.y] = s_n.[x, y]$, this reduces to showing that $s_n.\mathbf{e} \neq \mathbf{e}'$ eventually for any edges $\mathbf{e}, \mathbf{e}'$ in $\mathbf{T}$. Take $s \in \Gamma$ such that $s\mathbf{e} = \mathbf{e}'$. (If there is no such s, then we are already done.) Then $s_n.\mathbf{e} = \mathbf{e}'$ if and only if $s_n \in s\Gamma^{\mathbf{e}}$, where $\Gamma^{\mathbf{e}}$ is the edge stabilizer of $\mathbf{e}$. Hence $s_n.\mathbf{e} \neq \mathbf{e}'$ eventually, by assumption. □

Exercises

Exercise 5.2.1. Let X be a compact space which has no isolated points. Prove that the cardinality of X is at least c (cardinality of the continuum).

Exercise 5.2.2. Let X be a compact Γ-space whose cardinality is countable. Prove that there is a point $x \in X$ whose stabilizer subgroup Γ^x has finite index in Γ. (This explains the fact that Γ is exact if each stabilizer subgroup Γ^x is exact, which follows from Proposition 5.2.1 with $K = X$ and $\zeta\colon X \ni x \mapsto \delta_x \in \operatorname{Prob}(K)$.)

Exercise 5.2.3. Let $\mathbf{T}$ be a countable tree which has no infinite geodesic path. Prove that $\operatorname{Aut}(\mathbf{T})$ fixes either a point or an unoriented edge (i.e., a pair of points).

5.3. Hyperbolic groups

In this section we study an important class of graphs, namely those which are hyperbolic in the sense of Gromov. The main result is that groups which act properly on such graphs are exact.

Let K be a connected graph. We view K as a discrete metric space with the graph metric d (cf. Appendix E). As in the previous section, a geodesic path α is a sequence of vertices such that $d(\alpha(m), \alpha(n)) = |m - n|$ for every m and n. Since K is connected, for every pair $x, y \in K$, there exists a (not necessarily unique) geodesic path connecting x to y. Though not exactly well-defined, we often use $[x, y]$ to denote a geodesic path from x to y (multiple such geodesics may exist). For every subset $A \subset K$ and $r > 0$, we define

$$d(x, A) = \inf\{d(x, a) : a \in A\} \quad \text{and} \quad N_r(A) = \{x \in K : d(x, A) < r\}.$$

[8]That is, $\forall s, \forall \Lambda$ there exists n_0 such that $\forall n,\ n \geq n_0 \Rightarrow s_n \notin s\Lambda$.

The set $N_r(A)$ is called the r-tubular neighborhood of A in K. For subsets $A, B \subset K$, the *Hausdorff distance* between A and B is defined by

$$d_H(A, B) = \inf\{r : A \subset N_r(B) \text{ and } B \subset N_r(A)\}.$$

Definition 5.3.1. Let K be a connected graph. A *geodesic triangle* $\triangle$ in K consists of three points x, y, z in K and three geodesic paths $[x, y]$, $[y, z]$, $[z, x]$ connecting them.

Definition 5.3.2 (Hyperbolic graph)**.** For $\delta > 0$, we say a geodesic triangle $\triangle$ is *δ-slim* if each of its sides is contained in the open δ-tubular neighborhood of the union of the other two – i.e., $[x, y] \subset N_\delta([y, z] \cup [z, x])$ and similarly for the other two sides. We say that the graph K is *hyperbolic* if there exists $\delta > 0$ such that every geodesic triangle in K is δ-slim.

Note that hyperbolicity makes sense for any geodesic metric space (i.e., metric space where geodesics always exist). To get a feel for this concept, one should check that a tree is ε-hyperbolic (i.e., every geodesic triangle is ε-slim) for every $\varepsilon > 0$.

A *comparison tripod* is a geodesic triangle in a tree. It is not too hard to see that for every geodesic triangle $\triangle$ in a graph K there exist a unique tripod and a unique map f from $\triangle$ into the tripod that is isometric on all edges. Indeed, the lengths of the legs of the comparison tripod are determined by *the Gromov product*

$$\langle y, z\rangle_x = \frac{1}{2}(d(y, x) + d(z, x) - d(y, z)).$$[9]

Definition 5.3.3. For $\delta > 0$, we say that a geodesic triangle $\triangle$ is *δ-thin* if $u, v \in \triangle$ and $f(u) = f(v)$ imply that $d(u, v) < \delta$, where f is the unique map to $\triangle$'s comparison tripod.

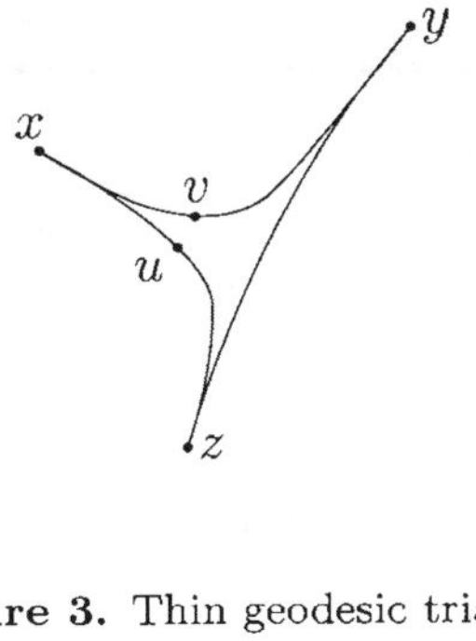

Figure 3. Thin geodesic triangle

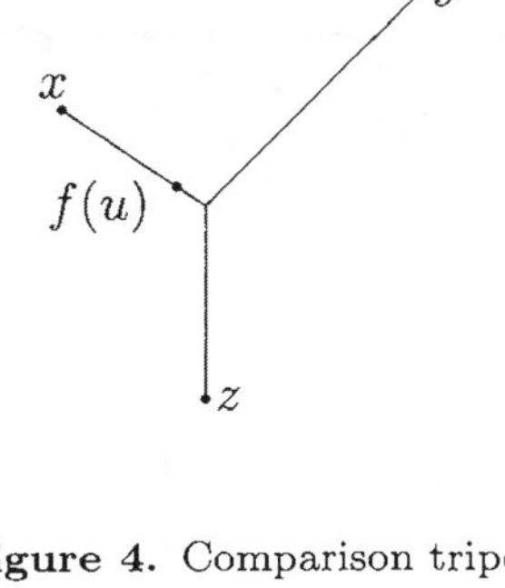

Figure 4. Comparison tripod

It is clear that any δ-thin geodesic triangle is δ-slim. The converse almost holds.

[9]This number is the distance from x to the intersection point in Figure 4. It is not an integer, in general, of course.

Proposition 5.3.4. *Let K be a hyperbolic graph. Then there exists $\delta > 0$ such that every geodesic triangle $\triangle$ is δ-thin.*

Proof. It will be convenient to think of K with the edges thrown in and each having length 1; that is, view K as a (continuous, rather than discrete) connected geodesic metric space. We will show that if every geodesic triangle in K is δ-slim, then they are all 4δ-thin.

Let $\triangle = [x, y] \cup [y, z] \cup [z, x]$ be a geodesic triangle and choose points u on $[x, y]$ and v on $[z, x]$ such that

$$d := d(u, x) = d(v, x) \leq \langle y, z \rangle_x.$$

By the intermediate value theorem, there is y' on $[x, y]$ such that $\langle y', z \rangle_x = d$. We note that u is on the subpath $[x, y']$ of $[x, y]$. Let $[y', z]$ be any geodesic connecting y' to z and let w be the point on $[y', z]$ such that $d(w, y') = d(u, y')$. It follows that $f(u) = f(v) = f(w)$ for the unique comparison map f from the geodesic triangle $\triangle' = [x, y'] \cup [y', z] \cup [z, x]$ onto its comparison tripod. Since $\triangle'$ is δ-slim, $u \in N_\delta([y'z] \cup [z, x])$ and $v \in N_\delta([x, y'] \cup [y'z])$. If $u \in N_\delta([z, x])$ or $v \in N_\delta([x, y'])$, then we must have $d(u, v) < 2\delta$ by the triangle inequality. Otherwise, we have $d(u, w) < 2\delta$ and $d(v, w) < 2\delta$. Therefore, we have $d(u, v) < 4\delta$ in either case. □

Now let Γ be a finitely generated group and $\mathcal{S}$ be a finite symmetric set of generators. We always equip the Cayley graph $\mathbf{X}(\Gamma, \mathcal{S})$ with the graph metric d (which is left invariant). Suppose that $\mathcal{S}'$ is another finite symmetric set of generators and let d' be the graph metric on $\mathbf{X}(\Gamma, \mathcal{S}')$. The vertex sets of $\mathbf{X}(\Gamma, \mathcal{S})$ and $\mathbf{X}(\Gamma, \mathcal{S}')$ are the same, of course; however their metric structures are different. But not that different. Indeed, if we choose $n \in \mathbb{N}$ so that $\mathcal{S} \subset (\mathcal{S}')^n = \{s_1 s_2 \cdots s_n : s_i \in \mathcal{S}'\}$ and $\mathcal{S}' \subset \mathcal{S}^n$, then it is readily seen that

$$n^{-1} d(x, y) \leq d'(x, y) \leq n d(x, y)$$

for every $x, y \in \Gamma$. Thus the formal identity from $\mathbf{X}(\Gamma, \mathcal{S})$ to $\mathbf{X}(\Gamma, \mathcal{S}')$ is *quasi-isometric*. More generally, we say that a map $f \colon (K, d) \to (K', d')$ between metric spaces is a *quasi-isometric embedding* if there exist $C \geq 1$ and $r > 0$ such that

$$C^{-1} d(x, y) - r \leq d'(f(x), f(y)) \leq C d(x, y) + r$$

for every $x, y \in K$. Thus, if Γ is finitely generated, its Cayley graph (with respect to a finite generating set) is unique up to quasi-isometry. Hence it is natural to look for properties which are quasi-isometry invariants, as they will provide invariants of groups.

Hyperbolicity turns out to be just such an invariant. This follows from the important fact that hyperbolic metric spaces enjoy *geodesic stability* – i.e., if a path is "close to being geodesic," then it is close (in Hausdorff distance) to an honest geodesic. To make this precise, we must define "close to being geodesic." For $C \geq 1$ and $r > 0$, we say that a finite sequence[10] α in K is *(C, r)-quasigeodesic* if

$$C^{-1}d(\alpha(m), \alpha(n)) - r \leq |m - n| \leq Cd(\alpha(m), \alpha(n)) + r$$

for every m, n.

Proposition 5.3.5. *Let K be a hyperbolic graph, $C \geq 1$ and $r > 0$. Then, there exists $D > 0$ with the following property: For any (C, r)-quasigeodesic sequence α and any geodesic path β having the same origin and terminal point as α, one has $d_H(\alpha, \beta) < D$.*

In particular, a graph is hyperbolic if it quasi-isometrically embeds into some hyperbolic graph.[11]

Proof. Let α and β be given. We set $D_0 = \max\{d(p, \alpha) : p \text{ on } \beta\}$ (hence β is contained in the D_0 tubular neighborhood of α).

Now suppose that q_0 is a point on α such that $d(q_0, \beta) \geq D_0$. By maximality of D_0, for every point u on β, there exists a point u' on α such that $d(u, u') \leq D_0$. Since the endpoints of α and β are the same, "the intermediate value theorem" implies the existence of consecutive u_0, u_1 on β such that u'_0 is on the origin (or "left") side and u'_1 is on the terminus (or "right") side of q_0. (That is, u' is on the left of q_0 when u is the starting point of β, and it is on the right when u is the endpoint – hence there is a place where u' jumps over q_0.) Since $d(u'_0, u'_1) \leq 2D_0 + 1$, the length of the subsequence of α from u'_0 to u'_1 is at most $C(2D_0 + 1) + r$. It follows that

$$d(u_0, q_0) \leq d(u_0, u'_0) + d(u'_0, q_0) \leq D_0 + C(C(2D_0 + 1) + 2r).$$

Therefore, we have $d_H(\alpha, \beta) \leq D$ for $D = D_0 + C(C(2D_0 + 1) + 2r)$. Hence, we must show that D_0 is bounded above by a function depending only on δ, C and r, where δ is the constant satisfying Definition 5.3.2.

Choose a point p_0 on β such that $d(p_0, \alpha) = D_0$. Choose two points b_0 and b_1 on β, one coming before p_0 and one after, such that $d(b_0, p_0) = 2D_0 = d(b_1, p_0)$ – or, if this isn't possible, take an endpoint of β. Let a_k, $k = 1, 2$, be points on α such that $d(b_k, a_k) = d(b_k, \alpha)$ and choose geodesic paths γ_0 and γ_1 connecting b_0 to a_0 and b_1 to a_1, respectively. (Note that if b_k is an endpoint, then $a_k = b_k$, so we take γ_k to be the single point $a_k = b_k$ in this case.) Maximality of D_0 implies that $d(b_k, a_k) \leq D_0$, and hence $d(p_0, \gamma_k) \geq D_0$. Let α' be the subsequence of α connecting a_0 to a_1. (It may flow backward.) Since

$$d(a_0, a_1) \leq d(a_0, b_0) + d(b_0, b_1) + d(b_1, a_1) \leq 6D_0$$

[10] It is important that we don't require α to be a path in this definition, because quasi-isometric embeddings don't always map paths to paths – i.e., neighbors need not map to neighbors.

[11] Think about the δ-slim condition and this is easily deduced.

and α is a (C, r)-quasigeodesic path, the length of α' is at most $6CD_0 + r$. By joining γ_0, α' and γ_1, we obtain a sequence γ connecting b_0 to b_1. For the reader's convenience, we list the properties of γ: it connects b_0 to b_1; $d(p_0, \gamma) \geq D_0$; the length $|\gamma|$ of the sequence γ is at most $(6C + 2)D_0 + r$; and $d(\gamma(k), \gamma(k+1)) \leq C(1 + r)$ for every k.

Now we apply the Weierstrass bisection process. Set $b_k^0 = b_k$, $p_0^0 = p_0$ and $\gamma^0 = \gamma$. Let c^0 be (one of) the midpoint(s) of γ^0 and consider a geodesic triangle $[b_0^0, c^0] \cup [c^0, b_1^0] \cup [b_1^0, b_0^0]$. Since K is hyperbolic, there exists p_0^1 in $[b_0^0, c^0] \cup [c^0, b_1^0]$ such that $d(p_0^0, p_0^1) \leq \delta$. If p_0^1 is on $[b_0^0, c^0]$, then we set $b_0^1 = b_0^0$ and $b_1^1 = c^0$ – otherwise let $b_0^1 = c^0$ and $b_1^1 = b_1^0$. Let γ^1 be the subsequence of γ connecting b_0^1 to b_1^1. We note that $|\gamma^1| \leq (2/3)|\gamma|$. Now, we continue this process by letting c^1 be (one of) the midpoint(s) of γ^1, and so on. This process terminates in l steps, with $l \leq \log|\gamma|/\log(3/2)$, and gives rise to p_0^l on $[b_0^l, b_1^l]$ such that b_0^l and b_1^l are consecutive points on γ. It follows that

$$D_0 \leq d(p_0, \gamma) \leq l\delta + d(b_0^l, b_1^l) \leq \delta \log(3/2)^{-1} \log\big((6C+2)D_0 + r\big) + C(1+r).$$

Since linear functions grow faster than logarithms, it follows that for each fixed C, r and δ, the numbers D_0 must be uniformly bounded (independent of α and β). □

Having established geodesic stability, the following definition is independent of the choice of generating set.

Definition 5.3.6 (Hyperbolic group). Let Γ be a finitely generated group. We say that Γ is *hyperbolic* if its Cayley graph is hyperbolic.

Remark 5.3.7. Since the Cayley graph of $\mathbb{F}_n$ is a tree, free groups are hyperbolic. Other examples include co-compact lattices in simple Lie groups of real rank one and the fundamental groups of compact Riemannian manifolds of negative sectional curvature (cf. [**70**]). Some of these groups have Kazhdan's property (T) (see [**15, 84**]).

Our next goal is to define the Gromov boundary; drawing lots of pictures will help.

Let K be a hyperbolic graph. We say that two infinite geodesic paths α and β in K are *equivalent* if

$$\liminf_{m,n\to\infty} \langle \alpha(m), \beta(n) \rangle_o = \infty$$

for some $o \in K$. Geometrically, this means α and β are pointing in the same direction. It is clear that the definition is independent of the choice of $o \in K$. It's not so clear that we have an equivalence relation, hence a lemma.

Lemma 5.3.8. *There exists a constant $C = C(K) > 0$ with the following property: For any two equivalent infinite geodesic paths α and β in K and any $m \geq d(\alpha(0), \beta(0))$, there exists n with $|m - n| \leq d(\alpha(0), \beta(0))$ such that $d(\alpha(m), \beta(n)) < C$.*

In particular, α and β are equivalent if and only if $\sup_m d(\alpha(m), \beta(m)) < \infty$ (and this is clearly an equivalence relation).

Proof. Choose $\delta > 0$ so that every geodesic triangle is δ-thin. Let $m \geq d(\alpha(0), \beta(0))$ be given. Let $o = \alpha(0)$ and find $m_1, n_1 \in \mathbb{N}$ such that $\langle \alpha(m_1), \beta(n_1) \rangle_o > m$. Choose any geodesic path $[o, \beta(n_1)]$ connecting o to $\beta(n_1)$. Let x be the vertex on $[o, \beta(n_1)]$ such that $d(o, x) = d(o, \alpha(m)) = m$. Since $\langle \alpha(m_1), \beta(n_1) \rangle_o > m$, we have $d(x, \alpha(m)) < \delta$. Let n be such that $n < n_1$ and $d(x, \beta(n_1)) = d(\beta(n), \beta(n_1))$. Since $d(o, x) \geq d(\alpha(0), \beta(0))$, such an n exists. Moreover $|m - n| \leq d(\alpha(0), \beta(0))$ and $d(x, \beta(n)) < \delta$. It follows that $d(\alpha(m), \beta(n)) < 2\delta$. This proves the first assertion.

For the second assertion, let equivalent geodesic paths α and β and $m \geq d(\alpha(0), \beta(0))$ be given. Then, by the first assertion, there is n with $|m - n| \leq d(\alpha(0), \beta(0))$ such that $d(\alpha(m), \beta(n)) < C$. Hence, $d(\alpha(m), \beta(m)) \leq d(\alpha(m), \beta(n)) + |m - n| \leq C + d(\alpha(0), \beta(0))$. Conversely, suppose $d_H(\alpha, \beta) < \infty$ and take $m_0, n_0 \geq 0$ such that $d(\alpha(m_0), \beta(n_0)) \leq d_H(\alpha, \beta) + 1$. Then, for any $m \geq m_0$ and $n \geq n_0$, one has

$$\begin{aligned}
2\langle \alpha(m), \beta(n) \rangle_o &= d(\alpha(m), o) + d(\beta(n), o) - d(\alpha(m), \beta(n)) \\
&\geq m - d(\alpha(0), o) + n - d(\beta(0), o) \\
&\qquad - ((m - m_0) + d(\alpha(m_0), \beta(n_0)) + (n - n_0)) \\
&\geq m_0 + n_0 - (d(\alpha(0), o) + d(\beta(0), o) + d_H(\alpha, \beta) + 1).
\end{aligned}$$

This proves $\liminf_{m,n\to\infty} \langle \alpha(m), \beta(n) \rangle_o = \infty$. □

Definition 5.3.9. We define the *Gromov boundary* ∂K of a hyperbolic graph K to be the set of all equivalence classes of infinite geodesic paths. We call $\bar{K} = K \cup \partial K$ the *Gromov compactification* of K (we soon describe the topology). For a hyperbolic group Γ, $\bar{\Gamma}$ denotes the Gromov compactification of its Cayley graph.

Definition 5.3.10. For a finite or infinite geodesic path $\alpha = x_0 x_1 \cdots$ in K, we denote by $\alpha_- = x_0$ its starting point and α_+ its terminal point (i.e., the boundary point α represents, in the infinite case). As before, we say that α *connects* α_- with α_+.

The Cayley graph of a finitely generated group is always *uniformly locally finite* (or has *bounded geometry*) – i.e., there is a uniform bound on the degree of vertices. This is a nice property for a graph to have, so *from now on, we assume that the hyperbolic graph K is uniformly locally finite* and, in particular, countable. We leave it as an exercise to check that every $x \in K$ can be connected to every $z \in \partial K$ via a geodesic path.

Although the topology on $\bar{K}$ can be defined like that of a tree, we give a different description. Fix a base point $o \in K$. For $z \in \partial K$ and $R > 0$, we set

$$U(z,R) = \{x \in \bar{K} : \exists \text{ geodesic paths } \alpha, \beta \text{ with } \alpha_+ = x,\ \beta_+ = z \text{ such that } \liminf_{m,n\to\infty} \langle \alpha(m), \beta(n)\rangle_o > R\},$$

where in the case $x \in K$, we choose the "geodesic" $\alpha(m) = x$ for all m. We also define

$$U'(z,R) = \{x \in \bar{K} : \forall \text{ geodesic paths } \alpha, \beta \text{ with } \alpha_+ = x,\ \beta_+ = z, \text{ we have } \liminf_{m,n\to\infty} \langle \alpha(m), \beta(n)\rangle_o > R\}.$$

It turns out these sets satisfy the axioms for a neighborhood basis. The resulting topology on $\bar{K}$ is as expected: $\bar{K}$ is compact and K is a dense open discrete subset. Let's prove this.

It is clear that $U'(z,R) \subset U(z,R)$. On the other hand, we have

Lemma 5.3.11. *There exists $C = C(K) > 0$ with the following property: If α, α' and β, β' are geodesic paths such that $\alpha_+ = \alpha'_+$ and $\beta_+ = \beta'_+$, then*

$$\liminf_{m,n\to\infty} \langle \alpha'(m), \beta'(n)\rangle_o \geq \liminf_{m,n\to\infty} \langle \alpha(m), \beta(n)\rangle_o - C.$$

In particular, $U'(z,R) \supset U(z,R+C)$ for every $z \in \partial K$ and $R > 0$.

Proof. This follows from Lemma 5.3.8 and the inequality

$$\langle \alpha'(m'), \beta'(n')\rangle_o \geq \langle \alpha(m), \beta(n)\rangle_o - (d(\alpha'(m'), \alpha(m)) + d(\beta'(n'), \beta(n))).$$

□

Lemma 5.3.12. *For any $R > 0$, there exists $S > 0$ with the following property: For any $y, z \in \partial K$ with $y \in U(z,S)$, we have $U(y,S) \subset U(z,R)$.*

Proof. Choose some $\delta > 0$ such that every geodesic triangle is δ-thin. By Lemma 5.3.11, it suffices to show that if $y \in U'(z,N)$ for $y, z \in \partial K$ and $N \in \mathbb{N}$, then $U'(y,N) \subset U(z,N-\delta)$. Let $x \in U'(y,N)$ and take geodesic paths α, β and γ connecting o to x, y and z, respectively. Since $\liminf \langle \gamma(n), \beta(m)\rangle_o > N$, we have $d(\gamma(N), \beta(N)) < \delta$. Similarly, $d(\beta(N), \alpha(N)) < \delta$ and hence $d(\gamma(N), \alpha(N)) < 2\delta$. It follows that for every $m, n \geq N$,

$$\begin{aligned} 2\langle \alpha(m), \gamma(n)\rangle_o &= m + n - d(\alpha(m), \gamma(n)) \\ &\geq m + n - (m - N + d(\alpha(N), \gamma(N)) + n - N) \\ &\geq 2N - 2\delta. \end{aligned}$$

This shows $x \in U(z, N-\delta)$. □

Now, it is easy to check that $\{U(z,R)\}_{R>0}$ defines a (not necessarily open) neighborhood basis and the resulting topology is Hausdorff.

Definition 5.3.13. We equip $\bar{K} = K \cup \partial K$ with a topology by declaring that a subset $O \subset \bar{K}$ is open if and only if for every $z \in \partial K \cap O$, there exists $R > 0$ such that $U(z,R) \subset O$. We note that for every $x \in K$, the singleton set $\{x\}$ is open in $\bar{K}$.

It is clear that this topology is independent of the choice of the base point o. Moreover, for a hyperbolic group Γ, the Gromov compactification $\bar{\Gamma}$ is independent of the choice of finite generating subset (thanks to Proposition 5.3.5).

Theorem 5.3.14. *Let K be a locally finite hyperbolic graph. Then the topological space $\bar{K}$ defined above is compact and contains K as a dense open subset. Every automorphism (i.e., isometric bijection) on K extends uniquely to a homeomorphism on $\bar{K}$.*

Proof. The proof is similar to that of Proposition 5.2.5. We only prove compactness; the rest is trivial. It suffices to show that an arbitrary net $(x_i)_{i\in I}$ in $\bar{K}$ has an accumulation point (Theorem A.8). For every i, choose a geodesic path α_i connecting o to x_i. For convenience, we set $\alpha_i(n) = x_i$ when $n \geq d(o,x_i)$. Let $\mathcal{U}$ be a cofinal ultrafilter on the directed set I. Since K is locally finite, for every n, there exists a unique point $\alpha(n) \in K$ such that $\{i : \alpha_i(n) = \alpha(n)\} \in \mathcal{U}$. Since each α_i is a geodesic path, α is also a geodesic path (or perhaps a path which is eventually constant). It is not too hard to see that $\alpha_+ \in \bar{K}$ is an accumulation point. □

Here is the exactness result we have been after.

Theorem 5.3.15. *Let K be a uniformly locally finite hyperbolic graph and Γ be a group acting properly*[12] *on it. Then the action of Γ on the Gromov compactification $\bar{K}$ is amenable. In particular, every hyperbolic group is exact (since it acts properly on its Cayley graph).*

Proof. For $x, y \in K$, we denote by $T(x,y)$ the set of $z \in \partial K$ such that there exists a geodesic path connecting x to z which passes through y. It is not hard to see that $T(x,y)$ is a closed subset of ∂K. For every $x \in K$, $z \in \partial K$ and integers l, k with $l \geq k$, we define a subset $S(x,z,l,k) \subset K$ by declaring

$$S(x,z,l,k) = \{\alpha(l) : \alpha \text{ a geodesic path in } K \text{ such that } d(\alpha_-, x) \leq k \text{ and } \alpha_+ = z\}.$$

[12] In this case, being proper is equivalent to saying every vertex stabilizer is finite.

Note that for every $x, y \in K$ and k, l, we have

$$\begin{aligned}&\{z \in \partial K : y \in S(x,z,l,k)\}\\&= \bigcup \{T(x',y) : x' \in K \text{ with } d(x',x) \leq k \text{ and } d(x',y) = l\}\end{aligned}$$

and hence the former set is Borel in $\bar{K}$. Also, note the inclusion $S(x,z,l,k) \subset S(x,z,l,k')$ whenever $k \leq k'$.

Let $C = C(K) > 0$ be the constant appearing in Lemma 5.3.8. Since K is uniformly locally finite, there exists $D > 0$ such that every ball in K of radius C contains at most $D/3$ points. By Lemma 5.3.8, the subset $S(x,z,l,k)$ is contained in the C-tubular neighborhood of a subpath $\alpha([l-k, l+k])$ of any geodesic path α connecting x to z. This implies that $|S(x,z,l,k)| \leq Dk$ for all x, z, k, l with $l \geq k$. For a finite subset $S \subset K$, we denote by $\chi_S \in \operatorname{Prob}(K)$ the normalized characteristic function on S. Define a sequence of Borel functions $\eta_n \colon K \times \partial K \to \operatorname{Prob}(K)$ by

$$\eta_n(x,z) = \frac{1}{n} \sum_{k=n+1}^{2n} \chi_{S(x,z,3n,k)}.$$

We claim that, for each $x, x' \in K$, we have

$$\lim_{n \to \infty} \sup_{z \in \partial K} \|\eta_n(x,z) - \eta_n(x',z)\| = 0.$$

Let $d = d(x,x')$. Fix $z \in \partial K$ and $n \geq d$ and set $S_k = S(x,z,3n,k)$ and $S'_k = S(x',z,3n,k)$. Then, we have $S_k \cup S'_k \subset S_{k+d}$ and $S_k \cap S'_k \supset S_{k-d}$ for every $n < k \leq 2n$. It follows that

$$\|\chi_{S_k} - \chi_{S'_k}\| = 2\left(1 - \frac{|S_k \cap S'_k|}{\max\{|S_k|, |S'_k|\}}\right) \leq 2\left(1 - \frac{|S_{k-d}|}{|S_{k+d}|}\right)$$

for $n < k \leq 2n$. Since $|S_k| \leq Dk$, we have

$$\begin{aligned}\|\eta_n(x,z) - \eta_n(x',z)\| &\leq \frac{1}{n} \sum_{k=n+1}^{2n} \|\chi_{S_k} - \chi_{S'_k}\|\\&\leq 2\left(1 - \frac{1}{n} \sum_{k=n+1}^{2n} \frac{|S_{k-d}|}{|S_{k+d}|}\right)\\&\leq 2\left(1 - \left(\prod_{k=n+1}^{2n} \frac{|S_{k-d}|}{|S_{k+d}|}\right)^{1/n}\right)\\&= 2\left(1 - \left(\frac{\prod_{k=n+1-d}^{n+d} |S_k|}{\prod_{k=2n+1-d}^{2n+d} |S_k|}\right)^{1/n}\right)\\&\leq 2\left(1 - (D(2n+d))^{-2d/n}\right).\end{aligned}$$

Since $(D(2n+d))^{-2d/n} \to 1$ as $n \to \infty$, this proves the claim.

Now we fix a base point $o \in K$, set $\zeta_n^z = \eta_n(o,z)$ and observe that the maps $\zeta_n \colon \partial K \to \mathrm{Prob}(K)$ are Borel. Since we have $(s.\eta_n)(x,z) = \eta_n(s.x, s.z)$ for every $s \in \Gamma$ and $(x,z) \in K \times \partial K$, it follows that

$$\lim_{n\to\infty} \sup_{z\in\partial K} \|s.\zeta_z^n - \zeta_{s.z}^n\| = \lim_{n\to\infty} \sup_{z\in\partial K} \|\eta_n(s.o, s.z) - \eta(o, s.z)\| = 0.$$

Finally, for $x \in K$ we set $\zeta_n^x = \delta_x$, and one can check that the Borel maps ζ_n satisfy the hypotheses of Proposition 5.2.1 – hence the action of Γ on $\bar{K}$ is amenable. □

We close this section with a few results which will be extremely important for later applications to von Neumann algebras (cf. Chapter 15).

Definition 5.3.16. A *compactification* of a group Γ is a compact topological space $\bar{\Gamma} = \Gamma \cup \partial\Gamma$ containing Γ as an open dense subset. We assume that a compactification is (left) *equivariant* in the sense that the left translation action of Γ on Γ extends to a continuous action on $\bar{\Gamma}$. The compactification $\bar{\Gamma}$ is said to be *small at infinity* if for every net $\{s_n\} \subset \Gamma$ converging to a boundary point $x \in \partial\Gamma$ and every $t \in \Gamma$, one has that $s_n t \to x$.

By Gelfand duality, there is a one-to-one correspondence between compactifications $\bar{\Gamma}$ and C*-algebras $C(\bar{\Gamma})$, where $c_0(\Gamma) \subset C(\bar{\Gamma}) \subset \ell^\infty(\Gamma)$ is left-translation invariant. The proof of the following lemma is a good exercise.

Lemma 5.3.17. *Let Γ be a group and $\bar{\Gamma} = \Gamma \cup \partial\Gamma$ be a compactification. The following are equivalent:*

(1) *the compactification $\bar{\Gamma}$ is small at infinity;*

(2) *the right translation action extends to a continuous action on $\bar{\Gamma}$ in such a way that it is trivial on $\partial\Gamma$;*

(3) *one has $f^t - f \in c_0(\Gamma)$ for every $f \in C(\bar{\Gamma})$ and $t \in \Gamma$, where $f^t(s) = f(st^{-1})$ for $f \in \ell^\infty(\Gamma)$.*

Proposition 5.3.18. *For any hyperbolic group Γ, the Gromov compactification $\bar{\Gamma}$ is small at infinity.*

Proof. Let a sequence $\{s_n\}$ converging to a boundary point $x \in \partial\Gamma$ and $t \in \Gamma$ be given. Let β be a geodesic path converging to x. Since $d(s_n t, s_n) = d(t,e)$ for every n, we have

$$\liminf_{m,n\to\infty} \langle s_m t, \beta(n)\rangle_e \geq \liminf_{m,n\to\infty} \langle s_m, \beta(n)\rangle_e - d(t,e) = \infty.$$

This means that $s_n t \to x$. □

We've seen that the left-translation action of a hyperbolic group Γ on $\ell^\infty(\Gamma)$ is amenable, but much more is true: the action of $\Gamma \times \Gamma$ on $\ell^\infty(\Gamma)$ (given by the left and right translations) is amenable mod $c_0(\Gamma)$.

Corollary 5.3.19. *If Γ is hyperbolic, then $\Gamma \times \Gamma$ acts amenably on the quotient algebra $\ell^\infty(\Gamma)/c_0(\Gamma)$.*

Proof. The previous proposition ensures that we can find a $(\Gamma\times\Gamma)$-invariant subalgebra $A \subset \ell^\infty(\Gamma)/c_0(\Gamma)$ such that the restriction of the $\Gamma \times \Gamma$ action to A is amenable on $\Gamma \times \{e\}$ and trivial on $\{e\} \times \Gamma$ (just let A be the image of $C(\bar{\Gamma})$ under the quotient map). By symmetry, we can also find $B \subset \ell^\infty(\Gamma)/c_0(\Gamma)$ such that the restriction of the $\Gamma\times\Gamma$ action to B is trivial on $\Gamma \times \{e\}$ and amenable on $\{e\} \times \Gamma$. The result now follows from Exercise 4.3.1. □

For free groups, the following fact was discovered by Akemann and Ostrand in [**2**].

Corollary 5.3.20. *Let Γ be hyperbolic, λ and ρ be the left and, respectively, right regular representations and $\pi\colon \mathbb{B}(\ell^2(\Gamma)) \to \mathbb{B}(\ell^2(\Gamma))/\mathbb{K}(\ell^2(\Gamma))$ be the quotient map. Then, the $*$-homomorphism*

$$C^*_\lambda(\Gamma) \odot C^*_\rho(\Gamma) \ni \sum_k a_k \otimes x_k \mapsto \pi(\sum_k a_k x_k) \in \mathbb{B}(\ell^2(\Gamma))/\mathbb{K}(\ell^2(\Gamma))$$

is continuous with respect to the minimal tensor norm.

Proof. This is in fact an immediate corollary of Corollary 5.3.19 and Theorem 4.3.4, but we give a different proof here. It suffices to show that there exists a nuclear C*-algebra $A \subset \mathbb{B}(\ell^2(\Gamma))$ such that $C^*_\lambda(\Gamma) \subset A$ and $\pi(A)$ commutes with $\pi(C^*_\rho(\Gamma))$. Indeed, if such A exists, then we have an inclusion $C^*_\lambda(\Gamma)\otimes C^*_\rho(\Gamma) \subset A\otimes C^*_\rho(\Gamma) = A\otimes_{\max} C^*_\rho(\Gamma)$ and a natural $*$-homomorphism $A \otimes_{\max} C^*_\rho(\Gamma) \to \mathbb{B}(\ell^2(\Gamma))/\mathbb{K}(\ell^2(\Gamma))$.

So, let $\bar{\Gamma} = \Gamma \cup \partial\Gamma$ be the Gromov compactification and embed $C(\bar{\Gamma}) \subset \ell^\infty(\Gamma)$ as above. By Theorems 5.3.15 and 4.4.3, the C*-subalgebra A of the uniform Roe algebra generated by $C(\bar{\Gamma})$ and $C^*_\lambda(\Gamma)$ is nuclear. (It is $*$-isomorphic to $C(\bar{\Gamma}) \rtimes_r \Gamma$ by Proposition 5.1.3.) By Proposition 5.3.18, we have

$$\rho_t^* f \rho_t - f = f^t - f \in c_0(\Gamma) \subset \mathbb{K}(\ell^2(\Gamma))$$

for any $f \in C(\bar{\Gamma})$ and any $t \in \Gamma$, which implies that $\pi(A)$ commutes with $\pi(C^*_\rho(\Gamma))$, as desired. □

5.4. Subgroups of Lie groups

Discrete subgroups of Lie groups provide lots of examples of amenable actions (hence exact groups). For example, let Γ be a discrete subgroup in the special linear group $\mathrm{SL}(n, \mathbb{R})$ (e.g., $\mathrm{SL}(n, \mathbb{Z})$) and let $P \subset \mathrm{SL}(n, \mathbb{R})$ be the closed subgroup of upper triangular matrices with positive diagonals. It turns out the left multiplication action of Γ on the compact homogeneous space $\mathrm{SL}(n, \mathbb{R})/P$ is amenable. (Note that $\mathrm{SL}(n, \mathbb{R})/P$ is compact since it is homeomorphic to $\mathrm{SO}(n, \mathbb{R})$.)

More generally, let G be a real semisimple Lie group, or any other (second countable) locally compact group which admits an Iwasawa decomposition $G = KAN$ into closed subgroups with K compact, A abelian and N nilpotent, such that A normalizes N. (When $G = SL(n, \mathbb{R})$, one can take $K = \mathrm{SO}(n, \mathbb{R})$, A to be the diagonal matrices with positive entries and determinant 1, and N to be the upper triangular matrices with 1's on the diagonal.) The closed subgroup $P = AN$ is solvable and hence is amenable (as a locally compact group[13]). Then, as we will prove below, any discrete subgroup $\Gamma \leq G$ (i.e., Γ is discrete in the relative topology) acts amenably on the compact homogeneous space $X = G/P$; this result is essentially due to Connes.

Theorem 5.4.1. *Let G be a second countable locally compact group, $\Gamma \leq G$ be a discrete subgroup and $P \subset G$ be a closed amenable subgroup such that $X = G/P$ is compact. Then, the left multiplication action of Γ on X is amenable.*

Proof. It is well known ([**10**, Theorem 3.4.1] applied to a compact subset of G which surjects onto X) that there exists a regular Borel cross section $\sigma\colon X \to G$; that is, σ is a Borel map such that $\sigma(x)P = x$ for all $x \in X$ and $\sigma(X)$ is pre-compact. Likewise, there is a Borel fundamental domain $Y \subset G$ for the Γ action - i.e., Y is Borel and G decomposes into the disjoint union $G = \bigsqcup_{s\in\Gamma} sY$. It follows that

$$\mathrm{Prob}(G) \ni \tilde{\mu} \mapsto (\tilde{\mu}(sY))_{s\in\Gamma} \in \mathrm{Prob}(\Gamma)$$

is a Γ-equivariant continuous map. Now, let a finite subset $E \subset \Gamma$ and $\varepsilon > 0$ be given. Then,

$$\tilde{E} = \{\sigma(sx)^{-1}s\sigma(x) : s \in E,\ x \in X\}$$

is a pre-compact subset in P. Since P is amenable, there exists $\nu \in \mathrm{Prob}(P)$ such that $\|g.\nu - \nu\| < \varepsilon$ for all $g \in \tilde{E}$. We regard ν as a measure on G and

[13] For information on amenability of locally compact groups, see [**139**].

define a Borel map $\tilde{\mu}\colon X \to \mathrm{Prob}(G)$ by $\tilde{\mu}_x = \sigma(x)\nu$. One checks that

$$\max_{s\in E}\sup_{x\in X}\|s\tilde{\mu}_x - \tilde{\mu}_{sx}\| = \max_{s\in E}\sup_{x\in X}\|\sigma(sx)^{-1}s\sigma(x)\nu - \nu\| < \varepsilon.$$

Composing $\tilde{\mu}$ with $\mathrm{Prob}(G) \to \mathrm{Prob}(\Gamma)$, we are done. □

5.5. Coarse metric spaces

Let (X,d) be a metric space. The associated topology on X reflects the "small scale structure" of the metric space – i.e., what X looks like under a microscope. In this section we take a step back...way back. Coarse geometry is the study of "large scale structure" of a (metric) space. For simplicity, we confine ourselves to countable discrete spaces X equipped with a proper metric d. (Recall that a metric d is said to be proper if any closed bounded subset of X is compact, i.e., finite because X is a discrete space.) We first consider a countable group Γ equipped with a proper metric d which is right invariant, meaning $d(s,t)$ depends only on st^{-1}. We will show that such a metric is intrinsic to Γ, but first we need a suitable notion of equivalence.

Definition 5.5.1. Let (X, d_X) and (Y, d_Y) be metric spaces. We say a map $f\colon X \to Y$ is *coarse* if the inverse image under f of any bounded subset in Y is bounded in X and if for any $R > 0$, there exists $S > 0$ such that $d_X(x,x') < R$ implies $d_Y(f(x), f(x')) < S$ for any $x, x' \in X$.

Let $f'\colon X \to Y$ be another map. We say f is *close* to f' if $d_Y(f(x), f'(x))$ is bounded on X. (You may want to verify that if f is close to a coarse map, then f is also coarse.) A map f is a *coarse isomorphism* if there exists a coarse map $g\colon Y \to X$ such that $g\circ f$ and $f\circ g$ are close to the identity maps on X and Y, respectively. Metric spaces (X, d_X) and (Y, d_Y) are *coarsely isomorphic* if there exists a coarse isomorphism $f\colon X \to Y$. Finally, we say metrics d and d' on X are *coarsely equivalent* if the identity map from (X,d) to (X,d') is a coarse isomorphism.

We loosely refer to a space X as a *coarse space* when a distinguished coarse equivalence class of metrics on X is chosen. The following result says that the coarse space structure of a countable discrete group Γ is unique.

Proposition 5.5.2. *Let Γ be a countable discrete group. Then, there exists a proper right invariant metric d on Γ. Moreover, such a metric d is unique up to coarse equivalence.*

Proof. Since Γ is countable, there exists a sequence

$$\{e\} = E_0 \subset E_1 \subset E_2 \subset \cdots \subset \Gamma$$

of finite symmetric subsets of Γ such that $\bigcup E_n = \Gamma$ and $E_n E_m \subset E_{n+m}$ for every n, m. It is not hard to see that the function d on $\Gamma \times \Gamma$ defined by

$$d(s,t) = \min\{n : st^{-1} \in E_n\}$$

is a proper right invariant metric on Γ. The second assertion is a nice exercise. □

Let $B_R(x) = \{y \in X : d(y,x) < R\}$ be the ball with center x and radius R. Recall that a (discrete) metric space (X,d) has *bounded geometry* if $\sup_{x\in X} |B_R(x)| < \infty$ for every $R > 0$.

Definition 5.5.3. Let (X,d) be a metric space with bounded geometry. For every $S > 0$, we set $T_S(X) = \{(x,y) \in X \times X : d(x,y) < S\}$. An operator $a \in \mathbb{B}(\ell^2(X))$ is said to have *finite propagation* if the kernel of a is supported on $T_S(X)$ for some S (i.e., $\langle a\delta_y, \delta_x\rangle \neq 0$ only if $d(x,y) < S$). The *translation algebra* $A(X)$ is the $*$-algebra of all operators with finite propagation. The closure of the translation algebra $A(X)$ in $\mathbb{B}(\ell^2(X))$ is called the *uniform Roe algebra* and is denoted by $C^*_u(X)$. Clearly, another metric d' on X which is coarsely equivalent to d gives rise to the same algebras $A(X)$ and $C^*_u(X)$.

Remark 5.5.4. If Γ is a countable group, $C^*_u(\Gamma)$ is easily seen to be the same as the Roe algebra defined in Section 5.1 (which Proposition 5.1.3 identifies with $\ell^\infty(\Gamma) \rtimes_r \Gamma$).

There is a notion of amenability for a coarse space defined via Følner sets, and a coarse space X with bounded geometry is amenable if and only if $C^*_u(X)$ has a tracial state. However, since every coarse metric space is embeddable into an amenable coarse metric space, amenability does not say much about the total structure of the space (unless the space is "coarsely homogeneous"). But, there is an important invariant, defined by Yu, which does pass to subspaces.

Definition 5.5.5 (Property A). We say a metric space (X,d) has *property* A if for any $R > 0$ and $\varepsilon > 0$, there exist a Hilbert space $\mathcal{H}$, a map $\xi\colon X \to \mathcal{H}$ and a number S such that

(1) $\|\xi_x\| = 1$ for every $x \in X$,
(2) if $d(x,y) < R$, then $\|\xi_x - \xi_y\| < \varepsilon$,
(3) and if $d(x,y) \geq S$, then $\langle \xi_y, \xi_x\rangle = 0$.

Just as for groups, we say a bounded kernel $k\colon X \times X \to \mathbb{C}$ is *positive definite* if the matrix $[k(x,y)]_{x,y\in\mathfrak{F}}$ is positive for every finite set $\mathfrak{F} \subset X$. The following lemma is very similar to our previous work on groups.

Lemma 5.5.6. *Let (X,d) be a discrete metric space with bounded geometry. The following are equivalent:*

(1) *X has property* A;

(2) *for any $R > 0$ and $\varepsilon > 0$, there exists a positive definite kernel k on X such that $k(x,x) = 1$ for every $x \in X$, $|1 - k(x,y)| < \varepsilon$ if $(x,y) \in T_R(X)$, and* $\operatorname{supp} k \subset T_S(X)$ *for some $S > 0$;*

(3) *for any $R > 0$ and $\varepsilon > 0$, there exist a map $\zeta\colon X \to \ell^2(X)$ and $S > 0$ such that $\|\zeta_x\| = 1$ for all x, $\|\zeta_x - \zeta_y\| < \varepsilon$ for $(x,y) \in T_R(X)$, and* $\operatorname{supp} \zeta_x \subset B_S(x)$ *for every x.*

Proof. (1) $\Rightarrow$ (2): Set $k(x,y) = \langle \xi_y, \xi_x \rangle$.

(2) $\Rightarrow$ (3): Since k has finite propagation, we may regard k as a positive bounded operator a_k on $\ell^2(X)$ which belongs to the translation algebra $A(X)$. Since the operator $a_k^{1/2}$ is in $C_u^*(X)$, the closure of $A(X)$, there is an operator $b \in A(X)$ such that $\|a_k - b^*b\| < \varepsilon$. Let $\eta_x = b\delta_x$, where $\{\delta_x\}_{x\in X}$ is the canonical orthonormal basis of $\ell^2(X)$. Since b has finite propagation, there exists $S > 0$ such that $\operatorname{supp} \eta_x \subset B_S(x)$ for every x. Moreover, we have

$$\langle \eta_y, \eta_x \rangle = \langle b^*b\delta_y, \delta_x \rangle \approx_\varepsilon \langle a_k\delta_y, \delta_x \rangle = k(x,y).$$

Hence, defining $\zeta\colon X \to \ell^2(X)$ by $\zeta_x = \|\eta_x\|^{-1}\eta_x$, we see that $\langle \zeta_y, \zeta_x \rangle \approx \langle \eta_y, \eta_x \rangle$ and thus ζ satisfies condition (3), modulo a change of ε (since $\|\eta_x\| \approx 1$, for all $x \in X$).

(3) $\Rightarrow$ (1) is obvious. □

Theorem 5.5.7. *Let Γ be a countable discrete group. Then Γ has property* A *if and only if Γ is exact. More generally, for a discrete metric space X with bounded geometry, property* A *is equivalent to the uniform Roe algebra $C_u^*(X)$ being nuclear.*

Proof. The first part follows from the previous lemma and Theorem 5.1.6.

For the second part, first suppose that X has property A. To show $C_u^*(X)$ is nuclear, it suffices to show that the identity map on $C_u^*(X)$ is approximated by u.c.p. maps which factor through nuclear C*-algebras. Let a finite subset $\mathfrak{F} \subset A(X)$ and $\varepsilon > 0$ be given. Choose some large $R > 0$ such that the support of a is contained in $T_R(X)$ for every $a \in \mathfrak{F}$. By assumption, there exist $\zeta\colon X \to \ell^2(X)$ and $S > 0$ satisfying condition (3) in Lemma 5.5.6 for $R > 0$ and $(\max_{a\in\mathfrak{F}} \|a\| \sup_x |B_R(x)|)^{-1}\varepsilon$. We define a u.c.p. map $\varphi\colon C_u^*(X) \to \prod_{x\in X} \mathbb{B}(\ell^2(B_S(x)))$ by

$$\varphi(a) = (\varphi_x(a))_{x\in X},$$

where $\varphi_x\colon \mathbb{B}(\ell^2(X)) \to \mathbb{B}(\ell^2(B_S(x)))$ is the standard compression map. Recall that there is an identification $\prod_{n\in\mathbb{N}} \mathbb{M}_k(\mathbb{C}) \cong \ell^\infty(\mathbb{N}) \otimes \mathbb{M}_k(\mathbb{C})$. Since X has bounded geometry, there is a uniform bound on the dimensions of the Hilbert spaces $\ell^2(B_S(x))$; thus $\prod_{x\in X} \mathbb{B}(\ell^2(B_S(x)))$ is nuclear, being a finite direct sum of the nuclear C*-algebras of the form $\ell^\infty \otimes \mathbb{M}_k(\mathbb{C})$. For each $x \in X$, we define $V_x \in \mathbb{B}(\ell^2(X), \ell^2(B_S(x)))$ by $V_x\delta_y = \zeta_y(x)\delta_y$ for every $y \in X$. It is not hard to check that $\sum_{x\in X} V_x^* V_x = 1$ in the strong operator topology. Hence, we get a u.c.p. map $\psi\colon \prod_{x\in X} \mathbb{B}(\ell^2(B_S(x))) \to \mathbb{B}(\ell^2(X))$ by defining

$$\psi((b_x)_{x\in X}) = \sum_{x\in X} V_x^* b_x V_x,$$

where convergence is in the strong operator topology. Since the support of $V_x^* b V_x$ is contained in $T_{2S}(X)$ for any $x \in X$ and $b \in \mathbb{B}(\ell^2(B_S(x)))$, the range of ψ is actually contained in $A(X)$. Define a positive definite kernel by $k(x, y) = \langle \zeta_y, \zeta_x\rangle$ and denote by m_k the corresponding Schur multiplier – i.e., $m_k([a_{x,y}]_{x,y\in X}) = [k(x, y)a_{x,y}]_{x,y\in X}$ (cf. Appendix D). Then, for any $a \in C_u^*(X)$, one has

$$\begin{aligned}\langle \psi \circ \varphi(a)\delta_y, \delta_x\rangle &= \sum_{z\in X} \langle V_z^* a V_z \delta_y, \delta_x\rangle \\ &= \langle a\delta_y, \delta_x\rangle \sum_z \zeta_y(z)\bar{\zeta}_x(z) = \langle m_k(a)\delta_y, \delta_x\rangle\end{aligned}$$

and hence $\psi \circ \varphi = m_k$. Therefore, for $a \in \mathfrak{F}$, we have

$$\|(\psi \circ \varphi)(a) - a\| \le \|a\| \sup_{(x,y)\in T_R(X)} |1 - k(x, y)| \sup_x |B_R(x)| < \varepsilon.$$

This proves the nuclearity of $C_u^*(X)$.

Now assume that $C_u^*(X)$ is nuclear and let $R > 0$ and $\varepsilon > 0$ be given. Since X has bounded geometry, one can find a finite set $\mathfrak{F}$ of partial isometries in $A(X)$ with the property that for every $(x, y) \in T_R(X)$, there exists $v \in \mathfrak{F}$ such that $v\delta_x = \delta_y$. Since $C_u^*(X)$ is nuclear, there exist u.c.p. maps $\varphi\colon C_u^*(X) \to \mathbb{M}_n(\mathbb{C})$ and $\psi\colon \mathbb{M}_n(\mathbb{C}) \to C_u^*(X)$ such that $\|(\psi\circ\varphi)(v) - v\| < \varepsilon$ for every $v \in \mathfrak{F}$. We set $\mathcal{H} = \ell_n^2 \otimes \ell_n^2$ and consider the Hilbert $C_u^*(X)$-module $\mathcal{H} \otimes C_u^*(X)$. (For the rest of this proof, inner products are linear in the second variable.) Note that $\mathbb{M}_n(\mathbb{C}) \otimes \mathbb{M}_n(\mathbb{C})$ naturally acts on $\mathcal{H} \otimes C_u^*(X)$ from the left. Let $\{e_{ij}\}$ be matrix units for $\mathbb{M}_n(\mathbb{C})$ and observe that $[\psi(e_{ij})]$ is positive in $\mathbb{M}_n(C_u^*(X))$, by complete positivity of ψ (see Proposition 1.5.12). Let $[b_{ij}] = [\psi(e_{ij})]^{1/2} \in \mathbb{M}_n(C_u^*(X))$ and

$$\xi_\psi = \sum_{j,k} \zeta_j \otimes \zeta_k \otimes b_{kj} \in \mathcal{H} \otimes C_u^*(X),$$

where ζ_j denotes the standard basis of ℓ_n^2. It is routine to check that $\psi(\alpha) = \langle \xi_\psi, (\alpha \otimes 1)\xi_\psi\rangle$ for every $\alpha \in \mathbb{M}_n(\mathbb{C})$. Perturbing ξ_ψ, we may assume that $\xi_\psi \in \mathcal{H} \otimes A(X)$. We write $\xi_\psi = \sum_l \xi_l \otimes a_l$ and set

$$\zeta_x(z) = \|\sum_l \xi_l \langle \delta_z, a_l\delta_x\rangle_{\ell^2(X)}\|_{\mathcal{H}}$$

for $x, z \in X$. Then, for some $S > 0$, we have $\operatorname{supp} \zeta_x \subset B_S(x)$ for all $x \in X$. Moreover, $\|\zeta_x\| = 1$ for all $x \in X$. Indeed,

$$\begin{aligned}
\|\zeta_x\|^2 &= \sum_z \|\sum_l \xi_l \langle \delta_z, a_l \delta_x \rangle \|_{\mathcal{H}}^2 \\
&= \sum_{l,m} \sum_z \langle \xi_l, \xi_m \rangle_{\mathcal{H}} \overline{\langle \delta_z, a_l \delta_x \rangle_{\ell^2(X)}} \langle \delta_z, a_m \delta_x \rangle_{\ell^2(X)} \\
&= \sum_{l,m} \langle \xi_l, \xi_m \rangle_{\mathcal{H}} \langle a_l \delta_x, a_m \delta_x \rangle_{\ell^2(X)} \\
&= \langle \delta_x, \langle \xi_\psi, \xi_\psi \rangle_{\mathcal{H} \otimes C_u^*(X)} \delta_x \rangle_{\ell^2(X)} = 1.
\end{aligned}$$

Finally, let $(x, y) \in T_R(X)$ and choose $v \in \mathfrak{F}$ so that $v\delta_x = \delta_y$. Since $\varphi(v)$ is a contraction in $\mathbb{M}_n(\mathbb{C})$, we have

$$\begin{aligned}
\langle \zeta_y, \zeta_x \rangle_{\ell^2(X)} &= \sum_z \|\sum_l \xi_l \langle \delta_z, a_l \delta_y \rangle \|_{\mathcal{H}} \|\sum_m \xi_m \langle \delta_z, a_m \delta_x \rangle \|_{\mathcal{H}} \\
&\geq \sum_z \|\sum_l \xi_l \langle \delta_z, a_l \delta_y \rangle \|_{\mathcal{H}} \|(\varphi(v) \otimes 1) \sum_m \xi_m \langle \delta_z, a_m \delta_x \rangle \|_{\mathcal{H}} \\
&\geq \Big|\sum_{l,m} \sum_z \langle \xi_l, (\varphi(v) \otimes 1) \xi_m \rangle_{\mathcal{H}} \overline{\langle \delta_z, a_l \delta_y \rangle_{\ell^2(X)}} \langle \delta_z, a_m \delta_x \rangle_{\ell^2(X)} \Big| \\
&= \big| \langle \delta_y, \langle \xi_\psi, (\varphi(v) \otimes 1) \xi_\psi \rangle_{\mathcal{H} \otimes C_u^*(X)} \delta_x \rangle_{\ell^2(X)} \big| \\
&= \big| \langle \delta_y, (\psi \circ \varphi)(v) \delta_x \rangle_{\ell^2(X)} \big| \approx 1.
\end{aligned}$$

This implies $\|\zeta_y - \zeta_x\|$ is small and hence X has property A. □

We finish off this section with two important facts. First, property A implies uniform embeddability into Hilbert space – this has important consequences for the Baum-Connes conjecture. The second follows from the first: there are coarse spaces which don't have property A.

Lemma 5.5.8. *A metric space X with property* A *is coarsely isomorphic to a subset of a Hilbert space.*

Proof. For each n, we find $\xi^n \colon X \to \mathcal{H}_n$ and numbers S_n satisfying Definition 5.5.5 for $R = n$ and $\varepsilon = 2^{-n}$. Define $\mathcal{H} = \bigoplus \mathcal{H}_n$ and $f \colon X \to \mathcal{H}$ by

$$f(x) = \bigoplus (\xi_x^n - \xi_o^n) \in \mathcal{H},$$

where $o \in X$ is a fixed base point. It follows that $f(o) = 0$ and

$$\|f(x) - f(y)\| \leq 2d(x, y) + \sum_{n \geq d(x,y)} \|\xi_x^k - \xi_y^k\| \leq 2d(x, y) + 1.$$

In particular, we have $\|f(x)\| < \infty$. On the other hand, if $d(x, y) \geq \max_{k \leq n} S_k$, then

$$\|f(x) - f(y)\| \geq \|\bigoplus_{k=1}^{n}(\xi_x^k - \xi_y^k)\| = \sqrt{2n}.$$

This implies that f is a coarse embedding. □

Recall from Appendix E that a *sequence of expanders* is a sequence $\{\mathbf{X}_n\}$ of finite connected d-regular graphs (with d fixed) such that $\inf \lambda_1(\mathbf{X}_n) > 0$ (the first nonzero eigenvalues of the Laplacians) and $\sup |\mathbf{X}_n| = \infty$. Here, we identify the graph $\mathbf{X}_n$ with the finite metric space of its vertices. We can glue the $\mathbf{X}_n$'s together to form a graph metric space $\mathbf{X}$, with bounded geometry, which contains the $\mathbf{X}_n$'s isometrically: choose a base point $x_n \in \mathbf{X}_n$, for every n, and put an edge between x_n and x_{n+1}. We note that the degrees of the vertices of $\mathbf{X}$ are at most $d+2$ and every ball of radius S has at most $(d+2)^S$ elements. A map g from a graph metric space X into any metric space Y is 1*-Lipschitz* if $d_Y(g(x), g(y)) \leq 1$ for every adjacent pair x and y in X.

Proposition 5.5.9. *Let X be a metric space with bounded geometry. Assume that there exists a sequence $\mathbf{X}_n$ of expanders with* 1*-Lipschitz maps $g_n \colon \mathbf{X}_n \to X$ such that*

$$\lim_{n\to\infty} \sup_{x\in X} \frac{|g_n^{-1}(B_S(x))|}{|\mathbf{X}_n|} = 0$$

for every $S > 0$. Then X is not coarsely embeddable into a Hilbert space. In particular, X does not have property A.

Proof. Let $\lambda_1 = \inf \lambda_1(\mathbf{X}_n) > 0$. Suppose that there exists a coarse embedding f of X into a Hilbert space $\mathcal{H}$. We have

$$C = \lambda_1^{-1/2} \sup_{d_X(x,y)\leq 1} \|f(x) - f(y)\| < \infty.$$

Since f is a coarse embedding, there exists $S > 0$ such that $\|f(x) - f(y)\| \leq 2C$ implies $d_X(x, y) < S$. Now, let n be fixed. Applying Lemma E.5 to $f_n = f \circ g_n \colon \mathbf{X}_n \to \mathcal{H}$, we have

$$\frac{1}{|\mathbf{X}_n|^2} \sum_{a,b\in\mathbf{X}_n} \|f_n(a) - f_n(b)\|^2 \leq \frac{1}{|\mathbf{E}(\mathbf{X}_n)|} \sum_{(a,b)\in\mathbf{E}(\mathbf{X}_n)} \frac{\|f_n(a) - f_n(b)\|^2}{\lambda_1} \leq C^2.$$

This implies the existence of a point $a \in \mathbf{X}_n$ such that $|\mathbf{X}_n|^{-1} \sum_{b\in\mathbf{X}_n} \|f_n(a) - f_n(b)\|^2 \leq C^2$. This further implies that the cardinality of the set $\{b \in \mathbf{X}_n : \|f_n(a) - f_n(b)\| \leq 2C\}$ is greater than $3|\mathbf{X}_n|/4$. It follows that $|g_n^{-1}(B_S(g_n(a)))| \geq 3|\mathbf{X}_n|/4$. Since n was arbitrary, this provides the desired contradiction. □

Remark 5.5.10. Gromov's infamous construction of nonexact groups yields examples whose Cayley graphs satisfy the hypotheses of the previous result – thus they aren't coarsely embeddable into Hilbert space and consequently can't be exact (cf. [**71**]).

Exercise

Exercise 5.5.1. Formulate and prove an analogue of Lemma 5.5.6 for spaces which are coarsely embeddable into Hilbert space. (Hint: The "supported in tubes" condition should be replaced with a "c_0-off the diagonal" condition.)

5.6. Groupoids

Groupoids (and their associated C*-algebras) generalize almost everything we have discussed so far: topological spaces, groups, group actions on spaces, coarse metric spaces, graphs, and many other things. Thus they provide a unifying framework for everything in this chapter.

To a groupoid G one can associate a reduced groupoid C*-algebra $C^*_\lambda(G)$. There is a natural notion of amenability for groupoids and the main result of this section states that $C^*_\lambda(G)$ is nuclear if and only if G is amenable.

Definition 5.6.1. A *groupoid* is a small category, in which every morphism is invertible. More specifically, a groupoid consists of a set G of morphisms and a distinguished subset $G^{(0)} \subset G$ of objects (often called *units*), together with source and range maps $s, r\colon G \to G^{(0)}$, and a composition map

$$G^{(2)} = \{(\alpha, \beta) \in G \times G : s(\alpha) = r(\beta)\} \ni (\alpha, \beta) \mapsto \alpha\beta \in G,$$

such that

(1) $s(\alpha\beta) = s(\beta)$ and $r(\alpha\beta) = r(\alpha)$ for every $(\alpha, \beta) \in G^{(2)}$,
(2) $s(x) = x = r(x)$ for every $x \in G^{(0)}$,
(3) $\gamma s(\gamma) = \gamma = r(\gamma)\gamma$ for every $\gamma \in G$,
(4) $(\alpha\beta)\gamma = \alpha(\beta\gamma)$, and
(5) every γ has an inverse γ^{-1}, with $\gamma\gamma^{-1} = r(\gamma)$ and $\gamma^{-1}\gamma = s(\gamma)$.

A topological groupoid G is a groupoid together with a topology on G such that all structure maps are continuous. We note that $G^{(2)}$ inherits the relative topology from $G \times G$.

A topological groupoid G is said to be *étale* (or *r-discrete*) if s and r are local homeomorphisms.[14]

We only consider (étale) locally compact groupoids and always assume that the Hausdorff property is included in the definition of local compactness

[14] Actually, if s is locally homeomorphic, then so is r.

(although there are important examples of non-Hausdorff locally compact groupoids). By the Hausdorff property, $G^{(0)}$ is closed.

Lemma 5.6.2. *Let G be an étale groupoid. Then, $G^{(0)} \subset G$ is a clopen subset.*

Proof. To show that $G^{(0)}$ is open in G, let $x \in G^{(0)}$ be given and take an open neighborhood U of x in G such that s is homeomorphic on U. Since $G^{(0)} \cap U$ is an open subset of $G^{(0)}$, $V = U \cap s^{-1}(G^{(0)} \cap U)$ is an open neighborhood of x in G. We claim that $V \subset G^{(0)}$. Indeed, for every $\gamma \in V$, we have $s(\gamma) \in U$. Since s is injective on U, $s(\gamma) = s(s(\gamma))$ implies that $\gamma = s(\gamma)$. □

Example 5.6.3 (Groups, spaces and actions on spaces)**.** Let Γ be a group. Then $G^{(0)} = \{e\} \subset G = \Gamma$ is a groupoid. Let X be a locally compact space. Then, $G^{(0)} = G = X$ is a groupoid. Now, let Γ act on X. Then,

$$X \rtimes \Gamma = \{(x, s, y) \in X \times \Gamma \times X : x = s.y\}$$

is a groupoid with $s(x, s, y) = (y, e, y)$, $r(x, s, y) = (x, e, x)$ and, finally, $(x, s, y)(y, t, z) = (x, st, z)$. The third component is often omitted because y is uniquely determined by (x, s). The topology on $X \rtimes \Gamma$ is nothing but that coming from $X \times \Gamma$, where Γ is viewed as a discrete space. With this topology, $X \rtimes \Gamma$ is an étale locally compact groupoid, which is called a *transformation groupoid.* The groupoid $X \rtimes \Gamma$ is a special case of the groupoid semidirect product.

Definition 5.6.4. A *partial homeomorphism* on X is a homeomorphism $f\colon U \to V$ between open subsets of X.[15] We denote by $\mathrm{dom}(f) = U$ the domain of f. For partial homeomorphisms $f\colon U \to V$ and $g\colon V' \to W$, we simply write $g \circ f$ for $(g|_{V\cap V'}) \circ (f|_{f^{-1}(V')})$. We note that $f|_{\mathrm{dom}(f)\cap U} = f \circ \mathrm{id}_U$. A *pseudogroup* $\mathcal{G}$ on X is a collection of partial homeomorphisms with the following properties:

(1) (units) the identity id_U on every open subset $U \subset X$ is in $\mathcal{G}$;

(2) (composition) if $f, g \in \mathcal{G}$, then $g \circ f \in \mathcal{G}$;

(3) (inverse) if $f\colon U \to V$ is in $\mathcal{G}$, then $f^{-1}\colon V \to U$ is in $\mathcal{G}$.

Sometimes in the literature the following is also required:

(4) (extension) if f is a partial homeomorphism with an open covering $\{U_i\}_i$ of $\mathrm{dom}(f)$ such that $f \circ \mathrm{id}_{U_i} \in \mathcal{G}$, then $f \in \mathcal{G}$.

Example 5.6.5. If X is a smooth (or Riemannian, etc.) manifold, then the collection of all partial diffeomorphisms (or partial isometries, etc.) is

[15]We allow the *empty map* which is the unique partial homeomorphism between empty subsets.

a pseudogroup. If $\sim$ is an equivalence relation on X, then the collection of all partial homeomorphisms f with $f(x) \sim x$ for all $x \in \text{dom}(f)$ is a pseudogroup.

Example 5.6.6 (Groupoid of germs). Let $\mathcal{G}$ be a pseudogroup on a locally compact space X. For $f \in \mathcal{G}$ and $x \in \text{dom}(f)$, the *germ* f_x of f at x consists of all $f' \in \mathcal{G}$ which agree with f on some neighborhood of x. For $f, g \in \mathcal{G}$ and $x \in \text{dom}(f)$ with $f(x) \in \text{dom}(g)$, we define the composition $g_{f(x)} \circ f_x$ of germs in the obvious way. The *groupoid of germs* is the set G of all germs equipped with its natural groupoid structure. We introduce the germ topology on G by declaring that

$$O(f) = \{f_x \in G : x \in \text{dom}(f)\}$$

is open for every $f \in \mathcal{G}$. This topology *may not be Hausdorff* (and we ignore such examples in this book). One can check that a neighborhood basis of $f_x \in G$ is given by $\{O(f \circ \text{id}_U)\}_U$ for a neighborhood basis U of x and that G is an étale locally compact groupoid.

Example 5.6.7 (Holonomy groupoid of a foliation). A foliation $(M, \mathcal{F})$ of codimension q can be defined by a Haefliger cocycle $\{s_i \colon U_i \to \mathbb{R}^q\}$, where $\{U_i\}$ is an open covering of M and the s_i's are submersions from U_i onto open subsets of $\mathbb{R}^q$ such that there are (necessarily unique) diffeomorphisms $f_{ij} \colon s_j(U_i \cap U_j) \to s_i(U_i \cap U_j)$ with $f_{ij} \circ s_j = s_i$ for every i, j. We assume that the fibers of the submersions are connected. Two cocycles define the same foliation on M if their common refinement is a cocycle. (See [**125**] for more on this subject.)

Let $\{s_i \colon U_i \to \mathbb{R}^q\}$ be a Haefliger cocycle. We define X to be the disjoint union $\bigsqcup s_i(U_i)$ and $\mathcal{G}$ to be the pseudogroup on X generated by the partial homeomorphisms f_{ij}. The *holonomy groupoid* of the Haefliger cocycle is the groupoid G of germs of $\mathcal{G}$.

The holonomy groupoid G defined above depends on the choice of the Haefliger cocycle, but equivalent Haefliger cocycles define *Morita equivalent* groupoids. Hence the Morita equivalence class of the holonomy groupoids is an invariant of the foliation $(M, \mathcal{F})$. Before introducing the notion of Morita equivalence for étale groupoids, we need one more example.

Example 5.6.8. Let G be a groupoid and $\mathcal{U} = \{U_i : i \in I\}$ be an open covering of $G^{(0)} = \bigcup U_i$. (Recall that $G^{(0)}$ is clopen in G.) The *localization of G over $\mathcal{U}$* is the groupoid

$$G_{\mathcal{U}} = \{(i, \gamma, j) \in I \times G \times I : s(\gamma) \in U_j \text{ and } r(\gamma) \in U_i\}$$

with $G_{\mathcal{U}}^{(0)} = \{(i, x, i) : i \in I,\ x \in U_i\} \cong \bigsqcup_i U_i$ (disjoint union) and with $s(i, \gamma, j) = (j, s(\gamma), j)$, $r(i, \gamma, j) = (i, r(\gamma), i)$ and $(i, \alpha, j)(j, \beta, k) = (i, \alpha\beta, k)$.

The topology of $G_{\mathcal{U}}$ is the relative product topology, where I is discrete. The groupoid $G_{\mathcal{U}}$ is an étale locally compact groupoid if G is.

Definition 5.6.9. Two étale locally compact groupoids G and G' are said to be *Morita equivalent* if there are localizations $G_{\mathcal{U}}$ and $G'_{\mathcal{U}'}$, respectively, which are isomorphic as topological groupoids. (NB: This definition is adapted for étale groupoids.)

Example 5.6.10 (Groupoids and coarse spaces)**.** To a coarse space X, one can associate a *translation groupoid* $G(X)$ (roughly, a groupoid of partial bijections on X). The reduced C*-algebra of $G(X)$ turns out to be isomorphic to the uniform Roe algebra $C^*_u(X)$. See [**167**, Chapter 10] for the construction of $G(X)$ and the C*-isomorphism result.

Example 5.6.11 (Graph groupoids)**.** There is a way to associate a groupoid to a directed graph, but the construction is technical and authors grow tired; see [**112, 140**].

Now we turn to the construction of C*-algebras associated to groupoids. Let G be an étale locally compact groupoid. Thus,

$$G_x = \{\gamma \in G : s(\gamma) = x\} \qquad \text{and} \qquad G^x = \{\gamma \in G : r(\gamma) = x\}$$

are discrete in G for every $x \in G^{(0)}$. The groupoid algebra $C_c(G)$ is the $*$-algebra of all continuous compactly supported functions $f\colon G \to \mathbb{C}$ with composition and $*$-operation given by

$$(f*g)(\gamma) = \sum_{\alpha\beta=\gamma} f(\alpha)g(\beta) = \sum_{\beta\in G_{s(\gamma)}} f(\gamma\beta^{-1})g(\beta) \qquad \text{and} \qquad f^*(\gamma) = \overline{f(\gamma^{-1})}.$$

If $K \subset G$ is a compact subset, we can cover it with finitely many open subsets on which s is a homeomorphism, and hence we have $\sup_x |K \cap G_x| < \infty$. It follows that $f * g$ indeed belongs to $C_c(G)$. For the convenience of the reader, we note that

$$(f^* * g)(\gamma) = \sum_{\beta\in G_{r(\gamma)}} \overline{f(\beta)}g(\beta\gamma).$$

The groupoid algebra $C_c(G)$ is naturally a right $C_0(G^{(0)})$-module with right action and $C_0(G^{(0)})$-valued inner product defined by

$$(\xi \cdot f)(\gamma) = \xi(\gamma)f(s(\gamma)) \qquad \text{and} \qquad \langle \xi, \eta\rangle(x) = \sum_{\gamma\in G_x} \overline{\xi(\gamma)}\eta(\gamma)$$

for $\xi, \eta \in C_c(G)$ and $f \in C_0(G^{(0)})$. Observe that $\langle \xi, \eta \rangle \in C_c(G^{(0)})$ is the restriction of $\xi^* * \eta \in C_c(G)$ to $G^{(0)}$. Denote by $L^2(G)$ the Hilbert $C_0(G^{(0)})$-module[16] arising from the completion of $C_c(G)$. The *left regular representation* $\lambda\colon C_c(G) \to \mathbb{B}(L^2(G))$ is defined by

$$\lambda(f)\xi = f * \xi$$

for $f, \xi \in C_c(G)$. It is not hard to see that $\|\lambda(f)\| < \infty$ and $\lambda(f)^* = \lambda(f^*)$. The *reduced groupoid* C*-*algebra* $C^*_\lambda(G)$ is the norm closure of $\lambda(C_c(G)) \subset \mathbb{B}(L^2(G))$. Since $G^{(0)}$ is clopen in G, we have $C_c(G^{(0)}) \subset C_c(G)$ and the restriction map $C_c(G) \to C_c(G^{(0)})$ gives rise to a conditional expectation E from $C^*_\lambda(G)$ onto $C_0(G^{(0)}) \subset C^*_\lambda(G)$, which implements the isomorphism $L^2(G) \cong L^2(C^*_\lambda(G), E)$ (see Section 4.6). For every $x \in G^{(0)}$, we define a $*$-representation λ_x of $C_c(G)$ on $\ell^2(G_x)$ by

$$(\lambda_x(f)\xi)(\gamma) = \sum_{\beta \in G_x} f(\gamma\beta^{-1})\xi(\beta).$$

Lemma 5.6.12. *For $f \in C_c(G)$, let*

$$\|f\|_{I,s} = \sup_{x \in G^{(0)}} \sum_{\gamma \in G_x} |f(\gamma)| \quad \text{and} \quad \|f\|_I = \max\{\|f\|_{I,s}, \|f^*\|_{I,s}\}.$$

Then, for every $f \in C_c(G)$, we have

$$\|f\|_\infty \le \|\lambda(f)\| = \sup_{x \in G^{(0)}} \|\lambda_x(f)\| \le \|f\|_I.$$

Proof. We only prove $\sup_{x \in G^{(0)}} \|\lambda_x(f)\| \le \|f\|_I$; the rest is trivial. For $f \in C_c(G)$ and $\xi, \eta \in \ell^2(G_x)$, we have

$$\begin{aligned} |\langle \xi, \lambda_x(f)\eta \rangle|^2 &= \Big| \sum_{\alpha,\beta \in G_x} \overline{\xi(\alpha)} f(\alpha\beta^{-1})\eta(\beta)\Big|^2 \\ &\le \sum_{\alpha,\beta \in G_x} |\xi(\alpha)|^2 |f(\alpha\beta^{-1})| \sum_{\alpha,\beta \in G_x} |f(\alpha\beta^{-1})|\,|\eta(\beta)|^2 \\ &\le \|\xi\|^2 \|f^*\|_{I,s} \|f\|_{I,s} \|\eta\|^2. \end{aligned}$$

Hence $\|\lambda_x(f)\| \le \|f^*\|_{I,s}^{1/2} \|f\|_{I,s}^{1/2} \le \|f\|_I$. □

The *full groupoid* C*-*algebra* $C^*(G)$ of the étale locally compact groupoid G is defined as the norm-completion of $C_c(G)$ with respect to the norm

$$\|f\|_{C^*(G)} = \sup \|\pi(f)\|,$$

where the supremum is taken over all (cyclic) $*$-representations π of $C_c(G)$ on Hilbert spaces, which are bounded on the commutative $*$-subalgebra

[16] If one wants an ordinary Hilbert space, choose a regular Borel measure μ on $G^{(0)}$ with full support and set $\langle \xi, \eta \rangle_\mu = \int_{G^{(0)}} \langle \xi, \eta \rangle(x)\, d\mu(x)$. Then we obtain a Hilbert space $L^2(G, \mu) \cong L^2(G) \otimes_{C_0(G^{(0)})} L^2(G^{(0)}, \mu)$.

$C_c(G^{(0)})$. Let $U \subset G$ be an open subset on which both s and r are homeomorphisms. Then, for every $f \in C_c(G)$ with $\operatorname{supp} f \subset U$, we have $f^* * f \in C_c(G^{(0)})$ with $(f^* * f)(s(\gamma)) = |f(\gamma)|^2$ for $\gamma \in U$. It follows that $\|f\|_{C^*(G)} = \|f\|_\infty$ for such f. Thus, by a partition of unity argument, for every compact subset $K \subset G$, there exists a constant $C_K > 0$ such that $\|g\|_{C^*(G)} \le C_K \|g\|_\infty$ for every $g \in C(K) = \{g \in C_c(G) : \operatorname{supp} g \subset K\}$. In particular, $\|g\|_{C^*(G)} < \infty$ for any $g \in C_c(G)$.[17] In passing, we observe that $C(K)$ is closed in $C^*(G)$. (Do you see the point?)

Let $G = X \rtimes \Gamma$ be a transformation groupoid. We leave it to the reader to check that $C^*_\lambda(G) \cong C(X) \rtimes_r \Gamma$ and $C^*(G) \cong C(X) \rtimes \Gamma$. It is a good exercise to check that the action is amenable if and only if there exists a net of compactly supported nonnegative functions $\mu_i \colon G \to \mathbb{C}$ such that $\sum_t \mu_i(x,t) \to 1$ and $\sum_t |\mu_i(x,t) - \mu_i(sx, st)| \to 0$ for $(x,s) \in G$ uniformly on compact subsets of G. Noticing that $(sx, st) = (sx, s)(x,t)$ in G, we are led to the definition of amenability for general étale locally compact groupoids (though multiplication on the left gets moved to the right for a technical reason).

Definition 5.6.13. An étale locally compact groupoid G is said to be *amenable* if there exists a net of compactly supported nonnegative functions $\mu_i \colon G \to \mathbb{C}$ such that

$$\sum_{\beta \in G_{r(\gamma)}} \mu_i(\beta) \to 1 \qquad \text{and} \qquad \sum_{\beta \in G_{r(\gamma)}} |\mu_i(\beta) - \mu_i(\beta\gamma)| \to 0,$$

for $\gamma \in G$, uniformly on compact subsets of G.

As usual, we have an L^2-characterization of amenability.

Lemma 5.6.14. *An étale locally compact groupoid G is amenable if and only if there exists a net $\zeta_i \in C_c(G)$ such that $\|\zeta_i\|_{L^2(G)} \le 1$ and $(\zeta_i^* * \zeta_i)(\gamma) \to 1$, for $\gamma \in G$, uniformly on compact subsets of G.*

Proof. First we'll prove the "only if" direction. Let μ_i be as in the definition of amenability and set $f_i(x) = \max\{\langle \mu_i^{1/2}, \mu_i^{1/2}\rangle(x), 1\}$. Then the functions $\zeta_i(\gamma) = \mu_i(\gamma)^{1/2} f_i(s(\gamma))^{-1/2}$ satisfy the desired properties. For the "if" direction, define $\mu_i(\gamma) = |\zeta_i(\gamma)|^2$ and do a computation. □

The functions $\zeta_i^* * \zeta_i$ are examples of positive-type functions.

Definition 5.6.15. A function $h \colon G \to \mathbb{C}$ on an étale locally compact groupoid G is said to be of *positive type* if $[h(\alpha\beta^{-1})]_{\alpha,\beta\in\mathfrak{F}}$ is positive definite for every $x \in G^{(0)}$ and every finite subset $\mathfrak{F} \subset G_x$.

---

[17] We actually have $\|g\|_{C^*(G)} \le \|g\|_I$, where $\|g\|_I$ is defined in Lemma 5.6.12. See page 43 in [**140**].

We note that $\sup_{\gamma\in G}|h(\gamma)| = \sup_{x\in G^{(0)}} h(x)$ for a positive-type function h. Also, $h \in C_c(G)$ is of positive type if and only if $\lambda(h)$ is positive in $C^*_\lambda(G)$.

Proposition 5.6.16. *Let $h\colon G \to \mathbb{C}$ be a continuous positive-type function with $\sup|h(\gamma)| \le 1$. Then, the multiplier map*

$$m_h\colon C_c(G) \ni f \mapsto hf \in C_c(G)$$

extends to a c.c.p. map on $C^*_\lambda(G)$ and on $C^*(G)$.

Proof. We first deal with $C^*_\lambda(G)$. For every $x \in G^{(0)}$, let m_h^x be the Schur multiplier of $[h(\alpha\beta^{-1})]_{\alpha,\beta\in G_x}$ on $\mathbb{B}(\ell^2(G_x))$. Since h is of positive type, m_h^x is a c.c.p. map (Theorem D.3). For every $f \in C_c(G)$ and $x \in G^{(0)}$, we have $\lambda_x(hf) = m_h^x(\lambda_x(f))$. Since $\{\lambda_x\}_{x\in G^{(0)}}$ is a faithful family, we are done.

We give an *ad hoc* proof for $C^*(G)$ assuming that $h = \zeta^* * \zeta$ for some $\zeta \in C_c(G)$ with $\|\zeta\|_{L^2(G)} \le 1$. (This case is good enough for these notes. The general case requires representation theory of groupoids [**140**].) Take a partition of unity $\{p_i\} \subset C_c(G)$ such that the source map is homeomorphic on $\operatorname{supp} p_i$. We define $\zeta_i \in C_c(G^{(0)})$ by the formula $\zeta_i(s(\gamma)) = \zeta(\gamma)p_i(\gamma)$ so that $\zeta(\gamma) = \sum_i \zeta_i(s(\gamma))$. We may assume that all but finitely many ζ_i are identically zero. Observe that

$$h(\gamma) = (\zeta^* * \zeta)(\gamma) = \sum_{\beta\in G_{r(\gamma)}} \overline{\zeta(\beta)}\zeta(\beta\gamma) = \sum_{i,j} \overline{\zeta_i(r(\gamma))}\zeta_j(s(\gamma)).$$

An element in $\mathbb{M}_n(C_c(G))$ is called *algebraically positive* if it is a finite sum of matrices of the form $[g_p^* * g_q]_{p,q}$. (By Lemma 4.2.1, any positive element in $\mathbb{M}_n(C^*(G))$ is approximated by algebraically positive elements.) We claim that m_h is completely algebraically positive – i.e., $\mathrm{id} \otimes m_h\colon \mathbb{M}_n(C_c(G)) \to \mathbb{M}_n(C_c(G))$ maps algebraically positive elements to algebraically positive elements, for all $n \in \mathbb{N}$. Let $f \in C_c(G)$ be given and define $\tilde{f} \in C_c(G)$ by $\tilde{f} = \sum f_i$, where $f_i(\gamma) = f(\gamma)\zeta_i(s(\gamma))$. Then,

$$\begin{aligned}
(\tilde{f}^* * \tilde{f})(\gamma) &= \sum_{i,j} \sum_{\beta\in G_{r(\gamma)}} \overline{f_i(\beta)} f_j(\beta\gamma) \\
&= \sum_{i,j} \sum_{\beta\in G_{r(\gamma)}} \overline{f(\beta)\zeta_i(r(\gamma))} f(\beta\gamma)\zeta_j(s(\gamma)) \\
&= h(\gamma) \sum_{\beta\in G_{r(\gamma)}} \overline{f(\beta)} f(\beta\gamma) = m_h(f^* * f)(\gamma).
\end{aligned}$$

It follows that m_h is algebraically positive. The proof of complete algebraic positivity is similar. We observe that for any $f \in C_c(G^{(0)})$ with $\|f\|_\infty \le 1$ and any $g_1, \ldots, g_n \in C_c(G)$, the element

$$[g_p^* * g_q]_{p,q} - [g_p^* * f^* * f * g_q]_{p,q} = [\tilde{g}_p^* * \tilde{g}_q]_{p,q} \in \mathbb{M}_n(C_c(G))$$

is algebraically positive, where $\tilde{g}_p(\gamma) = (1 - |f(r(\gamma))|^2)^{1/2} g_p(\gamma)$. Take an element $e \in C_c(G^{(0)})$ such that $0 \le e \le 1$ and $e(s(\gamma)) = 1 = e(r(\gamma))$ for every $\gamma \in \operatorname{supp} h$. Then, $m_h(e^* * f * e) = m_h(f)$ for every $f \in C_c(G)$. Now, it is not too hard to see that Stinespring's dilation of $m_h\colon C_c(G) \to C^*(G)$ gives rise to a continuous $*$-homomorphism on $C^*(G)$, and hence m_h is a c.c.p. map on $C^*(G)$. (Let e play the role of the identity 1.) □

Corollary 5.6.17. *Let G be an étale locally compact groupoid. If G is amenable, then $C^*(G) = C^*_\lambda(G)$ canonically.*

Proof. By Lemma 5.6.14, there exists a net $\zeta_i \in C_c(G)$ such that $\|\zeta_i\|_{L^2(G)} \le 1$ and $h_i = \zeta_i^* * \zeta_i \to 1$ uniformly on compact subsets of G. By Proposition 5.6.16, m_{h_i} is c.c.p. on $C^*(G)$ and on $C^*_\lambda(G)$. By the remarks following the definition of the full groupoid C*-algebra norm, we have $m_{h_i}(a) \to a$ in $C^*(G)$ for every $a \in C_c(G)$ and hence for every $a \in C^*(G)$. Moreover, we observe that $m_{h_i}(a) \in C_c(G)$ for every $a \in C^*(G)$. Now, we denote by $\pi\colon C^*(G) \to C^*_\lambda(G)$ the canonical quotient map and take $a \in \ker \pi$. Then, we have $(\pi \circ m_{h_i})(a) = (m_{h_i} \circ \pi)(a) = 0$. Since π is injective on $C_c(G)$, this implies that $m_{h_i}(a) = 0$ for all i, and hence $a = 0$. □

At the time of this writing, the converse (whether $C^*(G) = C^*_\lambda(G)$ implies amenability of G) is not known.

Theorem 5.6.18. *Let G be an étale locally compact groupoid. The following are equivalent:*

(1) *G is amenable;*

(2) *there exists a net of positive-type functions $h_i \in C_c(G)$ which converges to 1 uniformly on compact subsets of G;*

(3) *$C^*_\lambda(G)$ is nuclear.*

Proof. (1) ⇔ (2): Let $h \in C_c(G)$ be a positive-type function and let $\varepsilon > 0$. Since $\lambda(h) \ge 0$ in $C^*_\lambda(G)$, there exists $\zeta \in C_c(G)$ such that $\|\lambda(h - \zeta^* * \zeta)\| < \varepsilon$. By approximation, we are done.

(2) ⇒ (3): Let A be a C*-algebra and $Q\colon C^*_\lambda(G) \otimes_{\max} A \to C^*_\lambda(G) \otimes A$ be the quotient map. We will prove that Q is injective. We first claim that for every compact subset $K \subset G$, there exists a constant $C_K > 0$ such that

$$\|\sum_{j=1}^n \lambda(f_j) \otimes a_j\|_{C^*_\lambda(G) \otimes_{\max} A} \le C_K \sup_{\gamma \in K} \|\sum_{j=1}^n f_j(\gamma) a_j\|$$

for every $a_j \in A$ and $f_j \in C_c(G)$ with $\operatorname{supp} f_j \subset K$. Since K is compact, by a partition of unity reduction argument, we may assume that both s and r are homeomorphisms on K. It follows that $\operatorname{supp}(f_i^* * f_j) \subset G^{(0)}$ for every i, j.

Thus, writing $z = \sum \lambda(f_j) \otimes a_j$, we have $z^*z \in C_0(G^{(0)}) \odot A \subset C^*_\lambda(G) \otimes_{\max} A$. Since $C_0(G^{(0)})$ is nuclear, we have

$$\begin{aligned}
\|z^*z\|_{C^*_\lambda(G)\otimes_{\max} A} &= \sup_{x \in G^{(0)}} \|\sum_{i,j} \sum_{\gamma \in G_x} \overline{f_i(\gamma)} f_j(\gamma) a_i^* a_j\|_A \\
&= \sup_{\gamma \in K} \|\sum_{i,j} \overline{f_i(\gamma)} f_j(\gamma) a_i^* a_j\|_A \\
&= \sup_{\gamma \in K} \|\sum_j f_j(\gamma) a_j\|_A
\end{aligned}$$

as desired.

For a compact subset $K \subset G$, we denote by $C(K, A)$ the set of all A-valued continuous functions on G whose supports are contained in K. It is clear that $C(K, A)$ is a Banach space under the supremum norm. Let $\iota_K \colon C(K, A) \to C^*_\lambda(G) \otimes_{\max} A$ be the natural embedding. By the result of the previous paragraph, we have $\|\iota_K\| \leq C_K$, while it is trivial to see that $\|(Q \circ \iota_K)(f)\|_{C^*_\lambda(G)\otimes A} \geq \|f\|_\infty$ for every $f \in C(K, A)$. This means that Q is injective on $\iota_K(C(K, A))$ for every compact subset $K \subset G$.

Now take $h_i \in C_c(G)$ as in condition (2) and let $K_i = \operatorname{supp} h_i$. We may assume that $\sup |h_i(\gamma)| \leq 1$. By Proposition 5.6.16, the multiplier maps m_{h_i} are c.c.p. on $C^*_\lambda(G)$ and each $m_{h_i} \otimes_{\max} \mathrm{id}_A$ maps $C^*_\lambda(G) \otimes_{\max} A$ into $\iota_{K_i}(C(K_i, A))$ because $\iota_{K_i}(C(K_i, A))$ is closed. Since $m_{h_i} \to \mathrm{id}$ in the point-norm topology by Lemma 5.6.12, the net $m_{h_i} \otimes_{\max} \mathrm{id}_A$ converges to the identity on $C^*_\lambda(G) \otimes_{\max} A$. Let $x \in \ker Q$ be given. Then we have

$$Q\big((m_{h_i} \otimes_{\max} \mathrm{id}_A)(x)\big) = (m_{h_i} \otimes \mathrm{id}_A)(Q(x)) = 0.$$

This implies $(m_{h_i} \otimes_{\max} \mathrm{id}_A)(x) = 0$ for all i, so $x = \lim (m_{h_i} \otimes_{\max} \mathrm{id}_A)(x) = 0$.

(3) $\Rightarrow$ (1): Fix $\varepsilon > 0$ and a compact subset $K \subset G$. Let $U_1, \ldots, U_m$ be a relatively compact open covering of K such that, for every l, both s and r are homeomorphic on some neighborhood V_l of $\overline{U_l}$. Let $f_l \in C_c(G)$ be a function with $0 \leq f_l \leq 1$ such that $f_l = 1$ on U_l and $f_l = 0$ off V_l. Then $\|\lambda(f_l)\| \leq 1$ and $(f_l^* * f_l)(x) = 1$ for $x \in s(U_l)$. Let $\psi \colon C^*_\lambda(G) \to \mathbb{M}_n(\mathbb{C})$ and $\varphi \colon \mathbb{M}_n(\mathbb{C}) \to C^*_\lambda(G)$ be c.c.p. maps such that

$$\|(\varphi \circ \psi)(\lambda(f_l)) - \lambda(f_l)\| < \varepsilon \quad \text{and} \quad \|(\varphi \circ \psi)(\lambda(f_l^* * f_l)) - \lambda(f_l^* * f_l)\| < \varepsilon$$

for all l. Let $[b_{i,j}] = [\varphi(e_{i,j})]^{1/2} \in \mathbb{M}_n(C^*_\lambda(G))$ and set

$$\eta_\varphi = \sum_{j,k} \xi_j \otimes \xi_k \otimes b_{k,j} \in \ell^2_n \otimes \ell^2_n \otimes C^*_\lambda(G),$$

where $\{\xi_i\}_1^n$ is the standard basis of ℓ^2_n. We view $\ell^2_n \otimes \ell^2_n \otimes C^*_\lambda(G)$ as a Hilbert $C^*_\lambda(G)$-module. Then we have $\varphi(a) = \langle \eta_\varphi, (a \otimes 1 \otimes 1)\eta_\varphi \rangle$ for every

$a \in \mathbb{M}_n(\mathbb{C})$. For simplicity, let $a_l = \psi(\lambda(f_l))$. Then every a_l is a contraction such that

$$\begin{aligned}
&\langle (a_l \otimes 1 \otimes 1)\eta_\varphi - \eta_\varphi \lambda(f_l), (a_l \otimes 1 \otimes 1)\eta_\varphi - \eta_\varphi \lambda(f_l)\rangle \\
&\quad = \varphi(a_l^* a_l) + \lambda(f_l)^* \varphi(1)\lambda(f_l) - \varphi(a_l)^*\lambda(f_l) - \lambda(f_l)^*\varphi(a_l) \\
&\quad \le (\varphi\circ\psi)(\lambda(f_l^* f_l)) + \lambda(f_l^* f_l) - (\varphi\circ\psi)(\lambda(f_l))^*\lambda(f_l) \\
&\qquad - \lambda(f_l)^*(\varphi\circ\psi)(\lambda(f_l)) \\
&\quad \le 3\varepsilon.
\end{aligned}$$

This implies that $(a_l \otimes 1 \otimes 1)\eta_\varphi \approx \eta_\varphi \lambda(f_l)$ for every l. We approximate η_φ by $\eta'_\varphi \in \ell^2_n \otimes \ell^2_n \otimes C_c(G)$ of norm one such that $\|\eta_\varphi - \eta'_\varphi\| < \varepsilon$ and view η'_φ as an element in $C_c(G, \ell^2_n \otimes \ell^2_n)$. Now we define $\zeta \in C_c(G)$ by $\zeta(\beta) = \|\eta'_\varphi(\beta)\|_{\ell^2_n \otimes \ell^2_n}$. Thus, if $\eta'_\varphi = \sum_{j,k} \xi_j \otimes \xi_k \otimes \zeta_{k,j}$, then $\zeta(\beta)^2 = \sum_{j,k} |\zeta_{k,j}(\beta)|^2$. It follows that

$$\langle \zeta, \zeta\rangle_{L^2(G)}(x) = \sum_{\beta\in G_x} \zeta(\beta)^2 = \sum_{j,k} (\zeta_{k,j}^* * \zeta_{k,j})(x) = \langle \eta'_\varphi, \eta'_\varphi\rangle(x) \le 1$$

for every $x \in G^{(0)}$. Fix $\gamma \in U_l$ and let $x = s(\gamma)$. Then,

$$1 = (f_l^* * f_l)(x) \approx \langle (a_l \otimes 1 \otimes 1)\eta'_\varphi, \eta'_\varphi \lambda(f_l)\rangle(x)$$

since $\|f\|_\infty \le \|\lambda(f)\|$ for all $f \in C_c(G)$. For $\beta \in G_{s(\gamma)}$, we have

$$(\eta'_\varphi \lambda(f_l))(\beta) = \sum_{\alpha \in G_{s(\gamma)}} \eta'_\varphi(\beta\alpha^{-1}) f_l(\alpha) = \eta'_\varphi(\beta\gamma^{-1}) \in \ell^2_n \otimes \ell^2_n.$$

Therefore,

$$\begin{aligned}
1 &\approx |\langle (a_l \otimes 1 \otimes 1)\eta'_\varphi, \eta'_\varphi \lambda(f_l)\rangle(x)| \\
&= |\sum_{\beta\in G_x} \langle (a_l \otimes 1)\eta'_\varphi(\beta), (\eta'_\varphi\lambda(f_l))(\beta)\rangle_{\ell^2_n\otimes\ell^2_n}| \\
&\le \sum_{\beta\in G_x} \|\eta'_\varphi(\beta)\|_{\ell^2_n\otimes\ell^2_n} \|(\eta'_\varphi\lambda(f_l))(\beta)\|_{\ell^2_n\otimes\ell^2_n} \\
&= \sum_{\beta\in G_x} \zeta(\beta)\zeta(\beta\gamma^{-1}) = (\zeta^* * \zeta)(\gamma^{-1}).
\end{aligned}$$

Since $\|\zeta\|_{L^2(G)} \le 1$, this implies that $(\zeta^* * \zeta)(\gamma^{-1}) \approx 1$ for all $\gamma \in K$. □

5.7. References

Theorem 5.1.6 can be found in [**5**], [**74**] and [**130**]. Proposition 5.2.1 comes from [**136**]. Exactness of amalgamated free products follows from Dykema's C*-theorem (Corollary 4.8.3); however, geometric proofs – using Yu's property A – can be found in [**16**] and [**187**]. The compactification of a tree described here is a special case of the Bowditch compactification (unpublished). Theorem 5.4.1 follows easily from general results on Morita equivalence; see

[**6**] for more. The Gromov boundary of a hyperbolic space was defined by – care to guess? – Gromov. Theorem 5.3.15 is due to Adams, but our proof is taken from [**96**]. Proposition 5.3.18 and its Akemann-Ostrand corollary are explicit in [**85**], though previously known to Germain, Skandalis and perhaps others. Roe's lecture notes [**167**] on coarse geometry contain far more information than covered here (including connections with Baum-Connes conjectures). The fact that nuclearity of the uniform Roe algebra implies property A was proved in [**177**], using different methods than those in these notes. Finally, see the books [**6**], [**140**] and [**164**] for all things groupoid.

Chapter 6

Amenable Traces and Kirchberg's Factorization Property

In this chapter we explore another duality between finite-dimensional approximation properties and tensor products. This time the context is tracial states, but the end results have the same flavor as those characterizing nuclear or exact C*-algebras.

The first section contains some classical work of Murray and von Neumann. Section 6.2 contains the main result, Theorem 6.2.7, which connects tensor products with approximation properties of traces. The third section, perhaps out of place, discusses motivation and examples related to Theorem 6.2.7. In the final section we discuss Kirchberg's factorization property and see what this means for groups with Kazhdan's property (T).

6.1. Traces and the right regular representation

A classical theorem of Murray and von Neumann asserts that the commutant of the left regular representation of a group is just the weak closure of the right regular representation. Let's mimic their proof in the more general context of tracial GNS representations.

Definition 6.1.1. The *opposite algebra* of a C*-algebra A, denoted A^{op}, is simply A with reversed multiplication. That is, as a normed, involutive linear space $A^{\mathrm{op}} = A$, but we define a new multiplication by $a \cdot b = ba$.

Let τ be a tracial state on A, $L^2(A,\tau)$ be the GNS space, $\hat{A} \subset L^2(A,\tau)$ be the canonical image of A and $\pi_\tau \colon A \to \mathbb{B}(L^2(A,\tau))$ be the GNS representation. We will often refer to π_τ as the *left regular representation.*

One defines a $*$-representation $\pi_\tau^{\mathrm{op}} \colon A^{\mathrm{op}} \to \mathbb{B}(L^2(A,\tau))$ by

$$\pi_\tau^{\mathrm{op}}(a)\hat{b} = \widehat{ba}$$

for all $a \in A$ and $\hat{b} \in \hat{A} \subset L^2(A,\tau)$.[1] We will refer to this representation of A^{op} as the *right regular representation.*

Proposition 6.1.2. *For a C*-algebra A and tracial state τ the following assertions hold:*

(1) *the left and right regular representations commute – i.e., $\pi_\tau(A)' \supset \pi_\tau^{\mathrm{op}}(A^{\mathrm{op}})$;*

(2) *there is a unique conjugate linear*[2] *isometry $J \colon L^2(A,\tau) \to L^2(A,\tau)$ such that $J(\hat{b}) = \widehat{b^*}$ for all $\hat{b} \in \hat{A} \subset L^2(A,\tau)$ (also note that $J^2 = 1$);*

(3) *for all $a \in A$ we have $J\pi_\tau(a) = \pi_\tau^{\mathrm{op}}(a^*)J$ and $J\pi_\tau(a)J = \pi_\tau^{\mathrm{op}}(a^*)$. In particular, $J\pi_\tau(A)J = \pi_\tau^{\mathrm{op}}(A^{\mathrm{op}})$.*

Trivial calculations yield a proof. The next lemma is only slightly harder.

Lemma 6.1.3. *For a C*-algebra A and tracial state τ the following assertions hold:*

(1) *for all vectors $v, w \in L^2(A,\tau)$ we have $\langle Jv, w\rangle = \langle Jw, v\rangle$;*

(2) *for arbitrary $y \in \pi_\tau(A)'$ we have $Jy\hat{1} = y^*\hat{1}$, where $\hat{1} \in L^2(A,\tau)$ is the canonical cyclic vector (though we don't assume A is unital).*

Proof. The first assertion is easily verified on vectors of the form $v = \hat{a}$ and $w = \hat{b}$ so a simple approximation argument handles the general case. The second statement follows from the first and a little calculation:

$$\begin{aligned}\langle Jy\hat{1}, \hat{b}\rangle &= \langle J\hat{b}, y\hat{1}\rangle \\ &= \langle y^*\pi_\tau(b^*)\hat{1}, \hat{1}\rangle \\ &= \langle \pi_\tau(b^*)y^*\hat{1}, \hat{1}\rangle \\ &= \langle y^*\hat{1}, \hat{b}\rangle.\end{aligned}$$

Since this holds for arbitrary $b \in A$, we are finished. □

Since $J^2 = 1$, it is apparent that the map $T \mapsto JTJ$ is a conjugate linear $*$-isomorphism of $\mathbb{B}(L^2(A,\tau))$. In particular, this morphism preserves commutants in the sense that for each set $\mathcal{S} \subset \mathbb{B}(L^2(A,\tau))$ we have $J\mathcal{S}'J =$

[1] It's worth checking that this really works! For example, even the fact that $\pi_\tau^{\mathrm{op}}(a)$ is well-defined depends on τ being a trace.

[2] This means $J(\lambda v) = \bar{\lambda}v$ for all $\lambda \in \mathbb{C}$ and $v \in L^2(A,\tau)$.

$(JSJ)'$. With these simple observations and the technical calculations of the previous lemma, we can demonstrate the result of Murray and von Neumann mentioned earlier.

Theorem 6.1.4. *If τ is a tracial state on A, then $\pi_\tau(A)'' = \pi_\tau^{\mathrm{op}}(A^{\mathrm{op}})'$ and (hence) $\pi_\tau(A)' = \pi_\tau^{\mathrm{op}}(A^{\mathrm{op}})''$.*

Proof. Since $\pi_\tau(A)'' \subset \pi_\tau^{\mathrm{op}}(A^{\mathrm{op}})'$, we only have to observe the opposite inclusion. But $J\pi_\tau(A)J = \pi_\tau^{\mathrm{op}}(A^{\mathrm{op}})$ and hence

$$\pi_\tau^{\mathrm{op}}(A^{\mathrm{op}})' = J(\pi_\tau(A)')J.$$

Thus we will be finished once we observe that $J(\pi_\tau(A)')J$ and $\pi_\tau(A)'$ are commuting sets of operators. For any $x, y, z \in \pi_\tau(A)'$, we have by Lemma 6.1.3 that

$$JxJy(z\hat{1}) = Jxz^*y^*\hat{1} = yzx^*\hat{1} = yJxz^*\hat{1} = yJxJ(z\hat{1}).$$

Since $\hat{1}$ is $\pi_\tau(A)'$-cyclic and z was arbitrary, it follows that $JxJy = yJxJ$. □

Exercises

Exercise 6.1.1. A bijective, adjoint-preserving, linear map $\pi\colon A \to B$ is called an *anti-isomorphism* if $\pi(ab) = \pi(b)\pi(a)$. Prove that A is anti-isomorphic to itself if and only if it is isomorphic to A^{op}.

Exercise 6.1.2. The *conjugate* algebra of A, denoted $\overline{A} = \{\bar{a} : a \in A\}$, is just A as an involutive ring but it has the conjugate vector space structure – i.e., $\lambda\bar{a} = \overline{\lambda a}$ for all $\lambda \in \mathbb{C}$ and $a \in A$. Show that A^{op} is isomorphic to $\overline{A}$ via the adjoint map.

Exercise 6.1.3. Let $A = \mathbb{M}_2(\mathbb{C})$. Give an example of a state φ on A such that the obvious definition of the right regular representation fails (e.g., arrange that $\pi_\varphi^{\mathrm{op}}(a)\hat{b} = \widehat{ba}$ is not even well-defined).

Exercise 6.1.4. For a group Γ, let Γ^{op} denote the opposite group – i.e., $\Gamma^{\mathrm{op}} = \Gamma$ as sets but multiplication in Γ^{op} gets reversed ($g \cdot h = hg$). Show that $g \mapsto g^{-1}$ gives an isomorphism $\Gamma \to \Gamma^{\mathrm{op}}$. Prove that $C^*(\Gamma)^{\mathrm{op}} \cong C^*(\Gamma^{\mathrm{op}}) \cong C^*(\Gamma)$ and $C^*_\lambda(\Gamma)^{\mathrm{op}} \cong C^*_\lambda(\Gamma^{\mathrm{op}}) \cong C^*_\lambda(\Gamma)$. In other words, group C*-algebras are always anti-isomorphic to themselves.

Exercise 6.1.5. Show that the right regular representation, as defined in this section, is really the same (after identifying group C*-algebras with their opposite algebras) as the usual right regular representation of a group.

Exercise 6.1.6. Let tr be the unique tracial state on $\mathbb{M}_n(\mathbb{C})$. Show that the product map

$$\mathbb{M}_n(\mathbb{C}) \otimes \mathbb{M}_n(\mathbb{C})^{\mathrm{op}} \to \mathbb{B}(L^2(\mathbb{M}_n(\mathbb{C}), \mathrm{tr})),$$

arising from the left and right regular representations, is an isomorphism.

Exercise 6.1.7. Show that if $\varphi\colon A \to B$ is u.c.p., then it is still u.c.p. when regarded as a map of opposite algebras.

Exercise 6.1.8. Show that if A is nuclear (resp. exact), then A^{op} is nuclear (resp. exact).

6.2. Amenable traces

In 1975 Alain Connes proved a remarkable theorem which nearly completed the classification of injective factors with separable predual (cf. [**41**] – the one case not covered took a decade, and Uffe Haagerup, to resolve [**79**]). Connes's proof was a typical display of deep insight combined with remarkable technical strength and, in our opinion, ranks among the greatest achievements in the history of operator algebras. In this section we will explore one of the ingredients that went into the proof. The main conceptual idea is that certain traces on C*-algebras are analogous to invariant means on groups. When viewed this way, some ideas from the theory of amenable groups can be translated into a more general operator algebraic framework.

Unfortunately having the proper insight is only half of the battle; the proofs are still very technical and delicate. The main result of this section, Theorem 6.2.7, is essentially due to Connes in the unique trace case and Kirchberg in general. We have taken the quickest, most direct route (that we are aware of) to this result, postponing motivation and examples until the next section. For some it will be more palatable to skip ahead, take in the conceptual landscape and then come back for the gory details.

Definition 6.2.1. Let $A \subset \mathbb{B}(\mathcal{H})$ be a concretely represented unital C*-algebra. A state τ on A is called an *amenable trace*[3] if there exists a state φ on $\mathbb{B}(\mathcal{H})$ such that (1) $\varphi|_A = \tau$ and (2) $\varphi(uTu^*) = \varphi(T)$ for every unitary $u \in A$ and $T \in \mathbb{B}(\mathcal{H})$.[4]

Note that an amenable trace really is tracial. Indeed, if $T \in \mathbb{B}(\mathcal{H})$ is arbitrary and $u \in A$ is a unitary, then

$$\varphi(Tu) = \varphi(u(Tu)u^*) = \varphi(uT).$$

[3]Beware: Lots of different terminology can be found in the literature. Also, one can extend the notion to nonunital algebras, but we'll have no need for this.

[4]The state φ is sometimes called a "hypertrace" or an "A-central state."

It follows that if b is a linear combination of unitaries in A, then $\varphi(Tb) = \varphi(bT)$ and, since unitaries span A, this shows τ is a trace.

One should ask whether or not the definition of amenable trace depends on the choice of embedding $A \subset \mathbb{B}(\mathcal{H})$. Luckily, it doesn't.

Proposition 6.2.2. *Assume $A \subset \mathbb{B}(\mathcal{H})$ and τ is an amenable trace. For every faithful representation $\pi\colon A \to \mathbb{B}(\mathcal{K})$, there exists a state $\psi_{\mathcal{K}}$ on $\mathbb{B}(\mathcal{K})$ such that $\psi_{\mathcal{K}}(\pi(u)T\pi(u^*)) = \psi_{\mathcal{K}}(T)$, for all $T \in \mathbb{B}(\mathcal{K})$ and unitaries $u \in A$, and $\psi_{\mathcal{K}} \circ \pi = \tau$.*

Proof. By Arveson's Extension Theorem, there is a u.c.p. map $\Phi\colon \mathbb{B}(\mathcal{K}) \to \mathbb{B}(\mathcal{H})$ which extends the inverse map $\pi^{-1}\colon \pi(A) \to A \subset \mathbb{B}(\mathcal{H})$. Note that $\pi(A)$ is in the multiplicative domain of Φ. Defining $\psi_{\mathcal{K}} = \psi \circ \Phi$, where ψ is any state on $\mathbb{B}(\mathcal{H})$ as in the definition of amenable trace, we get a state on $\mathbb{B}(\mathcal{K})$. Evidently we have $\psi_{\mathcal{K}} \circ \pi = \tau$ and the rest of the proof is a simple multiplicative domain argument. Indeed, for $T \in \mathbb{B}(\mathcal{K})$ and a unitary $u \in A$ we have

$$\psi_{\mathcal{K}}(\pi(u)T\pi(u^*)) = \psi(u\Phi(T)u^*) = \psi_{\mathcal{K}}(T),$$

as desired. □

Recall that $\mathcal{S}_1, \mathcal{S}_2 \subset \mathbb{B}(\mathcal{H})$ denote the trace class and Hilbert-Schmidt operators, respectively. An easy, but important, fact is that both the L^1-norm $\|T\|_1 = \mathrm{Tr}(|T|)$ and L^2-norm $\|T\|_2 = \sqrt{\mathrm{Tr}(|T|^2)}$ are *unitarily invariant*, meaning that $\|uT\|_1 = \|Tu\|_1 = \|T\|_1$ for all unitary operators $u \in \mathbb{B}(\mathcal{H})$ (same for the L^2-norm).

Lemma 6.2.3. *For any finite-rank $h, k \in \mathbb{B}(\mathcal{H})$, arbitrary $x \in \mathbb{B}(\mathcal{H})$ and $p \in \{1, 2\}$,*[5] *one has $\|h^*\|_p = \|h\|_p$, $\|xh\|_p \leq \|x\|\|h\|_p$, $\|hx\|_p \leq \|x\|\|h\|_p$. Moreover, $\|hk\|_1 \leq \|h\|_2\|k\|_2$ and*

$$\|h\|_1 = \sup\{|\mathrm{Tr}(hy)| : y \in \mathbb{B}(\mathcal{H}), \|y\| \leq 1\}.$$

Proof. The last equation is well known ($\mathbb{B}(\mathcal{H})$ is the dual of $\mathcal{S}_1(\mathcal{H})$) and easily implies $\|h^*\|_1 = \|h\|_1$; that $\|h^*\|_2 = \|h\|_2$ is trivial. The inequality

$$\|xh\|_2^2 = \mathrm{Tr}(h^*x^*xh) \leq \|x\|^2\,\mathrm{Tr}(h^*h) = \|x\|^2\|h\|_2^2$$

also implies $\|hx\|_2 \leq \|x\|\|h\|_2$, after taking adjoints. Operator monotonicity of the square root function (applied to $|xh|^2 \leq \|x\|^2|h|^2$) implies $|xh| \leq \|x\||h|$; thus $\|xh\|_1 \leq \|x\|\|h\|_1$.

The Cauchy-Schwarz inequality says that

$$|\mathrm{Tr}(hk)| \leq \|h\|_2\|k\|_2$$

[5] In fact, this holds for any $1 \leq p < \infty$.

for all finite-rank h, k. Letting $hk = u|hk|$ be the polar decomposition, one checks that

$$\mathrm{Tr}(|hk|) = \mathrm{Tr}(u^*hk) \leq \|u^*h\|_2\|k\|_2 \leq \|h\|_2\|k\|_2,$$

which completes the proof. □

We will need the *Powers-Størmer inequality*:

Proposition 6.2.4. *For any $0 \leq h, k \in \mathcal{S}_2(\mathcal{H})$, one has*

$$\|h-k\|_2^2 \leq \|h^2-k^2\|_1 \leq \|h+k\|_2\|h-k\|_2.$$

In particular, if $u \in \mathbb{B}(\mathcal{H})$ is a unitary and $h \geq 0$ has finite rank, then

$$\|uh^{1/2} - h^{1/2}u\|_2 = \|uh^{1/2}u^* - h^{1/2}\|_2 \leq \|uhu^* - h\|_1^{1/2}.$$

Proof. We may assume that h and k are finite-rank. (Why?) The right inequality follows from the previous lemma and the identity

$$h^2 - k^2 = \big((h+k)(h-k) + (h-k)(h+k)\big)/2.$$

If h and k commute, then the operator inequality $|h-k|^2 \leq |h^2-k^2|$ implies the left inequality. Unfortunately, this operator inequality doesn't hold in the noncommutative case. To circumvent this problem, we exploit the fact that $\mathrm{Tr}(xy) \geq 0$ for any $x, y \geq 0$, so long as either x or y is finite-rank. Now, let $e = \chi_{[0,\infty)}(h-k)$ be the spectral projection and let $e^\perp = 1-e$. Since $(h-k)e \geq 0$ and $(k-h)e^\perp \geq 0$, one has

$$\begin{aligned}
\mathrm{Tr}((h-k)^2) &= \mathrm{Tr}((h-k)(h-k)e + (k-h)(k-h)e^\perp) \\
&\leq \mathrm{Tr}((h+k)(h-k)e + (k+h)(k-h)e^\perp) \\
&= \mathrm{Tr}((h^2-k^2)e + (k^2-h^2)e^\perp) \\
&\leq \mathrm{Tr}(|h^2-k^2|e + |k^2-h^2|e^\perp) \\
&= \mathrm{Tr}(|h^2-k^2|),
\end{aligned}$$

where the third line equality follows from

$$\mathrm{Tr}((h+k)(h-k)e) = \mathrm{Tr}((h+k)e(h-k)) = \mathrm{Tr}((h-k)(h+k)e)$$

and a similar identity for $e^\perp$.[6] □

Our next lemma is the main technical result of this section – you'll need lots of scratch paper. Seriously, lots.

[6]We thank Sorin Popa for sharing this elegant proof with us.

Lemma 6.2.5. *Let $h \in \mathbb{B}(\mathcal{H})$ be a positive, finite-rank operator with rational eigenvalues and such that $\operatorname{Tr}(h) = 1$. Then there exists a u.c.p. map $\varphi\colon \mathbb{B}(\mathcal{H}) \to \mathbb{M}_q(\mathbb{C})$ such that $\operatorname{tr}(\varphi(T)) = \operatorname{Tr}(hT)$ for all $T \in \mathbb{B}(\mathcal{H})$ and $|\operatorname{tr}(\varphi(uu^*) - \varphi(u)\varphi(u^*))| \leq 2\|uhu^* - h\|_1^{1/2}$ for every unitary operator $u \in \mathbb{B}(\mathcal{H})$.*

Proof. First we write

$$h = \frac{p_1}{q}Q_1 \oplus \frac{p_2}{q}Q_2 \oplus \cdots \oplus \frac{p_k}{q}Q_k$$

where $\frac{p_1}{q} < \frac{p_2}{q} < \cdots < \frac{p_k}{q}$ are the nonzero eigenvalues of h and $Q_1, \ldots, Q_k$ are the corresponding spectral projections. Notice that

$$\sum_{i=1}^{k} p_i \operatorname{Tr}(Q_i) = q$$

since $\operatorname{Tr}(h) = 1$.

Let $\mathcal{K}$ be a Hilbert space and $P_1 \leq P_2 \leq \cdots \leq P_k$ be projections such that $\operatorname{rank}(P_i) = p_i$ for all i. Since the Q_i's are orthogonal, we can define a projection $P \in \mathbb{B}(\mathcal{H} \otimes \mathcal{K})$ by

$$P = Q_1 \otimes P_1 + Q_2 \otimes P_2 + \cdots + Q_k \otimes P_k.$$

Note that

$$\operatorname{Tr}(P) = \sum_{i=1}^{k} \operatorname{Tr}(Q_i) \operatorname{Tr}(P_i) = q.$$

It turns out that the u.c.p. map $\varphi\colon \mathbb{B}(\mathcal{H}) \to \mathbb{M}_q(\mathbb{C})$ we are after is simply compression by P, i.e., $\varphi(T) = P(T \otimes 1_{\mathcal{K}})P$ for all $T \in \mathbb{B}(\mathcal{H})$.

We first check the trace-preserving condition. For $T \in \mathbb{B}(\mathcal{H})$ we have

$$\begin{aligned}\operatorname{tr}(\varphi(T)) &= \frac{\operatorname{Tr}(P(T \otimes 1)P)}{\operatorname{Tr}(P)} \\ &= \frac{\sum_{i=1}^{k} \operatorname{Tr}(Q_i T Q_i) \operatorname{Tr}(P_i)}{q} \\ &= \sum_{i=1}^{k} \frac{p_i}{q} \operatorname{Tr}(Q_i T) = \operatorname{Tr}(hT),\end{aligned}$$

as desired.

For a unitary operator $u \in \mathbb{B}(\mathcal{H})$, we now compare $\operatorname{tr}(\varphi(u)\varphi(u^*))$ with $\operatorname{Tr}(h^{\frac{1}{2}} u h^{\frac{1}{2}} u^*)$. Note that

$$\operatorname{Tr}(h^{\frac{1}{2}} u h^{\frac{1}{2}} u^*) = \sum_{i,j=1}^{k} \frac{\sqrt{p_i p_j}}{q} \operatorname{Tr}(Q_i u Q_j u^*),$$

and, by the Powers-Størmer inequality,

$$1-\operatorname{Tr}(h^{\frac{1}{2}}uh^{\frac{1}{2}}u^*)=\frac{1}{2}\|h^{1/2}-uh^{\frac{1}{2}}u^*\|_2^2\leq\frac{1}{2}\|h-uhu^*\|_1.$$

We use here and there the fact that $\operatorname{Tr}(ST)=\operatorname{Tr}(S^{1/2}TS^{1/2})\geq 0$ for any positive operators S,T with S finite-rank. Now, let's compute $\operatorname{tr}(\varphi(u)\varphi(u^*))$:

$$\begin{aligned}\operatorname{tr}(\varphi(u)\varphi(u^*))&=\frac{\operatorname{Tr}(P(u\otimes 1)P(u^*\otimes 1)P)}{\operatorname{Tr}(P)}\\&=\frac{\sum_{i,j=1}^k\operatorname{Tr}(Q_iuQ_ju^*)\operatorname{Tr}(P_iP_j)}{q}\\&=\sum_{i,j=1}^k\frac{\min\{p_i,p_j\}}{q}\operatorname{Tr}(Q_iuQ_ju^*)\\&\geq\operatorname{Tr}(h^{\frac{1}{2}}uh^{\frac{1}{2}}u^*)-\sum_{i,j=1}^k\frac{|p_i-p_j|}{2q}\operatorname{Tr}(Q_iuQ_ju^*)\end{aligned}$$

since $\min\{p_i,p_j\}=\frac{1}{2}(p_i+p_j-|p_i-p_j|)\geq\sqrt{p_ip_j}-\frac{1}{2}|p_i-p_j|$. We have further that

$$\sum_{i,j=1}^k\frac{p_i}{q}\operatorname{Tr}(Q_iuQ_ju^*)=\sum_{i=1}^k\frac{p_i}{q}\operatorname{Tr}(Q_iu(\sum_{j=1}^kQ_j)u^*)\leq\sum_{i=1}^k\frac{p_i}{q}\operatorname{Tr}(Q_i)=1$$

and hence

$$\begin{aligned}&\sum_{i,j=1}^k\frac{|p_i-p_j|}{2q}\operatorname{Tr}(Q_iuQ_ju^*)\\&=\sum_{i,j=1}^k\left(\frac{|p_i^{\frac{1}{2}}+p_j^{\frac{1}{2}}|}{2\sqrt{q}}\operatorname{Tr}(Q_iuQ_ju^*)^{\frac{1}{2}}\right)\left(\frac{|p_i^{\frac{1}{2}}-p_j^{\frac{1}{2}}|}{\sqrt{q}}\operatorname{Tr}(Q_iuQ_ju^*)^{\frac{1}{2}}\right)\\&\leq\left(\sum_{i,j=1}^k\frac{|p_i^{\frac{1}{2}}+p_j^{\frac{1}{2}}|^2}{4q}\operatorname{Tr}(Q_iuQ_ju^*)\right)^{\frac{1}{2}}\left(\sum_{i,j=1}^k\frac{|p_i^{\frac{1}{2}}-p_j^{\frac{1}{2}}|^2}{q}\operatorname{Tr}(Q_iuQ_ju^*)\right)^{\frac{1}{2}}\\&\leq\left(\frac{2+2\operatorname{Tr}(h^{\frac{1}{2}}uh^{\frac{1}{2}}u^*)}{4}\right)^{\frac{1}{2}}\left(2-2\operatorname{Tr}(h^{\frac{1}{2}}uh^{\frac{1}{2}}u^*)\right)^{\frac{1}{2}}\\&\leq\|h-uhu^*\|_1^{1/2},\end{aligned}$$

by the previous computations. Finally, combining everything, we get

$$\operatorname{tr}(\varphi(uu^*)-\varphi(u)\varphi(u^*))\leq 1-\operatorname{Tr}(h^{1/2}uh^{1/2}u^*)+\sum_{i,j=1}^k\frac{|p_i-p_j|}{2q}\operatorname{Tr}(Q_iuQ_ju^*)$$

$$\leq \frac{1}{2}\|h - uhu^*\|_1 + \|h - uhu^*\|_1^{1/2},$$

so the proof is complete. □

If ω is a state on B, then we define a seminorm $\|\cdot\|_{2,\omega}$ on B by $\|b\|_{2,\omega} = \sqrt{\omega(b^*b)}$.

Lemma 6.2.6. *If $\varphi\colon A \to B$ is a u.c.p. map and ω is a state on B, then*

$$\|\varphi(ab) - \varphi(a)\varphi(b)\|_{2,\omega} \leq \|a\|\omega\big(\varphi(b^*b) - \varphi(b^*)\varphi(b)\big)^{1/2},$$

for all $a, b \in A$.

Proof. Applying GNS, we may assume $B \subset \mathbb{B}(\mathcal{H})$ and $\omega = \langle \cdot\,\xi, \xi\rangle$ for some $\xi \in \mathcal{H}$. Let $\pi\colon A \to \mathbb{B}(\mathcal{K})$ be the Stinespring dilation of φ, with isometry $V\colon \mathcal{H} \to \mathcal{K}$. Then, for all $a, b \in A$, we have

$$\begin{aligned}
\|\varphi(ab) - \varphi(a)\varphi(b)\|_{2,\omega} &= \|(\varphi(ab) - \varphi(a)\varphi(b))\xi\| \\
&= \|V^*\pi(a)(1 - VV^*)\pi(b)V\xi\| \\
&\leq \|V^*\pi(a)(1 - VV^*)^{1/2}\|\|(1 - VV^*)^{1/2}\pi(b)V\xi\| \\
&\leq \|a\|\langle V^*\pi(b^*)(1 - VV^*)\pi(b)V\xi, \xi\rangle^{1/2} \\
&= \|a\|\omega\big(\varphi(b^*b) - \varphi(b^*)\varphi(b)\big)^{1/2}.
\end{aligned}$$

□

We are almost ready for the main theorem, just a little more notation. For a tracial state τ on A we consider the product $*$-homomorphism arising from the left and right regular representations:

$$\pi_\tau \times \pi_\tau^{\mathrm{op}}\colon A \odot A^{\mathrm{op}} \to \mathbb{B}(L^2(A, \tau)).$$

Composing this representation with the vector state $x \mapsto \langle x\hat{1}, \hat{1}\rangle$, where $\hat{1}$ denotes the natural image of the unit of A, we get a positive linear functional μ_τ on $A \odot A^{\mathrm{op}}$. Note that for an elementary tensor $a \otimes b \in A \odot A^{\mathrm{op}}$ we have

$$\mu_\tau(a \otimes b) = \langle \pi_\tau(a)\pi_\tau^{\mathrm{op}}(b)\hat{1}, \hat{1}\rangle = \langle \pi_\tau(a)\pi_\tau(b)\hat{1}, \hat{1}\rangle = \tau(ab),$$

since $\pi_\tau^{\mathrm{op}}(b)\hat{1} = \hat{b} = \pi_\tau(b)\hat{1}$. In the context of residually finite groups, we already used this functional (see the proof of Proposition 3.7.11); it will again play a crucial role.

Theorem 6.2.7. *Let A be unital with tracial state τ. The following are equivalent:*

(1) *τ is amenable;*

(2) *there exists a net of u.c.p. maps $\varphi_n\colon A \to \mathbb{M}_{k(n)}(\mathbb{C})$ such that $\tau(a) = \lim \mathrm{tr} \circ \varphi_n(a)$ and $\|\varphi_n(ab) - \varphi_n(a)\varphi_n(b)\|_{2,\mathrm{tr}} \to 0$, for all $a, b \in A$;*

(3) *the positive linear functional* $\mu_\tau\colon A \odot A^{\mathrm{op}} \to \mathbb{C}$ *is continuous with respect to the minimal tensor product norm;*

(4) *the product morphism* $\pi_\tau \times \pi_\tau^{\mathrm{op}}\colon A \odot A^{\mathrm{op}} \to \mathbb{B}(L^2(A,\tau))$ *is continuous with respect to the minimal tensor product norm;*

(5) *for any faithful representation* $A \subset \mathbb{B}(\mathcal{H})$ *there exists a u.c.p. map* $\Phi\colon \mathbb{B}(\mathcal{H}) \to \pi_\tau(A)''$ *such that* $\Phi(a) = \pi_\tau(a)$ *for all* $a \in A$.[7]

Proof. (1) $\Rightarrow$ (2): Let $A \subset \mathbb{B}(\mathcal{H})$ be a faithful representation. Since τ is an amenable trace, we can find a state ψ on $\mathbb{B}(\mathcal{H})$ which extends τ and such that $\psi(uTu^*) = \psi(T)$ for all unitaries $u \in A$ and operators $T \in \mathbb{B}(\mathcal{H})$. Since the normal states are weak-$*$ dense in the set of all states on $\mathbb{B}(\mathcal{H})$, we can find a net of positive operators $h_\lambda \in \mathcal{S}_1$ such that $\mathrm{Tr}(h_\lambda T) \to \psi(T)$ for all $T \in \mathbb{B}(\mathcal{H})$. Since $\psi(u^*Tu) = \psi(T)$, it follows that $\mathrm{Tr}(h_\lambda T) - \mathrm{Tr}((uh_\lambda u^*)T) \to 0$ for every $T \in \mathbb{B}(\mathcal{H})$ and unitary $u \in A$. In other words, for a fixed unitary $u \in A$, $h_\lambda - uh_\lambda u^* \to 0$ in the weak topology on $\mathcal{S}_1$. Hence, by the Hahn-Banach Theorem, some convex combinations of the elements $\{h_\lambda - uh_\lambda u^*\}$ will tend to zero in the L^1-norm.

As we saw in the proof of Lemma 2.3.4, a simple direct sum trick implies that we can extend this to a finite set of unitaries in A.[8] That is, if $\mathfrak{F} \subset A$ is a finite set of unitaries, then for every $\epsilon > 0$ we can find a positive trace class operator $h \in \mathcal{S}_1$ such that $\mathrm{Tr}(h) = 1$, $|\mathrm{Tr}(uh) - \tau(u)| < \epsilon$ and $\|h - uhu^*\|_1 < \epsilon$ for all $u \in \mathfrak{F}$. Since finite-rank operators are norm dense in $\mathcal{S}_1$, we may further assume that h is finite-rank with rational eigenvalues.

Applying Lemma 6.2.5 to bigger and bigger finite sets of unitaries and smaller and smaller ε's, one constructs a net of u.c.p. maps $\varphi_n\colon \mathbb{B}(\mathcal{H}) \to \mathbb{M}_{k(n)}(\mathbb{C})$ such that $\mathrm{tr}(\varphi_n(u)) \to \tau(u)$ and $|\,\mathrm{tr}(\varphi_n(uu^*) - \varphi_n(u)\varphi_n(u^*))| \to 0$ for every unitary element $u \in A$. From Lemma 6.2.6 it follows that

$$\|\varphi_n(au) - \varphi_n(a)\varphi_n(u)\|_{2,\mathrm{tr}} \to 0$$

for every unitary element $u \in A$ and we are finished.

(2) $\Rightarrow$ (3): Since we must show that for every $x \in A \odot A^{\mathrm{op}}$,

$$|\mu_\tau(x)| \le \|x\|_{\min},$$

it will suffice to exhibit a net of states which converge to μ_τ pointwise and are defined on all of $A \otimes A^{\mathrm{op}}$.

Let $\varphi_n\colon A \to \mathbb{M}_{k(n)}(\mathbb{C})$ be a net of u.c.p. maps as in condition (2). Note that we can also regard these maps as sending A^{op} to $\mathbb{M}_{k(n)}(\mathbb{C})^{\mathrm{op}}$

[7] This is also equivalent to the existence of a single representation with this property.

[8] More precisely, one considers the n-fold direct sum of $\mathcal{S}_1$ and, for a fixed n-tuple of unitaries $u_1, \dots, u_n \in A$, one shows that $(u_1h_\lambda u_1^* - h_\lambda, \dots, u_nh_\lambda u_n^* - h_\lambda)$ is converging weakly to zero. Hence a convex combination of the h_λ's will simultaneously bring all the norms $\|u_ih_\lambda u_i^* - h_\lambda\|_1$, $i = 1, \dots, n$, close to zero.

and they are still u.c.p. (Exercise 6.1.7). For notational convenience, we let $\varphi_n^{\rm op} : A^{\rm op} \to \mathbb{M}_{k(n)}(\mathbb{C})^{\rm op}$ be the "opposite" maps. Since u.c.p. maps induce u.c.p. maps on minimal tensor products, we may consider

$$\varphi_n \otimes \varphi_n^{\rm op} \colon A \otimes A^{\rm op} \to \mathbb{M}_{k(n)}(\mathbb{C}) \otimes \mathbb{M}_{k(n)}(\mathbb{C})^{\rm op}.$$

We'll now construct states μ_n on $\mathbb{M}_{k(n)}(\mathbb{C}) \otimes \mathbb{M}_{k(n)}(\mathbb{C})^{\rm op}$ such that

$$\mu_n \circ (\varphi_n \otimes \varphi_n^{\rm op})(x) \to \mu_\tau(x)$$

for all $x \in A \odot A^{\rm op}$.

The product map induced by the left and right regular representations gives an identification (Exercise 6.1.6)

$$\mathbb{M}_{k(n)}(\mathbb{C}) \otimes \mathbb{M}_{k(n)}(\mathbb{C})^{\rm op} \cong \mathbb{B}(L^2(\mathbb{M}_{k(n)}(\mathbb{C}), {\rm tr})).$$

As we have done for A, define a state μ_n on $\mathbb{M}_{k(n)}(\mathbb{C}) \otimes \mathbb{M}_{k(n)}(\mathbb{C})^{\rm op}$ by

$$T \otimes S \mapsto \langle \pi_{\rm tr}(T)\pi_{\rm tr}^{\rm op}(S)\hat{1}, \hat{1}\rangle = {\rm tr}(TS).$$

It follows that

$$\mu_n(\varphi_n \otimes \varphi_n^{\rm op}(a \otimes b)) = {\rm tr}(\varphi_n(a)\varphi_n(b)),$$

for all $a \in A, b \in A^{\rm op}$. The Cauchy-Schwarz inequality shows that for all $x \in \mathbb{M}_{k(n)}(\mathbb{C})$, $|\,{\rm tr}(x)| \le \|x\|_{2,{\rm tr}}$ and hence

$$|\,{\rm tr}(\varphi_n(a)\varphi_n(b)) - {\rm tr}(\varphi_n(ab))| \to 0.$$

Thus we see that

$$\mu_n(\varphi_n \otimes \varphi_n^{\rm op}(a \otimes b)) \to \tau(ab) = \mu_\tau(a \otimes b).$$

Since the linear span of elementary tensors is dense in $A \otimes A^{\rm op}$, we are done.

(3) $\Rightarrow$ (4) follows from uniqueness of GNS representations since the product map $A \odot A^{\rm op} \to \mathbb{B}(L^2(A,\tau))$ has a cyclic vector which implements μ_τ.

(4) $\Rightarrow$ (5): Apply The Trick to the inclusion $A \otimes A^{\rm op} \subset \mathbb{B}(\mathcal{H}) \otimes A^{\rm op}$ and use Theorem 6.1.4 to control the range of the extending u.c.p. map.

(5) $\Rightarrow$ (1) is very similar to the proof of Proposition 6.2.2. For every $T \in \mathbb{B}(\mathcal{H})$ and unitary $u \in A$ we have

$$\langle \Phi(uTu^*)\hat{1}, \hat{1}\rangle = \langle \pi_\tau(u)\Phi(T)\pi_\tau(u^*)\hat{1}, \hat{1}\rangle = \langle \Phi(T)\hat{1}, \hat{1}\rangle,$$

since u is in the multiplicative domain of Φ, $\Phi(T) \in \pi_\tau(A)''$ and $\langle \,\cdot\, \hat{1}, \hat{1}\rangle$ is a trace on $\pi_\tau(A)''$. Hence $T \mapsto \langle \Phi(T)\hat{1}, \hat{1}\rangle$ is the desired extending state on $\mathbb{B}(\mathcal{H})$. □

Remark 6.2.8. Of course, if A is a separable C*-algebra, then the net in part (2) of Theorem 6.2.7 can be replaced with a sequence. Moreover, the same thing is true if A is a von Neumann algebra with separable predual and τ is normal. Indeed, if h_λ is a net as in the proof of (1) $\Rightarrow$ (2) and $\psi_\lambda(\,\cdot\,) =$

$\mathrm{Tr}(h_\lambda\,\cdot\,)$ are the corresponding states, then the net $(\psi_\lambda)|_M$ converges to τ in the weak topology on M_*. Hence, by the Hahn-Banach Theorem, we may assume that $\|(\psi_\lambda)|_M - \tau\|_{M_*} \to 0$ and $\|h_\lambda - uh_\lambda u^*\|_{1,\mathrm{Tr}} \to 0$ for $u \in \mathcal{U}(A)$. It follows that the net φ_n in condition (2) can be taken so that $\|\tau - \mathrm{tr}\circ\varphi_n\|_{M_*} \to 0$. (In fact, one can arrange $\tau = \mathrm{tr}\circ\varphi_n$ with sufficient effort.) For a countable ultraweakly dense subset S in the unit ball of A, we can find a subsequence $\varphi_{n(k)}$ of φ_n such that $\|\varphi_{n(k)}(ab) - \varphi_{n(k)}(a)\varphi_{n(k)}(b)\|_{2,\mathrm{tr}} \to 0$, for all $a, b \in S$ and *a fortiori* for all $a, b \in A$.

Exercises

Exercise 6.2.1. Observe that every trace on an injective von Neumann algebra is amenable.

Exercise 6.2.2. Let $B = \mathbb{M}_{n(1)}(\mathbb{C}) \oplus \mathbb{M}_{n(2)}(\mathbb{C})$ and define a trace on B by

$$\tau(T \oplus S) = \frac{p}{q}\,\mathrm{tr}(S) + \frac{q-p}{q}\,\mathrm{tr}(T),$$

where $p < q \in \mathbb{N}$. Show that there exists a unital embedding $B \subset \mathbb{M}_{n(3)}(\mathbb{C})$ such that tr (on $\mathbb{M}_{n(3)}(\mathbb{C})$) restricts to τ (on B). In other words, there is a trace-preserving embedding $(B, \tau) \subset (\mathbb{M}_{n(3)}(\mathbb{C}), \mathrm{tr})$.

Exercise 6.2.3. Prove that if B is any finite-dimensional C*-algebra and τ is a trace on B, then for every $\varepsilon > 0$ there exists a matrix algebra $\mathbb{M}_n(\mathbb{C})$ and a unital $*$-homomorphism $\pi\colon B \to \mathbb{M}_n(\mathbb{C})$ such that

$$|\tau(b) - \mathrm{tr}(\pi(b))| \le \varepsilon\|b\|,$$

for all $b \in B$. Hence, there is nothing gained by replacing matrix algebras in part (2) of Theorem 6.2.7 with general finite-dimensional C*-algebras.[9]

Kirchberg first used the terminology "liftable" in reference to what we've called amenable traces. The following exercise explains his point of view. (See Appendix A for the ultraproduct R^ω of the hyperfinite II$_1$-factor R.)

Exercise 6.2.4. Show that a trace τ is amenable if and only if there exists an embedding

$$\pi_\tau(A)'' \subset R^\omega$$

such that the $*$-homomorphism $\pi_\tau\colon A \to \pi_\tau(A)'' \subset R^\omega$ has a u.c.p. lifting $A \to \ell^\infty(R)$. There are a couple of different proofs that one could give. Perhaps the most elegant is to use injectivity of $\ell^\infty(R)$ and the last condition in Theorem 6.2.7.

[9] On the other hand, it can be convenient to use finite-dimensional algebras in proofs and this exercise implies that this is legal.

In [**41**] another important characterization of amenable traces was given; it also illustrates the connection with classical amenability. In the literature this is sometimes called a *Følner condition* and has been very useful. For example, Sorin Popa used it to significantly simplify the proof of "injective implies hyperfinite" for von Neumann algebras with a faithful tracial state.

Exercise 6.2.5. Let $A \subset \mathbb{B}(\mathcal{H})$ be a representation which contains no nonzero compact operators and let τ be a tracial state on A. Show that τ is amenable if and only if there exist finite-rank projections $P_n \in \mathbb{B}(\mathcal{H})$ such that

$$\frac{\|[a, P_n]\|_2}{\|P_n\|_2} \to 0$$

and

$$\frac{\mathrm{Tr}(aP_n)}{\mathrm{Tr}(P_n)} \to \tau(a),$$

for all $a \in A$. (Hint: For the "only if" part we can use Voiculescu's Theorem to pull the u.c.p. maps appearing in statement (2) of Theorem 6.2.7 back to the given representation $A \subset \mathbb{B}(\mathcal{H})$, since we assumed A contains no compacts.)

6.3. Some motivation and examples

The first point which should be clarified is why amenable traces are analogues of invariant means on groups. Recall that a state φ on $\ell^\infty(\Gamma)$ is called an *invariant mean* if it is left translation invariant – i.e., $\varphi(s.f) = \varphi(f)$ for all $s \in \Gamma$ and $f \in \ell^\infty(\Gamma)$. In other words, invariant means are just states on a von Neumann algebra which are invariant under a particular group action.

Fix a concrete representation $A \subset \mathbb{B}(\mathcal{H})$ of a unital C*-algebra and let $\mathcal{U}(A)$ denote the unitary group of A (with the discrete topology). There is a natural action of $\mathcal{U}(A)$ on $\mathbb{B}(\mathcal{H})$ given by conjugation: $T \mapsto uTu^*$, for all $T \in \mathbb{B}(\mathcal{H})$ and $u \in \mathcal{U}(A)$. Hence the following formulation of amenable trace is equivalent to Definition 6.2.1.

Definition 6.3.1. Given $A \subset \mathbb{B}(\mathcal{H})$, an *amenable trace* on A is the restriction (to A) of a state on $\mathbb{B}(\mathcal{H})$ which is invariant under the conjugation action of $\mathcal{U}(A)$ on $\mathbb{B}(\mathcal{H})$.

Unlike the case of groups, there is really no canonical faithful embedding $A \subset \mathbb{B}(\mathcal{H})$ (unless one goes to the nonseparable universal representation). However, Proposition 6.2.2 states that being an amenable trace is independent of the choice of faithful representation and hence the lack of a canonical action of $\mathcal{U}(A)$ on $\mathbb{B}(\mathcal{H})$ doesn't cause problems with the notion of amenable trace.

As usual, group C*-algebras provide some insightful examples. Recall that the left translation action of Γ on $\ell^\infty(\Gamma)$ is spatially implemented by the left regular representation (see Section 2.5).

Proposition 6.3.2. *Let Γ be a discrete group. Then $l^\infty(\Gamma)$ has an invariant mean (i.e., Γ is amenable) if and only if $C^*_\lambda(\Gamma)$ has an amenable trace.*

Proof. ($\Rightarrow$) Assume ψ is an invariant mean on $\ell^\infty(\Gamma)$. Consider the conditional expectation $\Phi\colon \mathbb{B}(\ell^2(\Gamma)) \to \ell^\infty(\Gamma)$ defined by

$$\Phi(T) = \sum_{g\in\Gamma} e_{g,g} T e_{g,g},$$

where $e_{g,g}$ is the rank-one projection onto $\delta_g \in l^2(\Gamma)$ and the sum is taken in the strong operator topology. In the matrix view, Φ simply maps T to its diagonal (which, of course, is an element in $\ell^\infty(\Gamma)$). A straightforward calculation shows that

$$\Phi(\lambda_s T \lambda_s^*) = \lambda_s \Phi(T) \lambda_s^* = s.\Phi(T)$$

for all $T \in \mathbb{B}(\ell^2(\Gamma))$ and $s \in \Gamma$. Define a state φ on $\mathbb{B}(\ell^2(\Gamma))$ by

$$\varphi = \psi \circ \Phi$$

and we find that

$$\varphi(\lambda_s T \lambda_s^*) = \psi(s.\Phi(T)) = \psi(\Phi(T)) = \varphi(T).$$

It follows that $\varphi(xT) = \varphi(Tx)$ for all $x \in \mathbb{C}[\Gamma]$ and density of the group algebra in $C^*_\lambda(\Gamma)$ implies that $\varphi(uTu^*) = \varphi(T)$ for all unitaries $u \in C^*_\lambda(\Gamma)$.

($\Leftarrow$) The converse is even easier. Assume $C^*_\lambda(\Gamma)$ has an amenable trace and let φ be a state on $\mathbb{B}(l^2(\Gamma))$ such that

$$\varphi(uTu^*) = \varphi(T)$$

for all unitaries $u \in C^*_\lambda(\Gamma)$ and $T \in \mathbb{B}(\ell^2(\Gamma))$. In particular, if $f \in l^\infty(\Gamma)$, then

$$\varphi(s.f) = \varphi(\lambda_s f \lambda_s^*) = \varphi(f)$$

and thus the restriction of φ to $l^\infty(\Gamma)$ is an invariant mean. □

Here is a more general result which shows that for *reduced* group C*-algebras there is an "all or nothing" principle: Either every trace is amenable or none are.

Proposition 6.3.3. *Let Γ be a discrete group. Then the following are equivalent:*

(1) *Γ is amenable;*

(2) *$C^*_\lambda(\Gamma)$ has an amenable trace;*

(3) *every trace on $C^*_\lambda(\Gamma)$ is amenable.*

Proof. Thanks to the previous result, we only have to observe (1) $\Rightarrow$ (3). However, if Γ is amenable, then $C^*_\lambda(\Gamma)$ is nuclear (Theorem 2.6.8) and thus for every trace τ on $C^*_\lambda(\Gamma)$, the product map

$$\pi_\tau \times \pi_\tau^{\mathrm{op}} \colon C^*_\lambda(\Gamma) \odot C^*_\lambda(\Gamma)^{\mathrm{op}} \to \mathbb{B}(L^2(C^*_\lambda(\Gamma), \tau))$$

is continuous with respect to the spatial tensor product norm (Proposition 3.6.12). Thus Theorem 6.2.7 applies and the proof is complete. □

The next result is also an immediate consequence of Proposition 3.6.12 and Theorem 6.2.7.

Proposition 6.3.4. *Every trace on a nuclear* C*-*algebra is amenable.*

Let's turn to some easy permanence-type questions.

Proposition 6.3.5. *The following assertions hold:*

(1) *if $B \subset A$ and τ is an amenable trace on A, then $\tau|_B$ is amenable;*

(2) *if $J \triangleleft A$ is an ideal and τ is an amenable trace on A/J, then the induced trace on A is also amenable;*

(3) *if τ_A and τ_B are amenable traces on A and B, respectively, then the product trace $\tau_A \otimes \tau_B \colon A \otimes B \to \mathbb{C}$ is amenable (hence it is amenable on $A \otimes_{\max} B$ as well);*

(4) *if $0 \to J \to A \to A/J \to 0$ is a locally split extension and τ is an amenable trace on A such that $\tau|_J = 0$, then τ drops to an amenable trace on A/J.*

Proof. The first three statements are trivial and are left to the reader. There are various ways to prove the last fact. Here is a tensor product argument. (You may want to give another proof using approximation properties and Arveson's Extension Theorem.)

By Proposition 3.7.6, the sequence

$$0 \to J \otimes A^{\mathrm{op}} \to A \otimes A^{\mathrm{op}} \to A/J \otimes A^{\mathrm{op}} \to 0$$

is exact. But the product map

$$A \otimes A^{\mathrm{op}} \to \mathbb{B}(L^2(A, \tau))$$

evidently kills $J \otimes A^{\mathrm{op}}$ and thus we have a product map

$$A/J \otimes A^{\mathrm{op}} \to \mathbb{B}(L^2(A, \tau)).$$

Letting $A/J \subset \mathbb{B}(\mathcal{H})$ be any faithful representation, we can apply The Trick to the inclusion $A/J \otimes A^{\mathrm{op}} \subset \mathbb{B}(\mathcal{H}) \otimes A^{\mathrm{op}}$ and appeal to the last condition in Theorem 6.2.7. □

Statement (4) in the previous proposition reveals a subtlety which one should be aware of: *the notion of amenable trace depends on the domain.* Indeed, it can easily happen that τ is amenable as a trace on A, but is not amenable on the quotient A/J (when the extension is not locally split). Natural examples of this phenomenon will be explored in the next section (see Theorem 6.4.3 and the paragraph which follows it). On the other hand, this problem never occurs for exact C*-algebras.

Proposition 6.3.6. *If A is exact and τ is a trace on A/J which is amenable when regarded as a trace on A, then it is also amenable on A/J.*

Proof. We can use the same tensor product argument as above. Indeed, since A^{op} is exact (Exercise 6.1.8), the sequence

$$0 \to J \otimes A^{\mathrm{op}} \to A \otimes A^{\mathrm{op}} \to A/J \otimes A^{\mathrm{op}} \to 0$$

is exact and this is all we needed. □

We close this section with the observation that amenable traces are always a nice subset of the tracial space.

Proposition 6.3.7. *The set of amenable traces on A is always a weak-* closed, convex face in the space of tracial states on A.*

Proof. The finite-dimensional approximation property which characterizes amenable traces can be used to show that amenable traces are closed – just think about it.

The facial property is a simple application of Proposition 3.8.3. Indeed, if $\tau = \lambda\tau_1 + (1-\lambda)\tau_2$, where $0 < \lambda < 1$, then in the GNS representation for τ we can find a subrepresentation which is unitarily equivalent to the GNS representation with respect to τ_1.[10] Thus the last statement in Theorem 6.2.7 shows that if τ is amenable, then so is τ_1.

Convexity is also very simple. If $A \subset \mathbb{B}(\mathcal{H})$, τ_1 and τ_2 are amenable traces with corresponding states φ_1 and, respectively, φ_2 on $\mathbb{B}(\mathcal{H})$ (as in Definition 6.2.1) and $0 < \lambda < 1$, then $\lambda\varphi_1 + (1-\lambda)\varphi_2$ is a state on $\mathbb{B}(\mathcal{H})$ (which evidently extends $\lambda\tau_1 + (1-\lambda)\tau_2$) and is easily seen to be A-central. □

Exercises

Exercise 6.3.1. Observe that every trace on a C*-algebra with Lance's WEP (Definition 3.6.7) is amenable.

Exercise 6.3.2. Observe that the trace on $C^*(\Gamma)$ coming from the trivial representation is always amenable.

[10] If $y_{\tau_1} \in \pi_\tau(A)'$ is such that $\tau_1(a) = \langle \pi_\tau(a) y_{\tau_1} \hat{1}, \hat{1} \rangle$, then the invariant subspace generated by $y_{\tau_1}\hat{1}$ will do.

Exercise 6.3.3. Prove that Γ is amenable if and only if $C^*_\lambda(\Gamma)$ has a finite-dimensional representation.

6.4. The factorization property and Kazhdan's property (T)

Before defining Kirchberg's factorization property, let's make some observations which help motivate the concept.

In Exercise 3.6.3 we observed that a discrete group Γ is amenable if and only if the product map,

$$C^*_\lambda(\Gamma) \odot C^*_\rho(\Gamma) \to \mathbb{B}(\ell^2(\Gamma)),$$

induced by the left and right regular representations is min-continuous. One might wonder what would happen if the reduced group C*-algebra were replaced by the universal one.

Proposition 6.4.1. *If Γ is a discrete group, the following are equivalent:*

(1) *Γ is amenable;*
(2) *the product $*$-homomorphism $C^*_\lambda(\Gamma) \odot C^*_\rho(\Gamma) \to \mathbb{B}(\ell^2(\Gamma))$ is min-continuous;*
(3) *the product $*$-homomorphism $C^*(\Gamma) \odot C^*_\rho(\Gamma) \to \mathbb{B}(\ell^2(\Gamma))$ is min-continuous;*
(4) *the product $*$-homomorphism $C^*_\lambda(\Gamma) \odot C^*(\Gamma) \to \mathbb{B}(\ell^2(\Gamma))$ is min-continuous.*

Proof. We only need to show (3) $\Rightarrow$ (1) and (4) $\Rightarrow$ (1), but they are similar so let's do (3) $\Rightarrow$ (1). Applying The Trick to the inclusion $C^*(\Gamma) \otimes C^*_\rho(\Gamma) \subset C^*(\Gamma) \otimes \mathbb{B}(\ell^2(\Gamma))$, we get a u.c.p. map from $\mathbb{B}(\ell^2(\Gamma))$ into the von Neumann algebra generated by the right regular representation which restricts to the identity on the image of the right regular representation. As we have seen (in the proof of Theorem 2.6.8), this implies the existence of an amenable trace on $C^*_\lambda(\Gamma)$, so the proof is complete. □

Of course, if spatial tensor products are replaced by maximal ones, then continuity is never an issue. Hence, if we want a nontrivial condition which doesn't imply amenability, then we are forced to consider the *spatial* tensor product with the *universal* group C*-algebra in *both* variables.

Definition 6.4.2. A discrete group Γ has Kirchberg's *factorization property* if the product map,

$$C^*(\Gamma) \odot C^*(\Gamma) \to \mathbb{B}(\ell^2(\Gamma))$$

induced by the left and right regular representations is min-continuous.[11]

[11] In other words, the product map $C^*(\Gamma) \otimes_{\max} C^*(\Gamma) \to \mathbb{B}(\ell^2(\Gamma))$ factors through the spatial tensor product.

With Theorem 6.2.7 at our disposal, the following fact is immediate. (See Exercise 6.1.4 for the identification with opposite algebras.)

Theorem 6.4.3. *A discrete group Γ has the factorization property if and only if the trace τ_λ on $C^*(\Gamma)$ coming from the left regular representation is amenable.*

Though it is a natural notion, relatively little is known about the factorization property. It is clear that amenable groups enjoy this property and Proposition 3.7.10 says that residually finite groups do as well.[12] Since many classical groups are residually finite, this shows that we have lots of examples with the factorization property. Unfortunately, there is only one other nontrivial fact which is known about this concept, namely, that for groups with Kazhdan's property (T), residual finiteness and the factorization property are equivalent. This result, due to Kirchberg, will be our focus for the remainder of this section.

Definition 6.4.4. Γ has *Kazhdan's property* (T) if every unitary representation which has almost invariant vectors actually has invariant vectors. More precisely, if $\pi\colon \Gamma \to \mathbb{B}(\mathcal{H})$ is a unitary representation and $v_i \in \mathcal{H}$ is a net of unit vectors such that $\|\pi_g(v_i) - v_i\| \to 0$, for all $g \in \Gamma$, then there exists $0 \neq v \in \mathcal{H}$ such that $\pi_g(v) = v$ for all $g \in \Gamma$.

Proposition 6.4.5. *For a discrete group Γ the following statements are equivalent:*

(1) *Γ has Kazhdan's property* (T)*;*

(2) *(critical sets) there exists a finite set $F \subset \Gamma$ and $\kappa > 0$ such that if $\pi\colon \Gamma \to \mathbb{B}(\mathcal{H})$ is a representation and $w \in \mathcal{H}$ is a unit vector such that $\|\pi_g(w) - w\| < \kappa$ for all $g \in F$, then there exists a nonzero vector $v_0 \in \mathcal{H}$ such that $\pi_s(v_0) = v_0$ for all $s \in \Gamma$;*

(3) *(quantitative version) there exists a finite set $F \subset \Gamma$ and $\kappa > 0$ such that for every $\epsilon > 0$ and representation $\pi\colon \Gamma \to \mathbb{B}(\mathcal{H})$, if $w \in \mathcal{H}$ is a vector such that $\|\pi_g(w) - w\| < \epsilon$ for all $g \in F$, then there exists a vector $v_0 \in \mathcal{H}$ such that $\pi_s(v_0) = v_0$, for all $s \in \Gamma$, and $\|w - v_0\| \leq \frac{\epsilon}{\kappa}$.*

Proof. (1) $\Rightarrow$ (2): Proving the contrapositive, we assume that for each finite set $F \subset \Gamma$ and $\kappa > 0$ there exists a representation $\pi\colon \Gamma \to \mathbb{B}(\mathcal{H})$ such that π has *no fixed vectors* but there is a unit vector $v \in \mathcal{H}$ such that $\|\pi_g(v) - v\| < \kappa$ for all $g \in F$. Applying this assumption to larger and larger finite sets and smaller and smaller κ's, we can find a collection of

[12] Hence if Γ is a residually finite, nonamenable group, then the canonical trace on $C^*(\Gamma)$ is amenable, but it is not amenable when regarded as a trace on $C^*_\lambda(\Gamma)$.

representations $\pi^{(i)} : \Gamma \to \mathbb{B}(\mathcal{H}_i)$ such that each $\pi^{(i)}$ has no fixed vectors but there is a net of unit vectors $v_i \in \mathcal{H}_i$ such that $\|\pi_g^{(i)}(v_i) - v_i\| \to 0$ for all $g \in \Gamma$. Taking the direct sum of the $\pi^{(i)}$'s, we get a representation with no fixed vectors but which does contain almost invariant vectors.

(2) $\Rightarrow$ (3): Let $F \subset \Gamma$ and $\kappa > 0$ be as in the second statement above. For an arbitrary representation $\pi : \Gamma \to \mathbb{B}(\mathcal{H})$ we let $\mathcal{H}_0 \subset \mathcal{H}$ be the (closed subspace) of all fixed vectors (it is possible that $\mathcal{H}_0 = \{0\}$) and we let $\mathcal{K} \subset \mathcal{H}$ be its orthogonal complement. Evidently $\mathcal{H}_0$ is an invariant subspace and hence $\mathcal{K}$ is as well. Since $\mathcal{K}$ has no fixed vectors, it follows that for each $v \in \mathcal{K}$ there exists $g \in F$ such that $\|\pi_g(v) - v\| \geq \kappa\|v\|$.

Let $w \in \mathcal{H}$ be arbitrary and write $w = v_0 \oplus v \in \mathcal{H}_0 \oplus \mathcal{K} = \mathcal{H}$. Then

$$\|w - v_0\| = \|v\| \leq \frac{1}{\kappa}\|\pi_g(v) - v\| = \frac{1}{\kappa}\|\pi_g(w) - w\|,$$

for some $g \in F$.

(3) $\Rightarrow$ (1) is obvious. □

Remark 6.4.6. A finite set F as in statement (2) above is called a *critical set*. The constant κ is called a *Kazhdan constant* for (Γ, F). The pair (F, κ) is called a *Kazhdan pair*. See [**15, 84**] for much more on property (T). (See also Section 12.1.)

Corollary 6.4.7. *Property* (T) *groups are finitely generated.*

Proof. Let $F \subset \Gamma$ and $\kappa > 0$ be as in statement (3) of the last proposition and let $\Gamma_0 \subset \Gamma$ be the subgroup generated by F. Consider the left translation action of Γ on the ℓ^2 space of left cosets – i.e., $s.(t\Gamma_0) = st\Gamma_0$. Since every element of F leaves the trivial coset invariant, it follows that every group element leaves Γ_0 invariant; hence $\Gamma_0 = \Gamma$. □

With the help of Stinespring's Theorem, it is not too difficult to extend the representation-theoretic rigidity of property (T) groups from one-dimensional subspaces to finite-dimensional u.c.p. maps. In C*-language, rigidity means that a u.c.p. map which is nearly multiplicative on a critical set is close (in trace) to an honest homomorphism.

Proposition 6.4.8. *Let $F \subset \Gamma$ be a critical set and κ be some Kazhdan constant. If $\varphi : C^*(\Gamma) \to \mathbb{M}_n(\mathbb{C})$ is a u.c.p. map such that*

$$\operatorname{tr}(1 - \varphi(g)\varphi(g^*)) < \frac{1}{2}\varepsilon^2\kappa^2$$

for some $\varepsilon > 0$ and all $g \in F$, then there exists an integer m and a unital $*$-homomorphism $\pi : C^*(\Gamma) \to \mathbb{M}_m(\mathbb{C})$ such that $|\operatorname{tr} \circ \varphi(x) - \operatorname{tr} \circ \pi(x)| < 5\varepsilon\|x\|$ for all $x \in C^*(\Gamma)$.

Proof. We can find a representation $\sigma\colon C^*(\Gamma) \to \mathbb{B}(\mathcal{H})$ and a finite-rank projection $P \in \mathbb{B}(\mathcal{H})$ such that φ may be identified with $x \mapsto P\sigma(x)P$. It follows that

$$\begin{aligned}\frac{\|\sigma(g)P\sigma(g)^* - P\|_2}{\|P\|_2} &= \left(2 - 2\frac{\mathrm{Tr}(P\sigma(g)P\sigma(g)^*)}{\mathrm{Tr}(P)}\right)^{1/2} \\ &= (2\,\mathrm{tr}(1 - \varphi(g)\varphi(g^*)))^{1/2} \\ &\leq \varepsilon\kappa\end{aligned}$$

for all $g \in \Gamma$. Hence the Hilbert-Schmidt unit vector $\frac{1}{\|P\|_2}P$ is almost invariant under the conjugation action $T \mapsto \sigma(g)T\sigma(g)^*$. By the quantitative statement above, we can thus find a Hilbert-Schmidt operator R such that $\|R\|_2 = 1$, $\sigma(g)R\sigma(g)^* = R$ for all $g \in \Gamma$, and

$$\|R - \frac{1}{\|P\|_2}P\|_2 < 2\varepsilon.$$

In other words, $R \in \sigma(C^*(\Gamma))'$ and it is close to $\frac{1}{\|P\|_2}P$. Observe that

$$\begin{aligned}&|\,\mathrm{Tr}(\sigma(x)RR^*) - \mathrm{tr}(\varphi(x))| \\ &= |\,\mathrm{Tr}(R^*\sigma(x)R) - \frac{\mathrm{Tr}(P\sigma(x)P)}{\mathrm{Tr}(P)}| \\ &= \left|\mathrm{Tr}\left(R^*\sigma(x)(R - \frac{P}{\|P\|_2})\right) + \mathrm{Tr}\left((R^* - \frac{P}{\|P\|_2})\sigma(x)\frac{P}{\|P\|_2}\right)\right| \\ &\leq 2\varepsilon\|\sigma(x^*)R\|_2 + 2\varepsilon\|\sigma(x)\frac{P}{\|P\|_2}\|_2 \\ &\leq 4\varepsilon\|x\|,\end{aligned}$$

for all $x \in C^*(\Gamma)$. Since RR^* commutes with $\sigma(C^*(\Gamma))$, so do all its spectral projections and hence we can find a finite-rank positive operator h *which commutes with* $\sigma(C^*(\Gamma))$, has rational eigenvalues, $\mathrm{Tr}(h) = 1$ and $\|RR^* - h\|_1 < \varepsilon$. For such R we have

$$|\,\mathrm{Tr}(\sigma(x)RR^*) - \mathrm{Tr}(\sigma(x)h)| \leq \|\sigma(x)\|\|RR^* - h\|_1 < \varepsilon\|x\|$$

and thus

$$|\,\mathrm{Tr}(\sigma(x)h) - \mathrm{tr}(\varphi(x))| < 5\varepsilon\|x\|,$$

for all $x \in C^*(\Gamma)$. By Lemma 6.2.5, there is a u.c.p. map $\pi'\colon \mathbb{B}(\mathcal{H}) \to \mathbb{M}_m(\mathbb{C})$ such that

$$\mathrm{tr}(\pi'(\sigma(x))) = \mathrm{Tr}(\sigma(x)h) \approx \mathrm{tr}(\varphi(x))$$

and

$$\mathrm{tr}\left(1 - \pi'(\sigma(g))\pi'(\sigma(g^*))\right) = 0$$

for all $g \in \Gamma$ (since $h \in \sigma(C^*(\Gamma))'$). Since tr is faithful, it follows that $\pi'(\sigma(g))$ is a unitary element for every g and hence all of $\sigma(C^*(\Gamma))$ falls in the multiplicative domain of π'. Defining $\pi = \pi' \circ \sigma$, the proof is complete. □

Remark 6.4.9. It is possible to avoid Lemma 6.2.5 in the proof above, as follows. Write

$$h = \frac{p_1}{q} Q_1 \oplus \frac{p_2}{q} Q_2 \oplus \cdots \oplus \frac{p_k}{q} Q_k$$

where $\frac{p_1}{q} < \frac{p_2}{q} < \cdots < \frac{p_k}{q}$ are the nonzero eigenvalues of h and $Q_1, \ldots, Q_k$ are the corresponding spectral projections. Define a $*$-homomorphism

$$\pi' : C^*(\Gamma) \to Q_1 \mathbb{B}(\mathcal{H}) Q_1 \oplus Q_2 \mathbb{B}(\mathcal{H}) Q_2 \oplus \cdots \oplus Q_k \mathbb{B}(\mathcal{H}) Q_k$$

by

$$\pi'(x) = Q_1 \sigma(x) Q_1 \oplus \cdots \oplus Q_k \sigma(x) Q_k.$$

There is a trace τ on $Q_1 \mathbb{B}(\mathcal{H}) Q_1 \oplus \cdots \oplus Q_k \mathbb{B}(\mathcal{H}) Q_k$ such that $\tau(\pi'(x)) = \mathrm{Tr}(\sigma(x)h)$ (just think about it) and hence we can apply Exercise 6.2.3 and replace the finite-dimensional algebra $Q_1 \mathbb{B}(\mathcal{H}) Q_1 \oplus \cdots \oplus Q_k \mathbb{B}(\mathcal{H}) Q_k$ with a full matrix algebra.

The following corollary is immediate from the previous result and Theorem 6.2.7.

Corollary 6.4.10. *Assume Γ has property* (T) *and τ is a tracial state on $C^*(\Gamma)$. Then τ is amenable if and only if there exist $*$-homomorphisms $\pi_n : C^*(\Gamma) \to \mathbb{M}_{k(n)}(\mathbb{C})$ such that* $\lim \mathrm{tr}(\pi_n(x)) = \tau(x)$ *for all $x \in C^*(\Gamma)$.*

Remark 6.4.11. In most instances, the previous corollary is what we need, but note that more is true. Namely, if $J \triangleleft C^*(\Gamma)$ is an ideal and τ is an amenable trace on the quotient $C^*(\Gamma)/J$, then there exist $*$-homomorphisms

$$\pi_n : C^*(\Gamma)/J \to \mathbb{M}_{k(n)}(\mathbb{C})$$

such that $\lim \mathrm{tr}(\pi_n(x)) = \tau(x)$ for all $x \in C^*(\Gamma)/J$. This is because Proposition 6.4.8 also holds for $C^*(\Gamma)/J$ (the proof amounts to changing a bit of notation).

We are now very close to Kirchberg's result asserting property (T) groups with the factorization property are residually finite; we just need an important theorem of Malcev on the structure of finitely generated linear groups.

Theorem 6.4.12. *Let K be a field which is finitely generated as a ring. Then K is finite.*

Though elementary and well known, this result doesn't appear in many basic algebra texts; hence we include the proof at the end of this section. Deducing Malcev's result from it, however, is easy.

Theorem 6.4.13 (Malcev). *Let K be a field. Every finitely generated subgroup of $GL(n, K)$ is residually finite.*

Proof. Let $T_1, \ldots, T_n \in GL(n, K)$ be given and let Γ be the group they generate. Let $R \subset K$ be the ring generated by all the entries of the matrices $T_1, \ldots, T_n, T_1^{-1}, \ldots, T_n^{-1}$. Evidently R is finitely generated, abelian and $\Gamma \subset GL(n, R)$. It suffices to show that R is residually finite; that is, there exist ideals $I_n \lhd R$ such that $|R/I_n| < \infty$, for all n, and $\bigcap I_n = 0$. Indeed, if this is true, then the maps $GL(n, R) \to GL(n, R/I_n)$ will provide a separating family of finite quotients of Γ.

So, fix a finite set $F \subset R$. Let $\tilde{R} \subset K$ be the (finitely generated, abelian) ring generated by R and $\{(s-t)^{-1} : s, t \in F, s \neq t\}$. Let $M \lhd \tilde{R}$ be a (nontrivial) maximal ideal; hence, $\tilde{R}/M$ is a field, which is finitely generated as a ring; hence, it is a finite field, by the previous theorem. Note that M can't contain any of the elements $\{s - t : s, t \in F, s \neq t\}$ since they are all invertible in $\tilde{R}$ (and $M \neq \tilde{R}$). Thus $I = R \cap M$ is a finite-index ideal which separates points in F. □

Theorem 6.4.14. *Let Γ be a property (T) group. Then Γ has the factorization property if and only if Γ is residually finite.*

Proof. Let τ_λ be the trace on $C^*(\Gamma)$ coming from the left regular representation. Since we already know residually finite groups have the factorization property (Proposition 3.7.10), we may assume that τ_λ is amenable (cf. Theorem 6.4.3). By Corollary 6.4.10, there exist representations $\pi_n : C^*(\Gamma) \to \mathbb{M}_{k(n)}(\mathbb{C})$ such that for each nonneutral element $g \in \Gamma$ we have

$$\mathrm{tr}(\pi_n(g)) \to \tau_\lambda(g) = 0.$$

In particular, π_n is a separating family of representations and thus Malcev's theorem implies that Γ is residually finite. □

Remark 6.4.15. In [**70**] it was shown that there are infinite property (T) groups which are simple. Such groups are the only known examples which fail the factorization property.

In [**129**] it was shown that every hyperbolic group can be embedded into a hyperbolic group with property (T). Since residual finiteness obviously passes to subgroups, we get the following corollary.

Corollary 6.4.16. *Every hyperbolic group is residually finite if and only if every hyperbolic group has the factorization property.*

Whether or not every hyperbolic group is residually finite has been a long-standing open problem in geometric group theory. The corollary above says that this is the case if and only if the natural $*$-homomorphism

$$C^*(\Gamma) \odot C^*(\Gamma) \to \mathbb{B}(\ell^2(\Gamma))$$

is continuous with respect to the minimal norm for all hyperbolic groups Γ. However, if one passes to the Calkin algebra (instead of stopping in $\mathbb{B}(\ell^2(\Gamma))$), then the map is min-continuous for all hyperbolic groups (Corollary 5.3.20). This remark is unlikely to help resolve the question of residual finiteness for hyperbolic groups – it's just a cute observation.

As promised, we now give the proof of Theorem 6.4.12.[13] We need two simple lemmas.

Lemma 6.4.17. *Let R be a unique factorization domain with infinitely many distinct irreducibles. Then R's quotient field can't be finitely generated as a ring.*

In particular, $K(x_1, x_2, \ldots)$ – the field of rational functions in (possibly infinitely many) indeterminants $x_1, x_2, \ldots$ over a field K – is not finitely generated. Neither are the rational numbers $\mathbb{Q}$.

Proof. Every element in the quotient field F of R has an essentially unique representation as a fraction

$$\frac{p_1 p_2 \cdots p_s}{q_1 q_2 \cdots q_t},$$

for some irreducibles $p_1, \ldots, p_s$ and $q_1, \ldots, q_t$. Let $\{r_1, r_2, \ldots, r_n\} \subset F$ be a finite set. One easily checks that if an irreducible $q \in R$ does not appear in the denominator of any of the r_i's, then $\frac{1}{q}$ will not belong to the ring generated by $r_1, \ldots, r_n$. Hence F can't be finitely generated.

Euclid proved that $\mathbb{Z}$ has infinitely many primes (= irreducibles); the same proof shows the polynomial ring $K[x_1, x_2, \ldots]$ does too. □

Lemma 6.4.18. *Assume K is a finite field extension of another field F and K is finitely generated as a ring. Then so is F.*

Proof. By induction, we may assume that $K = \langle F, x \rangle$ is a simple extension – i.e., generated by F and one element $x \notin F$. Since $[K : F] < \infty$, x must be algebraic over F. Let $P(X) = X^d + f_{d-1}X^{d-1} + \cdots + f_0$, $f_i \in F$, be its minimal polynomial – hence $\{1, x, x^2, \ldots, x^{d-1}\}$ is a basis for K as a vector space over F ([**87**, Theorem V.1.6]). Let $k_1, \ldots, k_m$ be a finite generating set of K (as a ring). Find coefficients $g_{i,j} \in F$ such that $k_i = \sum_{j=0}^{d-1} g_{i,j} x^j$ and let R be the subring of F generated by the f_i's and $g_{i,j}$'s. We will show $R = F$.

[13]Our proof is borrowed from lecture notes for a geometric group theory course (Math 257) taught by Stallings at Berkeley in the fall of 2000. (Notes are available at Stallings's website.) He, in turn, credits the argument to Shalen.

Given $z \in F$, we can write z as an integer polynomial in the k_i's. Hence, substituting in our decompositions of the k_i's,

$$z = \sum_{s=0}^{n} r_n x^n$$

for some r_j's coming from R. If $n < d$, then linear independence of the elements $\{1, x, x^2, \dots, x^{d-1}\}$ implies $z = r_0$. If $n \geq d$, we can apply Euclid's algorithm and divide by the minimal polynomial P. Since the coefficients of P belong to R, we can rewrite z as a polynomial of degree $< d$ with coefficients in R, and this completes the proof. (If $T = QP + T'$ are polynomials and all coefficients of T, Q and P come from R – as they do in this case, since Q will be a polynomial in $k_1, \dots, k_m$ – then the coefficients of T' also come from R.) □

Now we can complete the proof of Theorem 6.4.12.

Proof of Theorem 6.4.12. Let K be a finitely generated field and P be its prime subfield. We will show $[K : P] < \infty$ and P is a finite field; evidently this implies the theorem.

Let $z_1, z_2, \cdots$ be a (possibly finite, maybe even empty) transcendence basis for K over P. The field generated by P and $\{z_1, z_2, \cdots\}$ is isomorphic to the field of rational functions $P(x_1, x_2, \dots)$ ([**87**, Theorem VI.1.2]). Every element of K is algebraic over $P(z_1, z_2, \dots)$ (by maximality of a transcendence basis); in particular, every element in a generating set of K is algebraic over $P(z_1, z_2, \dots)$, which implies $[K : P(z_1, z_2, \dots)] < \infty$. Thus Lemma 6.4.18 implies $P(z_1, z_2, \dots)$ is finitely generated. Hence, by Lemma 6.4.17, the transcendence basis must be empty – that is, $[K : P] < \infty$. Since P is either the rational numbers or the field of p elements for some prime number p, similar reasoning shows P is a finite field. □

Exercises

Exercise 6.4.1. Show that a discrete group Γ has the factorization property if and only if every finitely generated subgroup does. (Hint: You will need the fact that if $\Gamma_0 \subset \Gamma$ is a subgroup, then $C^*(\Gamma_0) \subset C^*(\Gamma)$, by Proposition 2.5.8.)

Exercise 6.4.2. A group is called *maximally almost periodic* if it has a separating family of finite-dimensional unitary representations (this is equivalent to being a subgroup of a compact group). Show that every maximally almost periodic group has the factorization property.

Exercise 6.4.3. Let Γ be a discrete group with property (T). Show that Γ is residually finite if and only if there exists an embedding of Γ into the unitary group of the hyperfinite II_1-factor R.[14]

6.5. References

Theorem 6.2.7 comes from [**105**], though our proof is found in [**134**]. The main results in Section 6.4 are also due to Kirchberg [**105**].

[14] Robertson has generalized this result (of Kirchberg) by replacing R with any II_1-factor with the Haagerup approximation property ([**166**]).

Chapter 7

Quasidiagonal C*-Algebras

In this chapter we present the basics of a large and intriguing class of C*-algebras. Quasidiagonality can be formulated in many ways, but the abundance of characterizations and natural examples should not lead one to believe that these algebras are easily understood. On the contrary, they can be deceptively difficult.

Most of the material in this chapter is relatively easy, but there are two important theorems which deserve recognition. The first is Voiculescu's homotopy invariance theorem, Theorem 7.3.6, which shows (among other things) that the natural K-theoretic operation of suspension always produces quasidiagonal C*-algebras – i.e., for arbitrary A, $SA = C_0(0,1) \otimes A$ is quasidiagonal. The other important result, due to Dadarlat, is a strong approximation theorem in the presence of exactness (Theorem 7.5.7).

7.1. The definition, easy examples and obstructions

As with nuclear and exact C*-algebras, we will develop the theory of quasidiagonality in reverse chronological order. Our definition is in terms of finite-dimensional approximations; the next section returns to the original representation-theoretic notion.

Definition 7.1.1. A C*-algebra A is called *quasidiagonal* (QD) if there exists a net of c.c.p. maps $\varphi_n \colon A \to \mathbb{M}_{k(n)}(\mathbb{C})$ which is both asymptotically multiplicative (i.e., $\|\varphi_n(ab) - \varphi_n(a)\varphi_n(b)\| \to 0$ for all $a, b \in A$) and asymptotically isometric (i.e., $\|a\| = \lim_{n\to\infty} \|\varphi_n(a)\|$ for all $a \in A$).

Remark 7.1.2. Completely positive maps respect linear, involutive and order structures; the definition above requires that all other C*-structures (i.e., norm and multiplication) be asymptotically preserved. Hence, one can think of QD C*-algebras as those which admit matrix models approximately recapturing *all* C*-structures.[1]

Just as for exactness and nuclearity, this is really a local property (hence, there is no substantial difference between the separable and nonseparable settings). Obviously we can localize as follows:

Lemma 7.1.3. *A is QD if and only if for each finite set $\mathfrak{F} \subset A$ and $\varepsilon > 0$ there exists a c.c.p. map $\varphi\colon A \to \mathbb{M}_n(\mathbb{C})$ such that*

$$\|\varphi(ab) - \varphi(a)\varphi(b)\| < \varepsilon$$

and

$$\|\varphi(a)\| > \|a\| - \varepsilon$$

for all $a, b \in \mathfrak{F}$.

As always, there are some nonunital questions that need to be resolved.

Lemma 7.1.4. *If A is unital and QD, then there exist u.c.p. maps $\varphi_n\colon A \to \mathbb{M}_{k(n)}(\mathbb{C})$ which are both asymptotically multiplicative and asymptotically isometric.*

Proof. Let $\varphi_n'\colon A \to \mathbb{M}_{l(n)}(\mathbb{C})$ be asymptotically multiplicative and isometric c.c.p. maps. Functional calculus shows that the spectra of the matrices $\varphi_n'(1_A)$ are contained in sets of the form $[0, \varepsilon_n) \cup (\varepsilon_n, 1]$, where $\varepsilon_n \to 0$, and hence

$$\|\varphi_n'(1_A) - P_n\| \to 0,$$

where $P_n \in \mathbb{M}_{l(n)}(\mathbb{C})$ are the spectral projections of $\varphi_n'(1_A)$ corresponding to $[1/2, 1]$.

Thus $\varphi_n'(1_A)P_n$ is an invertible element in $P_n\mathbb{M}_{l(n)}(\mathbb{C})P_n$ and more functional calculus shows

$$\|(\varphi_n'(1_A)P_n)^{-\frac{1}{2}} - P_n\| \to 0$$

as well. If $k(n)$ is the rank of P_n, then we get the desired u.c.p. maps $\varphi_n\colon A \to \mathbb{M}_{k(n)}(\mathbb{C})$ by defining

$$\varphi_n(a) = (\varphi_n'(1_A)P_n)^{-\frac{1}{2}}\varphi_n'(a)(\varphi_n'(1_A)P_n)^{-\frac{1}{2}}.$$

□

[1] This sentence better describes MF algebras (see Definition 11.1.6). For those who care, the thing which distinguishes QD algebras from MF algebras is the existence of c.c.p. maps connecting the algebra with its matrix models.

Another thing one should check is whether quasidiagonality passes to unitizations. It does (Exercise 7.1.2); thus most quasidiagonal questions can be reduced to the unital separable case.

How about some simple examples and permanence properties?

Proposition 7.1.5. *Every abelian* C*-*algebra is QD.*

Proof. Use direct sums of point evaluations to construct finite-dimensional $*$-homomorphisms. □

Abelian algebras are particular examples of *residually finite-dimensional* C*-algebras. Many QD questions can be reduced to these natural analogues of block diagonal operators on a Hilbert space.

Definition 7.1.6. A C*-algebra A is called *residually finite-dimensional* (RFD) if there exist finite-dimensional $*$-homomorphisms $\pi_n \colon A \to \mathbb{M}_{k(n)}(\mathbb{C})$ such that $\bigoplus \pi_n \colon A \to \prod \mathbb{M}_{k(n)}(\mathbb{C})$ is faithful.[2]

The following fact should be obvious.

Proposition 7.1.7. *Residually finite-dimensional* C*-*algebras are QD.*

Here is an often forgotten class of RFD C*-algebras.

Proposition 7.1.8. *Every type* I C*-*algebra with a* faithful *tracial state is RFD (hence QD).*

Proof. Let τ be a faithful trace on a type I algebra A and let $\pi_\tau \colon A \to \mathbb{B}(L^2(A,\tau))$ be the associated GNS representation. Since τ is faithful, so is π_τ.

By the structure theory of type I von Neumann algebras we have

$$\pi_\tau(A)'' \cong \prod_n L^\infty(X_n, \mu_n) \bar{\otimes} \mathbb{B}(\mathcal{H}_n).$$

But $\pi_\tau(A)''$ has a faithful trace (the vector state in the GNS representation is tracial and faithful) and hence the dimensions of all the $\mathcal{H}_n$'s must be finite. Since it is simple to show that $C(X) \otimes \mathbb{M}_k(\mathbb{C})$ is RFD, the remainder of the proof is straightforward. □

Of course, finite-dimensional algebras are QD, so the next proposition implies the same for AF algebras. In fact, since subhomogeneous algebras (Definition 2.7.6) are RFD – the direct sum of all irreducible representations is always faithful – it follows that inductive limits of subhomogeneous algebras (aka ASH algebras) are also QD.

[2]In the nonseparable setting one may need uncountably many representations, of course.

Proposition 7.1.9. *Quasidiagonality passes to inductive limits, so long as the connecting maps are injective.*[3]

Proof. Assume $A = \overline{\bigcup A_n}$ where $A_n \subset A_{n+1}$ are QD subalgebras of A. Apply Arveson's Extension Theorem to all c.c.p. maps $A_n \to \mathbb{M}_{k(n)}(\mathbb{C})$. □

Here are two trivial permanence properties (proofs are left to the reader).

Proposition 7.1.10. *Subalgebras of QD algebras are also QD.*

Proposition 7.1.11. *If each A_n is QD, then so is $\prod_n A_n$ and (hence) $\bigoplus_n A_n$.*

The next permanence property is a little harder.[4]

Proposition 7.1.12. *The minimal tensor product of two QD C*-algebras is again QD.*

Proof. Let $\varphi_i\colon A \to \mathbb{M}_{k(i)}(\mathbb{C})$ and $\psi_j\colon B \to \mathbb{M}_{l(j)}(\mathbb{C})$ be c.c.p. maps which are asymptotically multiplicative and isometric. The obvious thing to do is to consider

$$\varphi_i \otimes \psi_j\colon A \otimes B \to \mathbb{M}_{k(i)}(\mathbb{C}) \otimes \mathbb{M}_{l(j)}(\mathbb{C})$$

and hope that we get an asymptotically isometric sequence of maps (asymptotic multiplicativity is clear). This is essentially true but requires proof.

Let

$$\Phi\colon A \to \frac{\prod \mathbb{M}_{k(i)}(\mathbb{C})}{\bigoplus \mathbb{M}_{k(i)}(\mathbb{C})}$$

and

$$\Psi\colon B \to \frac{\prod \mathbb{M}_{l(j)}(\mathbb{C})}{\bigoplus \mathbb{M}_{l(j)}(\mathbb{C})}$$

be the maps obtained by composing $\bigoplus \varphi_i\colon A \to \prod \mathbb{M}_{k(i)}(\mathbb{C})$ and, respectively, $\bigoplus \psi_j\colon B \to \prod \mathbb{M}_{l(j)}(\mathbb{C})$ with the quotient maps

$$\prod \mathbb{M}_{k(i)}(\mathbb{C}) \to \frac{\prod \mathbb{M}_{k(i)}(\mathbb{C})}{\bigoplus \mathbb{M}_{k(i)}(\mathbb{C})} \text{ and, respectively, } \prod \mathbb{M}_{l(j)}(\mathbb{C}) \to \frac{\prod \mathbb{M}_{l(j)}(\mathbb{C})}{\bigoplus \mathbb{M}_{l(j)}(\mathbb{C})}.$$

Note that Φ and Ψ are injective $*$-homomorphisms. Hence the tensor product homomorphism

$$\Phi \otimes \Psi\colon A \otimes B \to \left(\frac{\prod \mathbb{M}_{k(i)}(\mathbb{C})}{\bigoplus \mathbb{M}_{k(i)}(\mathbb{C})}\right) \otimes \left(\frac{\prod \mathbb{M}_{l(j)}(\mathbb{C})}{\bigoplus \mathbb{M}_{l(j)}(\mathbb{C})}\right)$$

---

[3]The assumption of injective connecting maps is necessary – see Remark 17.3.3.

[4]A two-line proof can be given once we know the representation theorem from Section 7.2. However, the present argument uses some nice ideas which get used all of the time.

is also injective. Since injective representations of C*-algebras are always isometric, it follows that the (c.c.p.) tensor product map

$$(\bigoplus \varphi_i) \otimes (\bigoplus \psi_j) \colon A \otimes B \to \left(\prod \mathbb{M}_{k(i)}(\mathbb{C})\right) \otimes \left(\prod \mathbb{M}_{l(j)}(\mathbb{C})\right)$$

is also isometric. Using the identification

$$\left(\prod \mathbb{M}_{k(i)}(\mathbb{C})\right) \otimes \left(\prod \mathbb{M}_{l(j)}(\mathbb{C})\right) \cong \prod_{i,j} \mathbb{M}_{k(i)}(\mathbb{C}) \otimes \mathbb{M}_{l(j)}(\mathbb{C}),$$

one can find an asymptotically isometric sequence by taking direct sums of maps of the form $\varphi_i \otimes \psi_j$. □

Remark 7.1.13. It isn't known if the maximal tensor product of QD algebras is again QD.

We will see more examples and permanence properties in the exercises and later sections of this chapter. However, we now wish to discuss the only two known obstructions to quasidiagonality (which will give us lots of examples of nonquasidiagonal C*-algebras).

Obstruction 1: Every QD C*-algebra is stably finite.

Recall that an isometry s is called *proper* if $1 - ss^* \neq 0$.

Definition 7.1.14. A unital C*-algebra A is *stably finite* if $\mathbb{M}_n(\mathbb{C}) \otimes A$ contains no proper isometries, for every $n \in \mathbb{N}$. A nonunital algebra is stably finite if its unitization is.

Proposition 7.1.15. *Every QD C*-algebra is stably finite.*

Proof. The proof boils down to the following routine exercise: If $T_n \in \mathbb{M}_{k(n)}(\mathbb{C})$ and $\|T_n^*T_n - 1_{k(n)}\| \to 0$, then $\|T_nT_n^* - 1_{k(n)}\| \to 0$ too.

Hoping for a contradiction, we assume A is quasidiagonal and that it contains a proper isometry $s \in A$. (Since $\mathbb{M}_n(A)$ is QD whenever A is, we may assume A contains the proper isometry.) Let $\varphi_n \colon A \to \mathbb{M}_{k(n)}(\mathbb{C})$ be asymptotically multiplicative, asymptotically isometric u.c.p. maps. Then $1 = \varphi_n(s^*s) \approx \varphi_n(s^*)\varphi_n(s)$ while $\varphi_n(s)\varphi_n(s)^*$ can't get close to 1 since $1 - ss^* \neq 0$. This contradicts the exercise above. □

It follows that many standard examples of C*-algebras are *not* QD: the Toeplitz algebra, $\mathbb{B}(\mathcal{H})$, Cuntz algebras, and many others (cf. Sections 4.5 and 4.6).

One may wonder whether or not there is a converse to the previous proposition and the answer is: "No; yes; we don't know; maybe." More precisely, in general it is not true that every stably finite C*-algebra is QD –

we will see counterexamples after describing the other obstruction to quasidiagonality. In some cases (e.g., crossed products of either abelian or AF algebras by $\mathbb{Z}$) it is true that stable finiteness implies QD (though this is not easy to prove). In some cases there are questions which may turn out to have positive or negative answers and it is not clear which way things should go. For example, a question of Blackadar and Kirchberg asks whether every stably finite *nuclear* C*-algebra is QD. There is some evidence suggesting a positive answer, but we feel it is rather thin and we would not be surprised by counterexamples. Even the case of stably finite type I algebras is not presently known, as pointed out to us by David Kerr.

Obstruction 2: Every *unital* QD C*-algebra has an amenable trace.

Since amenable traces are characterized by a finite-dimensional approximation property (Theorem 6.2.7), this is a very simple observation indeed.

Proposition 7.1.16. *Every unital QD* C**-algebra has an amenable trace.*[5]

Proof. Let $\varphi_n \colon A \to \mathbb{M}_{k(n)}(\mathbb{C})$ be a sequence of u.c.p. maps which are asymptotically multiplicative. If tr is the unique tracial state on $\mathbb{M}_{k(n)}(\mathbb{C})$, then the functionals $\mathrm{tr} \circ \varphi_n$ are states on A, any cluster point of which is easily seen to be an amenable trace. □

Combining with Proposition 6.3.3, we get the following result.

Corollary 7.1.17. *If Γ is a discrete group and $C^*_\lambda(\Gamma)$ is QD, then Γ is amenable.*

Jonathan Rosenberg first observed the previous corollary and conjectures the converse. The problem appears to be very hard, however.

Here are some nonquasidiagonal, yet stably finite, examples.

Corollary 7.1.18. *If Γ is any discrete, nonamenable group, then $C^*_\lambda(\Gamma)$ is not QD. In particular, $C^*_\lambda(\mathbb{F}_n)$ is stably finite (since it has a faithful trace) and exact, but not QD.*

Exercises

Exercise 7.1.1. Show that A is QD if and only if all its separable subalgebras are QD.

Exercise 7.1.2. If A is nonunital and QD, then the unitization $\tilde{A}$ is also QD. (Use Proposition 2.2.1 and check that the unitized maps are still asymptotically multiplicative and isometric.)

[5]The assumption of a unit is necessary. The compact operators, being AF, are QD but have no tracial states.

Exercise 7.1.3. Show that A is QD if and only if there exists an injective $*$-homomorphism

$$A \to \frac{\prod_{n\in\mathbb{N}} \mathbb{M}_n(\mathbb{C})}{\bigoplus_{n\in\mathbb{N}} \mathbb{M}_n(\mathbb{C})}$$

which admits a c.c.p. splitting $A \to \prod_{n\in\mathbb{N}} \mathbb{M}_n(\mathbb{C})$.

Exercise 7.1.4. Let $P_1 \leq P_2 \leq \cdots \in \mathbb{B}(\mathcal{H})$ be finite-rank projections which converge strongly to the identity and denote by C the subset of

$$\prod_{n\in\mathbb{N}} P_n \mathbb{B}(\mathcal{H}) P_n$$

consisting of those sequences (T_n) for which there exists $T \in \mathbb{B}(\mathcal{H})$ such that $T_n \to T$ in the strong-$*$ topology (i.e., $T_n \to T$ and $T_n^* \to T^*$ in the strong operator topology). Prove that C is a C*-subalgebra of $\prod_{n\in\mathbb{N}} P_n \mathbb{B}(\mathcal{H}) P_n$. Show that $\mathbb{B}(\mathcal{H})$ is a quotient of C. Since C is clearly RFD, it follows that every separable C*-algebra is a quotient of an RFD algebra.

Definition 7.1.19. An extension $0 \to J \to A \to A/J \to 0$ is called *quasidiagonal* if J contains a quasicentral approximate unit consisting of projections.

Exercise 7.1.5. Show that if $0 \to J \to A \to A/J \to 0$ is a quasidiagonal extension, A is QD, *and* the extension is locally split, then A/J is also QD. (Hint: If $\{p_n\} \in J$ is a quasicentral approximate unit of projections and $\Phi\colon A/J \to A$ is a u.c.p. splitting, then the nonunital maps $\Phi_n(b) = (1 - p_n)\Phi(b)(1 - p_n)$ are asymptotically multiplicative. Now localize this fact.)

Exercise 7.1.6. Show that if $0 \to J \to A \to A/J \to 0$ is a quasidiagonal extension and both J and A/J are QD, then so is A.

7.2. The representation theorem

Though Definition 7.1.1 is perfectly natural from the C*-perspective, quasidiagonality was originally imported from single operator theory and formulated in terms of representation theory. The interested reader can find more on the operator theory origins in Chapter 16 but in this section we restrict our attention to the C*-ideas and prove the fundamental representation theorem.

Definition 7.2.1. Let $\Omega \subset \mathbb{B}(\mathcal{H})$ be an arbitrary collection of operators. Then Ω is called a *quasidiagonal set* if for each finite set $\mathfrak{F} \subset \Omega$, each finite set $\chi \subset \mathcal{H}$ and each $\varepsilon > 0$ there exists a finite-rank projection $P \in \mathbb{B}(\mathcal{H})$ such that

$$\|PT - TP\| < \varepsilon$$

for all $T \in \mathfrak{F}$ and

$$\|Pv - v\| < \varepsilon$$

for all $v \in \chi$.

As usual, this local formulation is handy when trying to verify quasidiagonality for a particular set and not so useful when proving something about a known quasidiagonal set. To get a better characterization, we need a lemma on perturbing projections.

Lemma 7.2.2. *Let A be a unital* C*-*algebra and let $p, q \in A$ be projections.*

(1) *If $\|q - p\| < 1$, then there is a unitary $u \in A$ with $uqu^* = p$ and $\|1 - u\| \le 4\|p - q\|$.*

(2) *If $\|q - pq\| < 1/4$, then there is a unitary element $u \in A$ such that $uqu^* \le p$ (with equality whenever $\|q - p\| < 1$) and $\|1 - u\| \le 10\|q - pq\|$.*

Proof. First assume $\|q - p\| < 1$, consider $x = pq$ and observe that

$$\|x^*x - q\| = \|q(p - q)q\| < 1;$$

hence x^*x is invertible in qAq. Letting $|x|_q^{-1} \in qAq$ be such that $|x|_q^{-1}|x| = q$ and $v = x|x|_q^{-1}$, we see that $x = v|x|$ is the polar decomposition. In particular, $v^*v = q$ and $p_0 = vv^* \le p$. But, xx^* is invertible in pAp ($xx^* - p = p(q - p)p$) and so the range projection of x must be p – i.e., $p_0 = vv^* = p$. We also have that

$$\|q - v\| \le \|q - x\| + \|v|x| - v\| \le \|(q - p)q\| + \|v(|x| - q)\| \le 2\|q - p\|$$

since $\||x| - q\| \le \|x^*x - q\| \le \|p - q\|$. Applying the same argument to orthogonal projections we find a partial isometry $w \in A$ such that $q^\perp = w^*w$, $p^\perp = ww^*$ and $\|q^\perp - w\| \le 2\|q - p\|$. Defining our unitary $u = v + w$ completes the proof of the first assertion.

Now suppose that $\varepsilon = \|q - pq\| < 1/4$. If $x = pq$, we have $\|q - x^*x\| \le \varepsilon$; hence $|x|$ is invertible in qAq. As above, put $v = x|x|_q^{-1}$ and evidently $q = v^*v$ and $p_0 = vv^* \le p$ (with equality whenever $\|q - p\| < 1$). Also, just as above, $\|q - v\| \le \|q - x\| + \|v|x| - v\| \le 2\varepsilon$, which implies $\|q - p_0\| \le 4\varepsilon$. Now, since $\|q^\perp - p_0^\perp\| \le 4\varepsilon < 1$, the first assertion yields a partial isometry $w \in A$ such that $q^\perp = w^*w$, $p_0^\perp = ww^*$ and $\|q^\perp - w\| \le 8\varepsilon$. It follows that again $u = v + w$ is a unitary element with the right properties. □

Proposition 7.2.3. *Let $\Omega \subset \mathbb{B}(\mathcal{H})$ be a norm separable, quasidiagonal set of operators on a separable Hilbert space.*[6] *There exists an increasing sequence of finite-rank projections $P_1 \le P_2 \le \cdots$ converging strongly to the identity and such that $\|[P_n, T]\| \to 0$ for all $T \in \Omega$.*

[6]We leave it to the reader to formulate and prove a nonseparable version.

Proof. With the help of the previous lemma we'll show that Definition 7.2.1 implies a stronger statement: For each finite set $\mathfrak{F} \subset \Omega$, $\chi \subset \mathcal{H}$ and $\varepsilon > 0$ there exists a finite-rank projection $P \in \mathbb{B}(\mathcal{H})$ such that $\|PT - TP\| < \varepsilon$ for all $T \in \mathfrak{F}$ and (here's the point) $P(v) = v$ for all $v \in \chi$. Once established, the remainder of the proof is routine since this stronger property evidently allows one to construct the desired increasing projections.

So, let $\mathfrak{F} \subset \Omega$, $\chi \subset \mathcal{H}$ and $\varepsilon > 0$ be given. If Q is the orthogonal projection onto the span of χ, then, by norm-compactness of the unit ball of $Q\mathcal{H}$, we can find a larger finite set $\tilde{\chi} \subset Q\mathcal{H}$ with the property that $\|\tilde{P}Q - Q\| < 3\delta$, whenever $\tilde{P}$ is a finite-rank projection satisfying $\|\tilde{P}(v) - v\| < \delta$ for all $v \in \tilde{\chi}$. Since Ω is a quasidiagonal set, we can find such a projection $\tilde{P}$ which also approximately commutes with $\mathfrak{F}$. Then, if δ is small, we can apply the previous lemma to find a unitary $U \in \mathbb{B}(\mathcal{H})$ such that $Q \leq U\tilde{P}U^*$ and $\|U - 1\| \leq 30\delta$; hence, $\|\tilde{P} - U\tilde{P}U^*\| \leq 60\delta$. Defining $P = U\tilde{P}U^*$, we have constructed a projection which dominates Q and almost commutes with $\mathfrak{F}$ (if δ is small enough) since it is near $\tilde{P}$ and $\tilde{P}$ almost commutes with $\mathfrak{F}$. □

Definition 7.2.4. Let $\pi\colon A \to \mathbb{B}(\mathcal{H})$ be a $*$-homomorphism. Then π is called a *quasidiagonal representation* if $\pi(A)$ is a quasidiagonal set of operators.[7]

Historically quasidiagonality for C*-algebras was defined in terms of the existence of a faithful quasidiagonal representation. The following result of Voiculescu – which is the representation theorem referred to in the title of this section – shows that the two definitions agree in the separable unital case.[8]

Theorem 7.2.5 (Voiculescu). *For a separable unital* C*-*algebra A, the following statements are equivalent:*

(1) *A is QD;*

(2) *A has a faithful quasidiagonal representation (on a separable Hilbert space);*

(3) *every faithful unital essential representation of A (on a separable Hilbert space) is quasidiagonal.*

Proof. (1) $\Rightarrow$ (3): Let $\varphi_n\colon A \to \mathbb{M}_{k(n)}(\mathbb{C})$ be u.c.p. maps which are asymptotically multiplicative and isometric. Let $\pi\colon A \to \mathbb{B}(\mathcal{H})$ be a faithful essential representation. We must show that if finite sets $\mathfrak{F} \subset A$, $\chi \subset \mathcal{H}$ and

[7]Warning: This is *not* equivalent to saying that $\pi(A)$ is a QD C*-algebra! Sad, but true (see Remark 7.5.3).

[8]They agree in full generality, but this case is sufficient for our purposes.

$\varepsilon > 0$ are given, then one can find a finite-rank projection which almost commutes (up to ε) with $\mathfrak{F}$ and almost leaves χ fixed.

The trick is simply to apply the technical version of Voiculescu's Theorem (version 1.7.6) to an appropriate faithful $*$-homomorphism modulo the compacts. The right thing to consider is the u.c.p. map

$$\Phi\colon A \to \prod_{n=1}^{\infty} \mathbb{M}_{k(n)}(\mathbb{C}) \subset \mathbb{B}(\bigoplus_{n=1}^{\infty} \ell^2_{k(n)}), \ \Phi(a) = \bigoplus_{n=1}^{\infty} \varphi_n(a).$$

As in the previous section, Φ is a faithful representation modulo the compacts. Moreover, if we first fix a finite set $\mathfrak{F} \subset A$ and $\varepsilon > 0$, then (going out far enough in the sequence) we may assume that each of the u.c.p. maps φ_n is almost multiplicative on $\mathfrak{F}$. In other words, we may assume that $\eta_\Phi(a) < \varepsilon$ for all $a \in \mathfrak{F}$ (notation as in the statement of Theorem 1.7.6). Hence we can find a unitary $U\colon \mathcal{H} \to \bigoplus_1^\infty \ell^2_{k(n)}$ such that $\|\pi(a) - U^*\Phi(a)U\| < \varepsilon$ for all $a \in \mathfrak{F}$. It should now be clear how to complete the proof since Φ obviously has finite-rank projections which commute with its image and tend strongly to one (which we can transfer over to $\mathcal{H}$ by $U^* \cdot U$).

(3) $\Rightarrow$ (2) is immediate so let's show (2) $\Rightarrow$ (1). Let $\pi\colon A \to \mathbb{B}(\mathcal{H})$ be a faithful quasidiagonal representation and $P_1 \leq P_2 \leq \cdots$ be finite-rank projections which converge to the identity strongly and asymptotically commute with $\pi(A)$. A straightforward bit of estimating shows that $\varphi_n\colon A \to P_n\mathbb{B}(\mathcal{H})P_n \cong \mathbb{M}_{k(n)}(\mathbb{C})$ defined by $\varphi_n(a) = P_n\pi(a)P_n$ is an asymptotically multiplicative and isometric sequence of u.c.p. maps. □

In most instances, the representation theorem above gets used in the following form.

Corollary 7.2.6. *Assume A is unital separable and QD. For each faithful essential representation $\pi\colon A \to \mathbb{B}(\mathcal{H})$, with $\mathcal{H}$ separable, there exist finite-rank projections $P_1 \leq P_2 \leq \cdots$, converging strongly to 1 and such that $\|[P_n, \pi(a)]\| \to 0$ for all $a \in A$.*

Exercises

For the following exercises you should assume that everything in sight is separable.

Exercise 7.2.1. Show that every representation of a simple unital QD algebra is quasidiagonal.

Exercise 7.2.2. Show that every representation of an AF algebra is quasidiagonal.

Exercise 7.2.3. Observe that if $A \subset \mathbb{B}(\mathcal{H})$ is a quasidiagonal set of operators, then so is $A + \mathbb{K}(\mathcal{H})$.

Exercise 7.2.4. Use the spectral theorem to show that $\{N\} \subset \mathbb{B}(\mathcal{H})$ is a quasidiagonal set, for every normal operator N. Hence $C^*(N) + \mathbb{K}(\mathcal{H})$ is as well.

Exercise 7.2.5. Let $A \subset \mathbb{B}(\mathcal{H})$ be a concrete C*-algebra. Show that A is a quasidiagonal set if and only if the extension

$$0 \to \mathbb{K}(\mathcal{H}) \to A + \mathbb{K}(\mathcal{H}) \to A/(A \cap \mathbb{K}(\mathcal{H})) \to 0$$

is quasidiagonal (in the sense of Exercise 7.1.5).

Exercise 7.2.6. Use the representation theorem to give a simple proof of Proposition 7.1.12 (in the unital separable case).

Exercise 7.2.7. Formulate and prove a nonunital version of the representation theorem.

7.3. Homotopy invariance

Quasidiagonality is not particularly well understood. This is probably due to its topological nature – a point of view first suggested by Voiculescu. In this section we prove an important theorem which both illustrates this topological nature and gives us some surprising examples of QD C*-algebras.

We need two quasicentral-approximate-unit facts.

Lemma 7.3.1. *Let $J \lhd A$ be a separable ideal. Then there exists a quasicentral approximate unit $\{e_j\} \subset J$ such that $e_{j+1}e_j = e_j$ for all $j \in \mathbb{N}$.*

Proof. Since J is separable, it contains a *strictly positive* element h (i.e., $\varphi(h) > 0$ for all states φ on J – let $h = \sum \frac{1}{2^j} e_j$ where $\{e_j\}$ is any approximate unit). For each n let $f_n \in C_0(0,1]$ be the function which is zero on the interval $(0, \frac{1}{2^n}]$, one on the interval $[\frac{1}{2^{n-1}}, 1]$ and linear in between. Evidently we have $f_{n+1}(h)f_n(h) = f_n(h)$, for all n, and it is also readily seen that $\varphi(f_n(h)) \to 1$ for all states φ on J. The usual Hahn-Banach convexity argument allows us to extract a quasicentral approximate unit from the convex hull of $\{f_n(h)\}$, so we leave the remaining details to you. □

Lemma 7.3.2. *Let $\varepsilon > 0$ and a continuous function $f \in C_0(0,1]$ be given. There exists a $\delta > 0$ such that for every C*-algebra A and pair of elements $e, a \in A$ in the unit ball of A, with $e \geq 0$, we have*

$$\|[e,a]\| < \delta \Longrightarrow \|[f(e),a]\| < \varepsilon.$$

Proof. By a standard approximation argument we may assume that f is a polynomial of the form

$$f(s) = t_1 s + t_2 s^2 + \cdots + t_n s^n.$$

Given $a \in A$, of norm one, we consider the inner derivation

$$D_a(b) = ba - ab, \;\; b \in A.$$

Assuming b also comes from the unit ball, one uses induction on the inequality

$$\|D_a(b^{i+1})\| = \|bD_a(b^i) + D_a(b)b^i\| \leq \|D_a(b^i)\| + \|D_a(b)\|$$

to deduce that

$$\|D_a(b^j)\| \leq j\|D_a(b)\|$$

for all natural numbers j.

So, let

$$\delta = \frac{\varepsilon}{\sum_{i=1}^n i|t_i|}$$

and assume that $\|D_a(e)\| = \|[e,a]\| < \delta$ for some positive element e in the unit ball of A. Then we have

$$\|[f(e),a]\| = \|t_1 D_a(e) + \cdots + t_n D_a(e^n)\| \leq \sum_{i=1}^n |t_i|(i\|D_a(e)\|) < \varepsilon$$

as desired. □

The notions of homotopic maps and homotopy equivalence of spaces are fundamental in topology. Here is how these ideas translate into C*-lingo.

Definition 7.3.3. Let $\sigma_0 \colon A \to B$ and $\sigma_1 \colon A \to B$ be $*$-homomorphisms. Then we say σ_0 and σ_1 are *homotopic* if there exist $*$-homomorphisms $\tilde{\sigma}_t \colon A \to B$, $0 \leq t \leq 1$, such that $\tilde{\sigma}_0 = \sigma_0$, $\tilde{\sigma}_1 = \sigma_1$ and for every fixed $a \in A$ the map $[0,1] \to B$, $t \mapsto \tilde{\sigma}_t(a)$, is continuous from the usual topology on $[0,1]$ to the norm topology on B – in other words, there should exist $*$-homomorphisms which give norm continuous paths from $\sigma_0(a)$ to $\sigma_1(a)$, for every $a \in A$. If you prefer, this is equivalent to asking for a $*$-homomorphism $A \to C[0,1] \otimes B$ which is σ_0 at the left endpoint and σ_1 at the right.

Definition 7.3.4. A C*-algebra A is said to *homotopically dominate* another C*-algebra B if there are $*$-homomorphisms $\pi \colon B \to A$ and $\sigma \colon A \to B$ such that $\sigma \circ \pi$ is homotopic to id_B. The algebras A and B are *homotopy equivalent* if there exist $*$-homomorphisms $\pi \colon B \to A$ and $\sigma \colon A \to B$ such that $\sigma \circ \pi$ is homotopic to id_B and $\pi \circ \sigma$ is homotopic to id_A.

We will show that if two C*-algebras are homotopy equivalent and one is QD, then so is the other. The proof is much easier if we isolate a preliminary step.

Proposition 7.3.5. *Let $\sigma_0, \sigma_1 \colon B \to C$ be homotopic $*$-homomorphisms such that σ_0 is injective and $\sigma_1(B)$ is a QD subalgebra of C. Then B is QD.*

Proof. Let $C \subset \mathbb{B}(\mathcal{K})$ be a faithful, essential representation and let $\sigma_t \colon B \to C$ be a homotopy of $*$-homomorphisms between σ_0 and σ_1. It suffices to show that for each finite set $\mathfrak{F} \subset B$ of the unit ball and $\varepsilon > 0$ there is a $*$-homomorphism $\pi \colon B \to \mathbb{B}(\mathcal{H})$ and a finite-rank projection $P \in \mathbb{B}(\mathcal{H})$ such that

$$\|[P, \pi(b)]\| \leq \varepsilon$$

and

$$\|P\pi(b)P\| \geq \|b\| - \varepsilon,$$

for all $b \in \mathfrak{F}$.

For any $\delta > 0$ (to be specified later) we can, by norm continuity of the homotopy and compactness of $[0,1]$, find a large n with the property that

$$\|\sigma_{\frac{j}{n}}(b) - \sigma_{\frac{j+1}{n}}(b)\| \leq \delta$$

for all $b \in \mathfrak{F}$ and $0 \leq j \leq n-1$. Since σ_0 is injective, we can find a finite-rank projection $Q \in \mathbb{B}(\mathcal{K})$ such that

$$\|Q\sigma_0(b)Q\| \geq \|b\| - \varepsilon$$

for all $b \in \mathfrak{F}$. Applying the existence of quasicentral approximate units to the ideal of compact operators, we can find positive, norm one, finite-rank operators

$$Q \leq F_0 \leq F_1 \leq F_2 \leq \cdots \leq F_n \leq 1$$

such that $F_{j+1}F_j = F_j$ and

$$\|[F_j, \sigma_{\frac{j}{n}}(b)]\| \leq \delta$$

for all $0 \leq j \leq n$ and $b \in \mathfrak{F}$. (Since any approximate unit of finite-rank operators will eventually *almost* dominate Q, we can perturb a little to *actually* dominate Q.)

Here is a crucial remark: Since $\sigma_1(B)$ is QD and $C \subset \mathbb{B}(\mathcal{K})$ is essential, we may assume that F_n is a projection! Indeed, the representation theorem says we can find a finite-rank projection which almost commutes with $\sigma_1(\mathfrak{F})$ and is almost the identity on a prescribed finite set of vectors (e.g., a basis for the range of whatever finite-rank operator F_n we are originally given). The proof of Proposition 7.2.3 allows us to perturb and *actually* dominate F_n and then replace F_n by such a projection.

Now comes the crucial trick: Consider the operator $V \colon \mathcal{K} \to \bigoplus_0^n \mathcal{K}$ defined by

$$Vh = G_0^{1/2}h \oplus G_1^{1/2}h \oplus \cdots \oplus G_n^{1/2}h,$$

where $G_0 = F_0$, $G_1 = F_1 - F_0, \ldots, G_n = F_n - F_{n-1}$. A calculation shows that

$$V^*V = F_n \in \mathbb{B}(\mathcal{K})$$

and hence V is a partial isometry. Thus VV^* is a finite-rank projection in $\mathbb{B}(\bigoplus_0^n \mathcal{K})$. A slightly unpleasant calculation, using the fact that $F_{k+1}F_k = F_k$, shows that the matrix of VV^* is given by

$$\begin{bmatrix} G_0 & G_0^{1/2}G_1^{1/2} & 0 & \cdots & 0 \\ G_1^{1/2}G_0^{1/2} & G_1 & G_1^{1/2}G_2^{1/2} & \cdots & 0 \\ 0 & G_2^{1/2}G_1^{1/2} & G_2 & \cdots & 0 \\ \vdots & \vdots & \vdots & \ddots & \vdots \\ 0 & 0 & 0 & \cdots & G_n \end{bmatrix}.$$

We let $P = VV^* \in \mathbb{B}(\bigoplus_0^n \mathcal{K})$ and define a representation of B on $\bigoplus_0^n \mathcal{K}$ by

$$\pi(b) = \sigma_0(b) \oplus \sigma_{\frac{1}{n}}(b) \oplus \sigma_{\frac{2}{n}}(b) \oplus \cdots \oplus \sigma_1(b).$$

Since $F_0 \geq Q$, it is not hard to see that

$$\|P\pi(b)P\| \geq \|b\| - \varepsilon,$$

for all $b \in \mathfrak{F}$, and hence we only have to show that P almost commutes with $\pi(\mathfrak{F})$.

For this it is helpful to look at the matrix description of P given above and to write $P = LD + D + UD$ where D is the diagonal of the matrix, while LD and UD denote the lower and upper diagonals, respectively. Evidently, it suffices to show that the three commutators $[LD, \pi(b)]$, $[D, \pi(b)]$ and $[UD, \pi(b)]$ all have small norms and this will follow from the fact that we arranged

$$\|[F_j, \sigma_{\frac{j}{n}}(b)]\| \leq \delta$$

and

$$\|\sigma_{\frac{j}{n}}(b) - \sigma_{\frac{j+1}{n}}(b)\| \leq \delta,$$

for all $0 \leq j \leq n$ and $b \in \mathfrak{F}$, at the very beginning. Indeed, it is not hard to see that

$$\|[D, \pi(b)]\| \leq 4\delta.$$

The harder estimates are the off-diagonal parts and here is where we use the fact that δ has not yet been prescribed.

Since ε is fixed, we can, by Lemma 7.3.2, find a δ such that $\|[X^{1/2}, b]\| \leq \varepsilon$ whenever $\|[X, b]\| \leq \delta$, $0 \leq X \leq 1$ and $\|b\| \leq 1$. Hence, taking δ small enough, we can ensure that

$$\|[\sigma_{\frac{j}{n}}(b), G_k^{1/2}]\| \leq \varepsilon$$

whenever $|j - k| \leq 1$. Using this, it is not too hard to show that the norms of $[LD, \pi(b)]$ and $[UD, \pi(b)]$ are also small for all $b \in \mathfrak{F}$. Indeed, if you write down the commutator in matrix form, you will see that one must estimate the norms of elements of the form

$$\sigma_{\frac{j}{n}}(b)G_j^{1/2}G_{j+1}^{1/2} - G_j^{1/2}G_{j+1}^{1/2}\sigma_{\frac{j+1}{n}}(b).$$

However, this operator has small norm since $\sigma_{\frac{j}{n}}(b)$ almost commutes with $G_j^{1/2}$ and $G_{j+1}^{1/2}$ and, furthermore, the norm of $\sigma_{\frac{j}{n}}(b) - \sigma_{\frac{j+1}{n}}(b)$ is small. □

Theorem 7.3.6 (Voiculescu). *If A homotopically dominates B and A is QD, then B is also QD. In particular, quasidiagonality is a homotopy-equivalence invariant.*

Proof. Let $\pi\colon B \to A$ and $\sigma\colon A \to B$ be $*$-homomorphisms such that $\sigma \circ \pi$ is homotopic to id_B. Define a $*$-homomorphism $\eta\colon B \to B \oplus \pi(B)$ by

$$\eta(b) = \sigma \circ \pi(b) \oplus \pi(b).$$

Evidently this $*$-homomorphism is homotopic to $\mathrm{id}_B \oplus \pi$ (which is faithful). Hence the proof will be complete, appealing to the previous proposition, once we observe that $\eta(B)$ is QD. But $\eta(B) \cong \pi(B) \subset A$ and A is QD by assumption. □

With this theorem in hand we can give some new examples of QD C*-algebras.

Corollary 7.3.7. *For any C*-algebra A, both the cone over A,*

$$CA = C_0(0,1] \otimes A,$$

and the suspension of A,

$$SA = C_0(0,1) \otimes A,$$

are QD.

Proof. Since $SA \lhd CA$, it suffices to show that CA is homotopic to zero. This is familiar to the K-theory crowd, but let's recall the proof. Define a family of $*$-homomorphisms $\sigma_t : C_0(0,1] \to C_0(0,1]$ by

$$\sigma_t(f)(s) = f(ts),$$

for $0 \le t \le 1$. One checks that each σ_t is a $*$-homomorphism and that this family defines a homotopy between the zero map and the identity. Tensoring with id_A shows that CA is homotopic to zero. □

We will actually use the corollary above in the nonseparable setting, so you may want to convince yourself that this case can be deduced from the separable one.

From the Bott periodicity theorem it follows that the K-theory of an arbitrary A is isomorphic to that of a QD algebra (i.e., S^2A). Of course, one loses the order structure (in the K_0 case) and we aren't suggesting that this is an enlightening reduction; it's just another cute observation.

In Exercise 7.1.4 we pointed out that every separable C*-algebra is a quotient of an RFD algebra. We can now improve that result in the cases of nuclear and exact C*-algebras.

Corollary 7.3.8. *Let A be a separable nuclear (resp. exact) C*-algebra. There is a nuclear (resp. exact) RFD C*-algebra B such that A is a quotient of B.*

Proof. Every algebra is a quotient of its cone, so we may assume that A is QD. Exercise 7.1.3 then provides us with an RFD algebra B with ideal J such that

$$0 \to J \to B \to A \to 0$$

is short exact and has a c.c.p. splitting. We can now appeal to Exercises 3.8.1 and 3.9.8 to conclude that B is nuclear (resp. exact) whenever A is nuclear (resp. exact). □

Exercise

Exercise 7.3.1. Given A, show that every faithful representation of either CA or SA is quasidiagonal. (Hint: CA has no nonzero projections.)

7.4. Two more examples

We've already seen that quasidiagonality of a *reduced* group C*-algebra implies amenability. But what about the universal algebras? As you might expect, things are very different (at least in the free group case).

Since $C^*(\Gamma)$ is defined by a representation-theoretic universal property, one might expect that it is residually finite-dimensional (Definition 7.1.6) whenever Γ is residually finite. This turns out to be false, in general, but true for free groups.[9] However, it is not residual finiteness that explains this phenomenon; it is universality of free groups.

Theorem 7.4.1 (Choi). *The full group C*-algebra $C^*(\mathbb{F}_n)$ is RFD for every $n = \infty, 1, 2, 3, \ldots$.*

[9] Bekka has shown that many residually finite groups do not have residually finite-dimensional universal algebras ([**14**]). The proof is not so simple, though, and it would be nice to have an elementary argument.

Proof. We first recall a basic operator theory fact: If $T \in \mathbb{B}(\mathcal{H})$ is a contraction, then

$$U = \begin{bmatrix} T & \sqrt{1-TT^*} \\ \sqrt{1-T^*T} & -T^* \end{bmatrix}$$

is a unitary operator on $\mathcal{H} \oplus \mathcal{H}$.[10]

Since $C^*(\mathbb{F}_\infty) \subset C^*(\mathbb{F}_2)$, it suffices to prove the theorem when $n = 2$. So, fix a faithful representation $C^*(\mathbb{F}_2) \subset \mathbb{B}(\mathcal{H})$ on a separable Hilbert space and let $P_k \leq P_{k+1}$ be finite-rank projections converging strongly to the identity. If u_1, u_2 are the canonical free generators of $\mathbb{F}_2$, then for each k we can dilate the contractions $P_k u_1 P_k, P_k u_2 P_k$, as above, to get unitaries $v_1^{(k)}, v_2^{(k)}$ acting on the finite-dimensional Hilbert space $P_k\mathcal{H} \oplus P_k\mathcal{H}$. Note that for each j,

$$v_j^{(k)} \to \begin{bmatrix} u_j & 0 \\ 0 & -u_j^* \end{bmatrix},$$

as $k \to \infty$, where convergence is in the strong* operator topology. It follows that for every noncommutative $*$-polynomial p of two variables, we have

$$p(v_1^{(k)}, v_2^{(k)}) \to \begin{bmatrix} p(u_1, u_2) & 0 \\ 0 & p(-u_1^*, -u_2^*) \end{bmatrix},$$

where convergence is in the strong operator topology, as $k \to \infty$.

By universality there exist $*$-homomorphisms $\pi_k \colon C^*(\mathbb{F}_2) \to \mathbb{B}(P_k\mathcal{H} \oplus P_k\mathcal{H})$ such that $\pi_k(u_j) = v_j^{(k)}$. The previous paragraph implies that for every element $x \in \mathbb{C}[\mathbb{F}_2]$ in the group algebra (i.e., $x = p(u_1, u_2)$ for some p) we have the inequality

$$\|x\|_{C^*(\mathbb{F}_2)} \leq \left\| \begin{bmatrix} x & 0 \\ 0 & p(-u_1^*, -u_2^*) \end{bmatrix} \right\| \leq \liminf_{k\to\infty} \|\pi_k(x)\|.$$

It follows that

$$\bigoplus_k \pi_k \colon C^*(\mathbb{F}_2) \to \prod_k \mathbb{B}(P_k\mathcal{H} \oplus P_k\mathcal{H})$$

is *isometric* on a dense subspace and hence must be injective on the whole C*-algebra. □

Remark 7.4.2. At first glance it may seem that RFD algebras can't be too exotic due to the abundance of finite-dimensional representations. However, nothing could be further from the truth, as $C^*(\mathbb{F}_n)$ is, in many ways, as exotic as they come. Every single separable unital C*-algebra arises as a quotient of this beast (if $n = \infty$) and hence there is no hope of understanding it at a very deep level.

---

[10]Hint: Observe that if p is any polynomial, then $T^*p(TT^*) = p(T^*T)T^*$ and hence the same relation passes to continuous functional calculus.

Remark 7.4.3. Free groups actually enjoy a stronger property: Every representation can be approximated in the Fell topology ([**15**]) by finite representations (i.e., representations factoring through finite quotients of $\mathbb{F}_n$); see [**123**].

Though it seems uninteresting at first, a major open problem is whether or not Choi's result can be extended to $\mathbb{F}_n \times \mathbb{F}_n$ (since it's equivalent to Connes's embedding problem).

Proposition 7.4.4. *The following statements are equivalent:*

(1) *every finite von Neumann algebra with separable predual can be embedded into the ultraproduct of the hyperfinite* II_1*-factor (this is Connes's embedding problem);*

(2) *there is a unique* C*-*norm on* $C^*(\mathbb{F}_n) \odot C^*(\mathbb{F}_n)$;

(3) $C^*(\mathbb{F}_n \times \mathbb{F}_n)$ *is residually finite-dimensional.*

Proof. For the equivalence of (1) and (2) see Theorem 13.3.1. Since they both satisfy the same universal property, it is easy to see that

$$C^*(\mathbb{F}_n \times \mathbb{F}_n) \cong C^*(\mathbb{F}_n) \otimes_{\max} C^*(\mathbb{F}_n).$$

Thus (2) $\Rightarrow$ (3) is easy (since the minimal tensor product of RFD algebras is easily seen to be RFD). So, assume $C^*(\mathbb{F}_n \times \mathbb{F}_n)$ is residually finite-dimensional and let $\pi_i\colon C^*(\mathbb{F}_n) \otimes_{\max} C^*(\mathbb{F}_n) \to \mathbb{M}_{k(i)}(\mathbb{C})$ be a separating family of representations. By finite-dimensionality, each of the π_i's must factor through the minimal tensor product, and hence the quotient mapping $C^*(\mathbb{F}_n) \otimes_{\max} C^*(\mathbb{F}_n) \to C^*(\mathbb{F}_n) \otimes C^*(\mathbb{F}_n)$ must be injective. □

Though residual finite-dimensionality remains open, it is not too hard to show quasidiagonality of $C^*(\mathbb{F}_n \times \mathbb{F}_n)$.

Proposition 7.4.5. *The full group* C*-*algebra* $C^*(\mathbb{F}_n \times \mathbb{F}_n)$ *is QD.*

Proof. Let $C^*(\mathbb{F}_n \times \mathbb{F}_n) \subset \mathbb{B}(\mathcal{H})$ be a faithful representation. We only consider the case $n = 2$; the general case follows since $C^*(\mathbb{F}_\infty \times \mathbb{F}_\infty) \subset C^*(\mathbb{F}_2 \times \mathbb{F}_2)$.

Let $U_1, V_1 \in \mathbb{B}(\mathcal{H})$ be the image of the unitaries which generate the "left" copy of $\mathbb{F}_2$ and let $U_2, V_2 \in \mathbb{B}(\mathcal{H})$ be the image of the unitaries which generate the "right" copy of $\mathbb{F}_2$.

Since the unitary group of $\mathbb{B}(\mathcal{H})$ is connected (Borel functional calculus), we can find a norm continuous path from any unitary to the identity. That is, we can find norm continuous maps $u_i : [0,1] \to \mathcal{U}(\mathcal{H})$ and $v_i : [0,1] \to \mathcal{U}(\mathcal{H})$, where $\mathcal{U}(\mathcal{H})$ is the unitary group of $\mathbb{B}(\mathcal{H})$ and $i = 1, 2$, such that

$$u_i(0) = 1, \ u_i(1) = U_i, \ v_i(0) = 1 \ \text{ and } \ v_i(1) = V_i.$$

A crucial remark is that we can even arrange a certain amount of commutativity. Namely, since the von Neumann algebras generated by $\{U_1, V_1\}$ and $\{U_2, V_2\}$ commute, we may choose the paths of unitaries to live inside these respective von Neumann algebras; hence *we may assume that the operators* $\{u_1(t), v_1(t)\}$ *commute with* $\{u_2(t), v_2(t)\}$ *for all* $t \in [0,1]$. It follows that the identity representation $C^*(\mathbb{F}_n \times \mathbb{F}_n) \hookrightarrow \mathbb{B}(\mathcal{H})$ is homotopic to the trivial representation into $\mathbb{C}1_{\mathcal{H}}$ and we are done, by Proposition 7.3.5. □

7.5. External approximation

In this section we exhibit some external approximation properties enjoyed by QD C*-algebras. All of the results obtained are, in one way or another, inspired by single operator theory. While this has certainly led to a better understanding of QD C*-algebras, it has also given back to single operator theory, as we will see in Chapter 16.

Our first result is a simple generalization of an argument which Halmos used in the single operator case.

Theorem 7.5.1 (Approximation by RFD algebras). *Let* $\mathcal{H}$ *be separable and* $A \subset \mathbb{B}(\mathcal{H})$ *be a separable* C*-*algebra. If* A *is a quasidiagonal set of operators, then for each finite set* $\mathfrak{F} \subset A$ *and* $\varepsilon > 0$ *there exists a residually finite-dimensional* C*-*algebra* $B \subset \mathbb{B}(\mathcal{H})$ *and a u.c.p. map* $\Phi\colon A \to B$ *such that*

$$\Phi(a) - a \in \mathbb{K}(\mathcal{H})$$

for all $a \in A$ *and*

$$\|\Phi(a) - a\| < \varepsilon$$

for all $a \in \mathfrak{F}$.

Proof. Let $a_1, a_2, \cdots$ be dense in A and, by Proposition 7.2.3, find increasing finite-rank projections, tending strongly to the identity, such that

$$\|[P_k, a_j]\| < \frac{1}{2^k}$$

for all $k \in \mathbb{N}$ and $1 \le j \le k$. If $\varepsilon > 0$ and a finite set $\mathfrak{F} \subset A$ from the unit ball are also given, we may assume that $\|[P_k, a]\| < \frac{\varepsilon}{2^k}$ for all k and $a \in \mathfrak{F}$.

Consider the orthogonal finite-rank projections

$$Q_1 = P_1, Q_2 = P_2 - P_1, Q_3 = P_3 - P_2, \ldots.$$

Since $\sum Q_i = 1$, we get a well-defined u.c.p. map by defining

$$\Phi(a) = \sum_{i=1}^{\infty} Q_i a Q_i,$$

for all $a \in A$. Letting $B = C^*(\Phi(A))$, it is clear that B is RFD. The remainder of the proof is contained in the following calculation:

$$\begin{aligned}\Phi(a) - a &= \sum_{i=1}^{\infty} Q_i a Q_i - (\sum_{i=1}^{\infty} Q_i) a \\ &= \sum_{i=1}^{\infty} (Q_i a Q_i - Q_i a) \\ &= \sum_{i=1}^{\infty} Q_i (a Q_i - Q_i a).\end{aligned}$$

Indeed, for arbitrary $a \in A$ all of the series above converge in the strong operator topology. But for each a_j the last sum is a norm convergent series of finite-rank operators and thus, by continuity, $\Phi(a) - a \in \mathbb{K}(\mathcal{H})$ for all $a \in A$. □

Note that $A + \mathbb{K}(\mathcal{H}) = B + \mathbb{K}(\mathcal{H})$ (since $\Phi(a) - a \in \mathbb{K}(\mathcal{H})$ for all $a \in A$) and hence the images of A and B down in the Calkin algebra agree. In particular, if $A \cap \mathbb{K}(\mathcal{H}) = \{0\}$, then A is a quotient of B.

Corollary 7.5.2. *Let A be separable, QD and let $\pi\colon A \to \mathbb{B}(\mathcal{H})$ be a faithful essential representation. Then, there exists an RFD algebra $B \subset \pi(A) + \mathbb{K}(\mathcal{H})$ and a short exact sequence*

$$0 \to \mathbb{K}(\mathcal{H}) \to B + \mathbb{K}(\mathcal{H}) \to A \to 0$$

such that π provides a ($*$-homomorphic) splitting.

Remark 7.5.3 (Fredholm index obstruction)**.** Assume $T \in \mathbb{B}(\mathcal{H})$ is a Fredholm, quasidiagonal operator. Then $\mathrm{Ind}(T) = 0$. Hence a Fredholm operator with nontrivial index can't be quasidiagonal. To see this, we use the approximation result above. For operators on a finite-dimensional space, the dimension of the kernel and cokernel are always equal; hence the same holds for any Fredholm, block diagonal operator $S = S_1 \oplus S_2 \oplus \cdots$. But if T is quasidiagonal, then $C^*(T) \subset \mathbb{B}(\mathcal{H})$ is a quasidiagonal set of operators, so $T = \Phi(T) + k$, where k is compact and $\Phi(T)$ is block diagonal. Since $\mathrm{Ind}(\,\cdot\,)$ is invariant under compact perturbations, the claim is proved.

Remark 7.5.4 (Nonquasidiagonal representations)**.** The previous remark implies that if A is QD and $\pi\colon A \to \mathbb{B}(\mathcal{H})$ is a faithful representation such that $\mathrm{Ind}(\pi(a)) \neq 0$, for some $a \in A$, then π can't be a quasidiagonal representation. With this observation it is easy to construct examples of QD C*-algebras with nonquasidiagonal faithful representations. For example, let $S \in \mathbb{B}(\ell^2(\mathbb{N}))$ be the unilateral shift and let $A = C^*(S^* \oplus S) \subset \mathbb{B}(\ell^2(\mathbb{N}) \oplus \ell^2(\mathbb{N}))$. The operator $S^* \oplus S$ is a compact (actually, rank one) perturbation of the (unitary) bilateral shift on $\mathbb{B}(\ell^2(\mathbb{Z}))$; hence A is QD (see

Exercise 7.2.4). However, A is evidently isomorphic to $C^*(S^* \oplus S \oplus S) \subset \mathbb{B}(\ell^2(\mathbb{N}) \oplus \ell^2(\mathbb{N}) \oplus \ell^2(\mathbb{N}))$. The resulting representation of A can't be quasi-diagonal since $\mathrm{Ind}(S^* \oplus S \oplus S) = -1$.

When we further assume that A is exact, a much stronger approximation is possible – by finite-dimensional algebras.[11] Unfortunately, the proof is not so easy. We will need a technical dilation lemma which is directly inspired by years of hard work in Elliott's classification program. Indeed, several variations of the next result can be found in the classification literature where they are used to prove "uniqueness" theorems for morphisms of various kinds.

Lemma 7.5.5. *Assume that A is unital exact and QD. For every finite set $\mathfrak{F} \subset A$ and $\varepsilon > 0$ there is an essential faithful nonunital representation $\pi\colon A \to \mathbb{B}(\mathcal{K})$, a c.p. map $\Phi\colon A \to \mathbb{B}(\mathcal{K})$ such that $\pi(1)$ and $\Phi(1)$ are orthogonal projections with $1_{\mathcal{K}} = \pi(1) + \Phi(1)$, and a unital finite-dimensional subalgebra $B \subset \mathbb{B}(\mathcal{K})$ together with a u.c.p. map $\gamma\colon A \to B$ such that*

$$\|\gamma(a) - (\pi(a) \oplus \Phi(a))\| < \varepsilon$$

and

$$\|\gamma(ab) - \gamma(a)\gamma(b)\| < \varepsilon$$

for all $a, b \in \mathfrak{F}$.

Proof. Assume that $A \subset \mathbb{B}(\mathcal{H})$ is a faithful essential representation. Fix a finite set $\mathfrak{F} \subset A$ from the unit ball of A. Enlarging, if necessary, we may assume that $\mathfrak{F}$ is self-adjoint and contains the unit of A. Define a larger finite set

$$\tilde{\mathfrak{F}} = \{ab : a, b \in \mathfrak{F}\}.$$

Note that $\mathfrak{F} \subset \tilde{\mathfrak{F}}$.

Since A is QD, the representation theorem (Corollary 7.2.6) provides us with increasing finite-rank projections $P_n \le P_{n+1}$ which asymptotically commute (in norm) with A and converge to the identity in the strong operator topology. Define $\varphi_n\colon A \to P_n\mathbb{B}(\mathcal{H})P_n \cong \mathbb{M}_{k(n)}(\mathbb{C})$ by $\varphi_n(a) = P_n a P_n$. By exactness (Exercise 3.9.5) we can find n large enough that there exists a u.c.p. map $\psi_n\colon \mathbb{M}_{k(n)}(\mathbb{C}) \to \mathbb{B}(\mathcal{H})$ such that

$$\|a - \psi_n(\varphi_n(a))\| < \varepsilon$$

for all $a \in \tilde{\mathfrak{F}}$. We may also assume that

$$\|\varphi_n(ab) - \varphi_n(a)\varphi_n(b)\| < \varepsilon$$

[11]Though the approximants are no longer compact perturbations; they can't be, as this would imply AF.

for all $a, b \in \mathfrak{F}$. It follows that ψ_n is almost multiplicative on $\varphi_n(\mathfrak{F})$. More precisely,

$$\|\psi_n(\varphi_n(a)\varphi_n(b)) - \psi_n(\varphi_n(a))\psi_n(\varphi_n(b))\|$$

is bounded above by

$$\|\psi_n(\varphi_n(a)\varphi_n(b)) - \psi_n(\varphi_n(ab))\| + \|\psi_n(\varphi_n(ab)) - ab\| \\ + \|ab - \psi_n(\varphi_n(a))\psi_n(\varphi_n(b))\|$$

which, in turn, is bounded above by 4ε whenever $a, b \in \mathfrak{F}$. To ease notation, we now fix n (large enough) and just write $\varphi\colon A \to \mathbb{M}_k(\mathbb{C})$ and $\psi\colon \mathbb{M}_k(\mathbb{C}) \to \mathbb{B}(\mathcal{H})$ instead of φ_n and ψ_n.

Let $\sigma\colon \mathbb{M}_k(\mathbb{C}) \to \mathbb{B}(\mathcal{K})$ be the Stinespring dilation of ψ and $V\colon \mathcal{H} \to \mathcal{K}$ be the isometry such that

$$\psi(x) = V^*\sigma(x)V$$

for all $x \in \mathbb{M}_k(\mathbb{C})$. Since we have arranged that ψ is 4ε-multiplicative on $\varphi(\mathfrak{F})$, it follows that the Stinespring projection $Q = VV^*$ almost commutes with $\sigma(\varphi(\mathfrak{F}))$. More precisely, we have

$$\|Q\sigma(\varphi(a))Q^{\perp}\|^2 = \|\psi(\varphi(a)\varphi(a)^*) - \psi(\varphi(a))\psi(\varphi(a))^*\| \leq 4\varepsilon$$

and similarly $\|Q^{\perp}\sigma(\varphi(a))Q\|^2 \leq 4\varepsilon$, for all $a \in \mathfrak{F}$. It follows that

$$\|\sigma(\varphi(a)) - \big(Q\sigma(\varphi(a))Q \oplus Q^{\perp}\sigma(\varphi(a))Q^{\perp}\big)\| \leq 2\sqrt{\varepsilon}$$

for all $a \in \mathfrak{F}$. But

$$Q\sigma(\varphi(a))Q = V\psi(\varphi(a))V^*$$

and hence

$$\|\sigma(\varphi(a)) - \big(VaV^* \oplus Q^{\perp}\sigma(\varphi(a))Q^{\perp}\big)\| \leq 2\sqrt{\varepsilon} + \varepsilon$$

for all $a \in \mathfrak{F}$. We complete the proof by defining $\pi\colon A \to \mathbb{B}(\mathcal{K})$ by $\pi(a) = VaV^*$, $\Phi\colon A \to \mathbb{B}(\mathcal{K})$ by $\Phi(a) = Q^{\perp}\sigma(\varphi(a))Q^{\perp}$ and $\gamma\colon A \to \sigma(\mathbb{M}_n(\mathbb{C})) \subset \mathbb{B}(\mathcal{K})$ by $\gamma(a) = \sigma(\varphi(a))$. □

Remark 7.5.6. Note that if we have $[P_n, a] = 0$ for all $a \in A$ (i.e., A is RFD and the inclusion $A \subset \mathbb{B}(\mathcal{H})$ is block diagonal), then the u.c.p. map γ is a $*$-homomorphism.

Let's take a moment and understand what the previous lemma really says: In the presence of exactness and quasidiagonality we can *dilate* a representation to make it close to something finite-dimensional. This may remind you of Arveson's proof of Voiculescu's Theorem (cf. [**11**]) where one of the major steps was dilating something to be block diagonal. We now recycle that idea to push our approximation results further.

It will be convenient to introduce some notation. Let $\varphi\colon A \to \mathbb{B}(\mathcal{H})$ and $\psi\colon A \to \mathbb{B}(\mathcal{K})$ be u.c.p. maps and suppose that a finite set $\mathfrak{F} \subset A$ and $\varepsilon > 0$ are given. Then we write

$$\varphi \overset{(\mathfrak{F},\varepsilon)}{\approx} \psi$$

if there is a unitary operator $U\colon \mathcal{H} \to \mathcal{K}$ such that $\|\varphi(a) - U^*\psi(a)U\| < \varepsilon$ for all $a \in \mathfrak{F}$. Note that if $\varphi \overset{(\mathfrak{F},\varepsilon)}{\approx} \psi$ and $\psi \overset{(\mathfrak{F},\delta)}{\approx} \gamma$, then $\varphi \overset{(\mathfrak{F},\varepsilon+\delta)}{\approx} \gamma$.

Theorem 7.5.7 (Dadarlat). *Let A be an exact QD* C**-algebra and $\pi\colon A \to \mathbb{B}(\mathcal{H})$ be a faithful essential representation. For each finite set $\mathfrak{F} \subset A$ and $\varepsilon > 0$, there exists a finite-dimensional* C**-algebra $B \subset \mathbb{B}(\mathcal{H})$ such that for each $a \in \mathfrak{F}$ there exists $b \in B$ with $\|a - b\| < \varepsilon$.*

Proof. We may assume A is unital and that a finite set $\mathfrak{F} \subset A$ and $\varepsilon > 0$ are given. Since all faithful essential representations are approximately unitarily equivalent, it suffices to show that *some* representation has the property stated in the theorem. By the previous lemma we can find an essential faithful nonunital representation $\pi\colon A \to \mathbb{B}(\mathcal{K})$, a c.p. map $\Phi\colon A \to \mathbb{B}(\mathcal{K})$ such that $\pi(1)$ and $\Phi(1)$ are orthogonal projections with $1_{\mathcal{K}} = \pi(1) + \Phi(1)$, and a unital finite-dimensional subalgebra $B \subset \mathbb{B}(\mathcal{K})$ together with a u.c.p. map $\gamma\colon A \to B$ such that

$$\|\gamma(a) - (\pi(a) \oplus \Phi(a))\| < \varepsilon$$

and

$$\|\gamma(ab) - \gamma(a)\gamma(b)\| < \varepsilon$$

for all $a, b \in \mathfrak{F}$. We now apply Arveson's $\infty + 1 = \infty$ trickery:

$$\begin{aligned}\gamma &\overset{(\mathfrak{F},\varepsilon)}{\approx} \pi \oplus \Phi \\ &\overset{(\mathfrak{F},\varepsilon)}{\approx} (\pi \oplus \pi) \oplus \Phi \\ &= \pi \oplus (\pi \oplus \Phi) \\ &\overset{(\mathfrak{F},\varepsilon)}{\approx} \pi \oplus \gamma \\ &\overset{(\mathfrak{F},2\sqrt{\varepsilon})}{\approx} \pi.\end{aligned}$$

If you believe each step above, then the proof is complete since

$$\gamma \overset{(\mathfrak{F},3\varepsilon+2\sqrt{\varepsilon})}{\approx} \pi$$

and the range of γ is contained in a finite-dimensional C*-algebra. So let's justify each step (after the first, which is immediate). The second line follows from Voiculescu's Theorem since π and $\pi \oplus \pi$ are both faithful, essential representations. The third is trivial while the fourth line follows from the first. Finally the last line also follows from Voiculescu's Theorem (Corollary 1.7.7) since we assumed that γ is ε-multiplicative on $\mathfrak{F}$. □

In view of Remark 7.5.6, the proof above implies the following corollary.

Corollary 7.5.8. *Let $A \subset \mathbb{B}(\mathcal{H})$ be an essential representation of an exact RFD* C*-*algebra. For each finite set $\mathfrak{F} \subset A$ and $\varepsilon > 0$ there exist a full matrix algebra $\mathbb{M}_k(\mathbb{C}) \subset \mathbb{B}(\mathcal{H})$ and a $*$-homomorphism $\gamma\colon A \to \mathbb{M}_k(C)$ such that $\|a - \gamma(a)\| < \varepsilon$ for all $a \in \mathfrak{F}$.*

Proof. Since A is RFD, we may start with an embedding

$$A \subset \prod_{n\in\mathbb{N}} \mathbb{M}_{k(n)}(\mathbb{C}) \subset \mathbb{B}(\bigoplus_{n\in\mathbb{N}} \ell^2_{k(n)}).$$

Now reread the proofs of the previous two results. □

Exercise

Exercise 7.5.1. Prove the converse of Theorem 7.5.7. That is, if $A \subset \mathbb{B}(\mathcal{H})$ and every finite subset of A is close to some finite-dimensional subalgebra of $\mathbb{B}(\mathcal{H})$, then A is exact and QD. (This fact was first observed by Voiculescu. It was a crucial ingredient in the resolution of Herrero's approximation problem.)

7.6. References

The representation theorem (Theorem 7.2.5) and the homotopy invariance theorem (Theorem 7.3.6) come from [**191**]. Choi proved his residual finite-dimensionality result in [**37**]. The results of Section 7.5 are taken from [**49**].

Chapter 8

AF Embeddability

Deciding when a particular C*-algebra is isomorphic to a subalgebra of some AF algebra – i.e., when it is AF embeddable – can be a very challenging problem. The seminal result in this direction is due to Pimsner and Voiculescu who used an AF embedding to compute the range of the unique trace on the irrational rotation algebras. Since then a number of authors have studied AF embeddability; this chapter introduces some of the techniques used and, at the end, gives a quick survey of the current state of affairs.

We have no intention of giving a comprehensive treatment, as the proofs of AF embeddability tend to be quite difficult and different situations require different techniques. On the other hand, there is one reasonably accessible case which demonstrates some common themes used in other cases and, most importantly, has a useful application to exactness, namely, the fact, due to Dadarlat, that the cone over a separable exact RFD algebra is always AF embeddable. In Section 8.3 we'll prove a more general result, but the RFD case is much easier to digest, so we start there.

Throughout this chapter A and B will denote separable C*-algebras.

8.1. Stable uniqueness and asymptotically commuting diagrams

Let's think about what it takes to embed something in an AF algebra. Lacking an abstract characterization of subalgebras of AF algebras, there is really no way to avoid a construction of some sort. In this section we introduce some common techniques, starting off gently and progressively generalizing. In one way or another, most AF-embedding results depend on the following simple fact.

Proposition 8.1.1 (Approximately commuting diagrams). *Given a* C*-*algebra* A, *assume there exist finite-dimensional* C*-*algebras* B_i, *injective* $*$-*homomorphisms* $\pi_i\colon B_i \to B_{i+1}$ *and u.c.p. maps* $\sigma_i\colon A \to B_i$ *such that*

(1) *the* σ_i*'s are asymptotically isometric and asymptotically multiplicative;*

(2) *there exists a set* $\mathfrak{S} \subset A$ *with dense linear span, such that for all* $a \in \mathfrak{S}$,

$$\sum_{i=i}^{\infty} \|\pi_i(\sigma_i(a)) - \sigma_{i+1}(a)\| < \infty.$$

Then A *is AF embeddable.*

Proof. Here's a sketch. First, let B be the inductive limit of the sequence

$$B_1 \xrightarrow{\pi_1} B_2 \xrightarrow{\pi_2} B_3 \xrightarrow{\pi_3} \cdots.$$

Since the maps π_i are assumed injective, we may (and will) identify each B_i with a subalgebra of B. Then we can regard each map σ_i as taking values in B and one uses hypothesis (2) to show that the sequence $\{\sigma_i(a)\} \subset B$ is Cauchy for every $a \in \mathfrak{S}$.

Letting $X \subset A$ be the linear span of $\mathfrak{S}$ (which is assumed dense in A), it follows that $\{\sigma_i(x)\} \subset B$ is Cauchy for every $x \in X$. Thus we can define $\sigma : X \to B$ by

$$\sigma(x) = \lim_{i\to\infty} \sigma_i(x).$$

Now one checks that σ is an isometric linear map; hence it has a unique (isometric) extension to all of A. Finally one checks that this extension has no choice but to be a $*$-homomorphism, thereby giving a C*-algebraic embedding of A into B. (There are a number of things to check, but they're all easy.) □

Sometimes variations of the previous result get used in providing AF embeddings. For example in the Pimsner-Voiculescu AF embedding of irrational rotation algebras there are no maps σ_i. However, the core idea is the same as they construct matrix models which converge in norm and asymptotically satisfy the defining relations of an irrational rotation algebra; the embedding then comes from universality. However, we'll use the previous result explicitly; hence we'll need an appropriate "uniqueness" result in order to invoke it. The following notation will be handy and used repeatedly.

Definition 8.1.2. Given $\sigma\colon A \to \mathbb{M}_k(\mathbb{C})$ and a positive integer N,

$$N\sigma\colon A \to \mathbb{M}_{Nk}(\mathbb{C})$$

denotes the N-fold direct sum of σ. We will often identify $N\sigma$ with the map

$$\sigma \otimes 1_N : A \to \mathbb{M}_k(\mathbb{C}) \otimes \mathbb{M}_N(\mathbb{C}).$$

To get a feel for the type of thing we're after, here's a very important first step.

Proposition 8.1.3. *Assume A is RFD and that it has the following "stable uniqueness" property: For each pair of $*$-homomorphisms $\sigma_1, \sigma_2 \colon A \to \mathbb{M}_k(\mathbb{C})$ there exists an integer N such that $(N+1)\sigma_1$ is unitarily equivalent to $\sigma_2 \oplus N\sigma_1$. Then A is AF embeddable.*

Proof. Fix a sequence of $*$-homomorphisms $\rho_i \colon A \to \mathbb{M}_{k(i)}(\mathbb{C})$ with the property that

$$\|a\| = \lim_{i\to\infty} \|\rho_i(a)\|$$

for all $a \in A$. (Hint: Take direct sums of a separating sequence of representations.) Our goal is to use the assumed stable uniqueness property to construct finite-dimensional C*-algebras B_i, injective $*$-homomorphisms $\pi_i \colon B_i \to B_{i+1}$ and $*$-homomorphisms $\sigma_i \colon A \to B_i$ such that each σ_i contains ρ_i as a direct summand (which ensures that the σ_i's are asymptotically isometric) and

$$\sigma_{i+1} = \pi_i \circ \sigma_i$$

for all i. This is evidently more than enough to invoke the previous proposition.

The construction is recursive so let's define $B_1 = \mathbb{M}_{k(1)}(\mathbb{C})$ and $\sigma_1 = \rho_1 \colon A \to B_1$. Now consider $B_1 \otimes \mathbb{M}_{k(2)}(\mathbb{C})$ and the pair of $*$-homomorphisms

$$\sigma_1 \otimes 1_{k(2)} = k(2)\sigma_1 \colon A \to B_1 \otimes \mathbb{M}_{k(2)}(\mathbb{C})$$

and

$$1_{B_1} \otimes \rho_2 \colon A \to B_1 \otimes \mathbb{M}_{k(2)}(\mathbb{C}).$$

Since these maps take values in the same matrix algebra, we can apply stable uniqueness to find an integer N_1 such that

$$(N_1+1)k(2)\sigma_1 \colon A \to B_1 \otimes \mathbb{M}_{k(2)(N_1+1)}(\mathbb{C})$$

and

$$\big(1_{B_1} \otimes \rho_2\big) \oplus N_1 k(2)\sigma_1 \colon A \to B_1 \otimes \mathbb{M}_{k(2)(N_1+1)}(\mathbb{C})$$

are unitarily equivalent. So let $u \in B_1 \otimes \mathbb{M}_{k(2)(N_1+1)}(\mathbb{C})$ be a unitary such that

$$u(N_1+1)k(2)\sigma_1(a)u^* = \big(1_{B_1} \otimes \rho_2(a)\big) \oplus N_1 k(2)\sigma_1(a)$$

for all $a \in A$.

We are ready for the next step: Let $B_2 = B_1 \otimes \mathbb{M}_{k(2)(N_1+1)}(\mathbb{C})$, $\pi_1 \colon B_1 \to B_2$ be defined by

$$\pi_1(T) = u(T \otimes 1_{k(2)(N_1+1)})u^*$$

and $\sigma_2 \colon A \to B_2$ be given by

$$\sigma_2(a) = \big(1_{B_1} \otimes \rho_2(a)\big) \oplus N_1 k(2)\sigma_1(a).$$

By construction we have $\pi_1 \circ \sigma_1 = \sigma_2$.

If you really understand the argument so far, the rest of the proof is routine. □

Actually, the previous proposition is completely useless (see Exercise 8.1.1). But, *the proof* is very important. In fact, one should absorb it completely before proceeding.

Not yet knowing what the right stable uniqueness property is, a clue is provided by homotopic $*$-homomorphisms.

Lemma 8.1.4. *Let $\sigma_0, \sigma_1 \colon A \to \mathbb{M}_k(\mathbb{C})$ be homotopic $*$-homomorphisms (Definition 7.3.3). Then for each finite set $\mathfrak{F} \subset A$ and $\varepsilon > 0$ there exists an integer N, a $*$-homomorphism $\rho \colon A \to \mathbb{M}_N(\mathbb{C})$ and a unitary $u \in \mathbb{M}_{k+N}(\mathbb{C})$ such that for all $a \in \mathfrak{F}$,*

$$\|u\big(\sigma_0(a) \oplus \rho(a)\big)u^* - \sigma_1(a) \oplus \rho(a)\| < \varepsilon.$$

Or, in the notation prior to Theorem 7.5.7,

$$\sigma_0 \oplus \rho \overset{(\mathfrak{F},\varepsilon)}{\approx} \sigma_1 \oplus \rho.$$

Proof. The proof amounts to a simple, but useful, trick. Let $\sigma_t \colon A \to \mathbb{M}_k(\mathbb{C})$ be a continuous family of homomorphisms connecting σ_0 and σ_1. Since $t \mapsto \sigma_t(a)$ is norm continuous for each $a \in A$, uniform continuity provides an integer M with the property that

$$\|\sigma_{\frac{i}{M}}(a) - \sigma_{\frac{i+1}{M}}(a)\| < \varepsilon$$

for all $a \in \mathfrak{F}$ and every $0 \le i \le M-1$. One then defines $N = k(M-1)$ and

$$\rho = \sigma_{\frac{1}{M}} \oplus \sigma_{\frac{2}{M}} \oplus \cdots \oplus \sigma_{\frac{M-1}{M}}.$$

Note that $k+N = kM$, so we can identify $\mathbb{M}_{k+N}(\mathbb{C})$ with $\mathbb{M}_k(\mathbb{C}) \otimes \mathbb{M}_M(\mathbb{C})$.

Finally, let $U \in M_M(\mathbb{C})$ be the cyclic shift unitary of order M and define the unitary we're really after by

$$u = 1_k \otimes U \in \mathbb{M}_k(\mathbb{C}) \otimes \mathbb{M}_M(\mathbb{C}) = M_{k+N}(\mathbb{C}).$$

A straightforward calculation completes the proof. □

Remark 8.1.5. Let A be a unital residually finite-dimensional C*-algebra with the property that any two unital $*$-representations on the same Hilbert space are homotopic (e.g., the unitized cone over an RFD algebra). In this case one always has the stable uniqueness property above to work with. Try to mimic the proof of Proposition 8.1.3 and deduce AF embeddability of such algebras. Don't spend too much time though – it's impossible! The point here is to identify where the proof of Proposition 8.1.3 breaks down so we can head toward a better stable uniqueness result.

Lemma 8.1.6. *Let $\sigma_0, \sigma_1 \colon A \to \mathbb{M}_k(\mathbb{C})$ be homotopic $*$-homomorphisms and $\pi \colon A \to \mathbb{B}(\mathcal{H})$ be any faithful essential representation. Then for each finite set $\mathfrak{F} \subset A$ and $\varepsilon > 0$ there exists a unitary*

$$U \in \mathbb{C}1_{\ell_k^2 \oplus \mathcal{H}} + \mathbb{K}(\ell_k^2 \oplus \mathcal{H})$$

such that

$$\|U(\sigma_0(a) \oplus \pi(a))U^* - \sigma_1(a) \oplus \pi(a)\| < \varepsilon$$

for all $a \in \mathfrak{F}$.

Proof. Note that both $\sigma_0 \oplus \pi$ and $\sigma_1 \oplus \pi$ are faithful essential representations. So the point is not that we can conjugate one to the other – that's just Voiculescu's Theorem – but rather the placement of the unitary U. Note also that the $\mathbb{C}1_{\ell_k^2 \oplus \mathcal{H}} + \mathbb{K}(\ell_k^2 \oplus \mathcal{H}) \subset \mathbb{B}(\ell_k^2 \oplus \mathcal{H})$ is invariant under conjugation by unitaries.

Here is the proof:

$$\sigma_0 \oplus \pi \sim \sigma_0 \oplus (\rho \oplus \pi) = (\sigma_0 \oplus \rho) \oplus \pi \sim (\sigma_1 \oplus \rho) \oplus \pi = \sigma_1 \oplus (\rho \oplus \pi) \sim \sigma_1 \oplus \pi.$$

Perhaps a few more details are in order? By the previous lemma we can find an integer N, a $*$-homomorphism $\rho \colon A \to \mathbb{M}_N(\mathbb{C})$ and a unitary $u \in \mathbb{B}(\ell_k^2 \oplus \ell_N^2)$ such that

$$\|u(\sigma_0(a) \oplus \rho(a))u^* - \sigma_1(a) \oplus \rho(a)\| < \varepsilon$$

for all $a \in \mathfrak{F}$.

By Voiculescu's Theorem (applied to $\rho \oplus \pi$ and π) we can find a unitary $v \colon \mathcal{H} \to \ell_N^2 \oplus \mathcal{H}$ such that

$$\|v\pi(a)v^* - \rho(a) \oplus \pi(a)\| < \varepsilon$$

for all $a \in \mathfrak{F}$.

Now define $U = (1_k \oplus v)^*(u \oplus 1_{\mathcal{H}})(1_k \oplus v)$. Since the unitary $u \oplus 1_{\mathcal{H}}$ is a compact perturbation of the identity, it follows that

$$U \in \mathbb{C}1_{\ell_k^2 \oplus \mathcal{H}} + \mathbb{K}(\ell_k^2 \oplus \mathcal{H})$$

and a simple calculation shows that the desired estimate holds. □

We are almost ready for the right stable uniqueness result. We just need a definition which brings exactness into the picture.

Definition 8.1.7. Let $\mathfrak{F} \subset A$, a finite set, and $\varepsilon > 0$ be given. We say that a finite-dimensional representation $\sigma \colon A \to \mathbb{M}_k(\mathbb{C})$ is $(\mathfrak{F}, \varepsilon)$-*admissible* if there exists a faithful essential representation $\pi \colon A \to \mathbb{B}(\mathcal{H})$ such that

$$\pi \overset{(\mathfrak{F},\varepsilon)}{\approx} \sigma_\infty$$

where $\sigma_\infty = \bigoplus_1^\infty \sigma \colon A \to \mathbb{B}(\bigoplus_1^\infty \ell_k^2)$ is the infinite amplification of σ.

Theorem 8.1.8. *Let $\sigma_0, \sigma_1 \colon A \to \mathbb{M}_k(\mathbb{C})$ be homotopic $*$-homomorphisms and assume that σ_0 is $(\mathfrak{F}, \varepsilon)$-admissible. Then there exists an integer N such that*

$$(N+1)\sigma_0 \overset{(\mathfrak{F},3\varepsilon)}{\approx} \sigma_1 \oplus N\sigma_0.$$

Proof. In a line, here is the proof:

$$(\sigma_0)_\infty \sim \sigma_0 \oplus \pi \sim \sigma_1 \oplus \pi \sim \sigma_1 \oplus (\sigma_0)_\infty.$$

More precisely, since σ_0 is $(\mathfrak{F}, \varepsilon)$-admissible, we can find a faithful essential representation $\pi \colon A \to \mathbb{B}(\bigoplus_1^\infty \ell^2_k)$ such that

$$\|(\sigma_0)_\infty(a) - \pi(a)\| < \varepsilon$$

for all $a \in \mathfrak{F}$. (Just replace whatever faithful essential representation the definition gives you by a unitarily conjugate representation on the Hilbert space $\bigoplus_1^\infty \ell^2_k$.) By the previous lemma there exists a unitary

$$U \in \mathbb{C}1_{\ell^2_k \oplus [\bigoplus_1^\infty \ell^2_k]} + \mathbb{K}(\ell^2_k \oplus [\bigoplus_1^\infty \ell^2_k])$$

such that

$$\|U\big(\sigma_0(a) \oplus \pi(a)\big)U^* - \sigma_1(a) \oplus \pi(a)\| < \varepsilon$$

for all $a \in \mathfrak{F}$. Hence the triangle inequality implies

$$\|U\big(\sigma_0(a) \oplus (\sigma_0)_\infty(a)\big)U^* - \sigma_1(a) \oplus (\sigma_0)_\infty(a)\| < 3\varepsilon$$

for all $a \in \mathfrak{F}$.

But why the big deal about our unitary being a compact perturbation of a scalar? Well, letting $P_M \in \mathbb{B}(\ell^2_k \oplus [\bigoplus_1^\infty \ell^2_k])$ be the orthogonal projection onto the subspace

$$\ell^2_k \oplus [\bigoplus_1^M \ell^2_k],$$

it follows that

$$\|[P_m, U]\| \to 0$$

as $M \to \infty$. Hence, by standard perturbation theory, we can find unitaries $U_M \in \mathbb{B}(\ell^2_k \oplus [\bigoplus_1^M \ell^2_k])$ such that $\|U_M - P_M U P_M\| \to 0$ as $M \to \infty$. It follows that

$$\limsup_{M\to\infty} \|U_M\big((M+1)\sigma_0(a)\big)U_M^* - \sigma_1(a) \oplus M\sigma_0(a)\| \le 3\varepsilon$$

for all $a \in \mathfrak{F}$, so the proof is complete. □

Exercise

Exercise 8.1.1. Show that the hypotheses of Proposition 8.1.3 imply that A is a finite-dimensional matrix algebra.

8.2. Cones over exact RFD algebras

A few more fairly simple facts and we'll be able to AF embed the cone over an exact RFD algebra. The first step is to observe the *existence* of $(\mathfrak{F}, \varepsilon)$-admissible representations.

Proposition 8.2.1. *Let A be exact and residually finite-dimensional. For each finite set $\mathfrak{F} \subset A$ and $\varepsilon > 0$ there exists an $(\mathfrak{F}, \varepsilon)$-admissible representation $\sigma\colon A \to \mathbb{M}_k(\mathbb{C})$.*

Proof. This is an immediate consequence of Corollary 7.5.8 since there is only one way to put a matrix algebra (unitally) inside of $\mathbb{B}(\mathcal{H})$. □

We now list, without proof, two trivial facts which are important below.

Lemma 8.2.2. *Let $\sigma\colon A \to \mathbb{M}_k(\mathbb{C})$ be an $(\mathfrak{F}, \varepsilon)$-admissible representation and $\pi\colon A \to \mathbb{M}_N(\mathbb{C})$ be any other representation. Then $\sigma \oplus \pi$ is also $(\mathfrak{F}, \varepsilon)$-admissible.*

Lemma 8.2.3. *Let $\tilde{CA}$ denote the unitization of the cone over some C*-algebra A. Any two unital $*$-representations of $\tilde{CA}$ on the same Hilbert space are homotopic.*

Theorem 8.2.4 (Dadarlat). *Let A be a separable exact residually finite-dimensional C*-algebra. Then the cone over A is AF embeddable.*[1]

Proof. The proof amounts to reviewing everything we have done so far.

Let $\tilde{CA}$ be the unitization of the cone over A and $\mathfrak{F}_n \subset \tilde{CA}$ be an increasing sequence of finite subsets whose union has dense linear span. Let $\varepsilon_n = \frac{1}{2^n}$ and let $\rho_n\colon \tilde{CA} \to \mathbb{M}_{k(n)}(\mathbb{C})$ be a sequence of $(\mathfrak{F}_n, \varepsilon_n)$-admissible representations. Note that the ρ_n's are automatically asymptotically isometric.

We now mimic the proof of Proposition 8.1.3 to construct a sequence of finite-dimensional C*-algebras B_i, injective $*$-homomorphisms $\pi_i\colon B_i \to B_{i+1}$ and $*$-homomorphisms $\sigma_i\colon \tilde{CA} \to B_i$ such that each σ_i contains ρ_i as a direct summand and

$$\sum_{i=1}^{\infty} \|\sigma_{i+1}(a) - \pi_i \circ \sigma_i(a)\| < \infty$$

for every $a \in \bigcup \mathfrak{F}_n$. As before, this will imply AF embeddability.

[1]One can even embed it into the UHF algebra of type 2^∞, but we won't prove this. (Our construction only gives a UHF embedding.) It requires a bit more work, keeping in mind that the unitized cone has a one-dimensional representation that can be used to modify dimension – i.e., increase the dimension of a given representation by taking direct sums with one-dimensional representations.

So, let $B_1 = \mathbb{M}_{k(1)}(\mathbb{C})$ and $\sigma_1 = \rho_1 \colon \tilde{CA} \to B_1$. Now consider $B_1 \otimes \mathbb{M}_{k(2)}(\mathbb{C})$ and the pair of $*$-homomorphisms

$$\sigma_1 \otimes 1_{k(2)} = k(2)\sigma_1 \colon \tilde{CA} \to B_1 \otimes \mathbb{M}_{k(2)}(\mathbb{C})$$

and

$$1_{B_1} \otimes \rho_2 \colon \tilde{CA} \to B_1 \otimes \mathbb{M}_{k(2)}(\mathbb{C}).$$

Note that $\sigma_1 \otimes 1_{k(2)}$ is still $(\mathfrak{F}_1, \varepsilon_1)$-admissible and homotopic to $1_{B_1} \otimes \rho_2$. Hence we can apply Theorem 8.1.8 to find an integer N_1 such that

$$(N_1+1)(\sigma_1 \otimes 1_{k(2)}) \overset{(\mathfrak{F}_1, 3\varepsilon_1)}{\approx} \big(1_{B_1} \otimes \rho_2\big) \oplus N(\sigma_1 \otimes 1_{k(2)}).$$

The remainder of the proof is very similar to that of Proposition 8.1.3; however one important point should be mentioned: *In order to repeat the recursive procedure, you have to know that* $\big(1_{B_1} \otimes \rho_2\big) \oplus N(\sigma_1 \otimes 1_{k(2)})$ *is* $(\mathfrak{F}_2, \varepsilon_2)$*-admissible!* In other words, adding on σ_1 does not *decrease* the level of admissibility. Luckily, this is the content of Lemma 8.2.2. □

The following corollary will play a crucial role in showing that exactness passes to quotients. It is due to Kirchberg, though his proof was quite different. We say C is a *subquotient* of A if there exists a subalgebra $B \subset A$ with ideal $J \lhd B$, such that $C = B/J$.

Corollary 8.2.5. *Every separable exact* C**-algebra is a subquotient of a UHF algebra.*

Proof. Since every algebra is a quotient of its cone, this follows immediately from Dadarlat's embedding theorem together with Corollary 7.3.8. □

8.3. Cones over general exact algebras

We now wish to remove the RFD hypothesis from Dadarlat's result. Though the main result of this section is a generalization of what we already know, one of the key steps in the proof is actually more elementary; the right stable uniqueness result is not very deep. However, setting everything up is far more complicated and the notation alone will tax one's memory. We start with an obvious adaptation of the admissible representations which were so important in the last section.

Definition 8.3.1. Let a finite set $\mathfrak{F} \subset A \subset \mathbb{B}(\mathcal{H})$ and $\varepsilon > 0$ be given. We say that a u.c.p. map $\varphi \colon A \to \mathbb{M}_k(\mathbb{C})$ is $(\mathfrak{F}, \varepsilon)$*-multiplicative* if

$$\max_{a \in \mathfrak{F}} \{\|\varphi(a^*a) - \varphi(a^*)\varphi(a)\|, \|\varphi(aa^*) - \varphi(a)\varphi(a^*)\|\} < \varepsilon.$$

We say φ is $(\mathfrak{F},\varepsilon)$-*suitable* if it is $(\mathfrak{F},\varepsilon)$-multiplicative and there is a u.c.p. map $\psi\colon \mathbb{M}_k(\mathbb{C}) \to \mathbb{B}(\mathcal{H})$ such that

$$\|\psi \circ \varphi(a) - a\| < \varepsilon$$

for all $a \in \mathfrak{F}$.

Remark 8.3.2. It is important (though trivial) to note that if φ is $(\mathfrak{F},\varepsilon)$-suitable and ψ is $(\mathfrak{F},\varepsilon)$-multiplicative, then $\varphi \oplus \psi$ is $(\mathfrak{F},\varepsilon)$-*suitable.* That is, adding multiplicative maps to suitable maps does not destroy suitability.

It follows easily from Arveson's Extension Theorem that the notion of $(\mathfrak{F},\varepsilon)$-suitability does not depend on the choice of representation $A \subset \mathbb{B}(\mathcal{H})$.

Here is the stable uniqueness result we need this time.

Lemma 8.3.3. *Let $\mathfrak{F} \subset A \subset \mathbb{B}(\mathcal{H})$ be a finite set of unitaries and let $\varepsilon > 0$. If $\varphi\colon A \to \mathbb{M}_p(\mathbb{C})$ is an $(\mathfrak{F},\varepsilon)$-suitable u.c.p. map and $\psi\colon A \to \mathbb{M}_q(\mathbb{C})$ is an $(\mathfrak{F},\varepsilon)$-multiplicative u.c.p. map, then there exists an integer $N \in \mathbb{N}$ and an $(\mathfrak{F}, 5\sqrt{\varepsilon})$-multiplicative u.c.p. map $\rho\colon A \to \mathbb{M}_{Np-q}(\mathbb{C})$ such that $N\varphi \overset{(\mathfrak{F},5\sqrt{\varepsilon})}{\approx} \psi \oplus \rho$.*

Proof. The proof uses some ideas and calculations from the proof of Lemma 7.5.5 so we describe how to get ρ and leave the estimates to the reader.

Let $\alpha\colon \mathbb{M}_p(\mathbb{C}) \to \mathbb{B}(\mathcal{H})$ be a u.c.p. map with $\|f - \alpha\circ\varphi(f)\| < \varepsilon$ for $f \in \mathfrak{F}$. We may assume that ψ is defined on all of $\mathbb{B}(\mathcal{H})$ and thus get a u.c.p. map $\psi \circ \alpha\colon \mathbb{M}_p(\mathbb{C}) \to \mathbb{M}_q(\mathbb{C})$ *which is almost multiplicative on* $\varphi(\mathfrak{F})$ (since α is necessarily close to multiplicative on this set and ψ is almost multiplicative on $\mathfrak{F}$). By Stinespring's Theorem, we get a $*$-homomorphism $\pi\colon \mathbb{M}_p(\mathbb{C}) \to \mathbb{B}(\mathcal{K})$ and a projection $P \in \mathbb{B}(\mathcal{K})$ such that $\psi \circ \alpha$ can be identified with compression by P. In this case $\mathcal{K}$ is a finite-dimensional Hilbert space; hence there is an integer N such that $\mathbb{B}(\mathcal{K}) \cong \mathbb{M}_p(\mathbb{C})\otimes\mathbb{M}_N(\mathbb{C})$ and we may identify $\pi\colon \mathbb{M}_p(\mathbb{C}) \to \mathbb{B}(\mathcal{K})$ with the canonical embedding $\mathbb{M}_p(\mathbb{C}) \to \mathbb{M}_p(\mathbb{C})\otimes\mathbb{M}_N(\mathbb{C})$, $T \mapsto T \otimes 1$. Thus $\psi \circ \alpha$ can be identified with the map $T \mapsto P(T \otimes 1)P$.

As we have seen, almost multiplicativity implies P almost commutes with $\varphi(\mathfrak{F}) \otimes 1 \subset \mathbb{M}_p(\mathbb{C}) \otimes \mathbb{M}_N(\mathbb{C})$ and hence, for every $a \in \mathfrak{F}$,

$$\begin{aligned} N\varphi(a) = \varphi(a) \otimes 1 &\approx P(\varphi(a) \otimes 1)P \oplus P^{\perp}(\varphi(a) \otimes 1)P^{\perp} \\ &\approx \psi(a) \oplus P^{\perp}(\varphi(a) \otimes 1)P^{\perp}. \end{aligned}$$

The desired map $\rho\colon A \to \mathbb{M}_{Np-q}(\mathbb{C})$ is defined by $\rho(a) = P^{\perp}(\varphi(a) \otimes 1)P^{\perp}$. □

Note that Theorem 7.5.7 implies the existence of $(\mathfrak{F},\varepsilon)$-admissible maps whenever A is exact and QD. Hence the stable uniqueness result above suggests the possibility of adapting Proposition 8.1.3 to deduce AF embeddability of all such algebras. You may want to figure out where the proof breaks down; it isn't necessary, but it will help motivate our next step for coned algebras.

For a unital separable exact C*-algebra A and $t \in [0,1]$ we let

$$C(t) = \{f \in C([t,1], A) : f(1) \in \mathbb{C}1_A\}.$$

Note that $C := C(0)$ is just the unitization of the cone CA over A and $C(1) = \mathbb{C}$.[2] We denote by $\pi(t)\colon C \to C(t)$ the natural restriction homomorphism from C onto $C(t)$ and let $\tilde{\sigma}(t)\colon C(t) \to C$ be the $*$-homomorphism given by

$$(\tilde{\sigma}(t)(f))(s) = \begin{cases} f(t) & \text{if } s < t, \\ f(s) & \text{if } s \geq t \end{cases}$$

for $f \in C$. Clearly $\sigma(t) := \tilde{\sigma}(t)\pi(t)$ is a continuous path of endomorphisms on C with $\sigma(0) = \mathrm{id}_C$ and $\sigma(1)(C) = \mathbb{C}1_C$. We will need to smear some functions around, so if $\mathfrak{F} \subset C$ and $\mathcal{P} \subset [0,1]$ are given finite sets, we let

$$\mathfrak{F}^{\mathcal{P}} = \{\sigma(p)(f) : f \in \mathfrak{F},\ p \in \mathcal{P}\} \subset C.$$

Note that $\sigma(p)(\mathfrak{F}^{\mathcal{P}}) \subset \mathfrak{F}^{\mathcal{P}}$ for every $p \in \mathcal{P}$.

The next lemma is the key technical fact. The proof resembles your worst nightmare, but the main point is this: Uniqueness of unital maps from $\mathbb{C}$ to a matrix algebra is automatic! Very roughly, we'll slowly work our way down the cone, applying stable uniqueness over and over, until we arrive at $\mathbb{C}$ where automatic uniqueness saves the day.

Lemma 8.3.4. *Let C be as above and suppose we are given*

(1) *a finite set $\mathfrak{F}_0$ of unitaries in C;*

(2) $\varepsilon_0 > 0$;

(3) *a finite set $\mathcal{P} = \{p_0, p_1, \ldots, p_N\} \subset [0,1]$ such that $p_i \leq p_{i+1}$, $p_0 = 0$, $p_N = 1$ and $\|f(s) - f(t)\| < \varepsilon_0$ for all $f \in \mathfrak{F}_0$ and $p_i \leq s, t \leq p_{i+1}$, $i = 1, \ldots, N-1$.*

Then there is a $\delta_0 > 0$ with the following property: For every

- *collection of $(\pi(p_i)(\mathfrak{F}_0^{\mathcal{P}}), \delta_0)$-suitable u.c.p. maps $\tilde{\theta}_0(i)\colon C(p_i) \to D_0(i)$, where $D_0(i)$ is a matrix algebra and $i = 1, \ldots, N$,*
- *finite set of unitaries $\mathfrak{F}_1 \subset C$ and*
- $\delta_1 > 0$,

[2] We have reversed things and unitized $C([0,1), A)$.

there is an $(\mathfrak{F}_1, \delta_1)$-suitable u.c.p. map θ_1 from C into a full matrix algebra D_1 and positive integers $n_0(i) \in \mathbb{N}$ $(i = 1, \ldots, N)$ such that

$$\theta_1 \overset{(\mathfrak{F}_0^{\mathcal{P}}, 4\varepsilon_0)}{\approx} \bigoplus_{i=1}^{N} n_0(i)\theta_0(i),$$

where $\theta_0(i) = \tilde{\theta}_0(i) \circ \pi(p_i)$ for every i.

Proof. Let $h(s) = 5\sqrt{s}$ and $h^j = h \circ \cdots \circ h$ be the composition of h with itself j times. We can take any δ_0 such that $h^N(\delta_0) < \varepsilon_0$.

Let the $\tilde{\theta}_0(i)$'s, $\mathfrak{F}_1$ and δ_1 be given. We may assume that $\mathfrak{F}_1 \supset \mathfrak{F}_0^{\mathcal{P}}$ and $\delta_1 < \delta_0$. Since each $C(p_i)$ is exact and QD, we can find $(\pi(p_i)(\mathfrak{F}_1), \delta_1)$-suitable maps $\tilde{\nu}(i)$ from $C(p_i)$ to matrix algebras, for $i = 0, 1, \ldots, N-1$. Let $\nu(i) = \tilde{\nu}(i) \circ \pi(p_i)$ and note that $\nu(0)$ is an $(\mathfrak{F}_1, \delta_1)$-suitable map. Hence it will suffice to construct θ_1 of the form

$$\theta_1 = \nu(0) \oplus m(1)\nu(1) \oplus \cdots \oplus m(N-1)\nu(N-1),$$

for some positive numbers $m(1), \ldots, m(N-1)$.

Applying Lemma 8.3.3 to the $(\pi(p_1)(\mathfrak{F}_0^{\mathcal{P}}), \delta_0)$-suitable map $\tilde{\theta}_0(1)$ and the $(\pi(p_1)(\mathfrak{F}_0^{\mathcal{P}}), \delta_0)$-multiplicative map $\nu(0) \circ \tilde{\sigma}(p_1)$, we find an integer $n_0(1)$ such that

$$n_0(1)\tilde{\theta}_0(1) \overset{(\pi(p_1)(\mathfrak{F}_0^{\mathcal{P}}), h(\delta_0))}{\approx} \nu(0) \circ \tilde{\sigma}(p_1) \oplus \tilde{\rho}(1),$$

for some $(\pi(p_1)(\mathfrak{F}_0^{\mathcal{P}}), h(\delta_0))$-multiplicative map $\tilde{\rho}(1)$ from $C(p_1)$ to a matrix algebra. Evidently this implies

$$n_0(1)\theta_0(1) \overset{(\mathfrak{F}_0^{\mathcal{P}}, h(\delta_0))}{\approx} \nu(0) \circ \sigma(p_1) \oplus \rho(1),$$

where $\rho(1) = \tilde{\rho}(1) \circ \pi(p_1)$.

Now one should apply Lemma 8.3.3 to the $(\pi(p_1)(\mathfrak{F}_0^{\mathcal{P}}), \delta_0)$-suitable map $\tilde{\nu}(1)$ and the $(\pi(p_1)(\mathfrak{F}_0^{\mathcal{P}}), h(\delta_0))$-multiplicative map $\tilde{\rho}(1)$ to find $m(1)$ such that

$$m(1)\tilde{\nu}(1) \overset{(\pi(p_1)(\mathfrak{F}_0^{\mathcal{P}}), h^2(\delta_0))}{\approx} \tilde{\rho}(1) \oplus \tilde{\mu}(1),$$

for some $(\pi(p_1)(\mathfrak{F}_0^{\mathcal{P}}), h^2(\delta_0))$-multiplicative map $\tilde{\mu}(1)$. Thus

$$m(1)\nu(1) \overset{(\mathfrak{F}_0^{\mathcal{P}}, h^2(\delta_0))}{\approx} \rho(1) \oplus \mu(1),$$

where $\mu(1) = \tilde{\mu}(1) \circ \pi(p_1)$.

Let's take a moment to absorb what we've done. Using the fact that

$$\nu(0) \overset{(\mathfrak{F}_0^{\mathcal{P}}, \varepsilon_0)}{\approx} \nu(0) \circ \sigma(p_1),$$

we see that

$$\nu(0) \oplus m(1)\nu(1) \overset{(\mathfrak{F}_0^{\mathcal{P}}, \varepsilon_0)}{\approx} \nu(0) \circ \sigma(p_1) \oplus \big(\rho(1) \oplus \mu(1)\big)$$

$$= \big(\nu(0) \circ \sigma(p_1) \oplus \rho(1)\big) \oplus \mu(1)$$
$$\overset{(\mathfrak{F}_0^{\mathcal{P}}, h(\delta_0))}{\approx} n_0(1)\theta_0(1) \oplus \mu(1).$$

Note that we've taken a step to the right in the sense that the "remainder" term $\mu(1)$ factors through $C(p_1)$.

Another application of Lemma 8.3.3 to the $(\pi(p_2)(\mathfrak{F}_0^{\mathcal{P}}), \delta_0)$-suitable map $\tilde{\theta}_0(2)$ and the $(\pi(p_2)(\mathfrak{F}_0^{\mathcal{P}}), h^2(\delta_0))$-multiplicative map $\mu(1) \circ \tilde{\sigma}(p_2)$ yields an integer $n_0(2)$ such that

$$n_0(2)\theta_0(2) \overset{(\mathfrak{F}_0^{\mathcal{P}}, h^3(\delta_0))}{\approx} \mu(1) \circ \sigma(p_2) \oplus \rho(2),$$

where $\rho(2) = \tilde{\rho}(2) \circ \pi(p_2)$ for some $(\pi(p_2)(\mathfrak{F}_0^{\mathcal{P}}), h^3(\delta_0))$-multiplicative map $\tilde{\rho}(2)$. It follows that

$$\begin{aligned} n_0(1)\theta_0(1) &\oplus n_0(2)\theta_0(2) \\ &\overset{(\mathfrak{F}_0^{\mathcal{P}}, h^3(\delta_0))}{\approx} \Big(\nu(0) \circ \sigma(p_1) \oplus \rho(1)\Big) \oplus \Big(\mu(1) \circ \sigma(p_2) \oplus \rho(2)\Big) \\ &= \nu(0) \circ \sigma(p_1) \oplus \Big(\rho(1) \oplus \mu(1) \circ \sigma(p_2)\Big) \oplus \rho(2) \\ &\overset{(\mathfrak{F}_0^{\mathcal{P}}, \varepsilon_0)}{\approx} \nu(0) \oplus \Big(\rho(1) \oplus \mu(1)\Big) \oplus \rho(2) \\ &\overset{(\mathfrak{F}_0^{\mathcal{P}}, h^2(\delta_0))}{\approx} \nu(0) \oplus m(1)\nu(1) \oplus \rho(2). \end{aligned}$$

Combining these estimates, we have

$$\nu(0) \oplus m(1)\nu(1) \oplus \rho(2) \overset{(\mathfrak{F}_0^{\mathcal{P}}, 3\varepsilon_0)}{\approx} n_0(1)\theta_0(1) \oplus n_0(2)\theta_0(2),$$

where the term $\rho(2)$ factors through $C(p_2)$; thus we've moved one step further to the right.

Now we repeat this procedure *ad nauseam*: Absorb $\rho(2)$ into a multiple of $\nu(2)$ with a remainder $\mu(2)$, which factors through $C(p_2)$, and then check that

$$\nu(0) \oplus m(1)\nu(1) \oplus m(2)\nu(2) \overset{(\mathfrak{F}_0^{\mathcal{P}}, 3\varepsilon_0)}{\approx} \Big(\bigoplus_{i=1}^{2} n_0(i)\theta_0(i)\Big) \oplus \mu(2).$$

Then absorb $\mu(2) \circ \sigma(p_3)$ into a multiple of $\theta_0(3)$ and find that

$$\nu(0) \oplus m(1)\nu(1) \oplus m(2)\nu(2) \oplus \rho(3) \overset{(\mathfrak{F}_0^{\mathcal{P}}, 3\varepsilon_0)}{\approx} \bigoplus_{i=1}^{3} n_0(i)\theta_0(i),$$

with a remainder $\rho(3)$ which now factors through $C(p_3)$, and so on.

Go back and forth until the $N-1$ step and you have

$$\nu(0)\oplus m(1)\nu(1)\oplus\cdots\oplus m(N-1)\nu(N-1) \overset{(\mathfrak{F}_0^{\mathcal{P}},3\varepsilon_0)}{\approx} \left(\bigoplus_{i=1}^{N-1} n_0(i)\theta_0(i)\right)\oplus\mu(N-1),$$

where $\mu(N-1)$ is a map which factors through $C(p_{N-1})$. Finally, we come to the end because there is an integer $n_0(N)$ such that

$$\mu(N-1) \overset{(\mathfrak{F}_0^{\mathcal{P}},\varepsilon_0)}{\approx} n_0(N)\theta_0(N).$$

Indeed, $\theta_0(N)$ factors through $C(p_N) = C(1) = \mathbb{C}$; hence we may assume its range is just $\mathbb{C}$ – and $\mu(N-1)$ factors through $C(p_{N-1})$, which means it maps every element of $\mathfrak{F}_0^{\mathcal{P}}$ to a matrix which is within ε_0 of a scalar multiple of the identity. Thus we have

$$\nu(0)\oplus m(1)\nu(1)\oplus\cdots\oplus m(N-1)\nu(N-1) \overset{(\mathfrak{F}_0^{\mathcal{P}},4\varepsilon_0)}{\approx} \bigoplus_{i=1}^{N} n_0(i)\theta_0(i)$$

and the proof is complete. □

Theorem 8.3.5. *The cone over every separable exact* C*-*algebra* A *is AF embeddable.*

Proof. As before C denotes the unitization of CA. It suffices to construct finite-dimensional C*-algebras $B_0, B_1, \ldots$, embeddings $B_i \hookrightarrow B_{i+1}$ and u.c.p. maps $\sigma_i : C \to B_i$ which satisfy the hypotheses of Proposition 8.1.1.

Fix an increasing sequence of finite sets of unitaries $\mathfrak{F}_0 \subset \mathfrak{F}_1 \subset \mathfrak{F}_2 \subset \cdots$ such that the linear span of their union is dense in C. Define $\varepsilon_n = 1/2^n$ and choose some finite sets $\mathcal{P}_n \subset [0,1]$ which satisfy condition (3) of Lemma 8.3.4 for $\mathfrak{F}_n$ and ε_n. Let $\delta_n < \varepsilon_n$ have the property asserted in Lemma 8.3.4 (applied to $\mathfrak{F}_n$, ε_n and $\mathcal{P}_n$).

To put the machine in motion, we choose any collection of $(\pi(p)(\mathfrak{F}_0^{\mathcal{P}_0}), \delta_0)$-suitable u.c.p. maps $\tilde{\theta}_0(p)\colon C(p) \to D_0(p)$, where p runs through the points in $\mathcal{P}_0$ and the $D_0(p)$'s are matrix algebras. We then let

$$B_0 = \bigoplus_{p\in\mathcal{P}_0} D_0(p)$$

and

$$\sigma_0 = \bigoplus_{p\in\mathcal{P}_0} \theta_0(p)\colon C \to B_0,$$

where $\theta_0(p) = \tilde{\theta}_0(p)\circ\pi(p)$.

Here is how we get B_1: For each $q \in \mathcal{P}_1$ we want to construct a $(\pi(q)(\mathfrak{F}_1^{\mathcal{P}_1}), \delta_1)$-suitable u.c.p. map $\tilde{\theta}_1(q)\colon C(q) \to D_1(q)$ and a nice connecting map $B_0 \to D_1(q)$ (which won't be injective for all q's but will be for the smallest $q = 0 \in \mathcal{P}_1$). *The key remark is that $C(q)$ is isomorphic to C* and hence we can apply Lemma 8.3.4 to $C(q)$ and get a $(\pi(q)(\mathfrak{F}_1^{\mathcal{P}_1}), \delta_1)$-suitable u.c.p. map $\tilde{\theta}_1(q)\colon C(q) \to D_1(q)$ and integers $n_{0,q}(p)$ such that

$$\tilde{\theta}_1(q) \overset{(\pi(q)(\mathfrak{F}_0),4\varepsilon_0)}{\approx} \bigoplus_{p\in\mathcal{P}_0, p\geq q} n_{0,q}(p)\tilde{\theta}_{0,q}(p),$$

where $\tilde{\theta}_{0,q}(p) = \tilde{\theta}_0(p)\circ\pi(p)\circ\tilde{\sigma}(q)$. Thus we can use the integers $n_{0,q}(p)$ to get a connecting map $B_0 \to D_1(q)$ which almost intertwines σ_0 and $\tilde{\theta}_1(q)\circ\pi(q)$ (up to $4\varepsilon_0$ on $\mathfrak{F}_0$) by simply cutting off the summands of B_0 which correspond to those points in $\mathcal{P}_0$ which are strictly less than q.

Define

$$B_1 = \bigoplus_{q\in\mathcal{P}_1} D_1(q)$$

and

$$\sigma_1 = \bigoplus_{q\in\mathcal{P}_1} \theta_1(q)\colon C \to B_1,$$

where $\theta_1(q) = \tilde{\theta}_1(q)\circ\pi(q)$. Since we know how to get a map from B_0 to each summand in B_1 (which almost intertwines σ_0 and σ_1), we can take a direct sum to get the desired embedding $B_0 \hookrightarrow B_1$ (which almost intertwines σ_0 and σ_1, as well). In other words, aside from the notational horrors, it is easy to see that we have constructed a diagram

$$\begin{array}{ccc} B_0 & \longrightarrow & B_1 \\ & \nwarrow_{\sigma_0} \quad {}^{\sigma_1}\nearrow & \\ & C & \end{array}$$

which $4\varepsilon_0$-commutes on $\mathfrak{F}_0$.

Now repeat.... □

8.4. Homotopy invariance

This section establishes the analogue of Voiculescu's homotopy invariance theorem (Theorem 7.3.6) for AF embeddability of exact C*-algebras. The proof requires Theorem 8.3.5 plus a number of other facts. The majority of the extra ingredients are elementary and complete proofs will be given. However, we need a nontrivial generalization of Voiculescu's Theorem, due to Kasparov, and refer the reader to [**97**] for a proof.

Definition 8.4.1. An ideal $I \lhd E$ is *essential* if it has "no orthogonal complement" – i.e.,

$$I^{\perp} := \{e \in E : ex = xe = 0, \text{ for all } x \in I\} = \{0\}.$$

It is a simple exercise to check that an ideal is essential if and only if every other nonzero ideal intersects it nontrivially.

Proposition 8.4.2. *Let I be a nonunital* C*-*algebra. Then there exists a unital* C*-*algebra $M(I)$, called* the multiplier algebra of I, *with the property that $M(I)$ is the largest* C*-*algebra containing I as an essential ideal – i.e., if $I \lhd A$ and I is essential, then there is a unique embedding $A \hookrightarrow M(I)$ which extends the canonical inclusion $I \subset M(I)$. Moreover, $M(I)$ is the unique (up to isomorphism) algebra with this property.*

Proof. Let $I \subset \mathbb{B}(\mathcal{H})$ be any nondegenerate representation and define

$$M(I) = \{T \in \mathbb{B}(\mathcal{H}) : Tx \in I \text{ and } xT \in I, \text{ for all } x \in I\}.$$

Since $I \subset \mathbb{B}(\mathcal{H})$ is nondegenerate, I sits in $M(I)$ as an essential ideal (an orthogonal element would give a subspace of $\mathcal{H}$ where I restricts to zero).

Now suppose $I \lhd A$. Then the map $I \hookrightarrow \mathbb{B}(\mathcal{H})$ extends uniquely to a $*$-homomorphism $\pi\colon A \to \mathbb{B}(\mathcal{H})$ by defining

$$\pi(a) = \lim ae_n$$

where $\{e_n\} \subset I$ is any approximate unit and the limit is taken in the strong operator topology (uniqueness of π is again due to the fact that $I \subset \mathbb{B}(\mathcal{H})$ is nondegenerate). If $I \lhd A$ is essential, then π must be injective – the kernel of π will be orthogonal to I – and hence we must show that $\pi(A) \subset M(I)$. However, if we take $\{e_n\} \subset I$ to be quasicentral, then this is easily verified. Indeed,

$$\pi(a)x = \lim(ae_n)x = \lim a(e_n x) = ax \in I$$

and multiplication on the other side is similar since $\{e_n\}$ is quasicentral.

Uniqueness of $M(I)$ follows easily from the universal property. □

Lemma 8.4.3. *Suppose $I \subset J$ is an inclusion of nonunital* C*-*algebras such that some approximate unit $\{e_n\} \subset I$ is also an approximate unit of J. Then we have a natural inclusion $M(I) \subset M(J)$ such that $M(I) \cap J = I$.*

Proof. Let $J \subset \mathbb{B}(\mathcal{H})$ be nondegenerate. Then $I \subset \mathbb{B}(\mathcal{H})$ is also nondegenerate and hence $I \subset M(I) \subset \mathbb{B}(\mathcal{H})$. Let $T \in M(I)$ and $x \in J$ be arbitrary. Then

$$Tx = \lim T(e_n x) = \lim(Te_n)x,$$

since $\{e_n\} \subset I$ is an approximate unit of J. But $Te_n \in I$ and thus $(Te_n)x \in J$ for all n – i.e., $T \in M(J)$. Now assume that $x \in M(I) \cap J$. Then $\|x - e_n x\| \to 0$ implies $x \in I$, since $e_n x \in I$ by assumption. □

Lemma 8.4.4. *Assume I is AF embeddable. Then there is an inclusion $I \subset J$ where J is AF and some approximate unit $\{e_n\} \subset I$ is also an approximate unit of J. (Hence $M(I) \subset M(J)$ and $M(I) \cap J = I$.)*

Proof. Let $I \subset B$ be an inclusion such that B is AF and let $\{e_n\} \subset I$ be an approximate unit of I. Let $J \subset B$ be the hereditary subalgebra generated by I (i.e., the norm closure of the algebras $e_n B e_n$). Evidently $\{e_n\}$ is also an approximate unit for J so we only need to recall that hereditary subalgebras of AF algebras are again AF.

This fact is standard fare so we only sketch the proof. The hardest part is to show that if $J \subset B$ is hereditary, then J has an approximate unit of projections. (The informed reader may recall that this fact even holds when B is only assumed to have real rank zero.) Once this is established, the proof is routine. Indeed, for any finite set $\mathfrak{F} \subset J$ and $\varepsilon > 0$ we can choose a projection $p \in J$ such that $\|px - x\| < \varepsilon$ for all $x \in \mathfrak{F}$. Then we can find a finite-dimensional subalgebra $C \subset B$ which almost contains $\{p\} \cup \mathfrak{F}$. Perturbation theory yields a projection $q \in C$ which is close to p and this implies the existence of a unitary $u \in B$ which is close to the identity and such that $uqu^* = p$. One then checks that $u(qCq)u^*$ is a finite-dimensional subalgebra of J which almost contains $\mathfrak{F}$. See [**64**] for the details. □

We now prove some results on the AF embeddability of extensions.

Lemma 8.4.5. *Let $0 \to I \to E \to B \to 0$ be exact where $I \lhd E$ is essential and I is AF embeddable. Then there exists a commutative diagram*

$$\begin{array}{ccccccccc} 0 & \longrightarrow & I & \longrightarrow & E & \longrightarrow & B & \longrightarrow & 0 \\ & & \downarrow & & \downarrow & & \downarrow{\scriptstyle \mathrm{id}_B} & & \\ 0 & \longrightarrow & J & \longrightarrow & F & \longrightarrow & B & \longrightarrow & 0 \end{array}$$

where J is AF and essential in F, the bottom row is exact, the vertical arrows on the left and in the middle are injective, and the vertical arrow on the right is the identity.

Proof. Apply the previous lemma and we have $E \subset M(I) \subset M(J)$. Since $M(I) \cap J = I$, it follows that $E \cap J = I$ and so we define $F = E + J$ to complete the proof. □

Here is (a special case of) the result of Kasparov that we will need (cf. [**97**]).

Theorem 8.4.6. *Let $I \subset \mathbb{B}(\mathcal{H})$ be a nondegenerate representation of a separable nuclear C*-algebra, B be separable and $\sigma\colon B \to M(\mathbb{K}(\mathcal{K}) \otimes I) \subset \mathbb{B}(\mathcal{K} \otimes \mathcal{H})$ be any $*$-homomorphism. If $\pi\colon B \to \mathbb{B}(\mathcal{K}) \otimes \mathbb{C}1 \subset M(\mathbb{K}(\mathcal{K}) \otimes I) \subset \mathbb{B}(\mathcal{K} \otimes \mathcal{H})$ is any faithful representation such that $\pi(B) \cap \mathbb{K}(\mathcal{K}) = 0$, then there exists a unitary operator $U\colon (\mathcal{K} \otimes \mathcal{H}) \oplus (\mathcal{K} \otimes \mathcal{H}) \to \mathcal{K} \otimes \mathcal{H}$ such that*

$$\pi(b) - U(\pi(b) \oplus \sigma(b))U^* \in \mathbb{K}(\mathcal{K}) \otimes I,$$

for all $b \in B$, *and*

$$\mathbb{K}(\mathcal{K}) \otimes I = U\mathbb{M}_2(\mathbb{K}(\mathcal{K}) \otimes I)U^*$$

where we identify $\mathbb{M}_2(\mathbb{K}(\mathcal{K}) \otimes I) \subset \mathbb{B}((\mathcal{K} \otimes \mathcal{H}) \oplus (\mathcal{K} \otimes \mathcal{H}))$ *canonically.*

Corollary 8.4.7. *If* $0 \to \mathbb{K} \otimes I \to E \to B \to 0$ *is exact, where* $\mathbb{K} \otimes I$ *is nuclear and essential in* E, *and there exists a* $*$*-homomorphic splitting* $\sigma\colon B \to E$, *then for any faithful essential representation* $\pi\colon B \to \mathbb{B}(\mathcal{K})$ *there is an embedding of* E *into the algebra*

$$\pi(B) \otimes \mathbb{C}1 + \mathbb{K}(\mathcal{K}) \otimes I.$$

Proof. Since $\mathbb{K} \otimes I$ is essential in E, we may identify $\mathbb{K} \otimes I \cong \mathbb{K}(\mathcal{K}) \otimes I \subset \mathbb{B}(\mathcal{K} \otimes \mathcal{H})$, for some nondegenerate inclusion $I \subset \mathbb{B}(\mathcal{H})$, and hence regard $E \subset M(\mathbb{K}(\mathcal{K}) \otimes I) \subset \mathbb{B}(\mathcal{K} \otimes \mathcal{H})$ and the splitting σ as taking values in $M(\mathbb{K}(\mathcal{K}) \otimes I)$.

Let $\iota_E\colon E \hookrightarrow \mathbb{B}(\mathcal{K} \otimes \mathcal{H})$ be the natural inclusion and extend the given map $\pi\colon B \to \mathbb{B}(\mathcal{K} \otimes \mathcal{H})$ to E by composing with the quotient map $E \to B$. Evidently

$$\iota_E \oplus \pi\colon E \to \mathbb{B}((\mathcal{K} \otimes \mathcal{H}) \oplus (\mathcal{K} \otimes \mathcal{H}))$$

is faithful and takes values in the C*-algebra

$$(\sigma \oplus \pi)(B) + \mathbb{M}_2(\mathbb{K}(\mathcal{K}) \otimes I).$$

But the unitary from the previous theorem twists this algebra (isomorphically) over to the algebra $\pi(B) \otimes \mathbb{C}1 + \mathbb{K}(\mathcal{K}) \otimes I$, so we're done. □

Here is a trivial case of a general result, due to Larry Brown, which asserts that extensions of AF algebras are always AF ([**53**, Theorem III.6.3]).

Lemma 8.4.8. *Assume* C *is AF and* $\pi\colon C \to \mathbb{B}(\mathcal{K})$ *is any representation. Then* $\pi(C) + \mathbb{K}(\mathcal{K})$ *is also AF.*

Proof. Let $\mathfrak{F} \subset \pi(C) + \mathbb{K}(\mathcal{K})$ be an arbitrary finite set and let $\varepsilon > 0$. By density we may assume that each $x \in \mathfrak{F}$ can be written as $x = \pi(c_x) + T_x$ where $T_x \in \mathbb{K}(\mathcal{K})$ is a finite-rank operator. In this situation, we can find a finite-rank projection $P \in \mathbb{K}(\mathcal{K})$ such that $PT_x = T_xP = T_x$ for all $x \in \mathfrak{F}$.

Since C is AF, we can find a finite-dimensional subalgebra $D \subset \pi(C)$ which contains, within ε, all of the elements $\pi(c_x)$, $x \in \mathfrak{F}$. Now let $Q \in \mathbb{K}(\mathcal{K})$ be the projection onto the subspace $DP\mathcal{K}$. Since this space is invariant for D, Q commutes with D. Hence $C^*(D, Q\mathbb{K}(\mathcal{K})Q) = D + Q\mathbb{K}(\mathcal{K})Q$ is a finite-dimensional subalgebra of $\pi(C) + \mathbb{K}(\mathcal{K})$ which nearly contains $\mathfrak{F}$. □

Putting things together, we arrive at (a special case of) a result of Spielberg (cf. [**179**]).

Proposition 8.4.9. *Assume* $0 \to I \to D \to B \to 0$ *is exact and there is a* $*$*-homomorphic splitting* $\sigma \colon B \to D$. *If* $I \lhd D$ *is essential and both* I *and* B *are AF embeddable, then* D *is also AF embeddable.*

Proof. By Lemma 8.4.5 we may assume that I is actually an AF algebra. (The larger exact sequence obviously splits since the smaller one does.) Tensoring everything with $\mathbb{K}$, if necessary, we may assume $I \cong \mathbb{K} \otimes I$. By Corollary 8.4.7, there is an embedding of D into the C*-algebra $\pi(B) \otimes \mathbb{C}1 + \mathbb{K}(\mathcal{K}) \otimes I$, where $\pi \colon B \to \mathbb{B}(\mathcal{K})$ is *any* faithful essential representation. But, since B embeds into some AF algebra C, we can take π to be a representation of C and then Lemma 8.4.8 ensures that

$$\big(\pi(C) + \mathbb{K}(\mathcal{K})\big) \otimes \big(\mathbb{C}1 + I\big)$$

is an AF algebra which contains D. □

After a few more remarks, we can harvest the homotopy invariance theorem.

Definition 8.4.10. If $\psi \colon B \to A$ is a $*$-homomorphism, then *the mapping cylinder of* ψ is the C*-algebra

$$Z_\psi := \{ f \oplus b \in (C[0,1] \otimes A) \oplus B : f(1) = \psi(b) \}.$$

Remark 8.4.11. We have a surjective coordinate mapping $Z_\psi \to B$ and evidently the kernel of this map can be identified with $CA = C[0,1) \otimes A$ – i.e., we have a natural short exact sequence

$$0 \to CA \to Z_\psi \to B \to 0$$

and this sequence has a $*$-homomorphic splitting defined by $b \mapsto \psi(b) \oplus b$ (where we think of the left side as being a constant function).

Remark 8.4.12. Note that in the case ψ is injective, the other coordinate mapping $Z_\psi \to C[0,1] \otimes A$ is an *isomorphism* onto the algebra

$$\{ f \in C[0,1] \otimes A : f(1) \in \psi(B) \}.$$

It is easily seen that CA will be an essential ideal in Z_ψ in this case – for any nonzero path we can construct an element of CA which won't multiply to zero – and hence the sequence

$$0 \to CA \to Z_\psi \to B \to 0$$

is essential with a $*$-homomorphic splitting.

Lemma 8.4.13. *Assume* $\psi \colon B \to A$ *is an injective* $*$*-homomorphism such that* B *is AF embeddable and* A *is exact. Then the mapping cylinder* Z_ψ *is also AF embeddable.*

Proof. This follows immediately from Remark 8.4.12 and Proposition 8.4.9, since Theorem 8.3.5 ensures that CA is AF embeddable. □

Theorem 8.4.14. *If A and B are separable exact homotopy equivalent (Definition 7.3.4) and B is AF embeddable, then A is also AF embeddable.*

Proof. As with quasidiagonality, it suffices to know that A is homotopically dominated by B. So assume $\varphi\colon A \to B$ and $\psi\colon B \to A$ are $*$-homomorphisms and $\gamma_t\colon A \to A$ is a path of $*$-homomorphisms such that $\gamma_0 = \mathrm{id}_A$ and $\gamma_1 = \psi \circ \varphi$.

By the previous lemma, the mapping cylinder of $\psi \oplus \mathrm{id}_B\colon B \to A \oplus B$ is AF embeddable and hence we only need to embed A into $Z_{\psi\oplus\mathrm{id}_B}$. For each $a \in A$ let $f_a\colon [0,1] \to A \oplus B$ be defined by

$$f_a(t) = \gamma_t(a) \oplus \varphi(a).$$

The mapping

$$A \to \Big(C[0,1] \otimes (A \oplus B)\Big) \oplus B, \ a \mapsto f_a \oplus \varphi(a)$$

is easily seen to give the desired $*$-monomorphism $A \hookrightarrow Z_{\psi\oplus\mathrm{id}_B}$. □

8.5. A survey

This section is a survey of results and problems; there are essentially no proofs, just definitions, statements and references.

Recall that a C*-algebra is *approximately subhomogeneous* (ASH) if it is an inductive limit of subhomogeneous algebras (Definition 2.7.6).

Proposition 8.5.1. *Every ASH algebra is AF embeddable.*

Proof. Assume $A = \overline{\bigcup A_n}$ where each A_n is subhomogeneous. Let $B \subset A^{**}$ be the *norm* closure of the union of the von Neumann algebras $A_n^{**} \subset A^{**}$. Each A_n^{**} is a (typically nonseparable) AF algebra (being isomorphic to a finite direct sum of algebras of the form $L^\infty(X) \otimes \mathbb{M}_n(\mathbb{C})$) and hence so is B. Evidently $A \subset B$, so we are done. □

In **[169]** Mikael Rørdam gave an alternate proof of Theorem 8.3.5 which is worth outlining. The main result, which is of independent interest, is this:

Theorem 8.5.2 (Rørdam). *There exists an ASH algebra B with the property that*

$$B \cong B \otimes \mathcal{O}_\infty.^3$$

---

[3]Actually, B is a nonunital AH algebra and the construction is due to Mortensen. What Rørdam shows is the tensorial absorption property.

Since $\mathcal{O}_\infty$ contains a copy of $C_0(0,1]$, so does B. Thus, if A is exact, we can invoke Kirchberg's embedding theorem (Theorem 10.2.2) to deduce

$$C_0(0,1] \otimes A \subset C_0(0,1] \otimes \mathcal{O}_\infty \subset B \otimes \mathcal{O}_\infty \cong B.$$

This yields Theorem 8.3.5, since B is AF embeddable.

As previously noted, the Pimsner-Voiculescu AF embedding of irrational rotation algebras first sparked interest in the subject. Hence it was natural to consider AF embeddability of more general crossed products and this has been studied by many authors. The results tend to be quite difficult and we really don't have a very good understanding of general crossed products at this point. However, for actions of $\mathbb{Z}$ there are two cases where the picture is clear. In [**145**] Pimsner proved the following theorem.

Theorem 8.5.3. *Let X be a compact metric space and $\alpha \in \mathrm{Aut}(C(X))$ be an automorphism. Then the following are equivalent:*

(1) *$C(X) \rtimes_\alpha \mathbb{Z}$ is AF embeddable;*
(2) *$C(X) \rtimes_\alpha \mathbb{Z}$ is QD;*
(3) *$C(X) \rtimes_\alpha \mathbb{Z}$ is stably finite;*
(4) *if $h_\alpha : X \to X$ is the homeomorphism corresponding to α, then "h_α compresses no open sets." That is, if $U \subset X$ is open and $h_\alpha(\overline{U}) \subset U$, then $U = h_\alpha(\overline{U})$, where $\overline{U}$ is the closure of U.*[4]

Related results for actions of other groups on abelian C*-algebras have been found by Pimsner (actions of $\mathbb{R}$ – see [**147**]) and Matui (actions of $\mathbb{Z}^2$ – see [**124**]). See also [**118**] and [**119**], and the references therein, for related results of Lin.

Voiculescu first studied the AF embeddability question for crossed products of AF algebras by the integers in [**188**]. His results, together with important Rohlin property contributions of Kishimoto and others, played an important role in the following ([**26**]):

Theorem 8.5.4. *Let A be an AF algebra and $\alpha \in \mathrm{Aut}(A)$ be an automorphism. Then the following are equivalent:*

(1) *$A \rtimes_\alpha \mathbb{Z}$ is AF embeddable;*
(2) *$A \rtimes_\alpha \mathbb{Z}$ is QD;*
(3) *$A \rtimes_\alpha \mathbb{Z}$ is stably finite;*

---

[4] A more dynamically pleasing formulation of this condition can be given in terms of "pseudoperiodic" points – namely, every point should be pseudoperiodic. By definition a point $x_0 \in X$ is *pseudoperiodic* if for every $\varepsilon > 0$ there exist points $x_1, \ldots, x_n$ such that $d(x_{i+1}, h_\alpha(x_i)) < \varepsilon$ (where d is a fixed metric on X) for $i \in \{0, \ldots, n-1\}$ and $d(x_0, h_\alpha(x_n)) < \varepsilon$ as well.

(4) *the induced map* $\alpha_*\colon K_0(A) \to K_0(A)$ *"compresses no element." In other words, if* $x \in K_0(A)$ *and* $\alpha_*(x) \leq x$ *(in the natural order on* $K_0(A)$*), then* $\alpha_*(x) = x$.

Katsura has studied the AF embeddability of Cuntz algebras by certain actions of $\mathbb{R}$ in [**98**].

Regarding the AF embeddability of general exact RFD algebras, we have the following striking result of Dadarlat ([**52**]).

Theorem 8.5.5. *Assume A is separable exact RFD and satisfies the universal coefficient theorem (UCT) of Rosenberg and Schochet ([**170**]). Then A is AF embeddable.*

In particular, every type I C*-algebra with a faithful tracial state is AF embeddable (since these are RFD by Proposition 7.1.8). A long-standing open question asks whether or not every nuclear C*-algebra satisfies the UCT. An affirmative answer would imply every nuclear RFD algebra is AF embeddable (or, if you prefer, the construction of an RFD nuclear algebra which is *not* AF embeddable would provide a counterexample to the UCT).

The last result we'll mention requires a definition.

Definition 8.5.6. A C*-algebra is called *residually finite* if every quotient is stably finite.

This notion is completely unrelated to residual finite-dimensionality (i.e., neither implies the other) and is much stronger than being stably finite. For example, the cone over any algebra is stably finite but often fails to be residually finite.

In [**179**] Spielberg proved the following:

Theorem 8.5.7. *Every residually finite type* I C*-*algebra is AF embeddable.*

A natural question, pointed out to us by Kerr, is whether the previous result can be extended to stably finite type I C*-algebras. Here are some related open problems.

Open Problems

(1) (Blackadar-Kirchberg) Is a stably finite nuclear C*-algebra necessarily QD?

(2) (Blackadar-Kirchberg) Is every nuclear QD C*-algebra AF embeddable?

(3) Does every exact QD C*-algebra embed into a nuclear QD C*-algebra?

(4) Is every exact QD C*-algebra AF embeddable? This, of course, is the ultimate goal as it would give an abstract characterization of subalgebras of AF algebras. Most days we conjecture this problem has an affirmative answer.

8.6. References

The main results of Sections 8.1 and 8.2 come from [**51**], while the results found in Sections 8.3 and 8.4 are from [**131**].

Chapter 9

Local Reflexivity and Other Tensor Product Conditions

This chapter introduces yet another finite-dimensional approximation property – local reflexivity – which is inspired by classical Banach space theory. Reformulating in terms of tensor products leads to three closely related notions: Properties C, C' and C''. It turns out that the first two are equivalent – and equivalent to exactness – and both imply property C'', which ends up being equivalent to local reflexivity. Consequently we arrive at Kirchberg's deep discovery that exactness implies local reflexivity, the main result of this chapter. It will follow that exactness passes to quotients, which might not sound exciting, but it is an important and very difficult permanence property.

These results are the culmination of years of hard work, by many able hands, and represent some of the deepest and most difficult theorems in C*-algebra theory (e.g., they depend on some of the deepest and most difficult theorems in von Neumann algebra theory). We have tried to keep the preceding chapters reasonably self-contained in the sense that theorems quoted without proof either could be found in introductory texts or are only used for results of peripheral interest. Unfortunately, that is not possible in the present chapter without adding many pages of hardcore W*-exposition. We will need the fact that the double dual of a nuclear C*-algebra is semidiscrete and presently there is no proof of this which avoids Connes's thesis (and the Tomita-Takesaki theory that preceded it). Hence we state, without proof, the existence of a modular automorphism and semifiniteness of the corresponding crossed product and then show how to derive what we need from these important facts.

9.1. Local reflexivity

In Banach space theory, the *principle of local reflexivity* refers to the following (cf. [**120**]):

Theorem 9.1.1. *Let X be a Banach space and $E \subset X^{**}$ be a finite-dimensional subspace. Then, there exists a net of contractions $\varphi_i \colon E \to X$ which converges to id_E in the point-weak* topology. Moreover, we may assume that the net $\{\varphi_i|_{E\cap X}\}$ converges in the point-norm topology to $\mathrm{id}_{E\cap X}$.*

Being Banach spaces, C*-algebras thus enjoy the (Banach space) principle of local reflexivity. But what if we require c.c.p. maps (instead of just contractions)?

Definition 9.1.2. A C*-algebra A is *locally reflexive* if for every finite-dimensional operator system $E \subset A^{**}$, there exists a net of c.c.p. maps $\varphi_i \colon E \to A$ which converges to id_E in the point-ultraweak topology.

As usual, it is often convenient to reduce to the unital case.

Lemma 9.1.3. *A nonunital C*-algebra A is locally reflexive if and only if its unitization $\tilde{A}$ is locally reflexive.*

Proof. This is a special case of the next proposition. □

Proposition 9.1.4. *Let $0 \to I \to A \to B \to 0$ be short exact with A unital. Then A is locally reflexive if and only if both I and B are locally reflexive* and *the extension is locally split.*

Proof. Let $\pi \colon A \to B$ denote the quotient map. Suppose first that A is locally reflexive and $E \subset B^{**}$ is a finite-dimensional operator system. By the natural identification $A^{**} = I^{**} \oplus B^{**}$, we may regard $E \subset A^{**}$. (If you want an operator system, just consider $\tilde{E} = E + \mathbb{C}1_{A^{**}}$.) Since A is locally reflexive, there exists a net of c.c.p. maps $\psi_i \colon E \to A$ which converges to id_E in the point-$\sigma(A^{**}, A^*)$ topology. Thus, the net $\pi \circ \psi_i \colon E \to B$ of c.c.p. maps converges to id_E in the point-$\sigma(B^{**}, B^*)$ topology. This proves that B is locally reflexive. We leave it to the reader to show the local reflexivity of I (Exercise 9.1.1). To prove local liftability of π, we start with $E \subset B$ (rather than $E \subset B^{**}$). Then the net $\pi \circ \psi_i$ of liftable c.c.p. maps will converge to id_E in the point-weak topology; taking convex combinations, we can assume norm convergence (see Lemma 2.3.4). Thus Arveson's Lemma C.2 ensures that id_E is liftable too.

Suppose now that both I and B are locally reflexive and the extension is locally split. Let a finite-dimensional operator system $E \subset A^{**}$ be given. Again writing $A^{**} = I^{**} \oplus B^{**}$, we may regard $E \subset E_I \oplus E_B \subset I^{**} \oplus B^{**}$, where E_I and E_B are finite-dimensional. Since I and B are locally reflexive, there exist nets of c.c.p. maps $\psi_i^I \colon E_I \to I$ and $\psi_i^B \colon E_B \to B$ which converge to the relevant identities. Let $\hat{\psi}_i^B \colon E_B \to A$ be a c.c.p. lifting of ψ_i^B. We denote the units of I^{**} and B^{**} in A^{**} by e_I and e_B, respectively. If (e_j) is an approximate unit for I, then $e_j^{1/2} \nearrow e_I$ and $(1-e_j)^{1/2} \searrow e_B$ ultrastrongly. Now define a net of c.c.p. maps $\varphi_{i,j} \colon E \to A$ by

$$\varphi_{i,j}(x) = e_j^{1/2}\psi_i^I(xe_I)e_j^{1/2} + (1-e_j)^{1/2}\hat{\psi}_i^B(xe_B)(1-e_j)^{1/2}.$$

Then, for every $x \in E$, we have

$$\lim_i \lim_j \varphi_{i,j}(x) = \lim_i e_I\psi_i^I(xe_I)e_I + e_B\hat{\psi}_i^B(xe_B)e_B = xe_I + xe_B = x$$

since $e_B\hat{\psi}_i^B(xe_B)e_B = \psi_i^B(xe_B)$ under the identification $A^{**} = I^{**} \oplus B^{**}$. □

Corollary 9.1.5. *Assume $I \triangleleft A$ is an ideal in a locally reflexive* C*-*algebra A. For every* C*-*algebra B the sequence*

$$0 \to I \otimes B \to A \otimes B \to (A/I) \otimes B \to 0$$

is exact.

Proof. We may assume A is unital, whence the result follows from Proposition 3.7.6. □

Corollary 9.1.6. *The* C*-*algebra $\mathbb{B}(\mathcal{H})$ is not locally reflexive. Neither is $C^*(\mathbb{F}_2)$.*

Proof. The previous corollary together with Exercise 3.9.7 implies that if $A = \mathbb{B}(\mathcal{H})$ is locally reflexive, then every C*-algebra is exact; hence $\mathbb{B}(\mathcal{H})$ is not locally reflexive. The free group case follows from Proposition 3.7.11, taking $A = B = C^*(\mathbb{F}_2)$. □

Exercise

Exercise 9.1.1. Use the definition to show local reflexivity passes to hereditary subalgebras.

9.2. Tensor product properties

To give tensor-product characterizations of local reflexivity, we must spend time with C*-tensor products of double duals; there are subtleties around every corner.

If M and N are von Neumann algebras, a $*$-homomorphism $M \odot N \to \mathbb{B}(\mathcal{H})$ is said to be *bi-normal* if both of the restriction maps $M = M \otimes \mathbb{C}1_N \to$

$\mathbb{B}(\mathcal{H})$ and $N = \mathbb{C}1_M \otimes N \to \mathbb{B}(\mathcal{H})$ are normal representations. Of course, we may replace $\mathbb{B}(\mathcal{H})$ with any other von Neumann algebra in this definition.

Proposition 9.2.1. *For every A and B, there is a canonical injective bi-normal map*

$$A^{**} \odot B^{**} \hookrightarrow (A \otimes B)^{**}.$$

Proof. Thanks to the existence of restriction maps (Theorem 3.2.6), the inclusion $A \odot B \hookrightarrow (A \otimes B)^{**}$ arises from commuting copies of A and B inside $(A \otimes B)^{**}$. Thus the weak closures of $A \subset (A \otimes B)^{**}$ and $B \subset (A \otimes B)^{**}$ also commute. Hence there is a bi-normal map

$$A^{**} \odot B^{**} \to (A \otimes B)^{**}.$$

To verify injectivity, one first recalls that functionals of the form $\varphi \odot \psi$, where φ (resp. ψ) is a linear functional on A (resp. B), separate the elements of $A^{**} \odot B^{**}$ (Exercise 3.1.5). By (the proof of) Takesaki's Theorem we can extend each $\varphi \odot \psi$ to a linear functional $\varphi \otimes \psi$ on $A \otimes B$ and then further extend to a normal functional on $(A \otimes B)^{**}$, also denoted by $\varphi \otimes \psi$. Now one must check that $\varphi \odot \psi$ is the same as

$$A^{**} \odot B^{**} \to (A \otimes B)^{**} \overset{\varphi \otimes \psi}{\to} \mathbb{C},$$

which implies injectivity, as desired. □

Since $A \subset A^{**}$, we also have a natural inclusion $A \odot B^{**} \hookrightarrow (A \otimes B)^{**}$. Though a bit strange at first, the following definitions are actually quite natural.

Definition 9.2.2. Let A be an arbitrary C*-algebra.

(1) A is said to have *property C* if

$$A^{**} \odot B^{**} \hookrightarrow (A \otimes B)^{**}$$

is min-continuous for every C*-algebra B.

(2) A is said to have *property C′* if

$$A \odot B^{**} \hookrightarrow (A \otimes B)^{**}$$

is min-continuous for every C*-algebra B.

(3) A is said to have *property C″* if

$$A^{**} \odot B \hookrightarrow (A \otimes B)^{**}$$

is min-continuous for every C*-algebra B.

To get acquainted with these notions, let's work out two simple permanence properties.

Proposition 9.2.3. *All three of these notions pass to subalgebras.*

Proof. Let $C \subset A$ be a C*-subalgebra. We only prove the case of property C as the other two are similar.

Since $C^{**} \subset A^{**}$ and $(C \otimes B)^{**} \subset (A \otimes B)^{**}$, we have a canonical commutative diagram

$$\begin{array}{ccc} C^{**} \odot B^{**} & \longrightarrow & (C \otimes B)^{**} \\ \downarrow & & \downarrow \\ A^{**} \odot B^{**} & \longrightarrow & (A \otimes B)^{**} \end{array}$$

for every C*-algebra B. Since $C^{**} \otimes B^{**} \subset A^{**} \otimes B^{**}$, min-continuity of the bottom arrow implies that of the top. □

Proposition 9.2.4. *If $I \lhd A$ is an ideal and A has property C (resp. C''), then A/I has property C (resp. C'').*

Proof. First assume that A has property C. Let $\pi \colon A \to A/I$ be the quotient map, π^{**} be its normal extension to double duals and $p \in A^{**}$ be the central support projection of π^{**} (i.e., $(A/I)^{**} = pA^{**}$). Recall that if $\{e_i\}$ is an approximate unit I, then $e_i \nearrow 1 - p$ ultraweakly.

Let θ be the (continuous) $*$-homomorphism defined via the following diagram:

$$\begin{array}{ccccc} (A/I)^{**} \otimes B^{**} & & \xrightarrow{\quad\quad\theta\quad\quad} & & ((A/I) \otimes B)^{**} \\ \downarrow \cong & & & & \uparrow (\pi \otimes \mathrm{id}_B)^{**} \\ pA^{**} \otimes B^{**} & \hookrightarrow & A^{**} \otimes B^{**} & \xrightarrow[\iota]{} & (A \otimes B)^{**}, \end{array}$$

where $\iota \colon A^{**} \otimes B^{**} \to (A \otimes B)^{**}$ is the inclusion coming from property C. We claim that θ restricted to $(A/I)^{**} \odot B^{**}$ coincides with the canonical inclusion $(A/I)^{**} \odot B^{**} \to ((A/I) \otimes B)^{**}$ (which evidently implies the result we are after). Indeed, for every $a \in A$ and $b \in B$, we have

$$\begin{aligned} \theta(\pi(a) \otimes b) &= (\pi \otimes \mathrm{id}_B)^{**} \circ \iota(pa \otimes b) \\ &= \lim_i (\pi \otimes \mathrm{id}_B)^{**} \circ \iota((1 - e_i)a \otimes b) \\ &= \lim_i \pi((1 - e_i)a) \otimes b \\ &= \pi(a) \otimes b \end{aligned}$$

and hence $\theta|_{(A/I) \odot B}$ coincides with the canonical inclusion. But θ is clearly bi-normal and this completes the proof.

We leave the case of property C'' to the reader as it is virtually identical. □

It turns out that property C' also passes to quotients (remarkably, properties C and C' are equivalent) but the proof above cannot be adapted to this case.

Coming back to finite-dimensional approximation properties, the next two results are very important. We will need to use operator space duality so the reader may wish to review Theorem B.13 before reading the proof of the next proposition.

Proposition 9.2.5. *A* C^**-algebra is locally reflexive if and only it has property* C''.

Proof. Suppose A is locally reflexive and let B be an arbitrary C^*-algebra. It suffices to show $\|z\|_{(A\otimes B)^{**}} \le \|z\|_{A^{**}\otimes B}$ for a given $z = \sum_{k=1}^n a_k \otimes b_k \in A^{**} \odot B$. Let E be the operator system in A^{**} spanned by $a_1, \ldots, a_n$. Since A is locally reflexive, there exists a net of c.c.p. maps $\varphi_i \colon E \to A$ which converge to id_E in the point-ultraweak topology. Because $A^{**} \odot B^{**} \to (A\otimes B)^{**}$ is bi-normal and multiplication by a fixed operator is ultraweakly continuous, it follows that for every $x \in E$ and $b \in B$ we have

$$\varphi_i(x) \otimes b \to xb \in (A \otimes B)^{**}$$

in the ultraweak topology on $(A \otimes B)^{**}$. This implies that

$$(\varphi_i \otimes \mathrm{id}_B)(z) \to z \in (A \otimes B)^{**}$$

in the ultraweak topology on $(A\otimes B)^{**}$. Hence we get the following inequalities:

$$\|z\|_{(A\otimes B)^{**}} \le \liminf_i \|(\varphi_i \otimes \mathrm{id}_B)(z)\|_{A\otimes B} \le \|z\|_{A^{**}\otimes B}.$$

We now prove the "if" direction, so assume A has property C''. We may assume that A is unital. Let $E \subset A^{**}$ be a finite-dimensional operator system. Invoking operator space duality (Theorem B.13), the inclusion $E \hookrightarrow A^{**}$ corresponds to an element $z \in E^* \otimes A^{**}$ with $\|z\| = 1$. We may assume that $E^* \subset \mathbb{B}(\mathcal{H})$ completely isometrically. Consider the following commutative diagram of canonical inclusions:

$$\begin{array}{ccc} E^* \otimes A^{**} & \longrightarrow & (E^* \otimes A)^{**} \\ \big\downarrow & & \big\downarrow \\ \mathbb{B}(\mathcal{H}) \otimes A^{**} & \hookrightarrow & (\mathbb{B}(\mathcal{H}) \otimes A)^{**}. \end{array}$$

Since the vertical inclusions and the bottom inclusion are all isometric, so is the top inclusion. Therefore, we have $\|z\|_{(E^*\otimes A)^{**}} = 1$. It follows that there exists a net $\{z_i\}$ in $E^* \otimes A$ such that $\sup_i \|z_i\| \le 1$ and the net $\{z_i\}$ converges to z in the weak*-topology on $(E^* \otimes A)^{**}$. Applying operator space duality again, each z_i corresponds to a c.c. map $\varphi_i \colon E \to A$ such that the net $\{\varphi_i\}$ converges to id_E in the point-ultraweak topology (see Remark

B.14). Passing to a convex combination if necessary, we may assume that the net $\{\varphi_i(1)\}$ converges in norm to the unit. Hence, we can perturb each φ_i to a u.c.p. map, thanks to Corollary B.11. □

Lemma 9.2.6. *If $I \lhd C$ and $x \in I^{**} \cap C$, then $x \in I$.*

Proof. The canonical map $C^{**} \to (C/I)^{**}$ restricted to C is just the quotient map $C \to C/I$. Hence, the set of elements in C belonging to the kernel is equal to I. On the other hand, the kernel of $C^{**} \to (C/I)^{**}$ is I^{**}, which evidently implies the lemma. □

Proposition 9.2.7. *A C*-algebra is exact if and only if it has property C'.*

Proof. We first assume A has property C' and let B be an arbitrary C*-algebra with ideal $J \lhd B$. Since $0 \to J^{**} \to B^{**} \to (B/J)^{**} \to 0$ splits, Proposition 3.7.6 implies that

$$0 \longrightarrow J^{**} \otimes A \longrightarrow B^{**} \otimes A \longrightarrow (B/J)^{**} \otimes A \longrightarrow 0$$

is exact. Since we have assumed property C', we get a natural commutative diagram

$$\begin{array}{ccccccccc}
0 & \longrightarrow & J \otimes A & \longrightarrow & B \otimes A & \longrightarrow & (B/J) \otimes A & \longrightarrow & 0 \\
 & & \downarrow & & \downarrow & & \downarrow & & \\
0 & \longrightarrow & J^{**} \otimes A & \longrightarrow & B^{**} \otimes A & \longrightarrow & (B/J)^{**} \otimes A & \longrightarrow & 0 \\
 & & \downarrow & & \downarrow & & \downarrow & & \\
0 & \longrightarrow & (J \otimes A)^{**} & \longrightarrow & (B \otimes A)^{**} & \longrightarrow & ((B/J) \otimes A)^{**} & \longrightarrow & 0.
\end{array}$$

It is important to remember that at this point we *do not* know if the bottom row is exact (since we don't know if the top row is exact). But, if we take $x \in B \otimes A$ with the property that x is in the kernel of the map $B \otimes A \to (B/J) \otimes A$, then exactness of the middle row implies that $x \in (J \otimes A)^{**} \cap B \otimes A = J \otimes A$ (by Lemma 9.2.6). This proves that A is exact.

For the converse, assume A is exact and B is arbitrary. For a directed set I, we let

$$B_I = \{(x(i))_{i \in I} \in \prod_{i \in I} B : \text{strong}^*\text{-}\lim_{i \in I} x(i) \text{ exists in } B^{**}\}.$$ [1]

It is easily checked that B_I is a C*-subalgebra of $\prod_{i \in I} B$ (multiplication on bounded sets is jointly strong*-continuous). Evidently we have a *-homomorphism $\sigma \colon B_I \to B^{**}$, given by

$$(x(i))_{i \in I} \mapsto \text{strong}^*\text{-}\lim_{i \in I} x(i) \in B^{**},$$

[1] This means the usual strong*-topology in the universal representation $B^{**} \subset \mathbb{B}(\mathcal{H}_u)$: $T_i \to T$ strong* if and only if $T_i \to T$ and $T_i^* \to T^*$ in the strong operator topology.

which is surjective (by Kaplansky's density theorem [**53**, Theorem I.7.3]) if we start with a large enough index set I (e.g., the directed set of finite subsets of B^*).

The key observation is that $\mathrm{id}_A \otimes \sigma \colon A \odot B_I \to A \odot B^{**} \subset (A \otimes B)^{**}$ is continuous with respect to the minimal norm since

$$\begin{aligned}\|(\mathrm{id}_A \otimes \sigma)(\sum_{k=1}^{n} a_k \otimes (x_k(i))_{i\in I})\|_{(A\otimes B)^{**}} & \\ &= \|\text{strong}^*\text{-}\lim_{i\in I} \sum_{k=1}^{n} a_k \otimes x_k(i)\|_{(A\otimes B)^{**}} \\ &\le \sup_{i\in I} \|\sum_{k=1}^{n} a_k \otimes x_k(i)\|_{A\otimes B} \\ &= \|\sum_{k=1}^{n} a_k \otimes (x_k(i))_{i\in I}\|_{A\otimes B_I}\end{aligned}$$

for every $\sum_{k=1}^{n} a_k \otimes (x_k(i))_{i\in I} \in A \odot B_I$. (The last equality follows from Lemma 3.9.4.) Letting $J \lhd B_I$ be the kernel of the map $B_I \to B^{**}$, it follows that $A \otimes J$ is contained in the kernel of

$$A \otimes B_I \to (A \otimes B)^{**}$$

and thus this $*$-homomorphism factors through

$$\frac{A \otimes B_I}{A \otimes J} = A \otimes (B_I/J) = A \otimes B^{**}$$

since we assumed A to be exact. The last step is to verify that the map we have constructed,

$$A \otimes B^{**} \to (A \otimes B)^{**},$$

agrees on elementary tensors with the canonical map $A \odot B^{**} \to (A \otimes B)^{**}$. But this is not hard, so we consider the proof complete. □

Though rarely needed, it is worth mentioning that property C'' is a local property in the sense that one can always reduce to the separable case.

Lemma 9.2.8. *A* C*-*algebra A is locally reflexive if and only if every separable* C*-*subalgebra of A is locally reflexive.*

Proof. Local reflexivity passes to subalgebras thanks to Propositions 9.2.5 and 9.2.3. Hence we must show the converse. We may assume that A is unital.

Let a finite-dimensional operator system $E \subset A^{**}$, a finite-dimensional subspace $F \subset A^*$ and $\varepsilon > 0$ be given. It suffices to find a u.c.p. map

$\varphi\colon E \to A$ such that for every $x \in E$ and $f \in F$,

$$|f(\varphi(x) - x)| < \varepsilon \|f\| \|x\|.$$

We regard $\mathbb{M}_n(E)$ as a finite-dimensional subspace of $\mathbb{M}_n(A)^{**}$ (via the isometric identification $\mathbb{M}_n(A^{**}) = \mathbb{M}_n(A)^{**}$). It follows from the principle of local reflexivity for Banach spaces (Theorem 9.1.1) that there exists a net of contractions from $\mathbb{M}_n(E)$ into $\mathbb{M}_n(A)$ which converges to the identity on $\mathbb{M}_n(E)$ in the point-weak* topology. Passing to convex combinations if necessary, we may assume that this net converges in the point-norm topology on $\mathbb{M}_n(\mathbb{C}1_A)$. Hence, we may find a contraction $\Psi_n\colon \mathbb{M}_n(E) \to \mathbb{M}_n(A)$ such that $\|\Psi_n|_{\mathbb{M}_n(\mathbb{C}1_A)} - \mathrm{id}_{\mathbb{M}_n(\mathbb{C}1_A)}\| < 1/n$ and

$$|(\omega \otimes f)(\Psi_n(x) - x)| < \frac{1}{n} \|\omega\| \|f\| \|x\|$$

for every $x \in \mathbb{M}_n(E)$, $\omega \in \mathbb{M}_n(\mathbb{C})^*$ and $f \in F$. We set

$$\tilde{\Psi}_n(x) = \iint_{U(n)\times U(n)} u^* \Psi_n(uxv^*) v \, du \, dv,$$

where $U(n)$ is the unitary group of $\mathbb{M}_n(\mathbb{C})$ (which is compact). It follows that $\tilde{\Psi}_n$ is an $\mathbb{M}_n(\mathbb{C})$-module map, or equivalently, there is $\psi_n\colon E \to A$ such that $\mathrm{id}_{\mathbb{M}_n(\mathbb{C})} \otimes \psi_n = \tilde{\Psi}_n$. It is not hard to check $\|\psi_n(1) - 1\| < 1/n$ and

$$|f(\psi_n(x) - x)| < \frac{1}{n} \|f\| \|x\|$$

for every $x \in E$ and $f \in F$. Let $A_0 \subset A$ be a separable C*-algebra containing all the subspaces $\psi_n(E)$, and let $\psi\colon E \to A_0^{**}$ be a cluster point of the sequence $\{\psi_n\}$. Then, ψ is unital and c.c. since $\|\mathrm{id}_{\mathbb{M}_k(\mathbb{C})} \otimes \psi_n\| \le \|\tilde{\Psi}_n\| \le 1$ for $n \ge k$. Moreover, $(f|_{A_0})(\psi(x)) = f(x)$ for every $x \in E$ and $f \in F$. Since A_0 is locally reflexive by assumption, the u.c.p. map $\psi\colon E \to A_0^{**}$ can be approximated by u.c.p. maps $\varphi\colon E \to A_0 \subset A$. □

We end this section with a technical result that will be needed in later chapters.

Lemma 9.2.9. *Let M and N be von Neumann algebras with weakly dense C*-subalgebras $B \subset M$ and $C \subset N$, respectively. Let $\Phi\colon M \odot N \to \mathbb{B}(\mathcal{H})$ be a bi-normal[2] u.c.p. map. If B has property C and Φ is* min*-continuous on $B \odot C$, then Φ is* min*-continuous on $M \odot N$.*

Proof. We may assume that B and C are unital. Since Φ is continuous on $B \otimes C$, it extends to a normal u.c.p. map $\tilde{\Phi}$ on $(B \otimes C)^{**}$. Since B has property C, the u.c.p. map $\tilde{\Phi}$ is bi-normal and min-continuous on $B^{**} \odot C^{**}$. Let $e \in B^{**}$ and $f \in C^{**}$ be the central projections such that $M = eB^{**}$

[2]As with homomorphisms, Φ is bi-normal if both $\Phi|_{M\otimes\mathbb{C}1}$ and $\Phi|_{\mathbb{C}1\otimes N}$ are normal.

and $N = fC^{**}$ (canonically). Then, $\tilde{\Phi}|_{eB^{**} \odot fC^{**}}$ is a bi-normal and min-continuous u.c.p. map on $M \odot N$. We claim that this u.c.p. map coincides with Φ. Let (b_i) be a bounded net in B such that $b_i \to e$ ultraweakly in B^{**}. Then, $b_i \to 1_M$ ultraweakly in M and hence

$$\tilde{\Phi}(e \otimes 1) = \lim_i \tilde{\Phi}(b_i \otimes 1) = \lim_i \Phi(b_i \otimes 1) = \Phi(1_M \otimes 1) = 1.$$

Likewise one has $\tilde{\Phi}(1 \otimes f) = 1$. It follows that $e \otimes f$ is in the multiplicative domain of $\tilde{\Phi}$ and $\tilde{\Phi}(eb \otimes fc) = \tilde{\Phi}(b \otimes c) = \Phi(b \otimes c)$ for every $b \in B$ and $c \in C$. Since both maps are bi-normal, this implies that $\Phi = \tilde{\Phi}|_{eB^{**} \odot fC^{**}}$. □

When dealing with double duals and tensor products, one cannot be too careful. As Pisier noted, "This subject is full of traps" ([**152**, page 309]). Below are a few to avoid.

Exercises

Exercise 9.2.1. Find an example of unital C*-algebras $B_0 \subset B$ and an ideal $J \lhd B$ such that the quotient map $\pi \colon B \to B/J$ is locally liftable but $\pi_0 = \pi|_{B_0} \colon B_0 \to \pi(B_0)$ is not. (Hint: B_0 should not contain J.)

Exercise 9.2.2. Here's a "proof" that π_0 (as in the previous exercise) is always locally liftable, whenever π is locally liftable. Where is the gap in the argument?

Proof. First observe that if $I \lhd C$, then there is a canonical embedding $C/I \subset (C/I)^{**} \subset C^{**}$ (since $C^{**} = (C/I)^{**} \oplus I^{**}$). By Theorem C.4, local liftability of π is equivalent to having a canonical identification

$$A \otimes B/J = (A \otimes B)/(A \otimes J),$$

for every A. Since $(A \otimes B)/(A \otimes J) \subset (A \otimes B)^{**}$, this in turn is equivalent to knowing that the canonical embedding

$$\theta \colon A \odot B/J \subset A \odot (B/J)^{**} \subset A \odot B^{**} \hookrightarrow (A \otimes B)^{**}$$

is min-continuous.

Now consider the commutative diagram

$$\begin{array}{ccc} A \otimes B/J & \overset{\theta}{\hookrightarrow} & (A \otimes B)^{**} \\ \uparrow & & \uparrow \\ A \otimes \pi_0(B_0) & \overset{\theta_0}{\hookrightarrow} & (A \otimes B_0)^{**}. \end{array}$$

Since the vertical inclusions are isometric, it follows that θ_0 is isometric whenever θ is isometric, and hence π_0 is locally liftable whenever π is. □

Exercise 9.2.3. Here is a "proof" of the *false* assertion that every C*-algebra A is locally reflexive. Where's the gap?

Proof. Let $\varphi\colon E \to A^{**}$ be a complete contraction with $\dim(E) < \infty$. By operator space duality, φ corresponds to a norm-one element $z \in E^* \otimes A^{**}$. Our task is to show $\|z\|_{(E^*\otimes A)^{**}} \leq 1$. (Note that since $\dim(E) < \infty$, we have a canonical algebraic identification $E^* \otimes A^{**} = (E^* \otimes A)^{**}$.)

Take a universal representation $A^{**} \subset \mathbb{B}(\mathcal{H})$ and let $B = A + \mathbb{K}(\mathcal{H})$. We regard φ as a map into $B'' \subset B^{**}$. Clearly, we can approximate φ by a net of complete contractions $\varphi_i\colon E \to B$ in the point-ultraweak topology. In other words, we have $\|z\|_{(E^*\otimes B)^{**}} \leq 1$. Since the embedding $(E^* \otimes A)^{**} \subset (E^* \otimes B)^{**}$ is isometric, we deduce that $\|z\|_{(E^*\otimes A)^{**}} \leq 1$. □

9.3. Equivalence of exactness and property C

This section contains the main result. We still have a lot of von Neumann algebra work to do, but after that only a few simple remarks remain. Why not proceed in reverse? The main theorem is

Theorem 9.3.1. *For a separable* C*-*algebra A, the following are equivalent:*

(1) *A is exact;*

(2) *A has property C;*

(3) *A has property C'.*

Since property C evidently implies property C'', it follows that every exact C*-*algebra is locally reflexive.*

Proof. Evidently property C implies property C'. Also, the equivalence of conditions (1) and (3) is the content of Proposition 9.2.7. Thus we must show (1) $\Rightarrow$ (2).

However, if A is exact, then it is isomorphic to a subquotient of a nuclear C*-algebra (Corollary 8.2.5). If we knew that every nuclear C*-algebra had property C, then we could appeal to Propositions 9.2.3 and 9.2.4 to complete the proof. Thus we devote the remainder of this section to proving nuclear C*-algebras have property C. □

There are two steps involved. The first is very easy.

Proposition 9.3.2. *If A^{**} is semidiscrete, then A has property C.*

Proof. Since A^{**} is semidiscrete, the map $A^{**} \to A^{**} \subset (A \otimes B)^{**} \subset \mathbb{B}(\mathcal{H})$ is weakly nuclear (into A^{**}) for any B and normal representation $(A\otimes B)^{**} \subset \mathbb{B}(\mathcal{H})$. Theorem 3.8.5 then implies that $A^{**} \odot B^{**} \to (A \otimes B)^{**} \subset \mathbb{B}(\mathcal{H})$ is min-continuous as desired. □

The second step is to show that the double dual of a nuclear C*-algebra is semidiscrete. However, a von Neumann algebra is semidiscrete if and only if its commutant is semidiscrete (Corollary 3.8.6) and we already know that if A is nuclear and $\pi\colon A \to \mathbb{B}(\mathcal{H})$ is any representation (e.g., the universal representation), then the commutant $\pi(A)'$ is an injective von Neumann algebra (Exercise 3.6.4). Hence we'll be finished once we establish the following remarkable theorem.

Theorem 9.3.3. *A von Neumann algebra is semidiscrete if it is injective.*[3]

We first prove this when M is semifinite – i.e., when it has no summand of type III – and then reduce the general case to the semifinite case via modular theory.

Theorem 9.3.4. *Every injective semifinite von Neumann algebra is semidiscrete.*

Proof. It is easily seen that if $M = M_1 \oplus M_2$, then M is injective (resp. semidiscrete) if and only if both M_1 and M_2 are injective (resp. semidiscrete). Since we already proved that type I von Neumann algebras are semidiscrete (Proposition 2.7.2), we may assume that $M = M_{\mathrm{II}_1} \oplus M_{\mathrm{II}_\infty}$, for some algebras of type II_1 and II_∞, respectively. But M_{II_∞} is an increasing union of type II_1 corners (which will be injective whenever M_{II_∞} is injective) and hence Lemma 2.7.1 allows us to assume that M is of type II_1 and has a normal faithful tracial state τ. Since M is injective, the tracial state τ is amenable (Exercise 6.2.1). Thus, by Theorem 6.2.7, the $*$-homomorphism $\pi_\tau \times \pi_\tau^{\mathrm{op}}\colon M \otimes M^{\mathrm{op}} \to \mathbb{B}(L^2(M,\tau))$ is continuous. We note that π_τ and π_τ^{op} are normal $*$-monomorphisms and in particular that $\pi_\tau(M)' = \pi_\tau^{\mathrm{op}}(M^{\mathrm{op}})$ (Theorem 6.1.4). Thus, by Theorem 3.8.5, the identity map on M is weakly nuclear – i.e., M is semidiscrete. □

Modular theory was one of the highlights of research in the 1970s. We won't prove the following theorem – it is deep and difficult – so consult [**184**, Theorem XII.1.1] if you must.

Theorem 9.3.5. *For any von Neumann algebra M there exists an action α of $\mathbb{R}$ on M (the so-called modular action) such that $M \rtimes_\alpha \mathbb{R}$ is a semifinite von Neumann algebra.*

Thus our work will be complete once we show that injectivity is preserved by $\mathbb{R}$-crossed products and that semidiscreteness passes from the crossed product back to M.

An $\mathbb{R}$-action is a group homomorphism $\alpha\colon \mathbb{R} \to \mathrm{Aut}(M)$ such that $t \mapsto \alpha_t(a)$ is ultraweakly continuous for every $a \in M$. We may assume that $M \subset \mathbb{B}(\mathcal{H})$ and the action α is implemented by a strongly continuous unitary

[3] Recall that the converse is easy – see Exercise 2.3.15.

representation v of $\mathbb{R}$ on $\mathbb{B}(\mathcal{H})$ – i.e., $v(t)av(t)^* = \alpha_t(a)$ for every $t \in \mathbb{R}$ and $a \in M$. We denote by λ the regular representation of $\mathbb{R}$ on $L^2(\mathbb{R})$, the Hilbert space of square integrable functions on $\mathbb{R}$ with respect to Lebesgue measure. The crossed product von Neumann algebra $M \rtimes_\alpha \mathbb{R}$ is (isomorphic to) the von Neumann subalgebra in $\mathbb{B}(\mathcal{H} \otimes L^2(\mathbb{R}))$ generated by $(v \otimes \lambda)(\mathbb{R})$ and $M \otimes \mathbb{C}1$.

Lemma 9.3.6. *There exists a conditional expectation E from $M \rtimes_\alpha \mathbb{R}$ onto M. Moreover, E is a point-ultraweak limit of normal u.c.p. maps from $M \rtimes_\alpha \mathbb{R}$ into M.*

Proof. For each n, fix a unit vector $f_n \in L^2(\mathbb{R})$ with $\operatorname{supp}(f_n) \subset [-1/n, 1/n]$ and define an isometry $V_n \colon \mathcal{H} \to L^2(\mathbb{R}, \mathcal{H}) \cong \mathcal{H} \otimes L^2(\mathbb{R})$ by

$$(V_n\xi)(t) = f_n(t)v(t)\xi$$

for $\xi \in \mathcal{H}$. We define a normal u.c.p. map $\varphi_n \colon M \rtimes_\alpha \mathbb{R} \to \mathbb{B}(\mathcal{H})$ by $\varphi_n(x) = V_n^* x V_n$. It is routine to check that

$$\langle \varphi_n((a \otimes 1)(v \otimes \lambda)(s))\eta, \xi\rangle = \int_{\mathbb{R}} \overline{f_n(t)} f_n(t-s) \langle \alpha_{-t}(a)\eta, \xi\rangle \, dt$$

for every $a \in M$ and every $s \in \mathbb{R}$. In particular,

$$\varphi_n(a \otimes 1) = \int_{\mathbb{R}} |f_n(t)|^2 \alpha_{-t}(a) \, dt$$

for $a \in M$. Hence, every φ_n maps $M \rtimes_\alpha \mathbb{R}$ into M and any cluster point of the sequence $\{\varphi_n\}$ is a conditional expectation onto M. □

The dual action $\hat{\alpha}$ of $\mathbb{R}$ on $M \rtimes_\alpha \mathbb{R}$ is implemented by $1 \otimes \mu$ on $\mathcal{H} \otimes L^2(\mathbb{R})$, where $\mu(s)$ is the unitary operator on $L^2(\mathbb{R})$ given by multiplication by e^{-ist}: $(\mu(s)\xi)(t) = e^{-ist}\xi(t)$ for every $s \in \mathbb{R}$ and $\xi \in L^2(\mathbb{R})$. This indeed gives rise to an $\mathbb{R}$-action as we have

$$(1 \otimes \mu)(s)(v \otimes \lambda)(t)(1 \otimes \mu)(s)^* = e^{-ist}(v \otimes \lambda)(t)$$

for every $s, t \in \mathbb{R}$ and

$$(1 \otimes \mu)(s)(a \otimes 1)(1 \otimes \mu)(s)^* = (a \otimes 1)$$

for every $s \in \mathbb{R}$ and $a \in M$. It follows that $M \rtimes_\alpha \mathbb{R} \rtimes_{\hat{\alpha}} \mathbb{R}$ is the von Neumann subalgebra of $\mathbb{B}(\mathcal{H} \otimes L^2(\mathbb{R}) \otimes L^2(\mathbb{R}))$ generated by

$$(1 \otimes \mu \otimes \lambda)(\mathbb{R}), \ (v \otimes \lambda \otimes 1)(\mathbb{R}) \text{ and } M \otimes \mathbb{C}1 \otimes \mathbb{C}1.$$

Takesaki's duality theorem states

Theorem 9.3.7. $M \rtimes_\alpha \mathbb{R} \rtimes_{\hat{\alpha}} \mathbb{R} \cong M \mathbin{\bar{\otimes}} \mathbb{B}(L^2(\mathbb{R}))$.

Proof. For simplicity, let $\tilde{M} = M \rtimes_\alpha \mathbb{R} \rtimes_{\hat{\alpha}} \mathbb{R}$. Since λ and μ are unitarily equivalent via the Fourier transform, $\tilde{M}$ is spatially isomorphic to the von Neumann algebra generated by

$$(1 \otimes \mu \otimes \mu)(\mathbb{R}),\ (v \otimes \lambda \otimes 1)(\mathbb{R}) \text{ and } M \otimes \mathbb{C}1 \otimes \mathbb{C}1.$$

If U is the unitary operator on $L^2(\mathbb{R}) \otimes L^2(\mathbb{R}) \cong L^2(\mathbb{R} \times \mathbb{R})$ given by $(U\xi)(s,t) = \xi(s-t,t)$, then one has $U(\mu \otimes \mu)(r)U^* = (\mu \otimes 1)(r)$ and $U(\lambda \otimes 1)(r)U^* = (\lambda \otimes 1)(r)$ for every $r \in \mathbb{R}$. Indeed,

$$\begin{aligned}(U(\mu \otimes \mu)(r)U^*\xi)(s,t) &= ((\mu \otimes \mu)(r)U^*\xi)(s-t,t)\\ &= e^{-ir(s-t)-irt}(U^*\xi)(s-t,t)\\ &= e^{-irs}\xi(s,t)\\ &= ((\mu \otimes 1)(r)\xi)(s,t)\end{aligned}$$

and

$$\begin{aligned}(U(\lambda \otimes 1)(r)U^*\xi)(s,t) &= ((\lambda \otimes 1)(r)U^*\xi)(s-t,t)\\ &= (U^*\xi)(s-t-r,t)\\ &= \xi(s-r,t)\\ &= ((\lambda \otimes 1)(r)\xi)(s,t).\end{aligned}$$

It follows that $\tilde{M}$ is spatially isomorphic to the von Neumann algebra generated by

$$(1 \otimes \mu \otimes 1)(\mathbb{R}),\ (v \otimes \lambda \otimes 1)(\mathbb{R}) \text{ and } M \otimes \mathbb{C}1 \otimes \mathbb{C}1.$$

We note that the von Neumann algebra generated by the first and third items is $M \mathbin{\bar{\otimes}} L^\infty(\mathbb{R}) \otimes \mathbb{C}1$. Let V be the unitary operator on $\mathcal{H} \otimes L^2(\mathbb{R})$ associated with v – i.e., $(V\xi)(t) = v(t)\xi(t)$ for $\xi \in L^2(\mathbb{R},\mathcal{H}) \cong \mathcal{H} \otimes L^2(\mathbb{R})$. It follows that $V^*(v \otimes \lambda)(r)V = (1 \otimes \lambda)(r)$ for $r \in \mathbb{R}$.

We claim that

$$V^*(M \mathbin{\bar{\otimes}} L^\infty(\mathbb{R}))V = M \mathbin{\bar{\otimes}} L^\infty(\mathbb{R}).$$

Since V acts "diagonally," V commutes with $\mathbb{C}1 \otimes L^\infty(\mathbb{R})$. On the other hand, $V^*(a \otimes 1)V = \pi_\alpha(a)$ for every $a \in M$, where $\pi_\alpha(a)$ is given by $(\pi_\alpha(a)\xi)(t) = \alpha_{-t}(a)\xi(t)$. It follows that $\pi_\alpha(M) \subset L^\infty(\mathbb{R}, M) \cong M \bar{\otimes} L^\infty(\mathbb{R})$. Hence, $V^*(M \mathbin{\bar{\otimes}} L^\infty(\mathbb{R}))V \subset M \mathbin{\bar{\otimes}} L^\infty(\mathbb{R})$. The converse inclusion follows from $V(M \otimes \mathbb{C}1)V^* \subset M \mathbin{\bar{\otimes}} L^\infty(\mathbb{R})$, which can be shown by a similar argument. This proves that $V^*(M \mathbin{\bar{\otimes}} L^\infty(\mathbb{R}))V = M \mathbin{\bar{\otimes}} L^\infty(\mathbb{R})$.

Consequently, $\tilde{M}$ is spatially isomorphic to the von Neumann algebra generated by

$$M \mathbin{\bar{\otimes}} L^\infty(\mathbb{R}) \otimes \mathbb{C}1 \text{ and } (1 \otimes \lambda \otimes 1)(\mathbb{R}).$$

Finally, we observe that the von Neumann algebra generated by $L^\infty(\mathbb{R})$ and $\lambda(\mathbb{R})$ is $\mathbb{B}(L^2(\mathbb{R}))$ since $L^\infty(\mathbb{R})' \cap \lambda(\mathbb{R})' = \mathbb{C}1$. □

Proof of Theorem 9.3.3. Let M be an injective von Neumann algebra and α be a modular action so that $M \rtimes_\alpha \mathbb{R}$ is semifinite (Theorem 9.3.5). Since $M \bar{\otimes} \mathbb{B}(L^2(\mathbb{R}))$ is injective (which isn't as obvious as you might think) and $M \rtimes_\alpha \mathbb{R}$ is the range of a conditional expectation from $M \bar{\otimes} \mathbb{B}(L^2(\mathbb{R}))$ by Theorem 9.3.7 and Lemma 9.3.6, $M \rtimes_\alpha \mathbb{R}$ is also injective. It follows from Theorem 9.3.4 that $M \rtimes_\alpha \mathbb{R}$ is semidiscrete. Let $\{\psi_i\}$ be a net of finite-rank u.c.p. maps on $M \rtimes_\alpha \mathbb{R}$ which converges to the identity in the point-ultraweak topology. By Lemma 9.3.6, there exists a net of normal u.c.p. maps $\varphi_n \colon M \rtimes_\alpha \mathbb{R} \to M$ which converges to the identity on M. It follows that $\varphi_n \circ \psi_i|_M$ are finite-rank u.c.p. maps on M such that

$$\text{ultraweak-}\lim_n \lim_i \varphi_n \circ \psi_i(a) = \text{ultraweak-}\lim_n \varphi_n(a) = a$$

for every $a \in M$. This proves semidiscreteness of M. □

Exercise

Exercise 9.3.1. Prove that A has property C if and only if A has both properties C' and C''. Use this fact, and Lemma 9.2.8, to remove the separability hypothesis in Theorem 9.3.1.

9.4. Corollaries

Corollary 9.4.1. *Exact* C**-algebras are locally reflexive.*

Proof. Assume A is separable and exact. Theorem 9.3.1 implies A has property C; evidently this implies property C''. Hence, by Proposition 9.2.5, A is locally reflexive. The nonseparable case follows, thanks to Lemma 9.2.8. □

Remark 9.4.2. It follows that quotient maps of exact C*-algebras are always locally liftable (Proposition 9.1.4).

Corollary 9.4.3. *Quotients of exact* C**-algebras are exact.*

Proof. Since an algebra is exact if and only if all its separable subalgebras are exact, we may assume separability. Now apply Theorem 9.3.1 and Proposition 9.2.4. □

Corollary 9.4.4. *Quotients of nuclear* C**-algebras are nuclear.*

Proof. Let A be nuclear and $J \triangleleft A$ be an ideal. It suffices to show $(A/J)^{**}$ is semidiscrete (Proposition 2.3.8). But $A^{**} = J^{**} \oplus (A/J)^{**}$, so it suffices to show A^{**} is semidiscrete (this clearly passes to direct summands). But we outlined this fact in the paragraph preceding Theorem 9.3.3. □

Combining with a deep result of Glimm, we recover a theorem of Blackadar ([**17**]).

Corollary 9.4.5. *A separable* C*-*algebra is type* I *if and only if every subalgebra is nuclear.*

Proof. Recall Glimm's Theorem: A separable C*-algebra A is *not* type I if and only if every UHF algebra arises as a subquotient of A (see [**142**, Section 6.8]).

This implies, first of all, that subalgebras of type I are again type I. Indeed, if a subalgebra were not type I, then it would have UHF subquotients and hence the larger algebra would too. Since type I C*-algebras are nuclear (Proposition 2.7.4), this evidently implies the "only if" direction.

For the opposite direction, we assume A is not type I and show that it has a nonnuclear subalgebra. By Glimm's Theorem, we can find a subalgebra $B \subset A$ which has a prescribed UHF algebra as a quotient. By Corollary 8.2.5, B must have a subquotient isomorphic to $C^*_\lambda(\mathbb{F}_2)$ (or any other exact nonnuclear C*-algebra). Since nuclearity passes to quotients, this implies B, hence A, contains a nonnuclear subalgebra. □

Finally, we present a folklore result (due to Tomiyama and Kirchberg) on the ideal structure of minimal tensor products (cf. [**186**]). It is a noncommutative analogue of the fact (definition) that rectangular open subsets form a topological basis in Cartesian products.

Corollary 9.4.6. *Let $I \lhd (A \otimes B)$ be an ideal. If A is exact, then*

$$\begin{aligned} I &= \overline{\operatorname{span}}\{I_A \odot I_B : I_A, I_B \text{ are ideals such that } I_A \odot I_B \subset I\} \\ &= \overline{\operatorname{span}}\{x \otimes y : x \otimes y \in I\}. \end{aligned}$$

Proof. First observe that the closed linear spans on the right hand side are equal and define a closed two-sided ideal, denoted by J, contained in I. We will show that if $z \in (A \otimes B) \setminus J$, then $z \notin I$. (You may want to review Exercise 3.4.2 before proceeding.)

Let $\sigma\colon A \otimes B \to \mathbb{B}(\mathcal{H})$ be an irreducible representation such that $J \subset \ker\sigma$ and $\sigma(z) \neq 0$. Letting σ_A and σ_B be the restrictions (Theorem 3.2.6), it is clear that $\ker\sigma_A \otimes B + A \otimes \ker\sigma_B \subset \ker\sigma$. Since A is exact, $A \otimes B/A \otimes \ker\sigma_B = A \otimes \sigma_B(B)$. Since A is locally reflexive, $A \otimes \sigma_B(B)/\ker\sigma_A \otimes \sigma_B(B) = \sigma_A(A) \otimes \sigma_B(B)$ (Corollary 9.1.5). It follows that

$$(A \otimes B)/(\ker\sigma_A \otimes B + A \otimes \ker\sigma_B) \cong \sigma_A(A) \otimes \sigma_B(B),$$

and hence σ factors through $\sigma_A(A) \otimes \sigma_B(B)$. In fact, σ drops to an isomorphism on $\sigma_A(A) \otimes \sigma_B(B)$. Indeed, since $\sigma_A(A)''$ and $\sigma_B(B)''$ are factors, we have $\sigma(A \odot B) \cong \sigma_A(A) \odot \sigma_B(B)$ and thus $\sigma(A \otimes B) \cong \sigma_A(A) \otimes \sigma_B(B)$ (cf. Exercise 3.4.1).

Hence we can find pure states φ on A and ψ on B such that $(\varphi \otimes \psi)|_{\ker \sigma} = 0$ and $(\varphi \otimes \psi)(z) \neq 0$. Let $a_i \in A$ excise φ and $b_i \in B$ excise ψ (Theorem 1.4.10). Since $(\varphi \otimes \psi)(a_i \otimes b_i) = 1$, one has $\|a_i \otimes b_i + J\| = 1$ for all i. We claim that $\|a_i \otimes b_i + I\| = 1$ as well (in particular, $\|a_i^2 \otimes b_i^2 + I\|$ is bounded away from zero). Once demonstrated, the proof will be complete because $(\varphi \otimes \psi)(z)(a_i^2 \otimes b_i^2 + I) \approx (a_i \otimes b_i)z(a_i \otimes b_i) + I$, by the excision property, and hence $z + I \neq 0$.

Suppose by contradiction that $\|a_i \otimes b_i + I\| = \delta < 1$. Consider the $*$-homomorphism $C[0,1] \otimes C[0,1] \to A \otimes B$ defined by $\sum_k f_k \otimes g_k \mapsto \sum_k f_k(a_i) \otimes g_k(b_i)$. Let $f_\delta \in C[0,1]$ be such that $0 \leq f_\delta \leq 1$, $f_\delta = 0$ on $[0, \delta^{1/2}]$ and $f_\delta = 1$ on $[\delta^{1/4}, 1]$. Then, $f_\delta(a_i) \otimes f_\delta(b_i) \in I$ and hence $f_\delta(a_i) \otimes f_\delta(b_i) \in J$, by definition. This implies that $\|a_i \otimes b_i + J\| \leq \delta^{1/4} < 1$ in contradiction to our observations above, so the proof is complete. □

9.5. References

Local reflexivity was imported to C*-algebras by Effros and Haagerup [**59**]. Propositions 9.1.4 and 9.2.5 appeared in that paper. Property C was introduced by Archbold and Batty in [**8**], where they showed property C implies exactness, it passes to subquotients, and it is enjoyed by nuclear C*-algebras. Theorem 9.3.1 is due to Kirchberg [**103**], [**106**]. That injectivity implies semidiscreteness (and even hyperfiniteness) for factors is Connes's remarkable discovery [**41**] (the general case is found in [**39**]). The proof given here is inspired by Wassermann's work in [**194**].

Chapter 10

Summary and Open Problems

Since a number of important characterizations and permanence properties are scattered throughout this text, we thought it may be useful to collect them in one place – hence this chapter. At the end, there are a few open problems too.

10.1. Nuclear C*-algebras

Here are the main characterizations of nuclear C*-algebras:

Theorem 10.1.1. *For a* C*-*algebra* A, *the following are equivalent:*

(1) *the identity map* $\mathrm{id}_A\colon A \to A$ *is nuclear (Definitions 2.1.1 and 2.3.1; see also Proposition 2.3.8 and Exercise 2.3.13);*

(2) *for every* C*-*algebra* B, $A \otimes_{\max} B = A \otimes B$ *(Theorem 3.8.7);*

(3) A^{**} *is injective (or semidiscrete; see Proposition 2.3.8 and Theorem 9.3.3).*

Subalgebras. In Remark 4.4.4 we saw that subalgebras of a nuclear C*-algebra *need not* be nuclear. However, there are two natural conditions which ensure that nuclearity is preserved.

Proposition 10.1.2. *A hereditary subalgebra of a nuclear* C*-*algebra is nuclear. If* A *is nuclear,* $B \subset A$ *and there exists a conditional expectation* $\Phi\colon A \to B$, *then* B *is also nuclear (Exercises 2.3.4 and 2.3.3).*

Extensions.

Proposition 10.1.3. *If* $0 \to I \to A \to B \to 0$ *is short exact and both* I *and* B *are nuclear, then so is* A *(Exercise 3.8.1).*

Quotients. At present, there is no elementary proof of the following fact.

Theorem 10.1.4. *If* $I \triangleleft A$ *and* A *is nuclear, then* A/I *is also nuclear (Corollary 9.4.4).*

Inductive limits. For inductive limits with injective connecting maps, the following result is easy (use Arveson's Extension Theorem). In general, it is not. But quotients of nuclear C*-algebras are nuclear, so everything reduces to the injective case.

Theorem 10.1.5. *An inductive limit of nuclear* C**-algebras is nuclear.*

Direct sums and products. Infinite direct products (aka ℓ^∞-direct sums) of nuclear algebras are rarely nuclear; $\prod_{n\in\mathbb{N}} \mathbb{M}_{k(n)}(\mathbb{C})$ is nuclear if and only if $\sup k(n) < \infty$ (Proposition 2.4.9). However, for sequences tending to zero in norm (aka c_0-direct sums), all is well.

Proposition 10.1.6. *Let* A_i*,* $i \in I$*, be a collection of* C**-algebras. Then* $\bigoplus_{i\in I} A_i$ *is nuclear if and only if each* A_i *is nuclear.*

The proof is trivial – take a (finite) direct sum of c.c.p. approximations on the A_i's.

Tensor products.

Proposition 10.1.7. *Let* A *and* B *be arbitrary* C**-algebras. The following statements are equivalent:*

(1) *both* A *and* B *are nuclear;*

(2) $A \otimes B$ *is nuclear;*

(3) $A \otimes_{\max} B$ *is nuclear.*

Proof. We only prove (1) $\Leftrightarrow$ (3) as the equivalence of (1) and (2) is similar.

First assume A and B are nuclear. Taking maximal tensor products of c.c.p. approximations, one easily checks that $A \otimes_{\max} B$ is also nuclear. For the other direction, choose a positive norm-one element $b \in B$ and a state φ on B such that $\varphi(b) = 1$. Then, for the c.c.p. embedding $\iota_b\colon A \to A \otimes_{\max} B$, defined by $\iota_b(a) = a \otimes b$, and the *slice map* $\mathrm{id}_A \otimes \varphi\colon A \otimes_{\max} B \to A \otimes \mathbb{C} \cong A$, we have $(\mathrm{id}_A \otimes \varphi) \circ \iota_b = \mathrm{id}_A$. Since $A \otimes_{\max} B$ is nuclear, A is nuclear, too. $\square$

Crossed products. For amenable groups everything is nice: The crossed product is nuclear if and only if the original algebra is too (Theorem 4.2.6). Otherwise, things get complicated. For example, $C(X) \rtimes_\alpha \Gamma$ is nuclear if and only if α is an amenable action (Theorem 4.4.3). For nonabelian algebras and exact groups, partial results can be formulated (e.g., Theorem 4.3.4), but they aren't very general.

Though unfairly neglected in this treatise, locally compact amenable groups also behave well with respect to crossed products: Full and reduced crossed products always agree, and the crossed product is nuclear whenever the algebra being acted upon is nuclear (see [**5**, Theorem 5.3] for a more general result).

Free products. The full free product (of unital algebras, with amalgamation over units)[1] of two nuclear C*-algebras is almost never nuclear ($C^*(\mathbb{F}_2) \cong C(\mathbb{T}) * C(\mathbb{T})$ is not even exact, by Proposition 3.7.11). Reduced free products are a little better.

Theorem 10.1.8. *The reduced (amalgamated) free product of two nuclear* C**-algebras is always exact (Corollary 4.8.3).*

In the pure-state case things work properly.

Theorem 10.1.9. *Let A and B be unital nuclear* C**-algebras with states φ and ψ, respectively. Assume φ is pure. Then*

$$(A, \varphi) * (B, \psi)$$

is nuclear (Theorem 4.8.7).

10.2. Exact C*-algebras

Here's the main theorem regarding exactness:

Theorem 10.2.1. *For a* C**-algebra A, the following are equivalent:*

(1) *there exists a faithful $*$-homomorphism $\pi\colon A \to \mathbb{B}(\mathcal{H})$ which is nuclear (Definitions 2.1.1 and 2.3.2; see Exercise 2.3.9 as well);*

(2) *for every short exact sequence $0 \to J \to B \to B/J \to 0$, the sequence*

$$0 \to J \otimes A \to B \otimes A \to (B/J) \otimes A \to 0$$

is also exact (Theorem 3.9.1 and Exercises 3.9.6 and 3.9.7).

[1] Which hasn't been defined, but satisfies the universal property that any pair of unital $*$-homomorphisms $A \to \mathbb{B}(\mathcal{H})$ and $B \to \mathbb{B}(\mathcal{H})$ extends uniquely to $A * B$.

We have seen that exactness is also equivalent to properties C and C' (Theorem 9.3.1). Finally, as mentioned in the preface, a fundamental theorem of Kirchberg is completely missing from these notes; at the very least, we should state it.

Theorem 10.2.2 ([**107**], [**168**]). *A separable* C**-algebra is exact if and only if it is isomorphic to a subalgebra of the Cuntz algebra* $\mathcal{O}_2$.

Subalgebras.

Proposition 10.2.3. *Exactness passes to subalgebras.*

This is immediate from the definition.

Extensions. We will show in Section 13.4 that extensions of exact C*-algebras need not be exact, in general. However, the following result is very useful.

Theorem 10.2.4. *Let* $0 \to I \to A \to B \to 0$ *be a short exact sequence where both* I *and* B *are exact and* A *is unital. Then* A *is exact if and only if the extension is locally split (Exercise 3.9.8).*

Quotients. At present, this is one of the hardest C*-results, ever.

Theorem 10.2.5. *A quotient of an exact* C**-algebra is again exact (Corollary 9.4.3).*

Inductive limits. Just like the nuclear case, the following fact is trivial when the connecting maps are injective and it is very hard in general.

Theorem 10.2.6. *An inductive limit of exact* C**-algebras is exact.*

Direct sums and products. This is identical to the nuclear case.

Tensor products. The proof of the following result is very similar to the nuclear case.

Proposition 10.2.7. *Let* A *and* B *be arbitrary* C**-algebras. The following are equivalent:*

(1) *both* A *and* B *are exact;*

(2) $A \otimes B$ *is exact.*

What happened to the maximal tensor product? Doesn't the nuclear proof carry over verbatim? Seriously, think about it for a minute. What goes wrong with the proof?

Proposition 10.2.8. *For any discrete group* Γ*, the* C*-*algebra*

$$C^*(\{\lambda_g \otimes \lambda_g : g \in \Gamma\}) \subset C^*_\lambda(\Gamma) \otimes_{\max} C^*_\lambda(\Gamma)$$

is isomorphic to $C^*(\Gamma)$. *In particular,* $C^*_\lambda(\mathbb{F}_2) \otimes_{\max} C^*_\lambda(\mathbb{F}_2)$ *is not exact since it contains a nonexact subalgebra.*

Proof. Let $\sigma\colon C^*(\Gamma) \to C^*_\lambda(\Gamma) \otimes_{\max} C^*_\lambda(\Gamma)$ be the $*$-homomorphism induced by the map $g \mapsto \lambda_g \otimes \lambda_g$. Fix a faithful representation $\pi\colon C^*(\Gamma) \to \mathbb{B}(\mathcal{H})$. Our goal is to show

$$\|\pi(x)\| \le \|\sigma(x)\|$$

for all $x \in C^*(\Gamma)$, as this implies σ is injective.

The trick is to construct a pair of commuting representations of Γ such that π is a subrepresentation of the product. Fell's absorbtion principle makes this straightforward. Indeed, we have a representation

$$C^*_\lambda(\Gamma) \to \mathbb{B}(\ell^2(\Gamma) \otimes \mathcal{H}), \ \lambda_g \mapsto \lambda_g \otimes \pi_g,$$

and a commuting representation given by

$$\lambda_g \mapsto \rho_g \otimes 1_{\mathcal{H}}.$$

Hence we have a product representation

$$\theta\colon C^*_\lambda(\Gamma) \otimes_{\max} C^*_\lambda(\Gamma) \to \mathbb{B}(\ell^2(\Gamma) \otimes \mathcal{H})$$

with the property that

$$\theta \circ \sigma(g) = \lambda_g \rho_g \otimes \pi_g.$$

The subspace $\delta_e \otimes \mathcal{H}$ is evidently invariant for each unitary $\theta \circ \sigma(g)$, and the restriction of $\theta \circ \sigma$ to $\delta_e \otimes \mathcal{H}$ is π. This implies the desired inequality. □

This result suggests the following conjecture: If A and B are exact C*-algebras, then $A \otimes_{\max} B$ is exact if and only if either A or B is nuclear.

Crossed products. There are many characterizations of exact groups – see Section 5.1. One of them is relevant to crossed products (Theorem 5.1.10).

Theorem 10.2.9. *Let* Γ *be an exact group and* A *be a* Γ-C*-*algebra. Then* $A \rtimes_r \Gamma$ *is exact if and only if* A *is exact.*

Proof. The "only if" direction is trivial. For the converse, let $0 \to J \to B \to (B/J) \to 0$ be short exact and note that $0 \to J \otimes A \to B \otimes A \to (B/J) \otimes A \to 0$ can be regarded as a short exact sequence of Γ-algebras by tensoring with the trivial action of Γ on B (compare with the first part of the proof of Theorem 5.1.10). Since $(B \otimes A) \rtimes_r \Gamma \cong B \otimes (A \rtimes_r \Gamma)$ in this case (Exercise 4.1.3), we have a commutative diagram

$$\begin{array}{ccccccccc} 0 & \longrightarrow & (J \otimes A) \rtimes_r \Gamma & \longrightarrow & (B \otimes A) \rtimes_r \Gamma & \longrightarrow & ((B/J) \otimes A) \rtimes_r \Gamma) & \longrightarrow & 0 \\ & & \cong\downarrow & & \cong\downarrow & & \cong\downarrow & & \\ 0 & \longrightarrow & J \otimes (A \rtimes_r \Gamma) & \longrightarrow & B \otimes (A \rtimes_r \Gamma) & \longrightarrow & (B/J) \otimes (A \rtimes_r \Gamma) & \longrightarrow & 0. \end{array}$$

Since the top row is exact, by Theorem 5.1.10, so is the bottom row. □

Free products. As in the nuclear case, full free products of exact C*-algebras are almost never exact. But reduced free products behave well.

Theorem 10.2.10. *The reduced free product (with arbitrary amalgamation) of exact* C*-*algebras is exact (Corollary 4.8.3).*

10.3. Quasidiagonal C*-algebras

Subalgebras. Quasidiagonality obviously passes to subalgebras.

Extensions. Extensions of QD C*-algebras need not be QD; the Toeplitz algebra is an elementary counterexample. Even for extensions which have a $*$-homomorphic splitting, it isn't known if quasidiagonality is preserved. However, there are some useful partial results. First we need a simple fact about multiplier algebras.

Proposition 10.3.1. *Let I be residually finite-dimensional. Then the multiplier algebra $M(I)$ is also RFD.*

Proof. Let $\pi_n \colon I \to A_n$ be *surjective* $*$-homomorphisms such that each A_n is unital and QD (e.g., finite-dimensional) and the map $\bigoplus \pi_n$ is faithful. Extend to $*$-homomorphisms $\tilde{\pi}_n \colon M(I) \to A_n$.[2]

In general, if $J \lhd A$ is an essential ideal and we have a $*$-homomorphism $\rho\colon A \to B$ with the property that $\rho|_J$ is injective, then ρ must be injective on all of A (since any nonzero ideal intersects J). Hence we see that the $*$-homomorphism

$$\bigoplus \tilde{\pi}_n \colon M(I) \to \prod A_n$$

is also injective. □

Proposition 10.3.2. *Assume $0 \to I \to A \xrightarrow{\pi} B \to 0$ is exact, where I is RFD and B is QD. Then A is QD.*

[2]One could quote the noncommutative Tietze extension theorem, but here it isn't necessary. For each n let $e_n \in I$ be a lift of the unit of A_n, and simply define $\tilde{\pi}_n(x) = \pi_n(e_n x e_n)$ for all $x \in M(I)$. This map might appear to be just u.c.p. but multiplicative domains suggest something more.

Proof. Let $\rho\colon A \to M(I)$ be the natural extension of the inclusion $I \hookrightarrow M(I)$ ([**198**, 2.2.14]). Then the map $\rho \oplus \pi\colon A \to M(I) \oplus B$ is injective. The previous proposition (plus the trivial fact that direct sums of QD C*-algebras are QD) now implies quasidiagonality. □

The proof shows that $M(I)$ is QD whenever I has a separating family of *unital* QD quotients. Hence the proposition above remains true for ideals of the form $I = J \otimes B$ where J is RFD and B is unital and QD. We still lack a description of those algebras whose multiplier algebras are QD.

Another instance where quasidiagonality passes to extensions is when the extension is quasidiagonal. Confusing, yes, but not a tautology.

Definition 10.3.3. Let $0 \to I \to A \xrightarrow{\pi} B \to 0$ be a short exact sequence. We call this a *quasidiagonal extension* if I has a quasicentral approximate unit *consisting of projections.*

It is important to note that in general an extension being quasidiagonal has nothing to do with whether or not the middle algebra A is QD – it is simply a statement about projections.

Proposition 10.3.4 (Exercise 7.1.6). *Let $0 \to I \to A \xrightarrow{\pi} B \to 0$ be a quasidiagonal extension where both I and B are QD. Then A is QD.*

Proof. Let $\{p_n\} \subset I$ be an approximate unit of projections which is quasicentral in A. Consider the c.c.p. maps $\varphi_n\colon A \to I \oplus B$, $\varphi_n(x) = p_n x p_n \oplus \pi(x)$. Evidently these maps are asymptotically multiplicative, so we must check that they are asymptotically isometric.

Let $P \in I^{**} \subset A^{**}$ be the (weak) limit of the p_n's. Then P is central in A^{**} and yields a decomposition $A^{**} = I^{**} \oplus B^{**}$. Hence for each $x \in A$ we have $\|x\| = \max\{\|PxP\|, \|(1-P)x(1-P)\|\}$. But $\|\pi(x)\| = \|(1-P)x(1-P)\|$ and $\|PxP\| \leq \liminf_n \|p_n x p_n\|$ since $p_n x p_n \to PxP$ in the strong operator topology. The other inequality is obvious. □

When all algebras in a sequence are nuclear and satisfy the UCT, K-theory can help determine quasidiagonality. See [**31**] for some partial results.

Quotients. Since $C^*(\mathbb{F}_\infty)$ is QD (even RFD – Theorem 7.4.1), it is clear that quasidiagonality need not pass to quotients. There is a simple sufficient condition, however.

Proposition 10.3.5 (Exercise 7.1.5). *Let $0 \to I \to A \xrightarrow{\pi} A/I \to 0$ be locally split and short exact. If A is QD and the extension is quasidiagonal, then A/I is QD.*

Proof. Let $\mathfrak{F} \subset A/I$ be a finite set and let $\varepsilon > 0$. Let $X \subset A/I$ be a finite-dimensional operator system which contains $\mathfrak{F}$ and $\{ab : a, b \in \mathfrak{F}\}$. Fix a u.c.p. splitting $\varphi\colon X \to A$. Now take a quasicentral approximate unit of projections, say $\{p_n\}$, and consider the (isometric, though no longer unital) completely positive splittings $\varphi_n(x) = (1 - p_n)\varphi(x)(1 - p_n)$. We claim that for sufficiently large n, these maps are ε-multiplicative on $\mathfrak{F}$; evidently this will imply the result.

For $a, b \in \mathfrak{F}$ we estimate

$$\begin{aligned}
&\|\varphi_n(ab) - \varphi_n(a)\varphi_n(b)\| \\
&\quad = \|(1 - p_n)\varphi(ab)(1 - p_n) - (1 - p_n)\varphi(a)(1 - p_n)\varphi(b)(1 - p_n)\| \\
&\quad \leq \|(1 - p_n)\Big(\varphi(ab) - \varphi(a)\varphi(b)\Big)(1 - p_n)\| \\
&\qquad + \|(1 - p_n)\Big((1 - p_n)\varphi(a) - \varphi(a)(1 - p_n)\Big)\varphi(b)(1 - p_n)\|.
\end{aligned}$$

Since φ is a splitting, $\|\pi(x)\| = \lim \|(1 - p_n)x(1 - p_n)\|$ and $\{p_n\}$ is quasicentral, the result follows easily. □

Note that local liftability is automatic whenever A is locally reflexive (e.g., exact).

Inductive limits. For inductive limits *with injective connecting maps*, a simple application of Arveson's Extension Theorem shows that quasidiagonality passes to inductive limits. For arbitrary inductive limits this is false; we will see a counterexample in Remark 17.3.3. However, in the presence of exactness nothing funny happens.

Proposition 10.3.6. *If $A = \varinjlim A_n$ and each A_n is QD and locally reflexive, then A is QD.*

Proof. We may assume everything is unital. Essentially by definition, A is a subalgebra of

$$\frac{\prod A_n}{\bigoplus A_n}.$$

Hence we have a quasidiagonal extension $0 \to \bigoplus A_n \to B \to A \to 0$. Moreover, B is clearly QD and the quotient mapping is locally liftable, since each A_n is locally reflexive. Thus quasidiagonality passes to the quotient A. □

Direct sums and products. Both c_0 and ℓ^∞-direct sums of QD algebras are again QD. This is easily deduced from the definition.

Tensor products. It isn't known if the maximal tensor product of QD C*-algebras is again QD. For spatial tensor products, everything is fine (see Proposition 7.1.12 and Exercise 7.2.6).

Proposition 10.3.7. *If A and B are QD, then so is $A \otimes B$.*

Crossed products. Quasidiagonality of crossed products is poorly understood. The only cases where we have complete information is $A \rtimes_\alpha \mathbb{Z}$ where A is either abelian or AF. (See Section 8.5 for precise statements.) Recall that Rosenberg's conjecture asks whether $C^*_\lambda(\Gamma) = \mathbb{C} \rtimes_r \Gamma$ is always QD for a discrete amenable group Γ. There are lots of open questions in this area.

Free products. The reduced free product of QD C*-algebras is almost never QD ($C^*_\lambda(\mathbb{F}_2) = C(\mathbb{T}) *_r C(\mathbb{T})$ is not QD, and many reduced free products contain this algebra as a subalgebra). A nice theorem of Florin Boca, however, reveals a different situation for full free products ([**23**]).

Theorem 10.3.8. *If A and B are unital QD C*-algebras, then the full free product $A * B$ (with amalgamation over $\mathbb{C}$) is also QD.*

It isn't known what happens with more general amalgamation.

10.4. Open problems

An affirmative answer to our first question would imply that every exact C*-algebra with the local lifting property is nuclear (it is known that $A \otimes_{\max} Q(\mathcal{H}) = A \otimes Q(\mathcal{H}) \Leftrightarrow A$ is both exact and has the local lifting property – see Section 13.1).

Problem 10.4.1. Let $Q(\mathcal{H}) = \mathbb{B}(\mathcal{H})/\mathbb{K}(\mathcal{H})$ be the Calkin algebra. Is it true that A is nuclear if and only if $A \otimes_{\max} Q(\mathcal{H}) = A \otimes Q(\mathcal{H})$?

The following problem appears tantalizing at first glance, but it's probably impossible.

Problem 10.4.2. Is there a simpler proof of the fact that the double dual of a nuclear C*-algebra is semidiscrete?

Problem 10.4.3. Let $N \subset M \subset \mathbb{B}(\mathcal{H})$ be von Neumann algebras and assume $\varphi\colon \mathbb{B}(\mathcal{H}) \to M$ is a u.c.p. map such that $\varphi|_N = \mathrm{id}_N$. Is the inclusion map $N \hookrightarrow M$ weakly nuclear?

Perhaps the most important problem in this section is

Problem 10.4.4. The following statements are equivalent ([**102**]). Are they true?

(1) (Connes's embedding problem) Every separable II_1-factor embeds into the ultraproduct R^ω of the hyperfinite II_1-factor.
(2) Every separable C*-algebra with the LLP also has the WEP (cf. Chapter 13).
(3) $C^*(\mathbb{F}_2 \times \mathbb{F}_2) = C^*(\mathbb{F}_2) \otimes_{\max} C^*(\mathbb{F}_2)$ is RFD.
(4) Every tracial state on $C^*(\mathbb{F}_\infty)$ is amenable.
(5) $C^*(\mathbb{F}_2) \otimes_{\max} C^*(\mathbb{F}_2) = C^*(\mathbb{F}_2) \otimes C^*(\mathbb{F}_2)$.
(6) $C^*(\mathbb{F}_2) \otimes_{\max} C^*(\mathbb{F}_2)$ has a faithful tracial state.
(7) Several others – see Section 13.3.

Group theory experts seem to agree: Gromov has shown that nonexact discrete groups exist (cf. [**71**]). We're willing to take their word for it, but we would love to understand it.

Problem 10.4.5. Is there a simple proof of the existence of nonexact groups? Is there an explicit example?

The fact that all linear groups are exact ([**73**]) suggests the following:

Problem 10.4.6. Is every residually finite group exact? Perhaps just coarsely embeddable into a Hilbert space?

An affirmative answer to the next question would imply the existence of hyperbolic groups which are not residually finite.

Problem 10.4.7. Does the factorization property pass to arbitrary inductive limits (i.e., allowing noninjective connecting maps) of discrete groups?

On the other hand, every hyperbolic group is residually finite if our next question has an affirmative answer.

Problem 10.4.8. Do hyperbolic groups have the factorization property?

More generally, it would be very nice to find geometric or dynamical proofs of the factorization property for particular examples.

Our final problems are for C*-fanatics; we aren't aware of any important applications, but they are basic questions which should be resolved.

Problem 10.4.9. Is the hyperfinite II_1-factor QD?[3]

Problem 10.4.10. Are separable exact QD C*-algebras AF embeddable?

Problem 10.4.11. Does local reflexivity imply exactness?

Kirchberg has suggested an affirmative answer to this last question.

---

[3]This seems unlikely, but a proof would be nice. On the other hand, if it is QD, then every simple unital stably finite nuclear C*-algebra would be QD (compare with Blackadar and Kirchberg's questions at the end of Section 8.5).

Part 2

Special Topics

Chapter 11

Simple C*-Algebras

This chapter contains a variety of results related to simple C*-algebras. Our first goal is to analyze some deep structural work of Blackadar and Kirchberg on simple nuclear QD C*-algebras. The generalized inductive limit theory necessary for this result is interesting in its own right, but our target is Corollary 11.3.10. Next, we discuss what Popa calls a "local quantization" technique; this gives an internal characterization of quasidiagonality for a large class of algebras and a simple proof of the fact that injective II_1-factors are AFD (Theorem 11.4.8). In the final section we present Connes's celebrated uniqueness theorem for injective II_1-factors.

11.1. Generalized inductive limits

Blackadar and Kirchberg introduced *generalized* inductive limits in hopes of broadening the scope of classification techniques. While we certainly follow many of their ideas here, we also draw inspiration from free probability theory and take a norm-microstate point of view whenever convenient.

Definition 11.1.1. Let $(A_m, \varphi_{n,m})$ be a sequence of C*-algebras and linear, adjoint-preserving[1] maps $\varphi_{n,m}\colon A_m \to A_n$, $m < n$, such that $\varphi_{s,n} \circ \varphi_{n,m} = \varphi_{s,m}$, whenever $m < n < s$. We say that $(A_m, \varphi_{n,m})$ is a *generalized inductive system* if

(1) $\sup_{n>m} \|\varphi_{n,m}(x)\| < \infty$ for all $m \in \mathbb{N}$ and $x \in A_m$ and

(2) for each k, any $a, b \in A_k$ and $\varepsilon > 0$ there is an M such that for all $M < m < n$ we have $\|\varphi_{n,m}(\varphi_{m,k}(a)\varphi_{m,k}(b)) - \varphi_{n,k}(a)\varphi_{n,k}(b)\| < \varepsilon$.

[1]This assumption is not necessary, but it can always be arranged, so there's no harm in making it.

To a generalized inductive system, we can associate a limit C*-algebra just as for usual inductive systems. However, some technicalities can be bypassed with the help of a corona algebra. That is, if $(A_m, \varphi_{n,m})$ is a generalized inductive system,

$$\pi\colon \prod_{n\in\mathbb{N}} A_n \to \frac{\prod_{n\in\mathbb{N}} A_n}{\bigoplus_{n\in\mathbb{N}} A_n}$$

is the canonical quotient map and, we define $\Phi_m = \bigoplus_{n>m} \varphi_{n,m}\colon A_m \to \prod_{n\in\mathbb{N}} A_n$ for each $m \in \mathbb{N}$ (with zeroes in the first m slots), then we can consider the adjoint-closed linear spaces

$$\pi \circ \Phi_m(A_m) \subset \frac{\prod_{n\in\mathbb{N}} A_n}{\bigoplus_{n\in\mathbb{N}} A_n}.$$

Note that the norm closure of the union $\bigcup_m \pi \circ \Phi_m(A_m)$ is a C*-algebra since the definition of a generalized inductive system implies that for each $a, b \in A_m$, the product $\pi(\Phi_m(a))\pi(\Phi_m(b))$ belongs to this closure (though it's *not* close to $\pi(\Phi_m(ab))$, in general).

Definition 11.1.2. If $(A_m, \varphi_{n,m})$ is a generalized inductive system, the *generalized inductive limit*, $A = \mathrm{g}\varinjlim(A_m, \varphi_{n,m})$, is the norm closure of

$$\bigcup_{m\in\mathbb{N}} \pi \circ \Phi_m(A_m) \subset \frac{\prod_{n\in\mathbb{N}} A_n}{\bigoplus_{n\in\mathbb{N}} A_n}.$$

Typically we forget about the map π and let $\Phi_m\colon A_m \to A$ denote the natural adjoint-preserving linear map.

We will primarily be concerned with the case where each A_m is a finite-dimensional C*-algebra. In this setting there is a related notion which arose from important work in the context of Voiculescu's free probability theory.

Definition 11.1.3. Let A be a C*-algebra with a finite generating set $\{x_k\}_{k=1}^s \subset A$ of self-adjoint elements. We say that A admits *norm microstates* if for every finite set $\mathfrak{F}$ of noncommutative polynomials in s variables and $\varepsilon > 0$ there exist $n \in \mathbb{N}$ and self-adjoint matrices $X_1, \ldots, X_s \in \mathbb{M}_n(\mathbb{C})$ such that for every $P \in \mathfrak{F}$ we have

$$\big| \|P(x_1, \ldots, x_s)\| - \|P(X_1, \ldots, X_s)\| \big| < \varepsilon.$$

The definition in the nonfinitely generated case is similar, except that one has to accommodate all noncommutative polynomials in any number of variables.

The proof of the following proposition is rather technical to write out precisely, though conceptually it is very easy. We leave the details to the reader, but the idea is to string together sequences of matrices, whose norms on polynomials provide better and better approximations, to get representing sequences in the quotient.

Proposition 11.1.4. *The* C*-*algebra* A *admits norm microstates if and only if there exist a sequence* $s(n)$ *in* $\mathbb{N}$ *and a* $*$-*homomorphic embedding*

$$A \subset \frac{\prod_{n\in\mathbb{N}} \mathbb{M}_{s(n)}(\mathbb{C})}{\bigoplus_{n\in\mathbb{N}} \mathbb{M}_{s(n)}(\mathbb{C})}.$$

Note that every QD C*-algebra admits norm microstates – this is immediate from the definition – but there are other examples as well (see Sections 13.5 and 17.3).

For a sequence $s(n) \in \mathbb{N}$, we still let $\pi\colon \prod_1^\infty \mathbb{M}_{s(n)}(\mathbb{C}) \to \frac{\prod_1^\infty \mathbb{M}_{s(n)}(\mathbb{C})}{\bigoplus_1^\infty \mathbb{M}_{s(n)}(\mathbb{C})}$ be the quotient map, but we also need the canonical projection maps

$$\rho_p^q\colon \prod_1^\infty \mathbb{M}_{s(n)}(\mathbb{C}) \to \prod_p^q \mathbb{M}_{s(n)}(\mathbb{C}).$$

Theorem 11.1.5. *Let* A *be a separable* C*-*algebra. The following statements are equivalent:*

(1) *A is isomorphic to a generalized inductive limit of finite-dimensional* C*-*algebras;*

(2) *there exist integers* $s(n) \in \mathbb{N}$ *and a* $*$-*homomorphic embedding*

$$A \subset \frac{\prod_{n\in\mathbb{N}} \mathbb{M}_{s(n)}(\mathbb{C})}{\bigoplus_{n\in\mathbb{N}} \mathbb{M}_{s(n)}(\mathbb{C})};$$

(3) *A admits norm microstates.*

Proof. Since every finite-dimensional C*-algebra embeds into a full matrix algebra, the implication (1) $\Rightarrow$ (2) is trivial; hence we only have to show the converse.

So, assume an embedding

$$A \subset \frac{\prod_{n\in\mathbb{N}} \mathbb{M}_{s(n)}(\mathbb{C})}{\bigoplus_{n\in\mathbb{N}} \mathbb{M}_{s(n)}(\mathbb{C})}$$

is given and fix a linear self-adjoint splitting $\sigma\colon A \to \prod_{n\in\mathbb{N}} \mathbb{M}_{s(n)}(\mathbb{C})$ (i.e., $\pi \circ \sigma = \mathrm{id}_A$).[2] Also fix some finite-dimensional self-adjoint subspaces

$$\mathcal{S}_1 \subset \mathcal{S}_2 \subset \cdots \subset A$$

whose union is dense in A and with the property that

$$\{ab : a, b \in \mathcal{S}_n\} \subset \mathcal{S}_{n+1}$$

[2]This map comes from linear algebra and isn't necessarily bounded. If σ doesn't preserve adjoints, then replace it with $\frac{1}{2}(\sigma + \sigma^*)$.

for all n. Let d_n be the dimension of $\mathcal{S}_n$. With a little work, one can find natural numbers $p_1 < q_1 < p_2 < q_2 < p_3 < q_3 < \cdots$ such that for

$$\alpha_m = \rho_{p_m}^{q_m} \circ \sigma \colon A \to A_m = \prod_{i=p_m}^{q_m} \mathbb{M}_{s(i)}(\mathbb{C}),$$

one has

(1) $(1 - 1/m)\|x\| < \|\alpha_m(x)\| < (1 + 1/m)\|x\|$ for all $x \in \mathcal{S}_m$ and

(2) $\|\alpha_m(xy) - \alpha_m(x)\alpha_m(y)\| < d_m^{-4}$ for all $x, y \in \mathcal{S}_m$ in the unit ball.

Note that each α_m is injective on $\mathcal{S}_m$ and hence $\alpha_m(\mathcal{S}_m)$ is a d_m-dimensional subspace of A_m. Using the Hahn-Banach Theorem, we can construct a self-adjoint projection $P_m \colon A_m \to \alpha_m(\mathcal{S}_m)$ of norm at most d_m (cf. Lemma B.10). Now define self-adjoint maps $\beta_m \colon A_m \to \mathcal{S}_m \subset A$ by

$$\beta_m = \alpha_m^{-1} \circ P_m.$$

The generalized inductive system we are after is given by the algebras A_m together with the connecting maps $\varphi_{n,m} = \alpha_n \circ \beta_m$. (The compatibility condition is satisfied since $\beta_j \circ \alpha_j|_{\mathcal{S}_j} = \mathrm{id}_{\mathcal{S}_j}$.) Note that

$$\|\varphi_{n,m}(x)\| \le 2\|\beta_m(x)\| \le 4d_m\|x\|$$

for $x \in A_m$ and all $n > m$; hence the first condition in Definition 11.1.1 is satisfied.

For the second condition we fix k, some elements $x, y \in A_k$ from the unit ball, and $\varepsilon > 0$. Then, for $n > m > k$, we have that

$$\begin{aligned}
&\|\varphi_{n,m}\big(\varphi_{m,k}(x)\varphi_{m,k}(y)\big) - \varphi_{n,m}\big(\alpha_m\big(\beta_k(x)\beta_k(y)\big)\big)\| \\
&\quad\le 4d_m\|\alpha_m(\beta_k(x))\alpha_m(\beta_k(y)) - \alpha_m\big(\beta_k(x)\beta_k(y)\big)\| \\
&\quad\le 4d_m(2d_k)^2 d_m^{-4} \\
&\quad\le 16d_m^{-1}
\end{aligned}$$

and that

$$\begin{aligned}
&\|\varphi_{n,m}\big(\alpha_m\big(\beta_k(x)\beta_k(y)\big)\big) - \varphi_{n,k}(x)\varphi_{n,k}(y)\| \\
&\quad= \|\alpha_n\big(\beta_k(x)\beta_k(y)\big) - \alpha_n(\beta_k(x))\alpha_n(\beta_k(y))\| \\
&\quad\le (2d_k)^2 d_n^{-4} \\
&\quad\le 4d_m^{-1}.
\end{aligned}$$

It follows that

$$\|\varphi_{n,m}\big(\varphi_{m,k}(x)\varphi_{m,k}(y)\big) - \varphi_{n,k}(x)\varphi_{n,k}(y)\| \le 20d_m^{-1},$$

which implies we have a generalized inductive system.

Showing that A is isomorphic to the induced limit is reasonably straightforward – one verifies directly that the α_n's give rise to an injective $*$-homomorphism from A onto $\mathrm{g}\varinjlim(A_m, \varphi_{n,m})$. Just string together the α_n's as in the definition of $\mathrm{g}\varinjlim(A_m, \varphi_{n,m})$. Everything is well-defined and isometric on the span of the $\mathcal{S}_n$'s; hence we can extend to an isometry on all of A and this extension has no choice but to be self-adjoint and multiplicative. We leave the details to the reader. □

Definition 11.1.6. A separable C*-algebra is called MF if it satisfies one of the equivalent conditions from Theorem 11.1.5.[3]

Taking the norm-microstate point of view, the following permanence properties are easy.

Proposition 11.1.7. *The following statements hold:*

(1) *Subalgebras of MF algebras are MF.*
(2) *Inductive limits (usual or even generalized) of MF algebras are again MF.*

Proof. Just think about it. □

We close this section with an inductive limit characterization. Let

$$A \subset \frac{\prod_{n\in\mathbb{N}} \mathbb{M}_{s(n)}(\mathbb{C})}{\bigoplus_{n\in\mathbb{N}} \mathbb{M}_{s(n)}(\mathbb{C})}$$

be given and $E \subset \prod \mathbb{M}_{s(n)}(\mathbb{C})$ be the pullback of A. If

$$P_n = 0 \oplus \ldots \oplus 0 \oplus 1_{s(n)} \oplus 1_{s(n+1)} \oplus \cdots \in \prod \mathbb{M}_{s(n)}(\mathbb{C})$$

and if we consider the inductive system $E \to P_1 E \to P_2 E \to \cdots$, it is readily seen that A is the corresponding inductive limit. Hence we have

Proposition 11.1.8. *A* C**-algebra is MF if and only if it is isomorphic to an (old-fashioned) inductive limit of RFD* C**-algebras.*

11.2. NF and strong NF algebras

We now restrict to the case of c.c.p. connecting maps. This turns out to describe an old friend: nuclear QD C*-algebras. Restricting further to connecting maps which are *complete order embeddings* is also interesting.

Definition 11.2.1. A u.c.p. map $\varphi\colon A \to B$ is called a *complete order embedding* if it is completely isometric.

[3] This terminology is due to Blackadar and Kirchberg and abbreviates "matricial field". We haven't presented the characterization which inspires the terminology, however.

Remark 11.2.2. If a u.c.p. map $\varphi\colon A \to B$ is a complete order embedding, then $\varphi^{-1}\colon \varphi(A) \to A$ is also completely positive – i.e., φ is an isomorphism, in the category of operator systems, onto its range (hence the terminology). Indeed, any unital completely contractive map is completely positive ([**63**, Corollary 5.1.2]), so in particular this is true for φ^{-1}.

However, more is true: φ^{-1} extends to a (surjective) $*$-homomorphism $C^*(\varphi(A)) \to A$. To see this, we assume $A \subset \mathbb{B}(\mathcal{H})$ and extend φ^{-1} to a u.c.p. map $B \to \mathbb{B}(\mathcal{H})$. The point is that $\varphi(A)$ is contained in the multiplicative domain of this map, since $1 \geq \varphi^{-1}(\varphi(u)^*\varphi(u)) \geq \varphi^{-1}(\varphi(u)^*)\varphi^{-1}(\varphi(u)) = 1$, for all unitaries $u \in A$ (and similarly with adjoints on the right).

Definition 11.2.3. Let $A = \mathrm{g}\varinjlim(A_m, \varphi_{n,m})$ be a generalized inductive limit where each A_m is finite-dimensional. If the connecting maps $\varphi_{n,m}$ are c.c.p. then we say A is an *NF algebra.*[4] If the $\varphi_{n,m}$'s are all complete order embeddings, then A is *strong NF.*

Though it isn't obvious, NF algebras are always nuclear.

Proposition 11.2.4. *NF algebras are nuclear and QD.*

Proof. First let $A = \mathrm{g}\varinjlim(A_m, \varphi_{n,m})$ be a *strong* NF algebra with connecting maps $\varphi_{n,m}$. If $\Phi_m\colon A_m \to A$ are the canonical maps, then evidently each Φ_m is a complete order embedding of A_m into A. Hence $\Phi_m^{-1}\colon \Phi_m(A_m) \to A_m$ can be extended to a u.c.p. map $A \to A_m$. Evidently this implies A is nuclear and QD.

For the general case, we may assume everything is unital. Indeed, if $A = \mathrm{g}\varinjlim(A_m, \varphi_{n,m})$ and the $\varphi_{n,m}$'s are not unital, then we adjoin new units everywhere and one checks that $\tilde{A} = \mathrm{g}\varinjlim(\tilde{A}_m, \tilde{\varphi}_{n,m})$. Since A is nuclear and QD whenever $\tilde{A}$ is, our reduction is complete.

So, assume $A = \mathrm{g}\varinjlim(A_m, \varphi_{n,m})$ with u.c.p. connecting maps. We have to use the strong NF case handled above. Indeed, we consider the generalized inductive system $(B_m, \psi_{n,m})$ where

$$B_m = A_1 \oplus \cdots \oplus A_m$$

and

$$\psi_{n,m}(a_1 \oplus \cdots \oplus a_m) = a_1 \oplus \cdots \oplus a_m \oplus \varphi_{m+1,m}(a_m) \oplus \cdots \oplus \varphi_{n,m}(a_m).$$

Since the $\psi_{n,m}$'s are complete order embeddings, $B = \mathrm{g}\varinjlim(B_m, \psi_{n,m})$ is nuclear. It is not hard to see that A is a quotient of B since we have a

[4]This abbreviates "nuclear and (stably) finite," even though it isn't known whether or not every nuclear stably finite C*-algebra is NF.

commutative diagram

$$\begin{array}{ccccccccc}
A_1 & \longrightarrow & A_1 \oplus A_2 & \longrightarrow & A_1 \oplus A_2 \oplus A_3 & \longrightarrow & \cdots & \longrightarrow & B \\
\downarrow & & \downarrow & & \downarrow & & & & \downarrow \\
A_1 & \longrightarrow & A_2 & \longrightarrow & A_3 & \longrightarrow & \cdots & \longrightarrow & A.
\end{array}$$

Since nuclearity passes to quotients, this implies that A is nuclear. That nuclear MF algebras are QD follows easily from the Choi-Effros Lifting Theorem (Theorem C.3) and Exercise 7.1.3. □

Now let's prove the converse.

Theorem 11.2.5. *For a separable* C*-*algebra A, the following statements are equivalent:*

(1) *A is an NF algebra;*

(2) *A is nuclear and isomorphic to a subalgebra of*

$$A \subset \frac{\prod_{n\in\mathbb{N}} \mathbb{M}_{s(n)}(\mathbb{C})}{\bigoplus_{n\in\mathbb{N}} \mathbb{M}_{s(n)}(\mathbb{C})}$$

for some sequence $s(n)$;

(3) *A is nuclear and QD;*

(4) *for every finite set $\mathfrak{F} \subset A$ and $\varepsilon > 0$ there exist c.c.p. maps $\alpha\colon A \to \mathbb{M}_n(\mathbb{C})$ and $\beta\colon \mathbb{M}_n(\mathbb{C}) \to A$ such that $\|\beta \circ \alpha(a) - a\| < \varepsilon$ and $\|\alpha(ab) - \alpha(a)\alpha(b)\| < \varepsilon$, for all $a, b \in \mathfrak{F}$. (In other words, the definitions of nuclearity and quasidiagonality can be incorporated into the same c.c.p. maps.)*

Proof. Thanks to the previous proposition, we only have to prove (3) ⇒ (4) and (4) ⇒ (1).

Assume $A \subset \mathbb{B}(\mathcal{H})$ is nuclear, QD and that it contains no nonzero compact operators. Fixing $\mathfrak{F} \subset A$ and $\varepsilon > 0$, we can apply Dadarlat's approximation theorem (Theorem 7.5.7) to find a finite-dimensional algebra $B \subset \mathbb{B}(\mathcal{H})$ which nearly contains $\mathfrak{F}$. We can also find c.c.p. maps $\tilde{\alpha}\colon A \to \mathbb{M}_n(\mathbb{C})$ and $\tilde{\beta}\colon \mathbb{M}_n(\mathbb{C}) \to A$ such that $\|\tilde{\beta} \circ \tilde{\alpha}(a) - a\| < \varepsilon$ for all $a \in \mathfrak{F}$. By Arveson's Extension Theorem we may assume that $\tilde{\alpha}$ is defined on all of $\mathbb{B}(\mathcal{H})$. Let $\varphi\colon \mathbb{B}(\mathcal{H}) \to B$ be a conditional expectation. One easily checks that $\alpha = \varphi|_A\colon A \to B$ and $\beta = \tilde{\beta} \circ \tilde{\alpha}|_B\colon B \to A$ are the desired maps, thus proving (4).

Now assume the approximation property in statement (4) and we will construct the right generalized inductive system. First, find some finite-dimensional self-adjoint subspaces

$$\mathcal{S}_1 \subset \mathcal{S}_2 \subset \cdots \subset A$$

and c.c.p. maps $\alpha_m \colon A \to \mathbb{M}_{s(m)}(\mathbb{C})$, $\beta_m \colon \mathbb{M}_{s(m)}(\mathbb{C}) \to A$ with all of the following properties:

(1) $\|\alpha_m(ab) - \alpha_m(a)\alpha_m(b)\| < \frac{1}{2^m}$ for all $a, b \in \mathcal{S}_m$ with $\|a\|, \|b\| \leq 1$;

(2) $\|\beta_m \circ \alpha_m(a) - a\| < \frac{1}{2^m}$ *and* $\|\beta_m \circ \alpha_m(ab) - ab\| < \frac{1}{2^m}$, for all $a, b \in \mathcal{S}_m$ with $\|a\|, \|b\| \leq 1$;

(3) $\mathcal{S}_{m+1}$ contains both of the following sets:

$$\{ab : a, b \in \mathcal{S}_m\} \text{ and } \{\beta_m(\alpha_m(a))\beta_m(\alpha_m(b)) : a, b \in \mathcal{S}_m\};$$

(4) the union of the $\mathcal{S}_m$'s is dense in A.

Just as in the proof of Lemma 7.5.5, the first two conditions ensure that β_m is almost multiplicative on the unit ball of $\alpha_m(\mathcal{S}_m)$. More precisely, for contractions $a, b \in \mathcal{S}_m$ we have

$$\|\beta_m(\alpha_m(a)\alpha_m(b)) - \beta_m(\alpha_m(a))\beta_m(\alpha_m(b))\| < \frac{1}{2^{m-2}}.$$

The generalized inductive system we are after is given by the matrix algebras $\mathbb{M}_{s(m)}(\mathbb{C})$ and connecting maps defined by

$$\varphi_{m+1,m} = \alpha_{m+1} \circ \beta_m$$

and

$$\varphi_{n,m} = \varphi_{n,n-1} \circ \cdots \circ \varphi_{m+1,m}$$

for all $n > m$. Since β_m is almost multiplicative on $\alpha_m(\mathcal{S}_m)$, our system is asymptotically multiplicative and, hence, an honest generalized inductive system. Indeed, also using the almost-multiplicativity of the α_m's, one finds that $\|\varphi_{n,m}(\varphi_{m,k}(x)\varphi_{m,k}(y)) - \varphi_{n,k}(x)\varphi_{n,k}(y)\|$ is bounded above by

$$\frac{1}{2^{m-2}} + \frac{1}{2^{m+1}} + \|\varphi_{n,m+1}(\varphi_{m+1,k}(x)\varphi_{m+1,k}(y)) - \varphi_{n,k}(x)\varphi_{n,k}(y)\|.$$

Repeating this inequality until we reach n, asymptotic multiplicativity is established.

Showing that A is isomorphic to the resulting generalized inductive limit C*-algebra is similar to Theorem 11.1.5 and will be left to the reader. □

Our next goal is a local characterization of strong NF algebras, but we first need a nontrivial perturbation lemma.

Lemma 11.2.6. *For any finite-dimensional* C**-algebra* A *and* $\varepsilon > 0$, *there exists* $\delta = \delta(A, \varepsilon) > 0$ *such that for every u.c.p. map* φ *from* A *into a finite-dimensional* C**-algebra* B *with*

$$\|\mathrm{id}_A \otimes \varphi^{-1}\| < 1 + \delta,$$

there exists a complete order embedding $\psi \colon A \to B$ *such that* $\|\psi - \varphi\| < \varepsilon$.

Proof. We will choose $\delta > 0$ later. Let $A = \bigoplus_{k=1}^{s} \mathbb{B}(\ell^2_{d(k)}) \subset \mathbb{B}(\ell^2_d)$ with $\bigoplus \ell^2_{d(k)} = \ell^2_d$, let $B = \bigoplus_{l=1}^{t} \mathbb{B}(\ell^2_{n(l)}) \subset \mathbb{B}(\ell^2_n)$ with $\bigoplus \ell^2_{n(l)} = \ell^2_n$, and let $\varphi = \bigoplus \varphi_l$ where $\varphi_l\colon A \to \mathbb{B}(\ell^2_{n(l)})$. (On a first reading, it will help to assume $s = t = 1$.) By Stinespring's Theorem, there are Hilbert spaces $\mathcal{H}_l$ and isometries $V_l\colon \ell^2_{n(l)} \to \ell^2_d \otimes \mathcal{H}_l$ such that

$$\varphi_l(x) = V_l^*(x \otimes 1)V_l$$

for $x \in A$. Let $\{\zeta_i^{(k)}\}_{i=1}^{d(k)}$ be the standard orthonormal basis for $\ell^2_{d(k)}$ and consider the unit vector

$$\xi_k = \frac{1}{\sqrt{d(k)}} \sum_{i=1}^{d(k)} \zeta_i^{(k)} \otimes \zeta_i^{(k)} \in \ell^2_{d(k)} \otimes \ell^2_{d(k)}.$$

If $\{e_{i,j}^{(k)}\}$ are the standard matrix units of $\mathbb{B}(\ell^2_{d(k)})$, then

$$E_k = d(k)^{-1} \sum_{i,j=1}^{d(k)} e_{i,j}^{(k)} \otimes e_{i,j}^{(k)} \in \mathbb{B}(\ell^2_{d(k)} \otimes \ell^2_{d(k)}) \subset A \otimes A$$

is the rank-one orthogonal projection onto $\mathbb{C}\xi_k$. Since

$$1 - \delta < \|(\mathrm{id}_A \otimes \varphi)(E_k)\|,$$

one can find a partition $\{K_l\}_{l=1}^{t}$ of the k's with the property that $k \in K_l$ implies $\|(\mathrm{id}_A \otimes \varphi_l)(E_k)\| > 1 - \delta$. Let $P_l = V_l V_l^*$. Then, for every $k \in K_l$, we have

$$\|(1_{\ell^2_d} \otimes P_l)(E_k \otimes 1_{\mathcal{H}_l})\|^2 = \|(\mathrm{id}_A \otimes \varphi_l)(E_k)\| > 1 - \delta.$$

Hence, there is a unit vector $\eta_k \in \mathcal{H}_l$ such that $\|(1 \otimes P_l)(\xi_k \otimes \eta_k)\|^2 > 1 - \delta$. Let $Q_k \in \mathbb{B}(\ell^2_d \otimes \mathcal{H}_l)$ be the orthogonal projection onto $\ell^2_{d(k)} \otimes \mathbb{C}\eta_k$; clearly the Q_k's are mutually orthogonal. Letting $R_l = \sum_{k \in K_l} Q_k$, we have

$$\begin{aligned}
\|(1 - P_l)R_l\|^2 &\le \|(1 - P_l)R_l\|_{2,\mathrm{Tr}}^2 \\
&= \sum_{k \in K_l} \sum_{i=1}^{d(k)} \|(1 - P_l)(\zeta_i^{(k)} \otimes \eta_k)\|^2 \\
&= \sum_{k \in K_l} \|(1 \otimes (1 - P_l))(\sum_{i=1}^{d(k)} \zeta_i^{(k)} \otimes \zeta_i^{(k)} \otimes \eta_k)\|^2 \\
&= \sum_{k \in K_l} d(k)(1 - \|(1 \otimes P_l)(\xi_k \otimes \eta_k)\|^2) < d\delta.
\end{aligned}$$

It follows that if we choose $\delta > 0$ small enough, then by Lemma 7.2.2 we may find a unitary operator U_l on $\ell^2_d \otimes \mathcal{H}_l$ such that $\|U_l - 1\| \le \varepsilon/2$ and $U_l^* R_l U_l \le P_l$. Defining $\psi = \bigoplus \psi_l$, where $\psi_l(x) = V_l^* U_l^*(x \otimes 1)U_l V_l$, we have that $\|\varphi - \psi\|_{\rm cb} = \max \|\varphi_l - \psi_l\|_{\rm cb} < \varepsilon$. Moreover, ψ is completely isometric since the map $A \to A$ defined by

$$A \ni x \mapsto \bigoplus_l \bigoplus_{k \in K_l} Q_k U_l V_l \psi(x) V_l^* U_l^* Q_k \in \bigoplus_l \bigoplus_{k \in K_l} \mathbb{B}(\ell^2_{d(k)} \otimes \mathbb{C}\eta_k) \cong A$$

can be shown to be a $*$-isomorphism. □

Theorem 11.2.7. *For a unital separable* C*-*algebra A, the following are equivalent:*

(1) *A is strong NF;*

(2) *for every finite set $\mathfrak{F} \subset A$ and $\varepsilon > 0$ there exists a finite-dimensional* C*-*algebra B and u.c.p. maps $\alpha\colon A \to B$, $\beta\colon B \to A$ such that $\|\beta(\alpha(a)) - a\| < \varepsilon$ for all $a \in \mathfrak{F}$ and β is a complete order embedding;*

(3) *for every finite set $\mathfrak{F} \subset A$ and $\varepsilon > 0$ there exist a finite-dimensional* C*-*algebra B and complete order embedding $\beta\colon B \to A$ such that $\mathfrak{F} \subset^{\varepsilon} \beta(B)$ (i.e., for each $a \in \mathfrak{F}$ there exists $b \in B$ such that $\|a - \beta(b)\| < \varepsilon$).*

Proof. Evidently we only have to prove (2) $\Rightarrow$ (1). But, with the previous perturbation result in hand, the proof is very similar to (4) $\Rightarrow$ (1) from Theorem 11.2.5, so we leave the details to the reader. □

While we can certainly give lots of examples of NF algebras (e.g., the cone over any nuclear algebra), we will have to wait until the next section to see examples of strong NF algebras; quite remarkably they are abundant. In the meantime, let's observe that they always admit a nice inductive limit structure.

Proposition 11.2.8. *If A is strong NF, then there exist subalgebras $A_1 \subset A_2 \subset \cdots \subset A$ such that each A_n is nuclear, RFD and the union of the A_n's is dense in A.*

Proof. If $A = \mathrm{g}\varinjlim(B_m, \varphi_{n,m})$ where each B_m is finite-dimensional and the connecting maps are all complete order embeddings, then it is not hard to see that the subalgebra $C^*(\Phi_m(B_m)) \subset A$ is residually finite-dimensional. (Any completely positive extensions of the maps $\Phi_n^{-1}\colon \Phi_n(B_n) \to B_n$, $n > m$, will contain $C^*(\Phi_m(B_m))$ in their multiplicative domains.) Unfortunately there is no reason to believe that $C^*(\Phi_m(B_m))$ is nuclear.

To fix this, we simply enlarge a bit. More precisely, for each k we recursively define finite-dimensional C*-algebras $C_{k+n} \subset B_{k+n}$, $n = 0, 1, \ldots$, by

$$C_k = B_k,\ C_{k+1} = C^*(\varphi_{k+1,k}(C_k)), \ldots, C_{k+n} = C^*(\varphi_{k+n,k+n-1}(C_{k+n-1})), \ldots$$

and use the $\varphi_{n,k}$, $n > k$, for connecting maps. Since we have defined C_{k+i+1} to be the algebra generated by the image of the previous algebra, we always have $*$-homomorphisms $C_{k+i+1} \to C_{k+i}$ which reverse the complete order embedding connecting maps. (This fact will be important below.)

The generalized inductive limit A_k of this sequence is obviously a strong NF (hence nuclear) subalgebra of A for each $k \in \mathbb{N}$. To prove residual finite-dimensionality of A_k, we note that for each n there is a commutative diagram

$$\begin{array}{ccccccccc} C_{k+n} & \longrightarrow & C_{k+n+1} & \longrightarrow & C_{k+n+2} & \longrightarrow & \cdots & \longrightarrow & A_k \\ \downarrow & & \downarrow & & \downarrow & & & & \downarrow \\ C_{k+n} & \longrightarrow & C_{k+n} & \longrightarrow & C_{k+n} & \longrightarrow & \cdots & \longrightarrow & C_{k+n} \end{array}$$

where the bottom arrows are all the identity and the vertical arrows are $*$-homomorphisms obtained by composing the $*$-homomorphisms $C_{k+n+i} \to C_{k+n+i-1} \to \cdots \to C_{k+n}$. This is enough to imply that each A_k is residually finite-dimensional, and it is clear that their union is dense since $\Phi_k(B_k) \subset A_k$. □

11.3. Inner quasidiagonality

This section is devoted to a natural subclass of QD algebras, first introduced by Blackadar and Kirchberg ([**19**]). Though interesting in their own right, we have a specific goal in mind and hence do not present everything that is known about inner QD C*-algebras.

Definition 11.3.1. A separable[5] C*-algebra A is *inner QD* if there exist projections $p_n \in A^{**}$ such that

(1) $\|[p_n, a]\| \to 0$ for all $a \in A \subset A^{**}$,
(2) $\|a\| = \lim \|p_n a p_n\|$ for all $a \in A$ and
(3) p_n is in the *socle*[6] of A^{**}, for every $n \in \mathbb{N}$.

Note that a nonunital algebra is inner QD if and only if its unitization is inner QD (since $(\tilde{A})^{**} = A^{**} \oplus \mathbb{C}$).

Remark 11.3.2. Every inner QD algebra is evidently QD, simply define $\varphi_n\colon A \to p_n A^{**} p_n$ by compression.

Remark 11.3.3. Every RFD C*-algebra A is inner QD, since the central covers $c(\pi) \in \mathcal{Z}(A^{**})$ of finite-dimensional representations always live in the socle of A^{**}.

[5]As usual, this is an assumption of convenience. The interested reader can work out the nonseparable case.

[6]This means $p_n A^{**} p_n$ is finite-dimensional.

At the other end of the spectrum, if A is simple and QD, then it is necessarily inner QD as well. Indeed, we can fix an irreducible representation $A \subset \mathbb{B}(\mathcal{H})$ – which is necessarily faithful – and then the representation theorem (Theorem 7.2.5) provides increasing, asymptotically commuting, finite-rank projections on $\mathcal{H}$.[7] Irreducibility ensures that these projections live in the weak closure of $A \subset \mathbb{B}(\mathcal{H})$ and thus give rise to projections in the socle of A^{**}.

Remark 11.3.4. Assume $p \in A^{**}$ is a projection in the socle and let $c(p)$ be the central cover. Then

$$c(p)A^{**} \cong \mathbb{B}(\mathcal{H}_1) \oplus \cdots \oplus \mathbb{B}(\mathcal{H}_n)$$

for some Hilbert spaces $\mathcal{H}_1, \ldots, \mathcal{H}_n$. That $c(p)A^{**}$ contains no summand of type II or III follows from the existence of minimal projections ($pA^{**}p$ contains minimal projections) and then the result follows from the structure theory of type I von Neumann algebras. (If $p \in L^\infty(X, \mu) \bar{\otimes} \mathbb{B}(\mathcal{H}) \cong L^\infty((X, \mu), \mathbb{B}(\mathcal{H}))$ has cental cover 1 and lives in the socle, then μ must be a finite sum of point masses.)

Though it isn't obvious, the benefit of inner quasidiagonality is that the c.p. maps in the definition of QD can be taken to have large multiplicative domains. This requires a lemma.

Lemma 11.3.5. *Let $p \in A^{**}$ be a projection in the socle and define*

$$A_p = \{a \in A : [a, p] = 0\}.$$

Then p belongs to the weak closure of A_p in A^{**} (i.e., A_p^{**}) and, hence, is the central cover of the $*$-representation $A_p \to pA_p$.

Proof. Fix an identification

$$c(p)A^{**} \cong \mathbb{B}(\mathcal{H}_1) \oplus \cdots \oplus \mathbb{B}(\mathcal{H}_n),$$

where $c(p)$ is the central cover (in A^{**}) of p. Applying (a tiny modification of the proof of) Corollary 1.4.8, we can find a net of self-adjoints $a_i \in A$ such that $a_i \to p$ ultraweakly and $a_i p = p$ for all i. Taking adjoints, it follows that $a_i p = pa_i$ and hence $p \in A_p^{**}$ as claimed.

It is immediate from the definition that p must be the central cover of the homomorphism $A_p \to pA_p$, so the proof is complete. □

Proposition 11.3.6. *Let $p \in A^{**}$ be in the socle and let $A_p = \{a \in A : [a, p] = 0\}$, as above. Then $d(a, A_p) = \|[a, p]\|$ for all $a \in A$.*[8]

---

[7] Note that if A is not isomorphic to $\mathbb{K}$, then the representation $A \subset \mathbb{B}(\mathcal{H})$ is necessarily essential – by simplicity – whereas the case $A = \mathbb{K}$ is trivial.

[8] Note that A_p is the multiplicative domain of the compression map $a \mapsto pap$.

Proof. First we must argue that A_p^{**} is equal to $pA^{**}p + (1-p)A^{**}(1-p)$. The inclusion $A_p^{**} \subset pA^{**}p + (1-p)A^{**}(1-p)$ is immediate.

Since p is the central cover of $A_p \to pA_p$, $1-p$ is open in A^{**} – i.e., there is an increasing net of positive elements $a_i \in A$ such that $a_i \to 1-p$ in the strong operator topology. Hence $(1-p)A(1-p) \cap A$ is weakly dense in $(1-p)A^{**}(1-p)$. Since $(1-p)A(1-p) \cap A \subset A_p$, we have $(1-p)A^{**}(1-p) \subset A_p^{**}$. Proving that $pA^{**}p \subset A_p^{**}$ is virtually identical to the proof of the previous lemma. Hence $A_p^{**} = pA^{**}p + (1-p)A^{**}(1-p)$, as desired.

Having identified the weak closure of A_p, the remainder of the proof is not hard. First, note that $d(a, A_p) = d(a, A_p^{**})$ (by convexity and the Hahn-Banach Theorem). Now define

$$x = pap + (1-p)a(1-p) \in pA^{**}p + (1-p)A^{**}(1-p) = A_p^{**}.$$

A simple calculation shows $\|a - x\| = \|[a,p]\|$ and thus $d(a, A_p) \leq \|[a,p]\|$. On the other hand, if $y \in A_p$, then

$$\|a-y\| \geq \max\{\|(1-p)(a-y)p\|, \|p(a-y)(1-p)\|\} = \|[a-y,p]\| = \|[a,p]\|,$$

which completes the proof. □

Corollary 11.3.7. *The* C*-*algebra* A *is inner QD if and only if there is a sequence of c.c.p. maps* $\varphi_n \colon A \to \mathbb{M}_{k(n)}(\mathbb{C})$ *such that* $\|a\| = \lim \|\varphi_n(a)\|$ *and* $d(a, A_{\varphi_n}) \to 0$ *for all* $a \in A$, *where* A_{φ_n} *is the multiplicative domain of* φ_n.

Proof. The "only if" direction follows from the previous result and Remark 11.3.2.

For the other implication we fix a finite set $\mathfrak{F} \subset A$ and $\varepsilon > 0$. Let $\varphi \colon A \to \mathbb{M}_n(\mathbb{C})$ be a c.c.p. map such that $d(a, A_\varphi) < \varepsilon$ and $\|\varphi(a)\| > \|a\| - \varepsilon$ for all $a \in \mathfrak{F}$. Define $B = \varphi(A_\varphi)$, decompose $B = \mathbb{B}(\mathcal{H}_1) \oplus \cdots \oplus \mathbb{B}(\mathcal{H}_k)$, for finite-dimensional Hilbert spaces $\mathcal{H}_i$, and consider the corresponding decomposition $\varphi|_{A_\varphi} = \pi_1 \oplus \cdots \oplus \pi_k$. Eliminating repetitive maps, if necessary, we may assume that the π_i's are pairwise disjoint representations of A_φ (that is, the corresponding projections in $\mathcal{Z}(A_\varphi^{**})$ are pairwise orthogonal).

Regarding each π_i as an irreducible representation, we can find irreducible representations $\tilde{\pi}_i \colon A \to \mathbb{B}(\tilde{\mathcal{H}}_i)$, $1 \leq i \leq k$, such that $\mathcal{H}_i \subset \tilde{\mathcal{H}}_i$ and if $P_i \colon \tilde{\mathcal{H}}_i \to \mathcal{H}_i$ is the orthogonal projection, then $P_i \tilde{\pi}_i(x) P_i = \pi_i(x)$ for all $x \in A_\varphi$. Sadly the $\tilde{\pi}_i$'s need not be pairwise disjoint, so we assume that $\tilde{\pi}_1, \ldots, \tilde{\pi}_j$ ($j \leq k$) are pairwise disjoint and each $\tilde{\pi}_i$ with $i > j$ is equivalent to one of these. Hence, for each $i > j$ there is a unitary conjugating the representation $\tilde{\pi}_i \colon A \to \mathbb{B}(\tilde{\mathcal{H}}_i)$ over to $\tilde{\pi}_m \colon A \to \mathbb{B}(\tilde{\mathcal{H}}_m)$, for some $m \leq j$, and this allows us to regard the projection P_i as acting on $\tilde{\mathcal{H}}_m$. Note that under this identification, P_i is orthogonal to P_m since the representations π_i and π_m are disjoint.

To finish the proof one considers the representation

$$\tilde{\pi} = \tilde{\pi}_1 \oplus \cdots \oplus \tilde{\pi}_j \colon A \to \mathbb{B}(\tilde{\mathcal{H}}_1 \oplus \cdots \oplus \tilde{\mathcal{H}}_j)$$

and the projection $P \in \mathbb{B}(\tilde{\mathcal{H}}_1 \oplus \cdots \oplus \tilde{\mathcal{H}}_j)$ obtained by adding up $P_1, \ldots, P_k$ (using the identifications $\mathbb{B}(\tilde{\mathcal{H}}_i) \cong \mathbb{B}(\tilde{\mathcal{H}}_m)$ for some $m \leq j$ whenever $i > j$). Since $\tilde{\pi}_1, \ldots, \tilde{\pi}_j$ are pairwise disjoint, $P \in \tilde{\pi}(A)''$. Finally, P commutes with $\tilde{\pi}(A_\varphi)$, compression by P recaptures $\varphi|_{A_\varphi}$, and this easily implies that P almost commutes with $\mathfrak{F}$ and approximately preserves the norms after compression. □

Here is the main technical result. It depends on Smith's lemma (Lemma B.4), which you may want to review before reading the proof.

Proposition 11.3.8. *The unital* C*-*algebra A is inner QD if and only if it satisfies the following approximate factorization property: For every finite set $\mathfrak{F} \subset A$, $\varepsilon > 0$ and every u.c.p. map $\varphi \colon A \to \mathbb{M}_k(\mathbb{C})$, there exist a projection p in the socle of A^{**} and a u.c.p. map $\psi \colon pA^{**}p \to \mathbb{M}_k(\mathbb{C})$ such that*

(1) $\|[a, p]\| < \varepsilon$ *and*

(2) $\|\psi(pap) - \varphi(a)\| < \varepsilon$ *for all* $a \in \mathfrak{F}$.

Proof. The "if" direction is trivial since one can always find u.c.p. maps $\varphi_n \colon A \to \mathbb{M}_{k(n)}(\mathbb{C})$ such that $\|a\| = \lim \|\varphi_n(a)\|$ for all $a \in A$ (just cut by finite-rank projections in some faithful representation).

To prove the "only if" direction, let's take p_n as in Definition 11.3.1 and denote by $\theta_n \colon A \to p_n A p_n$ the corresponding compressions. Since the θ_n's are asymptotically multiplicative and asymptotically isometric, one has

$$\lim_{n\to\infty} \|(\mathrm{id}_{\mathbb{M}_k(\mathbb{C})} \otimes \theta_n)(x)\| = \|x\|_{\mathbb{M}_k(\mathbb{C}) \otimes A}$$

for any fixed $k \in \mathbb{N}$ and $x \in \mathbb{M}_k(\mathbb{C}) \otimes A$. (Why?) Now suppose that a finite-dimensional operator system $E \subset A$, $\varepsilon > 0$ and a u.c.p. map $\varphi \colon A \to \mathbb{M}_k(\mathbb{C})$ are given. Note that $\theta_n|_E$ is a linear isomorphism for sufficiently large n, hence we can let $\Theta_n^{-1} \colon \theta_n(E) \to E$ denote the inverse. It follows from the above equality and Lemma B.4 that

$$\begin{aligned} \limsup_n \|\varphi \circ \Theta_n^{-1}\|_{\mathrm{cb}} &= \limsup_n \|\mathrm{id}_{\mathbb{M}_k(\mathbb{C})} \otimes (\varphi \circ \Theta_n^{-1})\| \\ &\leq \limsup_n \|(\mathrm{id}_{\mathbb{M}_k(\mathbb{C})} \otimes \theta_n|_E)^{-1}\| \\ &\leq 1. \end{aligned}$$

Hence, we can find n such that $\|\varphi \circ \Theta_n^{-1}\|_{\mathrm{cb}} < 1 + \varepsilon/2$. We may assume that $p = p_n$ satisfies condition (1). Since $\varphi \circ \Theta_n^{-1}$ is unital and self-adjoint, it can be perturbed to a u.c.p. map ψ by Corollary B.9. Extending ψ to all of $pA^{**}p$, by Arveson's Extension Theorem, we are done. □

Here is the main theorem of this section.

Theorem 11.3.9. *The unital* C*-*algebra A is strong NF if and only if it's separable unital nuclear and inner QD.*

Proof. If A is strong NF, then it is nuclear (Theorem 11.2.5) and an increasing union of RFD subalgebras (Proposition 11.2.8). Corollary 11.3.7 then implies inner QD because we can extend any homomorphism from an RFD subalgebra to a c.c.p. map on all of A, by Arveson's Extension Theorem.

For the other direction we will appeal to the third condition in Theorem 11.2.7. So fix a finite set $\mathfrak{F} \subset A$, $\varepsilon > 0$ and find some u.c.p. maps $\varphi\colon A \to \mathbb{M}_n(\mathbb{C})$ and $\tilde{\beta}\colon \mathbb{M}_n(\mathbb{C}) \to A$ such that $\|a - \tilde{\beta}(\varphi(a))\| < \varepsilon$ for all $a \in \mathfrak{F}$. Applying Proposition 11.3.8 to φ, we can find a projection p in the socle of A^{**} and a u.c.p. map $\psi\colon pA^{**}p \to \mathbb{M}_n(\mathbb{C})$ such that $\|[a,p]\| < \varepsilon$ and $\|\psi(pap) - \varphi(a)\| < \varepsilon$ for all $a \in \mathfrak{F}$.

Let $A_p = A \cap \{p\}'$ be the multiplicative domain of the compression map $A \to pA^{**}p$, let $J \lhd A_p$ be the kernel of this map (so $pA_p = A_p/J = pA^{**}p$) and let $\{e_i\} \subset J$ be a quasicentral approximate unit. Let $B = pA_p = pA^{**}p$ and $\gamma\colon B \to A_p$ be a u.c.p. splitting for the quotient map $A_p \to B$. It follows that γ is a complete order embedding. Thus the u.c.p. maps $\beta_i\colon B \to A$ defined by

$$\beta_i(b) = (1-e_i)^{\frac{1}{2}}\gamma(b)(1-e_i)^{\frac{1}{2}} + e_i^{\frac{1}{2}}\tilde{\beta}(\psi(b))e_i^{\frac{1}{2}}$$

are also complete order embeddings (since they're also splittings). We must show that for large i, the images of these maps almost contain $\mathfrak{F}$.

If $a \in \mathfrak{F}$, then we can find $c \in A_p$ such that $\|a - c\| < \varepsilon$. By quasicentrality we then have,

$$a \approx c \approx (1-e_i)^{\frac{1}{2}}c(1-e_i)^{\frac{1}{2}} + e_i^{\frac{1}{2}}ce_i^{\frac{1}{2}}.$$

We also have $\|c - \tilde{\beta}(\varphi(c))\| < 3\varepsilon$ which implies that the distance between $(1-e_i)^{\frac{1}{2}}c(1-e_i)^{\frac{1}{2}} + e_i^{\frac{1}{2}}ce_i^{\frac{1}{2}}$ and

$$(1-e_i)^{\frac{1}{2}}\gamma(pc)(1-e_i)^{\frac{1}{2}} + e_i^{\frac{1}{2}}\tilde{\beta}(\psi(pc))e_i^{\frac{1}{2}}$$

will be bounded above by 6ε, for all large i, since $\|\psi(pc) - \varphi(c)\| < 3\varepsilon$. As this latter element lives in the range of β, the proof is complete. □

We conclude with a striking decomposition theorem for simple nuclear QD C*-algebras.

Corollary 11.3.10. *If A is separable unital simple nuclear and QD, then it contains an increasing sequence of nuclear RFD subalgebras whose union is dense.*

Proof. Simple QD algebras are inner QD and hence the assumption of nuclearity implies that A must be strong NF. The conclusion then follows from Proposition 11.2.8. □

11.4. Excision and Popa's technique

Theorem 1.4.10 states that pure states can be excised. In this section we observe that a similar result holds for certain finite-dimensional u.c.p. maps, when the domain algebra has a sufficient supply of projections. We then use Popa's technique to derive some important corollaries.

Definition 11.4.1. Let A be a C*-algebra and $\varphi \in S(A)$ be a state on A. We say that φ *can be excised* if there exists a net of positive, norm-one elements $e_i \in A$ such that $\|e_i a e_i - \varphi(a)e_i^2\| \to 0$ for all $a \in A$.

If one can choose the e_i's to be projections, then we say φ *can be excised by projections.*

Recall that a C*-algebra A has *real rank zero* if every self-adjoint element in A can be approximated by a self-adjoint element with finite spectrum – in particular, real rank zero implies a rich supply of projections. The results of this section hold under weaker hypotheses regarding the abundance of projections, but we'll stick to the real rank-zero context as it is sufficient for most applications. (See [**159**] for more general results.)

Proposition 11.4.2. *Let A be a simple unital infinite-dimensional* C**-algebra with real rank zero. Then every state on A can be excised by projections.*

Proof. First let's observe that every state can be excised. For this, it suffices to show that the pure states are weak-$*$ dense in the state space of A. However, this fact follows easily from Glimm's Lemma (Lemma 1.4.11) and the simplicity of A. Indeed, if $A \subset \mathbb{B}(\mathcal{H})$ is an irreducible representation, then $A \cap \mathbb{K}(\mathcal{H}) = 0$, since A is unital and simple, and thus every state on A can be approximated by vector states (irreducibility implies vector states restrict to pure states on A); hence, every state on A can be excised.

We now use real rank zero to excise by projections. Fix a state φ on A and let e_i be a net of positive, norm-one elements in A such that $\|e_i a e_i - \varphi(a)e_i^2\| \to 0$ for all $a \in A$. Perturbing a little, we may assume that each e_i has finite spectrum and hence we can write

$$e_i = \sum_{j=1}^{k(i)} \alpha_j^{(i)} Q_j^{(i)},$$

where $\{Q_j^{(i)}\}$ are orthogonal projections and $1 = \alpha_1^{(i)} > \alpha_2^{(i)} > \cdots > \alpha_{k(i)}^{(i)} > 0$. The projections $Q_1^{(i)}$ also excise φ; indeed, we have the following inequality:

$$\begin{aligned}\|Q_1^{(i)} a Q_1^{(i)} - \varphi(a) Q_1^{(i)}\| &= \|Q_1^{(i)}(e_i a e_i - \varphi(a) e_i^2) Q_1^{(i)}\| \\ &\leq \|e_i a e_i - \varphi(a) e_i^2\|.\end{aligned}$$

□

Note that a state φ can be excised by projections $\{p_i\}$ if and only if there exists a net of $*$-monomorphisms $\rho_i \colon \mathbb{C} \hookrightarrow A$ (i.e., $\alpha \mapsto \alpha p_i$) such that

$$\|\rho_i(1) a \rho_i(1) - \rho_i(\varphi(a))\| \to 0,$$

for all $a \in A$. This somewhat awkward reformulation suggests how one should generalize to u.c.p. maps with finite-dimensional range. Before stating the result, however, we need a basic perturbation fact: approximate partial isometries with common support and orthogonal ranges can be perturbed to honest partial isometries with common support and orthogonal ranges.

Lemma 11.4.3. *For every $\varepsilon > 0$ and $n \in \mathbb{N}$ there exists $\delta = \delta(\varepsilon, n) > 0$ such that if A is a C^*-algebra, $p \in A$ is a projection and $\{w_1, \ldots, w_n\} \subset A$ are elements such that $\|w_j^* w_i - \delta_{i,j} p\| < \delta$ ($\delta_{i,j}$ is the Kronecker delta function), then there exist partial isometries $\{v_1, \ldots, v_n\} \subset A$ such that $v_j^* v_i = \delta_{i,j} p$ and $\|w_i - v_i\| < \varepsilon$, for $1 \leq i, j \leq n$.*

Proof. The proof is by induction so consider the case $n = 1$. Assume we have $w \in A$ such that $\|w^* w - p\| < \delta < 1$. Then $\|p w^* w p - p\| \leq \|w^* w - p\| < \delta$ and hence $p w^* w p$ is a positive invertible element in pAp. Applying functional calculus to $p w^* w p$ with $f(x) = x^{-1/2}$, we can find $0 \leq y \in pAp$ such that $y p w^* w p y = p$ and $\|y - p\| \leq \frac{1}{\sqrt{1-\delta}} - 1$. Defining $v = wpy$, we have

$$\begin{aligned}\|wpy - w\| &= \|wp(y - p) + (wp - w)\| \\ &\leq \|w\|\left(\frac{1}{\sqrt{1-\delta}} - 1\right) + \|(p-1) w^* w (p-1)\|^{1/2}.\end{aligned}$$

Since $\|(p-1) w^* w (p-1)\|^{1/2} = \|(p-1)(w^* w - p)(p-1)\|^{1/2} < \delta^{1/2}$ and $\|w\| \leq \sqrt{1+\delta}$, the right hand side of the inequality above goes to zero as $\delta \to 0$, and this evidently proves the $n = 1$ case of the lemma.

Now assume the lemma holds for $n - 1$, let $\varepsilon > 0$ be given and let $\{w_1, \ldots, w_n\} \subset A$ be such that $\|w_j^* w_i - \delta_{i,j} p\| < \delta$. If δ is sufficiently small, we can apply the induction hypothesis to find partial isometries $v_1, \ldots, v_{n-1}$

such that $\|w_i - v_i\| \leq \delta'$ (where δ' will be determined later) and $v_j^* v_i = \delta_{i,j} p$, for $1 \leq i, j \leq n-1$. Letting $Q = \sum_1^{n-1} v_i v_i^*$, one sees that

$$\begin{aligned}\|(1-Q)w_n - w_n\| &\leq \sum_1^{n-1} \|v_i^* w_n\| \leq \sum_1^{n-1} (\|v_i^* - w_i^*\| \|w_n\| + \delta) \\ &\leq (n-1)(\delta' \sqrt{1+\delta} + \delta).\end{aligned}$$

The remainder of the proof is bookkeeping, nothing more; the point is that $(1-Q)w_n$ is almost a partial isometry with support p (since it's close to w_n), so we can perturb it to an honest partial isometry with support p and range orthogonal to Q, by the (proof of the) $n=1$ case. We leave the details to the reader. □

We are now ready to excise certain u.c.p. maps. The excisable maps arise as compressions in the GNS constructions of nice states. More precisely, let $\varphi \in S(A)$ be a state which can be excised by projections and $\pi_\varphi \colon A \to \mathbb{B}(\mathcal{H}_\varphi)$ be the corresponding GNS representation. Let $\{y_i\}_{i=1}^m \subset A$ be such that $\varphi(y_j^* y_i) = \delta_{i,j}$ (i.e., in $\mathcal{H}_\varphi$, $\{\hat{y}_i\}_{i=1}^m$ is an orthonormal set of vectors), let $P \in \mathbb{B}(\mathcal{H}_\varphi)$ be the orthogonal projection onto the span of $\{\hat{y}_i\}_{i=1}^m$ and define a u.c.p. map $\Phi \colon A \to P\mathbb{B}(\mathcal{H}_\varphi)P$ by $\Phi(a) = P\pi_\varphi(a)P$.

Theorem 11.4.4 (Popa's local quantization). *With notation as above, the u.c.p. map Φ can be excised, meaning there exists a net of $*$-monomorphisms $\rho_i \colon P\mathbb{B}(\mathcal{H}_\varphi)P \hookrightarrow A$ such that*

$$\|\rho_i(P) a \rho_i(P) - \rho_i(\Phi(a))\| \to 0,$$

for all $a \in A$.

Proof. Let $\mathfrak{F} \subset A$ be a finite set of norm-one elements which contains the unit of A, and let $\varepsilon > 0$ be arbitrary. We must prove the existence of a $*$-monomorphism $\rho \colon P\mathbb{B}(\mathcal{H}_\varphi)P \hookrightarrow A$ such that $\|\rho(P)a\rho(P) - \rho(\Phi(a))\| < \varepsilon$, for all $a \in \mathfrak{F}$.

Since φ can be excised by projections, we can, for any $\delta > 0$, find a projection $p \in A$ such that $\|p(y_j^* x y_i)p - \varphi(y_j^* x y_i)p\| < \delta$ for all $x \in \mathfrak{F}$ and $1 \leq i, j \leq m$. In particular, note that $\|(y_j p)^*(y_i p) - \delta_{i,j} p\| < \delta$ for all $1 \leq i, j \leq m$.

In other words, $\{y_i p\}_{i=1}^m$ is almost a set of partial isometries with orthogonal ranges and common support p. Thus, if δ is sufficiently small, we can perturb them to honest partial isometries $\{v_i\}$ such that $v_j^* v_i = \delta_{i,j} p$ and $\|v_i - y_i p\| < \varepsilon/6m^2$ (δ depends on m and ε; however we may also assume that $\delta < \varepsilon/2m^2$). Hence if we define $f_{i,j} = v_i v_j^*$, then $\{f_{i,j}\}$ is a set of matrix units for an $m \times m$-matrix algebra. Moreover, if we let $q = \sum f_{i,i}$ be the unit of this matrix algebra, then cutting an element $x \in \mathfrak{F}$, we have

$$\begin{aligned} qxq &= \Big(\sum_{i=1}^{m} f_{i,i}\Big)x\Big(\sum_{j=1}^{m} f_{j,j}\Big) \\ &= \sum_{i,j=1}^{m} v_i v_i^* x v_j v_j^* \\ &\approx \sum_{i,j=1}^{m} v_i (y_i p)^* x (y_j p) v_j^* \\ &\approx \sum_{i,j=1}^{m} v_i (\varphi(y_i^* x y_j) p) v_j^* \\ &= \sum_{i,j=1}^{m} \varphi(y_i^* x y_j) f_{i,j}. \end{aligned}$$

In both of the approximations above, the norm difference is less than $\varepsilon/2$; hence

$$\|qxq - \sum_{i,j=1}^{m} \varphi(y_i^* x y_j) f_{i,j}\| < \varepsilon.$$

The only thing left to notice is that the matrix of $P\pi_\varphi(x)P$ with respect to the orthonormal basis $\{\hat{y}_i\}_{i=1}^m$ is just $[\langle \pi_\varphi(x)\hat{y}_j, \hat{y}_i\rangle]_{i,j} = [\varphi(y_i^* x y_j)]_{i,j}$. Hence we can identify $P\mathbb{B}(\mathcal{H}_\varphi)P$ with the matrix algebra $C^*(\{f_{i,j}\})$ in such a way that $P\pi_\varphi(x)P \mapsto \sum_{i,j=1}^m \varphi(y_i^* x y_j) f_{i,j}$, for all $x \in A$. This completes the proof. □

With a tiny approximation argument, the theorem above is easily seen to hold for arbitrary finite-rank projections P.

Remark 11.4.5 (Commutator estimates). An extremely important aspect of the result above is that the projections $\rho_i(P)$ almost commute with elements in A that almost commute with P. More precisely, we have

$$\|[u, \rho_i(P)]\|^2 \leq \|[P, \pi_\varphi(u)]\|^2 + 2\|\rho_i(P)u\rho_i(P) - \rho_i(\Phi(u))\|$$

and if φ is a trace,

$$\|[u, \rho_i(P)]\|_{2,\varphi}^2 \leq \|\rho_i(P)\|_{2,\varphi}^2 \left(\frac{\|[\pi_\varphi(u), P]\|_2^2}{\|P\|_2^2} + 4\|\rho_i(P)u\rho_i(P) - \rho_i(\Phi(u))\| \right),$$

for every unitary $u \in A$ (as usual, $\|x\|_{2,\varphi}^2 = \varphi(x^*x)$, and $\|\cdot\|_2$ is the Hilbert-Schmidt norm).

To prove the first inequality, we let $q = \rho_i(P)$, $u \in A$ be a unitary and compute

$$\begin{aligned}
\|[q,u]\|^2 &= \|quq^\perp - q^\perp uq\|^2\\
&= \max\{\|q^\perp u^*q\|^2, \|q^\perp uq\|^2\}\\
&= \max\{\|quq^\perp u^*q\|, \|qu^*q^\perp uq\|\}\\
&= \max\{\|q - ququ^*q\|, \|q - qu^*quq\|\}\\
&\leq \max\{\|q - \rho_i(\Phi(u)\Phi(u^*))\|, \|q - \rho_i(\Phi(u^*)\Phi(u))\|\}\\
&\quad + 2\|quq - \rho_i(\Phi(u))\|\\
&= \max\{\|P - P\pi_\varphi(u)P\pi_\varphi(u^*)P\|, \|P - P\pi_\varphi(u^*)P\pi_\varphi(u)P\|\}\\
&\quad + 2\|quq - \rho_i(\Phi(u))\|\\
&= \|[P,\pi_\varphi(u)]\|^2 + 2\|quq - \rho_i(\Phi(u))\|.
\end{aligned}$$

When φ is a trace, $\frac{1}{\varphi(q)}\varphi \circ \rho_i$ is the tracial state on $\mathbb{B}(P\mathcal{H}_\varphi)$ and the proof of the second inequality is similar:

$$\begin{aligned}
\|[q,u]\|_{2,\varphi}^2 &= \|q^\perp u^*q\|_{2,\varphi}^2 + \|q^\perp uq\|_{2,\varphi}^2\\
&= \varphi\big(q - \rho_i(\Phi(u)\Phi(u^*)) + q - \rho_i(\Phi(u^*)\Phi(u))\big)\\
&\quad + \varphi\big(\rho_i(\Phi(u)\Phi(u^*)) - ququ^*q + \rho_i(\Phi(u^*)\Phi(u)) - qu^*quq\big)\\
&= \varphi(q)\Big(\operatorname{tr}\big(P - P\pi_\varphi(u)P\pi_\varphi(u^*)P + P - P\pi_\varphi(u^*)P\pi_\varphi(u)P\big)\Big)\\
&\quad + \varphi\big(\rho_i(\Phi(u)\Phi(u^*)) - ququ^*q + \rho_i(\Phi(u^*)\Phi(u)) - qu^*quq\big)\\
&\leq \|q\|_{2,\varphi}^2\left(\frac{\|[\pi_\varphi(u),P]\|_2^2}{\|P\|_2^2} + 4\|quq - \rho_i(\Phi(u))\|\right),
\end{aligned}$$

where the last inequality uses the general Hölder-type inequality $|\varphi(ab)| \leq \varphi(|a|)\|b\|$.

Theorem 11.4.6. *Let A be simple unital quasidiagonal and let it have real rank zero. For every finite set $\mathfrak{F} \subset A$ and $\varepsilon > 0$ there exists a finite-dimensional subalgebra $B \subset A$ with unit Q such that $\|[a,Q]\| < \varepsilon$ and $d(QaQ, B) < \varepsilon$, for all $a \in \mathfrak{F}$.*

Proof. Let φ be any state on A. Since A is simple and unital, we have that the GNS representation π_φ is faithful and its image contains no nonzero compact operators. Since A is quasidiagonal, we can apply Theorem 7.2.5 to find a sequence of nonzero finite-rank projections $P_n \in \mathbb{B}(\mathcal{H}_\varphi)$ such that $\|[P_n, \pi_\varphi(a)]\| \to 0$ for all $a \in A$.

By Proposition 11.4.2, φ can be excised by projections; hence we can apply Popa's local quantization technique to excise the maps $a \mapsto P_n\pi_\varphi(a)P_n$. Since the P_n's asymptotically commute, the commutator estimates evidently imply the desired approximation property. □

A II_1-factor M is said to be *approximately finite-dimensional (AFD)* if the following holds: For every finite set $\mathfrak{F} \subset M$ there exists a finite-dimensional subalgebra $B \subset M$ such that $\mathfrak{F} \subset^{\varepsilon,2} B$ (i.e., for each $x \in \mathfrak{F}$ there exists $b \in B$ such that $\|x - b\|_2 < \varepsilon$, where $\|y\|_2^2 = \tau(y^*y)$ and τ is the trace on M).

Lemma 11.4.7. *Let M be an injective II_1-factor. Then, for every finite set $\mathfrak{F} \subset M$ and $\varepsilon > 0$ there exists a finite-dimensional subalgebra $B \subset M$ with unit q such that $\|[a, q]\|_2 < \varepsilon\|q\|_2$ and $d_2(qaq, B) < \varepsilon\|q\|_2$, for all $a \in \mathfrak{F}$.*

Proof. Let $M \subset \mathbb{B}(L^2(M))$ be in standard form. Since there is a conditional expectation $\mathbb{B}(L^2(M)) \to M$, the trace τ on M is amenable (Definition 6.2.1). By Theorem 6.2.7 and Voiculescu's Theorem, there exists a net of finite-rank projections $P_i \in \mathbb{B}(L^2(M))$ such that

$$\frac{\|[x, P_i]\|_2}{\|P_i\|_2} \to 0$$

and

$$\frac{\mathrm{Tr}(xP_i)}{\mathrm{Tr}(P_i)} \to \tau(x),$$

for all $x \in M$ (Exercise 6.2.5). Thus, as in the proof of the last theorem, Popa's technique together with the commutator estimates finish off the proof. □

Theorem 11.4.8. *Injective II_1-factors are AFD.*

Proof. Let M be an injective II_1-factor. As is easily verified, the previous lemma implies that M enjoys the following approximation property: For each finite set $\mathfrak{F} \subset M$ and $\varepsilon > 0$ there exists a finite-dimensional matrix algebra $B \subset M$ with unit $q \neq 0$ such that

$$\|E_B(qxq) - (x - q^\perp x q^\perp)\|_2^2 \leq \varepsilon^2\|q\|_2^2,$$

for all $x \in \mathfrak{F}$, where $E_B \colon qMq \to B$ is the unique trace-preserving conditional expectation. It is important to note that every corner of M is also injective (since $x \mapsto pxp$ defines a conditional expectation $M \to pMp$); hence it enjoys the same approximation property.

Now fix a finite set $\mathfrak{F} \subset M$, $\varepsilon > 0$ and consider the set $\mathcal{S}$ of all families of matrix subalgebras $\{B_i\}_{i \in I}$ of M such that their units q_i are mutually orthogonal and

$$\|E_{\prod_i B_i}(qxq) - (x - q^\perp x q^\perp)\|_2^2 \leq \varepsilon^2\|q\|_2^2,$$

where $q = \sum_i q_i$. The set $\mathcal{S}$ is nonempty, partially ordered by inclusion and it is easily seen that every increasing sequence in $\mathcal{S}$ has a least upper bound in $\mathcal{S}$ (by continuity, with respect to weak limits, of the quantities appearing in the relevant inequality). Hence there is a maximal family $\{C_i\}_{i\in I} \in \mathcal{S}$, with units $\{q_i\}$.

We show by contradiction that $q = \sum_i q_i = 1_M$. So, assume not and let $p = 1 - q \in M$. Since pMp is injective, we can find a matrix subalgebra $B \subset pMp$ with unit $e \neq 0$ such that

$$\|E_B(exe) - \big(pxp - (p-e)x(p-e)\big)\|_2^2 \leq \varepsilon^2 \|e\|_2^2.$$

To save space, we let $D = B \oplus (\prod_i C_i)$. Writing everything in matrix form with respect to the decomposition $1 = q + e + (p - e)$, it is routine to verify that

$$\begin{aligned}\|E_D((e+q)x(e+q)) - &\big(x - (p-e)x(p-e)\big)\|_2^2 \\ &= \|E_B(exe) - \big(pxp - (p-e)x(p-e)\big)\|_2^2 \\ &\quad + \|E_{\prod_i C_i}(qxq) - \big(x - (1-q)x(1-q)\big)\|_2^2 \\ &\leq \varepsilon^2\|e\|_2^2 + \varepsilon^2\|q\|_2^2 \\ &= \varepsilon^2\|e+q\|_2^2.\end{aligned}$$

Of course, this contradicts maximality of the family $\{C_i\}$, so $q = 1_M$.

Since $\sum_{i\in I}\tau(q_i) = 1$, we can find a finite subset $I_0 \subset I$ such that $\tau(q_0) > 1 - \varepsilon$ for $q_0 = \sum_{i\in I_0} q_i$. Finally, observe that $C = \mathbb{C}q_0^{\perp} \oplus \bigoplus_{i\in I_0} C_i$ is a finite-dimensional von Neumann subalgebra in M which almost contains $\mathfrak{F}$. This proves that M is AFD. □

Remark 11.4.9. For a while there was speculation that QD C*-algebras might always be nuclear. (There is even a "proof" of the false statement "QD implies exact" in the literature.) A less ambitious question asks whether every simple unital QD C*-algebra with real rank zero is nuclear; Popa's approximation property in Theorem 11.4.6 was viewed as evidence in favor of this due to its similarity with the approximation property characterizing the AFD II$_1$-factor (Lemma 11.4.7). It turns out, however, that quasidiagonality and nuclearity are completely unrelated, even in the presence of simplicity, real rank zero and almost any other assumptions one would like to make. The original counterexamples were constructed by Dadarlat ([**50**]); variations and refinements can be found in [**30**].

Also, we must remark that when the finite-dimensional algebras from Theorem 11.4.6 can be taken "large in trace", the algebra becomes amenable to classification. This fact was exploited in [**65**] and led Huaxin Lin to define tracially AF algebras ([**116**]) and classify them ([**117**]).

11.5. Connes's uniqueness theorem

We close this chapter with the celebrated uniqueness theorem for injective II_1-factors. Having reduced the problem to the AFD case in the previous section, we only have to present Murray and von Neumann's uniqueness theorem for AFD II_1-factors. The main ingredients in the proof are some nontrivial finite-dimensional perturbation lemmas. However, these preliminaries are well known and readily accessible in existing books ([**185**, Lemmas XIV.2.8 and XIV.2.10] or [**95**, Lemmas 12.2.3 – 12.2.6]), so we only sketch the main points. (The omitted details also make very good exercises.)

Lemma 11.5.1. *Let $\mathbb{M}_n(\mathbb{C}) \subset M$ be a unital inclusion, where M is a II_1-factor, and let $\{e_{i,j}\}$ be matrix units for $\mathbb{M}_n(\mathbb{C})$. For every $\varepsilon > 0$ there exists a $\delta = \delta(n, \varepsilon) > 0$ with the following property: If $1_M \in N \subset M$ is a subfactor of type I_{np} (i.e., $N = \mathbb{M}_{np}(\mathbb{C})$) with the property that for each i, j there is a contraction $n_{i,j} \in N$ such that $\|e_{i,j} - n_{i,j}\|_2 < \delta$, then there exists a unitary $u \in M$ such that $\|u - 1\|_2 < \varepsilon$ and $\mathbb{M}_n(\mathbb{C}) \subset u^*Nu$.*

Proof. The first step is to prove a version of this for orthogonal projections. More precisely, one shows that if $p, q \in M$ are orthogonal projections, $v \in M$ is a partial isometry with support p and range q, and $N \subset M$ is a type I subfactor which almost contains p, q and v (in 2-norm), then there is a partial isometry $s \in N$ such that $\|s^*s - p\|_2$, $\|ss^* - q\|_2$ and $\|s - v\|_2$ are all small.

With the projection version in hand one constructs matrices. That is, under the hypotheses of the lemma, we can recursively apply the projection result to find orthogonal projections $F_{i,i} \in N$ $(1 \le i \le n)$ and partial isometries $V_{i+1,i} \in N$ such that $\|F_{i,i} - e_{i,i}\|_2$ and $\|V_{i+1,i} - e_{i+1,i}\|_2$ are small and $V_{i+1,i}$ has support $F_{i,i}$ and range $F_{i+1,i+1}$. Sadly, the $n \times n$ matrix algebra generated by the $F_{i,i}$'s and $V_{i+1,i}$'s need not contain the unit of M, but this is easy to fix. Indeed, since N is of type I_{np} and $\tau(F_{i,i}) = \tau(F_{j,j}) = \frac{k}{np}$ for some $k \le p$, it follows that $\tau(1 - \sum_i F_{i,i}) = \frac{n(p-k)}{np}$; hence, $1 - \sum_i F_{i,i}$ is also the sum of n orthogonal, mutually equivalent projections (of trace $\frac{p-k}{np}$). Adding these projections and the corresponding partial isometries to the $F_{i,i}$'s and $V_{i+1,i}$'s gives a unital $n \times n$ matrix subalgebra of N which nearly contains the original copy of $\mathbb{M}_n(\mathbb{C}) \subset M$.

The upshot of the previous paragraph is that when attacking the general case, we may assume that N is also an $n \times n$ matrix algebra with matrix units $\{n_{i,j}\}$ that are close to the matrix units $\{e_{i,j}\}$. Constructing the right unitary is again piecemeal, via projection considerations. To start, if $v_{1,1}$ is the partial isometry in the polar decomposition of $n_{1,1}e_{1,1}$ and $w_{1,1}$ is any partial isometry with support $e_{1,1} - v_{1,1}^*v_{1,1}$ and range $n_{1,1} - v_{1,1}v_{1,1}^*$ (which exists, since $e_{1,1}$ and $n_{1,1}$ are equivalent), then $u_1 := v_{1,1} + w_{1,1}$ is a partial isometry with support $e_{1,1}$ and range $n_{1,1}$; moreover, it can be shown that $\|u_1 - e_{1,1}\|_2$ is small. Now, defining $u_i = n_{i,1}u_1e_{1,i}$, we get partial isometries from $e_{i,i}$ to $n_{i,i}$ with the property that $\|u_i - e_{i,i}\|_2$ is small. Hence, $u := \sum_i u_i$ is the unitary we want. □

Lemma 11.5.2. *For any finite-dimensional subalgebra C in a* II_1*-factor M and $\varepsilon > 0$ there exist an $n \in \mathbb{N}$ and a unital inclusion $\mathbb{M}_{2^n}(\mathbb{C}) \subset M$ such that for each $x \in C$ there exists $y \in \mathbb{M}_{2^n}(\mathbb{C})$ with $\|x - y\|_2 \leq \varepsilon\|x\|_2$.*

Proof. We may assume $1_M \in C$. Let $\{p_i\}_{1\leq i\leq k}$ be the minimal projections in the center of C (so $C = p_1C \oplus \cdots \oplus p_kC$, where each p_iC is a full matrix algebra) and let $\{e_{p,q}(i)\}$ be matrix units for p_iC. Assume first that there is an n with the property that for each $1 \leq i \leq k$ there is an integer s_i such that $\tau_M(e_{1,1}(i)) = \frac{s_i}{2^n}$. In this case standard manipulations, using the fact that equivalence of projections is determined by the trace, produce a unital inclusion $\mathbb{M}_{2^n}(\mathbb{C}) \subset M$ such that $C \subset \mathbb{M}_{2^n}(\mathbb{C})$.

The general case is conceptually similar: Approximate the traces of the $e_{1,1}(i)'s$ by appropriate dyadic rationals and then construct the approximating copy of $\mathbb{M}_{2^n}(\mathbb{C}) \subset M$. □

If M happens to be AFD in the previous lemma, then an improvement can be made: Any finite-dimensional sub*space* is almost contained in $\mathbb{M}_{2^n}(\mathbb{C}) \subset M$ (since they can first be approximated by finite-dimensional subalgebras in this case).

Lemma 11.5.3. *Let R be a separable AFD* II_1*-factor. For any finite-dimensional subspace $E \subset R$, $\varepsilon > 0$ and subfactor $1_R \in N \subset R$ of type* I_{2^n}*, there exists a subfactor $M \subset R$ of type* I_{2^m} *($m > n$) such that $N \subset M$ and for each $x \in E$ there exists $y \in M$ with $\|x - y\|_2 \leq \varepsilon\|x\|$.*

Proof. Using the previous lemma and the fact that R is AFD, we can find some large m and a unital inclusion $\mathbb{M}_{2^m}(\mathbb{C}) \subset R$ such that both E and N are nearly contained in $\mathbb{M}_{2^m}(\mathbb{C})$ (in 2-norm). Now apply Lemma 11.5.1 to find a unitary $u \in R$ which is close to 1 and such that $N \subset u\mathbb{M}_{2^m}(\mathbb{C})u^*$. □

Theorem 11.5.4. *Let R be a separable AFD* II_1*-factor and $M_{2^\infty} = \mathbb{M}_2(\mathbb{C}) \otimes \mathbb{M}_2(\mathbb{C}) \otimes \cdots$ be the UHF algebra of type 2^∞. Then $R \cong \pi_\tau(M_{2^\infty})''$, where τ is the unique trace on M_{2^∞} and π_τ is the GNS representation.*

In particular, all separable injective II_1*-factors are isomorphic.*

Proof. Applying Lemma 11.5.3 over and over, it is easy to see that R contains a weakly dense copy of M_{2^∞}. Since M_{2^∞} has a unique tracial state, uniqueness of GNS representations implies that $R \cong \pi_\tau(M_{2^\infty})''$, as desired. □

11.6. References

All the results in the first three sections come from [**18**] and [**19**]. Theorem 11.4.6 comes from [**159**], though our proof follows [**29**]. Theorem 11.4.8 is due to Connes ([**41**]), though the proof presented here is due to Popa (see [**155**]). Finally, the first part of Theorem 11.5.4 is due to Murray and von Neumann ([**128**]).

Chapter 12

Approximation Properties for Groups

This chapter contains a number of other approximation properties, as well as a fundamental notion of Kazhdan which typically prevents nice approximations from existing. Truth be told, we barely scratch the surface of any of these topics, but we hope our survey will help the reader tackle the literature.

12.1. Kazhdan's property (T)

Kazhdan's property (T) is extremely useful for constructing counterexamples and proving certain isomorphism theorems (among other things). Our goal in this section is quite modest: Discuss the basics and prove the very first application of this concept to operator algebras (Theorem 12.1.19).

Introduction to (relative) property (T).

Definition 12.1.1. Let Γ be a group and $(\pi, \mathcal{H})$ be a unitary representation of Γ. A vector $\xi \in \mathcal{H}$ is *Γ-invariant* if $\pi(s)\xi = \xi$ for all $s \in \Gamma$. A net (ξ_n) of unit vectors is *almost Γ-invariant* if $\lim \|\pi(s)\xi_n - \xi_n\| = 0$ for every $s \in \Gamma$. If $E \subset \Gamma$ and $\varepsilon > 0$, we say a (nonzero) vector $\xi \in \mathcal{H}$ is *(E, ε)-invariant* if

$$\sup_{s \in E} \|\pi(s)\xi - \xi\| < \varepsilon \|\xi\|.$$

Note that a unitary representation π has almost Γ-invariant vectors if and only if there exists a nonzero (E, ε)-invariant vector for any finite subset $E \subset \Gamma$ and $\varepsilon > 0$.[1]

Definition 12.1.2. Let $\Lambda \subset \Gamma$ be a subgroup. We say the inclusion $\Lambda \subset \Gamma$ has *relative property* (T) if any unitary representation $(\pi, \mathcal{H})$ of Γ which has almost Γ-invariant vectors has a nonzero Λ-invariant vector. We say Γ has *Kazhdan's property* (T) if the identity inclusion $\Gamma \subset \Gamma$ has relative property (T). A pair (E, κ), where $E \subset \Gamma$ and $\kappa > 0$, is called a *Kazhdan pair* for the inclusion $\Lambda \subset \Gamma$ (or for Γ, when $\Lambda = \Gamma$) if any unitary representation of Γ which has a nonzero (E, κ)-invariant vector has a nonzero Λ-invariant vector.

Kazhdan's motivation for considering property (T) was the following two facts: A group with property (T) is finitely generated (Corollary 6.4.7) and has finite abelianization (as explained below).

Example 12.1.3. Finite groups have property (T). It is a simple matter to prove that an amenable group with property (T) is finite (amenability provides almost invariant vectors in the left regular representation, so property (T) yields an honest invariant vector – which is impossible for an infinite group). Thus an amenable group has property (T) if and only if it is finite. Since property (T) clearly passes to quotients, it follows that the abelianization $\Gamma/[\Gamma, \Gamma]$ of a property (T) group is always finite.

The mere existence of an infinite group with property (T) is surprising, but it turns out that examples are abundant.

Example 12.1.4. Lattices in higher rank semisimple Lie groups, as well as lattices in $\mathrm{Sp}(1, n)$, have property (T). (See [**15**] – we will prove the case of $\mathrm{SL}(3, \mathbb{Z})$ shortly.)

Lemma 12.1.5. *For any group* Γ*, the pair* $(\Gamma, \sqrt{2})$ *is Kazhdan.*

Proof. Let a unitary representation $(\pi, \mathcal{H})$ and a nonzero $(\Gamma, \sqrt{2})$-invariant vector ξ be given. Then, the unique circumcenter ζ of the subset $\pi(\Gamma)\xi$ (see Exercise D.1) is Γ-invariant since $\pi(\Gamma)\xi$ is globally Γ-invariant. Letting $\Re$ denote the real part of a complex number, we have

$$\Re\langle \xi, \zeta \rangle \geq \inf_{s \in \Gamma} \Re\langle \xi, \pi(s)\xi \rangle = 1 - \frac{1}{2} \sup_{s \in \Gamma} \|\xi - \pi(s)\xi\|^2 > 0$$

and hence $\zeta \neq 0$. □

[1]In terms of the Fell topology ([**15**]), π has a nonzero invariant vector if and only if the trivial representation is contained in π, and π has almost invariant vectors if and only if the trivial representation is weakly contained in π.

Proposition 12.1.6. *Let Γ be a group, $\Lambda \lhd \Gamma$ be a normal subgroup and (E, κ) be a Kazhdan pair for the inclusion $\Lambda \subset \Gamma$. Let $(\pi, \mathcal{H})$ be a unitary representation of Γ and denote by P the orthogonal projection from $\mathcal{H}$ onto the subspace of all Λ-invariant vectors. Then, for every $\xi \in \mathcal{H}$, one has*

$$\|\xi - P\xi\| \leq \kappa^{-1} \sup_{s \in E} \|\pi(s)\xi - \xi\|.$$

Proof. Let $\xi_1 = \xi - P\xi$. Since Λ is normal in Γ, the subspace $P\mathcal{H}$ is globally Γ-invariant, and hence its orthogonal complement is too. Thus π restricts to a unitary representation π_1 on $\mathcal{H}_1 = \mathcal{H} \ominus P\mathcal{H}$. Since π_1 has no nonzero Λ-invariant vector, the assumption on (E, κ) implies that

$$\sup_{s \in E} \|\pi(s)\xi - \xi\| = \sup_{s \in E} \|\pi_1(s)\xi_1 - \xi_1\| \geq \kappa\|\xi_1\|,$$

as desired. □

Here are a few characterizations of relative property (T). (See Appendix D for more on 1-cocycles and isometric actions.)

Theorem 12.1.7. *Let Γ be a countable group and $\Lambda \subset \Gamma$ be a subgroup. The following are equivalent:*

(1) *the inclusion $\Lambda \subset \Gamma$ has relative property* (T)*;*

(2) *there exist a finite subset $E \subset \Gamma$ and $\kappa > 0$ with the following property:*[2] *If $(\pi, \mathcal{H})$ is a unitary representation of Γ and P is the orthogonal projection from $\mathcal{H}$ onto the subspace of all Λ-invariant vectors, then, for every $\xi \in \mathcal{H}$, one has*
$$\|\xi - P\xi\| \leq \kappa^{-1} \sup_{s \in E} \|\pi(s)\xi - \xi\|;$$

(3) *any sequence of positive definite functions on Γ that converges pointwise to the constant function 1 converges uniformly on Λ;*

(4) *every 1-cocycle $b\colon \Gamma \to \mathcal{H}$ of Γ is bounded on Λ;*

(5) *every action of Γ by affine isometries on a (real) Hilbert space has a Λ-fixed point.*

Moreover, if $\Lambda = \Gamma$, then the above conditions are equivalent to

(6) *the group Γ is finitely generated and for any generating subset $\mathcal{S} \subset \Gamma$, there exists $\kappa = \kappa(\Gamma, \mathcal{S}) > 0$ such that $(\mathcal{S}, \kappa)$ is a Kazhdan pair.*

Proof. (1) $\Rightarrow$ (4): We prove the contrapositive. Suppose that there exists a 1-cocycle b on Γ which is unbounded on Λ. By Theorem D.11 and the lemma following it, there exists a unitary representation $\pi = \bigoplus_n \pi^b_{1/n}$ which contains almost invariant vectors $(\xi^b_{1/n})_n$ but no nonzero Λ-invariant vectors.

[2] In particular, (E, κ) is a Kazhdan pair.

(4) $\Leftrightarrow$ (5) is a tautology.

(4) $\Rightarrow$ (2): We again prove the contrapositive. Let $E_1 \subset E_2 \subset \cdots \subset \Gamma$ be an increasing sequence of finite subsets with $\bigcup E_n = \Gamma$. If condition (2) does not hold, then for every n there exist a unitary representation $(\pi_n, \mathcal{H}_n)$ and a vector $\xi_n \in \mathcal{H}_n$ such that

$$4^{-n}\|\xi_n - P_n\xi_n\| > \sup_{s\in E_n} \|\xi_n - \pi_n(s)\xi_n\| =: \delta_n,$$

where P_n is the orthogonal projection from $\mathcal{H}_n$ onto the subspace of all Λ-invariant vectors. Define a map $\sigma\colon \Gamma \to \bigoplus \mathcal{H}_n$ by

$$\sigma(s) = \left(\frac{\xi_n - \pi_n(s)\xi_n}{2^n\delta_n}\right)_n \in \bigoplus_{n=1}^{\infty} \mathcal{H}_n.$$

The infinite series in the formula is convergent and σ is a 1-cocycle of Γ with coefficients in $(\bigoplus \pi_n, \bigoplus \mathcal{H}_n)$. By (the proof of) Lemma 12.1.5, for every n, there exists $s_n \in \Lambda$ such that $\|\xi_n - \pi_n(s_n)\xi_n\| \geq \|\xi_n - P_n\xi_n\|$. It follows that $\|\sigma(s_n)\| \geq 2^n$ and σ is unbounded on Λ.

(2) $\Rightarrow$ (3): Let (E, κ) be as in condition (2) and φ be any positive definite function on Γ with $\varphi(e) = 1$. Then, denoting by $(\pi_\varphi, \mathcal{H}_\varphi, \xi_\varphi)$ the GNS triplet, we have

$$\begin{aligned}\sup_{t\in\Lambda} |1 - \varphi(t)| &\leq \sup_{t\in\Lambda} \|\pi_\varphi(t)\xi_\varphi - \xi_\varphi\| \, \|\xi_\varphi\| \\ &= \sup_{t\in\Lambda} \|\pi_\varphi(t)P^\perp\xi_\varphi - P^\perp\xi_\varphi\| \\ &\leq 2\kappa^{-1} \max_{s\in E} \|\pi_\varphi(s)\xi_\varphi - \xi_\varphi\| \\ &= 2\kappa^{-1} \max_{s\in E} (2\Re(1 - \varphi(s)))^{1/2}.\end{aligned}$$

This clearly implies condition (3).

(3) $\Rightarrow$ (1): Let $(\pi, \mathcal{H})$ be a unitary representation of Γ which contains almost Γ-invariant unit vectors (ξ_n). Then, the sequence (φ_n) of the positive definite functions on Γ defined by $\varphi_n(s) = \langle \pi(s)\xi_n, \xi_n\rangle$, converges pointwise to 1. By assumption, convergence is uniform on Λ, which implies that ξ_n is $(\Lambda, 1)$-invariant for sufficiently large n. By Lemma 12.1.5, there exists a nonzero Λ-invariant vector.

(2) $\Rightarrow$ (6): See the proof of Corollary 6.4.7. Note that if (E, κ) is a Kazhdan pair for Γ, $\mathcal{S}$ is any generating subset of Γ and n is chosen such that $E \subset (\mathcal{S} \cup \mathcal{S}^{-1})^n$, then $(\mathcal{S}, \kappa/n)$ is also a Kazhdan pair. $\square$

Our next lemma gives a very useful spectral characterization of property (T). But first, note that the *uniform convexity*[3] of a Hilbert space $\mathcal{H}$ implies

[3] A Banach space X is *uniformly convex* if for each $\varepsilon > 0$, there exists $\delta > 0$ with the following property: For all $x, y \in X$ with $\|x\|, \|y\| \leq 1$ and $\|x - y\| \geq \varepsilon$, we have that $\|\frac{x+y}{2}\| \leq 1 - \delta$.

that if a convex combination $\xi = \sum_{k=1}^{n} \alpha_k \xi_k$ of unit vectors (ξ_k) in $\mathcal{H}$ is close to a unit vector η, then each ξ_k is close to η. More precisely, we have the estimate

$$\sum_{k=1}^{n} \alpha_k \|\eta - \xi_k\|^2 = 2 - 2\Re\langle \eta, \xi \rangle = 2\Re\langle \eta, \eta - \xi \rangle \leq 2\|\eta - \xi\|.$$

For convenience, we say a unitary representation $(\pi, \mathcal{H})$ of Γ is *universal* if the induced representation of the full group C*-algebra $C^*(\Gamma)$ is faithful (i.e., π weakly contains any other unitary representation). For instance, we can take $(\pi, \mathcal{H})$ to be the direct sum of all cyclic unitary representations of Γ.

Lemma 12.1.8. *Let Γ be a group which is generated by a finite symmetric set $\mathcal{S}$ and $\nu\colon \mathcal{S} \to \mathbb{R}$ be a strictly positive symmetric function (i.e., $\nu(s) = \nu(s^{-1}) > 0$ for every $s \in \mathcal{S}$). For a universal representation $(\pi, \mathcal{H})$ of Γ, set $|\nu| = \sum_{s \in \mathcal{S}} \nu(s)$ and*

$$h = \frac{1}{|\nu|} \sum_{s \in \mathcal{S}} \nu(s)\pi(s) \in \mathbb{B}(\mathcal{H}).$$

Then, Γ has property (T) *if and only if* 1 *is isolated in the spectrum $\sigma(h)$ of the self-adjoint operator h. Moreover, if $\sigma(h) \subset [-1, 1-\varepsilon] \cup \{1\}$, then $(\mathcal{S}, \sqrt{2\varepsilon})$ is a Kazhdan pair.*

Proof. First, observe the following consequences of uniform convexity: $\xi \in \mathcal{H}$ is Γ-invariant if and only if it is an eigenvector of h with eigenvalue 1, and a sequence (ξ_n) of unit vectors is almost Γ-invariant if and only if $\lim\langle h\xi_n, \xi_n \rangle = 1$.

Now, suppose Γ has property (T). Since a universal representation weakly contains the trivial representation, it must contain the trivial representation. Thus $1 \in \sigma(h)$ and we must show it's isolated. Letting $\mathcal{K} \subset \mathcal{H}$ be the orthogonal complement of the Γ-fixed vectors (which is a reducing subspace for h), it suffices to show 1 is not in the spectrum of $h|_{\mathcal{K}}$. But, if it were in the spectrum, we could find unit vectors $\xi_n \in \mathcal{K}$ such that $\lim\langle h\xi_n, \xi_n \rangle = 1$; this implies $\mathcal{K}$ has almost invariant vectors, which is impossible. Hence, 1 is isolated in $\sigma(h)$.

For the converse, assume the spectrum of h is contained in $[-1, 1-\varepsilon] \cup \{1\}$. Then the spectral projection P associated with the spectral subset $\{1\}$ coincides with the orthogonal projection onto the Γ-invariant vectors. Since $\varepsilon(1-P) \leq 1-h$, one has

$$\begin{aligned} 2\varepsilon\|\xi - P\xi\|^2 &\leq 2\langle (1-h)\xi, \xi \rangle \\ &= \frac{2}{|\nu|} \sum_{s \in \mathcal{S}} \nu(s)(\|\xi\|^2 - \Re\langle \pi(s)\xi, \xi \rangle) \end{aligned}$$

$$= \frac{1}{|\nu|} \sum_{s \in \mathcal{S}} \nu(s) \|\xi - \pi(s)\xi\|^2$$

$$\leq \max_{s \in \mathcal{S}} \|\pi(s)\xi - \xi\|^2$$

for every $\xi \in \mathcal{H}$. This implies that $(\mathcal{S}, \sqrt{2\varepsilon})$ is a Kazhdan pair. □

The following lemma will be needed in the next chapter. Its proof is similar to the proof of Lemma 12.1.8 and hence will be omitted.

Lemma 12.1.9. *Let Γ be a group and $\mathcal{S} \subset \Gamma$ be a finite generating subset which contains the unit. Then, Γ has property* (T) *if and only if the following is true: For any unitary representation $(\pi, \mathcal{H})$ without a nonzero invariant vector, one has*

$$\|\sum_{s \in \mathcal{S}} \pi(s)\| < |\mathcal{S}|.$$

Property (T) for $\mathrm{SL}(3, \mathbb{Z})$. Our next goal is to prove that $\mathrm{SL}(3, \mathbb{Z})$ has property (T), following ideas of Y. Shalom.[4] The proof takes several pages and consists of two parts. We first prove relative property (T) for certain subgroups of $\mathrm{SL}(3, \mathbb{Z})$ and then deduce property (T) from a bounded generation trick.

For $1 \leq i \neq j \leq n$, we denote by $E_{ij} \in \mathrm{SL}(n, \mathbb{Z})$ the matrix with 1's on the diagonal, 1 in the (i, j)-th entry, and 0's elsewhere. It is not hard to see that the set $\mathcal{S} = \{E_{ij} : i \neq j\}$ generates $\mathrm{SL}(3, \mathbb{Z})$. Let $G, H \subset \mathrm{SL}(3, \mathbb{Z})$ be the subgroups given by

$$G = \begin{bmatrix} * & * & 0 \\ * & * & 0 \\ 0 & 0 & 1 \end{bmatrix} \quad \text{and} \quad H = \begin{bmatrix} 1 & 0 & * \\ 0 & 1 & * \\ 0 & 0 & 1 \end{bmatrix}.$$

We note that $G \cong \mathrm{SL}(2, \mathbb{Z})$ normalizes $H \cong \mathbb{Z}^2$ and the subgroup GH is canonically isomorphic to the semidirect product $\mathbb{Z}^2 \rtimes \mathrm{SL}(2, \mathbb{Z})$, where $\mathrm{SL}(2, \mathbb{Z})$ acts on $\mathbb{Z}^2$ by linear transformations.

Theorem 12.1.10. *The inclusion $(\mathbb{Z}^2 \subset \mathbb{Z}^2 \rtimes \mathrm{SL}(2, \mathbb{Z}))$ has relative property* (T). *More precisely, if $\mathcal{S}_0 = \{E_{12}, E_{21}, E_{13}, E_{23}\}$, then $(\mathcal{S}_0, 10^{-1})$ is a Kazhdan pair for $(\mathbb{Z}^2 \subset \mathbb{Z}^2 \rtimes \mathrm{SL}(2, \mathbb{Z}))$.*

Proof. For notational simplicity, we set $g_1 = E_{12}$, $g_2 = E_{21}$, $h_1 = E_{13}$ and $h_2 = E_{23}$. Let $(\pi, \mathcal{H})$ be a unitary representation of $\mathbb{Z}^2 \rtimes \mathrm{SL}(2, \mathbb{Z})$ which does not have a nonzero invariant vector. Suppose by contradiction that there exists a unit vector $\xi \in \mathcal{H}$ such that

$$\delta = \max_{s \in \mathcal{S}_0} \|\pi(s)\xi - \xi\| < 10^{-1}.$$

[4]The argument works in greater generality, but we'll be content with this case since it's all we need for applications.

We identify the Pontryagin dual of $\mathbb{Z}$ with $\mathbb{T} = \mathbb{R}/\mathbb{Z}$ by the pairing $\mathbb{Z} \times \mathbb{T} \ni (n,t) \mapsto e^{2\pi \imath n t}$ (where $\pi = 3.14\cdots$ in the exponent has nothing to do with the representation π and where $\imath = \sqrt{-1}$). It follows that $C^*(\mathbb{Z}^2) \cong C(\mathbb{T}^2)$ and the representation $\pi|_{\mathbb{Z}^2}$ gives rise to a $*$-representation $\sigma\colon C(\mathbb{T}^2) \to \mathbb{B}(\mathcal{H})$ such that $\sigma(z_i) = \pi(h_i)$, where $z_i(t) = e^{2\pi \imath t_i}$ for $t = (t_1, t_2) \in \mathbb{T}^2$. We denote by μ the regular Borel probability measure on $\mathbb{T}^2$ defined by

$$\int_{\mathbb{T}^2} f\, d\mu = \langle \sigma(f)\xi, \xi\rangle.$$

Note that the formula makes sense for every bounded Borel function f on $\mathbb{T}^2$. Set $o = (0,0) \in \mathbb{T}^2$. Since $\sigma(\chi_{\{o\}})$ is the orthogonal projection onto the space of $\mathbb{Z}^2$-invariant vectors (which is zero by assumption), we have that $\mu(\{o\}) = \|\sigma(\chi_{\{o\}})\xi\|^2 = 0$. Setting $B_0 := \{t : \Re z_1(t) \le 0 \text{ or } \Re z_2(t) \le 0\}$, we claim that $\mu(B_0) \le \delta^2$. Indeed, for $i = 1, 2$, one has

$$\delta^2 \ge \|\pi(h_i)\xi - \xi\|^2 = \int |z_i(t) - 1|^2\, d\mu(t) \ge 2\mu(\{t : \Re z_i(t) \le 0\}).$$

The natural action of $\mathrm{SL}(2,\mathbb{Z})$ on $\mathbb{Z}^2$ gives rise to an action of $\mathrm{SL}(2,\mathbb{Z})$ on $\mathbb{T}^2$: For $g \in \mathrm{SL}(2,\mathbb{Z})$ and $f \in C(\mathbb{T}^2)$, one has

$$\pi(g)\sigma(f)\pi(g)^{-1} = \sigma(\hat{g} \cdot f),$$

where $\hat{g}_1(t_1, t_2) = (t_1, -t_1 + t_2)$ and $\hat{g}_2(t_1, t_2) = (t_1 - t_2, t_2)$. We claim that for $i = 1, 2$ and any Borel subset $B \subset \mathbb{T}^2$, one has $|\mu(\hat{g}_i B) - \mu(B)| < 2\delta$. Indeed, for any $f \in C(\mathbb{T}^2)$, one has

$$\left|\int (\hat{g}_i \cdot f - f)\, d\mu\right| = |\langle \sigma(f)\pi(g_i^{-1})\xi, \pi(g_i^{-1})\xi\rangle - \langle \sigma(f)\xi, \xi\rangle| \le 2\delta \|f\|_\infty.$$

It follows that $|\mu(\hat{g}_i B) - \mu(B)| = |\int (\hat{g}_i \cdot \chi_B - \chi_B)\, d\mu| \le 2\delta$.

Now consider the measurable partition

$$\mathbb{T}^2 = [-1/2, 1/2)^2 = \{o\} \sqcup \bigsqcup_{k=0}^{4} B_k$$

given by

$$\begin{aligned} B_0 &= \{(t_1,t_2) : |t_1| \ge 1/4 \text{ or } |t_2| \ge 1/4\},\\ B_1 &= \{(t_1,t_2) : |t_2| \le |t_1| < 1/4 \text{ and } t_1 t_2 > 0\},\\ B_2 &= \{(t_1,t_2) : |t_1| < |t_2| < 1/4 \text{ and } t_1 t_2 \ge 0\},\\ B_3 &= \{(t_1,t_2) : |t_1| \le |t_2| < 1/4 \text{ and } t_1 t_2 < 0\},\\ B_4 &= \{(t_1,t_2) : |t_2| < |t_1| < 1/4 \text{ and } t_1 t_2 \le 0\}. \end{aligned}$$

It is easy to check that

$$\hat{g}_1^{-1}(B_1 \cup B_2) \subset B_0 \cup B_2,$$

$$\hat{g}_2^{-1}(B_1 \cup B_2) \subset B_0 \cup B_1,$$
$$\hat{g}_1(B_3 \cup B_4) \subset B_0 \cup B_3,$$
$$\hat{g}_2(B_3 \cup B_4) \subset B_0 \cup B_4.$$

Therefore,

$$\mu(B_1) + \mu(B_2) = \mu(B_1 \cup B_2) \leq \mu(B_0 \cup B_2) + 2\delta \leq \mu(B_2) + \delta^2 + 2\delta$$

and hence $\mu(B_1) \leq \delta^2 + 2\delta$. The same inequality holds for the other $\mu(B_k)$'s. It follows that

$$1 = \mu(\mathbb{T}^2 \setminus \{o\}) = \sum_{k=0}^{4} \mu(B_k) \leq 5\delta^2 + 8\delta < 10\delta < 1,$$

which is a contradiction. □

Remark 12.1.11. Actually, one can show that whenever $\Gamma \subset \mathrm{SL}(2,\mathbb{Z})$ is nonamenable, the inclusion $(\mathbb{Z}^2 \subset \mathbb{Z}^2 \rtimes \Gamma)$ has relative property (T). Indeed, this follows from the following fact (combined with the proof above): There is no sequence (μ_n) of regular Borel probability measures on $\mathbb{T}^2$ with the following properties:

(1) $\mu_n(\{o\}) = 0$ for every n;

(2) the sequence (μ_n) converges weakly to the Dirac measure δ_o at o;

(3) the sequence (μ_n) is approximately Γ-invariant, i.e., for all $g \in \Gamma$

$$\limsup_n \{|\mu_n(\hat{g}B) - \mu_n(B)| : B \subset \mathbb{T}^2 \text{ a Borel subset}\} = 0.$$

To show that no such measures exist, we regard the μ_n's as probability measures on $\mathbb{T}^2 \setminus \{o\}$. After blowing up the hole and patching it with $\mathbb{R}P^1 \cong \mathbb{T}$, the resulting space $M = (\mathbb{T}^2 \setminus \{o\}) \cup \mathbb{R}P^1$ is a compact manifold. Moreover, the action of $\mathrm{SL}(2,\mathbb{Z})$ on $\mathbb{T}^2 \setminus \{o\}$ naturally extends to M. Now, μ_n is a probability measure on a compact space M and hence the sequence μ_n has a limit point μ. By assumption, μ is supported on $\mathbb{R}P^1$ and is Γ-invariant. However, this is impossible since Γ is nonamenable and $\mathbb{R}P^1$ is Γ-amenable (see Theorem 5.4.1 and Example E.10).

Now we turn to the bounded generation trick. We define subgroups $G, H_1, H_2 \subset \mathrm{SL}(3,\mathbb{Z})$ by

$$G = \begin{bmatrix} * & * & 0 \\ * & * & 0 \\ 0 & 0 & 1 \end{bmatrix}, \quad H_1 = \begin{bmatrix} 1 & 0 & * \\ 0 & 1 & * \\ 0 & 0 & 1 \end{bmatrix} \text{ and } H_2 = \begin{bmatrix} 1 & 0 & 0 \\ 0 & 1 & 0 \\ * & * & 1 \end{bmatrix}.$$

Note that G normalizes both H_1 and H_2.

Lemma 12.1.12. $\mathrm{SL}(3,\mathbb{Z}) = H_1 H_2 H_1 H_2 G$.

Proof. Let

$$s = \begin{bmatrix} * & * & * \\ * & * & * \\ x & y & z \end{bmatrix} \in \mathrm{SL}(3,\mathbb{Z})$$

be given. By the Chinese Remainder Theorem, there exists $m \in \mathbb{Z}$ such that $x+mz \equiv 1 \pmod{p}$ for all prime divisors p of y which do not divide z. Since $s \in \mathrm{SL}(3,\mathbb{Z})$, we have $\gcd(x,y,z)=1$ and thus no prime divisor of $\gcd(y,z)$ divides x. It follows that $\gcd(x+mz,y)=1$. Thus,

$$\begin{bmatrix} * & * & * \\ * & * & * \\ x & y & z \end{bmatrix}\begin{bmatrix} 1 & 0 & 0 \\ 0 & 1 & 0 \\ m & 0 & 1 \end{bmatrix} = \begin{bmatrix} * & * & * \\ * & * & * \\ x' & y & z \end{bmatrix}$$

with $\gcd(x',y)=1$. Hence, there exists $a,b \in \mathbb{Z}$ such that $ax'+by+z=1$, or equivalently,

$$\begin{bmatrix} * & * & * \\ * & * & * \\ x' & y & z \end{bmatrix}\begin{bmatrix} 1 & 0 & a \\ 0 & 1 & b \\ 0 & 0 & 1 \end{bmatrix} = \begin{bmatrix} * & * & u \\ * & * & v \\ x' & y & 1 \end{bmatrix}$$

for some $u,v \in \mathbb{Z}$. Finally, we have

$$\begin{bmatrix} 1 & 0 & -u \\ 0 & 1 & -v \\ 0 & 0 & 1 \end{bmatrix}\begin{bmatrix} * & * & u \\ * & * & v \\ x' & y & 1 \end{bmatrix}\begin{bmatrix} 1 & 0 & 0 \\ 0 & 1 & 0 \\ -x' & -y & 1 \end{bmatrix} = \begin{bmatrix} * & * & 0 \\ * & * & 0 \\ 0 & 0 & 1 \end{bmatrix} \in G.$$

Therefore, $s \in H_1GH_2H_1H_2 = H_1H_2H_1H_2G$. □

Definition 12.1.13. Let Γ be a group and $G, H_1, H_2 \subset \Gamma$ be subgroups. We say the subgroups satisfy the *Shalom property* if

(1) Γ is generated by H_1 and H_2;

(2) G normalizes both H_i;

(3) Γ is boundedly generated by G, H_1 and H_2, i.e., $\Gamma = H_{i(1)} \cdots H_{i(l)}G$;

(4) both $(H_i \subset \Gamma)$, $i=1,2$, have relative property (T).

We have seen that there is a system of subgroups $G, H_1, H_2 \subset \mathrm{SL}(3,\mathbb{Z})$ which satisfy the Shalom property. (That H_1 and H_2 generate Γ is a trivial exercise.)

Theorem 12.1.14. *Let Γ be a group which admits a system of subgroups with the Shalom property. Then Γ has property* (T). *In particular,* $\mathrm{SL}(3,\mathbb{Z})$ *has property* (T) *and* $(\mathcal{S}, 10^{-3})$ *is a Kazhdan pair, where* $\mathcal{S} = \{E_{ij} : i \neq j\}$.

Proof. By Theorem 12.1.7, there exist a finite subset $E \subset \Gamma$ and $\kappa > 0$ such that

$$\|\xi - P_i\xi\| \le \kappa^{-1} \sup_{s\in E} \|\pi(s)\xi - \xi\|$$

for any unitary representation $(\pi, \mathcal{H})$ of Γ and $\xi \in \mathcal{H}$, where P_i is the orthogonal projection onto the subspace of H_i-invariant vectors. Let l be such that $\Gamma = H_{i(1)} \cdots H_{i(l)} G$ and set $\varepsilon = \kappa/(7l+2)$. We will prove that (E, ε) is a Kazhdan pair for Γ. Suppose by contradiction that there exists a unitary representation $(\pi, \mathcal{H})$ of Γ having no nonzero invariant vector but having an (E, ε)-invariant vector $\xi_0 \in \mathcal{H}$.

Let P_i be the projection onto the H_i-invariant vectors and set $T = (P_1 + P_2)/2$. Observe that P_i and T commute with $\pi(G)$. Since $\|\xi_0 - P_i\xi_0\| \leq \varepsilon/\kappa$, we have

$$\langle T\xi_0, \xi_0 \rangle = \frac{1}{2}(\|P_1\xi_0\|^2 + \|P_2\xi_0\|^2) \geq 1 - (\varepsilon/\kappa)^2$$

and so the spectrum of T intersects $[1 - (\varepsilon/\kappa)^2, 1]$. But on the other hand, 1 is not in the point-spectrum of T, since Γ is generated by H_1 and H_2 and π does not have a nonzero invariant vector. It follows that there exists $0 < \delta < (\varepsilon/\kappa)^2$ such that the spectral projection $Q = \chi_{[1-(\varepsilon/\kappa)^2, 1-\delta]}(T)$ is nonzero.

We claim that $\|Q - \pi(h)Q\| \leq 2\sqrt{2}\varepsilon/\kappa$ for every $h \in H_1 \cup H_2$. Let $\xi \in Q\mathcal{H}$ be a unit vector. Then we have

$$1 - (\varepsilon/\kappa)^2 \leq \langle T\xi, \xi \rangle = \|P_1\xi\|^2/2 + \|P_2\xi\|^2/2 \leq 1/2 + \|P_i\xi\|^2/2,$$

and hence $\|P_i\xi\|^2 \geq 1 - 2(\varepsilon/\kappa)^2$. Since $\pi(H_i)$ commutes with P_i, this implies that

$$\forall h \in H_i, \quad \|\xi - \pi(h)\xi\| = \|P_i^{\perp}\xi - \pi(h)P_i^{\perp}\xi\| \leq 2\sqrt{2}\varepsilon/\kappa.$$

This proves the claim.

Note that $\pi(h)Q - Q\pi(h) = Q^{\perp}\pi(h)Q - Q\pi(h)Q^{\perp}$ and that $Q^{\perp}\pi(h)Q$ and $Q\pi(h)Q^{\perp}$ have mutually orthogonal domains and ranges. It follows from the inequality $\|Q^{\perp}\pi(h)Q\| \leq \|Q - \pi(h)Q\|$ that for all $h \in H_1 \cup H_2$ we have

$$\|\pi(h)Q - Q\pi(h)\| = \max\{\|Q^{\perp}\pi(h)Q\|, \|Q\pi(h)Q^{\perp}\|\} \leq 2\sqrt{2}\varepsilon/\kappa.$$

Since $\Gamma = H_{i(1)} \cdots H_{i(l)} G$ and $\pi(G)$ commutes with Q, it follows that

$$\sup_{s \in \Gamma} \|Q - \pi(s)Q\pi(s)^*\| \leq 2l\sqrt{2}\varepsilon/\kappa =: \varepsilon_0.$$

Now, let $\mathcal{C} = \overline{\mathrm{conv}}^w\{\pi(s)Q\pi(s)^* : s \in \Gamma\} \subset \mathbb{B}(\mathcal{H})$; this set is ultraweakly compact and Γ acts continuously on it by conjugation. By Zorn's Lemma, there exists a subset $\mathcal{C}_0$ of $\mathcal{C}$ which is minimal among the nonempty Γ-invariant ultraweakly-closed convex subsets of $\mathcal{C}$.[5] Since the Γ-action preserves the norm, all elements of $\mathcal{C}_0$ have the same norm, say c. Pick some $S \in \mathcal{C}_0$. We note that $\|Q - S\| \leq \varepsilon_0$ and for all $h \in H_1 \cup H_2$,

$$(*) \qquad \|S - \pi(h)S\| \leq 2\varepsilon_0 + \|Q - \pi(h)Q\| \leq (4l+2)\sqrt{2}\varepsilon/\kappa =: \varepsilon_1.$$

[5]If $\dim \mathcal{H} < \infty$, then $\mathcal{C}_0$ is a singleton, and we are essentially done here.

Set $S_0 = \sum 2^{-k}\pi(s_k)S\pi(s_k)^* \in \mathcal{C}_0$ for an enumeration $\Gamma = \{s_k\}_{k=1}^{\infty}$ and take a sequence (ζ_n) of unit vectors such that $\|c\zeta_n - S_0\zeta_n\| \to 0$. Since

$$\|\sum 2^{-k}\pi(s_k)S\pi(s_k)^*\| = c = \sum 2^{-k}\|\pi(s_k)S\pi(s_k)^*\|,$$

we have by convexity that

$$(**) \qquad \forall s \in \Gamma, \quad \lim_n \|c\zeta_n - \pi(s)S\pi(s)^*\zeta_n\| = 0.$$

We claim that the sequence (ζ_n) is almost $\pi(\Gamma)$-invariant. It suffices to show (modulo passing to a subsequence) that the vector $\zeta_\omega = (\zeta_n)_{n\to\omega} \in \mathcal{H}_\omega$ is $\pi_\omega(\Gamma)$-invariant. Here ω is a fixed free ultrafilter on $\mathbb{N}$ and π_ω is the ultrapower unitary representation of Γ on the ultrapower Hilbert space $\mathcal{H}_\omega$. This means that every vector ξ_ω in $\mathcal{H}_\omega$ is represented as a bounded sequence (ξ_n) in $\mathcal{H}$ in such a way that $\langle \eta_\omega, \xi_\omega\rangle = \lim_\omega \langle \eta_n, \xi_n\rangle$ and that $\pi_\omega(s)\xi_\omega = (\pi(s)\xi_n)_{n\to\omega}$. (See Appendix A for details.) Similarly, we consider $S_\omega \in \mathbb{B}(\mathcal{H}_\omega)$ given by the constant sequence S. Now the equations $(*)$ and $(**)$ read

$$\forall h \in H_1 \cup H_2, \quad \|S_\omega - \pi_\omega(h)S_\omega\| \le \varepsilon_1 \text{ and } \forall s \in \Gamma, \quad S_\omega\pi_\omega(s)\zeta_\omega = c\pi_\omega(s)\zeta_\omega.$$

Let $\mathcal{K} \subset \mathcal{H}_\omega$ be the $\pi_\omega(H_i)$-invariant Hilbert subspace spanned by $\pi_\omega(\Gamma)\zeta_\omega$. It follows from the above equations that $S_\omega|_{\mathcal{K}} = c1_{\mathcal{K}}$ and

$$\forall h \in H_1 \cup H_2, \quad \|1_{\mathcal{K}} - \pi_\omega(h)|_{\mathcal{K}}\| \le \frac{\varepsilon_1}{c} \le \frac{\varepsilon_1}{1-\varepsilon_0} =: \varepsilon_2.$$

Since we have chosen ε so that $\varepsilon_2 < \sqrt{2}$, this implies that $\pi_\omega(h)|_{\mathcal{K}} = 1_{\mathcal{K}}$ for all $h \in H_i$ by Lemma 12.1.5 (and the proposition following it). Since Γ is generated by H_1 and H_2, this proves the claim.

Since (ζ_n) is almost $\pi(\Gamma)$-invariant, we have $\lim_n \|\zeta_n - P_i\zeta_n\| = 0$ for each i, by relative property (T). This implies that $\lim_n \|\zeta_n - T\zeta_n\| = 0$. However, since

$$\lim_{n\to\omega} \|\zeta_n - Q\zeta_n\| \le 2\varepsilon_0 + \lim_{n\to\omega} \|c\zeta_n - S\zeta_n\| = 2\varepsilon_0 < 1$$

by the choice of ε, we have

$$\lim_{n\to\omega} \|\zeta_n - T\zeta_n\| \ge \lim_{n\to\omega} \|Q\zeta_n - TQ\zeta_n\| \ge \delta \lim_{n\to\omega} \|Q\zeta_n\| > 0.$$

(Recall that $Q = \chi_{[1-(\varepsilon/\kappa)^2, 1-\delta]}(T)$.) This is a contradiction.

Finally, that $(\mathcal{S}, 10^{-3})$ is a Kazhdan pair for $\mathrm{SL}(3,\mathbb{Z})$ follows from Theorem 12.1.10, Lemma 12.1.12 and the proof above. □

A spectral condition which ensures property (T). Let Γ be a discrete group which is generated by a finite symmetric set $\mathcal{S}$ with $e \notin \mathcal{S}$. The *link* of $(\Gamma, \mathcal{S})$ is a graph $L(\Gamma, \mathcal{S})$ whose vertex set is $\mathcal{S}$ and whose edge set is $\mathbf{E} = \{(s,t) \subset \mathcal{S}^2 : s^{-1}t \in \mathcal{S}\}$. Since $\mathcal{S}$ is symmetric, the link is a simple undirected graph. We denote by $\nu(s) = |\{t \in \mathcal{S} : (s,t) \in \mathbf{E}\}|$ the degree of

$s \in \mathcal{S}$, and we set $|\nu| = \sum \nu(s) = |\mathbf{E}|$. Let $L^2(\mathcal{S}, \nu)$ be the weighted L^2-space on $\mathcal{S}$ with inner product given by

$$\langle f, g\rangle_\nu = \frac{1}{|\nu|}\sum_{s\in\mathcal{S}} \nu(s)\, f(s)\overline{g(s)},$$

for functions f and g on $\mathcal{S}$.

Recall from Appendix E that the combinatorial Laplacian is defined as

$$\Delta = \frac{1}{2}d^*d \in \mathbb{B}(L^2(\mathcal{S}, \nu)),$$

where $d\colon L^2(\mathcal{S}, \nu) \to L^2(\mathbf{E})$ is given by $d(f)((s,t)) = f(t) - f(s)$. If the link $L(\Gamma, \mathcal{S})$ is connected, then 0 is a simple eigenvalue of Δ and the first nonzero eigenvalue of Δ is denoted by λ_1.

Theorem 12.1.15. *Let Γ be a discrete group which is generated by a finite symmetric set $\mathcal{S}$ with $e \notin \mathcal{S}$. Suppose that the link $L(\Gamma, \mathcal{S})$ is connected and $\lambda_1 > 1/2$. Then Γ has property* (T). *More precisely, $(\mathcal{S}, \sqrt{2(2 - \lambda_1^{-1})})$ is a Kazhdan pair.*

Proof. Let $(\pi, \mathcal{H})$ be a universal unitary representation and let

$$h = \frac{1}{|\nu|}\sum_{s\in\mathcal{S}} \nu(s)\pi(s).$$

Since $L(\Gamma, \mathcal{S})$ is connected, we have $\nu(s) = \nu(s^{-1}) > 0$ for every $s \in \mathcal{S}$. By Lemma 12.1.8, it suffices to show

$$\sigma(h) \subset [-1, \lambda_1^{-1} - 1] \cup \{1\},$$

where $\sigma(h)$ is the spectrum of h. Let $\xi \in \mathcal{H}$ be given and define $f\colon \mathcal{S} \to \mathcal{H}$ by $f(s) = \pi(s)\xi$. Notice that the mean of f is $h\xi$.

Now view f as an element of $L^2(\mathcal{S}, \nu)\otimes\mathcal{H}$. By a Poincaré-type inequality (Lemma E.5), we have

$$\lambda_1(\|f\|^2 - \|h\xi\|^2) \leq \langle(\Delta \otimes I_{\mathcal{H}})f, f\rangle = \frac{1}{2|\mathbf{E}|}\sum_{(s,t)\in\mathbf{E}} \|f(t) - f(s)\|^2.$$

Also note that $\|f\|^2 - \|h\xi\|^2 = \|\xi\|^2 - \|h\xi\|^2 = \langle(1 - h^2)\xi, \xi\rangle$. On the other hand, since $\nu(r) = |\{(s,t) \in \mathbf{E} : t^{-1}s = r\}|$ for every $r \in \mathcal{S}$, we have

$$\begin{aligned}
\frac{1}{2|\mathbf{E}|}\sum_{(s,t)\in\mathbf{E}} \|f(t) - f(s)\|^2 &= \frac{1}{2|\mathbf{E}|}\sum_{(s,t)\in\mathbf{E}} \|\pi(t)\xi - \pi(s)\xi\|^2 \\
&= \frac{1}{2|\nu|}\sum_{r\in\mathcal{S}} \nu(r)\|\xi - \pi(r)\xi\|^2 \\
&= \langle(1 - h)\xi, \xi\rangle.
\end{aligned}$$

It follows that $\lambda_1(1-h^2) \leq 1-h$, or equivalently, $\sigma(h)\cap(\lambda_1^{-1}-1, 1) = \emptyset$. $\square$

Unfortunately, providing examples to which this theorem applies is not so easy – though there are plenty! (See [**201**].)

Property (T) for von Neumann algebras.

Definition 12.1.16. Let $N \subset M$ be an inclusion of finite von Neumann algebras and τ be a faithful normal tracial state[6] on M. We say that an inclusion $N \subset M$ has *relative property* (T) if for any $\varepsilon > 0$, there exist a finite subset $\mathfrak{F} \subset M$ and $\delta > 0$ with the following property: If $\varphi\colon M \to M$ is a τ-preserving u.c.p. map such that $\|\varphi(x) - x\|_2 < \delta$ for every $x \in \mathfrak{F}$, then $\|\varphi(a) - a\|_2 < \varepsilon\|a\|$ for all $a \in N$.

We say M has property (T) if the identity inclusion $M \subset M$ has relative property (T).

Remark 12.1.17. The condition that φ be unital and τ-preserving can be relaxed: If $\varphi\colon M \to M$ is a c.p. map such that $\varphi(1) \leq 1$ and $\tau \circ \varphi \leq \tau$, then the map $\tilde{\varphi}\colon M \to M$ defined by

$$\tilde{\varphi}(x) = \varphi(x) + \frac{(\tau - \tau \circ \varphi)(x)}{1 - \tau \circ \varphi(1)}(1 - \varphi(1))$$

is a u.c.p. map such that $\tau \circ \tilde{\varphi} = \tilde{\varphi}$. We also note that $\tau \circ \varphi \leq \tau$ implies

$$\|\varphi(a)\|_2 = \tau(\varphi(a)^*\varphi(a))^{1/2} \leq \tau(\varphi(a^*a))^{1/2} \leq \tau(a^*a)^{1/2} = \|a\|_2$$

for every $a \in M$, by Proposition 1.5.7. In particular φ is normal.

Theorem 12.1.18. *Let $N \subset M$ be finite von Neumann algebras and τ be a faithful normal tracial state on M. The following are equivalent:*

(1) *the inclusion $N \subset M$ has relative property* (T)*;*

(2) *for any $\varepsilon > 0$, there exist a finite subset $\mathfrak{F} \subset M$ and $\delta > 0$ with the following property: If $\mathcal{H}$ is an M-M-bimodule and $\xi \in \mathcal{H}$ is a unit vector such that $\langle a\xi, \xi\rangle = \tau(a) = \langle \xi a, \xi\rangle$ for all $a \in M$ and such that $\|x\xi - \xi x\| < \delta$ for every $x \in \mathfrak{F}$, then there exists $\xi_0 \in \mathcal{H}$ with $\|\xi_0 - \xi\| < \varepsilon$ such that $a\xi_0 = \xi_0 a$ for all $a \in N$.*

Moreover, if $(N \subset M) = (L(\Lambda) \subset L(\Gamma))$ for an inclusion $\Lambda \subset \Gamma$ of groups, then the above conditions are equivalent to

(3) *the inclusion $\Lambda \subset \Gamma$ has relative property* (T).

Proof. (1) $\Rightarrow$ (2): Let $\varepsilon > 0$ be given and take $\mathfrak{F}$ and δ as in Definition 12.1.16. Let $\mathcal{H}$ be an M-M-bimodule and $\xi \in \mathcal{H}$ be a unit vector such that $\langle a\xi, \xi\rangle = \tau(a) = \langle \xi a, \xi\rangle$ for all $a \in M$ and $\|x\xi - \xi x\| < \delta$

[6] It turns out that this definition does not depend on the choice of faithful normal tracial state – but this isn't obvious. See [**144**] for a proof, and much more.

for every $x \in \mathfrak{F}$. We define a τ-preserving u.c.p. map $\varphi\colon M \to M$ by $\langle \varphi(a)\widehat{x}, \widehat{y}\rangle = \langle a\xi x, \xi y\rangle_{L^2(M)}$ (see Appendix F). Then we have

$$\begin{aligned}\|\varphi(x) - x\|_2^2 &= \|\varphi(x)\|_2^2 + \|x\|_2^2 - 2\Re\langle x\xi, \xi x\rangle \\ &\leq 2\|x\|_2^2 - 2\Re\langle x\xi, \xi x\rangle \\ &= \|x\xi - \xi x\|^2 \\ &< \delta^2\end{aligned}$$

for every $x \in \mathfrak{F}$. It follows that $\|\varphi(a) - a\|_2 < \varepsilon\|a\|$ for all $a \in N$. Hence, for every unitary element $u \in N$, we have

$$\|\xi - u\xi u^*\|_2^2 = 2 - 2\Re\tau(\varphi(u)u^*) < 2\varepsilon.$$

Let ξ_0 be the circumcenter of the set $\{u\xi u^* : u \in N \text{ unitary}\}$. By uniqueness of the circumcenter, we have $u\xi_0 u^* = \xi_0$ for every unitary element $u \in N$ and $\|\xi_0 - \xi\| \leq (2\varepsilon)^{1/2}$.

(2) $\Rightarrow$ (1): Let $\varepsilon > 0$ be given and take $\mathfrak{F}$ and δ as in condition (2). Let $\varphi\colon M \to M$ be a τ-preserving u.c.p. map such that $\|\varphi(x) - x\|_2 < (2\|x\|_2)^{-1}\delta^2$ for every $x \in \mathfrak{F}$. Let $(\mathcal{H}, \xi)$ be the M-M-bimodule arising from the minimal Stinespring dilation (see Appendix F). Then we have

$$\|x\xi - \xi x\|^2 = 2\|x\|_2^2 - 2\Re\tau(\varphi(x)x^*) < \delta^2$$

for every $x \in \mathfrak{F}$. It follows that there exists $\xi_0 \in \mathcal{H}$ with $\|\xi_0 - \xi\| < \varepsilon$ such that $a\xi_0 = \xi_0 a$ for all $a \in N$. Hence, for every $a \in N$ we have

$$\begin{aligned}\|\varphi(a) - a\|_2^2 &\leq 2\|a\xi\|^2 - 2\Re\langle a\xi, \xi a\rangle \\ &\leq 2\|a\xi\|\,\|a\xi - \xi a\| \\ &\leq 4\|a\|_2\|a\|\,\|\xi - \xi_0\| \\ &< 4\varepsilon\|a\|^2.\end{aligned}$$

Now, for the last assertion, we assume our inclusion arises from groups, i.e., $(N \subset M) = (L(\Lambda) \subset L(\Gamma))$. It is routine to check that relative property (T) of $(L(\Lambda) \subset L(\Gamma))$ implies condition (3) of Theorem 12.1.7; hence we only prove (3) $\Rightarrow$ (2). Let $\varepsilon > 0$ be given and take a finite subset $F \subset \Gamma$ and $\kappa > 0$ such that (F, κ) satisfies condition (2) of Theorem 12.1.7. We will show that $\lambda(F) \subset M$ and $\delta = \kappa\varepsilon > 0$ satisfy condition (2) of the present theorem. Let an M-M-bimodule $\mathcal{H}$ and a unit vector $\xi \in \mathcal{H}$ be given such that $\|\lambda(s)\xi - \xi\lambda(s)\| < \delta$ for every $s \in F$. Then, the unitary representation π of Γ on $\mathcal{H}$ given by

$$\pi(s)\zeta = \lambda(s)\zeta\lambda(s^{-1})$$

satisfies $\|\pi(s)\xi - \xi\| < \delta$ for every $s \in F$. It follows that there exists a $\pi(\Lambda)$-invariant vector $\xi_0 \in \mathcal{H}$ such that $\|\xi_0 - \xi\| \leq \kappa^{-1}\delta = \varepsilon$. Since ξ_0 is $\pi(\Lambda)$-invariant, we have $a\xi_0 = \xi_0 a$ for all $a \in L(\Lambda)$. □

For a II_1-factor M, we let $\mathrm{Aut}(M)$ denote the group of $*$-automorphisms of M, $\mathrm{Int}(M) \subset \mathrm{Aut}(M)$ denote the normal subgroup of inner automorphisms, and we let $\mathrm{Out}(M) = \mathrm{Aut}(M)/\mathrm{Int}(M)$ be the quotient group. It can be shown that the hyperfinite type II_1-factor and free group factors have uncountable outer-automorphism groups (see Exercises 12.1.2 and 12.1.3). In stark contrast, the following result of Connes was the first application of property (T) to operator algebras.

Theorem 12.1.19. *Let M be a II_1-factor with property* (T)*. Then* $\mathrm{Out}(M)$ *is countable.*

Proof. Take a finite subset $\mathfrak{F} \subset M$ and $\delta > 0$ as in Definition 12.1.16, for $\varepsilon = 1/2$. We claim that $\gamma \in \mathrm{Aut}(M)$, satisfying $\|\gamma(x) - x\|_2 < \delta$ for every $x \in \mathfrak{F}$, is inner. Indeed, relative property (T) implies that $\|\gamma(a) - a\|_2 < \varepsilon\|a\|$ for all $a \in M$. Let z be the circumcenter (see Exercise D.1) of $\{\gamma(u)u^* : u \in M \text{ unitary}\} \subset L^2(M, \tau)$. It follows that z is a nonzero element in M (why?) and $z = \gamma(u)zu^*$ for every unitary element $u \in M$. Let $z = v|z|$ be the polar decomposition of z. Since $|z|$ belongs to the center of M, it is a positive scalar and $\gamma = \mathrm{Ad}(v)$.

Suppose now that $\mathrm{Out}(M)$ is uncountable and choose a lift $\tilde{\alpha} \in \mathrm{Aut}(M)$ for each $\alpha \in \mathrm{Out}(M)$. Since M is separable in 2-norm (by property (T)), a simple cardinality argument shows that there exist two distinct lifts $\tilde{\alpha}$ and $\tilde{\beta}$ such that $\|\tilde{\alpha}(x) - \tilde{\beta}(x)\|_2 < \delta$ for every $x \in \mathfrak{F}$. Thus, $\gamma = \tilde{\beta}^{-1}\tilde{\alpha}$ is not inner, but it satisfies $\|\gamma(x) - x\|_2 < \delta$ for every $x \in \mathfrak{F}$: contradiction. □

A similar argument can be used to show that the fundamental group $\mathfrak{F}(M)$ of a II_1-factor with property (T) is countable. (Recall that $\mathfrak{F}(M)$ is the set of $t > 0$ such that $M_t \cong M$, where $M_t = P_t(\mathbb{B}(\mathcal{H}) \bar{\otimes} M)P_t$ for some projection P_t such that $(\mathrm{Tr} \otimes \tau)(P_t) = t$.)

Exercises

Exercise 12.1.1. Let Γ be a group and θ be an automorphism of Γ. Define $\tilde{\theta} \in \mathrm{Aut}(L(\Gamma))$ by $\tilde{\theta}(\lambda(s)) = \lambda(\theta(s))$. Prove that if $\tilde{\theta}$ is inner, then there exist $t \in \Gamma$ and a finite index normal subgroup $\Gamma_0 \subset \Gamma$ such that $\theta(s) = tst^{-1}$ for every $s \in \Gamma_0$.

Exercise 12.1.2. Let $\mathfrak{S}_\infty$ be the group of all finite permutations of $\mathbb{N}$. For any permutation φ of $\mathbb{N}$ which does not belong to $\mathfrak{S}_\infty$, we define an automorphism θ of $\mathfrak{S}_\infty$ by $\theta(\pi) = \varphi\pi\varphi^{-1}$. Prove that $\tilde{\theta} \in \mathrm{Aut}(L(\mathfrak{S}_\infty))$ is not inner.

Exercise 12.1.3. Let Γ be an ICC[7] group and let $\chi\colon \Gamma \to \{z : |z| = 1\}$ be a nontrivial character. Prove that $\theta_\chi \in \mathrm{Aut}(L(\Gamma))$, defined by $\theta_\chi(\lambda(s)) = \chi(s)\lambda(s)$, is not inner.

12.2. The Haagerup property

Haagerup's property is, in many ways, the opposite of property (T). Indeed, compare the definition below with condition (3) in Theorem 12.1.7.

Definition 12.2.1. A discrete group Γ has the *Haagerup property* if there exists a net (φ_i) of positive definite functions on Γ with $\varphi_i(e) = 1$, such that each φ_i vanishes at infinity (i.e., for any $\varepsilon > 0$, the set $\{s \in \Gamma : |\varphi_i(s)| > \varepsilon\}$ is finite) and $\varphi_i \to 1$ pointwise.

Example 12.2.2. Amenable groups have the Haagerup property (since the left regular representation weakly contains the trivial representation). We'll soon prove that free groups and $\mathrm{SL}(2, \mathbb{Z})$ also enjoy this property. In addition, the class of groups having the Haagerup property is closed under taking subgroups, direct products, free products and increasing unions. (Most of these are easy exercises; the proof of the free product result requires the fact that the free product of positive definite functions is positive definite – see Theorem 4.8.5.)

The next proposition is a trivial consequence of the definitions.

Proposition 12.2.3. *If Γ has the Haagerup property, then any infinite subgroup $\Lambda \subset \Gamma$ does not have relative property* (T). *In particular, a group with the Haagerup property and property* (T) *is finite.*

We say a 1-cocycle $b\colon \Gamma \to \mathcal{H}$ is (metrically) *proper* if for any $R > 0$, the set $\{s \in \Gamma : \|b(s)\| \leq R\}$ is finite (see Appendix D).

Theorem 12.2.4. *Let Γ be a countable discrete group. Then the following are equivalent:*

(1) Γ *has the Haagerup property;*

(2) Γ *admits a proper* 1*-cocycle;*

(3) Γ *admits a proper affine isometric action on a (real) Hilbert space.*

Proof. The equivalence (2) $\Leftrightarrow$ (3) is tautological (see Appendix D). The implication (2) $\Rightarrow$ (1) is a consequence of Schoenberg's Theorem D.11. To prove the converse, let φ_n be a sequence of positive definite functions on Γ satisfying the conditions in Definition 12.2.1. Let $\Gamma = \{s_n\}_{n=1}^\infty$ be an enumeration. Passing to a subsequence if necessary, we may assume that

[7] Meaning $\{sts^{-1} : s \in \Gamma\}$ is infinite for every nonneutral element $t \in \Gamma$.

$|1-\varphi_n(s_k)| < 2^{-n}$ for every k and $n \geq k$. By the GNS construction (see Theorem 2.5.11), for each n we have a unitary representation $(\pi_n, \mathcal{H}_n)$ and a unit vector $\xi_n \in \mathcal{H}_n$ such that $\varphi_n(s) = \langle \pi_n(s)\xi_n, \xi_n\rangle$. Let $\mathcal{H} = \bigoplus \mathcal{H}_n$ and $\pi = \bigoplus \pi_n$. We define a 1-cocycle b with coefficients in $(\pi, \mathcal{H})$ by

$$b\colon \Gamma \ni s \mapsto (\xi_n - \pi_n(s)\xi_n)_{n=1}^{\infty} \in \bigoplus_{n=1}^{\infty} \mathcal{H}_n = \mathcal{H}.$$

The infinite sum converges because

$$\|b(s)\|^2 = \sum_{n=1}^{\infty} \|\xi_n - \pi_n(s)\xi_n\|^2 = \sum_{n=1}^{\infty} 2\Re(1-\varphi_n(s))$$

for every $s \in \Gamma$. Moreover, this equation implies that b is proper because $\|b(s)\|^2 \leq N$ implies $|\varphi_n(s)| \geq 1/2$ for some $n \in \{1, \ldots, N\}$. □

In order to give examples of groups with the Haagerup property, we now show how to construct a 1-cocycle on a group which acts properly on a tree. (See Appendix E for our convention on trees.)

Let $\mathbf{T}$ be a tree and fix a base point o in $\mathbf{T}$. We view every edge $\mathbf{e} = (x, y) \in \mathbf{E}$ as a path of length one from x to y and denote its reverse path by $\bar{\mathbf{e}}$. For each vertex x in $\mathbf{T}$, we denote by $[o, x]$ the unique geodesic path that connects o to x. We define $b_o(x) \in \ell^2(\mathbf{E})$ by

$$b_o(x)(\mathbf{e}) = \begin{cases} 1 & \text{if } \mathbf{e} \text{ is on } [o,x], \\ -1 & \text{if } \bar{\mathbf{e}} \text{ is on } [o,x], \\ 0 & \text{otherwise.} \end{cases}$$

A computation confirms that

$$\|b_o(x) - b_o(y)\|^2 = 2d(x, y),$$

where d is the graph metric. It follows that the graph metric d on the tree $\mathbf{T}$ is a conditionally negative definite kernel. (See Section D.)

Now let Γ be a group which acts on $\mathbf{T}$. Denote by π the associated unitary representation of Γ on $\ell^2(\mathbf{E})$. We set $b(s) = b_o(so)$. It is not hard to check that b is a 1-cocycle on Γ with coefficients in $(\pi, \ell^2(\mathbf{E}))$ such that $\|b(s)\|^2 = 2d(o, so)$ (see Figure 1).

Since every finitely generated free group $\mathbb{F}_r$ acts properly on its Cayley graph (which is a tree), the associated 1-cocycle is proper on $\mathbb{F}_r$. Hence free groups have the Haagerup property. (Infinitely generated free groups also have the Haagerup property because they are increasing unions of finitely generated free groups.) The group $\mathrm{SL}(2,\mathbb{Z}) = (\mathbb{Z}/4\mathbb{Z}) *_{\mathbb{Z}/2\mathbb{Z}} (\mathbb{Z}/6\mathbb{Z})$ is another example which acts properly on a tree (being an amalgamated free product of finite groups, its action on the Bass-Serre tree is proper – see Appendix E).

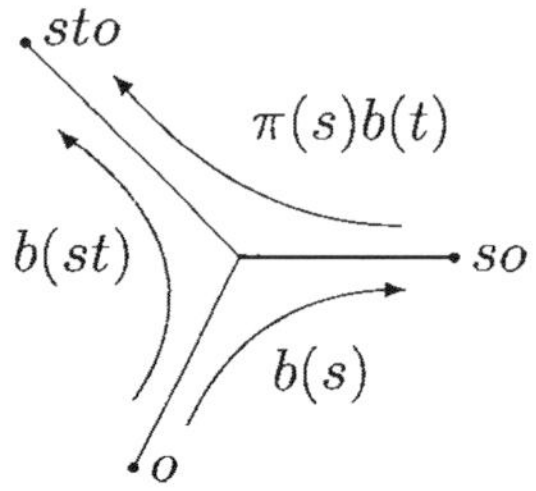

Figure 1. $b(st) = b(s) + \pi(s)b(t)$

In summary, we obtain the following theorem of Haagerup. (See Theorem D.11.)

Theorem 12.2.5. *The graph metric on a tree is conditionally negative definite. A group which acts properly on a tree has the Haagerup property. In particular, free groups and* $\mathrm{SL}(2,\mathbb{Z})$ *have the Haagerup property.*

A tree is an example of a space with walls.

Definition 12.2.6. A *space with walls* is a pair $(X, \mathcal{W})$ consisting of a set X and a family of partitions of X into two classes, called *walls*, such that the number $w(x, y)$ of walls separating x and y is finite for every $x, y \in X$. Here, a wall $w = \{H, H^c\}$ *separates* x and y if either $x \in H$ and $y \in H^c$ or $x \in H^c$ and $y \in H$.

Example 12.2.7. Let $\mathbf{T}$ be a tree, X be the vertex set and $\mathcal{W}$ be the edge set. Each edge $w \in \mathcal{W}$ of the tree $\mathbf{T}$ can be considered as a partition of X into two connected components. Hence, $(X, \mathcal{W})$ is a space with walls such that $w(x, y) = d(x, y)$, the graph metric.

Example 12.2.8. A finite Cartesian product of spaces with walls is naturally a space with walls. In particular, $(\mathbb{Z}^n, \text{hyperplanes})$ is a space with walls, such that w is again the graph metric on $\mathbb{Z}^n$.

We say a group Γ acts on a space $(X, \mathcal{W})$ with walls if Γ acts on X as permutations and preserves the wall structure. The action is said to be *proper* if $\lim_{s\to\infty} w(x, s.x) = \infty$ for every/some $x \in X$. The following result generalizes Theorem 12.2.5.

Theorem 12.2.9. *For any space* $(X, \mathcal{W})$ *with walls, the function* $w\colon X \times X \to \mathbb{Z}_{\geq 0}$ *is conditionally negative definite. A group which acts properly on a space with walls has the Haagerup property.*

Proof. Let $o \in X$ be a fixed base point. For every $x \in X$, we define $\zeta(x) \in \ell^2(\mathcal{W})$ to be the characteristic function of the set of walls separating

o and x. We leave it to the reader to check $w(x, y) = \|\zeta(y) - \zeta(x)\|^2$. Hence w is conditionally negative definite. The second assertion follows from Theorem D.11. □

Let us recall the definition of the wreath product $\Upsilon \wr \Lambda$ of a group Υ by another group Λ. To ease notation, denote by Υ_Λ the algebraic direct product group $\bigoplus_\Lambda \Upsilon$ and view an element $x \in \Upsilon_\Lambda$ as a finitely supported function $x\colon \Lambda \to \Upsilon$, where the support of x is $\mathrm{supp}(x) = \{p \in \Lambda : x(p) \neq e\}$. We note that $(xy)(p) = x(p)y(p) \in \Upsilon$ for $x, y \in \Upsilon_\Lambda$ and $p \in \Lambda$. Then, Λ acts on Υ_Λ by left translation: $\alpha_s(x)(p) = x(s^{-1}p)$. The *wreath product* $\Upsilon \wr \Lambda$ is defined to be the semidirect product $\Upsilon_\Lambda \rtimes_\alpha \Lambda$.

Definition 12.2.10. We say a group Γ has a *proper wall structure* if there is a family $\mathcal{W}$ of walls in Γ such that $(\Gamma, \mathcal{W})$ is a space with walls on which Γ acts properly by left multiplication.

Theorem 12.2.11. *Let $\Gamma = \Upsilon \wr \Lambda$ be the wreath product of Υ by Λ. If Υ is finite and Λ has a proper wall structure, then Γ has a proper wall structure. In particular, the group $(\mathbb{Z}/2\mathbb{Z}) \wr \mathbb{F}_2$ has the Haagerup property.*

Proof. Below, we'll let s and t denote elements of Λ, while x and y represent elements of Υ_Λ. Thus xs or yt will be generic elements of Γ.

Let $\mathcal{W}_\Lambda$ be a proper wall structure of Λ and $\mathcal{H}$ be the corresponding family of half spaces: $\mathcal{H} = \{H : \{H, H^c\} \in \mathcal{W}_\Lambda\}$. For $H \in \mathcal{H}$ and a finitely supported function $\mu\colon H^c \to \Upsilon$, we define

$$E(H, \mu) = \{xs \in \Gamma : s \in H \text{ and } x|_{H^c} = \mu\} \subset \Gamma.$$

Now, define a family $\mathcal{W}$ of walls in Γ by

$$\mathcal{W} = \{\{E(H, \mu), E(H, \mu)^c\} : H \in \mathcal{H},\ \mu\colon H^c \to \Upsilon \text{ finitely supported}\}.$$

We first claim that $(\Gamma, \mathcal{W})$ is a space with walls. Let $xs, yt \in \Gamma$ be given and we'll show that there are finitely many H's and μ's such that $xs \in E(H, \mu)$ and $yt \in E(H, \mu)^c$. Indeed, $xs \in E(H, \mu)$ means $s \in H$ and $x|_{H^c} = \mu$, while $yt \in E(H, \mu)^c$ means either $t \in H^c$ or $y|_{H^c} \neq \mu$. It follows that

$$H^c \cap (\{t\} \cup \mathrm{supp}(x^{-1}y)) \neq \emptyset$$

and hence the half space H separates s and some element in $\{t\} \cup \mathrm{supp}(x^{-1}y)$. Since the set $\{t\} \cup \mathrm{supp}(x^{-1}y)$ is finite, there are finitely many such H's. Note that $\mu = x|_{H^c}$ is uniquely determined by xs and H.

We next observe that Γ preserves the wall structure; that is, $sE(H, \mu) = E(sH, s.\mu)$ for $s \in \Lambda$ and $xE(H, \mu) = E(H, (x^{-1}|_{H^c})\mu)$ for $x \in \Upsilon_\Lambda$.

Finally, we check that the action of Γ on $(\Gamma, \mathcal{W})$ is proper. Let $w_\Gamma(xs)$ (resp. $w_\Lambda(s)$) be the number of walls in Γ (resp. in Λ) separating $xs \in \Gamma$ (resp.

$s \in \Lambda$) and the unit. We have to show that the set $\{xs \in \Gamma : w_\Gamma(xs) \leq N\}$ is finite for every $N \in \mathbb{N}$. For this, we claim that

$$w_\Gamma(xs) \leq N \Longrightarrow \{s\} \cup \operatorname{supp}(x) \subset \{t \in \Lambda : w_\Lambda(t) \leq N\} =: B_\Lambda(N).$$

Since $B_\Lambda(N)$ and Υ are finite by assumption, this suffices. Let $xs \in \Gamma$ and suppose by contrapositive that there is $t \in \{s\} \cup \operatorname{supp}(x)$ such that $w_\Lambda(t) > N$. Then there are $w_\Lambda(t)$ many half spaces $H \in \mathcal{H}$ such that $e \in H$ and $t \in H^c$, which implies $e \in E(H, 1_{H^c})$ and $xs \in E(H, 1_{H^c})^c$. (Here 1_{H^c} is the unit function.) It follows that $w_\Gamma(xs) > N$, so we're done. □

We now introduce the notion of a co-amenable subgroup – a convenient generalization of finite-index subgroups and normal subgroups with amenable quotients.

Definition 12.2.12. Let Λ be a subgroup of Γ. We say that Λ is *co-amenable* in Γ if there exists a left Γ-invariant mean μ on $\ell^\infty(\Gamma/\Lambda)$.

It can be shown that a subgroup Λ is co-amenable in Γ if and only if it satisfies a *co-Følner condition*: For any finite subset $E \subset \Gamma$ and any $\varepsilon > 0$, there exists a finite subset $F \subset \Gamma/\Lambda$ such that

$$\max_{s \in E} \frac{|sF \triangle F|}{|F|} < \varepsilon.$$

The proof of this fact is a verbatim translation of the proof of Theorem 2.6.8.

Proposition 12.2.13. *Let Λ be a co-amenable subgroup of Γ. If Λ has the Haagerup property, then Γ has the Haagerup property.*

Proof. Let (ψ_i) be a net of positive definite functions on Λ satisfying the conditions in Definition 12.2.1. We denote the induced positive definite functions by

$$\tilde{\psi}_i \colon \Gamma \ni s \mapsto \sum_{x \in \Gamma/\Lambda} \psi_i(\sigma(sx)^{-1} s \sigma(x)) e_{sx,x} \in \mathbb{B}(\ell^2(\Gamma/\Lambda)),$$

where $\sigma \colon \Gamma/\Lambda \to \Gamma$ is a fixed cross section. (See Lemma D.2.) For a finite subset $F \subset \Gamma/\Lambda$, let χ_F be the characteristic function on F and set

$$\tilde{\psi}_{i,F}(s) = \frac{1}{|F|} \langle \tilde{\psi}_i(s) \chi_F, \chi_F \rangle = \frac{1}{|F|} \sum_{x \in F \cap s^{-1}F} \psi_i(\sigma(sx)^{-1} s \sigma(x))$$

for $s \in \Gamma$. It follows that $\tilde{\psi}_{i,F}$ is positive definite and, for any $\varepsilon > 0$, the set

$$\{s \in \Gamma : |\tilde{\psi}_{i,F}(s)| > \varepsilon\} \subset \sigma(F)\{t \in \Lambda : |\psi_i(t)| > \varepsilon\}\sigma(F)^{-1}$$

is finite. Moreover, $\lim_j \lim_i \tilde{\psi}_{i,F_j} = 1$ pointwise for a Følner net $(F_j)_j$. □

Observe that the semidirect product $\mathbb{Z}^2 \rtimes \mathrm{SL}(2,\mathbb{Z})$ of $\mathbb{Z}^2$ by $\mathrm{SL}(2,\mathbb{Z})$ does not have the Haagerup property (by Theorem 12.1.10 and Proposition 12.2.3). Thus, an extension of groups with the Haagerup property need not have the Haagerup property. The group $\mathbb{Z}^2 \rtimes \mathrm{SL}(2,\mathbb{Z})$ also serves as a counterexample to preservation of the Haagerup property under amalgamated free products; indeed,

$$\mathbb{Z}^2 \rtimes \mathrm{SL}(2,\mathbb{Z}) \cong (\mathbb{Z}^2 \rtimes (\mathbb{Z}/4\mathbb{Z})) *_{\mathbb{Z}^2 \rtimes (\mathbb{Z}/2\mathbb{Z})} (\mathbb{Z}^2 \rtimes (\mathbb{Z}/6\mathbb{Z})).$$

The Haagerup property for von Neumann algebras. Let M be a von Neumann algebra with faithful normal tracial state τ and $\theta\colon M \to M$ be a τ-preserving u.c.p. map. Then θ extends to a contraction on $L^2(M,\tau)$.

Definition 12.2.14. Let M be a von Neumann algebra with a faithful normal tracial state τ.[8] We say that M has the *Haagerup property* if there exists a net of τ-preserving u.c.p. maps θ_i on M such that each θ_i extends to a compact operator on $L^2(M,\tau)$ and $\theta_i \to \mathrm{id}_M$ in the point-ultraweak topology.

Theorem 12.2.15. *A group Γ has the Haagerup property if and only if $L(\Gamma)$ has the Haagerup property.*

Proof. Let Γ be a group with the Haagerup property and take a net φ_i of positive definite functions as in Definition 12.2.1. Then the multipliers $\theta_i = m_{\varphi_i}$ (Theorem 2.5.11) are readily seen to satisfy the definition of the Haagerup property for $L(\Gamma)$.

To prove the converse, assume that $L(\Gamma)$ has the Haagerup property and take a net θ_i as in Definition 12.2.14. Let $V\colon \ell^2(\Gamma) \to \ell^2(\Gamma) \otimes \ell^2(\Gamma)$ be the isometry given by $V\delta_t = \delta_t \otimes \delta_t$ and

$$\sigma\colon L(\Gamma) \to L(\Gamma) \bar{\otimes} L(\Gamma) \subset \mathbb{B}(\ell^2(\Gamma) \otimes \ell^2(\Gamma))$$

be the normal $*$-homomorphism given by $\sigma(\lambda(s)) = \lambda(s) \otimes \lambda(s)$. It is easy to check that

$$V^*(\lambda(s) \otimes \lambda(t))V = \begin{cases} \lambda(s) & \text{if } s = t, \\ 0 & \text{if } s \neq t. \end{cases}$$

We set $\varphi_i(s) = \tau(\lambda(s)^*\theta_i(\lambda(s)))$ and observe that $m_{\varphi_i}(a) = V^*(\mathrm{id} \otimes \theta_i) \circ \sigma(a)V$. It follows that each φ_i is positive definite,

$$\limsup_{s\to\infty} |\varphi_i(s)| \leq \limsup_{s\to\infty} \|\theta_i(\lambda(s))\|_2 = 0,$$

and $\lim_i \varphi_i(s) = 1$ for every $s \in \Gamma$. This proves the Haagerup property for Γ. □

---

[8]As with property (T), the definition of the Haagerup property does not depend on the choice of the faithful normal tracial state – see [**90**] for more. Thus, for group von Neumann algebras we can (and will) always take the canonical trace.

In recent years Popa has exploited the Haagerup property in combination with relative property (T) – pitting one against the other, in some sense – with remarkable success. Here we content ourselves with one striking example of his work.

It is a fact that $L(\mathbb{Z}^2)$ is a Cartan subalgebra (Definition F.15) of $L(\mathbb{Z}^2 \rtimes \mathrm{SL}(2,\mathbb{Z}))$. Evidently the normalizer generates everything, so we only have to see why $L(\mathbb{Z}^2)$ is maximal. But in general, if $\Lambda \subset \Gamma$, $x \in L(\Gamma) \cap L(\Lambda)'$ and $x = \sum_{s\in\Gamma} x(s)\lambda(s)$ is the Fourier expansion of x, then $x(asa^{-1}) = x(s)$ for every $s \in \Gamma$ and $a \in \Lambda$ (since $[x, \lambda(a)] = 0$). As $x \in \ell^2(\Gamma)$, we conclude that

$$\operatorname{supp} x = \{s \in \Gamma : \text{the set } \{asa^{-1} : a \in \Lambda\} \text{ is finite}\}$$

for $x \in L(\Gamma) \cap L(\Lambda)'$. It is now easy to deduce that $L(\mathbb{Z}^2)$ is a maximal abelian subalgebra of $L(\mathbb{Z}^2 \rtimes \mathrm{SL}(2,\mathbb{Z}))$.

The inclusion $L(\mathbb{Z}^2) \subset L(\mathbb{Z}^2 \rtimes \mathrm{SL}(2,\mathbb{Z}))$ has relative property (T) by Theorem 12.1.10, while $\mathrm{SL}(2,\mathbb{Z})$ has the Haagerup property – this tension yields the following theorem of Popa.[9]

Theorem 12.2.16. *Let $\Gamma = \mathbb{Z}^2 \rtimes \mathrm{SL}(2,\mathbb{Z})$ and, to simplify notation, denote the inclusion $L(\mathbb{Z}^2) \subset L(\Gamma)$ by $A \subset M$. If $B \subset M$ is another Cartan subalgebra with relative property* (T)*, then there exists a unitary element $u \in M$ such that $u^*Bu = A$.*

Proof. We are going to use Theorem F.12 and Lemma F.17. Take a sequence φ_n of positive definite functions on $\mathrm{SL}(2,\mathbb{Z})$ as in Definition 12.2.1. It is not hard to check that the map $\theta_n \colon \mathbb{C}[\Gamma] \to \mathbb{C}[\Gamma]$, defined by

$$\theta_n(\lambda(as)) = \varphi_n(s)\lambda(as),$$

extends to a trace-preserving u.c.p. map on $L(\Gamma)$. Since $\theta_n \to \mathrm{id}$ in the point-ultraweak topology and $B \subset M$ has relative property (T), there exists n_0 such that $\|b - \theta_{n_0}(b)\|_2 < \|b\|/3$ for all $b \in B$. Let $T \in \mathbb{B}(L^2(M))$ be the positive contraction given by $T(\widehat{x}) = \widehat{\theta_{n_0}(x)}$ for $x \in M$ (or equivalently $T\delta_{as} = \varphi_{n_0}(s)\delta_{as}$ on $\ell^2(\Gamma)$) and $d = \chi_{[1/3,1]}(T)$. Note that T and d commute with the right A-action and hence $d \in \langle M, A\rangle$, where $\langle M, A\rangle$ is the basic construction (see Appendix F). Moreover, we have

$$\mathrm{Tr}(d) = |\{s \in \mathrm{SL}(2,\mathbb{Z}) : \varphi_{n_0}(s) \geq 1/3\}| < \infty$$

for the canonical trace Tr on $\langle M, A\rangle$. For any unitary element $w \in B$, we have

$$\|\widehat{1} - wdw^*\widehat{1}\| \leq 1/3 + \|\widehat{w^*} - T\widehat{w^*}\| = 1/3 + \|w^* - \theta_{n_0}(w^*)\|_2 \leq 2/3$$

[9] Combined with work of Gaboriau on ergodic theory, this result gives the first example of a type II_1-factor whose fundamental group is trivial ([**160**]).

and 0 is not contained in the ultraweakly closed convex hull of $\{wdw^* : w \in B \text{ unitary}\}$. By Theorem F.12 and Lemma F.17, we are done. □

Exercises

Exercise 12.2.1. Let Γ_i be groups with a common subgroup Λ and let $\Gamma = \Gamma_1 *_\Lambda \Gamma_2$ be the amalgamated free product. Let φ_i be a unital positive definite function on Γ_i which is bi-Λ-invariant: $\varphi_i(asb) = \varphi_i(s)$ for every $s \in \Gamma_i$ and $a, b \in \Lambda$. Prove that there is a unital positive definite function $(\varphi_1 * \varphi_2)$ on Γ such that

$$(\varphi_1 * \varphi_2)(as_1 \cdots s_n b) = \varphi_{i(1)}(s_1) \cdots \varphi_{i(n)}(s_n)$$

for $a, b \in \Lambda$ and $s_j \in \Gamma_{i(j)} \setminus \Lambda$ with $i(j) \neq i(j+1)$.

Exercise 12.2.2. Let Γ_i be groups with the Haagerup property and let $\Lambda \subset \Gamma_i$ be a finite common subgroup. Prove that the amalgamated free product $\Gamma_1 *_\Lambda \Gamma_2$ has the Haagerup property.

12.3. Weak amenability

Before proceeding, you may wish to review Appendix D on Schur and Herz-Schur multipliers.

Definition 12.3.1. A discrete group Γ is said to be *weakly amenable* if there exists a net (φ_i) of finitely supported functions on Γ such that $\varphi_i \to 1$ pointwise and $\limsup \|\varphi_i\|_{B_2} \leq C$.[10]

The *Cowling-Haagerup constant* $\Lambda_{\mathrm{cb}}(\Gamma)$ is the infimum of all C for which such a net (φ_i) exists. We set $\Lambda_{\mathrm{cb}}(\Gamma) = \infty$ if Γ is not weakly amenable.

Example 12.3.2. Amenable groups are weakly amenable. The class of weakly amenable groups is closed under taking subgroups and Cartesian products. If $\Gamma = \bigcup \Gamma_i$ is a directed union, then $\Lambda_{\mathrm{cb}}(\Gamma) = \sup \Lambda_{\mathrm{cb}}(\Gamma_i)$.

Let $\mathbf{T}$ be a tree (identified with the vertex set) and d be the graph metric. Let χ_n be the characteristic function on $\{(x, y) \in \mathbf{T}^2 : d(x, y) = n\}$ and $\chi_{\leq K} = \sum_{n=0}^{K} \chi_n$. For simplicity, we denote by $\|\theta\|_{\mathrm{cb}}$ the (cb-)norm of the corresponding Schur multiplier m_θ.

Theorem 12.3.3. *We have $\|\chi_n\|_{\mathrm{cb}} \leq 2n$ for every $n \geq 1$.*

Proof. Fix a geodesic ray ω in $\mathbf{T}$; that is, ω is any isometric function from $\mathbb{N}$ into $\mathbf{T}$.[11] For every $x \in \mathbf{T}$, there exists a unique geodesic ray ω_x which

---

[10] Recall that the Herz-Schur norm $\|\varphi\|_{B_2}$ is $\leq C$ if and only if there exist families of vectors $(\xi_s)_{s\in\Gamma}$ and $(\eta_t)_{t\in\Gamma}$ in a Hilbert space $\mathcal{H}$ such that $\varphi(st^{-1}) = \langle \eta_t, \xi_s \rangle$ for every $s, t \in \Gamma$ and $\sup_{s,t\in\Gamma} \|\xi_s\|\|\eta_t\| \leq C$.

[11] If there is no geodesic ray, just attach one to $\mathbf{T}$. This does not affect anything because the restriction to a rectangular subset does not increase the Schur multiplier norm.

starts at x and eventually flows into ω. It is not hard to see (cf. Figure 2) that

$$\psi_n(x,y) := \sum_{k=0}^{n} \langle \delta_{\omega_y(k)}, \delta_{\omega_x(n-k)} \rangle = \sum_{m=0}^{\lfloor n/2 \rfloor} \chi_{n-2m}(x,y)$$

for any $x, y \in \mathbf{T}$. In particular, $\chi_n = \psi_n - \psi_{n-2}$ for every $n \geq 2$. Since we have $\|\psi_n\|_{\mathrm{cb}} \leq n+1$ by Theorem D.4, we are done. □

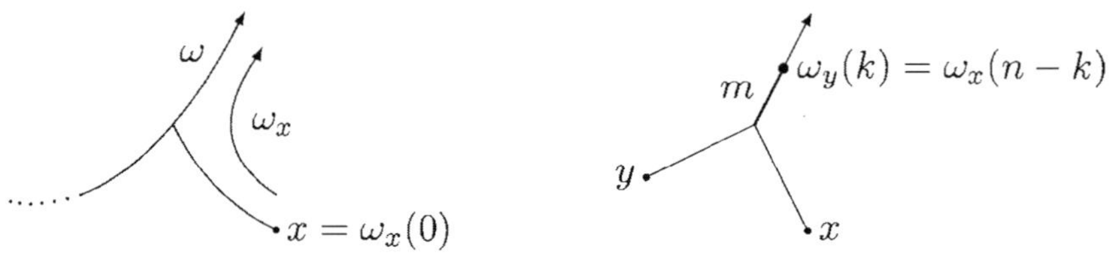

Figure 2

Corollary 12.3.4. *There is an increasing sequence $K_n \in \mathbb{N}$ such that the kernels*

$$\theta_n(x,y) = \chi_{\leq K_n}(x,y) \exp(-\frac{1}{n} d(x,y))$$

on the tree $\mathbf{T}$ *satisfy* $\limsup_{n\to\infty} \|\theta_n\|_{\mathrm{cb}} = 1$.

Proof. Let $\psi_n(x,y) = \exp(-n^{-1} d(x,y))$. By Theorems 12.2.5 and D.11, the kernel ψ_n is positive definite and hence $\|\psi_n\|_{\mathrm{cb}} = 1$. Since $\chi_k \psi_n = e^{-k/n} \chi_k$, Theorem 12.3.3 implies that $\|\chi_k \psi_n\|_{\mathrm{cb}} \leq 2k e^{-k/n}$ for every n and k. Therefore, for any integer K, one has

$$\|\chi_{\leq K} \psi_n\|_{\mathrm{cb}} \leq \|\psi_n\|_{\mathrm{cb}} + \|\sum_{k>K} \chi_k \psi_n\|_{\mathrm{cb}} \leq 1 + \sum_{k>K} 2k e^{-k/n}.$$

Thus, if K_n is chosen sufficiently large, we have $\|\theta_n\|_{\mathrm{cb}} < 1 + n^{-1}$. □

Corollary 12.3.5. *If a group Γ acts properly on a tree, then it is weakly amenable with $\Lambda_{\mathrm{cb}}(\Gamma) = 1$. In particular, free groups and* $\mathrm{SL}(2,\mathbb{Z})$ *are weakly amenable and their Cowling-Haagerup constants are* 1.

Proof. Let $\mathbf{T}$ be the tree on which Γ acts properly. Let θ_n be as in Corollary 12.3.4 and set $\varphi_n(s) = \theta_n(o, s^{-1}.o)$ for a fixed base point $o \in \mathbf{T}$. Then, $\varphi_n \to 1$ pointwise, and φ_n is finitely supported since the action is proper. Moreover, since $\varphi_n(st^{-1}) = \theta_n(s^{-1}.o, t^{-1}.o)$, we have $\|\varphi_n\|_{B_2} = \|\theta_n\|_{\mathrm{cb}} \to 1$. □

Here is another proof, using ideas from [**162**], of the corollary above, which simultaneously proves the Haagerup property.

Proof. For simplicity, we assume that $\mathbf{T}$ is uniformly locally finite. As in the proof of Theorem 12.3.3, we fix a geodesic ray ω in the tree $\mathbf{T}$ and, for every $x \in \mathbf{T}$, denote by ω_x the unique geodesic ray which starts at x and eventually flows into ω. For every $z \in \mathbb{D} = \{z \in \mathbb{C} : |z| < 1\}$, we define a function $\zeta_z : \mathbf{T} \to \ell^2(\mathbf{T})$ by

$$\zeta_z(x) = \sqrt{1-z^2} \sum_{k=0}^{\infty} z^k \delta_{\omega_x(k)},$$

where $\sqrt{1-z^2}$ denotes the principal branch of the square root. The series above converges absolutely in z and uniformly in x:

$$\|\zeta_z(x)\|^2 = |1-z^2| \sum_{k=0}^{\infty} |z|^{2k} = \frac{|1-z^2|}{1-|z|^2}.$$

In particular, the function $\mathbb{D} \ni z \mapsto \zeta_z \in \ell^\infty(\mathbf{T}, \ell^2(\mathbf{T}))$ is holomorphic. Hence the function $z \mapsto \theta_z$, defined by

$$\theta_z(x,y) = \langle \zeta_z(y), \overline{\zeta_z(x)} \rangle,$$

is also a holomorphic function into the Banach space of Schur multipliers. A calculation similar to that in the proof of Theorem 12.3.3 yields

$$\begin{aligned}
\theta_z(x,y) &= (1-z^2) \sum_{k,l=0}^{\infty} z^{k+l} \delta_{\omega_x(k),\omega_y(l)} \\
&= (1-z^2) \sum_{n=0}^{\infty} \sum_{k=0}^{n} z^n \delta_{\omega_x(k),\omega_y(n-k)} \\
&= (1-z^2) \sum_{m=0}^{\infty} z^{d(x,y)+2m} \\
&= z^{d(x,y)}.
\end{aligned}$$

Now, suppose that a group Γ acts properly on $\mathbf{T}$ and define $\varphi_z \in B_2(\Gamma)$ by $\varphi_z(s) = z^{d(s.o,o)}$, for a fixed base point $o \in \mathbf{T}$. It follows that the function $\mathbb{D} \ni z \mapsto \varphi_z \in B_2(\Gamma)$ is holomorphic. Since $\zeta_r = \overline{\zeta_r}$ for all $r \in \mathbb{R}$, φ_r is positive definite for $0 < r < 1$. Moreover, $\varphi_r \to 1$ as $r \to 1$ and hence it suffices to show that φ_r belongs to the Banach subspace $F \subset B_2(\Gamma)$ spanned by finitely supported functions. Since $\mathbf{T}$ is uniformly locally finite and the action of Γ is proper, there exists $C > 0$ such that $|\{s \in \Gamma : d(s.o,o) \leq n\}| \leq C^n$ for every n. It follows that $\varphi_z \in \ell^1(\Gamma) \subset F$ for $|z| < C^{-1}$. This implies that $\mathbb{D} \ni z \mapsto \varphi_z + F \in B_2(\Gamma)/F$ is a holomorphic function which is zero for $|z| < C^{-1}$. By uniqueness of holomorphic extensions, $\varphi_z \in F$ for all $z \in \mathbb{D}$. $\square$

Let $\Lambda \triangleleft \Gamma$ be a normal subgroup. Then, Γ acts on Λ by conjugation: $\mathrm{Ad}(s)(a) = sas^{-1}$ for $s \in \Gamma$ and $a \in \Lambda$. Abusing notation, we also denote by $\mathrm{Ad}(s)$ the action of $s \in \Gamma$ on $\ell^\infty(\Lambda)$, given by $\mathrm{Ad}(s)(f) = f \circ \mathrm{Ad}(s^{-1})$.

Theorem 12.3.6. *Let Γ be a group with $\Lambda_{\mathrm{cb}}(\Gamma) = 1$ and let $\Lambda \triangleleft \Gamma$ be a normal abelian subgroup. Then, there exists a Λ-invariant mean on $\ell^\infty(\Lambda)$ which is $\mathrm{Ad}(\Gamma)$-invariant.*

Proof. Let $\varphi_i \colon \Gamma \to \mathbb{C}$ be a net of finitely supported functions such that $\|\varphi_i\|_{B_2} \le 1$ and $\varphi_i \to 1$ pointwise. By Theorem D.4, there are Hilbert spaces $\mathcal{H}_i$ and unit vectors $\xi_i(s), \eta_i(t) \in \mathcal{H}$ such that $\varphi_i(st^{-1}) = \langle \eta_i(t), \xi_i(s)\rangle$ for every $s, t \in \Gamma$. For each $s \in \Gamma$, we have

$$\limsup_i \|\xi_i(st) - \xi_i(t)\| \le \limsup_i \|\xi_i(st) - \eta_i(t)\| + \limsup_i \|\xi_i(t) - \eta_i(t)\| = 0$$

uniformly for $t \in \Gamma$, since $\lim_i \varphi_i(s) = 1 = \lim_i \varphi_i(e)$. Likewise, $\lim_i \|\eta_i(st) - \eta_i(t)\| = 0$ uniformly for $t \in \Gamma$. It follows that $\|\varphi_i - \varphi_i \circ \mathrm{Ad}(s)\|_{B_2} \to 0$ for every $s \in \Gamma$.

Since Λ is amenable, there is a trivial character $\tau_0 \colon C^*_\lambda(\Lambda) \to \mathbb{C}$ (Theorem 2.6.8). Since the Herz-Schur multiplier m_{φ_i} maps $L(\Lambda)$ into $\mathbb{C}[\Lambda] \subset C^*_\lambda(\Lambda)$, we may define linear functionals ω_i on $L(\Lambda)$ by $\omega_i = \tau_0 \circ m_{\varphi_i}|_{L(\Lambda)}$. Observe that $\omega_i \in L(\Lambda)_*$ with $\|\omega_i\| \le \|\varphi_i\|_{B_2} \le 1$. Let $\widehat{\Lambda}$ be the Pontryagin dual of Λ and recall that $C^*_\lambda(\Lambda) \cong C(\widehat{\Lambda})$ and $L(\Lambda) \cong L^\infty(\widehat{\Lambda})$ via the Fourier transform $\ell^2(\Lambda) \cong L^2(\widehat{\Lambda})$. Also, note that $\mathrm{Ad}(s)$ acts naturally on $\widehat{\Lambda}$ and hence on $L^p(\widehat{\Lambda})$ for $1 \le p \le \infty$. We denote by $f_i \in L^1(\widehat{\Lambda})$ the element corresponding to $\omega_i \in L(\Lambda)_*$. Since $\|f_i\|_1 \le 1$ and $\int f_i = \omega_i(1) = \varphi_i(e) \to 1$, we have $\||f_i| - f_i\|_1 \to 0$. Let $\xi_i \in \ell^2(\Lambda)$ be the element corresponding to $|f_i|^{1/2} \in L^2(\widehat{\Lambda})$. Then, we have

$$\lim_i \langle \lambda(a)\xi_i, \xi_i\rangle = \lim_i |\omega_i|(\lambda(a)) = \lim_i \omega_i(\lambda(a)) = \lim_i \varphi_i(a) = 1$$

for every $a \in \Lambda$ and

$$\begin{aligned}
\lim_i \|\xi_i - \mathrm{Ad}(s)(\xi_i)\|_{\ell^2}^2 &= \lim_i \left\| |f_i|^{1/2} - \mathrm{Ad}(s)(|f_i|^{1/2}) \right\|_{L^2}^2 \\
&\le \lim_i \left\| |f_i| - \mathrm{Ad}(s)(|f_i|) \right\|_{L^1} \\
&= \lim_i \|f_i - \mathrm{Ad}(s)(f_i)\|_{L^1} \\
&\le \lim_i \|\varphi_i - \varphi_i \circ \mathrm{Ad}(s^{-1})\|_{\mathrm{cb}} \\
&= 0
\end{aligned}$$

for every $s \in \Gamma$. It follows that $\xi_i^2 \in \ell^1(\Lambda)$ is approximately Λ-invariant and approximately $\mathrm{Ad}(\Gamma)$-invariant. Taking any limit point of the net (ξ_i^2) in $\ell^\infty(\Lambda)^*$, we are done. $\square$

Corollary 12.3.7. *Let* $\Gamma = \Upsilon \wr \Lambda$ *be the wreath product of* Υ *by* Λ. *If* $|\Upsilon| > 1$ *and* Λ *is nonamenable, then* $\Lambda_{\mathrm{cb}}(\Gamma) > 1$.[12]

Proof. We assume that $\Lambda_{\mathrm{cb}}(\Gamma) = 1$ and derive a contradiction. Passing to a subgroup, we may assume that Υ is cyclic. Let $\Upsilon_\Lambda = \bigoplus_\Lambda \Upsilon$ be the canonical normal subgroup of $\Gamma = \Upsilon_\Lambda \rtimes \Lambda$. By Theorem 12.3.6, there exists a state μ on $\ell^\infty(\Upsilon_\Lambda)$ which is both Υ_Λ-invariant and $\mathrm{Ad}(\Lambda)$-invariant. Let $S \subset \Upsilon_\Lambda$ be a system of representatives of the Λ-orbits. Thus, $\Upsilon_\Lambda = \{e\} \sqcup \bigsqcup_{x \in S^0} \Lambda/\Lambda^x$ as a Λ-set, where $S^0 = S \setminus \{e\}$ and $\Lambda^x = \{s \in \Lambda : sxs^{-1} = x\}$ is the stabilizer subgroup of $x \in S^0$. Since Λ^x is finite for every $x \in S^0$, by averaging over the right Λ^x-action, we obtain a Λ-equivariant u.c.p. map from $\ell^\infty(\Lambda)$ into $\ell^\infty(\Lambda/\Lambda^x)$. Collecting them together, we obtain a Λ-equivariant u.c.p. map from $\ell^\infty(\Lambda)$ into $\ell^\infty(\Upsilon_\Lambda \setminus \{e\})$. Since Λ is nonamenable, the $\mathrm{Ad}(\Lambda)$-invariant mean μ must be concentrated on $\{e\}$. Such a μ cannot be Υ_Λ-invariant – a contradiction. □

There exist weakly amenable groups whose Cowling-Haagerup constants are greater than 1. Before stating the theorem, we should mention that the definition of weak amenability extends to locally compact groups G. Moreover, for a lattice $\Gamma \leq G$, it turns out that $\Lambda_{\mathrm{cb}}(G) = \Lambda_{\mathrm{cb}}(\Gamma)$. The proofs of these facts are beyond the scope of this book, but we summarize some important results below.

Theorem 12.3.8. *The following statements are true:*

(1) $\Lambda_{\mathrm{cb}}(\mathbb{Z}^2 \rtimes \mathrm{SL}(2, \mathbb{Z}^2)) = \infty$;
(2) $\Lambda_{\mathrm{cb}}(\mathrm{SO}(1,n)) = 1$ *and* $\Lambda_{\mathrm{cb}}(\mathrm{SU}(1,n)) = 1$;
(3) $\Lambda_{\mathrm{cb}}(\mathrm{Sp}(1,n)) = 2n - 1$ *and* $\Lambda_{\mathrm{cb}}(\mathrm{F}_{4(-20)}) = 21$;
(4) *if* G *is a simple Lie group with real rank greater than or equal to* 2, *then* $\Lambda_{\mathrm{cb}}(G) = \infty$.

Definition 12.3.9. We say a C*-algebra A has the *CBAP* (*completely bounded approximation property*) if there exists a net of finite-rank maps $\theta_i \colon A \to A$ such that $\theta_i \to \mathrm{id}_A$ in the point-norm topology and $\sup \|\theta_i\|_{\mathrm{cb}} \leq C$. The *Haagerup constant* $\Lambda_{\mathrm{cb}}(A)$ is the infimum of those C for which such a net (θ_i) exists. We set $\Lambda_{\mathrm{cb}}(A) = \infty$ if A does not have the CBAP.

We say a von Neumann algebra M has the *W* CBAP* (*weak* CBAP*) if there exists a net of ultraweakly-continuous finite-rank maps $\theta_i \colon M \to M$ such that $\theta_i \to \mathrm{id}_M$ in the point-ultraweak topology and $\sup \|\theta_i\|_{\mathrm{cb}} \leq C$. The Haagerup constant $\Lambda_{\mathrm{cb}}(M)$ is again the infimum of those C for which such a net (θ_i) exists, and $\Lambda_{\mathrm{cb}}(M) = \infty$ if M does not have the weak* CBAP.

[12] It is plausible that $\Lambda_{\mathrm{cb}}(\Gamma) = \infty$.

Though the context should always be clear, one must be careful not to mix up Λ_{cb} for C*-algebras and for von Neumann algebras.

Theorem 12.3.10. *Let Γ be a discrete group. Then*

$$\Lambda_{\text{cb}}(\Gamma) = \Lambda_{\text{cb}}(C^*_\lambda(\Gamma)) = \Lambda_{\text{cb}}(L(\Gamma)).$$

Proof. We trivially have $\Lambda_{\text{cb}}(\Gamma) \geq \Lambda_{\text{cb}}(C^*_\lambda(\Gamma))$ and $\Lambda_{\text{cb}}(\Gamma) \geq \Lambda_{\text{cb}}(L(\Gamma))$. To prove the reverse inequalities at once, let a finite subset $E \subset \Gamma$ and $\varepsilon > 0$ be given and choose a finite-rank map $\theta\colon C^*_\lambda(\Gamma) \to L(\Gamma)$ such that $\|\theta\|_{\text{cb}} = C$ and $|1 - \tau(\lambda(s)^*\theta(\lambda(s)))| < \varepsilon$ for $s \in E$. Since $\|\psi\|_{B_2} \leq \|\psi\|_{\ell^2}$ for any ψ, it suffices to show that the function $\varphi(s) = \tau(\lambda(s)^*\theta(\lambda(s)))$ is in $\ell^2(\Gamma)$ and $\|(m_\varphi)_{|C^*_\lambda(\Gamma)}\|_{\text{cb}} \leq C$. (See Proposition D.6.) Since θ has finite rank, there exist finite sequences $\omega_1, \ldots, \omega_n \in C^*_\lambda(\Gamma)^*$ and $x_1, \ldots, x_n \in L(\Gamma)$ such that $\theta(a) = \sum_{k=1}^n \omega_k(a)x_k$ for all $a \in C^*_\lambda(\Gamma)$. It follows that

$$\varphi(s) = \tau(\lambda(s)^*\theta(\lambda(s))) = \sum_{k=1}^{n} \omega_k(\lambda(s))\tau(\lambda(s)^*x_k).$$

Since $\sup_{s\in\Gamma} |\omega_k(\lambda(s))| \leq \|\omega_k\|$ and $\sum_{s\in\Gamma} |\tau(\lambda(s)^*x_k)|^2 = \|x_k\delta_e\|^2 < \infty$ for every k, the function φ is in $\ell^2(\Gamma)$. We denote by π the $*$-homomorphism from $C^*_\lambda(\Gamma)$ into $C^*_\lambda(\Gamma)\otimes C^*_\lambda(\Gamma)$ given by $\pi(\lambda(s)) = \lambda(s)\otimes\lambda(s)$ for every $s \in \Gamma$. Let V be the isometry from $\ell^2(\Gamma)$ into $\ell^2(\Gamma) \otimes \ell^2(\Gamma)$ given by $V\delta_s = \delta_s \otimes \delta_s$ for $s \in \Gamma$. It is not hard to check that

$$m_\varphi(a) = V^*(\text{id}_{C^*_\lambda(\Gamma)} \otimes \theta)(a)V$$

for $a \in C^*_\lambda(\Gamma)$, and hence $\|(m_\varphi)_{|C^*_\lambda(\Gamma)}\|_{\text{cb}} \leq \|\theta\|_{\text{cb}}$. □

Proposition 12.3.11. *Let Λ be a co-amenable subgroup of Γ. Then there exist nets of u.c.p. maps*

$$\varphi_i\colon C^*_\lambda(\Gamma) \to \mathbb{M}_{n(i)}(C^*_\lambda(\Lambda)) \text{ and } \psi_i\colon \mathbb{M}_{n(i)}(C^*_\lambda(\Lambda)) \to C^*_\lambda(\Gamma)$$

such that $\psi_i \circ \varphi_i \to 1$ in the point-norm topology.

Proof. Fix a cross section $\sigma\colon \Gamma/\Lambda \to \Gamma$ and identify $(\Gamma/\Lambda) \times \Lambda$ with Γ via $(p, a) \mapsto \sigma(p)a$. Then we have

$$\lambda_\Gamma(s) = \sum_{p\in\Gamma/\Lambda} e_{sp,p} \otimes \lambda_\Lambda(\sigma(sp)^{-1}s\sigma(p)) \in \mathbb{B}(\ell^2(\Gamma/\Lambda) \otimes \ell^2(\Lambda)),$$

where λ_Γ and λ_Λ are the left regular representations of Γ and Λ, respectively. For a finite subset $F \subset \Gamma/\Lambda$, let

$$\varphi_F\colon \mathbb{B}(\ell^2(\Gamma/\Lambda) \otimes \ell^2(\Lambda)) \to \mathbb{B}(\ell^2(F) \otimes \ell^2(\Lambda))$$

be the compression map; observe that $\varphi_F(C^*_\lambda(\Gamma)) \subset \mathbb{B}(\ell^2(F)) \otimes C^*_\lambda(\Lambda)$. Let

$$V_F\colon \ell^2(\Gamma) \to \ell^2(\Gamma/\Lambda) \otimes \ell^2(F) \otimes \ell^2(\Lambda)$$

be an isometry such that, for each $s \in \Gamma$,

$$V_F \delta_s = \frac{1}{\sqrt{|F|}} \sum_{p \in F} \delta_{s^{-1}p} \otimes \delta_p \otimes \delta_{\sigma(p)^{-1}s\sigma(s^{-1}p)}.$$

Setting $\psi_F(x) = V_F^*(1_{\ell^2(\Gamma/\Lambda)} \otimes x)V_F$ for $x \in \mathbb{B}(\ell^2(F)) \otimes C^*_\lambda(\Lambda)$, it is routine to check that

$$\psi_F(e_{p,q} \otimes \lambda_\Lambda(a)) = \frac{1}{|F|} \delta_{a\sigma(q)^{-1}q,\sigma(p)^{-1}p} \lambda(\sigma(p)a\sigma(q)^{-1}) \in C^*_\lambda(\Gamma)$$

for every $p, q \in F$ and $a \in \Lambda$. (As you probably guessed, $\delta_{a\sigma(q)^{-1}q,\sigma(p)^{-1}p}$ is the Kronecker delta.) Hence, the range of ψ_F is contained in $C^*_\lambda(\Gamma)$ and

$$\psi_F \circ \varphi_F(\lambda(s)) = \frac{|F \cap sF|}{|F|} \lambda(s)$$

for every $s \in \Gamma$. Taking F to be a suitable Følner subset, we are done. □

Corollary 12.3.12. *Let Λ be a co-amenable subgroup of Γ. Then $\Lambda_{\mathrm{cb}}(\Gamma) = \Lambda_{\mathrm{cb}}(\Lambda)$.*

Theorem 12.3.13. *For C*-algebras A_1 and A_2 we have $\Lambda_{\mathrm{cb}}(A_1 \otimes A_2) = \Lambda_{\mathrm{cb}}(A_1)\Lambda_{\mathrm{cb}}(A_2)$. The same is true for von Neumann algebras (replacing $\otimes$ with $\bar{\otimes}$). In particular, $\Lambda_{\mathrm{cb}}(\Gamma_1 \times \Gamma_2) = \Lambda_{\mathrm{cb}}(\Gamma_1)\Lambda_{\mathrm{cb}}(\Gamma_2)$, for discrete groups Γ_1 and Γ_2.*

This theorem holds for arbitrary operator spaces (with the obvious definition of Λ_{cb}). Hence the proof requires some operator space results, which we now establish.

Let A and E be operator spaces with $\dim(E) < \infty$. Then there is a natural inclusion $A \odot E^* \hookrightarrow \mathrm{CB}(A, E)^*$, where the duality between $\mathrm{CB}(A, E)$ and $A \odot E^*$ is given by $\langle \varphi, u \rangle = \sum_k f_k(\varphi(a_k))$ for $\varphi \in \mathrm{CB}(A, E)$ and $u = \sum_k a_k \otimes f_k \in A \odot E^*$. We denote by $A \widehat{\otimes} E^*$ the space $A \odot E^*$ equipped with the norm induced from $\mathrm{CB}(A, E)^*$. An element $f \in \mathbb{M}_n(E)^*$ is written as $f = [f_{i,j}] \in \mathbb{M}_n(E^*)$, where $\langle f, x \rangle = \sum f_{i,j}(x_{i,j})$ for $x = [x_{i,j}] \in \mathbb{M}_n(E)$. For $a = [a_{i,j}] \in \mathbb{M}_n(A)$ and $f = [f_{i,j}] \in \mathbb{M}_n(E)^*$, we define $a \boxtimes f = \sum a_{i,j} \otimes f_{i,j} \in A \widehat{\otimes} E^*$.

Lemma 12.3.14. *Let A and E be as above. Then, $(A \widehat{\otimes} E^*)^* = \mathrm{CB}(A, E)$ canonically isometrically and the set*

$$\{a \boxtimes f : n \in \mathbb{N},\ a \in \mathbb{M}_n(A),\ f \in \mathbb{M}_n(E)^* \text{ with } \|a\|_{\mathbb{M}_n(A)} < 1,\ \|f\|_{\mathbb{M}_n(E)^*} < 1\}$$

is dense in the unit ball of $A \widehat{\otimes} E^*$. In particular, the map

$$A_1 \widehat{\otimes} E_1^* \times A_2 \widehat{\otimes} E_2^* \ni (u_1, u_2) \mapsto u_1 \times u_2 \in (A_1 \otimes A_2) \widehat{\otimes} (E_1 \otimes E_2)^*,$$

defined by $(a_1 \otimes f_1) \times (a_2 \otimes f_2) = (a_1 \otimes a_2) \otimes (f_1 \otimes f_2)$ on elementary tensor products, satisfies $\|u_1 \times u_2\| \leq \|u_1\| \|u_2\|$.

Proof. It is not hard to see that $\Omega = \{a \boxtimes f : \|a\| < 1, \|f\| < 1\}$ is a convex subset of the unit ball of $A \widehat{\otimes} E^*$ and that $\|\varphi\|_{\mathrm{cb}} = \sup |\langle \varphi, a \boxtimes f \rangle|$ for every $\varphi \in \mathrm{CB}(A, E)$. It follows that the duality pairing from $\mathrm{CB}(A, E)$ into $(A \widehat{\otimes} E^*)^*$ is a surjective isometry and Ω is dense in the unit ball of $A \widehat{\otimes} E^*$, by the Hahn-Banach separation theorem. To prove the second part, we observe that the map $\times$ is continuous since each E_i is finite-dimensional. Therefore, it suffices to show $\|u_1 \times u_2\| < 1$ for $u_i = a_i \boxtimes f_i$, where $a_i \in \mathbb{M}_{n_i}(A_i)$ and $f_i \in \mathbb{M}_{n_i}(E_i)^*$ with $\|a_i\|\|f_i\| < 1$. In this case, $u_1 \times u_2 = a \boxtimes f$ for $a = a_1 \otimes a_2 \in \mathbb{M}_{n_1 n_2}(A_1 \otimes A_2)$ and $f = f_1 \otimes f_2 \in \mathbb{M}_{n_1 n_2}(E_1 \otimes E_2)^*$ (modulo shuffling of indices) and $\|u_1 \times u_2\| \le \|a\|\|f\| = \|a_1\|\|a_2\|\|f_1\|\|f_2\| < 1$. □

We will use repeatedly the following lemma, known as the small perturbation argument.

Lemma 12.3.15. *Let $F \subset A$ be a finite-dimensional subspace with basis $(x_k)_{k=1}^n \subset F$, and let $(x_k^*)_{k=1}^n \subset F^*$ be a dual basis: $x_j^*(x_k) = \delta_{j,k}$. Then, for any $(a_k)_{k=1}^n \subset A$, there is a map θ on A such that $\theta(x_k) = a_k$ for every k and*

$$\|\theta - \mathrm{id}_A\|_{\mathrm{cb}} \le \sum \|x_k^*\|\|a_k - x_k\|.$$

Proof. By the Hahn-Banach Theorem, we may assume that $x_k^* \in A^*$. The map θ defined by $\theta(a) = a + \sum x_k^*(a)(a_k - x_k)$ has the desired property. (Note that $\|x_k^*\|_{\mathrm{cb}} = \|x_k^*\|$.) □

It follows that $\Lambda_{\mathrm{cb}}(A) < C$ if and only if for any finite-dimensional subspace $F \subset A$ there is a finite-rank map φ on A such that $\varphi|_F = \mathrm{id}_F$ and $\|\varphi\|_{\mathrm{cb}} < C$.

Lemma 12.3.16. *Let $F \subset E$ be finite-dimensional subspaces of A and let $C > 0$. Then, there is a map $\varphi \in \mathrm{CB}(A, E)$ such that $\varphi|_F = \mathrm{id}_F$ and $\|\varphi\|_{\mathrm{cb}} \le C$ if and only if $|\mathrm{tr}(u)| \le C\|u\|_{A \widehat{\otimes} E^*}$ for every $u = \sum_k a_k \otimes f_k \in F \odot E^*$, where $\mathrm{tr}(u) = \sum_k f_k(a_k)$.*

Proof. If $\varphi \in \mathrm{CB}(A, E)$ is such that $\varphi|_F = \mathrm{id}_F$ and $\|\varphi\|_{\mathrm{cb}} \le C$, then

$$|\mathrm{tr}(u)| = |\langle \varphi, u \rangle| \le \|\varphi\|_{\mathrm{cb}}\|u\|_{A \widehat{\otimes} E^*} \le C\|u\|_{A \widehat{\otimes} E^*}$$

for every $u \in F \odot E^*$. Conversely, if $\|\mathrm{tr}|_{F \odot E^*}\| \le C$, then by the Hahn-Banach Theorem we can find $\varphi \in (A \widehat{\otimes} E^*)^* = \mathrm{CB}(A, E)$ which is a norm-preserving extension of $\mathrm{tr}|_{F \odot E^*}$. Then $\varphi|_F = \mathrm{id}_F$ and $\|\varphi\|_{\mathrm{cb}} = \|\mathrm{tr}|_{F \odot E^*}\| \le C$. □

Proof of Theorem 12.3.13. We only prove that

$$\Lambda_{\mathrm{cb}}(A_1 \otimes A_2) = \Lambda_{\mathrm{cb}}(A_1)\Lambda_{\mathrm{cb}}(A_2)$$

for C*-algebras A_1 and A_2. The inequality $\Lambda_{\text{cb}}(A_1 \otimes A_2) \leq \Lambda_{\text{cb}}(A_1)\Lambda_{\text{cb}}(A_2)$ is trivial. To prove the opposite inequality, suppose by contradiction that $\Lambda_{\text{cb}}(A_1 \otimes A_2) < \Lambda_{\text{cb}}(A_1)\Lambda_{\text{cb}}(A_2)$, and choose $0 < C_i < \Lambda_{\text{cb}}(A_i)$ such that $\Lambda_{\text{cb}}(A_1 \otimes A_2) < C_1 C_2$. It follows that there is a finite-dimensional subspace $F_i \subset A_i$ such that there is no finite-rank map φ_i on A_i with $\varphi_i|_{F_i} = \text{id}_{F_i}$ and $\|\varphi_i\|_{\text{cb}} \leq C_i$. Choose a finite-rank map φ on $A_1 \otimes A_2$ such that $\varphi|_{F_1 \otimes F_2} = \text{id}_{F_1 \otimes F_2}$ and $\|\varphi\|_{\text{cb}} < C_1 C_2$. By the small perturbation argument, we may assume that the range of φ is contained in $E_1 \otimes E_2$ for some finite-dimensional subspaces $E_i \subset A_i$. By Lemma 12.3.16, there is $u_i \in F_i \odot E_i^*$ such that $|\operatorname{tr}(u_i)| > C_i \|u_i\|_{A_i \widehat{\otimes} E_i^*}$. By Lemma 12.3.14, the element $u = u_1 \times u_2 \in (F_1 \otimes F_2) \odot (E_1 \otimes E_2)^*$ satisfies

$$\begin{aligned}
|\operatorname{tr}(u)| &= |\operatorname{tr}(u_1)\operatorname{tr}(u_2)| \\
&> C_1 \|u_1\|_{A_1 \widehat{\otimes} E_1^*} C_2 \|u_2\|_{A_2 \widehat{\otimes} E_2^*} \\
&\geq C_1 C_2 \|u\|_{(A_1 \otimes A_2) \widehat{\otimes} (E_1 \otimes E_2)^*}.
\end{aligned}$$

This is in contradiction to the fact that $\|\varphi\|_{\text{cb}} < C_1 C_2$, by Lemma 12.3.16. □

It is proved in [**165**] that the class of groups with Cowling-Haagerup constant equal to 1 is closed under free products (even allowing amalgamation over a finite subgroup), but it is not yet known whether weak amenability is closed under free products.

12.4. Another approximation property

Definition 12.4.1. Let A be a C*-algebra. We say that A has the *OAP* (*operator approximation property*) if there exists a net of finite-rank continuous linear maps φ_i on A which converges to id_A in the *stable point-norm topology*: $\varphi_i \otimes \text{id}_{\mathbb{K}(\ell^2)}$ converges to the identity map on $A \otimes \mathbb{K}(\ell^2)$ in the point-norm topology.[13] We say that A has the *SOAP* (*strong OAP*) if there exists a net of finite-rank continuous linear maps φ_i on A which converges to id_A in the *strong stable point-norm topology*: $\varphi_i \otimes \text{id}_{\mathbb{B}(\ell^2)}$ converges to the identity map on $A \otimes \mathbb{B}(\ell^2)$ in the point-norm topology.

Let M be a von Neumann algebra. We say that M has the *W*OAP* (*weak* OAP*) if there exists a net of finite-rank ultraweakly-continuous linear maps φ_i on M which converges to id_M in the *stable point-ultraweak topology*: $\varphi_i \otimes \text{id}_{\mathbb{B}(\ell^2)}$ converges to the identity map on $M \bar{\otimes} \mathbb{B}(\ell^2)$ in the point-ultraweak topology.

It is clear that CBAP⇒SOAP⇒OAP and that W*CBAP⇒W*OAP.

[13]The crucial point is that the net (φ_i) need not be uniformly bounded. The Principle of Uniform Boundedness implies that this net cannot be a sequence (unless A has the CBAP).

Remark 12.4.2. A Banach space X is said to have the AP (approximation property) if there exists a net of finite-rank continuous linear maps φ_i on X such that $\varphi_i \to \mathrm{id}_X$ uniformly on compact subsets. We note that $\varphi_i \to \mathrm{id}_X$ uniformly on compact subsets if and only if $\varphi_i \otimes \mathrm{id}_{c_0(\mathbb{N})}$ converges to the identity map on $X \check{\otimes} c_0 \cong c_0(\mathbb{N}; X)$ in the point-norm topology. Since $c_0(\mathbb{N}) \subset \mathbb{K}(\ell^2)$, the OAP implies the AP. Szankowski [**180**] proved that $\mathbb{B}(\ell^2)$ does not have the AP. It follows that $\mathbb{B}(\ell^2)$ does not have the OAP.

Though we've already encountered them, it seems fitting to properly introduce slice maps on tensor products: Given C*-algebras A and B and a linear functional $\omega \in A^*$, the map $\omega \otimes \mathrm{id}_B : A \otimes B \to \mathbb{C} \otimes B \cong B$ is called a *slice map*.

Definition 12.4.3. Let A and B be C*-algebras and $X \subset B$ be a closed subspace. We define a subset $F(A, B, X) \subset A \otimes B$ (F is for Fubini) by

$$F(A, B, X) = \{x \in A \otimes B : \forall \omega \in A^*, \quad (\omega \otimes \mathrm{id}_B)(x) \in X\}.$$

We say a triple (A, B, X) satisfies the *slice map property* if $F(A, B, X) = A \otimes X$, the norm closure of $A \odot X$ in $A \otimes B$.

Let M and N be von Neumann algebras and $X \subset N$ be an ultraweakly closed subspace. Define a subset $F_\sigma(M, N, X) \subset M \bar{\otimes} N$ by

$$F_\sigma(M, N, X) = \{x \in M \bar{\otimes} N : \forall \omega \in M_*, \quad (\omega \otimes \mathrm{id}_N)(x) \in X\}$$

and we'll say the triple (M, N, X) satisfies the *weak slice map property* if $F_\sigma(M, N, X)$ is the ultraweak closure of $M \odot X$ in $M \bar{\otimes} N$.

One should check that for an ideal J of B, the triple (A, B, J) satisfies the slice map property if and only if the sequence

$$0 \longrightarrow A \otimes J \longrightarrow A \otimes B \longrightarrow A \otimes (B/J) \longrightarrow 0$$

is exact.

Theorem 12.4.4. *The following are true:*

(1) *a* C*-*algebra A has the OAP if and only if $(A, \mathbb{K}(\ell^2), X)$ satisfies the slice map property for any closed subspace $X \subset \mathbb{K}(\ell^2)$;*

(2) *a* C*-*algebra A has the SOAP if and only if (A, B, X) satisfies the slice map property for any B and any closed subspace $X \subset B$;*

(3) *a von Neumann algebra M has the W*OAP if and only if (M, N, X) satisfies the weak slice map property for any N and any ultraweakly closed subspace $X \subset N$.*

In particular, the SOAP implies exactness.

Proof. We only prove assertion (2). Denote by $F(A)$ the set of all continuous finite-rank linear maps on A. Note that every $\varphi \in F(A)$ is of the form $\varphi(\,\cdot\,) = \sum_{k=1}^{n} \omega_k(\,\cdot\,)a_k$ for some $n \in \mathbb{N}$, $\omega_k \in A^*$ and $a_k \in A$. For $x \in A \otimes B$, we set $R_x = \{(\omega \otimes \mathrm{id}_B)(x) : \omega \in A^*\} \subset B$ (which may not be closed). It is easy to see that $\{(\varphi \otimes \mathrm{id}_B)(x) : \varphi \in F(A)\} = A \odot R_x$ and that $x \in F(A, B, X)$ if and only if $R_x \subset X$.

Now suppose that A has the SOAP and $x \in F(A, B, X)$ is given. We may assume that B is separable since for every $x \in A \otimes B$ one can find a separable C*-subalgebra $B_0 \subset B$ such that $x \in A \otimes B_0$. Let (φ_i) be a net of finite-rank continuous linear maps on A such that the net $(\varphi_i \otimes \mathrm{id}_{\mathbb{B}(\ell^2)})$ converges pointwise to the identity map on $A \otimes \mathbb{B}(\ell^2)$. Since $B \hookrightarrow \mathbb{B}(\ell^2)$, we have $x = \lim(\varphi_i \otimes \mathrm{id}_B)(x) \in A \otimes X$.

To prove the converse, let $x_1, \ldots, x_n \in A \otimes \mathbb{B}(\ell^2)$ and $\varepsilon > 0$ be given. Now set

$$x = \mathrm{diag}(x_1, \ldots, x_n) \in \bigoplus_{k=1}^{n}(A \otimes \mathbb{B}(\ell^2)) \subset A \otimes \mathbb{B}(\bigoplus_{k=1}^{n} \ell^2)$$

and let $X = \text{norm-cl}\, R_x$ be the norm closure of R_x. If the triple

$$(A, \mathbb{B}(\bigoplus \ell^2), X)$$

satisfies the slice map property, then

$$x \in F(A, \mathbb{B}(\bigoplus \ell^2), X) = A \otimes X = \text{norm-cl}\{(\varphi \otimes \mathrm{id}_{\mathbb{B}(\bigoplus \ell^2)})(x) : \varphi \in F(A)\}$$

and hence there exists $\varphi \in F(A)$ such that $\|x_k - (\varphi \otimes \mathrm{id}_{\mathbb{B}(\bigoplus \ell^2)})(x_k)\| < \varepsilon$ for every k. □

Remark 12.4.5. In the proof above, we have not used the C*-structure of A at all – i.e., the theorem, except the statement that SOAP implies exactness, holds for an arbitrary operator space.

We should also mention that Tomita's celebrated commutation theorem is equivalent to the validity of the weak slice map property for arbitrary von Neumann algebras (M, N, B). The standard formulation of Tomita's result ([**183**, Theorem IV.5.9]) is that $(M \bar{\otimes} N)' = M' \bar{\otimes} N'$ in $\mathbb{B}(\mathcal{H} \otimes \mathcal{K})$, for arbitrary von Neumann algebras $M \subset \mathbb{B}(\mathcal{H})$ and $N \subset \mathbb{B}(\mathcal{K})$. But it is easy to see that

$$\begin{aligned} F_\sigma(M, N, B) &\subset (M \bar{\otimes} \mathbb{B}(\mathcal{K})) \cap (\mathbb{B}(\mathcal{H}) \bar{\otimes} B) \\ &= (M' \bar{\otimes} \mathbb{C}1)' \cap (\mathbb{C}1 \bar{\otimes} B')' = (M' \bar{\otimes} B')'. \end{aligned}$$

Finally, it is a long-standing open problem whether or not there exists an exact C*-algebra without the OAP. The reduced group C*-algebra $C^*_\lambda(\mathrm{SL}(3, \mathbb{Z}))$ is a candidate [**80**]. We note that Haagerup and Kraus [**80**] and Kirchberg [**101**] independently proved that an exact (or locally reflexive) C*-algebra with the OAP has the SOAP.

Corollary 12.4.6. *The class of algebras with the OAP is closed under extensions.*

Proof. Let $0 \to J \to B \xrightarrow{\pi} A \to 0$ be a short exact sequence of C*-algebras with J and A having the OAP. By Theorem 12.4.4, to prove that B has the OAP, it suffices to check the slice map property for $(B, \mathbb{K}(\ell^2), X)$, where $X \subset \mathbb{K}(\ell^2)$ is an arbitrary closed subspace.

Let $x \in F(B, \mathbb{K}(\ell^2), X)$ be given. Since A has the OAP, it follows that $(\pi \otimes \mathrm{id})(x) \in F(A, \mathbb{K}(\ell^2), X) = A \otimes X$. Since X is $\otimes$-exact (as defined in Section 3.9), there is a canonical isometric isomorphism $A \otimes X = (B \otimes X)/(J \otimes X)$ and $(\pi \otimes \mathrm{id})(x)$ lifts to $y \in B \otimes X$. It follows that $x - y \in \ker(\pi \otimes \mathrm{id}_{\mathbb{K}(\ell^2)}) = J \otimes \mathbb{K}(\ell^2)$ and $x - y \in F(J, \mathbb{K}(\ell^2), X) = J \otimes X$. Consequently, $x \in B \otimes X$. □

Remark 12.4.7. Consider a nonsemisplit extension B of $\mathrm{Cone}(C^*_\lambda(\mathbb{F}_2))$ by $\mathbb{K}(\ell^2)$

$$0 \longrightarrow \mathbb{K}(\ell^2) \longrightarrow B \longrightarrow \mathrm{Cone}(C^*_\lambda(\mathbb{F}_2)) \longrightarrow 0$$

(which will be shown to exist in Section 13.4). The C*-algebra B is not exact, but it has the OAP by Corollary 12.4.6. It follows that the OAP does not imply the SOAP.

Definition 12.4.8. We say a group Γ has the *AP* (approximation property) if there exists a net[14] (φ_i) of finitely supported functions on Γ such that $\varphi_i \to 1$ weak* in $B_2(\Gamma)$ – i.e., $\omega(\varphi_i) \to \omega(1)$ for every $\omega \in Q(\Gamma)$. (See Appendix D.)

It should be clear that the AP passes to subgroups and increasing unions.

Theorem 12.4.9. *For a group Γ, the following are equivalent:*

(1) *the group Γ has the AP;*

(2) *the reduced group C*-algebra $C^*_\lambda(\Gamma)$ has the OAP;*

(3) *the reduced group C*-algebra $C^*_\lambda(\Gamma)$ has the SOAP;*

(4) *the group von Neumann algebra $L(\Gamma)$ has the W*OAP.*

Proof. We only prove the equivalence of (1) through (3). The implication (3) ⇒ (2) is trivial and the proof of (2) ⇒ (1) is similar to that of Theorem 12.3.10, thanks to Lemma D.9. Thus, we must prove (1) ⇒ (3).

[14]The Principle of Uniform Boundedness again implies that this net cannot be a sequence unless Γ is weakly amenable.

By Lemma D.7, if $\varphi_i \in \mathbb{C}[\Gamma]$ converges to 1 weak*, then $m_{\varphi_i} \otimes \mathrm{id}_{\mathbb{B}(\ell^2)} \to \mathrm{id}_{C^*_\lambda(\Gamma)\otimes\mathbb{B}(\ell^2)}$ in the point-weak topology. Hence, by the Hahn-Banach Theorem, there is a net of convex combinations of the m_{φ_i}'s which converges to $\mathrm{id}_{C^*_\lambda(\Gamma)}$ in the strong stable point-norm topology. □

It follows from this result that a nonexact group does not have the AP. The algebraic direct product of infinitely many copies of $(\mathbb{Z}/2\mathbb{Z}) \wr \mathbb{F}_2$ is not weakly amenable (Corollary 12.3.7 and Theorem 12.3.13), but it has the Haagerup property (Theorem 12.2.11) and the AP.

Proposition 12.4.10. *The class of groups with the AP is closed under extensions.*

Proof. Let $\{1\} \to \Lambda \to \Gamma \xrightarrow{\pi} \bar{\Gamma} \to \{1\}$ be a short exact sequence of groups with Λ and $\bar{\Gamma}$ having the AP. We denote by Z the weak*-closure of $\mathbb{C}[\Gamma]$ in $B_2(\Gamma)$. It suffices to show $1 \in Z$.

Consider the inclusion $\iota\colon B_2(\Lambda) \hookrightarrow B_2(\Gamma)$, where $\iota(\varphi)(s) = \varphi(s)$ for $s \in \Lambda$, and $\iota(\varphi) = 0$ off Λ. It is clear that ι is an isometry such that $\iota(\mathbb{C}[\Lambda]) \subset \mathbb{C}[\Gamma]$. We claim that ι is weak*-continuous. For this, it suffices to show $\iota^*(Q(\Gamma)) \subset Q(\Lambda)$, which in turn is equivalent to the trivial assertion that $\iota^*(\delta_s) \in Q(\Lambda)$ for every $s \in \Gamma$. Since Λ has the AP, we have $\chi_\Lambda = \iota(1) \in Z$. It also follows that $\chi_{s\Lambda} \in Z$ for every $s \in \Gamma$. Therefore, for the inclusion $\pi_*\colon B_2(\bar{\Gamma}) \hookrightarrow B_2(\Gamma)$ given by $\pi_*(\varphi) = \varphi \circ \pi$, we have $\pi_*(\mathbb{C}[\bar{\Gamma}]) \subset Z$. Since π_* is a weak*-continuous contraction (why?), the assertion $1 \in Z$ follows from the AP of $\bar{\Gamma}$. □

Exercises

Exercise 12.4.1. Check the details of the proof of (2) ⇒ (1) in Theorem 12.4.9.

Exercise 12.4.2. Let $C^*_u(\Gamma) = C^*(\lambda(\Gamma), \ell^\infty(\Gamma))$ be the uniform Roe algebra (Section 5.1). We say the group Γ satisfies the *invariant translation approximation property* (ITAP) if

$$C^*_u(\Gamma) \cap L(\Gamma) = C^*_\lambda(\Gamma).$$

Prove that if Γ has the AP, then Γ has the ITAP. (At present, it isn't known whether there exists a group without the ITAP – cf. [**167, 199**].)

Exercise 12.4.3. Let Γ be a group and $\Lambda \subset \Gamma$ be a co-amenable subgroup. Prove that Γ has the AP if Λ has the AP.

Exercise 12.4.4. Let Γ be a group acting on a locally finite tree. Prove that Γ has the AP if one of the vertex stabilizers has the AP.

Exercise 12.4.5. Let $\Gamma = \Gamma_1 * \Gamma_2$ be the free product of Γ_1 and Γ_2. Prove that Γ has the AP if each Γ_i has the AP. (Hint: Prove that, for every $i \in \{1, 2\}$ and $d, n \in \mathbb{N}$, there is a bounded map $T\colon B_2(\Gamma_i) \to B_2(\Gamma)$ such that $T(\varphi)(s) = \delta_{n,m}\delta_{i,i_d}\,\varphi(s_d)$ for $s = s_1 \cdots s_m$ with $s_j \in \Gamma_{i_j}$ and $i_j \neq i_{j+1}$.)

12.5. References

(Relative) property (T) was introduced by Kazhdan (and Margulis). Since then, several groups have been shown to have property (T) using formidable machinery from Lie group theory or spectral geometry. The first "elementary proof" of the fact that some infinite group (e.g., $\mathrm{SL}(3, \mathbb{Z})$) has property (T) was recently given by Shalom [**174**]. Theorem 12.1.7 (especially (1) $\Rightarrow$ (2)) is due to Jolissaint [**91**]. Lemma 12.1.8 is due to de la Harpe, Robertson and Valette [**83**]. Theorem 12.1.10 is due to Margulis, but its quantitative proof is due to Shalom [**173**]. Remark 12.1.11 is due to Burger [**32**]. The proof of Theorem 12.1.14 is based on Shalom's work [**174**]. Theorem 12.1.15 is due to Żuk [**201**]. The notion of property (T) for von Neumann algebras was introduced by Connes and Jones [**45**]. Theorem 12.1.18 is a routine modification of [**45**] to the relative property (T) context. We should mention that an analogue of Theorem 12.1.7 is true for von Neumann algebras [**144**]. Theorem 12.1.19 is due to Connes [**43**]. For a comprehensive treatment of property (T), we refer to the book of Bekka, de la Harpe and Valette [**15**].

Haagerup's property was introduced by, well, Haagerup; Theorems 12.2.5 and 12.2.4 are proved in his seminal paper [**75**]. Theorem 12.2.9 comes from [**33**]. Theorem 12.2.11 is due to de Cornulier, Stalder and Valette [**55**]. Theorem 12.2.15 was proved by Choda [**34**] (see also [**90**]), while Theorem 12.2.16 is due to Popa [**160**]. For a comprehensive treatment of the Haagerup property, we refer to the book of Cherix, Cowling, Jolissaint, Julg and Valette [**33**].

The CBAP was also introduced by Haagerup. Theorem 12.3.3 and its corollary were proved by Bożejko and Picardello [**25**]. The fact that $\Lambda_{\mathrm{cb}}(\mathbb{F}_n) = 1$ is due to De Cannière and Haagerup [**54**]. More general versions of Theorem 12.3.6 and its corollary were proved by Ozawa and Popa [**138**]. Theorem 12.3.8 combines work of De Cannière and Haagerup [**54**], Cowling [**46**], Haagerup [**78**] and Cowling and Haagerup [**47**]. See also [**56**] for a proof of nonweak amenability of $\mathbb{Z}^2 \rtimes \mathrm{SL}(2, \mathbb{Z})$ and other groups. Theorem 12.3.10 is due to Haagerup [**78**], while Theorem 12.3.13 was proved by Cowling and Haagerup [**47**] for group algebras and by Sinclair and Smith [**175**] in general.

Theorem 12.4.4 is due to Kraus [**111**]. Corollary 12.4.6 is work of Kirchberg [**101**]. Finally, Theorem 12.4.9 and Proposition 12.4.10 were proved by Haagerup and Kraus [**80**].

Chapter 13

Weak Expectation Property and Local Lifting Property

This chapter contains a potpourri of concepts and results which appear unrelated but turn out to have deep, harmonious connections. We begin with two technical C*-concepts (the local lifting property and the weak expectation property), proving tensorial characterizations in Corollary 13.2.5. The last three sections explore connections with Kirchberg's QWEP conjecture (see Section 13.3), surprising examples of nonsplit extensions, and the fundamental fact (due to Junge and Pisier) that $\mathbb{B}(\mathcal{H}) \odot \mathbb{B}(\mathcal{H})$ has more than one C*-norm.

13.1. The local lifting property

Definition 13.1.1. Let A be a C*-algebra, J be a closed two-sided ideal in a C*-algebra B and $\pi: B \to B/J$ be the quotient map. We say a c.c.p. map $\varphi: A \to B/J$ is *liftable* if there exists a c.c.p. map $\psi: A \to B$ such that $\pi \circ \psi = \varphi$. We say φ is *locally liftable* if for any finite-dimensional operator system $E \subset A$, there exists a c.c.p. map $\psi: E \to B$ such that $\pi \circ \psi = \varphi|_E$:

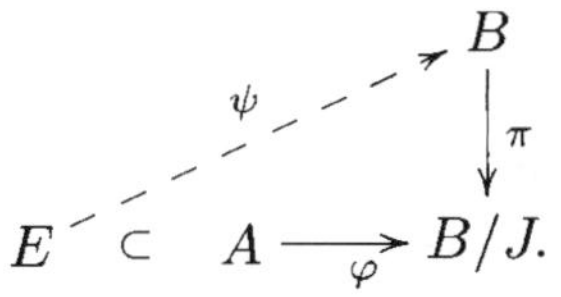

A unital C*-algebra A has the *lifting property* (LP) (resp. *local lifting property* (LLP)) if any c.c.p. map from A into a quotient C*-algebra B/J is liftable (resp. locally liftable). A nonunital C*-algebra A has the LP (resp. LLP) if its unitization has that property.

Not surprisingly, we can usually restrict our attention to u.c.p. maps.

Lemma 13.1.2. *A unital* C*-*algebra A has the LLP (resp. LP) if any u.c.p. map from A into a quotient* C*-*algebra B/J is locally liftable (resp. liftable and A is separable).*

Proof. Suppose that a c.c.p. map $\varphi\colon A \to B/J$ is given and let $c = \varphi(1) \in B/J$. Let $c_n = h_n(c)$ for the positive continuous function $h_n(t) = \max\{t, 1/n\}$. Fix a state ω on A and define u.c.p. maps $\varphi_n\colon A \to B/J$ by

$$\varphi_n(a) = c_n^{-1/2}\big(\varphi(a) + \omega(a)(c_n - c)\big)c_n^{-1/2}.$$

By assumption, φ_n has a (local) lifting ψ_n into B. Let $b_n \in B$ be any lifting of c_n with $0 \le b_n \le 1$. Then, one has

$$\pi(b_n^{1/2}\psi_n(a)b_n^{1/2}) = \varphi(a) + \omega(a)(c_n - c) \to \varphi(a)$$

for every $a \in A$. Hence, by Lemma C.2, φ is (locally) liftable. □

The Choi-Effros Lifting Theorem (Theorem C.3) implies that separable nuclear C*-algebras have the LP. Here is a nonnuclear example:

Theorem 13.1.3 (Kirchberg). *The full* C*-*algebra $C^*(\mathbb{F}_n)$ of a countable free group $\mathbb{F}_n$ has the LP.*

Proof. Assume $n = \infty$ (the finite case is identical). We first show that a $*$-homomorphism $\theta\colon C^*(\mathbb{F}_n) \to B/J$ is liftable. To this end, let $x_1, x_2, \ldots \in B$ be contractive liftings of $\theta(U_1), \theta(U_2), \ldots$, where $U_1, U_2, \ldots$ are the standard generators of $C^*(\mathbb{F}_n)$. Then, each x_n dilates to a unitary

$$\hat{x}_n = \begin{bmatrix} x_n & (1 - x_n x_n^*)^{1/2} \\ (1 - x_n^* x_n)^{1/2} & -x_n^* \end{bmatrix} \subset \mathbb{M}_2(B).$$

By universality, there is a unital $*$-homomorphism $\rho\colon C^*(\mathbb{F}_n) \to \mathbb{M}_2(B)$ with $\rho(U_n) = \hat{x}_n$. It is not hard to see that the $(1,1)$-corner of ρ is a u.c.p. lifting of θ, so the homomorphism case is complete.

Now, let $\varphi\colon C^*(\mathbb{F}_n) \to B/J$ be a u.c.p. map. Since $\mathbb{F}_n$ is countable, we may assume B/J is separable. By the Kasparov-Stinespring Dilation Theorem [**97**], there is a $*$-homomorphism $\theta\colon C^*(\mathbb{F}_n) \to M(\mathbb{K}\otimes(B/J))$ such that $\varphi(a) = \theta(a)_{11}$ for $a \in C^*(\mathbb{F}_n)$, where x_{11} is the $(1,1)$-entry of x in the multiplier algebra of $\mathbb{K} \otimes (B/J)$. By the noncommutative Tietze extension theorem (Proposition 6.8 in [**114**]), the surjective $*$-homomorphism $\mathrm{id} \otimes \pi$ from $\mathbb{K} \otimes B$ onto $\mathbb{K} \otimes (B/J)$ extends to a surjective $*$-homomorphism $\tilde{\pi}$

between their multiplier algebras. Hence, by the first part of this proof, there is a u.c.p. map $\rho\colon C^*(\mathbb{F}_n) \to M(\mathbb{K}\otimes B)$ such that $\theta = \tilde{\pi}\circ\rho$. The u.c.p. map $\psi\colon C^*(\mathbb{F}_n) \ni a \mapsto \rho(a)_{11} \in B$ is our desired lifting of φ. □

Drawing a diagram, one easily deduces the following fact.

Corollary 13.1.4. *Let A be a separable* C*-*algebra and J be a closed two-sided ideal in $C^*(\mathbb{F}_\infty)$ such that $A \cong C^*(\mathbb{F}_\infty)/J$. Then, A has the LP (resp. the LLP) if and only if* $\mathrm{id}_A\colon A \to C^*(\mathbb{F}_\infty)/J$ *is liftable (resp. locally liftable).*

Remark 13.1.5 (LP versus LLP)**.** The full C*-algebra of an uncountable free group has the LLP (but probably does not have the LP). Indeed, when $\mathbb{F}_I$ is the free group on an uncountable set I, every element x (resp. every separable C*-subalgebra A) in $C^*(\mathbb{F}_I)$ sits inside $C^*(\mathbb{F}_{I'})$ for some countable subset $I' \subset I$. Moreover, $C^*(\mathbb{F}_{I'}) \subset C^*(\mathbb{F}_I)$ is the range of a conditional expectation. Hence the LLP of $C^*(\mathbb{F}_I)$ follows from that of $C^*(\mathbb{F}_{I'})$.

There is a (nonseparable) C*-algebra which has the LLP but not the LP. Indeed, the commutative C*-algebra ℓ^∞/c_0 is such an example as there is no bounded linear lifting from ℓ^∞/c_0 into ℓ^∞ (Exercise 13.1.1). On the other hand, an affirmative answer to the QWEP conjecture (see Section 13.3) would imply that the LP and the LLP are equivalent for *separable* C*-algebras (cf. [**102**]).

Recall that in the particular case of quotients, Effros and Haagerup gave a tensorial characterization of local liftability (Theorem C.4):

Theorem 13.1.6. *Let J be a closed two-sided ideal in a unital* C*-*algebra B and let $\pi\colon B \to B/J$ be the quotient map. Then, the following are equivalent:*

(1) *for any* C*-*algebra A, the following sequence is exact:*

$$0 \longrightarrow A\otimes J \longrightarrow A\otimes B \longrightarrow A\otimes(B/J) \longrightarrow 0;$$

(2) *same as above but with $A = \mathbb{B}(\ell^2)$;*

(3) *the identity map on B/J is locally liftable.*

Note that condition (1) holds whenever $A \otimes_{\max} (B/J) = A \otimes (B/J)$ (Corollary 3.7.3). Thus, if $B/J \odot \mathbb{B}(\ell^2)$ has a unique C*-norm, it follows that $\mathrm{id}_{B/J}$ is locally liftable; Corollary 13.1.4 then implies that B/J has the LLP (since we may assume $B = C^*(\mathbb{F}_\infty)$). In the next section we prove the converse.

Exercise

Exercise 13.1.1. Prove that $c_0(\mathbb{R})$ (where $\mathbb{R}$ is just viewed as an uncountable set) is $*$-isomorphic to a C*-subalgebra of ℓ^∞/c_0. Conclude that there is no injective continuous linear map from ℓ^∞/c_0 into ℓ^∞.

13.2. Tensorial characterizations of the LLP and WEP

To get tensorial characterizations of the LLP and WEP, we will need another striking theorem of Kirchberg.

Theorem 13.2.1 (Kirchberg). *For any free group $\mathbb{F}$ and any Hilbert space $\mathcal{H}$, we have*

$$C^*(\mathbb{F}) \otimes_{\max} \mathbb{B}(\mathcal{H}) = C^*(\mathbb{F}) \otimes \mathbb{B}(\mathcal{H})$$

canonically.

A bit of thought reveals that one only needs to consider the case $\mathbb{F} = \mathbb{F}_2$ and $\mathcal{H} = \ell^2$. Our proof follows Pisier's beautiful ideas, and requires two lemmas.

Let E_n be the n-dimensional operator space in $C^*(\mathbb{F}_{n-1})$ spanned by the unit $U_0 = 1$ and the standard unitary generators $U_1, \dots, U_{n-1}$ of $C^*(\mathbb{F}_{n-1})$. It is not hard to see that E_n is canonically isometric to ℓ_n^1 (the n-dimensional ℓ^1 space), or equivalently,

$$\|\sum_{k=0}^{n-1} \alpha_k U_k\| = \sum_{k=0}^{n-1} |\alpha_k|,$$

for all $(\alpha_k)_{k=0}^{n-1} \in \mathbb{C}^n$. By duality, we have a one-to-one correspondence between elements $z = \sum_{k=0}^{n-1} U_k \otimes x_k \in E_n \otimes \mathbb{B}(\ell^2)$ and maps

$$T_z \colon \ell_n^\infty \ni (\alpha_k)_{k=0}^{n-1} \longmapsto \sum_{k=0}^{n-1} \alpha_k x_k \in \mathbb{B}(\ell^2).$$

Lemma 13.2.2. *The above operator space E_n is canonically completely isometrically isomorphic to the dual operator space $\ell_n^1 = (\ell_n^\infty)^*$, or equivalently, $\|z\|_{\min} = \|T_z\|_{\mathrm{cb}}$ for every $z \in E_n \otimes \mathbb{B}(\ell^2)$.*

Proof. Since $(U_k)_{k=0}^{n-1} \in E_n \otimes \ell_n^\infty$ is contractive and

$$z = (\mathrm{id}_{E_n} \otimes T_z)((U_k)_{k=0}^{n-1}) \in E_n \otimes \mathbb{B}(\ell^2),$$

we have $\|z\|_{\min} \leq \|T_z\|_{\mathrm{cb}}$. To prove the opposite inequality, we give ourselves contractions $a_0, \dots, a_{n-1} \in \mathbb{B}(\mathcal{H})$ and let $\hat{a}_k \in \mathbb{M}_2(\mathbb{B}(\mathcal{H}))$ be their unitary dilations (see the proof of Theorem 13.1.3). It follows that the map $\varphi \colon E_n \to \mathbb{M}_2(\mathbb{B}(\mathcal{H}))$ defined by $\varphi(U_k) = \hat{a}_0^{-1}\hat{a}_k$, $k = 0, \dots, n-1$, is c.c. since it extends to a $*$-homomorphism on $C^*(\mathbb{F}_{n-1})$. Hence, the map $\theta \colon E_n \to \mathbb{B}(\mathcal{H})$ defined by $\theta(U_k) = (\hat{a}_0\varphi(U_k))_{11} = a_k$, $k = 0, \dots, n-1$, is also c.c. Therefore, we have

$$\|(\mathrm{id}_{\mathbb{B}(\mathcal{H})} \otimes T_z)((a_k)_{k=0}^{n-1})\|_{\min} = \|(\theta \otimes \mathrm{id}_{\mathbb{B}(\ell_2)})(z)\|_{\min} \leq \|z\|_{\min}.$$

Since the contractive element $(a_k)_{k=0}^{n-1} \in \mathbb{B}(\mathcal{H}) \otimes \ell_\infty^n$ was arbitrary, $\|T_z\|_{\mathrm{cb}} \leq \|z\|_{\min}$ and we are done. $\square$

Lemma 13.2.3. *Let $X_i \subset \mathbb{B}(\mathcal{H}_i)$ $(i = 1, 2)$ be unital operator subspaces and let $\varphi\colon X_1 \to X_2$ be a unital complete isometry. Suppose that X_2 is spanned by unitary elements in $\mathbb{B}(\mathcal{H}_2)$. Then, φ uniquely extends to a $*$-homomorphism between the C*-subalgebras $C^*(X_i)$ generated by X_i in $\mathbb{B}(\mathcal{H}_i)$.*

Proof. By Arveson's Extension Theorem, φ extends to a c.c. map from $\mathbb{B}(\mathcal{H}_1)$ into $\mathbb{B}(\mathcal{H}_2)$, which we still denote by φ. Since φ is unital, it has to be a u.c.p. map. Since $\varphi|_{X_1}$ is isometric and X_2 is spanned by unitary elements, X_1 is contained in the multiplicative domain of φ. Hence, φ is a $*$-homomorphism on $C^*(X_1)$. □

Proof of Theorem 13.2.1. Thanks to Lemma 13.2.3, it suffices to show that the formal identity map from $E_n \otimes_{\min} \mathbb{B}(\ell^2)$ into $C^*(\mathbb{F}_{n-1}) \otimes_{\max} \mathbb{B}(\ell^2)$ is c.c. for every n (or just $n = 3$). We give ourselves $z = \sum_{k=0}^{n-1} U_k \otimes x_k \in E_n \otimes \mathbb{B}(\ell^2)$ with $\|z\|_{\min} = 1$. By Lemma 13.2.2, the corresponding map $T_z\colon \ell_n^\infty \to \mathbb{B}(\ell^2)$ is c.c. Hence, by the factorization theorem for completely bounded maps (Theorem B.7), there exist a Hilbert space $\mathcal{H}$, a $*$-homomorphism $\pi\colon \ell_n^\infty \to \mathbb{B}(\mathcal{H})$ and isometries $V, W \in \mathbb{B}(\ell^2, \mathcal{H})$ such that $T_z(f) = V^*\pi(f)W$ for $f \in \ell_n^\infty$. We may assume that $\mathcal{H} = \ell^2$. Then, $a_k = \pi(\delta_k)V$ and $b_k = \pi(\delta_k)W$ in $\mathbb{B}(\ell^2)$ are such that $x_k = a_k^* b_k$ for every k and $\sum_{k=0}^{n-1} a_k^* a_k = 1 = \sum_{k=0}^{n-1} b_k^* b_k$. It follows that

$$\begin{aligned}\|\sum_{k=0}^{n-1} U_k \otimes x_k\|_{C^*(\mathbb{F}_\infty)\otimes_{\max}\mathbb{B}(\ell^2)} &= \|\sum_{k=0}^{n-1}(1 \otimes a_k)^*(U_k \otimes b_k)\|_{\max} \\ &\le \|\sum_{k=0}^{n-1}(1 \otimes a_k)^*(1 \otimes a_k)\|_{\max}^{1/2}\|\sum_{k=0}^{n-1}(U_k \otimes b_k)^*(U_k \otimes b_k)\|_{\max}^{1/2} = 1.\end{aligned}$$

This shows that the formal identity from $E_n \otimes_{\min} \mathbb{B}(\ell^2)$ into $C^*(\mathbb{F}_\infty) \otimes_{\max} \mathbb{B}(\ell^2)$ is contractive. Since $\mathbb{B}(\ell^2)$ is stable, our formal identity is also c.c. □

Remark 13.2.4. Theorem 13.2.1 and a modification of Theorem 13.1.6 can be used to give an alternate proof of the fact that $C^*(\mathbb{F})$ has the LLP.

The C*-algebra $\mathbb{B}(\ell^2)$ is universal in the sense that it contains all separable C*-algebras and has the WEP (Definition 3.6.7), since it is injective. The full free group C*-algebra $C^*(\mathbb{F}_\infty)$ is universal in the sense that it has the LP and any separable unital C*-algebra is a quotient of it. With these simple observations we can now establish tensorial characterizations of the WEP and LLP.

Corollary 13.2.5. *For C*-algebras A and B, we have the following:*

(1) *$A \otimes_{\max} B = A \otimes B$ canonically if A has the LLP and B has the WEP;*

(2) *$A \otimes_{\max} \mathbb{B}(\ell^2) = A \otimes \mathbb{B}(\ell^2)$ canonically if and only if A has the LLP;*

(3) *$C^*(\mathbb{F}_\infty) \otimes_{\max} B = C^*(\mathbb{F}_\infty) \otimes B$ if and only if B has the WEP.*

Proof. Assertion (3) follows from Theorem 13.2.1 and Proposition 3.6.6. The "only if" part of (2) follows from Corollary 13.1.4 and Theorem 13.1.6. We will prove the "if" part of (2) and leave the proof of (1) to the reader.

Let A be a unital C*-algebra with the LLP. Let $\mathbb{F}$ be a free group of suitable cardinality and let $\pi\colon C^*(\mathbb{F}) \to A$ be a surjective *-homomorphism. By the LLP, for any finite-dimensional operator system E, there is a u.c.p. map $\psi\colon E \to C^*(\mathbb{F})$ such that $\pi \circ \psi = \mathrm{id}_E$. It follows that the formal identity

$$E \otimes \mathbb{B}(\ell^2) \xrightarrow{\psi\otimes\mathrm{id}} C^*(\mathbb{F}) \otimes \mathbb{B}(\ell^2) = C^*(\mathbb{F}) \otimes_{\max} \mathbb{B}(\ell^2) \xrightarrow{\pi\otimes\mathrm{id}} A \otimes_{\max} \mathbb{B}(\ell^2)$$

is contractive. Since E was arbitrary, we are done. □

Exercises

Exercise 13.2.1. Prove the following: If A is nuclear and B has the WEP, then $A \otimes B$ has the WEP; if A has the LLP and B is nuclear, then $A \otimes B$ has the LLP; if A is separable with the LP and B is separable and nuclear, then $A \otimes B$ has the LP.

Exercise 13.2.2. Let A be a C*-algebra with the WEP and let $B = C^*(\mathbb{F}_\infty)/J$. Prove that $A \otimes_{\max} B = (A \otimes C^*(\mathbb{F}_\infty))/(A \otimes J)$. Use tensor products to give another proof of the fact that a C*-algebra which is exact and has the WEP is nuclear (Exercise 2.3.14).

13.3. The QWEP conjecture

We say a C*-algebra A is *QWEP* if it is a quotient of a C*-algebra with the WEP. Kirchberg's QWEP conjecture asserts that every C*-algebra is QWEP, and it turns out to be equivalent to several seemingly unrelated open problems, including Connes's embedding problem (statement (3) below). Consequently, the QWEP conjecture is one of the most important open problems in the theory of operator algebras.

Theorem 13.3.1 (Kirchberg). *The following conjectures are equivalent:*

(1) *every C*-algebra is QWEP;*

(2) *$C^*(\mathbb{F}_\infty) \otimes_{\max} C^*(\mathbb{F}_\infty) = C^*(\mathbb{F}_\infty) \otimes C^*(\mathbb{F}_\infty)$ canonically;*

(3) *every type II_1-factor with separable predual is embeddable into the ultraproduct R^ω of the hyperfinite type II_1-factor R;*

(4) *the (separable) predual of any von Neumann algebra is isometrically isomorphic to a subspace of the Banach space ultraproduct $(S_1)_\omega$ of the predual S_1 of $\mathbb{B}(\ell^2)$.*

We will only prove the equivalence of (1), (2) and (3); see [**134**] for the last condition (and the exercises at the end of this section for a few more equivalent statements). We again need several lemmas. First, we add a local characterization of "relative weak injectivity" to Proposition 3.6.6.

Lemma 13.3.2. *Let $A \subset B$ be* C*-*algebras. The following are equivalent.*

(1) *the* C*-*algebra A is relatively weakly injective in B – i.e., there exists a c.c.p. map $\varphi\colon B \to A^{**}$ such that $\varphi|_A = \mathrm{id}_A$;*

(2) *for any finite-dimensional subspace $E \subset B$ and any $\varepsilon > 0$, there exists a contraction $\psi\colon E \to A$ such that $\|\psi|_{E\cap A} - \mathrm{id}_{E\cap A}\| < \varepsilon$.*

Proof. The implication (1) $\Rightarrow$ (2) follows from the principle of local reflexivity for Banach spaces (Theorem 9.1.1). To prove the converse, assume that for every finite-dimensional subspace $E \subset B$ and $\varepsilon > 0$, there exists a contraction $\psi_{E,\varepsilon}\colon E \to A$ such that $\|\psi|_{E\cap A} - \mathrm{id}_{E\cap A}\| < \varepsilon$. Let $\varphi\colon B \to A^{**}$ be a cluster point of $\psi_{E,\varepsilon}$ in the point-weak* topology. Then, φ is a contraction with $\varphi|_A = \mathrm{id}_A$. The weak*-continuous extension of φ on B^{**} is still denoted by φ. Then, $\varphi\colon B^{**} \to A^{**}$ is a contraction such that $\varphi|_{A^{**}} = \mathrm{id}_{A^{**}}$. By Tomiyama's Theorem (Theorem 1.5.10), φ is a conditional expectation, hence c.c.p. □

Lemma 13.3.3. *Let $\{A_i\}$ be a family of* C*-*algebras with the WEP. Then their product $\prod A_i$ also has the WEP.*

Proof. Let $A_i \subset \mathbb{B}(\mathcal{H}_i)$. Since $\prod \mathbb{B}(\mathcal{H}_i)$ is injective, it is enough to verify condition (2) in Lemma 13.3.2 for $\prod A_i \subset \prod \mathbb{B}(\mathcal{H}_i)$. Let $E \subset \prod \mathbb{B}(\mathcal{H}_i)$ be a finite-dimensional subspace and let $\varepsilon > 0$. Then, there exist finite-dimensional subspaces $E_i \subset \mathbb{B}(\mathcal{H}_i)$ such that $E \subset \prod E_i$. Since A_i has the WEP for each i, there exists a contraction $\psi_i\colon E_i \to A_i$ such that $\|\psi_i|_{E_i\cap A_i} - \mathrm{id}_{E_i\cap A_i}\| < \varepsilon$. We define a contraction $\psi\colon E \to \prod A_i$ by $\psi((x_i)_i) = (\psi_i(x_i))_i$ for $(x_i)_i \in E \subset \prod E_i$. Evidently we have $\|\psi|_{E\cap\prod A_i} - \mathrm{id}_{E\cap\prod A_i}\| < \varepsilon$. □

Lemma 13.3.4. *Let $A_0 \subset A$ be* C*-*algebras such that A_0 is relatively weakly injective in A. If $\pi\colon B \to A$ is a surjective $*$-homomorphism from a* C*-*algebra B, then $B_0 = \pi^{-1}(A_0)$ is relatively weakly injective in B. In particular, if A is QWEP, then so is A_0.*

Proof. Since A_0 is relatively weakly injective in A, there exists a (normal) conditional expectation ψ from A^{**} onto A_0^{**}. If we set $J = \ker \pi$, then

$$B_0^{**} \cong J^{**} \oplus A_0^{**} \subset J^{**} \oplus A^{**} \cong B^{**}.$$

Hence, $\mathrm{id}_{J^{**}} \oplus \psi$ is a conditional expectation from B^{**} onto B_0^{**}. If A is QWEP and B has the WEP, then B_0 has the WEP and A_0 is QWEP. □

Lemma 13.3.5. *Let A and B be unital* C*-*algebras. If B has the WEP and there exists a u.c.p. map $\varphi\colon B \to A$ which maps the closed unit ball of B onto the closed unit ball of A, then A is QWEP.*

Proof. Let $B_0 \subset B$ be the multiplicative domain of φ (Definition 1.5.8). Since φ maps the closed unit ball of B onto that of A, the restriction of φ to B_0 is a surjective $*$-homomorphism π onto A. We will prove that B_0 has the WEP. By Proposition 3.6.6, it suffices to show the canonical $*$-homomorphism $B_0 \otimes_{\max} C \to B \otimes_{\max} C$ is injective for every C*-algebra C. Let $J = \{x \in B : \varphi(x^*x) = 0 = \varphi(xx^*)\}$ be a hereditary C*-subalgebra in B. We observe that $J \subset B_0$ and in fact $J = \ker \pi$. Injectivity follows from the commutative diagram

$$\begin{array}{ccccccccc}
0 & \longrightarrow & J \otimes_{\max} C & \hookrightarrow & B_0 \otimes_{\max} C & \xrightarrow{\pi\otimes \mathrm{id}} & A \otimes_{\max} C & \longrightarrow & 0 \\
 & & \| & & \downarrow & & \| & & \\
 & & J \otimes_{\max} C & \hookrightarrow & B \otimes_{\max} C & \xrightarrow{\varphi\otimes \mathrm{id}} & A \otimes_{\max} C & &
\end{array}$$

since the top row is exact and $J \otimes_{\max} C \to B \otimes_{\max} C$ is injective (since J is hereditary). □

Lemma 13.3.6. *Let $\{A_i\}_{i\in I}$ be an increasing net of* C*-*subalgebras in $\mathbb{B}(\mathcal{H})$. If the A_i's are all QWEP, then so are $\overline{\bigcup A_i}$ and $(\bigcup A_i)''$.*

Proof. We only prove that the von Neumann algebra $M = (\bigcup A_i)''$ is QWEP. Adjoining the unit of $\mathbb{B}(\mathcal{H})$ if necessary, we may assume that all A_i's are unital. For each $i \in I$, fix a C*-algebra B_i with the WEP and a surjective $*$-homomorphism $\pi_i\colon B_i \to A_i$. Let J be a suitable directed set. It follows from Lemma 13.3.3 that $B = \prod_{(i,j)\in I\times J} B_i$ has the WEP. Fix a cofinal ultrafilter $\mathcal{U}$ (resp. $\mathcal{V}$) on the directed set I (resp. J) and define a u.c.p. map $\varphi\colon B \to M$ by

$$\varphi((x_{i,j})) = \lim_{j\to\mathcal{V}} \lim_{i\to\mathcal{U}} \pi_i(x_{i,j}),$$

where the limits are taken with respect to the ultraweak topology. If J is chosen large enough, then φ maps the closed unit ball of B onto that of M, by Kaplansky's density theorem. By Lemma 13.3.5, this implies that M is QWEP. □

Every C*-algebra A is obviously relatively weakly injective in its double dual A^{**}. Hence Lemmas 13.3.4 and 13.3.6 imply the useful fact that *A is QWEP if and only if A^{**} is QWEP.*

Proof of Theorem 13.3.1. (1) $\Rightarrow$ (2): It suffices to show that if a C*-algebra A is QWEP and has the LP, then A has the WEP. Let B be a C*-algebra with the WEP and let $\pi\colon B \to A$ be a surjective $*$-homomorphism.

Since A has the LP, there is a c.c.p. lifting $\psi\colon A \to B$. Let $A^{**} \subset \mathbb{B}(\mathcal{H})$ be a universal representation. Since B has the WEP, the c.c.p. map ψ extends to a c.c.p. map $\bar{\psi}\colon \mathbb{B}(\mathcal{H}) \to B^{**}$:

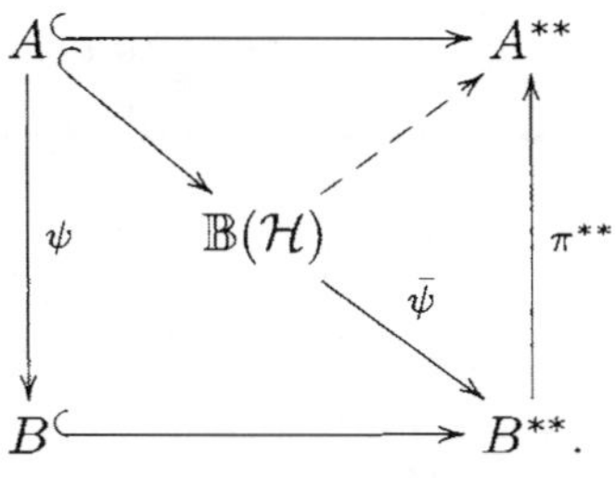

It follows that $\varphi = \pi^{**} \circ \bar{\psi}$ is a c.c.p. map such that $\varphi|_A = \pi \circ \psi = \mathrm{id}_A$.

(2) $\Rightarrow$ (3): Let M be a finite von Neumann algebra with separable predual and τ be a faithful normal trace on M. Let $\pi\colon C^*(\mathbb{F}_\infty) \to M$ be a $*$-homomorphism with ultraweakly dense range. By assumption and Theorem 6.2.7, the trace $\tau \circ \pi$ is an amenable trace on $C^*(\mathbb{F}_\infty)$. It follows that there is a $*$-homomorphism $\rho\colon C^*(\mathbb{F}_\infty) \to R^\omega$ such that $\tau_\omega \circ \rho = \tau \circ \pi$. It is easy to see that the von Neumann subalgebra in R^ω generated by $\rho(C^*(\mathbb{F}_\infty))$ is $*$-isomorphic to M (cf. Exercise 6.2.4).

(3) $\Rightarrow$ (1): The C*-algebra R^ω is QWEP by construction. Let M be a von Neumann subalgebra in R^ω. Since every von Neumann subalgebra in a finite von Neumann algebra is the range of a conditional expectation, M is also QWEP by Lemma 13.3.4. Since every finite von Neumann algebra with separable predual is embeddable into a type II_1-factor with separable predual (e.g., take a tracial free product with the hyperfinite II_1-factor R, [**154**]), every finite von Neumann algebra with separable predual is embeddable into R^ω by assumption and is QWEP. It follows from Lemma 13.3.6 that all semifinite von Neumann algebras are QWEP. By Takesaki's duality theorem (Theorems 9.3.5 and 9.3.7) we conclude that all von Neumann algebras are QWEP. In particular, all double duals are QWEP, so all C*-algebras are QWEP, too. □

Modifying the proof of (1) $\Rightarrow$ (2), it can be shown that a QWEP C*-algebra with the LLP also has the WEP. Hence, the QWEP conjecture is equivalent to saying that the LLP implies the WEP. There is no known example of a nonnuclear C*-algebra with both the WEP and the LLP.

From the proof of (3) $\Rightarrow$ (1), we see that a finite von Neumann algebra with separable predual is QWEP if it is embeddable into R^ω. (The converse is also true.) Hence we obtain the following corollary, the proof of which is left to the reader.

Corollary 13.3.7. *Let Γ be a group with Kirchberg's factorization property. Then, the group von Neumann algebra $L(\Gamma)$ is QWEP.*

Of course, if $C^*_\lambda(\Gamma)$ is QWEP, then so is $L(\Gamma)$, by Lemma 13.3.6. But typically it's very hard to show reduced group C*-algebras are QWEP. Here's one case where it can be done.

Proposition 13.3.8. *The reduced group* C*-*algebra* $C^*_\lambda(\mathbb{F}_r)$ *of the free group* $\mathbb{F}_r$ *is relatively weakly injective in the group von Neumann algebra* $L(\mathbb{F}_r)$. *In particular,* $C^*_\lambda(\mathbb{F}_r)$ *is QWEP.*

Proof. By Lemma 13.3.4, it suffices to show that $C^*_\lambda(\mathbb{F}_r)$ is relatively weakly injective in $L(\mathbb{F}_r)$. For this, we will verify condition (2) in Lemma 13.3.2. Recall that $\mathbb{F}_r$ is weakly amenable with $\Lambda_{\mathrm{cb}}(\mathbb{F}_r) = 1$ (Corollary 12.3.5). Hence, there exists a sequence of finitely supported functions φ_n on $\mathbb{F}_r$ which converge pointwise to 1 and such that their multipliers m_{φ_n} are (completely) contractive on $L(\mathbb{F}_r)$. Since each m_{φ_n} maps $L(\mathbb{F}_r)$ into $C^*_\lambda(\mathbb{F}_r)$ and since the sequence $\{m_{\varphi_n}\}$ converges to the identity on $C^*_\lambda(\mathbb{F}_r)$, we are done. □

A variant of the proof above shows that the reduced group C*-algebra $C^*_\lambda(\Gamma)$ is QWEP provided that Γ is residually finite and weakly amenable. However, it is not known whether or not $C^*_\lambda(\mathrm{SL}(3,\mathbb{Z}))$ is QWEP. Actually, the following question appears to be open: Does there exist a group Γ such that $C^*_\lambda(\Gamma)$ is *not* relatively weakly injective in $L(\Gamma)$?

Exercises

Exercise 13.3.1. Let A and B be unital C*-algebras and τ be a tracial state on $A \odot B$ such that $\pi_\tau(A \odot B)''$ is a factor. Prove that $\tau = (\tau|_A) \otimes (\tau|_B)$.

Exercise 13.3.2. Let C be a unital C*-algebra and T be the set of tracial states on C. Prove that $\tau \in T$ is an extreme point if and only if $\pi_\tau(C)''$ is a factor.

Exercise 13.3.3. Let A and B be unital C*-algebras and τ be a tracial state on $A \odot B$. Prove that τ is continuous on $A \otimes B$. (This is equivalent to asserting that any $*$-homomorphism from $A \odot B$ into a finite von Neumann algebra is continuous on $A \otimes B$.)

Exercise 13.3.4. Prove that $C^*(\mathbb{F}_\infty) \otimes_{\max} C^*(\mathbb{F}_\infty) = C^*(\mathbb{F}_\infty) \otimes C^*(\mathbb{F}_\infty)$ canonically if and only if $C^*(\mathbb{F}_\infty \times \mathbb{F}_\infty)$ has a faithful trace.

Exercise 13.3.5. Let C_i $(i = 1, 2)$ be C*-algebras with the LLP and J_i be a closed two-sided ideal in C_i. Prove that if $A_1 = C_1/J_1$ is QWEP, then

$$A_1 \otimes_{\max} A_2 = C_1 \otimes C_2/(J_1 \otimes C_2 + C_1 \otimes J_2)$$

canonically.

Exercise 13.3.6. Prove that a finite von Neumann algebra M with separable predual is embeddable into R^ω if it is QWEP. (Hint: Combine Exercise 13.3.5 and Theorem 6.2.7).

Exercise 13.3.7. Use the previous result and Exercise 6.2.4 to show that the QWEP conjecture is equivalent to knowing that every trace on $C^*(\mathbb{F}_\infty)$ is amenable.

Exercise 13.3.8. Prove that if Γ is a countable group with Kirchberg's factorization property, then $L(\Gamma)$ is embeddable into R^ω.

Exercise 13.3.9. Prove that a C*-algebra which is locally reflexive and which has the WEP is nuclear.

13.4. Nonsemisplit extensions

This section is devoted to the following theorem:

Theorem 13.4.1 (Kirchberg). *Let A be a separable QWEP C*-algebra and $CA = C_0(0,1] \otimes A$ be the cone over A. Then, there exists a quasidiagonal extension (Definition 10.3.3)*

$$0 \longrightarrow \mathbb{K}(\ell^2) \longrightarrow B \longrightarrow CA \longrightarrow 0$$

such that B has the WEP.

Remark 13.4.2. Arveson's Extension Theorem implies that if an extension $0 \to J \to C \to D \to 0$ is locally split and C has the WEP, then D also has the WEP. Since there is always a splitting for the quotient map $CA \to A$, it follows that if A in the theorem above does not have the WEP, then the sequence $0 \to \mathbb{K}(\ell^2) \to B \to CA \to 0$ can't be locally split. In the case $A = C^*_\lambda(\mathbb{F}_2)$, which is QWEP (Proposition 13.3.8) but does not have the WEP (see Exercise 13.2.2), we can apply the theorem and deduce that the resulting algebra B is not locally reflexive, hence not exact. Thus, *an extension of exact C*-algebras need not be exact* and exactness cannot be characterized in terms of double duals.[1]

It is not known whether or not in the theorem one can take B to have the LLP.

Turning to the proof of the theorem, let's ease notation by letting F denote the full free group C*-algebra $C^*(\mathbb{F}_\infty)$. Let A be a separable unital C*-algebra and J be a closed two-sided ideal in C such that $A = F/J$. We regard F as the subalgebra of constant functions in $C[0,1] \otimes F$ (which is in the multiplier algebra of the cone CF). Fix a quasicentral approximate unit $\{e_n\}$ of J in $C^*(\mathbb{F}_\infty)$ and define $*$-homomorphisms ρ_n by

$$\rho_n \colon C_0(0,1] \ni f \mapsto f(t \otimes (1 - e_n)) \in CF,$$

where $t \in C_0(0,1]$ is the identity function on $(0,1]$.

[1] Another example of a nonsemisplit extension of exact C*-algebras was given by Haagerup and Thorbjørnsen who proved that $C^*_\lambda(\mathbb{F}_2) \hookrightarrow \prod \mathbb{M}_n(\mathbb{C}) / \bigoplus \mathbb{M}_n(\mathbb{C})$.

Lemma 13.4.3. *For every $x \in J$, $a \in F$ and $f \in C_0(0,1]$, we have*

(1) $\lim_{n\to\infty} \|\rho_n(f)x\| = 0$,

(2) $\rho_n(f)a - fa \in CJ$ *for every* $n \in \mathbb{N}$, *and*

(3) $\lim_{n\to\infty} \|\rho_n(f)a - a\rho_n(f)\| = 0$.

Proof. These assertions are all trivial when f is a polynomial with $f(0) = 0$. Since such polynomials are dense in $C_0(0,1]$, we are done. □

For a C*-algebra D, we set $D_\infty = \prod_{n=1}^\infty D / \bigoplus_{n=1}^\infty D$ and define a $*$-homomorphism ρ_∞ by

$$\rho_\infty \colon C_0(0,1] \ni f \mapsto (\rho_n(f))_n + \bigoplus_{n=1}^\infty CF \in (CF)_\infty.$$

We regard $F \subset (C[0,1] \otimes F)_\infty$ as constant functions. Since the range of ρ_∞ commutes with F, part (3) of Lemma 13.4.3 implies that they give rise to a $*$-homomorphism $\varphi = \rho_\infty \times \mathrm{id}_F \colon CF \to (CF)_\infty$. Since the C*-algebra CF has the LP, φ has a c.c.p. lifting $\psi \colon CF \to \prod_{n=1}^\infty CF$. It is not hard to see that $\ker \varphi = CJ$ and hence φ induces an injective $*$-homomorphism $CA \hookrightarrow (CF)_\infty$. Indeed, we have the following stronger result.

Lemma 13.4.4. *For any* C*-*algebra* D, *the c.c.p. map* $\psi \otimes \mathrm{id}_D$ *induces an isometric* $*$-*homomorphism* θ:

$$\begin{array}{ccc} CF \otimes D & \xrightarrow{\psi\otimes\mathrm{id}} & \left(\prod_{n=1}^\infty CF\right) \otimes D \\ {\scriptstyle Q}\big\downarrow & & \big\downarrow \\ \dfrac{CF \otimes D}{CJ \otimes D} & \overset{\theta}{\hookrightarrow} & \dfrac{\left(\prod_{n=1}^\infty CF\right) \otimes D}{\left(\bigoplus_{n=1}^\infty CF\right) \otimes D}. \end{array}$$

Proof. For any $f \in C_0(0,1]$ and $a \in F$, we have $\psi(fa) = (\rho_n(f)a)_n$ modulo $\bigoplus_{n=1}^\infty CF$. Hence, ψ is multiplicative modulo $\bigoplus_{n=1}^\infty CF$ and ψ maps CJ into $\bigoplus_{n=1}^\infty CF$ by part (1) of Lemma 13.4.3. It follows that θ is a well-defined $*$-homomorphism. Moreover, for any $y = \sum_k f_k a_k \otimes x_k \in CF \otimes D$, we have

$$\|\theta(Q(y))\| = \limsup_{n\to\infty} \Big\|\sum_k \rho_n(f_k)a_k \otimes x_k\Big\| \geq \|Q(y)\|$$

by part (2) of Lemma 13.4.3. This implies that θ is isometric. □

Lemma 13.4.5. *For any separable* D, *there is a c.c.p. map* $\Psi \colon CF \to \prod_{n=1}^\infty \mathbb{M}_n(\mathbb{C})$ *such that* Ψ *is multiplicative modulo* $\bigoplus_{n=1}^\infty \mathbb{M}_n(\mathbb{C})$, Ψ *maps*

CJ into $\bigoplus_{n=1}^{\infty} \mathbb{M}_n(\mathbb{C})$, and the c.c.p. map $\Psi \otimes \mathrm{id}_D$ induces an isometric $*$-homomorphism π:

$$\begin{array}{ccc} CF \otimes D & \xrightarrow{\Psi\otimes\mathrm{id}} & \prod_{n=1}^{\infty} \mathbb{M}_n(\mathbb{C}) \otimes D \\ {\scriptstyle Q}\downarrow & & \downarrow \\ \dfrac{CF \otimes D}{CJ \otimes D} & \overset{\pi}{\hookrightarrow} & \dfrac{\left(\prod_{n=1}^{\infty} \mathbb{M}_n(\mathbb{C})\right) \otimes D}{\left(\bigoplus_{n=1}^{\infty} \mathbb{M}_n(\mathbb{C})\right) \otimes D}. \end{array}$$

Proof. Let ψ be as in Lemma 13.4.4 and $\psi_n : CF \to CF$ be the c.c.p. map such that $\psi = (\psi_n)$. Choose a dense sequence $\{y_j\}_{j=1}^{\infty}$ in $CF \otimes D$. Since CF is residually finite-dimensional by Theorem 7.4.1, for each n, there is a $*$-homomorphism $\sigma_n : CF \to \mathbb{M}_{k(n)}(\mathbb{C})$ such that

$$\|(\sigma_n \circ \psi_n \otimes \mathrm{id}_D)(y_j)\|_{\mathbb{M}_{k(n)}(\mathbb{C})\otimes D} \geq \|(\psi_n \otimes \mathrm{id}_D)(y_j)\|_{CF\otimes D} - \frac{1}{n}$$

for every $j = 1, \ldots, n$. We may assume that $\{k(n)\}$ is increasing and define $\Psi : CF \to \prod_{n=1}^{\infty} \mathbb{M}_{k(n)}(\mathbb{C})$ by $\Psi = (\sigma_n \circ \psi_n)_n$. It is not hard to see that Ψ is multiplicative modulo $\bigoplus_{n=1}^{\infty} \mathbb{M}_{k(n)}(\mathbb{C})$ and Ψ maps CJ into $\bigoplus_{n=1}^{\infty} \mathbb{M}_{k(n)}(\mathbb{C})$. We regard $\prod_{n=1}^{\infty} \mathbb{M}_{k(n)}(\mathbb{C})$ as a "lacunary" subalgebra in $\prod_{k=1}^{\infty} \mathbb{M}_k(\mathbb{C})$. Now the $*$-homomorphism π in the diagram is well-defined and, for every j, we have

$$\begin{aligned} \|\pi(Q(y_j))\| &= \limsup_{n\to\infty} \|(\sigma_n \circ \psi_n \otimes \mathrm{id}_D)(y_j)\| \\ &= \limsup_{n\to\infty} \|(\psi_n \otimes \mathrm{id}_D)(y_j)\| \\ &= \|Q(y_j)\| \end{aligned}$$

by Lemma 13.4.4. This implies that π is isometric. $\square$

We are now in a position to prove Theorem 13.4.1.

Proof of Theorem 13.4.1. Let $D = F$ and take $\Psi : CF \to \prod_{n=1}^{\infty} \mathbb{M}_n(\mathbb{C})$ as in Lemma 13.4.5. We embed $\prod_{n=1}^{\infty} \mathbb{M}_n(\mathbb{C})$ into $\mathbb{B}(\ell^2)$ via a unitary identification $\bigoplus_{n=1}^{\infty} \ell_n^2 \cong \ell^2$. Let $B \subset \mathbb{B}(\ell^2)$ be the (nonunital) C*-algebra generated by $\Psi(CF)$ and $\mathbb{K} = \mathbb{K}(\ell^2)$. It follows that $B/\mathbb{K}$ is $*$-isomorphic to $CF/CJ = CA$. Since A is QWEP, the left hand side of the bottom line of the diagram in Lemma 13.4.5 is canonically $*$-isomorphic to $CA \otimes_{\max} F$ (Exercise 13.3.5). It follows that the bottom row of the following commuting diagram is exact (the top row is also exact, by universality of the maximal tensor product):

$$\begin{array}{ccccccccc} 0 & \longrightarrow & \mathbb{K}\otimes_{\max} F & \longrightarrow & B\otimes_{\max} F & \longrightarrow & CA\otimes_{\max} F & \longrightarrow & 0 \\ & & \downarrow & & \downarrow & & \downarrow & & \\ 0 & \longrightarrow & \mathbb{K}\otimes F & \longrightarrow & B\otimes F & \longrightarrow & CA\otimes_{\max} F & \longrightarrow & 0. \end{array}$$

Since $\mathbb{K}$ is nuclear, we have $B\otimes_{\max} F = B\otimes F$ by the 5 Lemma. □

Remark 13.4.6. With additional effort, one can find B with the property that $B\otimes_{\max} B^{\mathrm{op}} = B\otimes B^{\mathrm{op}}$. We sketch the argument. In the proof of Lemma 13.4.5, we have

$$\limsup_{n\to\infty} \|(\psi_n\otimes\psi_n^{\mathrm{op}})(y)\| = \|y + (CJ\otimes CF^{\mathrm{op}} + CF\otimes CJ^{\mathrm{op}})\|$$

for every $y\in CF\otimes CF^{\mathrm{op}}$. Hence, we can take Ψ in Lemma 13.4.5 so that it satisfies

$$\begin{array}{ccc} CF\otimes CF^{\mathrm{op}} & \xrightarrow{\ \Psi\otimes\Psi^{\mathrm{op}}\ } & M\otimes M^{\mathrm{op}} \\ \downarrow & & \downarrow \\ \dfrac{CF\otimes CF^{\mathrm{op}}}{CJ\otimes CF^{\mathrm{op}} + CF\otimes CJ^{\mathrm{op}}} & \hookrightarrow & \dfrac{M\otimes M^{\mathrm{op}}}{K\otimes M^{\mathrm{op}} + M\otimes K^{\mathrm{op}}}, \end{array}$$

where $M = \prod \mathbb{M}_n(\mathbb{C})$ and $K = \bigoplus \mathbb{M}_n(\mathbb{C})$. The rest of the proof is similar to that of Theorem 13.4.1.

13.5. Norms on $\mathbb{B}(\ell^2)\odot\mathbb{B}(\ell^2)$

We close this chapter with another fundamental result.

Theorem 13.5.1 (Junge-Pisier). *The* C*-*algebra $\mathbb{B}(\ell^2)$ does not have the LLP; in other words,*

$$\mathbb{B}(\ell^2)\otimes_{\max}\mathbb{B}(\ell^2) \neq \mathbb{B}(\ell^2)\otimes\mathbb{B}(\ell^2).$$

For a Hilbert space $\mathcal{H}$, we denote by $\overline{\mathcal{H}}$ its complex conjugate Hilbert space. More precisely, $\mathcal{H}\ni\xi\mapsto\bar{\xi}\in\overline{\mathcal{H}}$ is an isomorphism as a real Hilbert space, but we have $\overline{\lambda\xi} = \bar{\lambda}\bar{\xi}$ and $\langle\bar{\xi},\bar{\eta}\rangle_{\overline{\mathcal{H}}} = \overline{\langle\xi,\eta\rangle_{\mathcal{H}}}$ for $\lambda\in\mathbb{C}$ and $\xi,\eta\in\mathcal{H}$. For $x\in\mathbb{B}(\mathcal{H})$, we denote by $\bar{x}$ the corresponding operator in $\mathbb{B}(\overline{\mathcal{H}})$. Finally, the complex conjugate of a C*-algebra $A\subset\mathbb{B}(\mathcal{H})$ is defined as $\overline{A} = \{\bar{x}\in\mathbb{B}(\overline{\mathcal{H}}) : x\in A\}$. We note that $\overline{A}$ is $*$-isomorphic to the opposite A^{op} via the adjoint map. The following lemma is well known, so we omit the proof.

Lemma 13.5.2. *The Hilbert space $\mathcal{H}\otimes\overline{\mathcal{K}}$ can be identified with the space $\mathcal{S}_2(\mathcal{K},\mathcal{H})$ of Hilbert-Schmidt class operators from $\mathcal{K}$ into $\mathcal{H}$ via the isomorphism $\theta\colon\mathcal{H}\otimes\overline{\mathcal{K}}\to\mathcal{S}_2(\mathcal{K},\mathcal{H})$ given by*

$$\theta(\sum\xi_i\otimes\bar{\eta}_i)\zeta = \sum\langle\zeta,\eta_i\rangle_{\mathcal{K}}\xi_i$$

for $\zeta \in \mathcal{K}$. *Under this identification,* $x \otimes \bar{y} \in \mathbb{B}(\mathcal{H} \otimes \overline{\mathcal{K}})$ *acts on* $\mathcal{S}_2(\mathcal{K}, \mathcal{H})$ *as*

$$\mathcal{S}_2(\mathcal{K}, \mathcal{H}) \ni h \mapsto xhy^* \in \mathcal{S}_2(\mathcal{K}, \mathcal{H}).$$

If A is a C*-algebra with a tracial state τ, then for any unitary elements $u_1, \ldots, u_k \in A$, we have

$$\|\sum_{i=1}^{k} u_i \otimes \overline{u_i}\|_{A \otimes_{\max} \overline{A}} = k$$

since the functional $A \otimes_{\max} \overline{A} \ni \sum x_i \otimes \overline{y_i} \mapsto \sum \tau(x_i y_i^*) \in \mathbb{C}$ is a continuous state (see Chapter 6). In particular, for any unitary matrices $u_1, \ldots, u_k \in \mathbb{M}_N(\mathbb{C})$, we have

$$\|\sum_{i=1}^{k} u_i \otimes \overline{u_i}\|_{\mathbb{M}_N(\mathbb{C}) \otimes \overline{\mathbb{M}_N(\mathbb{C})}} = k.$$

Definition 13.5.3. Let $(u_1(n), \ldots, u_k(n)) \in \mathbb{M}_{N(n)}(\mathbb{C})^k$ be a sequence of k-tuples of unitary matrices. We say the family $\{(u_i(n))_{i=1,\ldots,k} : n \in \mathbb{N}\}$ is *coding* if

$$\sup\{\|\sum_{i=1}^{k} u_i(m) \otimes \overline{u_i(n)}\|_{\mathbb{M}_{N(m)}(\mathbb{C}) \otimes \overline{\mathbb{M}_{N(n)}(\mathbb{C})}} : m \neq n\} < k.$$

Theorem 13.5.4 (Voiculescu). *There exists a coding family of unitary k-tuples, for every $k \geq 3$.*

Proof. To prove the theorem, we use the fact that there exists a residually finite group Γ with property (T) (e.g., we can take $\Gamma = \mathrm{SL}(3, \mathbb{Z})$; see Theorem 12.1.14).

First we must find an infinite sequence of finite-dimensional irreducible representations $\pi_n \colon \Gamma \to \mathbb{M}_{N(n)}(\mathbb{C})$ such that π_n is *not* equivalent to π_m whenever $n \neq m$. Using residual finiteness of Γ, this is easy. Indeed, if $(\Gamma_n)_{n \in \mathbb{N}}$ is a sequence of finite quotients of Γ, then the left regular representations $\Gamma_n \to \mathbb{B}(\ell^2(\Gamma_n))$ extend to representations $\Gamma \to \mathbb{B}(\ell^2(\Gamma_n))$. Each such representation is a direct sum of irreducible representations, and infinitely many of these irreducible components must be mutually inequivalent (since Γ is infinite). Hence we can find representations π_n with the desired properties.

Schur's Lemma (Lemma 17.2.1) implies for $m \neq n$ the unitary representation $\pi_m \otimes \overline{\pi_n}$ on $\mathcal{H}_m \otimes \overline{\mathcal{H}_n}$ does not have a nonzero invariant vector (since, by Lemma 13.5.2, an invariant vector can be identified with a Hilbert-Schmidt operator $T \colon \mathcal{H}_n \to \mathcal{H}_m$ which intertwines π_m and π_n). Let $\mathcal{S} \subset \Gamma$ be a finite subset of generators that contains the unit e. (We remark that $\mathrm{SL}(3, \mathbb{Z})$ is generated by two elements and $|\mathcal{S}|$ can be 3.) Since $\bigoplus_{m\neq n}\pi_m\otimes\overline{\pi_n}$ does not have a nonzero invariant vector, we have

$$\sup\{\|\sum_{s\in\mathcal{S}}\pi_m(s)\otimes\overline{\pi_n(s)}\| : m\neq n\} < |\mathcal{S}|,$$

by Lemma 12.1.9. □

Remark 13.5.5. Here is another proof. A deep theorem of Haagerup and Thorbjørnsen [**81**] states that there is an embedding

$$\pi\colon C^*_\lambda(\mathbb{F}_k)\hookrightarrow\prod_{n=1}^\infty\mathbb{M}_n(\mathbb{C})/\bigoplus_{n=1}^\infty\mathbb{M}_n(\mathbb{C}).$$

Let $g_1,\ldots,g_k$ ($k\geq 2$) be the free generators of $\mathbb{F}_k$ and take for each i a lifting $(u_i(n))_{n=1}^\infty\in\prod_{n=1}^\infty\mathbb{M}_n(\mathbb{C})$ of $\pi(g_i)$ such that $u_i(n)$ are unitary for all n. We set $u_{-i}(n)=u_i(n)^*$. Then, by Fell's absorption principle, we have

$$\begin{aligned}\limsup_{m\to\infty}\|\sum_{i=\pm1}^{\pm k}u_i(m)\otimes\overline{u_i(n)}\| &= \|\sum_{i=\pm1}^{\pm k}\lambda(g_i)\otimes\overline{u_i(n)}\|\\ &= \|\sum_{i=\pm1}^{\pm k}\lambda(g_i)\|\\ &< 2k,\end{aligned}$$

since $\mathbb{F}_k$ is nonamenable (Theorem 2.6.8). Therefore, passing to an appropriate subsequence, we obtain a coding family of unitary matrices.

Remark 13.5.6. The reduced group C*-algebra $C^*_\lambda(\mathbb{F}_k)$ is a non-QD C*-subalgebra of $\prod_{n=1}^\infty\mathbb{M}_n(\mathbb{C})/\bigoplus_{n=1}^\infty\mathbb{M}_n(\mathbb{C})$. It turns out that any coding family of unitary k-tuples gives rise to such a non-QD subalgebra. This observation of S. Wassermann [**195**] will be revisited in Chapter 17, but let's sketch a proof now.

Let $\{(u_i(n))_{i=1,\ldots,k} : n\in\mathbb{N}\}$ be a coding family, $M=\prod_{n=1}^\infty\mathbb{M}_{N(n)}(\mathbb{C})$ and $u_i=(u_i(n))_{n=1}^\infty\in M$. Denote by $\pi\colon M\to M/K$ the quotient map, where $K=\bigoplus_{n=1}^\infty\mathbb{M}_{N(n)}(\mathbb{C})$ and set $v_i=\pi(u_i)$. Then we have

$$\begin{aligned}\|\sum_{i=1}^k v_i\otimes\overline{v_i}\|_{M/K\otimes\overline{M/K}} &\leq \|\sum_{i=1}^k u_i\otimes\overline{v_i}\|_{M\otimes\overline{M/K}}\\ &= \sup_{m\in\mathbb{N}}\|\sum_{i=1}^k u_i(m)\otimes\overline{v_i}\|_{\mathbb{M}_{N(m)}(\mathbb{C})\otimes\overline{M/K}}\\ &= \sup_{m\in\mathbb{N}}\limsup_{n\to\infty}\|\sum_{i=1}^k u_i(m)\otimes\overline{u_i(n)}\|_{\mathbb{M}_{N(m)}(\mathbb{C})\otimes\overline{\mathbb{M}_{N(n)}(\mathbb{C})}}\\ &< k.\end{aligned}$$

It follows that $C^*(\{v_i : i = 1, \ldots, k\}) \subset M/K$ cannot have an amenable tracial state (since, by Theorem 6.2.7, such a trace τ would give a state on $C^*(\{v_i : i = 1, \ldots, k\}) \otimes \overline{C^*(\{v_i : i = 1, \ldots, k\})}$ defined by $x \otimes \overline{y} \mapsto \tau(xy^*)$) and hence cannot be quasidiagonal (Proposition 7.1.16).

We now present Pisier's proof of Theorem 13.5.1.

Proof of Theorem 13.5.1. Let $\{(u_i(n))_{i=1,\ldots,k} \in \mathbb{M}_{N(n)}(\mathbb{C})^k : n \in \mathbb{N}\}$ be a coding family of unitary k-tuples. Let $\sigma_n \colon C^*(\mathbb{F}_k) \to \mathbb{M}_{N(n)}(\mathbb{C})$ be the $*$-homomorphism given by $\sigma_n(U_i) = u_i(n)$ for every $i = 1, \ldots, k$, where the U_i's are the free generators of $C^*(\mathbb{F}_k)$. Taking a subsequence, we may assume that the sequence $\mathrm{tr}_{N(n)}(\sigma_n(x))$ converges for every $x \in C^*(\mathbb{F}_k)$. For each $j = 1, 2$, let

$$M_j = \prod_{n \in \mathbb{N}} \mathbb{M}_{N(2n+j)}(\mathbb{C}) \quad \text{and} \quad u_i^{(j)} = (u_i(2n+j))_{n \in \mathbb{N}} \in M_j.$$

Let $A_j = C^*(\{u_i^{(j)} : i = 1, \ldots, k\}) \subset M_j$. Fix a free ultrafilter ω on $\mathbb{N}$ and consider the GNS representations π_j of the tracial states τ_j on M_j, which are given by $\tau_j((x_n)_n) = \lim_\omega \mathrm{tr}_{N(2n+j)}(x_n)$. Then, by assumption, we have an isomorphism from $\pi_1(A_1)$ onto $\pi_2(A_2)$ sending $\pi_1(u_i^{(1)})$ to $\pi_2(u_i^{(2)})$. We set $N = \pi_j(A_j)''$ and $v_i = \pi_j(u_i^{(j)}) \in N$. Since there exists a (trace-preserving) conditional expectation from $\pi_j(M_j)''$ onto $\pi_j(A_j)''$, there exist u.c.p. maps $\varphi_j \colon M_j \to N$ such that $\varphi_j(u_i^{(j)}) = v_i$ for $j = 1, 2$.

By the coding assumption, we have

$$\Big\| \sum_{i=1}^k u_i^{(1)} \otimes \overline{u_i^{(2)}} \Big\|_{M_1 \otimes \overline{M_2}} = \sup_{m,n} \Big\| \sum_{i=1}^k u_i(2m+1) \otimes \overline{u_i(2n+2)} \Big\| < k.$$

On the other hand, since N has a tracial state, we have

$$\Big\| \sum_{i=1}^k u_i^{(1)} \otimes \overline{u_i^{(2)}} \Big\|_{M_1 \otimes_{\max} \overline{M_2}} \geq \Big\| \sum_{i=1}^k v_i \otimes \overline{v_i} \Big\|_{N \otimes_{\max} \overline{N}} = k.$$

It follows that $M_1 \otimes_{\max} \overline{M_2} \neq M_1 \otimes \overline{M_2}$. Since $M_j \hookrightarrow \mathbb{B}(\ell^2)$ in such a way that M_j is the range of a conditional expectation, this proves the assertion. $\square$

13.6. References

Most of the results in this chapter are due to Kirchberg [**102**]. Notable exceptions are the proof of Theorem 13.2.1, which is taken from [**150**], and Junge and Pisier's Theorem 13.5.1, which is taken from [**93, 153**]. See [**134**] for more information on the QWEP conjecture.

Chapter 14

Weakly Exact von Neumann Algebras

As noted earlier, if one defines exact von Neumann algebras as those which have a weakly nuclear embedding into $\mathbb{B}(\mathcal{H})$ – i.e., the obvious adaptation of our C*-definition – one has defined nothing new (since every c.c.p. map into $\mathbb{B}(\mathcal{H})$ is weakly nuclear). However, there is a sensible alternative, based on tensor products, which we now briefly explore.

14.1. Definition and examples

Throughout this chapter, B will be an arbitrary unital C*-algebra and $J \lhd B$ will be an ideal. The canonical quotient map will be denoted by $Q\colon B \to B/J$. All C*-algebras, except ideals, are assumed to be unital.

Recall that a C*-algebra A is exact if there is a canonical identification

$$\frac{B \otimes A}{J \otimes A} \cong (B/J) \otimes A,$$

for any $J \lhd B$ (Theorem 3.9.1). One may rephrase this result as follows: A is exact if for any $J \lhd B$ and any $*$-representation $\pi\colon A \otimes B \to \mathbb{B}(\mathcal{H})$ with $A \otimes J \subset \ker \pi$, the induced representation $\tilde{\pi}\colon A \odot (B/J) \to \mathbb{B}(\mathcal{H})$ is min-continuous. Inserting the adjective *normal* in properly, we have our W*-definition.

Definition 14.1.1. A von Neumann algebra M is said to be *weakly exact* if for any $J \lhd B$ and any left normal $*$-representation[1] $\pi\colon M \otimes B \to \mathbb{B}(\mathcal{H})$

[1]Recall that π is left normal if $\pi|_{M \otimes \mathbb{C}1}$ is normal.

with $M \otimes J \subset \ker \pi$, the induced representation $\tilde{\pi}\colon M \odot (B/J) \to \mathbb{B}(\mathcal{H})$ is min-continuous.

Note that by Theorem 3.8.5, a left normal $*$-representation $\pi\colon M \odot B \to \mathbb{B}(\mathcal{H})$ is min-continuous if and only if the $*$-homomorphism $\pi|_B\colon B \to \pi(M \otimes \mathbb{C}1_B)'$ is weakly nuclear.

Theorem 14.1.2. *A von Neumann algebra M is weakly exact if it contains a weakly dense* C*-*algebra A which is exact.*

Proof. Let A be an exact C*-algebra such that $M = A''$. By Theorem 9.3.1, A is locally reflexive. Let $J \lhd B$, π and $\tilde{\pi}$ be given as in Definition 14.1.1. Since A is exact, $\tilde{\pi}$ is continuous on $A \otimes (B/J)$. Since local reflexivity is equivalent to property C'' (Proposition 9.2.5), $\tilde{\pi}$ extends to a left normal $*$-representation ρ on $A^{**} \otimes (B/J)$. Let $p \in A^{**}$ be the central support projection of $\rho|_{A^{**}}\colon A^{**} \to M$ (so that we may identify M with pA^{**}). We denote this identification by $\iota\colon M \to pA^{**} \subset A^{**}$. It follows that $\tilde{\pi}$ coincides with $\rho \circ (\iota \otimes \mathrm{id})$, which is min-continuous. □

Remark 14.1.3. Suppose that the second dual A^{**} of a C*-algebra A is weakly exact. Then, A is exact if and only if it is locally reflexive (Exercise 14.1.1). The assumption of local reflexivity is essential. Indeed, there exists a nonexact C*-algebra whose second dual is weakly exact. For instance, an extension of exact C*-algebras need not be exact (see Theorem 13.4.1), but it is easily checked that the direct sum of two weakly exact von Neumann algebras is again weakly exact. It is not known whether there exists a locally reflexive C*-algebra which is not exact.

Proposition 14.1.4. *Let M be a von Neumann algebra. Suppose that there exists a net M_i of weakly exact von Neumann algebras and nets of normal c.p. contractions $\varphi_i\colon M \to M_i$ and $\psi_i\colon M_i \to M$ such that the net $\psi_i \circ \varphi_i$ converges to the identity on M in the point-ultraweak topology. Then, M is weakly exact.*

Proof. Let $J \lhd B$ and a left normal $*$-representation $\pi\colon M \otimes B \to \mathbb{B}(\mathcal{H})$ with $M \otimes J \subset \ker \pi$ be given. We consider the left normal c.p. contraction

$$\Psi_i = \pi \circ (\psi_i \otimes \mathrm{id}_B)\colon M_i \otimes B \to \mathbb{B}(\mathcal{H}).$$

Since $M_i \ni a \mapsto \Psi_i(x(a \otimes 1_B)y) \in \mathbb{B}(\mathcal{H})$ is ultraweakly continuous for every $x, y \in M_i \otimes B$, the minimal Stinespring dilation of Ψ_i is a left normal $*$-representation which still vanishes on $M_i \otimes J$. (One should verify this.) Thus, it follows from weak exactness of M_i that the induced c.p. map $\tilde{\Psi}_i\colon M_i \odot (B/J) \to \mathbb{B}(\mathcal{H})$ is min-continuous. Since π is left normal, the net of c.p. contractions

$$\tilde{\Psi}_i \circ (\varphi_i \otimes \mathrm{id}_{B/J})\colon M \otimes (B/J) \to \mathbb{B}(\mathcal{H})$$

converges to $\tilde{\pi}$ in the point-ultraweak topology. This implies the continuity of $\tilde{\pi}$. □

Corollary 14.1.5. *All of the following statements are true.*

(1) *Every injective (semidiscrete) von Neumann algebra is weakly exact.*

(2) *If M is a weakly exact von Neumann algebra, then $M \bar{\otimes} \mathbb{B}(\mathcal{H})$ is weakly exact.*

(3) *A von Neumann algebra M is weakly exact if and only if its commutant M' is.*

(4) *If M is a weakly exact von Neumann algebra and G is an amenable (locally compact) group which acts on M, then $M \rtimes G$ is weakly exact.*

(5) *A von Neumann subalgebra $N \subset M$ is weakly exact provided that M is weakly exact and there exists a normal conditional expectation from M onto N. In particular, every von Neumann subalgebra of a finite weakly exact von Neumann algebra is weakly exact.*

The proofs of these facts make nice exercises, so we leave them to you. It is unknown whether or not every conditional expectation from a von Neumann algebra M onto a von Neumann subalgebra N can be approximated, in the point-ultraweak topology, by normal c.p. contractions from M into N. However, it is known to be true "up to Morita equivalence" [**4**]. Hence the last fact of the above corollary holds without the normality assumption on the conditional expectation.

If Γ is an exact discrete group, then the group von Neumann algebra $L(\Gamma)$ is weakly exact since it contains the weakly dense exact C*-algebra $C^*_\lambda(\Gamma)$. The converse is also true, though we postpone the proof until Section 14.2.

Theorem 14.1.6. *For a discrete group Γ, the group von Neumann algebra $L(\Gamma)$ is weakly exact if and only if Γ is exact.*

More generally, if α is a measure-preserving action of Γ on a probability space (X, μ), then the crossed product von Neumann algebra $L^\infty(X, \mu) \rtimes \Gamma$ is weakly exact if and only if Γ is exact. Indeed, if Γ is exact, then the C*-crossed product $L^\infty(X, \mu) \rtimes_r \Gamma$ is weakly dense in $L^\infty(X, \mu) \rtimes \Gamma$ and is exact since $L^\infty(X, \mu)$ is an exact (nuclear) C*-algebra; the converse follows from the previous theorem, together with the fact that weak exactness passes to subalgebras of finite von Neumann algebras. In particular, exactness is preserved under so-called *measure equivalence* (cf. [**185**, Proposition XIII.2.16] and [**67**]), as this implies stable isomorphism of the corresponding crossed products.

Remark 14.1.7. Since there exists a discrete group which is not exact [71] (see also Remark 5.5.10), there exists a von Neumann algebra which is not weakly exact. However, it would be very nice to have simpler examples; a prime candidate is R^ω.

A large number of von Neumann algebras have the W*OAP (see Section 12.4). We now show that the W*OAP implies weak exactness. Denote by $\mathrm{CB}(M)$ the Banach space of all completely bounded maps on M and define

$$CB(M)_* = \text{the closed linear span of } \{\omega_{a,f} : a \in M,\ f \in M_*\} \subset \mathrm{CB}(M)^*,$$

where $\omega_{a,f}(\varphi) = f(\varphi(a))$ for $\varphi \in \mathrm{CB}(M)$. For $a \in M \bar{\otimes} \mathbb{B}(\ell^2)$ and $f \in (M \bar{\otimes} \mathbb{B}(\ell^2))_*$, we define $\omega_{a,f} \in \mathrm{CB}(M)^*$ with $\|\omega_{a,f}\| \le \|a\|\|f\|$ by[2]

$$\omega_{a,f}(\varphi) = f(\varphi \otimes \mathrm{id}_{\mathbb{B}(\ell^2)}(a)).$$

The proof of the following proposition is very similar to that of Lemmas D.7–D.9, and hence it is omitted.

Proposition 14.1.8. *For a von Neumann algebra M, the following are true.*

(1) *One has $\omega_{a,f} \in \mathrm{CB}(M)_*$ for every $a \in M \bar{\otimes} \mathbb{B}(\ell^2)$ and $f \in (M \bar{\otimes} \mathbb{B}(\ell^2))_*$.*
(2) *There is a canonical isometric isomorphism $\mathrm{CB}(M) = (\mathrm{CB}(M)_*)^*$.*
(3) *Every element in $\mathrm{CB}(M)_*$ is of the form $\omega_{a,f}$ with $a \in M \otimes \mathbb{K}(\ell^2)$ and $f \in (M \bar{\otimes} \mathbb{B}(\ell^2))_*$.*

In particular, the stable point-ultraweak topology (Definition 12.4.1) coincides with the $\sigma(\mathrm{CB}(M), \mathrm{CB}(M)_*)$-topology.

Corollary 14.1.9. *Let B be a C*-algebra, $x \in M \otimes B$ and $f \in (M \otimes B)^*$. If $f(\,\cdot\, \otimes b) \in M_*$ for every $b \in B$, then $\omega_{x,f} \in \mathrm{CB}(M)_*$, where $\omega_{x,f}(\varphi) = f((\varphi \otimes \mathrm{id}_B)(x))$ for all $\varphi \in \mathrm{CB}(M)$.*

Proof. Let (x_n) be a sequence in $M \odot B$ which converges to x. It is clear that the sequence $(\omega_{x_n,f})$ is in $\mathrm{CB}(M)_*$ and converges to $\omega_{x,f}$. □

Proposition 14.1.10. *A von Neumann algebra M with the W*OAP is weakly exact.*

[2]The map $\varphi \otimes \mathrm{id}_{\mathbb{B}(\ell^2)}$ is defined as the point-ultraweak limit of $\varphi \otimes \Psi_n$, where Ψ_n is the compression onto $\mathbb{B}(\ell^2_n)$. It is uniquely determined by the identity $(\mathrm{id}_M \otimes g) \circ (\varphi \otimes \mathrm{id}_{\mathbb{B}(\ell^2)}) = \varphi \circ (\mathrm{id}_M \otimes g)$ for all $g \in \mathbb{B}(\ell^2)_*$.

Proof. Let M be a von Neumann algebra with the W*OAP and let φ_i be a net of finite-rank maps on M which converges to id_M in the stable point-ultraweak topology. Let $J \lhd B$, π and $\tilde{\pi}$ be given as in Definition 14.1.1. Recall that $Q\colon B \to B/J$ is the canonical quotient map. It follows from Corollary 14.1.9 that the net $\pi \circ (\varphi_i \otimes \mathrm{id}_B)(x)$ converges ultraweakly to $\pi(x)$ for every $x \in M \otimes B$. Let $x \in \ker(\mathrm{id}_M \otimes Q)$ be given. Since

$$(\mathrm{id}_M \otimes Q)((\varphi_i \otimes \mathrm{id}_B)(x)) = (\varphi_i \otimes \mathrm{id}_{B/J})(\mathrm{id}_M \otimes Q)(x) = 0$$

and since φ_i is of finite-rank, we have $(\varphi_i \otimes \mathrm{id}_B)(x) \in M \odot J$ for all i. It follows that

$$\pi(x) = \lim_i \pi \circ (\varphi_i \otimes \mathrm{id}_B)(x) = 0.$$

Therefore, $\ker(\mathrm{id}_M \otimes Q) \subset \ker \pi$ and $\tilde{\pi}$ is min-continuous. □

Exercises

Exercise 14.1.1. Let A be a C*-algebra such that A^{**} is weakly exact. Prove that A is exact if it is locally reflexive.

Exercise 14.1.2. Prove Corollary 14.1.5.

Exercise 14.1.3. Prove that a von Neumann algebra M is weakly exact if and only if for any $J \lhd B$ and any weakly nuclear u.c.p. map $\varphi\colon B \to M$ with $J \subset \ker \varphi$, the induced u.c.p. map $\tilde{\varphi}\colon B/J \to M$ is weakly nuclear.

Exercise 14.1.4. Let $M \subset N \subset \mathbb{B}(\mathcal{H})$ be von Neumann algebras with M weakly exact. Suppose that there exists a u.c.p. map $\Phi\colon \mathbb{B}(\mathcal{H}) \to N$ such that $\Phi|_M = \mathrm{id}_M$. Prove that the inclusion $M \hookrightarrow N$ is weakly nuclear. (It is not known whether the weak exactness assumption on M is necessary.)

14.2. Characterization of weak exactness

In this section, we will give an approximation characterization of weakly exact von Neumann algebras, which is analogous to that of exact C*-algebras. First, we need a von Neumann algebra analogue of Proposition 9.2.7. Recall that for von Neumann algebras M and N, a $*$-representation π of $M \odot N$ (or $M \otimes N$) is said to be *bi-normal* if both $\pi|_M$ and $\pi|_N$ are normal.

Proposition 14.2.1. *A von Neumann algebra M is weakly exact if and only if for any* C**-algebra B and any left normal representation $\pi\colon M \otimes B \to \mathbb{B}(\mathcal{H})$, the bi-normal extension $\tilde{\pi}\colon M \odot B^{**} \to \mathbb{B}(\mathcal{H})$ is* min*-continuous.*

Proof. We first prove the "if" part. Let $J \lhd B$ and $\pi\colon M \otimes B \to \mathbb{B}(\mathcal{H})$ be as in Definition 14.1.1. Let p be the central projection which supports the normal $*$-homomorphism $Q^{**}\colon B^{**} \to (B/J)^{**}$, and identify $(B/J)^{**}$ with pB^{**}. We denote the canonical inclusion by

$$\psi\colon B/J \to (B/J)^{**} = pB^{**} \subset B^{**}.$$

By assumption, π extends to a bi-normal $*$-homomorphism $\hat{\pi}$ on $M \otimes B^{**}$. Since $M \otimes J \subset \ker \pi$, we have $\hat{\pi}(1 \otimes p) = 1$. It follows that $\tilde{\pi}$ coincides with $\hat{\pi} \circ (\mathrm{id} \otimes \psi)$ and hence is min-continuous. Indeed, for any $a \in M$ and $x \in B$, we have

$$\hat{\pi} \circ (\mathrm{id} \otimes \psi)(a \otimes Q(x)) = \hat{\pi}(a \otimes px) = \pi(a \otimes x) = \tilde{\pi}(a \otimes Q(x)).$$

We next prove the "only if" direction; it's almost the same as Proposition 9.2.7. For a given B, let I, B_I and $\sigma\colon B_I \to B^{**}$ be as in the proof of Proposition 9.2.7. For a given left normal representation $\pi\colon M \otimes B \to \mathbb{B}(\mathcal{H})$, we consider the representation $\rho\colon M \otimes B_I \to \mathbb{B}(\mathcal{H})$ defined by

$$\rho(\sum_{k=1}^{n} a_k \otimes (x_k(i))_{i \in I}) = \text{strong}^*\text{-}\lim_i \pi(\sum_{k=1}^{n} a_k \otimes x_k(i)).$$

Since M is weakly exact, ρ is left normal and $M \otimes \ker \sigma \subset \ker \rho$, the induced representation $\tilde{\rho}$ of $M \odot B^{**}$ is min-continuous. Since $\hat{\pi} = \tilde{\rho}$, this completes the proof. □

The advantage of the above formulation is that it can be generalized to the case that B is not a C*-algebra. This fact is important, so let's be more precise. Let X be a C*-algebra or an operator space. Let $p \in M^{**}$ be the central support of the identity representation of M so that we may identify M with pM^{**}. Then, the canonical bi-normal embedding $M^{**} \odot X^{**} \subset (M \otimes X)^{**}$ gives rise to a (nonunital) bi-normal embedding

$$\Theta_X\colon M \odot X^{**} = pM^{**} \odot X^{**} \hookrightarrow (M \otimes X)^{**}.$$

Let $a \in M$ and $x \in X^{**}$ be given and take nets $p_i \in M$ and $x_j \in X$ such that $p_i \to p$ and $x_j \to x$ in the weak*-topology. Then, we have $p_i a \otimes x_j \in M \otimes X$ and

$$\Theta_X(a \otimes x) = \text{weak}^*\text{-}\lim_{i,j}(p_i a \otimes x_j) \in (M \otimes X)^{**}.$$

Now, assume that M is weakly exact and let B be a C*-algebra. Since Θ_B is a bi-normal $*$-homomorphism which is continuous on $M \otimes B$, it follows from Proposition 14.2.1 that Θ_B is continuous (and isometric) on $M \otimes B^{**}$. This implies that Θ_X is isometric on $M \otimes X^{**}$ for any operator space X. Indeed, when $X \subset B$, we have the commuting diagram

$$\begin{array}{ccc} M \otimes X^{**} & \xrightarrow{\Theta_X} & (M \otimes X)^{**} \\ \downarrow & & \downarrow \\ M \otimes B^{**} & \xrightarrow{\Theta_B} & (M \otimes B)^{**}, \end{array}$$

where the bottom and the vertical inclusions are all isometric. In summary, we have the following result.

Corollary 14.2.2. *Let M be a weakly exact von Neumann algebra and X be an operator space. Then, for any $z \in M \odot X^{**}$ with $\|z\|_{\min} \le 1$, there exists a net (z_i) in $M \odot X$ with $\|z_i\|_{\min} \le 1$ which converges to z in the $\sigma(M \otimes X^{**}, M_* \odot X^*)$-topology.*

Proof. Let $z \in M \odot X^{**}$ be such that $\|z\|_{\min} \le 1$. By the above discussion, we have $\|\Theta_X(z)\| \le 1$. Hence, there exists a net (z_i) in $M \odot X$ with $\|z_i\|_{\min} \le 1$ which converges to $\Theta_X(z)$ in the weak*-topology. We claim that the net (z_i) converges to z in the $\sigma(M \otimes X^{**}, M_* \odot X^*)$-topology. Let $f \in M_*$ and $g \in X^*$ be given. We'll write $\bar{f}$ for f when it is regarded as an element in M^*. Then, $\bar{f}(a) = f(\pi(a))$ for $a \in M^{**}$, where $\pi\colon M^{**} \to M$ is the normal extension of the identity on M. (We have identified $M = pM^{**}$ and $\pi(a) = pa$.) Now, $\bar{f} \otimes g \in (M \otimes X)^*$ and

$$(\bar{f} \otimes g)(\Theta_X(a \otimes x)) = \bar{f}(pa)g(x) = f(a)g(x) = (f \otimes g)(a \otimes x)$$

for every $a \in M$ and $x \in X^{**}$. Therefore, we have

$$(f \otimes g)(z) = (\bar{f} \otimes g)(\Theta_X(z)) = \lim_i (\bar{f} \otimes g)(z_i) = \lim_i (f \otimes g)(z_i)$$

as claimed. □

Operator space duality (Theorem B.13) immediately yields the following reformulation.

Corollary 14.2.3. *Let M be a weakly exact von Neumann algebra and X be an operator space. Then, for any finite-rank c.c. map $\varphi\colon X^* \to M$, there exists a net (φ_i) of weak*-continuous finite-rank c.c. maps $\varphi_i\colon X^* \to M$ which converges to φ in the point-ultraweak topology.*

We need the notion of exactness for operator systems. An operator system S is said to be exact (or more precisely 1-exact) if $(S \otimes B)/(S \otimes J) = S \otimes (B/J)$ canonically isometrically for any $J \triangleleft B$. All the characterizations of exactness for C*-algebras (Section 3.9) are still valid for operator systems, with obvious modifications (one can verify this or see [**61**]). Moreover, exact operator systems are locally reflexive in the sense of Definition 9.1.2 (cf. Theorem 9.3.1) – although this is not so easy to prove (see [**61**] for the details).

The following is a von Neumann algebra analogue of Theorem 3.9.1 and a partial converse to Theorem 14.1.2. The implication (1) ⇔ (2) holds without the separability assumption, by replacing sequences with nets.

Theorem 14.2.4. *Let M be a von Neumann algebra with separable predual. The following are equivalent:*

(1) *M is weakly exact;*

(2) *for any finite-dimensional operator system E in M, there exist sequences of u.c.p. maps $\varphi_i\colon E \to \mathbb{M}_{n(i)}(\mathbb{C})$ and $\psi_i\colon \varphi_i(E) \to M$ such that the net $(\psi_i \circ \varphi_i)$ converges to id_E in the point-ultraweak topology;*

(3) *there exist an exact operator system S and normal u.c.p. maps $\varphi\colon M \to S^{**}$, $\psi\colon S^{**} \to M$ such that $\psi \circ \varphi = \mathrm{id}_M$.*

Proof. (1) $\Rightarrow$ (2): Let M be a weakly exact von Neumann algebra acting on a Hilbert space $\mathcal{H}$ and let $E \subset M$ be a finite-dimensional operator system. For simplicity we assume $\mathcal{H} = \ell^2$ and we let $\Phi_n\colon E \to \mathbb{M}_n(\mathbb{C})$ be compression by the projection onto $\ell^2_n \subset \ell^2$. Let $E_n = \Phi_n(E)$ and $\Phi = \bigoplus \Phi_n\colon E \to \prod E_n$. We note that Φ is a complete isometry and $\prod E_n \subset \prod \mathbb{M}_n(\mathbb{C})$ is an ultraweakly closed operator subspace. A left inverse of Φ is given by

$$\Psi\colon \prod E_n \ni (x_n)_{n=1}^{\infty} \mapsto \lim_{n\to\omega} \Phi_n^{-1}(x_n) \in E \subset M,$$

where ω is a fixed free ultrafilter on $\mathbb{N}$. The limit exists since E is finite-dimensional and Φ_n is injective for large n. Moreover, we have

$$\|\mathrm{id}_{\mathbb{M}_k(\mathbb{C})} \otimes \Psi\| \leq \lim_{n\to\omega} \|(\mathrm{id}_{\mathbb{M}_k(\mathbb{C})} \otimes \Phi_n)^{-1}\| = 1$$

for every k. Hence Ψ is a complete contraction with $\Psi \circ \Phi = \mathrm{id}_E$. By Corollary 14.2.3, there exists a net of ultraweakly continuous c.c. maps $\Psi_i\colon \prod E_n \to M$ such that $\lim_i \Psi_i = \Psi$ in the point-ultraweak topology. Since each Ψ_i is ultraweakly continuous, if we set

$$\varphi_N = \bigoplus_{n=1}^{N} \Phi_n\colon E \to \bigoplus_{n=1}^{N} E_n \subset \prod_{n=1}^{\infty} E_n$$

and $\psi_{i,N} = \Psi_i|_{\varphi_N(E)}$, then we have

$$\lim_i \lim_{N\to\infty} \psi_{i,N} \circ \varphi_N = \lim_i \Psi_i \circ \Phi = \Psi \circ \Phi = \mathrm{id}_E.$$

Since $\varphi_N(E) \subset \bigoplus_{n=1}^{N} E_n \subset \bigoplus_{n=1}^{N} \mathbb{M}_n(\mathbb{C})$, this implies condition (2) except that the $\psi_{i,N}$'s may not be u.c.p. One can fix this problem as follows. First, passing to a convex combination, one may assume convergence in the point-ultrastrong topology. Then, arguing as in the proof of Proposition 3.8.2, one may further assume that $\psi_{i,N}$ is approximately unital in norm. Finally, one invokes Corollary B.11 to perturb $\psi_{i,N}$ to u.c.p. maps.

(2) $\Rightarrow$ (1): Let $J \triangleleft B$, π and $\tilde{\pi}$ be given as in Definition 14.1.1. Let $E \subset M$ be a finite-dimensional operator system and take nets (φ_i) and (ψ_i) as in condition (2). Since $E_i = \varphi_i(E) \subset \mathbb{M}_{n(i)}(\mathbb{C})$, we have $(E_i \otimes B)/(E_i \otimes J) = E_i \otimes (B/J)$ isometrically. It follows that $\tilde{\pi} \circ (\psi_i \otimes \mathrm{id}_{B/J})$ is contractive on $E_i \otimes (B/J)$. Since $\tilde{\pi}$ is left normal, the net $\tilde{\pi} \circ ((\psi_i \circ \varphi_i) \otimes \mathrm{id}_{B/J})$ of contractions converges to $\tilde{\pi}$ on $E \otimes (B/J)$. Since $E \subset M$ is arbitrary, $\tilde{\pi}$ is contractive on $M \otimes (B/J)$.

(3) $\Rightarrow$ (1): The proof is virtually identical to that of Theorem 14.1.2.

(2) $\Rightarrow$ (3): Let M be a weakly exact von Neumann algebra with separable predual and A be an ultraweakly-dense norm-separable C*-subalgebra in M. We denote by $\|\cdot\|_\sigma$ a norm which defines the ultraweak topology on the unit ball of M. Using condition (2) recursively, we can construct sequences of finite-dimensional operator systems and connecting u.c.p. maps to get a diagram

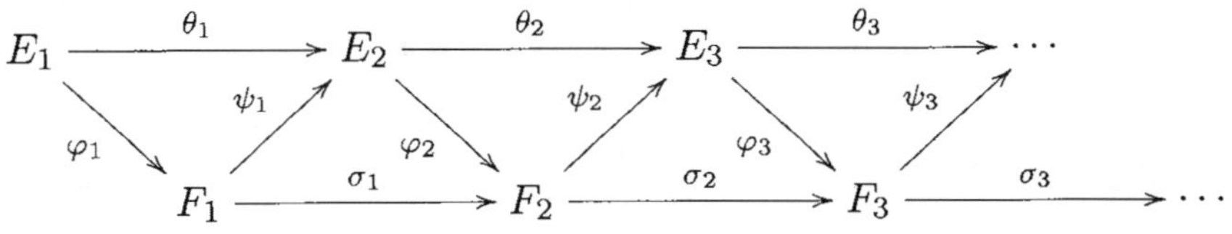

such that

(1) $E_1 \subset E_2 \subset \cdots \subset M$ and the norm closure of $\bigcup E_n$ contains A,

(2) $F_n \subset \mathbb{M}_{k(n)}(\mathbb{C})$ for some $k(n) \in \mathbb{N}$,

(3) the diagram commutes and $\|\theta_n(a) - a\|_\sigma < 2^{-n}\|a\|$ for $a \in E_n$.

We define the operator system S to be the inductive limit of (F_n, σ_n); more precisely, let $Q\colon \prod \mathbb{M}_{k(n)}(\mathbb{C}) \to \prod \mathbb{M}_{k(n)}(\mathbb{C})/\bigoplus \mathbb{M}_{k(n)}(\mathbb{C})$ be the quotient map and let $S = Q(\tilde{S})$, where $\tilde{S}$ is the norm closure of

$$\{(x_n)_n \in \prod_{n=1}^{\infty} \mathbb{M}_{k(n)}(\mathbb{C}) : \exists m \text{ with } x_m \in F_m \text{ and } x_{n+1} = \sigma_n(x_n) \text{ for } n \geq m\}$$
$$\cong \bigcup_{m\geq 1} \big(\mathbb{M}_{k(1)}(\mathbb{C}) \oplus \cdots \oplus \mathbb{M}_{k(m-1)}(\mathbb{C}) \oplus F_m\big).$$

Since the operator system $\tilde{S}$ is exact (and locally reflexive), so is the quotient $Q(\tilde{S})$ by the "ideal" $\bigoplus \mathbb{M}_{k(n)}(\mathbb{C})$. (That's an exercise.) Let $\Phi\colon \bigcup E_n \to S^{**}$ be a point-weak* cluster point of the sequence

$$\Phi_m\colon E_m \ni x \mapsto Q((\sigma_{n-1} \circ \cdots \circ \sigma_m \circ \varphi_m(x))_{n=m}^{\infty}) \in S \subset S^{**}$$

and let $\tilde{\psi}\colon \tilde{S} \to M$ be a point-weak* cluster point of the sequence

$$\tilde{\psi}_m\colon \tilde{S} \ni (x_n)_n \mapsto \psi_m(x_m) \in M.$$

It is not hard to see that $\tilde{\psi} = \psi \circ Q$ for some u.c.p. map $\psi\colon S \to M$. The unique weak*-continuous extension of ψ on S^{**} is still denoted by ψ. Then, for any $a \in \bigcup E_n$, we have that

$$\begin{aligned}\|a - (\psi \circ \Phi)(a)\|_\sigma &= \|a - \lim_m (\psi \circ \Phi_m)(a)\|_\sigma \\ &= \|a - \lim_m \lim_n (\psi_n \circ \sigma_{n-1} \circ \cdots \circ \sigma_m \circ \varphi_m)(a)\|_\sigma \\ &= \|a - \lim_m \lim_n (\theta_n \circ \cdots \circ \theta_m)(a)\|_\sigma\end{aligned}$$

$$\le \lim_m \lim_n \left(\frac{1}{2^n} + \frac{1}{2^{n-1}} + \cdots + \frac{1}{2^m}\right)\|a\|$$
$$= 0,$$

where the limits are taken along appropriate converging subnets. It follows that $\psi \circ \Phi$ is the identity map on $\bigcup E_n$. Now extend Φ to the norm closure of $\bigcup E_n$, by norm continuity, and then further extend $\Phi|_A$ to $\bar{\Phi}\colon A^{**} \to S^{**}$ by ultraweak continuity. Then, $\psi \circ \bar{\Phi}\colon A^{**} \to M$ is a normal u.c.p. map such that $\psi \circ \bar{\Phi}|_A = \mathrm{id}_A$. This means that $\psi \circ \bar{\Phi}$ is a normal $*$-homomorphism which is the extension of the identity representation $A \hookrightarrow M$. Hence, the restriction φ of $\bar{\Phi}$ to the (nonunital) von Neumann subalgebra $M \subset A^{**}$ satisfies $\psi \circ \varphi = \mathrm{id}_M$. □

It would be interesting to know whether or not one can choose S in condition (3) to be an exact C*-algebra (and, in addition, ψ to be a normal $*$-homomorphism).

Proof of Theorem 14.1.6. Let Γ be a discrete group whose group von Neumann algebra $L(\Gamma)$ is weakly exact. We will verify condition (2) of Theorem 5.1.6.

Let a finite subset $E \subset \Gamma$ and $\varepsilon > 0$ be given. By Theorem 14.2.4 and Arveson's Extension Theorem, there exist u.c.p. maps $\varphi\colon L(\Gamma) \to \mathbb{M}_n(\mathbb{C})$ and $\psi\colon \mathbb{M}_n(\mathbb{C}) \to \mathbb{B}(\ell^2(\Gamma))$ such that for $\theta = \psi \circ \varphi$, one has

$$\theta(\lambda(s)) \in L(\Gamma) \text{ and } \|\lambda(s) - \theta(\lambda(s))\|_2 < \varepsilon$$

for every $s \in E$. We may assume that $\varphi(\lambda(s)) = 0$ for all but finitely many $s \in \Gamma$. It follows that the kernel

$$u(s,t) = \langle \theta(\lambda(st^{-1}))\lambda_t\delta_e, \lambda_s\delta_e \rangle$$

is positive definite and satisfies $|u(s,t) - 1| < \varepsilon$ whenever $st^{-1} \in E$. This completes the proof. □

Corollary 14.2.5. *Let M_1 and M_2 be von Neumann algebras with separable preduals. If both M_i are weakly exact, then $M_1 \bar{\otimes} M_2$ is weakly exact.*

Proof. By Theorem 14.2.4, there exist exact operator spaces S_i and normal u.c.p. maps $\varphi_i\colon M_i \to S_i^{**}$ and $\psi_i\colon S_i^{**} \to M_i$ such that $\psi_i \circ \varphi_i = \mathrm{id}_{M_i}$. Since S_i is exact and has property C, the canonical bi-normal inclusion $\iota\colon S_1^{**} \odot S_2^{**} \hookrightarrow (S_1 \otimes S_2)^{**}$ is a continuous u.c.p. map on the minimal tensor product. We first consider $\psi = (\psi_1 \otimes \psi_2)|_{S_1 \otimes S_2}$ and extend it to a normal u.c.p. map, still denoted by ψ, from $(S_1 \otimes S_2)^{**}$ into $M_1 \bar{\otimes} M_2$. We have $\psi \circ \iota = \psi_1 \otimes \psi_2$ since both maps are bi-normal and agree on $S_1 \otimes S_2$. Let $\tilde{\varphi}\colon (M_1 \otimes M_2)^{**} \to (S_1 \otimes S_2)^{**}$ be the normal extension of the u.c.p. map $\iota \circ (\varphi_1 \otimes \varphi_2)\colon M_1 \otimes M_2 \to (S_1 \otimes S_2)^{**}$. Then, $\psi \circ \tilde{\varphi}\colon (M_1 \otimes M_2)^{**} \to M_1 \bar{\otimes} M_2$

is a normal u.c.p. map such that $(\psi \circ \tilde{\varphi})|_{M_1 \otimes M_2} = \mathrm{id}_{M_1 \otimes M_2}$. Hence the restriction φ of $\tilde{\varphi}$ to $M_1 \bar{\otimes} M_2 \subset (M_1 \otimes M_2)^{**}$ satisfies $\psi \circ \varphi = \mathrm{id}_{M_1 \bar{\otimes} M_2}$. □

Remark 14.2.6. The separability assumption in Corollary 14.2.5 can be dropped if one employs the following fact (which is proved using Takesaki's conditional expectation theorem – see Theorem IX.4.2 in [**184**]): For any von Neumann algebra M, there exists a net E_i of normal conditional expectations such that $E_i(M)$ are all von Neumann subalgebras with separable predual and $E_i \to \mathrm{id}_M$ in the point-ultraweak topology.

Exercises

Exercise 14.2.1. Let J be an ideal in a C*-algebra B and $Q\colon B \to B/J$ be the quotient map. Let $\tilde{S} \subset B$ be an exact (and hence locally reflexive) operator subsystem such that $J \subset \tilde{S}$. Prove that $S = Q(\tilde{S}) \subset B/J$ is exact.

Exercise 14.2.2. Let $S \subset A$ and $T \subset B$ be operator systems. Prove that there exists a bi-normal embedding $S^{**} \odot T^{**} \hookrightarrow (S \otimes T)^{**}$ which extends the canonical inclusion $S \odot T \subset (S \otimes T)^{**}$.

Exercise 14.2.3. Let S and T be operator systems and assume that S is exact (and hence locally reflexive). Prove that the bi-normal embedding $S^{**} \odot T^{**} \hookrightarrow (S \otimes T)^{**}$ is contractive with respect to the minimal norm.

14.3. References

The notion of weak exactness was introduced by Kirchberg in [**104**], where he proved all the results in Section 14.1, except for Theorem 14.1.6 and Proposition 14.1.8. Proposition 14.1.8 comes from [**63**] and [**80**], while Theorem 14.1.6 and Section 14.2 are adapted from [**132**].

Part 3

Applications

Chapter 15

Classification of Group von Neumann Algebras

In recent years, C*-tensor product theory and amenable actions of groups have joined forces to provide some deep results related to the classification of group von Neumann algebras. For example, it is now possible to recover Ge's celebrated result that free group factors are prime ([**68**]) from tensor product/amenable action techniques; in fact, this approach yields a vast generalization of Ge's result with no additional effort. Another surprising application deals with the classification of tensor products (or free products) of certain von Neumann algebras. Our goal is to explain these applications in a unified and succinct way.

Throughout this chapter, we make the blanket assumption that all groups are countable and von Neumann algebras have separable predual.

15.1. Subalgebras with noninjective relative commutants

Definition 15.1.1. Let Γ be a group and $\mathcal{G}$ be a family of subgroups of Γ. We say a subset Ω of Γ is *small relative to* $\mathcal{G}$ if it is contained in a finite union of $s\Lambda t$'s, where $s, t \in \Gamma$ and $\Lambda \in \mathcal{G}$. (Here $s\Lambda t = \{sat : a \in \Lambda\} \subset \Gamma$.)

Let $c_0(\Gamma; \mathcal{G}) \subset \ell^\infty(\Gamma)$ be the C*-subalgebra generated by functions whose supports are small relative to $\mathcal{G}$. More intuitively, for a net (s_i) in Γ, we write $s_i \to \infty/\mathcal{G}$ if $s_i \notin s\Lambda t$ eventually[1] for every $s, t \in \Gamma$ and $\Lambda \in \mathcal{G}$. Hence, for $f \in \ell^\infty(\Gamma)$ we have that $f \in c_0(\Gamma; \mathcal{G}) \Leftrightarrow \{x \in \Gamma : |f(x)| > \varepsilon\}$ is small relative to $\mathcal{G}$ for every $\varepsilon > 0 \Leftrightarrow \lim_{s\to\infty/\mathcal{G}} f(s) = 0$.

[1] $\forall s, t, \forall \Lambda, \exists i_0$ such that $\forall i$ we have the implication $i \geq i_0 \Rightarrow s_i \notin s\Lambda t$.

Definition 15.1.2. We say the group Γ is *bi-exact relative to* $\mathcal{G}$ if it is exact and there exists a map

$$\mu\colon \Gamma \to \operatorname{Prob}(\Gamma)$$

such that for every $s, t \in \Gamma$, one has

$$\lim_{x\to\infty/\mathcal{G}} \|\mu(sxt) - s.\mu(x)\| = 0.$$

Remark 15.1.3. A countable exact group Γ is bi-exact relative to $\mathcal{G}$ if for every finite subset $E \subset \Gamma$ and $\varepsilon > 0$, there exists $\mu\colon \Gamma \to \operatorname{Prob}(\Gamma)$ such that for every $s, t \in E$ the subset $\{x : \|\mu(sxt) - s.\mu(x)\| \geq \varepsilon\}$ is small relative to $\mathcal{G}$. Since we won't need this fact, we won't prove it. However, the main points are laid out in Exercise 15.1.1.

Let $\mathbb{K}(\Gamma;\mathcal{G})$ be the hereditary C*-subalgebra of $\mathbb{B}(\ell^2(\Gamma))$ generated by $c_0(\Gamma;\mathcal{G})$:

$$\mathbb{K}(\Gamma;\mathcal{G}) = \text{the norm closure of } c_0(\Gamma;\mathcal{G})\mathbb{B}(\ell^2(\Gamma))c_0(\Gamma;\mathcal{G}).$$

Since the left and right regular representations λ and, respectively, ρ normalize $c_0(\Gamma;\mathcal{G})$, the reduced group C*-algebras $C^*_\lambda(\Gamma)$ and $C^*_\rho(\Gamma)$ are in the multipliers of $\mathbb{K}(\Gamma;\mathcal{G})$.

Lemma 15.1.4. *Let Γ be an exact group and $\mathcal{G}$ be a nonempty family of subgroups of Γ. Then Γ is bi-exact relative to $\mathcal{G}$ if and only if there exists a u.c.p. map*

$$\theta\colon C^*_\lambda(\Gamma) \otimes C^*_\rho(\Gamma) \to \mathbb{B}(\ell^2(\Gamma))$$

such that $\theta(a \otimes b) - ab \in \mathbb{K}(\Gamma;\mathcal{G})$ for every $a \in C^*_\lambda(\Gamma)$ and $b \in C^*_\rho(\Gamma)$.

Proof. We first prove the "if" direction. Let θ be a u.c.p. map such that $\theta(a\otimes b) - ab \in \mathbb{K}(\Gamma;\mathcal{G})$. By Voiculescu's Theorem (Theorem 1.7.8), there is an isometry $V\colon \ell^2(\Gamma) \to \ell^2(\Gamma\times\Gamma)$ such that $\theta(a\otimes b) - V^*(a\otimes b)V \in \mathbb{K}(\ell^2(\Gamma))$ for every a and b. It follows that

$$V^*(\lambda(s)\otimes\rho(t))V - \lambda(s)\rho(t) \in \mathbb{K}(\Gamma;\mathcal{G})$$

for every $s, t \in \Gamma$. Define a map $\mu\colon \Gamma \to \operatorname{Prob}(\Gamma)$ by

$$\mu(x)(y) = \sum_{z\in\Gamma} |(V\delta_x)(y,z)|^2.$$

It follows that

$$\begin{aligned}\|\mu(sxt) - s.\mu(x)\|_1 &\leq \|\,|V\delta_{sxt}|^2 - |(\lambda(s)\otimes\rho(t))^{-1}V\delta_x|^2\|_1\\ &\leq 2\|V\delta_{sxt} - (\lambda(s)\otimes\rho(t))^{-1}V\delta_x\|_2 \to 0\end{aligned}$$

as $x \to \infty/\mathcal{G}$.

Now we prove the "only if" direction. Define a unitary operator U on $\ell^2(\Gamma)\otimes\ell^2(\Gamma)$ by $U(\delta_x\otimes\delta_y) = \delta_x\otimes\delta_{x^{-1}y}$, so that $U^*(\lambda(s)\otimes\rho(t))U =$

$(\lambda \otimes \lambda)(s)(1 \otimes \rho)(t)$ (cf. Fell's absorption principle). Let $\mu\colon \Gamma \to \mathrm{Prob}(\Gamma)$ be a map as in the definition of bi-exactness and define an isometry $V\colon \ell^2(\Gamma) \to \ell^2(\Gamma) \otimes \ell^2(\Gamma)$ by $V\delta_x = U(\mu(x)^{1/2} \otimes \delta_x)$. Then, it is routine to check that

$$V^*(\lambda(s) \otimes \rho(t))V\delta_x = \langle \lambda(s)(\mu(x)^{1/2}), \mu(sxt^{-1})^{1/2} \rangle \lambda(s)\rho(t)\delta_x.$$

Since

$$\lim_{x \to \infty/\mathcal{G}} \|\lambda(s)(\mu(x)^{1/2}) - \mu(sxt^{-1})^{1/2}\|_2^2 \le \lim_{x \to \infty/\mathcal{G}} \|s.\mu(x) - \mu(sxt^{-1})\|_1 = 0,$$

we have $V^*(\lambda(s) \otimes \rho(t))V - \lambda(s)\rho(t) \in \mathbb{K}(\Gamma; \mathcal{G})$ for every $s, t \in \Gamma$. □

Here is the main theorem of this section. (See Definition F.13 for the terminology "N embeds in $L(\Lambda)$ inside $L(\Gamma)$".)

Theorem 15.1.5. *Let Γ be a countable group and $\mathcal{G}$ be a countable family of subgroups of Γ. Assume that the group Γ is bi-exact relative to $\mathcal{G}$. Let $p \in L(\Gamma)$ be a projection and $N \subset pL(\Gamma)p$ be a von Neumann subalgebra. If the relative commutant $N' \cap pL(\Gamma)p$ is noninjective, then there exists $\Lambda \in \mathcal{G}$ such that N embeds in $L(\Lambda)$ inside $L(\Gamma)$.*

The proof of this result requires some preparation. Let $M \subset \mathbb{B}(\mathcal{H})$ be a von Neumann algebra and consider the $*$-homomorphism

$$\Phi_M\colon M \odot M' \ni \sum a_k \otimes b_k \mapsto \sum a_k b_k \in \mathbb{B}(\mathcal{H}).$$

We note that Φ_M is min-continuous if and only if M is injective (Theorem 9.3.3). We will need a refinement of this result for von Neumann subalgebras contained in corners $P \subset pMp$.

Proposition 15.1.6. *Let $M \subset \mathbb{B}(\mathcal{H})$ be a finite von Neumann algebra and $p \in M$ be a projection. Let $P \subset pMp$ be a von Neumann subalgebra and $E_P\colon pMp \to P$ be the trace-preserving conditional expectation. Consider the bi-normal u.c.p. map*

$$\Phi_P\colon M \odot M' \ni \sum_k a_k \otimes b_k \mapsto \sum_k E_P(pa_kp)b_kp \in \mathbb{B}(p\mathcal{H}).$$

Suppose that there are weakly dense C*-*subalgebras $C_l \subset M$ and $C_r \subset M'$ such that C_l is exact and Φ_P is* min-*continuous on $C_l \odot C_r$. Then P is injective.*

Proof. By Lemma 9.2.9, our assumptions imply that Φ_P is min-continuous on $M \odot M'$. By The Trick, $\Phi_P|_M$ extends to a u.c.p. map ψ from $\mathbb{B}(\mathcal{H})$ into $(pM')' = pMp$. (Note that the argument for The Trick only requires $\Phi_P|_{\mathbb{C}1 \otimes M'}$ to be $*$-homomorphic.) It follows that $E_P \circ \psi|_{\mathbb{B}(p\mathcal{H})}$ is a conditional expectation from $\mathbb{B}(p\mathcal{H})$ onto P. □

We primarily consider Φ_P in the case where $P = B' \cap pMp$ for a projection $p \in M$ and a *diffuse* abelian von Neumann subalgebra $B \subset pMp$ (meaning B has no nonzero minimal projections). Every diffuse abelian von Neumann algebra B with separable predual is $*$-isomorphic to $L^\infty[0,1]$ and hence is generated by a single unitary element $u_0 \in B$ (e.g., $u_0(t) = e^{2\pi i t}$). Fixing such a generator, we define a c.p. map Ψ_B from $\mathbb{B}(\mathcal{H})$ into $\mathbb{B}(p\mathcal{H})$ by

$$\Psi_B(x) = \text{ultraweak-}\lim_n \frac{1}{n}\sum_{k=1}^{n} u_0^k x u_0^{-k},$$

where the limit is taken along some fixed ultrafilter. It is not hard to see that Ψ_B is a (nonunital) conditional expectation onto $B' \cap \mathbb{B}(p\mathcal{H})$ and that $\Psi_B|_{pMp}$ is a trace-preserving conditional expectation from pMp onto $B' \cap pMp$. By uniqueness of the trace-preserving conditional expectation, one has $\Psi_B(a) = E_P(pap)$ for every $a \in M$. It follows that

$$\Psi_B(\sum_k a_k b_k) = \sum_k E_P(pa_kp)b_kp = \Phi_P(\sum_k a_k \otimes b_k)$$

for $a_k \in M$ and $b_k \in M'$.

Proof of Theorem 15.1.5. By contradiction, suppose that the conclusion of the theorem is not true. Then, by Corollary F.14, there is a diffuse abelian von Neumann subalgebra $B \subset N$ such that B does not embed in $L(\Lambda)$ inside $M = L(\Gamma)$ for any Λ. We will use Theorem F.12 with $A = L(\Lambda)$. For this, observe that $\chi_\Lambda \in \ell^\infty(\Gamma) \subset \mathbb{B}(L^2(M))$ is nothing but the orthogonal projection e_A onto $L^2(A)$ and hence $\chi_{s\Lambda} = \lambda(s)e_A\lambda(s)^* \in \langle M, A\rangle_+$ with $\mathrm{Tr}(\chi_{s\Lambda}) = 1$. It follows that $\Psi_B(\chi_{s\Lambda})$ is a positive element in $p\langle M, A\rangle p \cap B'$ such that $\mathrm{Tr}(\Psi_B(\chi_{s\Lambda})) \leq 1$. By assumption and Theorem F.12, $\Psi_B(\chi_{s\Lambda}) = 0$. Since $\rho(\Gamma)$ is in the multiplicative domain of Ψ_B, this implies that $\Psi_B(\chi_{s\Lambda t}) = 0$ for every $s, t \in \Gamma$ and $\Lambda \in \mathcal{G}$, or equivalently, $\mathbb{K}(\Gamma; \mathcal{G}) \subset \ker \Psi_B$. Hence, for the u.c.p. map θ given in Lemma 15.1.4, one has $\Phi_P = \Psi_B \circ \theta$ and Φ_P is min-continuous on $C^*_\lambda(\Gamma) \odot C^*_\rho(\Gamma)$. Injectivity of $P = B' \cap pMp$ now follows from Proposition 15.1.6. □

Exercise

Exercise 15.1.1. Prove the claim made in Remark 15.1.3. Here is a hint: Let $\{e\} = E_0 \subset E_1 \subset E_2 \cdots$ be an increasing sequence of finite symmetric subsets of Γ with $\bigcup E_n = \Gamma$. Find μ_n for E_n and $\varepsilon = 1/n$. Define relatively small sets Ω_n inductively by

$$\Omega_n = \bigcup_{s,t \in E_n} \{x : \|\mu_n(sxt) - s.\mu_n(x)\| \geq 1/n\} \cup E_n \Omega_{n-1} E_n.$$

Set $|x| = \min\{n : x \in \Omega_n\}$ and $\mu(x) = |x|^{-1} \sum_{n=1}^{|x|} \mu_n(x)$.

15.2. On bi-exactness

Definition 15.2.1. Let Γ be a group and $\mathcal{G}$ be a family of subgroups of Γ. For $f \in \ell^\infty(\Gamma)$ and $t \in \Gamma$, we define the right translation $f^t \in \ell^\infty(\Gamma)$ by $f^t(s) = f(st^{-1})$. Note that $(f^t)^{t'} = f^{tt'}$. Now define a compact space $\bar{\Gamma}^{\mathcal{G}}$ by

$$C(\bar{\Gamma}^{\mathcal{G}}) = \{f \in \ell^\infty(\Gamma) : f - f^t \in c_0(\Gamma; \mathcal{G}) \text{ for every } t \in \Gamma\}$$

and view it as a Γ-space, where Γ acts by *left* translation. We define another compact Γ-space $\Delta^{\mathcal{G}}\Gamma \subset \bar{\Gamma}^{\mathcal{G}}$ by

$$C(\Delta^{\mathcal{G}}\Gamma) = C(\bar{\Gamma}^{\mathcal{G}})/c_0(\Gamma; \mathcal{G})$$

and we call it the $\mathcal{G}$-boundary of Γ.

Remark 15.2.2. It is not hard to see that $x \in \bar{\Gamma}^{\mathcal{G}}$ belongs to $\Delta^{\mathcal{G}}$ if and only if there is a net (s_n) in Γ such that $s_n \to x$ and $s_n \to \infty/\mathcal{G}$.

It is possible that $\mathcal{G} = \emptyset$ and $c_0(\Gamma; \mathcal{G}) = \{0\}$, but otherwise we have $c_0(\Gamma) \subset c_0(\Gamma; \mathcal{G}) \subset C(\bar{\Gamma}^{\mathcal{G}})$ and $\bar{\Gamma}^{\mathcal{G}}$ is an equivariant compactification of Γ.[2] By Gelfand duality, there is a one-to-one correspondence between equivariant compactifications $\bar{\Gamma}$ of Γ and intermediate C*-subalgebras $c_0(\Gamma) \subset C(\bar{\Gamma}) \subset \ell^\infty(\Gamma)$ which are left translation invariant. It is possible that $\Gamma \in \mathcal{G}$ and $c_0(\Gamma; \mathcal{G}) = \ell^\infty(\Gamma)$ and $\Delta^{\mathcal{G}}\Gamma = \emptyset$.

Note that $f \in C(\bar{\Gamma}^{\mathcal{G}})$ if $f - f^t \in c_0(\Gamma; \mathcal{G})$ for all t in some generating subset of Γ, since $f - f^{tt'} = (f - f^{t'}) + (f - f^t)^{t'}$.

Proposition 15.2.3. *Let Γ be a countable group and $\mathcal{G}$ be a nonempty family of subgroups of Γ. Then the following are equivalent:*

(1) *Γ is bi-exact relative to $\mathcal{G}$;*
(2) *the $\mathcal{G}$-boundary $\Delta^{\mathcal{G}}\Gamma$ is amenable;*[3]
(3) *the Gelfand spectrum of $\ell^\infty(\Gamma)/c_0(\Gamma; \mathcal{G})$ is amenable as a $\Gamma \times \Gamma$-space (with the left-times-right translation action).*

Proof. Assume condition (1) and let $\mu \colon \Gamma \to \mathrm{Prob}(\Gamma)$ be a map as in Definition 15.1.2. Then, the u.c.p. map $\mu_* \colon \ell^\infty(\Gamma) \to \ell^\infty(\Gamma)$ defined by $\mu_*(f)(x) = \langle f, \mu(x)\rangle$ has the property that $\mu_*(s.f) - s.\mu_*(f)^t \in c_0(\Gamma; \mathcal{G})$. In particular, $\mu_*(f) \in C(\bar{\Gamma}^{\mathcal{G}})$. Let $Q \colon C(\bar{\Gamma}^{\mathcal{G}}) \to C(\Delta^{\mathcal{G}}\Gamma)$ be the quotient map. Then, $Q \circ \mu_*$ is a Γ-equivariant u.c.p. map from $\ell^\infty(\Gamma)$ into $C(\Delta^{\mathcal{G}}\Gamma)$. One can now deduce the amenability of $\Delta^{\mathcal{G}}\Gamma$ from Exercise 15.2.2.

Next, we assume condition (2) and let X denote the Gelfand spectrum of $\ell^\infty(\Gamma)/c_0(\Gamma; \mathcal{G})$. The inclusion $C(\bar{\Gamma}^{\mathcal{G}}) \subset \ell^\infty(\Gamma)$ induces a continuous map

---

[2] A *compactification* is a compact space $\bar{\Gamma}$ containing Γ as an open dense subset; it is *equivariant* if the left translation action of Γ on Γ extends continuously to $\bar{\Gamma}$. (This is the same as Definition 5.3.16, where equivariance was assumed.)

[3] By convention, we say that the empty Γ-space $\emptyset$ is amenable if Γ is exact.

$\varphi_l \colon X \to \Delta^{\mathcal{G}}\Gamma$ which is $\Gamma \times \Gamma$-equivariant, where the right action of Γ on $\Delta^{\mathcal{G}}\Gamma$ is trivial. By symmetry, there exists a continuous $\Gamma \times \Gamma$-equivariant map $\varphi_r \colon X \to \Delta_r^{\mathcal{G}}\Gamma$, where $\Delta_r^{\mathcal{G}}\Gamma$ is amenable as a $1 \times \Gamma$-space and is trivial as a $\Gamma \times 1$-space. Thus, the $\Gamma \times \Gamma$-space $\Delta^{\mathcal{G}}\Gamma \times \Delta_r^{\mathcal{G}}\Gamma$ is amenable and $\varphi_l \times \varphi_r$ is a $\Gamma \times \Gamma$-equivariant continuous map from X into it. Therefore, X is amenable.

Finally, assume condition (3) and define a C*-algebra D by

$$D = C^*(\lambda(\Gamma), \rho(\Gamma), \ell^\infty(\Gamma)) + \mathbb{K}(\Gamma; \mathcal{G}) \subset \mathbb{B}(\ell^2(\Gamma)).$$

It is not hard to see that $\mathbb{K}(\Gamma; \mathcal{G})$ is an ideal in D and $D/\mathbb{K}(\Gamma; \mathcal{G})$ is a quotient of the crossed product of $\ell^\infty(\Gamma)/c_0(\Gamma; \mathcal{G})$ by $\Gamma \times \Gamma$ (actually, it's isomorphic to this crossed product). By assumption, the canonical $*$-homomorphism $C^*_\lambda(\Gamma) \odot C^*_\rho(\Gamma) \to D/\mathbb{K}(\Gamma; \mathcal{G})$ is min-continuous and $D/\mathbb{K}(\Gamma; \mathcal{G})$ is nuclear. Hence, the quotient map from D to $D/\mathbb{K}(\Gamma; \mathcal{G})$ has a u.c.p. splitting on any separable C*-subalgebra, by the Choi-Effros Lifting Theorem. Thanks to Lemma 15.1.4, we are done. □

It will be more convenient to work with $\Delta^{\mathcal{G}}\Gamma$ than the original definition of bi-exactness. This allows us to exploit the technology developed in previous chapters.

Definition 15.2.4. Let $\bar{\Gamma}$ be an equivariant compactification of Γ. We say $\bar{\Gamma}$ is *small at infinity relative to* $\mathcal{G}$ if the following holds: If (s_n) is a net in Γ such that $s_n \to x \in \bar{\Gamma}$ and $s_n \to \infty/\mathcal{G}$, then $s_n t \to x$ for every $t \in \Gamma$.

One should check that an equivariant compactification $\bar{\Gamma}$ of Γ is small at infinity relative to $\mathcal{G}$ if and only if the identity map on Γ extends to a continuous map from $\bar{\Gamma}^{\mathcal{G}}$ onto $\bar{\Gamma}$. The image of $\Delta^{\mathcal{G}}\Gamma$ under this map is the set of $x \in \bar{\Gamma}$ such that there is a net (s_n) in Γ with the property that $s_n \to x$ and $s_n \to \infty/\mathcal{G}$.

Example 15.2.5. In the examples below, amenability of $\Delta^{\mathcal{G}}\Gamma$ follows from that of $\bar{\Gamma}^{\mathcal{G}}$.

(1) Let $\mathcal{G}$ be the empty family. Then, $c_0(\Gamma; \mathcal{G}) = \{0\}$ and $\bar{\Gamma}^{\mathcal{G}}$ is a one-point set. Hence $\bar{\Gamma}^{\mathcal{G}}$ is amenable if and only if Γ is amenable.

(2) Let $\mathcal{G} = \{\mathbf{1}\}$, where $\mathbf{1}$ is the trivial subgroup consisting of the neutral element. Then, $c_0(\Gamma; \mathcal{G}) = c_0(\Gamma)$ and $\bar{\Gamma}^{\mathcal{G}}$ is the universal compactification which is small at infinity. Recall from Section 5.3 that if Γ is a hyperbolic group, then there is a Γ-equivariant continuous map from $\bar{\Gamma}^{\mathcal{G}}$ onto the Gromov compactification – hence $\bar{\Gamma}^{\mathcal{G}}$ is amenable for a hyperbolic group (Corollary 5.3.19).

(3) Suppose $\Gamma \in \mathcal{G}$. Then $c_0(\Gamma; \mathcal{G}) = C(\bar{\Gamma}^{\mathcal{G}}) = \ell^\infty(\Gamma)$ and $\bar{\Gamma}^{\mathcal{G}} = \beta\Gamma$. Hence $\bar{\Gamma}^{\mathcal{G}}$ is amenable if and only if Γ is exact.

It is often useful to ignore amenable subgroups.

Lemma 15.2.6. *Let Γ be an exact group, $\Upsilon \subset \Gamma$ be an amenable subgroup and $\mathcal{G}$ be a family of subgroups of Γ. If there is a map*

$$\zeta\colon \Gamma \to \ell^1(\Gamma/\Upsilon)$$

such that

$$\lim_{x\to\infty/\mathcal{G}} \frac{\|\zeta(sxt) - s.\zeta(x)\|}{\|\zeta(x)\|} = 0$$

for every $s, t \in \Gamma$, then Γ is bi-exact relative to $\mathcal{G}$.

Proof. We define $\mu\colon \Gamma \to \mathrm{Prob}(\Gamma/\Upsilon)$ by $\mu(x) = \|\zeta(x)\|^{-1}|\zeta(x)|$. Then,

$$\begin{aligned}\|\mu(sxt) - s.\mu(x)\| &\le \left|1 - \frac{\|\zeta(sxt)\|}{\|\zeta(x)\|}\right| + \frac{\|\zeta(sxt) - s.\zeta(x)\|}{\|\zeta(x)\|}\\ &\le 2\frac{\|\zeta(sxt) - s.\zeta(x)\|}{\|\zeta(x)\|} \to 0 \quad \text{as } x \to \infty/\mathcal{G}\end{aligned}$$

for every $s, t \in \Gamma$. Let $\mu_*\colon \ell^\infty(\Gamma/\Upsilon) \to \ell^\infty(\Gamma)$ be the u.c.p. map defined by $\mu_*(f)(x) = \langle \mu(x), f\rangle$. It is not hard to see that $\mu_*(f) \in C(\bar{\Gamma}^{\mathcal{G}})$ and composed with the quotient map, it gives rise to a Γ-equivariant u.c.p. map from $\ell^\infty(\Gamma/\Upsilon)$ into $C(\Delta^{\mathcal{G}}\Gamma)$. We view $\ell^\infty(\Gamma/\Upsilon)$ as the C*-subalgebra of right Υ-invariant functions in $\ell^\infty(\Gamma)$. Since Υ is amenable, by taking an "average" over the right Υ-action, one can find a (left) Γ-equivariant conditional expectation from $\ell^\infty(\Gamma)$ onto $\ell^\infty(\Gamma/\Upsilon)$. Combining these two Γ-equivariant u.c.p. maps, we obtain a Γ-equivariant u.c.p. map from $\ell^\infty(\Gamma)$ into $C(\Delta^{\mathcal{G}}\Gamma)$. Amenability of $\Delta^{\mathcal{G}}\Gamma$ now follows from Exercise 15.2.2. □

Proposition 15.2.7. *Let Γ be a group. For families $\mathcal{G}$ and $\mathcal{G}'$ of subgroups of Γ, define*

$$\mathcal{G} \wedge \mathcal{G}' = \{\Lambda \cap s\Lambda' s^{-1} : \Lambda \in \mathcal{G},\ \Lambda' \in \mathcal{G}',\ s \in \Gamma\}.$$

If Γ is bi-exact relative to $\mathcal{G}$ and to $\mathcal{G}'$, then Γ is bi-exact relative to $\mathcal{G} \wedge \mathcal{G}'$.

Proof. For notational simplicity, set $\tilde{\Gamma} = \Gamma\times\Gamma$, $A = \ell^\infty(\Gamma)$ and $I = c_0(\Gamma;\mathcal{G})$, $I' = c_0(\Gamma;\mathcal{G}')$. Then, the natural short exact sequence

$$0 \longrightarrow (I/(I\cap I')) \rtimes \tilde{\Gamma} \longrightarrow (A/(I\cap I')) \rtimes \tilde{\Gamma} \longrightarrow (A/I) \rtimes \tilde{\Gamma} \longrightarrow 0$$

is exact. By assumption, $(I/(I\cap I')) \rtimes \tilde{\Gamma} \cong ((I + I')/I') \rtimes \tilde{\Gamma} \lhd (A/I') \rtimes \tilde{\Gamma}$ and $(A/I) \rtimes \tilde{\Gamma}$ are nuclear. Hence the middle algebra $(A/(I\cap I')) \rtimes \tilde{\Gamma}$ is also nuclear. Therefore, it suffices to show that $I \cap I' = c_0(\Gamma; \mathcal{G}\wedge\mathcal{G}')$. We may assume that $\mathcal{G}$ is *saturated* in the sense that $s\Lambda s^{-1} \in \mathcal{G}$ for any $\Lambda \in \mathcal{G}$ and $s \in \Gamma$, and likewise for $\mathcal{G}'$. It is not hard to see that $I \cap I'$ is generated by a function whose support is contained in $\Lambda t \cap \Lambda' t'$. Pick any $x \in \Lambda t \cap \Lambda' t'$ (unless it is empty) and observe that $\Lambda t \cap \Lambda' t' = (\Lambda \cap \Lambda')x$. This completes the proof. □

Exercises

Exercise 15.2.1. Let X be a compact Γ-space and $\mathrm{Prob}(X)$ be the state space of $C(X)$ equipped with the natural Γ-action. Prove that X is amenable if and only if $\mathrm{Prob}(X)$ is amenable.

Exercise 15.2.2. Let X be a compact Γ-space and assume there is a Γ-equivariant u.c.p. map from $\ell^\infty(\Gamma)$ into $C(X)$. Prove that X is amenable provided that Γ is exact.

15.3. Examples

Hyperbolic groups. We have already seen that $\bar{\Gamma}^{\mathcal{G}}$ and (hence) $\Delta^{\mathcal{G}}\Gamma$ are amenable for a hyperbolic group Γ and $\mathcal{G} = \{\mathbf{1}\}$.

Theorem 15.3.1. *Let Γ be a hyperbolic group and $B \subset L(\Gamma)$ be a diffuse von Neumann subalgebra. Then $B' \cap L(\Gamma)$ is injective.*

Proof. It follows from Theorem 15.1.5 that noninjectivity of $B' \cap L(\Gamma)$ implies that B embeds in $\mathbb{C}1$ inside $L(\Gamma)$, which means that B has a nonzero minimal projection. □

A type II_1-factor N is said to be *prime* if $N \cong N_1 \bar{\otimes} N_2$ implies that N_1 or N_2 is finite-dimensional.

Corollary 15.3.2. *Let Γ be a hyperbolic group and $N \subset L(\Gamma)$ be a subfactor. Then N is either injective or prime. In particular, the free group factors $L(\mathbb{F}_r)$ $(r \geq 2)$ are prime.*

Proof. Suppose $N_1 \bar{\otimes} N_2 \cong N \subset L(\Gamma)$ is noninjective. Then either N_1 or N_2 is noninjective and the other cannot be diffuse by the theorem above. We note that N_i is a factor since N is a factor and that a finite factor with a minimal projection is finite-dimensional. □

Direct product of hyperbolic groups.

Lemma 15.3.3. *Let $\Gamma_1, \ldots, \Gamma_n$ be groups and $\Gamma = \prod_{i=1}^n \Gamma_i$ be the direct product. Let $\mathcal{G}_i$ be a family of subgroups of Γ_i and define a family $\mathcal{G}$ of subgroups of Γ by*

$$\mathcal{G} = \bigcup_i \{\Lambda \times \prod_{j \neq i} \Gamma_j : \Lambda \in \mathcal{G}_i\}.$$

If each of Γ_i is bi-exact relative to $\mathcal{G}_i$, then Γ is bi-exact relative to $\mathcal{G}$.

We leave the proof as an exercise.

Theorem 15.3.4. *Let $\Gamma_1, \ldots, \Gamma_n$ be hyperbolic groups and $N_1, \ldots, N_m$ be noninjective II_1-factors. If there exists an embedding*

$$N_1 \bar{\otimes} \cdots \bar{\otimes} N_m \hookrightarrow pL(\Gamma_1 \times \cdots \times \Gamma_n)p,$$

for some projection $p \in L(\Gamma_1 \times \cdots \times \Gamma_n)$, then $m \leq n$.

Proof. By Theorem 15.1.5 and Lemma 15.3.3, after permuting indices, one has

$$e_1 N_1 e_1 \bar{\otimes} \cdots \bar{\otimes} N_{m-1} \hookrightarrow p_0 L(\Gamma_1 \times \cdots \times \Gamma_{n-1}) p_0$$

for some nonzero projections $e_1 \in N_1$ and $p_0 \in L(\Gamma_1 \times \cdots \times \Gamma_{n-1})$. By induction, we are done. □

Semidirect products and wreath products.

Lemma 15.3.5. *Let $\Gamma = \Upsilon \rtimes \Lambda$ be a semidirect product of discrete groups. Let $\mathcal{G}_\Lambda$ be a family of subgroups of Λ and set $\mathcal{G} = \{\Upsilon \rtimes \Lambda_0 : \Lambda_0 \in \mathcal{G}_\Lambda\}$. If Υ is amenable and Λ is bi-exact relative to $\mathcal{G}_\Lambda$, then Γ is bi-exact relative to $\mathcal{G}$.*

Proof. Let $\mu\colon \Lambda \to \mathrm{Prob}(\Lambda)$ be a map as in Definition 15.1.2. It is not hard to see that the composition of μ with the quotient $\Gamma \to \Lambda = \Gamma/\Upsilon$ satisfies the conditions of Lemma 15.2.6. □

This is not so interesting unless $\mathcal{G}_\Lambda$ is very small (e.g., if Λ is hyperbolic). So, we consider another example, namely the wreath product, which was defined right before Definition 12.2.10.

In what follows, we denote $\Upsilon \wr \Lambda$ by Γ and agree that p, s and t represent elements of Λ, while x, y and z represent elements of Υ_Λ (the group $\bigoplus_\Lambda \Upsilon$ of finitely supported functions from Λ into Υ). Hence a typical element of Γ will be denoted by xs or yt. In particular, $sx = \alpha_s(x)s$, where α is the left translation action of Λ on Υ_Λ.

Proposition 15.3.6. *Let $\Gamma = \Upsilon \wr \Lambda$ be the wreath product and let $\mathcal{G} = \{\Lambda\}$. If Υ is amenable and Λ is exact, then Γ is bi-exact relative to $\mathcal{G}$.*

The proof of this proposition requires several steps. We fix a proper length function $|\cdot|_\Lambda$ on Λ:

(1) $|s|_\Lambda = |s^{-1}|_\Lambda \in \mathbb{R}_{\geq 0}$ for $s \in \Lambda$ and $|s|_\Lambda = 0$ if and only if $s = e$;

(2) $|st|_\Lambda \leq |s|_\Lambda + |t|_\Lambda$ for every $s, t \in \Lambda$;

(3) the subset $B_\Lambda(R) = \{s \in \Lambda : |s|_\Lambda \leq R\}$ is finite for every $R > 0$.

(See Proposition 5.5.2 for the existence of such a function.) Likewise, fix a length function on Υ. For $yt \in \Gamma$, we define $\zeta(yt) \in \ell^1(\Lambda)$ by

$$\zeta(yt)(p) = \begin{cases} \min\{|p|_\Lambda, |t^{-1}p|_\Lambda\} + |y(p)|_\Upsilon & \text{if } p \in \mathrm{supp}(y), \\ 0 & \text{if } p \notin \mathrm{supp}(y). \end{cases}$$

Lemma 15.3.7. *For $\mathcal{G} = \{\Lambda\}$, one has*

$$\lim_{yt\to\infty/\mathcal{G}} \frac{|\operatorname{supp}(y)|}{\|\zeta(yt)\|} = 0.$$

Proof. We first claim that $\lim_{yt\to\infty/\mathcal{G}} \|\zeta(yt)\| = \infty$. Let $R > 0$ be given and suppose $yt \in \Gamma$ is such that $\|\zeta(yt)\| \leq R$. Then $\operatorname{supp}(y) \subset B_R(\Lambda) \cup tB_R(\Lambda)$ and $y(p) \in B_R(\Upsilon)$ for every $p \in \Lambda$. Define $y' \in \Upsilon_\Lambda$ by $y'(p) = y(p)$ for $p \in B_R(\Lambda)$ and $y'(p) = e$ for $p \notin B_R(\Lambda)$. Then, $y = y'y''$ with $\operatorname{supp}(y') \subset B_R(\Lambda)$ and $\operatorname{supp}(y'') \subset tB_R(\Lambda)$. Hence, for the finite subset

$$E = \{z \in \Upsilon_\Lambda : \operatorname{supp}(z) \subset B_R(\Lambda) \text{ and } z(p) \in B_R(\Upsilon) \text{ for every } p \in \Lambda\}$$

of Υ_Λ, we have

$$yt = y't\alpha_{t^{-1}}(y'') \in \bigcup_{z',z''\in E} z'\Lambda z''.$$

This means that the subset $\Omega_R = \{yt \in \Gamma : \|\zeta(yt)\| \leq R\}$ is small relative to $\mathcal{G}$ and the claim follows.

Let $C > 0$ be given and suppose $yt \in \Gamma$ is such that $\|\zeta(yt)\| \leq C|\operatorname{supp}(y)|$. Since $\zeta(yt)(p) \geq 2C$ for $p \in \operatorname{supp}(y) \setminus \big(B_{2C}(\Lambda) \cup tB_{2C}(\Lambda)\big)$, we have

$$|\operatorname{supp}(y) \setminus \big(B_{2C}(\Lambda) \cup tB_{2C}(\Lambda)\big)| \leq |\operatorname{supp}(y)|/2.$$

This implies $|\operatorname{supp}(y)| \leq 4|B_{2C}(\Lambda)|$ and $yt \in \Omega_R$ for $R = 4C|B_{2C}(\Lambda)|$. By the first part of the proof, $\{yt \in \Gamma : |\operatorname{supp}(y)|/\|\zeta(yt)\| \geq C^{-1}\}$ is small relative to $\mathcal{G}$. □

Lemma 15.3.8. *The following hold:*

1. $\|\zeta(xyt) - \zeta(yt)\| \leq \|\zeta(x)\|$ *for every* $x, y \in \Upsilon_\Lambda$ *and* $t \in \Lambda$;
2. $\|\zeta(syt) - s.\zeta(yt)\| \leq |s|_\Lambda |\operatorname{supp}(y)|$ *for every* $y \in \Upsilon_\Lambda$ *and* $s, t \in \Lambda$;
3. $\|\zeta(ytx) - \zeta(yt)\| \leq \|\zeta(x)\|$ *for every* $x, y \in \Upsilon_\Lambda$ *and* $t \in \Lambda$;
4. $\|\zeta(yts) - \zeta(yt)\| \leq |s|_\Lambda |\operatorname{supp}(y)|$ *for every* $y \in \Upsilon_\Lambda$ *and* $s, t \in \Lambda$.

Proof. Note that $\zeta(xyt)(p) - \zeta(yt)(p)$ is nonzero only if $p \in \operatorname{supp}(x)$. Also,

$$\begin{aligned}
|\zeta(xyt)(p) - \zeta(yt)(p)| &= \begin{cases} \big|\,|x(p)y(p)|_\Upsilon - |y(p)|_\Upsilon\big| & \text{if } p \in \operatorname{supp}(y) \cap \operatorname{supp}(xy), \\ \zeta(xt)(p) & \text{otherwise} \end{cases} \\
&\leq \zeta(x)(p)
\end{aligned}$$

for $p \in \operatorname{supp}(x)$. This yields the first assertion. For the second, observe that $\zeta(syt)(p)$ and $(s.\zeta(yt))(p)$ are nonzero only if $p \in s\operatorname{supp}(y)$ and that

$$\begin{aligned}
|\zeta(syt)(p) - (s.\zeta(yt))(p)| &= |\min\{|p|_\Lambda, |(st)^{-1}p|_\Lambda\} - \min\{|s^{-1}p|_\Lambda, |(st)^{-1}p|_\Lambda\}| \\
&\leq |s|.
\end{aligned}$$

This yields the second assertion. For the third assertion, we observe that $\zeta(ytx)(p)-\zeta(yt)(p)$ is nonzero only if $p \in t\,\mathrm{supp}(x)$ and that for $q \in \mathrm{supp}(x)$, one has

$$\zeta(ytx)(tq) = \begin{cases} \min\{|tq|_\Lambda, |q|_\Lambda\} + |y(tq)x(q)|_\Upsilon & \text{if } y(tq) \neq x(q)^{-1}, \\ 0 & \text{if } y(tq) = x(q)^{-1}, \end{cases}$$

$$\zeta(yt)(tq) = \begin{cases} \min\{|tq|_\Lambda, |q|_\Lambda\} + |y(tq)|_\Upsilon & \text{if } tq \in \mathrm{supp}(y), \\ 0 & \text{if } tq \notin \mathrm{supp}(y). \end{cases}$$

Hence for $q \in \mathrm{supp}(x)$, one has

$$|\zeta(ytx)(tq) - \zeta(yt)(tq)| \leq |q|_\Lambda + |x(q)|_\Upsilon = \zeta(x)(q)$$

and the third assertion follows. Finally, since $\zeta(yts)(p)$ and $\zeta(yt)(p)$ are nonzero only if $p \in \mathrm{supp}(y)$ and

$$|\zeta(yts)(p) - \zeta(yt)(p)| = |\min\{|p|_\Lambda, |s^{-1}t^{-1}p|_\Lambda\} - \min\{|p|_\Lambda, |t^{-1}p|_\Lambda\}| \leq |s|,$$

for $p \in \mathrm{supp}(y)$, the fourth assertion follows. □

Proof of Proposition 15.3.6. With Lemmas 15.3.7 and 15.3.8 in hand, it is easy to verify the condition of Lemma 15.2.6. Indeed, one just has to check the condition separately for $x \in \Upsilon_\Lambda$ and $s \in \Lambda$, acting from the left or the right. □

Corollary 15.3.9. *Let $\Gamma = \Upsilon \wr \Lambda$ be the wreath product. Suppose that Υ is amenable and Λ is bi-exact relative to $\{\mathbf{1}\}$ (e.g., if Λ is hyperbolic). Then, Γ is bi-exact relative to $\{\mathbf{1}\}$.*

Proof. Combine Lemma 15.3.5, Proposition 15.3.6 and Proposition 15.2.7. □

Theorem 15.3.10. *Let $\Gamma = \Upsilon \wr \Lambda$ be the wreath product of an amenable group Υ by an exact group Λ. If $N \subset pL(\Gamma)p$ is a von Neumann subalgebra with a noninjective relative commutant, then N embeds in $L(\Lambda)$ inside $L(\Gamma)$.*

Proof. Combine Theorem 15.1.5 and Proposition 15.3.6. □

Corollary 15.3.11. *Let $\Gamma = \Upsilon \wr \Lambda$ be the wreath product of an amenable group Υ by an exact group Λ. If $N \subset L(\Gamma)$ is a noninjective nonprime factor whose relative commutant $N' \cap L(\Gamma)$ is a factor, then there exists a unitary element $u \in L(\Gamma)$ such that $uNu^* \subset L(\Lambda)$.*

Proof. Write N as a tensor product $N = N_1 \bar{\otimes} N_2$ of type II_1-factors N_1 and N_2. Since N is noninjective, we may assume that N_2 is noninjective. By Theorem 15.3.10, N_1 embeds in $L(\Lambda)$ inside $L(\Gamma)$. By Lemma F.18 and Theorem F.20, we can find a unitary element $u \in L(\Gamma)$ such that $uN_1u^* \subset L(\Lambda)$. This implies $uN_2u^* \subset (uN_1u^*)' \cap L(\Gamma) \subset L(\Lambda)$, by Theorem F.20. Therefore, $uNu^* \subset L(\Lambda)$. □

Amalgamated free products.

Proposition 15.3.12. *Let $\Gamma = \Gamma_1 *_\Lambda \Gamma_2$ be an amalgamated free product and let $\mathcal{G} = \{\Gamma_1, \Gamma_2\}$. If both Γ_i are exact and Λ is amenable, then $\bar{\Gamma}^{\mathcal{G}}$ is amenable and, in particular, Γ is bi-exact relative to $\mathcal{G}$.*

Before giving the proof, we point out that the amenability assumption on Λ is essential. Indeed, if $\Gamma_i = \Gamma_i' \times \Lambda$ and Λ is nonamenable, then $\Gamma = (\Gamma_1' * \Gamma_2') \times \Lambda$ and thus $L(\Gamma_1' * \Gamma_2')$ has a noninjective commutant in $L(\Gamma)$. More generally, if $s_i \in \Gamma_i \setminus \Lambda$ normalize Λ and $s_1 a s_1^{-1} = s_2 a s_2^{-1}$ for all $a \in \Lambda$, then $s = s_1 s_2^{-1} \in \Gamma$ has infinite order and commutes with Λ.

Proof. We first prove that the Γ-space $\bar{\Gamma}^{\mathcal{G}}$ is amenable as a Γ_i-space. We prove this for $i = 1$. Let $A \subset \ell^\infty(\Gamma_1)$ be the C*-subalgebra of those functions f such that $f = f^t$ for all $t \in \Lambda$. Averaging over the right Λ-action, we obtain a (left) Γ_1-equivariant conditional expectation from $\ell^\infty(\Gamma_1)$ onto A. By Exercise 15.2.2, it suffices to find a Γ_1-equivariant $*$-homomorphism π from A into $C(\bar{\Gamma}^{\mathcal{G}})$. Fix a system $\{e\} \sqcup S_i^0 \subset \Gamma_i$ of representatives of $\Lambda \backslash \Gamma_i$, and set

$$\mathfrak{X} = \{e\} \sqcup S_2^0 \sqcup S_2^0 S_1^0 \sqcup S_2^0 S_1^0 S_2^0 \sqcup \cdots \subset \Gamma.$$

Then, every $s \in \Gamma$ can be uniquely written in the form $s = s_1 x$, where $s_1 \in \Gamma_1$ and $x \in \mathfrak{X}$ (cf. Appendix E). We define $\pi \colon A \to \ell^\infty(\Gamma)$ by $\pi(f)(s_1 x) = f(s_1)$ for $s_1 \in \Gamma_1$ and $x \in \mathfrak{X}$. Our task is to show $\pi(f) - \pi(f)^t \in c_0(\Gamma; \mathcal{G})$ for every $t \in \Gamma_1 \cup \Gamma_2$. Suppose first that $t \in \Gamma_1$. Then, for every $s_1 x \in \Gamma$, one has either $s_1 x t^{-1} = s_1 t^{-1}$ (if $x = e$) or $s_1 x t^{-1} = s_1 a y$ for some $a \in \Lambda$ and $y \in \mathfrak{X}$ (if $x \neq e$). Since f is right Λ-invariant, $\pi(f) - \pi(f)^t$ has support in Γ_1. It follows that $\pi(f) - \pi(f)^t \in c_0(\Gamma; \mathcal{G})$. Suppose next that $t \in \Gamma_2$. Then, one has $\pi(f) - \pi(f)^t = 0$ by similar reasoning. Altogether, this implies $\bar{\Gamma}^{\mathcal{G}}$ is amenable as a Γ_i-space.

Let $\mathbf{T} = \Gamma/\Gamma_1 \sqcup \Gamma/\Gamma_2$ be the Bass-Serre tree on which $\Gamma = \Gamma_1 *_\Lambda \Gamma_2$ acts and let $\bar{\mathbf{T}}$ be its compactification,[4] as defined in Section 5.2. We will find a Γ-equivariant continuous map from $\bar{\Gamma}^{\mathcal{G}}$ into $\bar{\mathbf{T}}$, which suffices to show the amenability of $\bar{\Gamma}^{\mathcal{G}}$ by Proposition 5.2.1 and Lemma 5.2.6. Choose a base point $o \in \mathbf{T}$ and define a Γ-equivariant $*$-homomorphism $\sigma \colon C(\bar{\mathbf{T}}) \to \ell^\infty(\Gamma)$ by $\sigma(f)(s) = f(so)$. We will show $\sigma(f) - \sigma(f)^t \in c_0(\Gamma; \{\Lambda\})$ for every $f \in C(\bar{\mathbf{T}})$ and $t \in \Gamma$. Suppose by contradiction that this is not the case. Then, there exists $\varepsilon > 0$ such that the set

$$\Omega = \{s \in \Gamma : |f(so) - f(st^{-1}o)| \geq \varepsilon\} \subset \Gamma$$

is not small relative to $\{\Lambda\}$. Hence, there exists a net (s_n) in Ω such that $s_n \to \infty/\{\Lambda\}$. We may assume that $s_n o \to z$ for some $z \in \bar{\mathbf{T}}$. Since every edge stabilizer of the Γ-action on the Bass-Serre tree is an inner conjugate

[4]Although we use the term "compactification", $\mathbf{T}$ is not open in $\bar{\mathbf{T}}$.

of Λ, we can apply Lemma 5.2.8 and deduce that $s_n t^{-1}o \to z$. Hence we obtain the contradiction

$$\varepsilon \leq \lim_n |f(s_n o) - f(s_n t^{-1}o)| = |f(z) - f(z)| = 0.$$

Therefore, $\sigma(f) - \sigma(f)^t \in c_0(\Gamma; \{\Lambda\}) \subset c_0(\Gamma; \mathcal{G})$ and we are done. □

Theorem 15.3.13. *Let $\Gamma = \Gamma_1 *_\Lambda \Gamma_2$ be an amalgamated free product such that both Γ_i are exact and Λ is amenable. If $N \subset L(\Gamma)$ is a von Neumann subalgebra with a noninjective relative commutant, then there exists i such that N embeds in $L(\Gamma_i)$ inside $L(\Gamma)$.*

Proof. Combine Theorem 15.1.5 and Proposition 15.3.12. □

Recall that a group Γ is said to have *infinite conjugacy classes (ICC)* if the sets $\{sts^{-1} : s \in \Gamma\}$ are infinite for every nonneutral element $t \in \Gamma$.

Corollary 15.3.14. *Let $\Gamma = \Gamma_1 * \Gamma_2$ be a free product of ICC exact groups. If $N \subset L(\Gamma)$ is a noninjective nonprime factor whose relative commutant $N' \cap L(\Gamma)$ is a factor, then there exist $i \in \{1, 2\}$ and a unitary element $u \in L(\Gamma)$ such that $uNu^* \subset L(\Gamma_i)$.*

We omit the proof of this corollary as it is very similar to the proof of Corollary 15.3.11.

We say Γ is a *product* group if it is isomorphic to a direct product of nontrivial groups. We note that if $\Gamma = \Gamma' \times \Gamma''$ is an ICC product group, then Γ' and Γ'' are also ICC and, in particular, infinite.

Corollary 15.3.15. *Let $\Gamma_1, \ldots, \Gamma_n$ and $\Lambda_1, \ldots, \Lambda_m$ be ICC nonamenable exact product groups. If*

$$M = L(\mathbb{F}_\infty * \Gamma_1 * \cdots * \Gamma_n) \cong L(\mathbb{F}_\infty * \Lambda_1 * \cdots * \Lambda_m),$$

then $n = m$ and, modulo permutation of indices, $L(\Gamma_i)$ is unitarily conjugated to $L(\Lambda_i)$ inside M for every $1 \leq i \leq n$.

Proof. It follows from Corollary 15.3.14 that there are maps $\imath, \jmath$ and unitary elements $u_1, \ldots, u_m$ and $v_1, \ldots v_n$ such that $u_j L(\Lambda_j) u_j^* \subset L(\Gamma_{\imath(j)})$ and $v_i L(\Gamma_i) v_i^* \subset L(\Lambda_{\jmath(i)})$. It follows that

$$v_{\imath(j)} u_j L(\Lambda_j) u_j^* v_{\imath(j)}^* \subset L(\Lambda_{\jmath(\imath(j))})$$

for every j. By Theorem F.20 and Exercise F.8, this implies $\jmath(\imath(j)) = j$ and $v_{\imath(j)} u_j \in L(\Lambda_j)$. In particular, the above inclusions are tight and $u_j L(\Lambda_j) u_j^* = L(\Gamma_{\imath(j)})$. Likewise, one has $\imath(\jmath(i)) = i$ for every $1 \leq i \leq n$. □

This corollary is an analogue of Kurosh's isomorphism theorem for groups and, like Kurosh's Theorem, it says almost nothing about the positions of the copies of $L(\mathbb{F}_\infty)$.

15.4. References

This chapter grew out of [**133, 137, 135**], which contain some additional results. It would be nice to eliminate the exactness assumption from Theorem 15.1.5. Theorem 15.3.10 and its corollaries were obtained by Popa [**161**], who used completely different methods which did not require exactness. Theorem 15.3.13 and its corollaries also have stronger versions, due to Ioana, Peterson and Popa [**89**] and Peterson [**143**], which do not require exactness.

Chapter 16

Herrero's Approximation Problem

Paul Halmos introduced the notion of a quasidiagonal operator (on a separable Hilbert space) over 30 years ago. The purpose of this chapter is to study approximation properties of such operators and to give a complete solution to a question of Domingo Herrero. At present there is no operator-theoretic solution – we will have to rely on the theory of exact C*-algebras.

16.1. Description of the problem

Definition 16.1.1. An operator $S \in \mathbb{B}(\mathcal{H})$ is called *block diagonal* if there exist finite-rank projections $P_n \leq P_{n+1}$ which converge strongly to the identity and such that $[T, P_n] = 0$ for all $n \in \mathbb{N}$. We denote the set of all such operators by $\mathcal{BD}(\mathcal{H})$.

Keep in mind the matrix picture of block diagonal operators. If we write

$$\mathcal{H} = P_1\mathcal{H} \oplus (P_2 - P_1)\mathcal{H} \oplus (P_3 - P_2)\mathcal{H} \oplus \cdots,$$

then the matrix of S with respect to this decomposition is really "block diagonal" with finite-dimensional blocks.

Definition 16.1.2. An operator $T \in \mathbb{B}(\mathcal{H})$ is called *quasidiagonal* if there exist finite-rank projections $P_n \leq P_{n+1}$ which converge strongly to the identity and such that $\|[T, P_n]\| \to 0$ for all $n \in \mathbb{N}$. We denote the set of all such operators by $\mathcal{QD}(\mathcal{H})$.

Remark 16.1.3. Note that Proposition 7.2.3 shows $T \in \mathcal{QD}(\mathcal{H})$ if and only if $\{T\}$ is a quasidiagonal set in the sense of Definition 7.2.1. From the local formulation of quasidiagonal sets, it is evident that $\mathcal{QD}(\mathcal{H})$ is a norm closed set of operators (which contains block diagonal operators).

We can now reformulate Theorem 7.5.1, which is really due to Halmos, as follows.

Theorem 16.1.4. *If $T \in \mathcal{QD}(\mathcal{H})$, then for every $\varepsilon > 0$ there exists $S \in \mathcal{BD}(\mathcal{H})$ such that $T - S \in \mathbb{K}(\mathcal{H})$ and $\|T - S\| < \varepsilon$. In particular,*

$$\mathcal{QD}(\mathcal{H}) = \overline{\mathcal{BD}(\mathcal{H})},$$

where the closure is taken in norm.

For reasons we will soon explain, the following subset of $\mathcal{QD}(\mathcal{H})$ is important.

Definition 16.1.5. An operator $S \in \mathbb{B}(\mathcal{H})$ is called *block diagonal with bounded blocks* if there exist finite-rank projections $P_n \le P_{n+1}$ which converge strongly to the identity, such that $[T, P_n] = 0$ for all $n \in \mathbb{N}$ and

$$\sup_n \operatorname{rank}(P_n - P_{n-1}) < \infty.$$

We denote the set of all such operators by $\mathcal{BD}_{bdd}(\mathcal{H})$.

Again, thinking of block matrices and the decomposition

$$\mathcal{H}P_1\mathcal{H} \oplus (P_2 - P_1)\mathcal{H} \oplus (P_3 - P_2)\mathcal{H} \oplus \cdots ,$$

this terminology should be clear.

The study of various examples suggested that the norm closures of $\mathcal{BD}(\mathcal{H})$ and $\mathcal{BD}_{bdd}(\mathcal{H})$ may coincide (i.e., for a while, all the known examples of quasidiagonal operators could be shown to lie in the norm closure of $\mathcal{BD}_{bdd}(\mathcal{H})$). So, Herrero asked if this was the case:

Is every quasidiagonal operator in the norm closure of $\mathcal{BD}_{bdd}(\mathcal{H})$?

It turns out that the answer is "no" and we will describe some explicit counterexamples in the last section of this chapter. However, knowing that $\mathcal{QD}(\mathcal{H})$ properly contains the norm closure of $\mathcal{BD}_{bdd}(\mathcal{H})$, the original question naturally evolved into *Herrero's approximation problem*, which asks:

What is the norm closure of $\mathcal{BD}_{bdd}(\mathcal{H})$?

The answer to this question is found in Theorem 16.3.3.

16.2. C*-preliminaries

To solve Herrero's problem, we will need a technical generalization of Dadarlat's approximation theorem for exact QD C*-algebras (compare Theorem 7.5.7 with Theorem 16.2.6 below). This, in turn, requires a few preliminary results.

Lemma 16.2.1. *If $A \subset \mathbb{B}(\mathcal{H})$ is a unital exact* C*-*algebra, then $A + \mathbb{K}(\mathcal{H})$ is also exact.*

Proof. Since exactness passes to quotients (Corollary 9.4.3) and locally split extensions of exact C*-algebras are again exact (Exercise 3.9.8), it suffices to show that the extension

$$0 \to \mathbb{K}(\mathcal{H}) \to A + \mathbb{K}(\mathcal{H}) \to A/(A \cap \mathbb{K}(\mathcal{H})) \to 0$$

is locally split. Since quotient mappings of exact C*-algebras are always locally liftable (Remark 9.4.2), the related extension

$$0 \to A \cap \mathbb{K}(\mathcal{H}) \to A \to A/(A \cap \mathbb{K}(\mathcal{H})) \to 0$$

is locally split. Evidently this implies what we want. □

The following corollary isn't really necessary, as all the arguments which follow can be localized to finite-dimensional operator subsystems. However, it makes things a little cleaner (and follows immediately from the previous fact and Remark C.5).

Corollary 16.2.2. *If $A \subset \mathbb{B}(\mathcal{H})$ is unital and exact, then there exists a u.c.p. splitting for the sequence*

$$0 \to \mathbb{K}(\mathcal{H}) \to A + \mathbb{K}(\mathcal{H}) \to A/(A \cap \mathbb{K}(\mathcal{H})) \to 0.$$

Here is another requisite application of local liftability.

Proposition 16.2.3. *If $A \subset \mathbb{B}(\mathcal{H})$ is a quasidiagonal set of operators and A is exact, then $\pi(A)$ is exact and QD, where $\pi\colon \mathbb{B}(\mathcal{H}) \to Q(\mathcal{H})$ is the quotient map to the Calkin algebra.*

Proof. We may assume A is unital. Since A is a quasidiagonal set of operators, the extension

$$0 \to \mathbb{K}(\mathcal{H}) \to A + \mathbb{K}(\mathcal{H}) \to A/(A \cap \mathbb{K}(\mathcal{H})) \to 0$$

is quasidiagonal (Exercise 7.2.5). Since $A + \mathbb{K}(\mathcal{H})$ is exact, the quotient mapping $A + \mathbb{K}(\mathcal{H}) \to A/(A \cap \mathbb{K}(\mathcal{H}))$ is liftable and hence Exercise 7.1.5 completes the proof.

The reader who didn't do those exercises is probably a bit annoyed right now. Since we actually need the proofs more than the results, here are the details.

To ease notation, let $B = \pi(A)$. We have a short exact sequence

$$0 \to \mathbb{K}(\mathcal{H}) \to A + \mathbb{K}(\mathcal{H}) \to B \to 0$$

and by the previous corollary we can find a u.c.p. map $\Phi\colon B \to A + \mathbb{K}(\mathcal{H})$ such that $\pi \circ \Phi = \mathrm{id}_B$. Since $A \subset \mathbb{B}(\mathcal{H})$ is a quasidiagonal set, we can find increasing finite-rank projections $P_n \leq P_{n+1}$ which tend to the identity and asymptotically commute with $A + \mathbb{K}(\mathcal{H})$. To show that B is QD, it suffices to find asymptotically multiplicative, asymptotically isometric contractive c.p. maps from B into $A + \mathbb{K}(\mathcal{H})$.

Let $P_n^{\perp} = 1 - P_n$ and consider the (nonunital) c.p. (isometric) maps

$$\Phi_n(b) = P_n^{\perp}\Phi(b)P_n^{\perp}.$$

Note that $\Phi_n(b) \in A + \mathbb{K}(\mathcal{H})$ for all n and $b \in B$, since $P_n \in A + \mathbb{K}(\mathcal{H})$. Thus the proof will be complete once we understand why

$$\|\Phi_n(bc) - \Phi_n(b)\Phi_n(c)\| \to 0$$

for all $b, c \in B$. But this follows from the fact that $\{P_n\}$ is an approximate unit for $\mathbb{K}(\mathcal{H})$ and quasicentral in $A + \mathbb{K}(\mathcal{H})$. □

Proposition 16.2.4. *With the same assumptions as the previous proposition, for each finite set $\mathfrak{F} \subset A$ and $\varepsilon > 0$ there exists a finite-dimensional C*-subalgebra $D \subset Q(\mathcal{H})$ such that $\pi(\mathfrak{F}) \subset^{\varepsilon} D$ – i.e., for each $a \in \mathfrak{F}$ there exists $d \in D$ such that $\|\pi(a) - d\| < \varepsilon$.*

Proof. Let $B = \pi(A)$ and $\Phi_n(b) = P_n^{\perp}\Phi(b)P_n^{\perp}$ as in the proof of the last proposition. Since each Φ_n is a splitting, we may regard Φ_n as a faithful $*$-homomorphism modulo the compacts (as in Theorem 1.7.6). To make Φ_n unital, we regard it as taking values in $\mathbb{B}(P_n^{\perp}\mathcal{H})$ (rather than $\mathbb{B}(\mathcal{H})$). Keep in mind that $P_n^{\perp}\mathcal{H}$ has finite codimension in $\mathcal{H}$ and hence the Calkin algebra can't see the difference between the two.

Let $\sigma\colon B \to \mathbb{B}(\mathcal{K})$ be a faithful essential representation. Since B is exact and QD it follows from Dadarlat's approximation theorem (Theorem 7.5.7) that we can find finite-dimensional subalgebras in $\mathbb{B}(\mathcal{K})$ which approximate prescribed finite subsets of $\sigma(B)$ arbitrarily well. Since the maps $\Phi_n\colon B \to \mathbb{B}(P_n^{\perp}\mathcal{H})$ are asymptotically multiplicative faithful $*$-homomorphisms modulo the compacts, we can, by Theorem 1.7.6, find unitary operators $U_n\colon \mathcal{K} \to P_n^{\perp}\mathcal{H}$ such that

$$\|\Phi_n(b) - U_n\sigma(b)U_n^*\| \to 0$$

for all $b \in B$. So, if a finite subset of B and $\varepsilon > 0$ are given, we can find finite-dimensional approximations inside $Q(\mathcal{H})$ by first approximating in the representation $\sigma\colon B \to \mathbb{B}(\mathcal{K})$, then conjugating over to $\Phi_n\colon B \to \mathbb{B}(P_n^{\perp}\mathcal{H})$ (for a sufficiently large n) and, finally, passing to $Q(P_n^{\perp}\mathcal{H})$. Since

P_n was finite-rank, we have $Q(P_n^\perp \mathcal{H}) = Q(\mathcal{H})$ (canonically) and the proof is complete. □

We need a well-known theorem of Larry Brown. (See [**53**, Theorem III.6.3] for a proof.)

Theorem 16.2.5. *Let $0 \to J \to A \to B \to 0$ be a short exact sequence where both J and B are AF algebras. Then A is also AF. In other words, extensions of AF algebras are AF.*

Finally, we present the generalization of Theorem 7.5.7 that we've been after.

Theorem 16.2.6. *Let A be an exact* C*-*algebra and $\sigma\colon A \to \mathbb{B}(\mathcal{H})$ be a quasidiagonal representation. For each finite set $\mathfrak{F} \subset A$ and $\varepsilon > 0$ there exists a finite-dimensional* C*-*subalgebra $B \subset \mathbb{B}(\mathcal{H})$ such that*

$$\sigma(\mathfrak{F}) \subset^\varepsilon B.$$

Proof. Fix a finite set $\mathfrak{F} \subset A$ and $\varepsilon > 0$. As before, $\pi\colon \mathbb{B}(\mathcal{H}) \to Q(\mathcal{H})$ is the canonical quotient map. Let $D \subset Q(\mathcal{H})$ be a finite-dimensional subalgebra such that

$$\pi \circ \sigma(\mathfrak{F}) \subset^\varepsilon D.$$

Let $C \subset \mathbb{B}(\mathcal{H})$ be the canonical extension of $\mathbb{K}(\mathcal{H})$ by D (i.e., the pullback of $D \subset Q(\mathcal{H})$ in $\mathbb{B}(\mathcal{H})$). Since C is AF, it suffices to show that

$$\sigma(\mathfrak{F}) \subset^\varepsilon C,$$

which is easy, since $\mathbb{K}(\mathcal{H}) \subset C$. □

16.3. Resolution of Herrero's problem

To resolve Herrero's problem, we must consider another natural class of operators.

Definition 16.3.1. An operator $S \in \mathbb{B}(\mathcal{H})$ is called *banded* if there exists an orthonormal basis $\{v_i\}$ of $\mathcal{H}$ such that the matrix of S with respect to $\{v_i\}$ is banded (meaning only a finite number of diagonals are nonzero – that is, there exists $N \in \mathbb{N}$ such that $\langle Sv_i, v_j\rangle = 0$ whenever $|i - j| > N$). We let $\mathrm{Band}(\mathcal{H})$ denote the set of banded operators on $\mathcal{H}$.[1]

In order to apply Theorem 16.2.6, we will need a simple observation.

Lemma 16.3.2. *If $T \in \overline{\mathrm{Band}(\mathcal{H})}$, then $C^*(T)$ is exact.*

---

[1]In our previous terminology, these are finite propagation operators, or operators supported in tubes.

Proof. By Exercise 2.3.10 we may assume that $T \in \mathrm{Band}(\mathcal{H})$. However, every banded operator can be identified with an element in $\ell^\infty(\mathbb{Z}) \rtimes \mathbb{Z}$, since this algebra is precisely the norm closure of the banded operators on $\ell^2(\mathbb{Z})$ (see Proposition 5.1.3) and we can always identify $\mathcal{H}$ with $\ell^2(\mathbb{N}) \subset \ell^2(\mathbb{Z})$ in such a way that T lands inside $\ell^\infty(\mathbb{Z}) \rtimes \mathbb{Z}$. Since exactness passes to subalgebras and $\ell^\infty(\mathbb{Z}) \rtimes \mathbb{Z}$ is nuclear (Theorem 4.2.4), the proof is complete. □

Here's the answer to Herrero's problem:

Theorem 16.3.3. $\overline{\mathcal{BD}_{bdd}(\mathcal{H})} = \mathcal{QD}(\mathcal{H}) \cap \overline{\mathrm{Band}(\mathcal{H})}$. *In other words, if there exists a banded sequence S_n such that $\|T - S_n\| \to 0$ and a block diagonal sequence U_n such that $\|T - U_n\| \to 0$, then there exists a sequence X_n which is simultaneously banded and block diagonal such that $\|T - X_n\| \to 0$.*

Proof. We already observed that $\mathcal{BD}_{bdd}(\mathcal{H}) \subset \mathcal{QD}(\mathcal{H}) \cap \mathrm{Band}(\mathcal{H})$ and hence the same inclusion with norm closures is immediate. Assume now that $T \in \mathcal{QD}(\mathcal{H}) \cap \overline{\mathrm{Band}(\mathcal{H})}$. Since T is quasidiagonal, one easily checks that $C^*(T) \subset \mathbb{B}(\mathcal{H})$ is a quasidiagonal set of operators. The previous lemma implies that $C^*(T)$ is also exact. Hence, by Theorem 16.2.6, there exist operators T_n such that $\|T - T_n\| \to 0$ and each of the C*-algebras $C^*(T_n)$ is finite-dimensional. However, standard representation theory of finite-dimensional C*-algebras shows that each T_n is both block diagonal and banded, and this completes the proof. □

16.4. Counterexamples

The first counterexamples to Herrero's original question were discovered by Szarek ([**181**]). His proof was nonconstructive, however, and in [**189**] Voiculescu gave the first explicit examples of quasidiagonal operators which don't generate exact C*-algebras. To do this, he constructed finitely generated QD C*-algebras which aren't exact. Passing to a single operator requires the following trick:

Proposition 16.4.1. *Let $A = C^*(b_1, \ldots, b_n)$ be a unital C*-algebra generated by n self-adjoint elements. Then $\mathbb{M}_n(A)$ is a singly generated C*-algebra.*

Proof. We only sketch the ingredients required. The first step is to show that we may assume that each $b_i \geq 0$, b_i is invertible and the spectra $\sigma(b_i^2)$ are pairwise disjoint. (Hint: replace b_i with $b_i + \lambda_i 1_A$ for appropriate positive constants λ_i. Invertibility implies the algebra generated by $b_i + \lambda_i 1_A$ contains b_i.)

After making this reduction, we consider the operator $T \in \mathbb{M}_n(A)$ defined as follows:

$$T = \begin{bmatrix} 0 & b_1 & 0 & \cdots & 0 & 0 \\ 0 & 0 & b_2 & \cdots & 0 & 0 \\ 0 & 0 & 0 & \cdots & 0 & 0 \\ \vdots & \vdots & \vdots & \ddots & \vdots & \vdots \\ 0 & 0 & 0 & \cdots & 0 & b_{n-1} \\ b_n & 0 & 0 & \cdots & 0 & 0 \end{bmatrix}.$$

Step 2 is to convince yourself that it suffices to show $C^*(T)$ contains the matrices

$$\begin{bmatrix} b_1 & 0 & \cdots & 0 \\ 0 & 0 & \cdots & 0 \\ \vdots & \vdots & \ddots & \vdots \\ 0 & 0 & \cdots & 0 \end{bmatrix}, \begin{bmatrix} 0 & 0 & \cdots & 0 \\ 0 & b_2 & \cdots & 0 \\ \vdots & \vdots & \ddots & \vdots \\ 0 & 0 & \cdots & 0 \end{bmatrix}, \dots, \begin{bmatrix} 0 & 0 & \cdots & 0 \\ 0 & 0 & \cdots & 0 \\ \vdots & \vdots & \ddots & \vdots \\ 0 & 0 & \cdots & b_n \end{bmatrix},$$

together with all of the matrix units of $\mathbb{M}_n(\mathbb{C}) \subset \mathbb{M}_n(A)$. If this is the case, then matrix unit manipulations show we can generate all of $\mathbb{M}_n(A)$.

Step 3 is to use functional calculus on TT^* (remember we separated the spectra of the b_i^2's) to show that $C^*(T)$ contains all the matrices

$$\begin{bmatrix} b_1^2 & 0 & \cdots & 0 \\ 0 & 0 & \cdots & 0 \\ \vdots & \vdots & \ddots & \vdots \\ 0 & 0 & \cdots & 0 \end{bmatrix}, \begin{bmatrix} 0 & 0 & \cdots & 0 \\ 0 & b_2^2 & \cdots & 0 \\ \vdots & \vdots & \ddots & \vdots \\ 0 & 0 & \cdots & 0 \end{bmatrix}, \dots, \begin{bmatrix} 0 & 0 & \cdots & 0 \\ 0 & 0 & \cdots & 0 \\ \vdots & \vdots & \ddots & \vdots \\ 0 & 0 & \cdots & b_n^2 \end{bmatrix}.$$

It follows that $C^*(T)$ contains all the matrices

$$\begin{bmatrix} b_1 & 0 & \cdots & 0 \\ 0 & 0 & \cdots & 0 \\ \vdots & \vdots & \ddots & \vdots \\ 0 & 0 & \cdots & 0 \end{bmatrix}, \begin{bmatrix} 0 & 0 & \cdots & 0 \\ 0 & b_2 & \cdots & 0 \\ \vdots & \vdots & \ddots & \vdots \\ 0 & 0 & \cdots & 0 \end{bmatrix}, \dots, \begin{bmatrix} 0 & 0 & \cdots & 0 \\ 0 & 0 & \cdots & 0 \\ \vdots & \vdots & \ddots & \vdots \\ 0 & 0 & \cdots & b_n \end{bmatrix}$$

and

$$\begin{bmatrix} b_1^{-1} & 0 & \cdots & 0 \\ 0 & 0 & \cdots & 0 \\ \vdots & \vdots & \ddots & \vdots \\ 0 & 0 & \cdots & 0 \end{bmatrix}, \begin{bmatrix} 0 & 0 & \cdots & 0 \\ 0 & b_2^{-1} & \cdots & 0 \\ \vdots & \vdots & \ddots & \vdots \\ 0 & 0 & \cdots & 0 \end{bmatrix}, \dots, \begin{bmatrix} 0 & 0 & \cdots & 0 \\ 0 & 0 & \cdots & 0 \\ \vdots & \vdots & \ddots & \vdots \\ 0 & 0 & \cdots & b_n^{-1} \end{bmatrix},$$

since b_i is the unique positive square root of b_i^2 and C*-algebras are inverse closed.

The final step is to play around with all of the matrices we now know belong to $C^*(T)$ and to show that all of the matrix units must also be in there. For example, multiplying

$$\begin{bmatrix} b_1^{-1} & 0 & \cdots & 0 \\ 0 & 0 & \cdots & 0 \\ \vdots & \vdots & \ddots & \vdots \\ 0 & 0 & \cdots & 0 \end{bmatrix}$$

by T gives a nondiagonal matrix unit and similar fiddling gives the rest. □

Remark 16.4.2. Though we are after explicit examples, the proposition above shows that starting with any finitely generated QD nonexact C*-algebra (e.g., $C^*(\mathbb{F}_n)$), we can construct a counterexample to Herrero's original problem by passing to matrices over the given algebra and representing them on a Hilbert space.

To get concrete examples, we start with a finitely generated residually finite nonamenable discrete group Γ. We fix a descending sequence of normal, finite-index subgroups Γ_k whose intersection is the neutral element. For many classical groups, like $\mathrm{SL}(n, \mathbb{Z})$, these subgroups and their quotients are easily described explicitly.

For each k we let

$$\pi_k \colon C^*(\Gamma) \to \mathbb{B}(\ell^2(\Gamma/\Gamma_k))$$

be the representation induced by the left regular representation of Γ/Γ_k; then we take their direct sum

$$\pi = \bigoplus_k \pi_k \colon C^*(\Gamma) \to \mathbb{B}(\bigoplus_k \ell^2(\Gamma/\Gamma_k)).$$

Theorem 16.4.3. *The algebra $\pi(C^*(\Gamma))$ is (obviously) RFD, but it is not exact.*[2] *Hence sufficiently large matrices over $\pi(C^*(\Gamma))$ yield a singly generated nonexact QD* C**-algebra, thus concrete counterexamples to Herrero's original approximation problem.*

Proof. For notational convenience let $A = \pi(C^*(\Gamma))$. Note that A has an amenable trace τ (Theorem 6.2.7) whose GNS representation is unitarily equivalent to the left regular representation of Γ. (Take any cluster point of the traces $A \to \mathbb{B}(\ell^2(\Gamma/\Gamma_k)) \xrightarrow{\mathrm{tr}} \mathbb{C}$.) Hence we have a continuous product map

$$A \otimes A^{\mathrm{op}} \to \mathbb{B}(\ell^2(\Gamma))$$

taking A^{op} to the right regular representation of Γ.

---

[2]Voiculescu used precisely this construction but assumed that Γ has property (T). That $\pi(C^*(\Gamma))$ is not exact in this more general context was pointed out to us by Marius Dadarlat.

The remainder of the proof is virtually identical to the proof of Proposition 3.7.11. Indeed, if A is exact, then so is A^{op} and hence exactness of the sequence

$$0 \to J \otimes A^{\mathrm{op}} \to A \otimes A^{\mathrm{op}} \to C^*_\lambda(\Gamma) \otimes A^{\mathrm{op}} \to 0$$

would imply that the product map

$$C^*_\lambda(\Gamma) \otimes A^{\mathrm{op}} \to \mathbb{B}(\ell^2(\Gamma))$$

induced by the left and right regular representations is also min-continuous. However, this implies Γ is amenable (Proposition 6.4.1), which is a contradiction. □

16.5. References

This chapter evolved out of [**27**].

Chapter 17

Counterexamples in K-Homology and K-Theory

In our final chapter, we present an application of finite-dimensional approximation properties to a natural problem in analytic K-homology. Shortly after the pioneering work of Brown, Douglas and Fillmore, Anderson constructed an example of a C*-algebra A for which the Ext semigroup is not a group ([**7**]). Since then a number of other counterexamples have been given.[1] Here we will present Simon Wassermann's examples [**195**], using residually finite property (T) groups.

In Section 17.2 we prove a structure theorem for $C^*(\Gamma)$, where Γ is a property (T) group. Using this result, it is easy to show that some natural extensions (actually, the same ones used by Voiculescu in the last chapter) are not invertible, and thus Ext need not be a group.

17.1. BDF preliminaries

The goal of BDF (Brown-Douglas-Fillmore) theory is to classify extensions of a fixed algebra A by the compact operators $\mathbb{K}$. The original motivation was a natural problem in single operator theory – the classification of essentially normal operators – and the solution required importing topological ideas into the C*-world. This section contains the necessary definitions and a few

[1]In fact, Sections 13.4 and 13.5 describe counterexamples of Kirchberg and of Haagerup and Thorbjørnsen, respectively.

fundamental facts, but it is really a bare minimum treatment of the subject (cf. [**86**]). We will only deal with *unital* C*-algebras and *unital* maps.

If A is a C*-algebra, then we define an *essential extension of* $\mathbb{K}$ *by* A to be a short exact sequence

$$0 \to \mathbb{K} \xrightarrow{\iota} \mathcal{E} \xrightarrow{\pi} A \to 0,$$

where $\mathcal{E}$ is a unital C*-algebra, both ι and π are $*$-homomorphisms and $\iota(\mathbb{K}) \lhd \mathcal{E}$ is an essential ideal (Definition 8.4.1). It is often more convenient to work with the *Busby invariant* of an extension which, by definition, is a $*$-homomorphism[2] into the Calkin algebra. It is clear that any injective $*$-homomorphism into the Calkin algebra gives rise to an essential extension – identify A with its image and let $\mathcal{E}$ be the pullback in $\mathbb{B}(\mathcal{H})$ – but the converse is also true. Indeed, suppose that

$$0 \to \mathbb{K} \xrightarrow{\iota} \mathcal{E} \xrightarrow{\pi} A \to 0$$

is an essential extension and $\mathbb{K}$ is given as the compact operators on some concrete Hilbert space $\mathcal{H}$. The map $\iota^{-1}\colon \iota(\mathbb{K}) \to \mathbb{K} \subset \mathbb{B}(\mathcal{H})$ extends to a $*$-homomorphism $\mathcal{E} \to \mathbb{B}(\mathcal{H})$ and this map is faithful, since we assumed that $\iota(\mathbb{K})$ is essential (if there were a kernel, it would be orthogonal to $\iota(\mathbb{K})$). Hence we get an induced $*$-homomorphism $\sigma\colon A \to Q(\mathcal{H})$ by composing the representation $\mathcal{E} \to \mathbb{B}(\mathcal{H})$ with the quotient map $\mathbb{B}(\mathcal{H}) \to Q(\mathcal{H})$. Note, however, that this procedure does not produce a unique map into the Calkin algebra, since there are infinitely many ways to identify an abstract copy of $\mathbb{K}$ with the (concrete) compact operators on a Hilbert space $\mathcal{H}$. Hence, at this point, it is an abuse of terminology to refer to "the" Busby invariant associated to an extension. Luckily, we can mod out by a natural equivalence relation and eliminate this ambiguity.

Definition 17.1.1. Let $U\colon \mathcal{H} \to \mathcal{K}$ be a unitary operator. We will let

$$\mathrm{Ad}_U\colon Q(\mathcal{H}) \to Q(\mathcal{K})$$

denote the isomorphism induced by the isomorphism $\mathbb{B}(\mathcal{H}) \to \mathbb{B}(\mathcal{K})$, $T \mapsto UTU^*$.

Definition 17.1.2. Two unital injective $*$-homomorphisms $\varphi\colon A \to Q(\mathcal{H})$ and $\psi\colon A \to Q(\mathcal{K})$ are called *equivalent* if there exists a unitary operator $U\colon \mathcal{H} \to \mathcal{K}$ such that

$$\mathrm{Ad}_U \circ \varphi = \psi.$$

This gives the appropriate equivalence relation on maps to the Calkin algebra and *we use the term Busby invariant to refer to the equivalence class of a map.* Now we define an equivalence relation on the essential extensions.

[2]Actually, an equivalence class – see Definition 17.1.2.

Definition 17.1.3. Two essential extensions, $0 \to \mathbb{K} \xrightarrow{\iota_1} \mathcal{E}_1 \xrightarrow{\pi_1} A \to 0$ and $0 \to \mathbb{K} \xrightarrow{\iota_2} \mathcal{E}_2 \xrightarrow{\pi_2} A \to 0$, are *equivalent* if there exists a $*$-isomorphism $\Phi\colon \mathcal{E}_1 \to \mathcal{E}_2$ such that $\Phi(\iota_1(\mathbb{K})) = \iota_2(\mathbb{K})$ and the induced isomorphism of quotients is the identity on A – i.e., there exists a commutative diagram

$$\begin{array}{ccccccccc} 0 & \longrightarrow & \mathbb{K} & \longrightarrow & \mathcal{E}_1 & \longrightarrow & A & \longrightarrow & 0 \\ & & \downarrow{\cong} & & \downarrow{\cong} & & \downarrow{\mathrm{id}_A} & & \\ 0 & \longrightarrow & \mathbb{K} & \longrightarrow & \mathcal{E}_2 & \longrightarrow & A & \longrightarrow & 0. \end{array}$$

If we identify equivalent extensions, then *there is a one-to-one correspondence between Busby invariants and essential extensions.* This is not immediate, though not too difficult, and we give some hints in Exercises 17.1.1 and 17.1.2. For our purposes, the Busby invariant will be more useful, so the reader can safely concentrate on maps to the Calkin algebra.

Having dispensed with the technicalities, here's the BDF extension semigroup.

Definition 17.1.4. Let A be a unital separable C*-algebra. Then $\mathrm{Ext}(A)$ denotes the set of (equivalence classes of) essential extensions of A by $\mathbb{K}$ – i.e., $\mathrm{Ext}(A)$ is just the set of Busby invariants. If $\varphi\colon A \to Q(\mathcal{H})$ is a unital $*$-monomorphism, then $[\varphi] \in \mathrm{Ext}(A)$ will denote its equivalence class.

As mentioned above, $\mathrm{Ext}(A)$ is always a semigroup: If $\varphi\colon A \to Q(\mathcal{H})$ and $\psi\colon A \to Q(\mathcal{K})$ are unital $*$-monomorphisms, there is a natural diagonal embedding of $Q(\mathcal{H}) \oplus Q(\mathcal{K})$ into $Q(\mathcal{H} \oplus \mathcal{K})$ and hence we get a unital $*$-monomorphism $\varphi \oplus \psi\colon A \to Q(\mathcal{H} \oplus \mathcal{K})$ via the embedding $Q(\mathcal{H}) \oplus Q(\mathcal{K}) \hookrightarrow Q(\mathcal{H} \oplus \mathcal{K})$; this allows us to define addition in $\mathrm{Ext}(A)$ as

$$[\varphi] + [\psi] = [\varphi \oplus \psi]$$

(you'll have the opportunity in Exercise 17.1.3 to check that this is well-defined and turns $\mathrm{Ext}(A)$ into an abelian semigroup).

Though it is far from obvious, it turns out that $\mathrm{Ext}(A)$ always has a neutral element, given by a trivial extension.

Definition 17.1.5. A unital $*$-monomorphism $\varphi\colon A \to Q(\mathcal{H})$ is called *trivial* if it has a unital $*$-homomorphic lifting – i.e., there exists a unital representation $\sigma\colon A \to \mathbb{B}(\mathcal{H})$ such that $\pi_{\mathcal{H}} \circ \sigma = \varphi$, where $\pi_{\mathcal{H}}\colon \mathbb{B}(\mathcal{H}) \to Q(\mathcal{H})$ is the quotient map.

Note that, since φ is injective, $\sigma\colon A \to \mathbb{B}(\mathcal{H})$ is necessarily faithful and essential. Hence Voiculescu's Theorem (Theorem 1.7.3) implies that *all trivial extensions give rise to the same element in* $\mathrm{Ext}(A)$. In fact, more is true.

Theorem 17.1.6 (Voiculescu). *If* $0 \in \mathrm{Ext}(A)$ *denotes the class of a trivial extension, then* 0 *is an additive identity for* $\mathrm{Ext}(A)$.

Proof. Let $\psi\colon A \to Q(\mathcal{H})$ be a unital $*$-monomorphism and $\sigma\colon A \to \mathbb{B}(\mathcal{K})$ be a unital faithful essential representation. Let $\mathcal{E} \subset \mathbb{B}(\mathcal{H})$ be the pullback of $\psi(A) \subset Q(\mathcal{H})$ and consider the representation

$$\sigma \circ \psi^{-1} \circ \pi_{\mathcal{H}}\colon \mathcal{E} \to \mathbb{B}(\mathcal{K}).$$

It suffices to show that the natural inclusion $\iota\colon \mathcal{E} \hookrightarrow \mathbb{B}(\mathcal{H})$ is approximately unitarily equivalent relative to the compacts to the map

$$\iota \oplus (\sigma \circ \psi^{-1} \circ \pi_{\mathcal{H}})\colon \mathcal{E} \to \mathbb{B}(\mathcal{H} \oplus \mathcal{K}).$$

This, however, is immediate from Theorem 1.7.3. □

It turns out that it is easier to characterize the invertible elements of Ext(A) than it is to decide whether or not every element is invertible, so let's tackle the easy question first.

Definition 17.1.7. A unital $*$-monomorphism $\varphi\colon A \to Q(\mathcal{H})$ is called *liftable* if there exists a u.c.p. map $\sigma\colon A \to \mathbb{B}(\mathcal{H})$ such that $\pi_{\mathcal{H}} \circ \sigma = \varphi$.

Proposition 17.1.8 (Invertible elements in Ext(A)). *$[\varphi] \in$ Ext(A) is invertible if and only if $\varphi\colon A \to Q(\mathcal{H})$ is liftable.*

Proof. We leave it to the reader to check that one representative of a Busby invariant is liftable if and only if all other representatives have this property.

First suppose that $[\varphi] \in$ Ext(A) is invertible. Then there exists an element $[\psi] \in$ Ext(A) such that $[\varphi \oplus \psi] = 0 \in$ Ext(A). In other words, there exists a unital faithful essential representation $\eta\colon A \to \mathbb{B}(\mathcal{H} \oplus \mathcal{K})$ such that

$$\pi_{\mathcal{H}\oplus\mathcal{K}} \circ \eta(a) = \varphi(a) \oplus \psi(a) \in Q(\mathcal{H} \oplus \mathcal{K}).$$

Let $P_{\mathcal{H}} \in \mathbb{B}(\mathcal{H} \oplus \mathcal{K})$ be the projection onto $\mathcal{H} \oplus 0$. One checks that our desired u.c.p. map $\sigma\colon A \to \mathbb{B}(\mathcal{H})$ is given by $\sigma(a) = P\eta(a)P$.

For the converse, assume that $\varphi\colon A \to Q(\mathcal{H})$ is liftable and let $\sigma\colon A \to \mathbb{B}(\mathcal{H})$ be a u.c.p. lifting. Let $\eta\colon A \to \mathbb{B}(\mathcal{K})$ be the Stinespring dilation of σ and $P \in \mathbb{B}(\mathcal{K})$ be the Stinespring projection (so we identify σ with $a \mapsto P\eta(a)P$). Define a u.c.p. map $\gamma\colon A \to \mathbb{B}(P^{\perp}\mathcal{K})$ by $\gamma(a) = P^{\perp}\eta(a)P^{\perp}$.

The key observation is *that γ is a $*$-homomorphism modulo the compacts* (i.e., $\gamma(ab) - \gamma(a)\gamma(b)$ is compact for all $a, b \in A$). To see this, observe that the $*$-homomorphism φ can be identified with $a \mapsto \pi_{\mathcal{K}}(P)\pi_{\mathcal{K}}(\eta(a))\pi_{\mathcal{K}}(P) \in Q(\mathcal{K})$. Thus, the calculation used to identify multiplicative domains of u.c.p. maps shows that $\pi_{\mathcal{K}}(P)$ commutes with $\pi_{\mathcal{K}}(\eta(A))$. Hence $[P, \eta(a)] \in \mathbb{K}(\mathcal{K})$, for all $a \in A$, and this implies γ is a $*$-homomorphism modulo the compacts.

If γ happens to be a *faithful* $*$-homomorphism modulo the compacts, then we are done, and if not we simply add on a faithful essential representation to complete the proof. □

In [**11**] Arveson proved the previous result and simplified the proof of the Choi-Effros Lifting Theorem (Theorem C.3). This latter result immediately implies

Theorem 17.1.9. *The semigroup* Ext(A) *is a group for every nuclear* C*-*algebra* A.

Exercises

Exercise 17.1.1. Prove that every isomorphism $\Phi\colon \mathbb{K}(\mathcal{H}) \to \mathbb{K}(\mathcal{K})$ is inner – i.e., there exists a unitary $U\colon \mathcal{H} \to \mathcal{K}$ such that $\Phi(T) = UTU^*$ for all $T \in \mathbb{K}(\mathcal{H})$. (Hint: Φ must map rank-one projections to rank-one projections and thus we can define U via the correspondence between orthonormal bases and orthogonal rank-one projections.)

Exercise 17.1.2. Let $\mathbb{K}(\mathcal{H}) \subset \mathcal{E}_1 \subset \mathbb{B}(\mathcal{H})$ and $\mathbb{K}(\mathcal{K}) \subset \mathcal{E}_2 \subset \mathbb{B}(\mathcal{K})$ be given and assume there is an isomorphism $\Phi\colon \mathcal{E}_1 \to \mathcal{E}_2$ such that $\Phi(\mathbb{K}(\mathcal{H})) = \mathbb{K}(\mathcal{K})$. Let $U\colon \mathcal{H} \to \mathcal{K}$ be the unitary from the previous exercise and show that $\Phi(T) = UTU^*$ for all $T \in \mathcal{E}_1$. Use this fact to show that there is a one-to-one correspondence between Busby invariants and equivalence classes of essential extensions. (Hint: If $\{p_n\} \in \mathbb{K}(\mathcal{H})$ is an approximate unit, then $\{\Phi(p_n)\}$ must be an approximate unit for $\mathbb{K}(\mathcal{K})$ and hence $\Phi(T)$ is the strong limit of $\Phi(p_n)\Phi(T) = (Up_nU^*)(UTU^*)$, for all $T \in \mathcal{E}_1$.)

Exercise 17.1.3. Prove that addition in Ext(A) is well-defined and abelian.

17.2. Property (T) and Kazhdan projections

This section contains an important structure theorem for the universal C*-algebras of discrete groups with property (T) (Definition 6.4.4). Essentially the result states that every finite-dimensional unitary representation of Γ arises from compression by a projection in $C^*(\Gamma)$. In particular, $C^*(\Gamma)$ has at least one nontrivial projection, coming from the trivial representation. (Compare with $C^*(\mathbb{F}_n)$, which contains no nontrivial projections, [**53**, Theorem VII.6.6]).

We'll need a well-known observation of Schur.

Lemma 17.2.1 (Schur). *Let* $\pi\colon A \to \mathbb{B}(\mathcal{H})$ *and* $\sigma\colon A \to \mathbb{B}(\mathcal{K})$ *be two* $*$-*representations which have a nontrivial intertwiner – i.e., assume there exists a bounded linear operator* $T\colon \mathcal{H} \to \mathcal{K}$ *such that* $T\pi(a) = \sigma(a)T$ *for all* $a \in A$. *If* π *is irreducible, then* π *is unitarily equivalent to a subrepresentation of* σ.[3]

---

[3]Recall that this means there is a subspace $\mathcal{L} \subset \mathcal{K}$ which is invariant for $\sigma(A)$ and such that π is unitarily equivalent to the restriction $\sigma(A)|_{\mathcal{L}}$.

Proof. Taking adjoints and replacing a with a^*, our intertwining assumption also implies

$$\pi(a)T^* = T^*\sigma(a).$$

Now observe that T^*T commutes with $\pi(A)$:

$$T^*T\pi(a) = T^*(\sigma(a)T) = (T^*\sigma(a))T = \pi(a)T^*T.$$

Since π is irreducible, the commutant of $\pi(A)$ consists of scalar multiples of the identity and hence $T^*T = \lambda 1_{\mathcal{H}}$ for some $\lambda \geq 0$. Replacing T by $\frac{1}{\sqrt{\lambda}}T$, we may assume that T is an isometry. Since T is a unitary operator onto its range, we will be done once we know that the range of T is an invariant subspace for $\sigma(A)$ – i.e., TT^* commutes with $\sigma(A)$. But this is trivial. □

One often sees property (T) defined as follows: *If a unitary representation weakly contains the trivial representation, then it has nonzero fixed vectors (i.e., actually contains the trivial representation).* This one-dimensional property implies a stronger version of itself.

Proposition 17.2.2. *Assume Γ has property* (T), $\pi: C^*(\Gamma) \to \mathbb{M}_n(\mathbb{C})$ *is an irreducible representation and* $\sigma: C^*(\Gamma) \to \mathbb{B}(\mathcal{K})$ *is any representation such that there exist isometries* $V_k: \ell^2_n \to \mathcal{K}$ *with* $\|\sigma(g)V_k - V_k\pi(g)\| \to 0$ *for all* $g \in \Gamma$.[4] *Then π is unitarily equivalent to a subrepresentation of σ.*

Proof. Let $\mathcal{S}_2(\ell^2_n, \mathcal{K})$ denote the Hilbert space of Hilbert-Schmidt class operators from ℓ^2_n to $\mathcal{K}$.[5] We define a unitary representation ρ of Γ on $\mathcal{S}_2(\ell^2_n, \mathcal{K})$ by

$$\rho(g): T \mapsto \sigma(g)T\pi(g)^*.$$

By Schur's Lemma, it suffices to show that this representation has a nonzero fixed vector (since this vector will be an intertwiner). Since Γ has property (T), we only need to observe the existence of a sequence of asymptotically invariant unit vectors. A routine calculation shows that the vectors $\frac{1}{\sqrt{n}}V_k \in \mathcal{S}_2(\ell^2_n, \mathcal{K})$ have this property. □

Though the result above is completely elementary, it has a striking consequence: The central projection in $C^*(\Gamma)^{**}$ corresponding to a finite-dimensional irreducible representation actually lives in $C^*(\Gamma)$. In this context, finite-dimensional central covers are often called something else.

[4] This is equivalent to requiring $\|V_k^*\sigma(g)V_k - \pi(g)\| \to 0$ for all $g \in \Gamma$.

[5] Recall that if $S, T \in \mathbb{B}(\ell^2_n, \mathcal{K})$, then their inner product is defined by

$$\langle S, T\rangle_{\mathcal{S}_2} = \sum_{i=1}^{n} \langle Se_i, Te_i\rangle,$$

where $\{e_i\}$ is your favorite orthonormal basis of ℓ^2_n.

Definition 17.2.3 (Kazhdan projections). If Γ has property (T) and the representation $\pi\colon C^*(\Gamma) \to \mathbb{M}_n(\mathbb{C})$ is irreducible, then the central cover $c(\pi)$ is also known as the *Kazhdan projection* associated with π.

Theorem 17.2.4 (Structure theorem for property (T) groups). *Let Γ be a discrete group with property* (T)*. For each finite-dimensional irreducible representation $\pi\colon C^*(\Gamma) \to \mathbb{M}_n(\mathbb{C})$, the Kazhdan projection $c(\pi)$ is an element of $C^*(\Gamma)$.*

Proof. Let $\sigma\colon C^*(\Gamma) \to \mathbb{B}(\mathcal{H})$ be an essential representation such that (1) $\pi \oplus \sigma\colon C^*(\Gamma) \to \mathbb{B}(\ell^2_n \oplus \mathcal{H})$ is faithful and (2) σ contains no subrepresentation which is unitarily equivalent to π.[6]

It suffices to show that σ can't be faithful. Indeed, if σ has a kernel $0 \neq J \lhd C^*(\Gamma)$, then $\pi|_J$ must be faithful (since $\pi \oplus \sigma$ is faithful and $\sigma|_J = 0$). Hence J is finite-dimensional and, as such, has a unit p_π which is a central projection in $C^*(\Gamma)$. Note that $\pi(p_\pi) = 1$, since π is irreducible, and thus we may identify π with the $*$-homomorphism $A \to p_\pi A$, $a \mapsto p_\pi a$ (i.e., π is an isomorphism on $p_\pi A$). Since p_π is also a central projection in $C^*(\Gamma)^{**}$, the representation $C^*(\Gamma)^{**} \to p_\pi C^*(\Gamma)^{**}$ is quasi-equivalent to π and hence has the same central cover as π – i.e., $p_\pi = c(\pi)$.

So, why isn't σ faithful? Well, if it were, then Voiculescu's Theorem would imply that σ is approximately unitarily equivalent to $\sigma \oplus \pi$. In other words, σ weakly contains π and thus, by Proposition 17.2.2, it actually contains π. This contradicts our assumption (2) above and the proof is complete. □

Here are two nice applications.

Corollary 17.2.5. *If Γ has property* (T)*, then it has at most countably many nonequivalent finite-dimensional representations.*

Proof. A separable C*-algebra (e.g., $C^*(\Gamma)$) can have at most countably many orthogonal projections (since it can be represented on a separable Hilbert space). □

Corollary 17.2.6. *Assume Γ has property* (T) *and let $J_f \lhd C^*(\Gamma)$ be the ideal generated by all of Γ's Kazhdan projections. Then*

$$C^*(\Gamma)/J_f$$

has no amenable traces.

---

[6]There are many ways to do this. Perhaps the simplest is to take a direct sum of all the irreducible representations except π and then inflate to force the essential part. However, one wants to stay on a separable Hilbert space so one should restrict to a countable set of such irreducible representations while preserving condition (1).

Proof. If $C^*(\Gamma)/J_f$ had an amenable trace τ, then it would also have a finite-dimensional quotient (Remark 6.4.11). Hence Γ would have a finite-dimensional irreducible representation which factored through $C^*(\Gamma)/J_f$ and this is impossible. □

17.3. Ext need not be a group

With the structure theorem for $C^*(\Gamma)$ in hand, there is little more to do. We start with an infinite residually finite discrete group Γ which has property (T) (say $\mathrm{SL}(3,\mathbb{Z})$). Then Γ necessarily has an infinite (but countable) number of nonequivalent finite-dimensional irreducible representations which we denote by $\pi_k\colon C^*(\Gamma) \to \mathbb{B}(\mathcal{H}_k)$, $k \in \mathbb{N}$. Let

$$\pi = \bigoplus_{k\in\mathbb{N}} \pi_k\colon C^*(\Gamma) \to \mathbb{B}(\bigoplus_{k\in\mathbb{N}} \mathcal{H}_k).$$

Here is the key observation/exercise: If $J_f \lhd C^*(\Gamma)$ is the ideal generated by all of the Kazhdan projections, then

$$\pi(J_f) = \bigoplus_{k\in\mathbb{N}} \mathbb{B}(\mathcal{H}_k) \subset \mathbb{K}(\bigoplus_{k\in\mathbb{N}} \mathcal{H}_k) \cap \prod_{k\in\mathbb{N}} \mathbb{B}(\mathcal{H}_k) \subset \mathbb{B}(\bigoplus_{k\in\mathbb{N}} \mathcal{H}_k).$$

Hence, passing to the Calkin algebra, we have an induced $*$-homomorphism

$$\varphi\colon C^*(\Gamma)/J_f \to Q(\bigoplus_{k\in\mathbb{N}} \mathcal{H}_k).$$

Theorem 17.3.1. *Let $C = \varphi(C^*(\Gamma)/J_f) \subset Q(\bigoplus_{k\in\mathbb{N}} \mathcal{H}_k)$. Then* $\mathrm{Ext}(C)$ *is not a group.*

Proof. We will show that if $\mathrm{Ext}(C)$ were a group, then C would have an amenable trace, which contradicts Corollary 17.2.6 (since C is a quotient of $C^*(\Gamma)/J_f$).

If there existed a u.c.p. lifting to the natural inclusion $C \subset Q(\bigoplus_{k\in\mathbb{N}} \mathcal{H}_k)$, then we could compose with a conditional expectation to force it to take values in $\prod_{k\in\mathbb{N}} \mathbb{B}(\mathcal{H}_k) \subset \mathbb{B}(\bigoplus_{k\in\mathbb{N}} \mathcal{H}_k)$. Evidently this implies C has an amenable trace, so we are finished. □

For some time, experts wondered whether a quasidiagonal set of operators would always descend to a QD C*-algebra in the Calkin algebra. Wassermann's work also answers that question, negatively, since the pull-back of C is obviously a quasidiagonal set of operators.

Corollary 17.3.2. *Since C has no amenable trace, it is not QD.*

Remark 17.3.3. Since C is MF, by definition, it follows that inductive limits of RFD algebras need not be QD (Proposition 11.1.8).

17.4. Topology on Ext

Finally, we would like to show that noninvertible elements in Ext are often abundant. This will follow from topological considerations and a result of Dadarlat (whom we thank for some helpful conversations regarding his work cited below).

First, though, we define a quotient of Ext by relaxing the equivalence relation on Busby invariants. Namely, we say two unital $*$-monomorphisms $\varphi, \psi\colon A \to Q(\mathcal{H})$ are *weakly equivalent* if there exists a unitary $u \in Q(\mathcal{H})$ such that $u\varphi(a)u^* = \psi(a)$ for all $a \in A$. Defining $\mathrm{Ext}_w(A)$ to be weak equivalence classes of unital $*$-monomorphisms into the Calkin algebra, it is obvious that $\mathrm{Ext}_w(A)$ is a quotient of $\mathrm{Ext}(A)$ and one can check that invertibility in $\mathrm{Ext}_w(A)$ is still characterized by the existence of a u.c.p. splitting (as in Proposition 17.1.8).

There is a natural topology on Ext_w: the quotient of the pointwise-convergence topology on the set of unital $*$-monomorphisms from A into $Q(\mathcal{H})$. This topology is not Hausdorff, but it is pseudometrizable since A is separable. To be more precise, we fix a dense sequence $x_1, x_2, \ldots$ in the unit ball of A and define a pseudometric d on $\mathrm{Ext}_w(A)$ by

$$d([\varphi],[\psi]) = \inf_u \sum_{i=1}^{\infty} \frac{1}{2^i} \|u\varphi(x_i)u^* - \psi(x_i)\|,$$

where $u \in Q(\mathcal{H})$ is a unitary. The topology on $\mathrm{Ext}_w(A)$ defined by this pseudometric is called the (Larry) Brown-Salinas topology. It is independent of the choice of dense sequence, of course. We note that the group $\mathrm{Ext}_w(A)^{-1}$ of invertible elements is closed in this topology, by Lemma C.2. It is not very difficult to show that this topology makes $\mathrm{Ext}_w(A)$ a topological semigroup: Addition and inversion are continuous operations. A theorem of Dadarlat (which depends on KK-theory; hence we won't prove it – see [**52**]) states that $\mathrm{Ext}_w(A)^{-1}$ is separable. In this section, we'll show that the set of quasidiagonal extensions is nonseparable in many cases; hence noninvertible elements abound.

Definition 17.4.1. A unital $*$-monomorphism $\varphi\colon A \to Q(\mathcal{H})$ is said to be *quasidiagonal* if $\pi^{-1}(\varphi(A)) \subset \mathbb{B}(\mathcal{H})$ is a quasidiagonal set of operators. We say $[\varphi] \in \mathrm{Ext}_w(A)$ is *quasidiagonal* if φ is.[7]

Lemma 17.4.2. *If $\varphi\colon A \to Q(\mathcal{H})$ is quasidiagonal and liftable, then $[\varphi]$ belongs to the closure of the neutral element in* $\mathrm{Ext}_w(A)$.

---

[7] It isn't entirely obvious, but this is well-defined, as one should check (or see [**171**, Corollary 2.6]).

Proof. Let $\tilde{\varphi}\colon A \to \mathbb{B}(\mathcal{H})$ be a u.c.p. lifting and $\sigma\colon A \to \mathbb{B}(\mathcal{H})$ be a faithful unital essential $*$-representation. Since $\tilde{\varphi}(A)$ is a quasidiagonal set, there is a sequence of finite-rank projections P_n which converges to 1 strongly and commutes with $\tilde{\varphi}(A)$ asymptotically. Consider the u.c.p. maps $\tilde{\varphi}_n\colon A \to \mathbb{B}(P_n^{\perp}\mathcal{H})$ obtained by compressing $\tilde{\varphi}$ to $P_n^{\perp}\mathcal{H}$. Clearly all the $\tilde{\varphi}_n$'s still represent $[\varphi]$ (after passing to the Calkin algebra) and it is easy to see that the sequence $(\tilde{\varphi}_n)$ is asymptotically multiplicative. Hence by Theorem 1.7.6, there are unitary operators $U_n\colon P_n^{\perp}\mathcal{H} \to \mathcal{K}$ such that

$$\lim_{n\to\infty} \| \operatorname{Ad}_{U_n} \circ \tilde{\varphi}_n(a) - \sigma(a)\| = 0$$

for all $a \in A$. This means that $d([\varphi], [\sigma]) = 0$. □

If you actually checked that our notion of quasidiagonality for elements of Ext_w is well-defined, you may also want to prove the converse of the previous lemma for QD C*-algebras ([**171**, Theorem 2.9]).

Let $(u_1(n), \ldots, u_k(n)) \in \mathbb{M}_{N(n)}(\mathbb{C})^k$ be a coding family of unitary k-tuples (Definition 13.5.3), which means that

$$\sup\{\|\sum_{i=1}^{k} u_i(m) \otimes \overline{u_i(n)}\|_{\mathbb{M}_{N(m)}(\mathbb{C})\otimes\overline{\mathbb{M}_{N(n)}(\mathbb{C})}} : m \neq n\} = k - \delta$$

for some $\delta > 0$. Let

$$v_i = (u_i(n))_{n=1}^{\infty} + \mathbb{K}(\bigoplus_{n=1}^{\infty} \ell^2_{N(n)}) \in Q(\bigoplus_{n=1}^{\infty} \ell^2_{N(n)})$$

and define A to be the C*-algebra generated by $\{v_1, \ldots, v_k\}$ in $Q(\bigoplus \ell^2_{N(n)})$.

Proposition 17.4.3. *Let A be as above. Then, the set of quasidiagonal elements in $\mathrm{Ext}_w(A)$ is nonseparable.*

Proof. Passing to a subsequence, we may assume that the sequence

$$\|f(u_1(n), \ldots, u_k(n))\|$$

is convergent for every $f \in \mathbb{C}[\mathbb{F}_k]$. For an infinite subset $\alpha \subset \mathbb{N}$, we identify $\prod_{n\in\alpha} \mathbb{M}_{N(n)}(\mathbb{C})$ with the corresponding block-diagonal subalgebra of $\mathbb{B}(\mathcal{H}_\alpha)$, where $\mathcal{H}_\alpha = \bigoplus_{n\in\alpha} \ell^2_{N(n)}$. Let

$$u_i^\alpha = (u_i(n))_{n\in\alpha} \in \mathbb{B}(\mathcal{H}_\alpha) \text{ and } v_i^\alpha = u_i^\alpha + \mathbb{K}(\mathcal{H}_\alpha) \in Q(\mathcal{H}_\alpha).$$

Observe that $\|f(v_1^\alpha, \ldots, v_k^\alpha)\| = \lim_n \|f(u_1(n), \ldots, u_k(n))\|$ for every $f \in \mathbb{C}[\mathbb{F}_k]$, and hence $\varphi_\alpha(v_i) = v_i^\alpha$ defines a $*$-monomorphism φ_α from A into $Q(\mathcal{H}_\alpha)$.

We redefine the pseudometric d on $\mathrm{Ext}_w(A)$ by

$$d([\varphi],[\psi]) = \inf_u \sum_{i=1}^{k} \|u\varphi(v_i)u^* - \psi(v_i)\|.$$

(Note that this metric is equivalent to the previous one, since A is generated by the v_i's.) As is well known (and a fun exercise), there is an uncountable family Ω of infinite subsets of $\mathbb{N}$ such that $|\alpha \cap \beta| < \infty$ for every $\alpha, \beta \in \Omega$ with $\alpha \neq \beta$. Thus, it suffices to show $|\alpha \cap \beta| < \infty$ implies $d([\varphi_\alpha],[\varphi_\beta]) \geq \delta$.

Suppose by contradiction that $|\alpha \cap \beta| < \infty$ and $d([\varphi_\alpha],[\varphi_\beta]) < \delta$. Then, there exists a contraction $U\colon \mathcal{H}_\alpha \to \mathcal{H}_\beta$ such that

$$\sum_{i=1}^{k} \|(\mathrm{Ad}_U(u_i^\alpha) - u_i^\beta) + \mathbb{K}(\mathcal{H}_\beta)\|_{Q(\mathcal{H}_\beta)} < \delta.$$

Thus one has

$$\begin{aligned} k &= \left\| \left(\sum_{i=1}^{k} u_i^\beta \otimes \overline{u_i^\beta} \right) + \mathbb{K}(\mathcal{H}_\beta) \otimes \overline{\mathbb{B}(\mathcal{H}_\beta)} \right\| \\ &< \delta + \left\| \left(\sum_{i=1}^{k} u_i^\alpha \otimes \overline{u_i^\beta} \right) + \mathbb{K}(\mathcal{H}_\alpha) \otimes \overline{\mathbb{B}(\mathcal{H}_\beta)} \right\| \\ &= \delta + \limsup_{\substack{m\in\alpha \\ m\to\infty}} \sup_{n\in\beta} \left\| \sum_{i=1}^{k} u_i(m) \otimes \overline{u_i(n)} \right\| \\ &\leq k, \end{aligned}$$

by the coding property, which is absurd. □

17.5. References

As mentioned already, the main result of this chapter is due to Wassermann, building on Voiculescu's counterexample to Herrero's problem. Our approach, however, follows [**30**]. For more on Kirchberg's and Haagerup and Thorbjørnsen's counterexamples see [**102**] and [**81**], respectively.

Part 4

Appendices

Appendix A

Ultrafilters and Ultraproducts

Since we need them throughout this book, here's a brief account of ultrafilters and ultraproducts.

Ultrafilters. An ultrafilter on a set labels every subset as "big" or "small" in such a way that a finite intersection of big subsets is big and a finite union of small sets is small.

Definition A.1. Let I be a set. A *filter* on the set I is a nonempty family $\mathcal{U}$ of subsets of I with the following properties:

(1) (nontriviality) $\emptyset \notin \mathcal{U}$;

(2) (finite intersection property) if $I_0, I_1 \in \mathcal{U}$, then there exists $J \in \mathcal{U}$ such that $J \subset I_0 \cap I_1$;

(3) (directedness) if $I_0 \in \mathcal{U}$ and $I_0 \subset I_1 \subset I$, then $I_1 \in \mathcal{U}$.

A nonempty family $\mathcal{U}'$ of subsets of I with properties (1) and (2) is called a *filter base*. Such a family can always be enlarged to a filter by including all subsets which contain a member of $\mathcal{U}'$. A filter $\mathcal{U}$ is called an *ultrafilter* if it satisfies

(4) (maximality) for any subset $I_0 \subset I$, either $I_0 \in \mathcal{U}$ or $I \setminus I_0 \in \mathcal{U}$.

It is easy to check that an ultrafilter $\mathcal{U}$ has the property

(5) if $\bigcup_{k=1}^{n} I_k \in \mathcal{U}$, then at least one of I_k's is in $\mathcal{U}$.

Example A.2. Let I be a set. The *principal ultrafilter* generated by $i_0 \in I$ is the family of all subsets which contain i_0.

Example A.3. Let I be an infinite set. The *Fréchet filter* on I is the family of all subsets whose complements are finite.

Example A.4. Let I be a directed set. The *cofinal filter base* on I is the family of all subsets of the forms $\{i \in I : i \geq i_0\}$ for some $i_0 \in I$.

Theorem A.5. *Let I be a set and $\mathcal{U}'$ be a filter base on I. Then, there exists an ultrafilter $\mathcal{U}$ which contains $\mathcal{U}'$.*

Proof. Let $\mathcal{U}'$ be a filter base on I. By Zorn's Lemma, there exists a maximal filter base $\mathcal{U}$ on I which contains $\mathcal{U}'$. By maximality, $\mathcal{U}$ is a filter. Now, let $I_0 \subset I$ be given. We claim that if $I_0 \notin \mathcal{U}$, then $I_0 \cap J = \emptyset$ for some $J \in \mathcal{U}$. Otherwise, the family $\mathcal{U} \cup \{I_0 \cap J : J \in \mathcal{U}\}$ would be a filter base and would contradict the maximality of $\mathcal{U}$. Since $J \subset I \setminus I_0$ and $J \in \mathcal{U}$, this means that $I_0 \notin \mathcal{U}$ implies $I \setminus I_0 \in \mathcal{U}$. □

Definition A.6. A *cofinal ultrafilter* $\mathcal{U}$ on a directed set I is an ultrafilter which contains the cofinal filter base. It is also called a *free* or *nonprincipal ultrafilter* when $I = \mathbb{N}$, the directed set of natural numbers.

The notion of filter comes from topology. It can be used to axiomatize topological spaces, but it lost out to the open set formalism. Still it has some advantages since it is easier to incorporate the Axiom of Choice (ultrafilter) into it.

Definition A.7. Let X be a topological space. A net $(x_i)_{i \in I}$ in X is said to *converge along* the filter $\mathcal{U}$ on I if there exists $x \in X$ such that for any open neighborhood O of x, the set $\{i \in I : x_i \in O\}$ belongs to $\mathcal{U}$. The limit point x is denoted by $\lim_{i \to \mathcal{U}} x_i$. Observe that the limit point x is unique provided that the topology is Hausdorff.

If $\mathcal{U}$ is a principal ultrafilter generated by i_0, then $\lim_{i \to \mathcal{U}} x_i = x_{i_0}$. If I is a directed set and $\mathcal{U}$ is a cofinal ultrafilter, then the limit point (if it exists) is a cluster point in the ordinary sense of topology. The following is the first application of ultrafilters. We leave the proof as an exercise.

Theorem A.8. *Let X be a Hausdorff topological space. The following are equivalent:*

(1) *X is compact;*

(2) *any net in X converges along any ultrafilter on the index set;*

(3) *any net in X converges along some cofinal ultrafilter on the index set.*

Let I be a set and $\mathcal{U}$ be an ultrafilter on I. Since bounded subsets of $\mathbb{C}$ are pre-compact, the above theorem allows us to define a map $\varphi \colon \ell^\infty(I) \to \mathbb{C}$

by

$$\varphi(f) = \lim_{i \to \mathcal{U}} f(i).$$

It is not too hard to check that φ is a character (i.e., a $*$-homomorphism) on $\ell^\infty(I)$. Conversely, if φ is a nonzero character on $\ell^\infty(I)$, then the family of subsets J such that $\varphi(\chi_J) = 1$ is an ultrafilter on I, which gives the character φ, whence there is a natural bijection between the set of ultrafilters on I and the set of characters on $\ell^\infty(I)$ (which, in turn, is isomorphic to the Stone-Čech compactification βI of I).

Now we assume that I is a directed set and $\mathcal{U}$ is a cofinal ultrafilter. Let X, Y be Banach spaces and $(T_i)_{\in I}$ be a bounded net in $\mathbb{B}(X, Y^*)$. Since bounded subsets of Y^* are pre-compact in the weak*-topology, we can define a map $T_{\mathcal{U}}\colon X \to Y^*$ by

$$T_{\mathcal{U}}(x) = \text{weak*-} \lim_{i \to \mathcal{U}} T_i(x).$$

This turns out to be a bounded linear map in the set of point-weak* cluster points of the net $(T_i)_{i \in I}$.

Ultraproducts. Let $\mathcal{U}$ be an ultrafilter on a set I. Let $(X_i)_{i \in I}$ be a net of Banach spaces. We denote by $\prod X_i$ the ℓ^∞-direct sum of the X_i's and let $N_{\mathcal{U}}$ be the Banach subspace of $\mathcal{U}$-null nets:

$$N_{\mathcal{U}} = \{(x_i)_{i \in I} \in \prod_{i \in I} X_i : \lim_{i \to \mathcal{U}} \|x_i\| = 0\}.$$

The *ultraproduct Banach space* of $(X_i)_{i \in I}$ is defined as $X_{\mathcal{U}} = (\prod X_i)/N_{\mathcal{U}}$. We write $x_{\mathcal{U}}$, or $(x_i)_{i \to \mathcal{U}}$, for the element represented by $(x_i)_{i \in I}$. It is a nice exercise to check that $\|x_{\mathcal{U}}\| = \lim_{\mathcal{U}} \|x_i\|$. If $A_i = X_i$ are all C*-algebras, then the ultraproduct $A_{\mathcal{U}}$ is again a C*-algebra. If $\mathcal{H}_i = X_i$ are all Hilbert spaces, then the ultraproduct $\mathcal{H}_{\mathcal{U}}$ is again a Hilbert space such that $\langle \eta_{\mathcal{U}}, \xi_{\mathcal{U}} \rangle = \lim_{\mathcal{U}} \langle \eta_i, \xi_i \rangle$. If $A_i \subset \mathbb{B}(\mathcal{H}_i)$, then $A_{\mathcal{U}} \subset \mathbb{B}(\mathcal{H}_{\mathcal{U}})$ with $a_{\mathcal{U}} \xi_{\mathcal{U}} = (a_i \xi_i)_{i \to \mathcal{U}}$.

Ultraproducts of von Neumann algebras are not quite as straightforward as the C*-case. To keep the notation consistent[1], let ω be a free ultrafilter on $\mathbb{N}$. Let (M, τ) be a von Neumann algebra with a faithful tracial state τ (or, more generally, a sequence of von Neumann algebras M_n with faithful tracial states τ_n). Let $N_\omega^{(2)}$ be the norm-closed ideal of $\prod M$, given by

$$N_\omega^{(2)} = \{(x_n)_{n \in \mathbb{N}} \in \prod_{n \in \mathbb{N}} M : \lim_{n \to \omega} \|x_n\|_2 = 0\},$$

where $\|x_n\|_2 = \tau(x_n^* x_n)^{1/2}$. The (tracial) ultraproduct of (M, τ) is defined to be $M^\omega = (\prod M)/N_\omega^{(2)}$; it has a faithful tracial state τ_ω given by $\tau_\omega(x_\omega) = \lim_\omega \tau(x_n)$. We note that $\|x_\omega\|_2 := \tau_\omega(x_\omega^* x_\omega)^{1/2} = \lim_{n \to \omega} \|x_n\|_2$.

[1] It is a lamentable habit, but operator algebraists tend to use the symbol ω for an ultrafilter on $\mathbb{N}$.

Lemma A.9. *The ultraproduct M^ω is a von Neumann algebra and the faithful tracial state τ_ω is normal.*

Proof. We prove that M^ω coincides with the von Neumann algebra generated by the GNS representation associated with τ_ω. It suffices to show that the closed unit ball Ω of M^ω is complete in the 2-norm – i.e., any Cauchy sequence $\{x_\omega^{(k)}\}_{k=1}^\infty$ in $(\Omega, \| \ \|_2)$ is convergent. We may assume that $\|x_\omega^{(k+1)} - x_\omega^{(k)}\|_2 < 2^{-k}$ for every k.

Claim. Let $k \in \mathbb{N}$ and suppose that $(x_n^{(k)})_n \in \prod M$ is a lifting of $x_\omega^{(k)}$ with $\|(x_n^{(k)})_n\| \leq 1$. Then, we can find a lifting $(x_n^{(k+1)})_n \in \prod M$ of $x_\omega^{(k+1)}$ such that $\|(x_n^{(k+1)})_n\| \leq 1$ and $\|x_n^{(k+1)} - x_n^{(k)}\|_2 < 2^{-k}$ for all n.

Indeed, let $(x_n^{(k+1)})_n \in \prod M$ be any lifting of $x_\omega^{(k+1)}$ with $\|(x_n^{(k+1)})_n\| \leq 1$. Then, we have $\{n \in \mathbb{N} : \|x_n^{(k+1)} - x_n^{(k)}\|_2 < 2^{-k}\} \in \omega$. Hence, we can replace $x_n^{(k+1)}$ with $x_n^{(k)}$ for all n outside of the above set, without affecting $x_\omega^{(k+1)}$.

Now, we inductively choose liftings $(x_n^{(k)})_n \in \prod M$ of $x_\omega^{(k)}$ as in the claim. Then, for $x_\omega = (x_n^{(n)})_{n\to\omega} \in M^\omega$, one immediately checks that

$$\|x_\omega - x_\omega^{(k)}\|_2 = \lim_{n\to\omega} \|x_n^{(n)} - x_n^{(k)}\|_2 \leq \sum_{n=k}^{\infty} 2^{-n} = 2^{-(k-1)}.$$

Hence the sequence $\{x_\omega^{(k)}\}_{k=1}^\infty$ converges to x_ω. □

Exercises

Exercise A.1. Prove Theorem A.8.

Exercise A.2. Using Theorem A.8, give a three line proof of Tychonoff's theorem: The product of compact spaces is compact.

Exercise A.3. Check that $\ell^\infty(I) \ni f \mapsto \lim_{i\to\mathcal{U}} f(i) \in \mathbb{C}$ is a character for every ultrafilter $\mathcal{U}$ on I.

Exercise A.4. Prove that every projection $e \in M^\omega$ lifts to a projection $(e_n)_n \in \prod M$ such that $\lim_{n\to\omega} \tau(e_n) = \tau_\omega(e)$.

Exercise A.5. Prove that if M is a factor (of type II_1), then so is M^ω. (Hint: Use the previous exercise.)

Operator Spaces, Completely Bounded Maps and Duality

Here is a crash course on the operator space theory needed in these notes.

Completely bounded maps and perturbation results.

Definition B.1. An operator space X is a closed subspace of a C*-algebra A. Note that for each $n \in \mathbb{N}$, $\mathbb{M}_n(X)$ inherits a norm from $\mathbb{M}_n(A)$. Let φ be a linear map from an operator space $X \subset A$ into an operator space $Y \subset B$. We say φ is *completely bounded* (abbreviated c.b.) if

$$\|\varphi\|_{\mathrm{cb}} := \sup_{n \in \mathbb{N}} \|\varphi_n \colon \mathbb{M}_n(X) \to \mathbb{M}_n(Y)\| < \infty,$$

where φ_n is defined by $\varphi_n([x_{i,j}]) = [\varphi(x_{i,j})]$ for $x = [x_{i,j}] \in \mathbb{M}_n(X)$. The number $\|\varphi\|_{\mathrm{cb}}$ is called the *cb norm* of φ. We say φ is *completely contractive* (c.c.) if $\|\varphi\|_{\mathrm{cb}} \leq 1$; φ is *completely isometric* (c.i.) if $\varphi_n \colon \mathbb{M}_n(X) \to \mathbb{M}_n(Y)$ is isometric for every n. We denote by $\mathrm{CB}(X, Y)$ the Banach space of all c.b. maps from X into Y, equipped with the cb norm.

Example B.2. A bounded linear functional φ on an operator space $X \subset A$ is c.b. with $\|\varphi\|_{\mathrm{cb}} = \|\varphi\|$. Indeed, for any $[x_{i,j}] \in \mathbb{M}_n(X)$, we have

$$\begin{aligned}\|\varphi_n([x_{i,j}])\|_{\mathbb{M}_n(\mathbb{C})} &= \sup\{|\langle \varphi_n([x_{i,j}])\eta, \xi\rangle| : \xi, \eta \in \ell_n^2,\ \|\xi\| = 1 = \|\eta\|\} \\ &= \sup\{|\varphi(\sum_{i,j} \overline{\xi_i}\eta_j x_{i,j})| : \sum_i |\xi_i|^2 = 1 = \sum_j |\eta_j|^2\}\end{aligned}$$

$$\leq \|\varphi\| \sup\{\| \sum_{i,j} \overline{\xi_i}\eta_j x_{i,j}\| : \sum_i |\xi_i|^2 = 1 = \sum_j |\eta_j|^2\}$$
$$\leq \|\varphi\| \|[x_{i,j}]\|.$$

If $X \subset A$ is an operator system and $\varphi\colon X \to B$ is c.p., then φ is c.b. and $\|\varphi\|_{\mathrm{cb}} = \|\varphi\| = \|\varphi(1)\|$. Indeed, this follows from Arveson's Extension Theorem and Stinespring's Dilation Theorem.

Remark B.3. If $\varphi\colon X \to Y$ is a c.b. map, then it is not hard to see that for any infinite-dimensional Hilbert space $\mathcal{H}$,

$$\|\varphi\|_{\mathrm{cb}} = \|\varphi \otimes \mathrm{id}_{\mathbb{B}(\mathcal{H})}\|.$$

Indeed, the inequality $\leq$ is immediate by compressing to finite-dimensional subspaces. For the other inequality, one shows that for $T \in X \odot \mathbb{B}(\mathcal{H})$, $\|\varphi \otimes \mathrm{id}_{\mathbb{B}(\mathcal{H})}(T)\|$ is the supremum over compressions to finite-dimensional subspaces of $\mathcal{H}$.

The following fact can be quite useful.

Lemma B.4 (Smith's lemma). *Let X be an operator space and $\varphi\colon X \to \mathbb{M}_n(\mathbb{C})$ be a bounded linear map. Then,*

$$\|\varphi\|_{\mathrm{cb}} = \|\mathrm{id}_{\mathbb{M}_n(\mathbb{C})} \otimes \varphi\|.$$

Proof. Since $\|\varphi\|_{\mathrm{cb}} = \|\mathrm{id}_{\mathbb{B}(\ell^2)} \otimes \varphi\|$ by Remark B.3, for any $\varepsilon > 0$, there is a contraction $x \in \mathbb{B}(\ell^2) \otimes X$ and unit vectors $\xi, \eta \in \ell^2 \otimes \ell^2_n$ such that

$$|\langle (\mathrm{id}_{\mathbb{B}(\ell^2)} \otimes \varphi)(x)\eta, \xi\rangle| > \|\varphi\|_{\mathrm{cb}} - \varepsilon.$$

Note that there are n-dimensional Hilbert subspaces $\mathcal{H}, \mathcal{K} \subset \ell^2$ such that $\xi \in \mathcal{H} \otimes \ell^2_n$ and $\eta \in \mathcal{K} \otimes \ell^2_n$. It follows that $\|\mathrm{id}_{\mathbb{B}(\mathcal{K},\mathcal{H})} \otimes \varphi\| > \|\varphi\|_{\mathrm{cb}} - \varepsilon$. Since $\mathbb{B}(\mathcal{K}, \mathcal{H})$ is spatially isomorphic to $\mathbb{M}_n(\mathbb{C})$ (i.e., there are unitaries $U\colon \mathcal{K} \to \ell^2_n$ and $V\colon \mathcal{H} \to \ell^2_n$ such that $\mathbb{M}_n(\mathbb{C}) \cong V\mathbb{B}(\mathcal{K}, \mathcal{H})U^*$ completely isometrically) and $\varepsilon > 0$ was arbitrary, we are done. □

Many facts about c.p. maps can be transferred to results about c.b. maps via an indispensable trick of Paulsen. For an operator space X in a unital C*-algebra A, we define an operator system $S_X \subset \mathbb{M}_2(A)$ by

$$S_X = \{\begin{bmatrix} \lambda 1_A & x \\ y^* & \mu 1_A \end{bmatrix} : \lambda, \mu \in \mathbb{C},\ x, y \in X\}.$$

For a map φ from X into a unital C*-algebra B, we define a map $S_\varphi\colon S_X \to \mathbb{M}_2(B)$ by

$$S_\varphi(\begin{bmatrix} \lambda 1_A & x \\ y^* & \mu 1_A \end{bmatrix}) = \begin{bmatrix} \lambda 1_B & \varphi(x) \\ \varphi(y)^* & \mu 1_B \end{bmatrix}.$$

Theorem B.5 (Paulsen). *The map $S_\varphi\colon S_X \to \mathbb{M}_2(B)$ is u.c.p. if and only if the map $\varphi\colon X \to B$ is c.c.*

For the proof, we need the following lemma.

Lemma B.6. *For any operators $a, b \geq 0$ and x, we have*

$$\begin{bmatrix} a & x \\ x^* & b \end{bmatrix} \geq 0 \iff \forall \varepsilon > 0 \quad \|(a+\varepsilon 1)^{-1/2} x (b+\varepsilon 1)^{-1/2}\| \leq 1.$$

Proof. Let $z(\varepsilon) = (a+\varepsilon 1)^{-1/2} x (b+\varepsilon 1)^{-1/2}$. The follow facts all make good exercises:

$$\begin{aligned} \begin{bmatrix} a & x \\ x^* & b \end{bmatrix} \geq 0 &\iff \forall \varepsilon > 0 \quad \begin{bmatrix} a+\varepsilon 1 & x \\ x^* & b+\varepsilon 1 \end{bmatrix} \geq 0 \\ &\iff \forall \varepsilon > 0 \quad \begin{bmatrix} 1 & z(\varepsilon) \\ z(\varepsilon)^* & 1 \end{bmatrix} \geq 0 \\ &\iff \forall \varepsilon > 0 \quad \|z(\varepsilon)\| \leq 1. \end{aligned}$$

□

Proof of Theorem B.5. The 'only if' part is trivial. To prove the other direction, let $T = [T_{i,j}] \in \mathbb{M}_n(S_X)$ be a given positive element, where

$$T_{i,j} = \begin{bmatrix} \lambda_{i,j} & x_{i,j} \\ y_{i,j}^* & \mu_{i,j} \end{bmatrix} \in S_X \subset \mathbb{M}_2(A).$$

We canonically identify $\mathbb{M}_n(\mathbb{M}_2(A))$ with $\mathbb{M}_2(\mathbb{M}_n(A))$ and $\mathbb{M}_n(\mathbb{M}_2(B))$ with $\mathbb{M}_2(\mathbb{M}_n(B))$. Let $\lambda = [\lambda_{i,j}]$, $\mu = [\mu_{i,j}]$, $x = [x_{i,j}]$ and $y = [y_{i,j}]$. Under the above identification, T is of the form

$$T = \begin{bmatrix} \lambda & x \\ y* & \mu \end{bmatrix} \in \mathbb{M}_2(\mathbb{M}_n(A)).$$

Since T is positive, λ and μ are positive and $y = x$. It follows from Lemma B.6 that $\|(\lambda+\varepsilon 1)^{-1/2} x (\mu+\varepsilon)^{-1/2}\| \leq 1$ for any $\varepsilon > 0$. Since φ is c.c.,

$$\|(\lambda+\varepsilon 1)^{-1/2} \varphi_n(x) (\mu+\varepsilon 1)^{-1/2}\| = \|\varphi_n((\lambda+\varepsilon 1)^{-1/2} x (\mu+\varepsilon 1)^{-1/2})\| \leq 1$$

for any $\varepsilon > 0$. This means

$$(S_\varphi)_n(T) = \begin{bmatrix} \lambda & \varphi_n(x) \\ \varphi_n(y)* & \mu \end{bmatrix} \in \mathbb{M}_2(\mathbb{M}_n(B))$$

is positive. Hence, S_φ is c.p. □

Here is an important generalization of Stinespring's Theorem.

Theorem B.7 (Haagerup, Paulsen, Wittstock). *Let $X \subset A$ be an operator space and $\varphi\colon X \to \mathbb{B}(\mathcal{H})$ be a c.c. map. Then, there exist a Hilbert space $\widehat{\mathcal{H}}$, a $*$-representation $\pi\colon A \to \mathbb{B}(\widehat{\mathcal{H}})$ and isometries $V, W\colon \mathcal{H} \to \widehat{\mathcal{H}}$ such that*

$$\varphi(x) = V^* \pi(x) W$$

for every $x \in X$. In particular, φ extends to a c.c. map on A.

Proof. Without loss of generality, we may assume that A is unital. Since $\varphi\colon X \to \mathbb{B}(\mathcal{H})$ is c.c., it follows from Theorem B.5 that $S_\varphi\colon S_X \to \mathbb{M}_2(\mathbb{B}(\mathcal{H}))$ is u.c.p. Let $(\tilde{\pi}, \tilde{\mathcal{H}}, \tilde{V})$ be a Stinespring triplet of a u.c.p. extension of S_φ to $\mathbb{M}_2(A)$. We observe that $\tilde{\mathcal{H}}$ can be decomposed as $\tilde{\mathcal{H}} = \widehat{\mathcal{H}} \oplus \widehat{\mathcal{H}}$ in such a way that $\mathbb{B}(\tilde{\mathcal{H}}) = \mathbb{M}_2(\mathbb{B}(\widehat{\mathcal{H}}))$ and $\tilde{\pi}\colon \mathbb{M}_2(A) \to \mathbb{M}_2(\mathbb{B}(\widehat{\mathcal{H}}))$ is of the form $\mathrm{id}_2 \otimes \pi$, for some $*$-representation $\pi\colon A \to \mathbb{B}(\mathcal{H})$. Under this identification, $\tilde{V} = [V_{i,j}] \in \mathbb{M}_2(\mathbb{B}(\mathcal{H}, \widehat{\mathcal{H}}))$. But since

$$V^* \begin{bmatrix} 1 & 0 \\ 0 & 0 \end{bmatrix} V = \begin{bmatrix} 1 & 0 \\ 0 & 0 \end{bmatrix} \text{ and } V^* \begin{bmatrix} 0 & 0 \\ 0 & 1 \end{bmatrix} V = \begin{bmatrix} 0 & 0 \\ 0 & 1 \end{bmatrix},$$

$V := V_{1,1}$ and $W := V_{2,2}$ are isometries from $\mathcal{H}$ into $\widehat{\mathcal{H}}$ (and $V_{1,2} = 0 = V_{2,1}$). Hence,

$$\begin{aligned} V^*\pi(x)W &= \begin{bmatrix} 1 & 0 \end{bmatrix} \begin{bmatrix} V & 0 \\ 0 & W \end{bmatrix}^* \begin{bmatrix} 0 & \pi(x) \\ 0 & 0 \end{bmatrix} \begin{bmatrix} V & 0 \\ 0 & W \end{bmatrix} \begin{bmatrix} 0 \\ 1 \end{bmatrix} \\ &= \begin{bmatrix} 1 & 0 \end{bmatrix} S_\varphi(\begin{bmatrix} 0 & x \\ 0 & 0 \end{bmatrix}) \begin{bmatrix} 0 \\ 1 \end{bmatrix} \\ &= \varphi(x) \end{aligned}$$

for every $x \in A$. □

Let $X \subset A$ and $Y \subset B$ be operator spaces. We define the minimal tensor product $X \otimes Y$ of X and Y to be the norm closure of $X \odot Y$ in $A \otimes B$.

The proof of the following corollary is very similar to C*-results in the body of this text (cf. Theorem 3.5.3 and the proof of Proposition 3.3.11).

Corollary B.8. *Let $X_i \subset A_i$ and $Y_i \subset B_i$ be operator spaces ($i = 1, 2$) and let $\varphi\colon X_1 \to X_2$ and $\psi\colon Y_1 \to Y_2$ be c.c. maps. Then,*

$$\varphi \otimes \psi\colon X_1 \otimes Y_1 \to X_2 \otimes Y_2$$

is a c.c. map. For $z \in X \otimes Y$, one has

$$\|z\|_{\min} = \sup \|(\varphi \otimes \psi)(z)\|_{\mathbb{M}_m(\mathbb{C}) \otimes \mathbb{M}_n(\mathbb{C})}$$

where the supremum is taken over all $m, n \in \mathbb{N}$ and c.c. maps $\varphi\colon X \to \mathbb{M}_m(\mathbb{C})$ and $\psi\colon Y \to \mathbb{M}_n(\mathbb{C})$.

Though we won't prove it, we remind you that for C*-algebras A, B, C and a c.c. map $\varphi\colon A \to B$, the map $\varphi \otimes \mathrm{id}_C\colon A \odot C \to B \odot C$ need not be continuous with respect to the maximal tensor norm ([**88**]).

It follows from Theorem B.7 that if φ is unital and c.c., then $V = W$ and φ is automatically c.p. More generally, we have the following perturbation fact:

Corollary B.9. *Let $E \subset A$ be an operator system and $\varphi\colon E \to \mathbb{B}(\mathcal{H})$ be a unital self-adjoint map (i.e., $\varphi(1) = 1$ and $\varphi(a^*)^* = \varphi(a)$ for $a \in E$). Then, there exists a u.c.p. map $\psi\colon E \to \mathbb{B}(\mathcal{H})$ with $\|\varphi - \psi\|_{\mathrm{cb}} \le 2(\|\varphi\|_{\mathrm{cb}} - 1)$.*

Proof. Let $\lambda = \|\varphi\|_{\mathrm{cb}}$. By Theorem B.7, there exist a $*$-representation $A \to \mathbb{B}(\widehat{\mathcal{H}})$ and isometries $V, W\colon \mathcal{H} \to \widehat{\mathcal{H}}$ such that

$$\varphi(a) = \lambda V^*\pi(a)W = \lambda W^*\pi(a)V$$

for $a \in E$. Let $\psi\colon E \to \mathbb{B}(\mathcal{H})$ be the u.c.p. map defined by

$$\psi(a) = \frac{1}{2}(V^*\pi(a)V + W^*\pi(a)W)$$

for $a \in E$. Then,

$$\lambda\psi(a) - \varphi(a) = \frac{1}{2}\lambda(V - W)^*\pi(a)(V - W).$$

Since φ is unital, we have $\lambda V^*W = 1$ and

$$\|\varphi - \psi\|_{\mathrm{cb}} \le \frac{1}{2}\lambda\|(V - W)^*(V - W)\| + (\lambda - 1) = 2(\lambda - 1).$$

□

Let E be an n-dimensional linear space with a basis $\{x_i\}_{i=1}^n$. Then, its dual basis, denoted by $\{\widehat{x}_i\}_{i=1}^n \subset E^*$, is defined by the relation $\widehat{x}_i(x_j) = \delta_{i,j}$.

Lemma B.10. *Let E be an n-dimensional operator system. Then, there exists a basis $\{x_i\}_{i=1}^n$ consisting of self-adjoint elements such that $\|x_i\| = 1 = \|\widehat{x}_i\|$ for every i.*

Proof. Fix a basis $\{z_i\}_{i=1}^n$ for E and consider the multilinear function

$$\Phi\colon E^n \ni (y_1, \ldots, y_n) \longmapsto \det[\widehat{z}_i(y_j)] \in \mathbb{C}.$$

We denote by $\mathrm{Ball}(E)_{\mathrm{sa}}$ the self-adjoint part of the closed unit ball of E. Since $\mathrm{Ball}(E)_{\mathrm{sa}}^n$ is compact, there exists $(x_1, \ldots, x_n) \in \mathrm{Ball}(E)_{\mathrm{sa}}^n$ which attains the maximum of the absolute value of Φ on $\mathrm{Ball}(E)_{\mathrm{sa}}^n$. It is not hard to see that $\{x_j\}_{j=1}^n$ is a basis for E and $\{\widehat{x}_i\}_{i=1}^n \subset E^*$ given by

$$\widehat{x}_i(y) = \Phi(x_1, \ldots, x_n)^{-1}\Phi(x_1, \ldots, x_{i-1}, y, x_{i+1}, \ldots, x_n)$$

forms the dual basis. Since $\widehat{x}_i$ is self-adjoint, the maximality assumption on $\{x_j\}_{j=1}^n$ yields that $\|\widehat{x}_i\| = 1$ for every i. □

The following is a variant of Corollary B.9, which works for maps into a unital C*-algebra B rather than into $\mathbb{B}(\mathcal{H})$.

Corollary B.11. *Let E be a finite-dimensional operator system, B be a unital C*-algebra and $\varphi\colon E \to B$ be unital self-adjoint map. Then, there exists a u.c.p. map $\psi\colon E \to B$ with $\|\varphi - \psi\|_{\mathrm{cb}} \le 2\dim(E)(\|\varphi\|_{\mathrm{cb}} - 1)$.*

Proof. Let $B \subset \mathbb{B}(\mathcal{H})$. Then, Corollary B.9 provides a u.c.p. map $\psi'\colon E \to \mathbb{B}(\mathcal{H})$ with $\|\varphi - \psi'\|_{\mathrm{cb}} \le 2(\|\varphi\|_{\mathrm{cb}} - 1)$. Let $\sigma = \psi' - \varphi$ and $n = \dim(E)$. We claim that there is a positive linear functional θ on E, with $\|\theta\| \le n\|\sigma\|$, such that $\theta - \sigma$ is c.p. If this is the case, then defining $\psi = \varphi + \theta$, we are done. Fix a basis $\{x_i\}_{i=1}^n$ for E as in Lemma B.10 and let $\theta = \|\sigma\| \sum_{i=1}^n |\widehat{x}_i|$. For every $a \ge 0$, we have

$$\sigma(a) = \sum_{i=1}^n \widehat{x}_i(a)\sigma(x_i) \le \sum_{i=1}^n |\widehat{x}_i|(a)\|\sigma(x_i)\| \le \|\sigma\| \sum_{i=1}^n |\widehat{x}_i|(a) = \theta(a)$$

and hence $\theta - \sigma$ is positive. The proof of complete positivity is similar. □

We close this section with a c.b. variant of Corollary 1.6.3, though we leave the proof to the reader. (Hint: Theorem B.5.)

Lemma B.12. *Let $X \subset \mathbb{B}(\mathcal{H})$ be an ultraweakly closed operator space and let $\varphi\colon X \to \mathbb{M}_n(\mathbb{C})$ be a c.c. map. Then, there exists a net of ultraweakly continuous c.c. maps $\varphi_\lambda\colon X \to \mathbb{M}_n(\mathbb{C})$ which converges to φ in the point-norm topology.*

Operator space duality. Let X be an operator subspace of a C*-algebra A. In many cases, the position of X inside A does not matter. What matters is the *operator space structure* of X – i.e., the norms on $\mathbb{M}_n(X)$, $n \in \mathbb{N}$, inherited from $\mathbb{M}_n(X) \subset \mathbb{M}_n(A)$. Hence, we often identify the operator space X with another operator space which is completely isometrically isomorphic. In other words, we consider an operator space as an abstract Banach (or normed) space equipped with a distinguished family of matrix norms on $\mathbb{M}_n(X)$ (satisfying some axioms) and view $X \subset A$ as a concrete realization of X. This viewpoint is very similar to that for C*-algebras and their representations. Like the Gelfand-Naimark Theorem for C*-algebras, there is Ruan's Theorem which gives an axiomatic characterization of operator spaces, but we won't need it (we just had to mention it – see [**63**]).

Recall that for operator spaces $X \subset A$ and $Y \subset B$, we have $X \otimes Y \subset A \otimes B$ and the operator space structure of $X \otimes Y$ depends only on those of X and Y (cf. Corollary B.8).

There is a very important notion of dual operator spaces which was introduced in [**21, 22, 62**]. Let $X \subset \mathbb{B}(\mathcal{H})$ be an operator space and let X^* be its dual Banach space. For $x = [x_{k,l}] \in \mathbb{M}_m(X)$, we define $\theta_x\colon X^* \ni \varphi \mapsto [\varphi(x_{k,l})] \in \mathbb{M}_m(\mathbb{C})$. Denote by $\mathrm{Ball}_m(X)$ the closed unit ball of $\mathbb{M}_m(X)$. One puts an operator space structure on X^* by the isometric inclusion

$$\Theta\colon X^* \ni \varphi \longmapsto (\theta_x(\varphi))_x \in \prod_{m\in\mathbb{N}} \prod_{x\in\mathrm{Ball}_m(X)} \mathbb{M}_m(\mathbb{C}) \subset \mathbb{B}(\mathcal{H}),$$

where $\mathcal{H} = \bigoplus_{m\in\mathbb{N}} \bigoplus_{x\in\mathrm{Ball}_m(X)} \ell_m^2$. Unless otherwise stated, we always assume that the dual space X^* is equipped with this operator space structure. We observe that the weak*-topology of X^* agrees with the ultraweak topology of $\mathbb{B}(\mathcal{H})$ and in particular that if a net $\varphi_\lambda \in \mathbb{M}_n(X^*)$ converges to an element $\varphi \in \mathbb{M}_n(X^*)$ in the entrywise weak*-topology, then $\|\varphi\|_{\mathbb{M}_n(X^*)} \le \liminf_\lambda \|\varphi_\lambda\|_{\mathbb{M}_n(X^*)}$.

For $\varphi = [\varphi_{i,j}] \in \mathbb{M}_n(X^*)$, we define $T_\varphi\colon X \ni x \mapsto [\varphi_{i,j}(x)] \in \mathbb{M}_n(\mathbb{C})$. It follows from the definitions that

$$\begin{aligned}\|\varphi\|_{\mathbb{M}_n(X^*)} &= \sup\{\|(\theta_x)_n(\varphi)\|_{\mathbb{M}_n(\mathbb{M}_m(\mathbb{C}))} : m \in \mathbb{N},\ x \in \mathrm{Ball}_m(X)\}\\ &= \sup\{\|(T_\varphi)_m(x)\|_{\mathbb{M}_m(\mathbb{M}_n(\mathbb{C}))} : m \in \mathbb{N},\ x \in \mathrm{Ball}_m(X)\}\\ &= \|T_\varphi\|_{\mathrm{cb}},\end{aligned}$$

where we used the standard identification $\mathbb{M}_n(\mathbb{M}_m(\mathbb{C})) = \mathbb{M}_m(\mathbb{M}_n(\mathbb{C}))$. This implies that the bijection

$$X^* \otimes \mathbb{M}_n(\mathbb{C}) \cong \mathbb{M}_n(X^*) \ni \varphi \mapsto T_\varphi \in \mathrm{CB}(X, \mathbb{M}_n(\mathbb{C}))$$

is isometric. Hence, we sometimes identify φ with T_φ (and $\mathbb{M}_n(X^*)$ with $\mathrm{CB}(X, \mathbb{M}_n(\mathbb{C}))$). This identification can be generalized as follows.

Theorem B.13. *Let X and Y be operator spaces. For $z = \sum_k \varphi_k \otimes y_k \in X^* \odot Y$, we define a finite-rank map $T_z\colon X \to Y$ by*

$$T_z(x) = \sum_k \varphi_k(x) y_k.$$

This correspondence is isometric, meaning $\|z\|_{X^*\otimes Y} = \|T_z\|_{\mathrm{cb}}$, and hence extends to a canonical (isometric) inclusion

$$X^* \otimes Y \subset \mathrm{CB}(X, Y).$$

Moreover, the inclusion is bijective if either X or Y is finite-dimensional.

Proof. Let $Y \subset \mathbb{B}(\mathcal{H})$. For simplicity, we assume that $\mathcal{H}$ is separable and let $P_1 \le P_2 \le \cdots$ be an increasing sequence of finite-rank projections in $\mathbb{B}(\mathcal{H})$ with $\mathrm{rank}(P_n) = n$ and $P_n \nearrow 1$ in the strong operator topology. We denote by $\Phi_n\colon \mathbb{B}(\mathcal{H}) \to \mathbb{B}(P_n\mathcal{H}) \cong \mathbb{M}_n(\mathbb{C})$ the compression by P_n. Fix $z \in X^* \odot Y$. Since $T_{(\mathrm{id}\otimes\Phi_n)(z)} = \Phi_n \circ T_z$ in $\mathrm{CB}(X, \mathbb{M}_n(\mathbb{C}))$, we have

$$\begin{aligned}\|z\|_{\min} &= \sup_n \|(\mathrm{id} \otimes \Phi_n)(z)\|_{X^*\otimes\mathbb{M}_n(\mathbb{C})}\\ &= \sup_n \|\Phi_n \circ T_z\|_{\mathrm{cb}}\\ &= \|T_z\|_{\mathrm{cb}}.\end{aligned}$$

The rest of the proof is trivial. ∎

Remark B.14 (Technical observation). Let E be a finite-dimensional operator space and A be a C*-algebra. Assume we are given elements $z \in E^* \otimes A^{**}$ and $z_i \in E^* \otimes A$. Let $T_z\colon E \to A^{**}$ and $T_{z_i}\colon E \to A$ be the corresponding c.b. maps. Note that there is an algebraic (though not necessarily isometric) linear isomorphism $E^* \otimes A^{**} \cong (E^* \otimes A)^{**}$ (Proposition 9.2.1). We claim that if $z_i \to z$ in the weak*-topology coming from $(E^* \otimes A)^{**}$, then $T_{z_i} \to T_z$ in the point-ultraweak topology.

The point is that finite-dimensionality of E gives an algebraic identification $(E^* \otimes A)^* = E \otimes A^*$. Hence, any $\xi \in (E^* \otimes A)^*$ can be written

$$\xi = \sum e_j \otimes \eta_j,$$

where $e_j \in E$ and $\eta_j \in A^*$. Straightforward calculations, using nothing but the definitions, show convergence works as asserted.

Suppose that X and Y are operator spaces and $\varphi\colon X^* \to Y$ is a finite-rank weak*-continuous c.b. map. By the above theorem, φ corresponds to an element $z \in X \odot Y$ (rather than $z \in X^{**} \odot Y$) with $\|\varphi\|_{\mathrm{cb}} = \|z\|_{X^{**} \otimes Y}$. It is natural to ask whether $\|z\|_{X^{**} \otimes Y} = \|z\|_{X \otimes Y}$ for $z \in X \odot Y$. Fortunately, this is true as the canonical inclusion of X into X^{**} is completely isometric for any operator space X.

Lemma B.15. *For any operator space X, the canonical inclusion $X \subset X^{**}$ is completely isometric.*

Proof. For $x = [x_{i,j}] \in \mathbb{M}_n(X)$, we have

$$\begin{aligned}\|x\|_{\mathbb{M}_n(X)} &= \sup\{\|\varphi_n(x)\|_{\mathbb{M}_n(\mathbb{M}_m(\mathbb{C}))} : m \in \mathbb{N},\ \varphi\colon X \to \mathbb{M}_m(\mathbb{C}),\ \|\varphi\|_{\mathrm{cb}} \leq 1\} \\ &= \sup\{\|(\theta_x)_m(\varphi)\|_{\mathbb{M}_m(\mathbb{M}_n(\mathbb{C}))} : m \in \mathbb{N},\ \varphi \in \mathbb{M}_m(X^*),\ \|\varphi\| \leq 1\} \\ &= \|\theta_x\|_{\mathrm{CB}(X^*, \mathbb{M}_n(\mathbb{C}))} = \|x\|_{\mathbb{M}_n(X^{**})}\end{aligned}$$

since $\mathrm{CB}(X^*, \mathbb{M}_n(\mathbb{C})) = \mathbb{M}_n(X^{**})$ by definition. □

Let X be an operator space. There is a canonical algebraic identification $\mathbb{M}_n(X^{**}) = \mathbb{M}_n(X)^{**}$. As one might expect, under this identification a net in $\mathbb{M}_n(X)^{**}$ converges in the $\sigma(\mathbb{M}_n(X)^{**}, \mathbb{M}_n(X)^*)$-topology if and only if each matrix entry converges in the $\sigma(X^{**}, X^*)$-topology. Moreover, $\mathbb{M}_n(X^{**}) = \mathbb{M}_n(X)^{**}$ is a completely isometric identification.

Proposition B.16. *Let X be an operator space. The canonical isomorphism $\mathbb{M}_n(X^{**}) \cong \mathbb{M}_n(X)^{**}$ is (completely) isometric. In particular, if A is a C*-algebra, then the operator space structure of A^{**} as a second dual operator space agrees with the von Neumann algebra structure.*

Proof. We will show $\|x\|_{\mathbb{M}_n(X^{**})} = \|x\|_{\mathbb{M}_n(X)^{**}}$ for every $x = [x_{i,j}] \in \mathbb{M}_n(X^{**})$. Suppose first that $\|x\|_{\mathbb{M}_n(X)^{**}} \leq 1$. By the bi-polar theorem, there exists a net $\{x_\lambda\}_\lambda$ of norm-one elements in $\mathbb{M}_n(X)$ which converges to x in the $\sigma(\mathbb{M}_n(X)^{**}, \mathbb{M}_n(X)^*)$-topology. But since $X \subset X^{**}$ completely isometrically, by Lemma B.15, one has $\|x_\lambda\|_{\mathbb{M}_n(X^{**})} \leq 1$ for all λ. Since each entry of x_λ converges in the weak*-topology to the corresponding entry of x, one has $\|x\|_{\mathbb{M}_n(X^{**})} \leq 1$. For the converse inequality, suppose this time that $\|x\|_{\mathbb{M}_n(X^{**})} \leq 1$. Then the corresponding map $T_x \colon X^* \to \mathbb{M}_n(\mathbb{C})$ has $\|T_x\|_{\mathrm{cb}} \leq 1$. It follows from Lemma B.12 that T_x can be approximated in the point-norm topology by weak*-continuous complete contractions $T_\lambda \colon X^* \to \mathbb{M}_n(\mathbb{C})$. Then, the corresponding element $x_\lambda \in \mathbb{M}_n(X)$ satisfies $\|x_\lambda\| = \|T_\lambda\|_{\mathrm{cb}}$ for every λ. Since the net $(x_\lambda)_\lambda$ converges to x in $\sigma(\mathbb{M}_n(X)^{**}, \mathbb{M}_n(X)^*)$, we have $\|x\|_{\mathbb{M}_n(X)^{**}} \leq 1$. This completes the proof. □

If X is a dual operator space, then by definition there is a weak*-ultraweak homeomorphic complete isometry from X onto an ultraweakly closed subspace of $\mathbb{B}(\mathcal{H})$. On the other hand, if $X \subset \mathbb{B}(\mathcal{H})$ is an ultraweakly closed operator subspace, then the space X_* of ultraweakly continuous linear functionals on X is a predual of X, i.e., $X = (X_*)^*$ canonically. Since $X_* \subset X^*$ as a Banach space and X^* is a (dual) operator space, we can equip the predual X_* with an operator space structure via $X_* \subset X^*$. Fortunately, the equality $X = (X_*)^*$ holds completely isometrically as usual. (One should verify this.) We note that even if an operator space X happens to have a predual X_* as a Banach space, it does not mean X_* has an operator space structure with $X = (X_*)^*$ completely isometrically. This is because such X need not have a weak*-ultraweak homeomorphic c.i. representation [**115**].

References. The theory of operator spaces took off after Ruan's axiomatic formalization, although many notions had appeared earlier in papers of Arveson, of Choi and Effros, etc. The notion of duality was introduced by Blecher [**21**], Blecher and Paulsen [**22**] and Effros and Ruan [**62**]. Theorems B.5 and B.7 are taken from [**141**], and Corollary B.11 is taken from [**59**]. Standard text books for operator spaces and completely bounded maps are [**63, 152, 141**].

Appendix C

Lifting Theorems

Throughout this appendix, we assume that J is a closed two-sided ideal in a unital C*-algebra B and $\pi\colon B \to B/J$ is the quotient map.

Definition C.1. Let E be an operator system (or a C*-algebra). A c.c.p. map $\varphi\colon E \to B/J$ is said to be *liftable* if there exists a c.c.p. map $\psi\colon E \to B$ such that $\pi \circ \psi = \varphi$. It is *locally liftable* if for every finite-dimensional operator system $F \subset E$, the c.c.p. map $\varphi|_F$ is liftable.

If φ is a liftable u.c.p. map, then we can take a unital lifting ψ. Indeed, one just has to choose a unital positive linear functional θ on E and replace ψ with $\psi + (1 - \psi(1))\theta$.

The following result of Arveson is very important – so is its proof.

Lemma C.2. *Let J be a closed two-sided ideal in a C*-algebra B and let E be a separable operator system (or a separable C*-algebra). Then, the set of c.c.p. maps from E into B/J which are liftable is closed in the point-norm topology.*

Proof. Let $\varphi\colon E \to B/J$ be a c.c.p. map and let $\psi_n'\colon E \to B$ be c.c.p. maps such that the sequence $\pi\circ\psi_n'$ converges to φ in the point-norm topology. Fix a dense sequence $\{x_k\}_k$ in E. Passing to a subsequence if necessary, we may assume that $\|\pi \circ \psi_n'(x_k) - \varphi(x_k)\| < 1/2^n$ for $k \leq n$. We claim that there exists a sequence of c.c.p. maps $\psi_n\colon E \to B$ such that $\|\pi\circ\psi_n(x_k) - \varphi(x_k)\| < 1/2^n$ and $\|\psi_{n+1}(x_k) - \psi_n(x_k)\| < 1/2^{n-1}$ for $k \leq n$. Once this is established, the proof is essentially complete. Indeed, since the sequence $\{\psi_n\}_n$ of c.c.p. maps converges point-norm on a dense subset $\{x_k\}_k$, it converges everywhere to a c.c.p. map $\psi\colon E \to B$, which is clearly a lifting of φ.

To prove the claim, we proceed by induction. First, set $\psi_1 = \psi_1'$. Suppose now that we have constructed $\psi_1, \ldots, \psi_n$ with the desired property. Let $(e_\lambda)_\lambda$ be a quasicentral approximate unit of J in B. Then, for $k \le n$,

$$\lim_\lambda \|(1-e_\lambda)^{1/2}\psi_n(x_k)(1-e_\lambda)^{1/2} + e_\lambda^{1/2}\psi_n(x_k)e_\lambda^{1/2} - \psi_n(x_k)\| = 0$$

and for $b_k = \psi_{n+1}'(x_k) - \psi_n(x_k)$ with $k \le n$,

$$\lim_\lambda \|(1-e_\lambda)^{1/2}b_k(1-e_\lambda)^{1/2}\| = \|\pi(b_k)\| < 3/2^{n+1}.$$

Hence, we can take $e = e_\lambda \in J$ so that for every $k \le n$

$$\|(1-e)^{1/2}\psi_n(x_k)(1-e)^{1/2} + e^{1/2}\psi_n(x_k)e^{1/2} - \psi_n(x_k)\| < 1/2^{n+1}$$

and

$$\|(1-e)^{1/2}b_k(1-e)^{1/2}\| < 3/2^{n+1}.$$

Then the c.c.p. map $\psi_{n+1}\colon E \to B$ defined by

$$\psi_{n+1}(x) = (1-e)^{1/2}\psi_{n+1}'(x)(1-e)^{1/2} + e^{1/2}\psi_n(x)e^{1/2}$$

satisfies the desired property. (Note that $\pi \circ \psi_{n+1} = \pi \circ \psi_{n+1}'$.) □

We observe that if a c.c.p. map $\varphi\colon E \to B/J$ has a c.p. lifting, then it actually has a c.c.p. lifting. Indeed, if E is unital and $\psi\colon E \to B$ is a c.p. lifting of φ, then for an approximate unit $(e_\lambda)_\lambda$ in J, the c.c.p. maps $\psi_\lambda(\,\cdot\,) = (1-e_\lambda)\psi(\,\cdot\,)(1-e_\lambda)$ are c.p. liftings of φ with $\lim_\lambda \|\psi_\lambda\| = 1$. When E is a nonunital C*-algebra, the proof is more cumbersome and we leave it to the reader.

The following result is due to Choi and Effros.

Theorem C.3. *Every nuclear c.c.p. map from a separable* C**-algebra A into a quotient* C**-algebra B/J is liftable. In particular, every c.c.p. map from a separable nuclear* C**-algebra is liftable.*

Proof. By Lemma C.2, the set of liftable c.c.p. maps is closed in the point-norm topology. Since a nuclear c.c.p. map is approximated by c.c.p. maps which factor through full matrix algebras, it suffices to show that every c.c.p. map φ from a full matrix algebra $\mathbb{M}_n(\mathbb{C})$ into B/J is liftable. We use Proposition 1.5.12. Let $a = [\varphi(e_{i,j})]$ be a positive element in $\mathbb{M}_n(B/J)$. Since $\pi_n\colon \mathbb{M}_n(B) \to \mathbb{M}_n(B/J)$ is a surjective $*$-homomorphism, the positive element a lifts to a positive element $b = [b_{i,j}] \in \mathbb{M}_n(B)$ and the corresponding map $\psi'\colon \mathbb{M}_n \to B$ is a c.p. lifting. □

The separability assumption in Theorem C.3 is essential. Indeed, there is no bounded linear lifting from ℓ^∞/c_0 to ℓ^∞ (see Exercise 13.1.1).

Here is a celebrated lifting theorem due to Effros and Haagerup.

Theorem C.4. *Let J be a closed two-sided ideal in a unital* C**-algebra B and $\pi\colon B \to B/J$ be the quotient map. Then, the following are equivalent:*

(1) *for any* C**-algebra A, the following sequence is exact:*

$$0 \longrightarrow A \otimes J \longrightarrow A \otimes B \longrightarrow A \otimes (B/J) \longrightarrow 0;$$

(2) *same as above but with $A = \mathbb{B}(\ell^2)$;*

(3) *for any finite-dimensional operator system $E \subset B/J$, the inclusion of E into B/J is liftable.*

Proof. Obviously, (1) implies (2). Let's prove (3) implies (1). Given $z \in \ker(\mathrm{id}_A \otimes \pi) \subset A \otimes B$, we must show $z \in A \otimes J$. For any $\varepsilon > 0$, there is $y = \sum_i a_i \otimes x_i \in A \odot B$ with $\|y - z\| < \varepsilon$. Let $E \subset B/J$ be a finite-dimensional operator system containing the $\pi(x_i)$'s and let $\psi\colon E \to B$ be a u.c.p. lifting. It follows that $y - (\mathrm{id}_A \otimes \psi \circ \pi)(y) \in A \odot J$ and

$$\|(\mathrm{id}_A \otimes \psi \circ \pi)(y)\| \le \|(\mathrm{id}_A \otimes \pi)(y)\| \le \|y - z\| < \varepsilon.$$

Therefore,

$$\mathrm{dist}(z, A \otimes J) \le \|z - (y - (\mathrm{id}_A \otimes \psi \circ \pi)(y))\| < 2\varepsilon.$$

Since $\varepsilon > 0$ was arbitrary, $z \in A \otimes J$ as desired.

We now prove that (2) implies (3). Let $E \subset B/J$ be given. By operator space duality, the inclusion $E \subset B/J$ corresponds to an element $z \in E^* \otimes (B/J)$ with $\|z\| = 1$. We may assume $E^* \subset \mathbb{B}(\ell^2)$. Since

$$(E^* \otimes B)/(E^* \otimes J) \subset (\mathbb{B}(\ell^2) \otimes B)/(\mathbb{B}(\ell^2) \otimes J) = \mathbb{B}(\ell_2) \otimes (B/J)$$

isometrically (Proposition 3.7.2), we have $E^* \otimes (B/J) = (E^* \otimes B)/(E^* \otimes J)$ isometrically. Hence, for any $\varepsilon > 0$, one can lift z to an element $\tilde{z} \in E^* \otimes B$ with $\|\tilde{z}\| < 1 + \varepsilon$. Then, the map $\psi'\colon E \to B$ corresponding to $\tilde{z}$ is a lifting of φ with $\|\psi'\|_{\mathrm{cb}} < 1 + \varepsilon$. We may assume that ψ' is self-adjoint. Since $\psi'(1) - 1 \in J$, we can find $e \in J$ with $0 \le e \le 1$ such that

$$\|(1-e)^{1/2}(\psi'(1) - 1)(1-e)^{1/2}\| < \varepsilon.$$

Choose a unital positive linear functional θ on E and let $\psi''\colon E \to B$ be defined by

$$\begin{aligned}\psi''(x) &= (1-e)^{1/2}\Big(\psi'(x) - \theta(x)(\psi'(1) - 1)\Big)(1-e)^{1/2} + \theta(x)e \\ &= \begin{bmatrix}(1-e)^{1/2} & e^{1/2}\end{bmatrix} \begin{bmatrix}\psi'(x) & 0 \\ 0 & \theta(x)\end{bmatrix} \begin{bmatrix}(1-e)^{1/2} \\ e^{1/2}\end{bmatrix} \\ &\quad + \theta(x)(1-e)^{1/2}(\psi'(1) - 1)(1-e)^{1/2}.\end{aligned}$$

It follows that ψ'' is a lifting of φ with $\psi''(1) = 1$ and $\|\psi''\|_{\mathrm{cb}} < 1 + 2\varepsilon$. By Corollary B.11, there exists a u.c.p. map $\psi\colon E \to B$ such that $\|\psi - \psi''\|_{\mathrm{cb}} < 4\dim(E)\varepsilon$. Since $\varepsilon > 0$ was arbitrary, we are done by Lemma C.2. □

Remark C.5. Though one can almost always get around it by working locally, a convenient generalization is available when the ideal is nuclear. Namely, we can take $E = B/J$ in (3) provided that B/J is separable and J is nuclear. The proof of this needs a variant of Lemma C.2 and is found in [**59**].

References. Theorem C.3 was proved in [**38**], but our proof follows [**11**]. Theorem C.4 is from [**59**].

Appendix D

Positive Definite Functions, Cocycles and Schoenberg's Theorem

Unitary representations. As in Section 2.5, Γ denotes a discrete group, and $\mathbb{C}[\Gamma]$ is its complex group algebra. Let $(\pi, \mathcal{H})$ and $(\pi', \mathcal{H}')$ be two unitary representations of Γ. Recall that π and π' are *unitarily equivalent* if there exists a unitary $U\colon \mathcal{H} \to \mathcal{H}'$ such that $\pi'(s) = U\pi(s)U^*$ for every $s \in \Gamma$. We say π' is *contained* in π if π' is equivalent to a subrepresentation of π; π' is *weakly contained* in π if the map $\pi(s) \mapsto \pi'(s)$ extends to a (continuous) $*$-homomorphism from $C^*(\pi(\Gamma))$ into $C^*(\pi'(\Gamma))$; π and π' are *weakly equivalent* if the map $\pi(s) \mapsto \pi'(s)$ extends to a $*$-isomorphism from $C^*(\pi(\Gamma))$ onto $C^*(\pi'(\Gamma))$.

Generalizing Definition 2.5.6, positive definite functions make perfectly good sense for operator-valued functions.

Definition D.1. We say a function $\varphi\colon \Gamma \to \mathbb{B}(\mathcal{H})$ is *positive definite* if for any finite sequence $s_1, \ldots, s_n \in \Gamma$, the operator matrix $[\varphi(s_i^{-1}s_j)]_{i,j} \in \mathbb{M}_n(\mathbb{B}(\mathcal{H}))$ is positive.

One should check that Theorem 2.5.11 holds *mutatis mutandis* for an operator-valued function. In particular, $\varphi\colon \Gamma \to \mathbb{B}(\mathcal{H})$ with $\varphi(e) = 1$ is positive definite if and only if there exist a Hilbert space $\widehat{\mathcal{H}} \supset \mathcal{H}$ and a unitary representation π of Γ on $\widehat{\mathcal{H}}$ such that $\varphi(s) = P_{\mathcal{H}}\pi(s)|_{\mathcal{H}}$ for every $s \in \Gamma$. Therefore, we may induce positive definite functions.

Lemma D.2. *Let Γ be a discrete group and $\Lambda \leq \Gamma$ be a subgroup. If $\varphi\colon \Lambda \to \mathbb{B}(\mathcal{H})$ is a positive definite function, then the function*

$$\mathrm{Ind}_\Lambda^\Gamma \varphi\colon \Gamma \ni s \mapsto \sum_{x\in\Gamma/\Lambda} e_{sx,x} \otimes \varphi(\sigma(sx)^{-1}s\sigma(x)) \in \mathbb{B}(\ell^2(\Gamma/\Lambda)\otimes\mathcal{H})$$

is positive definite, where $\sigma\colon \Gamma/\Lambda \to \Gamma$ is a fixed cross section.

Schur multipliers. Let Γ be an index set. A *kernel* is a function $k\colon \Gamma\times\Gamma \to \mathbb{C}$. The kernel k is *positive definite* if for any finite sequence $s_1, \ldots, s_n \in \Gamma$, the matrix $[k(s_i, s_j)]_{i,j} \in \mathbb{M}_n(\mathbb{C})$ is positive. (Sometimes this property is called positive semidefinite.) When Γ is a group, a function φ on Γ is positive definite if and only if the kernel $k(s,t) = \varphi(s^{-1}t)$ is positive definite. Every $x \in \mathbb{B}(\ell^2(\Gamma))$ can be represented as a $\Gamma\times\Gamma$ matrix: $x = [x_{s,t}]_{s,t\in\Gamma}$, where $x_{s,t} = \langle x\delta_t, \delta_s\rangle$ for $s,t \in \Gamma$. The *Schur multiplier* $m_k\colon \mathbb{B}(\ell^2(\Gamma)) \to \mathbb{B}(\ell^2(\Gamma))$ is defined by $m_k([x_{s,t}]) = [k(s,t)x_{s,t}]$. This may not be well-defined because $[k(s,t)x_{s,t}]$ may not belong to $\mathbb{B}(\ell^2(\Gamma))$.

Theorem D.3. *Let $k\colon \Gamma\times\Gamma \to \mathbb{C}$ be a kernel with $k(s,s) = 1$ for all $s \in \Gamma$. The following are equivalent:*

(1) *the kernel k is positive definite;*

(2) *there exist a Hilbert space $\mathcal{H}$ and unit vectors $\xi_s \in \mathcal{H}$ such that $k(s,t) = \langle \xi_t, \xi_s\rangle$ for every $s,t \in \Gamma$;*

(3) *the multiplier m_k is a (continuous) u.c.p. map on $\mathbb{B}(\ell^2(\Gamma))$.*

Proof. (1)$\Rightarrow$(2): This follows from a standard GNS construction. (Mimic the Hilbert space construction following Definition 2.5.6 with a one-variable positive definite function replaced by k.)

(2)$\Rightarrow$(3): Let $V\colon \ell^2(\Gamma) \to \ell^2(\Gamma)\otimes\mathcal{H}$ be the isometry such that $V\delta_t = \delta_t \otimes \xi_t$ for $t \in \Gamma$. It is not hard to see that $m_k(x) = V^*(x\otimes 1)V$ for $x \in \mathbb{B}(\ell^2(\Gamma))$.

(3)$\Rightarrow$(1): For any finite subset $E \subset \Gamma$, let $x \in \mathbb{B}(\ell^2(\Gamma))$ be such that $x_{s,t}$ is the characteristic function of $E\times E$. Since x is positive, so is $m_k(x)$. This shows that $[k(s,t)]_{s,t\in E}$ is positive for any finite subset $E \subset \Gamma$ and hence k is positive definite. □

Theorem D.4. *Let $k\colon \Gamma\times\Gamma \to \mathbb{C}$ be a kernel. The multiplier m_k is completely bounded with $\|m_k\|_{\mathrm{cb}} \leq 1$ if and only if there exist a Hilbert space $\mathcal{H}$ and vectors $\xi_s, \eta_t \in \mathcal{H}$, with $\|\xi_s\|, \|\eta_t\| \leq 1$, such that $k(s,t) = \langle \eta_t, \xi_s\rangle$ for every $s,t \in \Gamma$.*

Proof. First suppose that $\|m_k\|_{\mathrm{cb}} \le 1$. By Wittstock's factorization theorem for completely bounded maps (Theorem B.7), there exist a Hilbert space $\mathcal{H}$, a $*$-representation $\pi\colon \mathbb{B}(\ell^2(\Gamma)) \to \mathbb{B}(\mathcal{H})$ and isometries $V, W\colon \ell^2(\Gamma) \to \mathcal{H}$ such that $m_k(x) = V^*\pi(x)W$ for every $x \in \mathbb{B}(\ell^2(\Gamma))$. Fix $r \in \Gamma$ and set $\xi_s = \pi(e_{r,s})V\delta_s$ and $\eta_t = \pi(e_{r,t})W\delta_t$. It follows that

$$\langle \eta_t, \xi_s\rangle = \langle V^*\pi(e_{s,t})W\delta_t, \delta_s\rangle = m_k(e_{s,t})_{s,t} = k(s,t).$$

Conversely, if $k(s,t) = \langle \eta_t, \xi_s\rangle$ for some $\xi_s, \eta_t \in \mathcal{H}$ with $\|\xi_s\|, \|\eta_t\| \le 1$, then we define contractions $V, W\colon \ell^2(\Gamma) \to \ell^2(\Gamma) \otimes \mathcal{H}$ by $V\delta_s = \delta_s \otimes \xi_s$ and $W\delta_t = \delta_t \otimes \eta_t$. Now check that $m_k(x) = V^*(x \otimes 1)W$ for $x \in \mathbb{B}(\ell^2(\Gamma))$. □

If $k\colon \Gamma_1 \times \Gamma_1 \to \mathbb{C}$ is a kernel and $f\colon \Gamma_2 \to \Gamma_1$ is a map, then for the kernel $k_f(s,t) = k(f(s), f(t))$ on Γ_2, we have $\|m_{k_f}\|_{\mathrm{cb}} \le \|m_k\|_{\mathrm{cb}}$.

It turns out that a bounded linear map $\theta\colon \mathbb{B}(\ell^2(\Gamma)) \to \mathbb{B}(\ell^2(\Gamma))$ is a Schur multiplier if and only if it is an $\ell^\infty(\Gamma)$-module map – i.e., $\theta(fxg) = f\theta(x)g$ for $f, g \in \ell^\infty(\Gamma)$ and $x \in \mathbb{B}(\ell^2(\Gamma))$. Since $\ell^\infty(\Gamma)$ has a cyclic vector (when Γ is countable), any bounded Schur multiplier θ is automatically completely bounded and $\|\theta\|_{\mathrm{cb}} = \|\theta\|$. The result below comes from [**178**], inspired by work of Christensen and Haagerup.

Proposition D.5. *Let $A \subset \mathbb{B}(\mathcal{H})$ be a C*-subalgebra with a cyclic vector. Let $X \subset \mathbb{B}(\mathcal{H})$ be an operator space such that $AXA \subset X$. If $\varphi\colon X \to \mathbb{B}(\mathcal{H})$ is a bounded A-A-bimodule map, then it is c.b. with $\|\varphi\|_{\mathrm{cb}} = \|\varphi\|$.*

Proof. The proof is by contradiction. Suppose φ is contractive, but there are $n \in \mathbb{N}$ and $[x_{ij}] \in \mathbb{M}_n(X)$ with $\|[x_{ij}]\| = 1$ such that $\|[\varphi(x_{ij})]\| > 1$. Then there are $\varepsilon > 0$, $\xi_1, \ldots, \xi_n, \eta_1, \ldots, \eta_n \in \mathcal{H}$ such that $\sum \|\xi_i\|^2 = \sum \|\eta_j\|^2 = 1 - \varepsilon$ and $|\sum \langle \varphi(x_{ij})\eta_j, \xi_i\rangle| > 1$. Let ζ be an A-cyclic unit vector. We may assume that $\xi_i = a_i\zeta$ and $\eta_j = b_j\zeta$. For $a = \varepsilon + \sum a_i^* a_i$, we have $\|a^{1/2}\zeta\|^2 = \varepsilon + \sum \|\xi_i\|^2 = 1$; similarly for $b = \varepsilon + \sum b_j^* b_j$. Let $c_i = a_i a^{-1/2}$ and $d_j = b_j b^{-1/2}$. We have $\sum c_i^* c_i \le 1$ and $\sum d_j^* d_j \le 1$. It follows that

$$\begin{aligned}
|\sum_{i,j} \langle \varphi(x_{ij})\eta_j, \xi_i\rangle| &= |\sum_{i,j} \langle \varphi(x_{ij})d_j b^{1/2}\zeta, c_i a^{1/2}\zeta\rangle| \\
&= |\langle \varphi(\sum_{i,j} c_i^* x_{ij} d_j) b^{1/2}\zeta, a^{1/2}\zeta\rangle| \\
&\le \|\sum_{i,j} c_i^* x_{ij} d_j\|_X \\
&\le \|\sum_i c_i^* c_i\|^{1/2} \, \|[x_{ij}]\|_{\mathbb{M}_n(X)} \, \|\sum_j d_j^* d_j\|^{1/2} \\
&\le \|[x_{ij}]\|_{\mathbb{M}_n(X)} = 1.
\end{aligned}$$

This is a contradiction. □

Let Γ be a discrete group and $\varphi\colon \Gamma \to \mathbb{C}$ be a function. Abusing notation, we denote by m_φ the Schur multiplier on $\mathbb{B}(\ell^2(\Gamma))$ that is associated with the kernel $(s,t) \mapsto \varphi(st^{-1})$. This notation is compatible with Definition 2.5.10: $m_\varphi(\lambda(s)) = \varphi(s)\lambda(s)$ for every $s \in \Gamma$.

Proposition D.6. *Let Γ be a discrete group and $\varphi\colon \Gamma \to \mathbb{C}$ be a function. The multiplier m_φ on $\mathbb{B}(\ell^2(\Gamma))$ is completely bounded if and only if it is completely bounded on $C^*_\lambda(\Gamma)$. Moreover, their cb-norms coincide.*

Proof. Suppose that m_φ is completely contractive on $C^*_\lambda(\Gamma)$. Let U be the unitary operator on $\ell^2(\Gamma) \otimes \ell^2(\Gamma)$ given by $U\delta_s \otimes \delta_t = \delta_s \otimes \delta_{st}$ for $s,t \in \Gamma$. An easy computation shows

$$U(e_{s,t} \otimes 1)U^* = e_{s,t} \otimes \lambda(st^{-1}) \in \mathbb{B}(\ell^2(\Gamma)) \otimes C^*_\lambda(\Gamma)$$

for every $s,t \in \Gamma$. It follows that

$$U(m_\varphi(x) \otimes 1)U^* = (\mathrm{id}_{\mathbb{B}(\ell^2(\Gamma))} \otimes (m_\varphi)_{|C^*_\lambda(\Gamma)})(U(x \otimes 1)U^*)$$

for any $x \in \mathbb{K}(\ell^2(\Gamma))$. Hence, m_φ is completely contractive on $\mathbb{K}(\ell^2(\Gamma))$. By ultraweak continuity, this implies that m_φ is completely contractive on $\mathbb{B}(\ell^2(\Gamma))$. □

We note that the norm of m_φ on $C^*_\lambda(\Gamma)$ does not coincide with the cb-norm in general.

A function φ on a group Γ is called a *Herz-Schur multiplier* if m_φ is completely bounded. (Sometimes one also refers to m_φ as a Herz-Schur multiplier.) We denote by $B_2(\Gamma)$ the Banach space of Herz-Schur multipliers equipped with the Herz-Schur norm defined by $\|\varphi\|_{B_2} = \|m_\varphi\|_{\mathrm{cb}}$. Let $\varphi\colon \Gamma \to \mathbb{C}$ be a function and ω_φ be the corresponding linear functional on $\mathbb{C}[\Gamma]$ (cf. Definition 2.5.10). If ω_φ is bounded on $C^*(\Gamma)$, then φ is a Herz-Schur multiplier with $\|\varphi\|_{B_2} \le \|\omega\|$ (cf. proof of Theorem 2.5.11, (3) $\Rightarrow$ (4)). In particular, $\|\varphi\|_{B_2} \le \|\varphi\|_2$ for every $\varphi \in \ell^2(\Gamma)$. (This fact also follows easily from Theorem D.4.) It follows that the norm closure of $\mathbb{C}[\Gamma]$ in $B_2(\Gamma)$ contains $\ell^2(\Gamma)$.

A finitely supported function $\omega \in \mathbb{C}[\Gamma]$ defines an element of $B_2(\Gamma)^*$ by the formula $\omega(\varphi) = \sum_{s\in\Gamma} \omega(s)\varphi(s)$. We denote by $Q(\Gamma)$ the norm closure of $\mathbb{C}[\Gamma]$ in $B_2(\Gamma)^*$. For $a \in C^*_\lambda(\Gamma) \otimes \mathbb{B}(\ell^2)$ and $f \in (C^*_\lambda(\Gamma) \otimes \mathbb{B}(\ell^2))^*$ or for $a \in L(\Gamma) \bar{\otimes} \mathbb{B}(\ell^2)$ and $f \in (L(\Gamma) \bar{\otimes} \mathbb{B}(\ell^2))_*$, we define $\omega_{a,f} \in B_2(\Gamma)^*$ by

$$\omega_{a,f}(\varphi) = f\big(m_\varphi \otimes \mathrm{id}(a)\big).$$

It is clear that $\|\omega_{a,f}\|_{B_2^*} \le \|a\|\|f\|$.

Lemma D.7. *For every a and f as above, $\omega_{a,f} \in Q(\Gamma)$.*

Proof. We first prove the assertion for $a \in C^*_\lambda(\Gamma) \otimes \mathbb{B}(\ell^2)$ and $f \in (C^*_\lambda(\Gamma) \otimes \mathbb{B}(\ell^2))^*$. Since $\|\omega_{a,f}\|_{B_2^*} \leq \|a\|\|f\|$, taking an approximation, we may assume that $a = \sum_{s\in\Gamma} \lambda(s) \otimes a(s)$, where only finitely many $a(s) \in \mathbb{B}(\ell^2)$ are nonzero. Then,

$$\omega_{a,f}(\varphi) = f\big(\sum_{s\in\Gamma} \varphi(s)\lambda(s) \otimes a(s)\big) = \sum_{s\in\Gamma} f\big(\lambda(s) \otimes a(s)\big)\,\varphi(s)$$

and $\omega_{a,f} \in \mathbb{C}[\Gamma] \subset Q(\Gamma)$. The other case is similar – one approximates f instead of a. □

Lemma D.8. *There is a canonical isometric isomorphism $B_2(\Gamma) = Q(\Gamma)^*$.*

Proof. Since $Q(\Gamma) \subset B_2(\Gamma)^*$, we have a canonical contraction $B_2(\Gamma) \to Q(\Gamma)^*$. We need to show it is a surjective isometry. Since

$$\|\varphi\|_{B_2} = \sup\{|\omega_{a,f}(\varphi)| : \|a\| \leq 1,\ \|f\| \leq 1\},$$

it follows that $B_2(\Gamma) \to Q(\Gamma)^*$ is an isometry. Let $\psi \in Q(\Gamma)^*$ be given and define $\varphi \in \ell^\infty(\Gamma)$ by $\varphi(s) = \langle \psi, \delta_s \rangle$. Since $\varphi = \psi$ on $\mathbb{C}[\Gamma]$, one sees that $\varphi \in B_2(\Gamma)$ and $\varphi = \psi$ on $Q(\Gamma)$. □

Lemma D.9. *Every $\omega \in Q(\Gamma)$ is of the form $\omega = \omega_{a,f}$ for some $a \in C^*_\lambda(\Gamma) \otimes \mathbb{K}(\ell^2)$ and $f \in (L(\Gamma) \bar{\otimes} \mathbb{B}(\ell^2))_*$ (with $\|a\|\|f\|$ arbitrarily close to $\|\omega\|$).*

Proof. Consider the set

$$S = \{\omega_{a,f} : a \in C^*_\lambda(\Gamma)\otimes\mathbb{K}(\ell^2),\ f \in (L(\Gamma)\bar{\otimes}\mathbb{B}(\ell^2))_* \text{ with } \|a\| \leq 1,\ \|f\| \leq 1\}.$$

Since $\ell^2 \oplus \ell^2 \cong \ell^2$, the set S is a convex subset of the closed unit ball of $Q(\Gamma)$ such that

$$\|\varphi\|_{B_2} = \sup\{|\omega_{a,f}(\varphi)| : \omega_{a,f} \in S\},$$

for every $\varphi \in B_2(\Gamma)$. By the Hahn-Banach separation theorem, S is norm-dense in the closed unit ball of $Q(\Gamma)$. Hence, for every norm-one element $\omega \in Q(\Gamma)$, there exists a sequence $(\omega_{a_n,f_n})_{n=0}^\infty$ in S such that $\omega = \sum_{n=0}^\infty 4^{-n}\omega_{a_n,f_n}$. We will view elements in $C^*_\lambda(\Gamma) \otimes \mathbb{K}(\ell^2 \otimes \ell^2)$ (resp. $L(\Gamma) \bar{\otimes} \mathbb{B}(\ell^2 \otimes \ell^2)$) as $\infty \times \infty$ matrices with entries in $C^*_\lambda(\Gamma) \otimes \mathbb{K}(\ell^2)$ (resp. $L(\Gamma) \bar{\otimes} \mathbb{B}(\ell_2)$). It follows that

$$a := \operatorname{diag}(2^{-n}a_n) \in C^*_\lambda(\Gamma) \otimes \mathbb{K}(\ell^2 \otimes \ell^2)$$

and $f \in (L(\Gamma) \bar{\otimes} \mathbb{B}(\ell^2 \otimes \ell^2))_*$, defined by

$$f\colon L(\Gamma) \bar{\otimes} \mathbb{B}(\ell^2 \otimes \ell^2) \ni [x_{ij}] \mapsto \sum_{n=0}^{\infty} 2^{-n} f_n(x_{nn}) \in \mathbb{C},$$

satisfy $\omega_{a,f} = \sum_{n=0}^\infty 4^{-n}\omega_{a_n,f_n} = \omega$. This completes the proof. □

Cocycles of unitary representations. Let $(\pi, \mathcal{H})$ be a unitary representation of Γ. Then, a 1-*cocycle* on Γ (with coefficients in $(\pi, \mathcal{H})$) is a function $b\colon \Gamma \to \mathcal{H}$ such that

$$b(st) = b(s) + \pi(s)b(t)$$

for every $s, t \in \Gamma$. We note that $b(e) = 0$ and that the closed linear span of $b(\Gamma)$ is a $\pi(\Gamma)$-invariant subspace of $\mathcal{H}$. We often do not bother to mention the unitary representation $(\pi, \mathcal{H})$. Let b_i, $i = 1, 2$, be 1-cocycles on Γ with coefficients in $(\pi_i, \mathcal{H}_i)$ such that $b_i(\Gamma)$ have dense linear span in $\mathcal{H}_i$. If there exists a unitary operator u from $\mathcal{H}_1$ onto $\mathcal{H}_2$ such that $ub_1(s) = b_2(s)$ for all $s \in \Gamma$, then $u\pi_1(s)u^* = \pi_2(s)$ for all $s \in \Gamma$.

We observe that a function $b\colon \Gamma \to \mathcal{H}$ is a 1-cocycle if and only if the map θ from Γ into the group of affine isometries on $\mathcal{H}$, defined by

$$\theta(s)\colon \mathcal{H} \ni \xi \mapsto \pi(s)\xi + b(s) \in \mathcal{H}$$

for $s \in \Gamma$ and $\xi \in \mathcal{H}$, is a group homomorphism. Note that any group homomorphism from Γ into the group of (affine) isometries on a real Hilbert space $\mathcal{H}_{\mathbb{R}}$ is of the above form, where π is an orthogonal representation of Γ on $\mathcal{H}_{\mathbb{R}}$.

A 1-cocycle b on Γ is called a 1-*coboundary* if there exists $\xi \in \mathcal{H}$ such that $b(s) = \xi - \pi(s)\xi$ for all $s \in \Gamma$ (or equivalently $\theta(s)\xi = \xi$ for all $s \in \Gamma$).

Lemma D.10. *A 1-cocycle is a 1-coboundary if and only if it is bounded.*

Proof. Let $b\colon \Gamma \to \mathcal{H}$ be a bounded 1-cocycle. Since $b(\Gamma)$ is a bounded subset in a Hilbert space, there exists a unique circumcenter $\xi \in \mathcal{H}$ of $b(\Gamma)$ (see Exercise D.1). Since $b(\Gamma)$ is invariant under the affine isometry $\theta(s)$, the vector ξ is invariant under $\theta(s)$ for every $s \in \Gamma$. It follows that $b(s) = \xi - \pi(s)\xi$. □

We observe that if b is a 1-cocycle on a group Γ, then

$$\|b(s) - b(t)\| = \| - \pi(s)b(s^{-1}t)\| = \|b(s^{-1}t)\|$$

for every $s, t \in \Gamma$. Let us forget for a moment the group structure of Γ. We say a kernel $k\colon \Gamma \times \Gamma \to \mathbb{R}$ is *conditionally negative definite* if there exists a function b from Γ into a Hilbert space $\mathcal{H}$ such that $k(s, t) = \|b(s) - b(t)\|^2$. It is well known and not hard to see that a real-valued kernel k is conditionally negative definite if and only if k is symmetric (i.e., $k(s, t) = k(t, s)$), $k(s, s) = 0$ for every $s \in \Gamma$ and $\sum_{i,j=1}^{n} k(s_i, s_j)\alpha_i\alpha_j \leq 0$ for any finite sequences $s_1, \ldots, s_n \in \Gamma$ and $\alpha_1, \ldots, \alpha_n \in \mathbb{R}$ with $\sum_{i=1}^{n} \alpha_i = 0$.

Theorem D.11 (Schoenberg). *Let k be a conditionally negative definite kernel. Then the kernel $\Gamma \times \Gamma \ni (s, t) \mapsto \exp(-k(s, t)) \in \mathbb{R}$ is positive* *definite. In particular, for any* 1*-cocycle* b *on a group* Γ *and* $\gamma > 0$*, the function* φ_γ^b *on* Γ*, defined by*

$$\varphi_\gamma^b(s) = \exp(-\gamma\|b(s)\|^2),$$

is positive definite.

Proof. Let $k(s,t) = \|b(s) - b(t)\|^2$. A computation shows

$$\exp(-k(s,t)) = \exp(-\|b(s)\|^2)\exp(-\|b(t)\|^2)\exp(2\Re\langle b(s), b(t)\rangle).$$

It is easy to see that $(s,t) \mapsto \exp(-\|b(s)\|^2)\exp(-\|b(t)\|^2)$ is a positive definite kernel. On the other hand,

$$\exp(2\Re\langle b(s), b(t)\rangle) = \sum_{n\geq 0} \frac{(2\Re\langle b(s), b(t)\rangle)^n}{n!}$$

and it is easy to see that $\Re\langle b(s), b(t)\rangle$ is positive definite and so are the iterated Schur products $(\Re\langle b(s), b(t)\rangle)^n$. Hence, $(s,t) \mapsto \exp(-k(s,t))$ is positive definite since it is a Schur product of positive definite kernels. □

Let b be a 1-cocycle on Γ. We note that $\varphi_\gamma^b(e) = 1$ and $\varphi_\gamma^b \to 1$ pointwise as $\gamma \to 0$. On the other hand, $\varphi_\gamma^b \to 1$ uniformly as $\gamma \to 0$ if and only if b is bounded.

Lemma D.12. *Let* b *be a* 1*-cocycle of a group* Γ *and let* $\gamma > 0$*. Let* $(\pi_\gamma^b, \mathcal{H}_\gamma^b, \xi_\gamma^b)$ *be the GNS triplet of the positive definite function* φ_γ^b *on* Γ *given in Theorem D.11. Suppose that* b *is unbounded on a subgroup* Λ *of* Γ*. Then,* $(\pi_\gamma^b, \mathcal{H}_\gamma^b)$ *has no nonzero* Λ*-invariant vector.*

Proof. Suppose that b is unbounded on a subgroup Λ and choose a sequence (s_n) in Λ such that $\|b(s_n)\| \to \infty$. Since $\|b(ts_nt')\| \geq \|b(s_n)\| - (\|b(t)\| + \|b(t')\|)$, we have

$$\limsup_{n\to\infty} \varphi_\gamma^b(ts_nt') = \limsup_{n\to\infty} \exp(-\gamma\|b(ts_nt')\|^2) = 0$$

for every $t, t' \in \Gamma$. Hence, for any $\zeta = \sum_{i=1}^n \alpha_i\pi_\gamma^b(t_i)\xi \in \mathcal{H}_\gamma^b$, we have

$$\limsup_{n\to\infty} |\langle\pi_\gamma^b(s_n)\zeta, \zeta\rangle| = \limsup_{n\to\infty} |\sum_{i,j} \overline{\alpha_i}\alpha_j\, \varphi_\gamma^b(t_i^{-1}s_nt_j)| = 0.$$

By continuity, $\lim\langle\pi_\gamma^b(s_n)\zeta, \zeta\rangle = 0$ for every $\zeta \in \mathcal{H}_\gamma^b$ (i.e., $\pi_\gamma^b|_\Lambda$ is *weakly mixing*). Clearly, there is no nonzero Λ-invariant vector. □

Exercises

Exercise D.1. Let V be a bounded subset of a Hilbert space and let

$$r_0 = \inf\{r \geq 0 : V \text{ is contained in a closed ball of radius } r\}.$$

Prove that there exists a unique vector ζ, called the *circumcenter* of V, such that V is contained in the closed ball with center ζ and radius r_0. Prove also that the circumcenter ζ is in the closed convex hull of V.

Exercise D.2. The complexification of a real Hilbert space $\mathcal{H}_{\mathbb{R}}$ is $\mathcal{H}_{\mathbb{C}} = \mathcal{H}_{\mathbb{R}} + \sqrt{-1}\mathcal{H}_{\mathbb{R}}$ with the obvious complex inner product. Observe that one can complexify an orthogonal representation of a group on $\mathcal{H}_{\mathbb{R}}$. Conversely, the "realification" of a complex Hilbert space $\mathcal{H}$ is the real Hilbert space $\mathcal{H}$ with the real inner product $\Re\langle\,\cdot\,,\,\cdot\,\rangle$. Observe that one can "realify" a unitary representation of a group on $\mathcal{H}$.

We outline below another proof of Schoenberg's Theorem.

Exercise D.3. For a real Hilbert space $\mathcal{H}$ and arbitrary vector $\xi \in \mathcal{H}$, we define

$$\mathrm{EXP}(\mathcal{H}) = \mathbb{C}\Omega \oplus \bigoplus_{n\geq 1} \mathcal{H}^{\otimes n} \quad \text{and} \quad \mathrm{EXP}(\xi) = \Omega + \sum_{n\geq 1} \frac{\xi^{\otimes n}}{\sqrt{n!}},$$

where Ω is a fixed unit vector. Observe that $\langle \mathrm{EXP}(\eta), \mathrm{EXP}(\xi)\rangle = \exp\langle \eta, \xi\rangle$. Check that if we define a unit vector

$$\xi_\gamma = \exp(-\frac{1}{2}\gamma^2\|\xi\|^2)\,\mathrm{EXP}(\gamma\xi) \in \mathrm{EXP}(\mathcal{H})$$

for $\gamma > 0$ and $\xi \in \mathcal{H}$, then we have $\langle \eta_\gamma, \xi_\gamma\rangle = \exp(-\gamma^2\|\eta - \xi\|^2/2)$.

Suppose that θ is an action of Γ on $\mathcal{H}$ by affine isometries. We define a representation σ of Γ on $\mathcal{K} = \overline{\mathrm{span}}\{\mathrm{EXP}(\xi) : \xi \in \mathcal{H}\} \subset \mathrm{EXP}(\mathcal{H})$ by

$$\sigma(s)\sum_i \lambda_i \,\mathrm{EXP}(\xi_i) = \sum_i \lambda_i \exp(\frac{1}{2}\|\xi_i\|^2 - \frac{1}{2}\|\theta(s)\xi_i\|^2)\,\mathrm{EXP}(\theta(s)\xi_i).$$

Prove σ is a well-defined orthogonal representation such that $\langle \sigma(s)\Omega, \Omega\rangle = \exp(-\|b(s)\|^2/2)$, where $b(s) \in \mathcal{H}$ is the translation part of the affine isometry $\theta(s)$.

References. Most of the results in this appendix are classical. Exceptions are Proposition D.6, which comes from [**24**], and Lemmas D.7–D.9, which come from [**80**]. Many other important results on Herz-Schur multipliers are found in [**47, 54, 78, 80**]; see also [**151**].

Appendix E

Groups and Graphs

Definition of graphs.

Definition E.1. A *graph* $\mathbf{X}$ consists of a *vertex set* $\mathbf{V} = \mathbf{V}(\mathbf{X})$ and an *edge set* $\mathbf{E} = \mathbf{E}(\mathbf{X})$, equipped with two maps $s, r \colon \mathbf{E} \to \mathbf{V}$, called the *source map* and the *range map* ($s(\mathbf{e})$ is the vertex at which an edge begins, while $r(\mathbf{e})$ is the vertex at which it ends). A *simple* graph is a graph without loops or multiple edges, i.e., $s(\mathbf{e}) \neq r(\mathbf{e})$ for every $\mathbf{e} \in \mathbf{E}$ and the map $\mathbf{E} \ni \mathbf{e} \mapsto (s(\mathbf{e}), r(\mathbf{e})) \in \mathbf{V}^2$ is injective. When dealing with a simple graph, we identify the edge $\mathbf{e}$ with $(s(\mathbf{e}), r(\mathbf{e})) \in \mathbf{V}^2$. A simple graph is *undirected* if $(x, y) \in \mathbf{E}$ implies $(y, x) \in \mathbf{E}$.

Except for those used in graph C*-algebras, all graphs are assumed to be simple and undirected. We say two vertices $x, y \in \mathbf{V}$ are *adjacent* if $(x, y) \in \mathbf{E}$. We often identify the graph $\mathbf{X}$ with its vertex set $\mathbf{V}$. A *path* α of length n (possibly $n = \infty$) in $\mathbf{X}$ is a sequence $x_0 x_1 \cdots x_n$ of vertices such that x_{k+1} is adjacent to x_k for every $0 \leq k < n$. We say the path α *connects* x_0 to x_n. It's convenient to assume the graph $\mathbf{X}$ is *connected*: For any pair of vertices, there exists a path which connects one to the other. The (graph) metric on the vertex set $\mathbf{V}$ is

$$d(x, y) = \min\{n : \exists \text{ a path of length } n \text{ which connects } x \text{ to } y\}.$$

A path $x_0 x_1 \cdots$ is called a *geodesic* if $d(x_n, x_m) = |n - m|$ for all n, m.

Lemma E.2. *Let $\alpha = x_0 x_1 \cdots$ be an infinite geodesic path in $\mathbf{X}$ and $y \in \mathbf{X}$ be a vertex. Then, there is a geodesic path $\beta = y_0 y_1 \cdots$ starting at $y = y_0$ and which eventually flows into α, i.e., there exists $k \in \mathbb{Z}$ such that $y_n = x_{k+n}$ for all large enough n.*

Proof. For m, n with $m \leq n$, we define $h(m,n) = d(y, x_m) + d(x_m, x_n) - d(y, x_n)$. Note that h decreases (resp. increases) as m (resp. n) increases and that $h(m,k) = h(m,n) + h(n,k)$ for every $m \leq n \leq k$. Since $0 \leq h(m,n) \leq 2d(y, x_0)$, $h(m) = \lim_k h(m,k)$ exists. We note that $h(m) = h(m,n) + h(n)$ for every $m \leq n$. It follows that for m_0 such that $h(m_0) = \inf h(m)$, we have $h(m_0, n) = 0$ for every $n \geq m_0$. Therefore, letting β be the concatenation of a geodesic path connecting y to x_{m_0} and the subpath $x_{m_0} x_{m_0+1} \cdots$ of α, we are done. □

Definition E.3. A (simple and connected) graph $\mathbf{X}$ is called a *tree* if there is no nontrivial loop, i.e., there is no path $x_0 x_1 \cdots x_n$ such that $x_n = x_0$, $n > 2$, and $x_1, \ldots, x_n$ are all distinct.

Example E.4. Let Γ be a group and $\mathcal{S}$ be a set of generators such that $e \notin \mathcal{S}$. Then, the Cayley graph $\mathbf{X} = \mathbf{X}(\Gamma, \mathcal{S})$ of Γ with respect to $\mathcal{S}$ is the graph whose vertex set is Γ and whose edge set is $\mathbf{E} = \{(s,t) : s^{-1}t \in \mathcal{S} \cup \mathcal{S}^{-1}\}$. For example, if $\Gamma = \mathbb{F}_2$ is freely generated by $\mathcal{S} = \{a, b\}$, then the corresponding Cayley graph is a tree (of degree four).

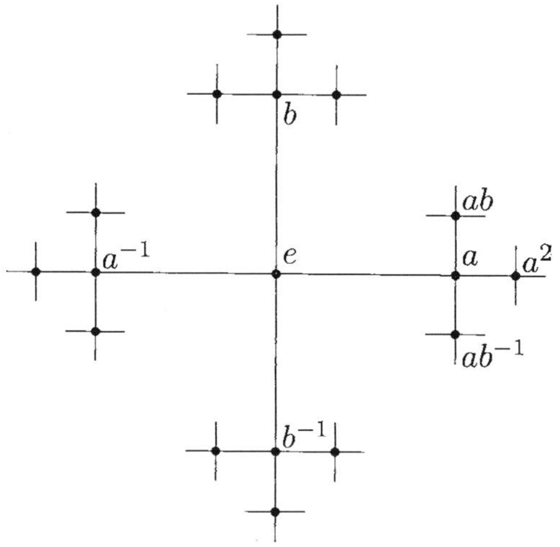

Figure 1. The Cayley graph of $\mathbb{F}_2$

The group of automorphisms on $\mathbf{X}$ (i.e., isometric bijections on $\mathbf{V}$) is denoted by $\mathrm{Aut}(\mathbf{X})$. An *action* of a group Γ on the graph $\mathbf{X}$ is a group homomorphism from Γ into $\mathrm{Aut}(\mathbf{X})$. The action is said to be *(metrically) proper* if for every $F \subset \mathbf{V}$ of finite diameter, the set $|\{s \in \Gamma : sF \cap F \neq \emptyset\}|$ is finite. The *stabilizer* of a vertex $x \in \mathbf{V}$ is the subgroup

$$\Gamma^x = \{s \in \Gamma : s.x = x\} \subset \Gamma.$$

We note that $s\Gamma^x s^{-1} = \Gamma^{s.x}$. The stabilizer of an edge $(x,y) \in \mathbf{E}$ is $\Gamma^{(x,y)} = \Gamma^x \cap \Gamma^y$. Every group acts by left multiplication on its Cayley graph $\mathbf{X}(\Gamma, \mathcal{S})$; note that this action is *free* (i.e., all stabilizers are trivial).

Expanders. In this section, we consider a finite connected graph $\mathbf{X}$. For each vertex $x \in \mathbf{V}$, the *degree* of x is defined as

$$\nu(x) = |\{y \in \mathbf{V} : (x,y) \in \mathbf{E}\}|.$$

A graph $\mathbf{X}$ is said to be *d-regular* if $\nu(x) = d$ for all x. We set $|\nu| = \sum_{x\in\mathbf{V}} \nu(x) = |\mathbf{E}|$ and denote by $L^2(\mathbf{V},\nu)$ the weighted L^2-space on $\mathbf{V}$: For functions f and g on $\mathbf{V}$, we have

$$\langle f, g\rangle_\nu = \frac{1}{|\nu|}\sum_{x\in\mathbf{V}} \nu(x) f(x)\overline{g(x)}.$$

We define $L^2(\mathbf{E})$ to be the L^2-space on $\mathbf{E}$ with the uniform weight. One defines the "differential operator" $d\colon L^2(\mathbf{V},\nu) \to L^2(\mathbf{E})$ by $(df)(x,y) = f(y) - f(x)$. Then the *combinatorial Laplacian* $\Delta = \Delta_{\mathbf{X}}$ is defined as the positive operator $d^*d/2$ on $L^2(\mathbf{V},\nu)$. One checks that

$$\begin{aligned}
\langle \Delta f, f\rangle_\nu = \frac{1}{2}\|df\|^2 &= \frac{1}{2|\mathbf{E}|}\sum_{(x,y)\in\mathbf{E}} |f(x)-f(y)|^2 \\
&= \frac{1}{|\nu|}\sum_{x\in\mathbf{V}}\sum_{y;\,(x,y)\in\mathbf{E}} \left(|f(x)|^2 - f(y)\overline{f(x)}\right) \\
&= \frac{1}{|\nu|}\sum_{x\in\mathbf{V}} \nu(x)\Big(f(x) - \frac{1}{\nu(x)}\sum_{y;\,(x,y)\in\mathbf{E}} f(y)\Big)\overline{f(x)}.
\end{aligned}$$

It follows that

$$(\Delta f)(x) = f(x) - \frac{1}{\nu(x)}\sum_{y;\,(x,y)\in\mathbf{E}} f(y).$$

Since the graph $\mathbf{X}$ is connected, eigenvectors of the eigenvalue zero are constant functions. We denote by $\lambda_1 = \lambda_1(\mathbf{X})$ the first nonzero positive eigenvalue of Δ.

Let $\mathcal{H}$ be a Hilbert space and $f\colon \mathbf{V} \to \mathcal{H}$ be a function. We view f as an element of the Hilbert space $L^2(\mathbf{V},\nu)\otimes\mathcal{H}$. Here's a Poincaré-type inequality.

Lemma E.5. *Let $\mathbf{X}$ be a finite connected graph. Then, for any map f from the vertex set $\mathbf{V}$ into a Hilbert space $\mathcal{H}$, we have*

$$\frac{\lambda_1(\mathbf{X})}{2|\nu|^2}\sum_{x,y\in\mathbf{V}} \nu(x)\nu(y)\|f(x)-f(y)\|^2 = \lambda_1(\mathbf{X})\left(\|f\|^2_{L^2(\mathbf{V},\nu)\otimes\mathcal{H}} - \|m\|^2_{\mathcal{H}}\right)$$

$$\leq \frac{1}{2|\mathbf{E}|} \sum_{(x,y)\in\mathbf{E}} \|f(x) - f(y)\|^2,$$

where $m = |\nu|^{-1} \sum_{x\in\mathbf{V}} \nu(x) f(x) \in \mathcal{H}$ *is the mean of* f.

Proof. The first equality is routine to check. Let $\mathbb{E}$ be the orthogonal projection from $L^2(\mathbf{V}, \nu)$ onto the subspace $\mathbb{C}1$ of constant functions. It is easy to see $(\mathbb{E} \otimes 1)(f) = m$, where $m \in \mathcal{H}$ is viewed as a constant function in $L^2(\mathbf{V}) \otimes \mathcal{H}$. It follows that

$$\begin{aligned}\lambda_1(\mathbf{X})(\|f\|^2_{L^2(\mathbf{V},\nu)\otimes\mathcal{H}} - \|m\|^2_{\mathcal{H}}) &= \lambda_1(\mathbf{X})\|f - (\mathbb{E}\otimes 1)(f)\|^2_{L^2(\mathbf{V})\otimes\mathcal{H}} \\ &\leq \langle(\Delta \otimes I_{\mathcal{H}})f, f\rangle_{L^2(\mathbf{V})\otimes\mathcal{H}} \\ &= \frac{1}{2|\mathbf{E}|} \sum_{(x,y)\in\mathbf{E}} \|f(x) - f(y)\|^2.\end{aligned}$$

□

Definition E.6. A sequence $\{\mathbf{X}_n\}$ of finite connected graphs is called a *sequence of expanders* if $\lim |\mathbf{V}(\mathbf{X}_n)| = \infty$ and $\inf \lambda_1(\mathbf{X}_n) > 0$.

Usually, a sequence of expanders is required to have uniformly bounded degrees. By (discrete) probabilistic methods, it is relatively easy to show that "most" d-regular graphs have "large" λ_1. However, the first explicit construction of a sequence of expanders was given by Margulis using (relative) property (T). Recall that a group Γ has property (T) if the following is true: If a unitary representation $(\pi, \mathcal{H})$ of Γ has a sequence ξ_n of unit vectors such that $\lim \|\xi_n - \pi(s)\xi_n\| = 0$ for every $s \in \Gamma$, then there exists a unit vector $\xi \in \mathcal{H}$ such that $\pi(s)\xi = \xi$ for all $s \in \Gamma$. Recall that $\mathrm{SL}(3,\mathbb{Z})$ has property (T) (see Sections 6.4 and 12.1 for more information).

Theorem E.7. *Let* Γ *be a residually finite group with property* (T) *(e.g.,* $\Gamma = \mathrm{SL}(3,\mathbb{Z})$*) and* $\mathcal{S}$ *be a finite symmetric generating subset without the unit. Let* $\{\Gamma_n\}$ *be a sequence of finite quotients of* Γ *such that the image* $\mathcal{S}_n$ *of* $\mathcal{S}$ *in* Γ_n *does not contain the unit. Then, the Cayley graphs* $\mathbf{X}(\Gamma_n, \mathcal{S}_n)$ *form a sequence of expanders.*

Proof. First note that the Cayley graphs $\mathbf{X}_n = \mathbf{X}(\Gamma_n, \mathcal{S}_n)$ are all connected and $|\mathcal{S}|$-regular. Composing the left regular representation of Γ_n with the quotient homomorphism, we obtain a unitary representation $(\pi_n, \ell^2(\Gamma_n))$ of Γ. We observe that

$$\Delta_n = 1_{\ell^2(\Gamma_n)} - \frac{1}{|\mathcal{S}|} \sum_{s\in\mathcal{S}} \pi_n(s) \in \mathbb{B}(\ell^2(\Gamma_n))$$

is the combinatorial Laplacian of $\mathbf{X}_n$ (modulo identification of $\ell^2(\Gamma_n)$ and $L^2(\Gamma_n)$). Let $\ell^2(\Gamma_n)^0 \subset \ell^2(\Gamma_n)$ be the orthogonal complement of the constant functions and π_n^0 be the restriction of π_n to the π_n-invariant subspace $\ell^2(\Gamma_n)^0$. We set $\pi^0 = \bigoplus \pi_n^0$ and observe that π does not have a nonzero invariant vector. Suppose by contradiction that $\inf \lambda_1(\mathbf{X}_n) = 0$. Let $\xi_n \in \ell^2(\Gamma_n)^0$ be a unit eigenvector of Δ_n with eigenvalue $\lambda_1(\mathbf{X}_n)$. Passing to a subsequence, we may assume that $\lim \|\Delta_n \xi_n\| = 0$. By uniform convexity of Hilbert spaces, this implies that $\lim \|\xi_n - \pi_n(s)\xi_n\| = 0$ for every $s \in \mathcal{S}$ and hence for all $s \in \Gamma$. This contradicts property (T). For a qualitative proof, refer to Lemma 12.1.8. □

Amalgamated free products. In this section, we state the definitions of amalgamated free products and HNN extensions (that they exist is a theorem which we do not prove).

Definition E.8 (Amalgamated free products). Let Γ_i, $i = 1, 2$, be groups and Λ be a common subgroup (i.e., Λ comes with an injective homomorphism into each Γ_i). Then, the amalgamated free product $\Gamma = \Gamma_1 *_\Lambda \Gamma_2$ is the group satisfying the following properties:

(1) Γ contains Γ_1 and Γ_2 as subgroups and Γ is generated by Γ_1 and Γ_2;

(2) $\Gamma_1 \cap \Gamma_2 = \Lambda$ in Γ;

(3) $s_1 \cdots s_n a \neq e$ whenever $n \geq 1$, $a \in \Lambda$ and $s_k \in \Gamma_{i_k} \setminus \Lambda$ with $i_k \neq i_{k+1}$ for $1 \leq k < n$;

(4) if we choose systems S_i of representatives of Γ_i/Λ and let $S_i^0 = S_i \setminus \{e\}$ (we always assume that the representative of the coset Λ is e), then any element s in Γ can be uniquely written as

$$s = s_1 \cdots s_n a,$$

where $a \in \Lambda$ and $s_k \in S_{i_k}^0$ such that $i_k \neq i_{k+1}$ for $1 \leq k < n$.

The expression in condition (4) is called the *normal form* of $s \in \Gamma$. By definition, for any homomorphisms φ_i from Γ_i into a group Γ' which agree on Λ, there exists a unique homomorphism $\varphi\colon \Gamma \to \Gamma'$ such that $\varphi_{|\Gamma_i} = \varphi_i$.

Since we do not use it, we do not give the normal form theorem for an HNN extension. But here's the definition.

Definition E.9 (HNN extension). Let Γ be a group, $\Lambda \leq \Gamma$ be a subgroup and $\theta\colon \Lambda \to \Gamma$ be an injective homomorphism. Then, the HNN extension

$$\Gamma^* = \langle \Gamma, z \mid z^{-1} a z = \theta(a) \text{ for } a \in \Lambda \rangle$$

is the group satisfying the following properties:

(1) Γ^* contains Γ as a subgroup and Γ^* is generated by Γ and a distinguished element z of infinite order;

(2) for every $a \in \Lambda$, we have $z^{-1}az = \theta(a)$ in Γ^*;

(3) if $s_0 z^{d_1} s_1 \cdots z^{d_n} s_n = e$ for some $n \geq 1$, $s_k \in \Gamma$ and $d_k \in \mathbb{Z} \setminus \{0\}$, then there exists $k \geq 1$ such that either $s_k \in \Lambda$ and $d_k < 0 < d_{k+1}$ or $s_k \in \theta(\Lambda)$ and $d_k > 0 > d_{k+1}$.

The HNN-extension $\Gamma^* = \langle \Gamma, z \mid z^{-1}az = \theta(a) \text{ for } a \in \Lambda \rangle$ can also be described in the following manner. Let $\Gamma^{(n)}$, $n \in \mathbb{Z}$, be copies of Γ and write $s^{(n)}$ for the element in $\Gamma^{(n)}$ corresponding to s in Γ. Let $\Lambda^{(n)} \leq \Gamma^{(n)}$ be the subgroup corresponding to Λ and consider the embedding

$$\theta^{(n)} \colon \Lambda^{(n)} \ni a^{(n)} \mapsto \theta(a)^{(n+1)} \in \Gamma^{(n+1)}.$$

We define Γ_∞ as the iterated amalgamated product

$$\begin{aligned}\Gamma_\infty &= \cdots * \Gamma^{(n-1)} *_{\Lambda^{(n-1)} = \theta(\Lambda)^{(n)}} \Gamma^{(n)} *_{\Lambda^{(n)} = \theta(\Lambda)^{(n+1)}} \Gamma^{(n+1)} * \cdots \\ &= \langle \bigcup \Gamma^{(n)} \mid a^{(n)} = \theta(a)^{(n+1)} \text{ for } a \in \Lambda \text{ and } n \in \mathbb{Z} \rangle.\end{aligned}$$

By universality of the amalgamated free product, the "shift"

$$\Gamma^{(n)} \ni s^{(n)} \mapsto s^{(n-1)} \in \Gamma^{(n-1)} \subset \Gamma_\infty$$

extends to an automorphism of Γ_∞. Then, the HNN-extension Γ^* is known to be isomorphic to the subgroup of the semidirect product $\Gamma_\infty \rtimes \mathbb{Z}$, generated by $\Gamma^{(0)}$ and z, where z is the element in $\Gamma_\infty \rtimes \mathbb{Z}$ that implements the shift automorphism.

Example E.10. Recall that

$$\mathrm{SL}(2,\mathbb{Z}) = \{ \begin{bmatrix} a & b \\ c & d \end{bmatrix} : a, b, c, d \in \mathbb{Z} \text{ and } ad - bc = 1 \}$$

and that $\mathrm{PSL}(2,\mathbb{Z}) = \mathrm{SL}(2,\mathbb{Z})/\{\pm I\}$, where I is the identity matrix. We will show that

$$s = \begin{bmatrix} 0 & 1 \\ -1 & 0 \end{bmatrix} \text{ and } t = \begin{bmatrix} 0 & -1 \\ 1 & 1 \end{bmatrix}$$

freely generate $\mathrm{PSL}(2,\mathbb{Z})$ and hence $\mathrm{PSL}(2,\mathbb{Z}) \cong (\mathbb{Z}/2\mathbb{Z}) * (\mathbb{Z}/3\mathbb{Z})$. (Note that $\langle s \rangle \cong \mathbb{Z}/2\mathbb{Z}$ and $\langle t \rangle \cong \mathbb{Z}/3\mathbb{Z}$ in $\mathrm{PSL}(2,\mathbb{Z})$.) This also implies that $\mathrm{SL}(2,\mathbb{Z}) \cong (\mathbb{Z}/4\mathbb{Z}) *_{\mathbb{Z}/2\mathbb{Z}} (\mathbb{Z}/6\mathbb{Z})$. Consider the action of $\mathrm{PSL}(2,\mathbb{Z})$ on the projective line $\mathbb{R}P^1 = \mathbb{R} \cup \{\infty\}$:

$$\begin{bmatrix} a & b \\ c & d \end{bmatrix} : \mathbb{R} \cup \{\infty\} \ni x \mapsto \frac{ax+b}{cx+d} \in \mathbb{R} \cup \{\infty\}.$$

We observe that $s(-\infty, 0) = (0, \infty)$ and $t^d(0,\infty) \subset (-\infty, 0)$ for $d = 1, 2$. Let w be a reduced word in $\langle s \rangle$ and $\langle t \rangle$. By conjugating t or t^2 if necessary, we may assume that both the first and the last letters of w come from $\langle t \rangle$

– i.e., $w = t^{d_1} s t^{d_2} s \cdots s t^{d_n}$ for some $d_1, \ldots, d_n \in \{1, 2\}$. It follows that $w(0, \infty) \subset (-\infty, 0)$ and hence $w \neq 1$ in $\mathrm{PSL}(2, \mathbb{Z})$. (This is the so-called "table tennis lemma" of Klein.) One can now check that s and t generate $\mathrm{SL}(2, \mathbb{Z})$. We note that $\left[\begin{smallmatrix} 1 & 2 \\ 0 & 1 \end{smallmatrix}\right]$ and $\left[\begin{smallmatrix} 1 & 0 \\ 2 & 1 \end{smallmatrix}\right]$ generate a copy of the free group $\mathbb{F}_2$ (which has index 12 in $\mathrm{SL}(2, \mathbb{Z})$).

Note that $\mathrm{SL}(2, \mathbb{R})/P \cong \mathbb{R}P^1$ as an $\mathrm{SL}(2, \mathbb{R})$-space, where P is the subgroup of upper triangular matrices. Indeed, $\mathrm{SL}(2, \mathbb{R})$ acts transitively on the projective line $\mathbb{R}P^1$ and P coincides with the stabilizer of $\infty \in \mathbb{R}P^1$.

Example E.11. Let $p, q > 1$ be a relatively prime pair of natural numbers. The Baumslag-Solitar group

$$\Gamma_{p,q} = \langle a, z \mid z^{-1} a^p z = a^q \rangle$$

is isomorphic to the HNN-extension associated with $p\mathbb{Z}, q\mathbb{Z} \subset \mathbb{Z}$. It contains $T_{p,q} = \langle x, y \mid x^p = y^q \rangle \cong \mathbb{Z} *_{p\mathbb{Z}=q\mathbb{Z}} \mathbb{Z}$ – the *torus knot group* – which is nonamenable (because $p, q \neq 1$ and $(p, q) \neq (2, 2)$) and has an infinite center (generated by x^p). The group $\Gamma_{p,q}$ has several interesting properties. Let $\varphi \colon \Gamma \to \Gamma$ be the endomorphism given by

$$\varphi(a) = a^p \text{ and } \varphi(z) = z.$$

The endomorphism φ is surjective since $p, q > 1$ are relatively prime, but it is not injective since the commutator $[a, z^{-1} a z] \neq e$ is in $\ker \varphi$. (A group which admits a surjective and noninjective endomorphism is said to be non-Hopfian.) This implies that $\Gamma_{p,q}$ is not residually finite. Indeed, for any finite-index subgroup $\Lambda \leq \Gamma_{p,q}$, the subgroup $\varphi^{-1}(\Lambda)$ has the same index in $\Gamma_{p,q}$ as Λ. But since $\Gamma_{p,q}$ is finitely generated, the set of subgroups of index n is finite for every n. It follows that the surjective endomorphism φ acts bijectively on the set of subgroups of index n, and thus $\ker \varphi$ is contained in the intersection of all the finite-index subgroups.

Let $N_k = \ker \varphi^k$ be an increasing sequence of normal subgroups in $\Gamma_{p,q}$, and let $N_\infty = \bigcup N_k$. Then, $\Gamma_{p,q}/N_k \cong \Gamma_{p,q}$ for every k. Let us show that Γ/N_∞ is meta-abelian. Let $\sigma \colon \Gamma_{p,q} \to \langle z \rangle \cong \mathbb{Z}$ be a homomorphism given by $\sigma(a) = e$ and $\sigma(z) = z$. It is not hard to see that σ factors through $\Gamma_{p,q}/N_\infty$. We observe that $\ker \sigma$ is generated as a group by $\{z^{-n} a z^n : n \in \mathbb{Z}\}$. But, $z^{-n} a z^n = a^{q^n}$ in $\Gamma_{p,q}/N_{p^n}$; hence $\ker \sigma = \langle a \rangle$ in $\Gamma_{p,q}/N_\infty$.

We present a result from [**12**]. Let $\mathcal{S} = \{a, a^{-1}, z, z^{-1}\}$ be the standard set of generators of $\Gamma_{p,q}$. Consider the Cayley graphs $\mathbf{X}_k = \mathbf{X}(\Gamma_{p,q}, \varphi^k(\mathcal{S}))$ and let $R > 0$ be arbitrary. Then, one can find $K > 0$ such that

$$\mathcal{S}^{2R} \cap N_\infty = \mathcal{S}^{2R} \cap \bigcup_{k \geq 1} N_k = \mathcal{S}^{2R} \cap N_K.$$

It follows that for all $k \geq K$ the balls $B_R(\mathbf{X}_k)$ of radius R in $\mathbf{X}_k$ are identical to $B_R(\mathbf{X}_\infty)$, where $\mathbf{X}_\infty = \mathbf{X}(\Gamma_{p,q}/N_\infty, \mathcal{S})$. Since $\Gamma_{p,q}/N_\infty$ is amenable, we conclude that

$$\lim_{k\to\infty} \|\frac{1}{|S|} \sum_{s\in\varphi^k(S)} \lambda(s)\|_{\mathbb{B}(\ell^2(\Gamma_{p,q}))} = 1.$$

That is, the Baumslag-Solitar group $\Gamma_{p,q}$ is not "uniformly nonamenable."

Groups acting on trees.

Example E.12. Let $\Gamma = \Gamma_1 *_\Lambda \Gamma_2$ be the amalgamated free product of Γ_1 and Γ_2 with a common subgroup Λ. Then the associated Bass-Serre tree is the graph $\mathbf{T}$ whose vertex set is $\Gamma/\Gamma_1 \sqcup \Gamma/\Gamma_2$ and whose edge set is Γ/Λ, where $s\Lambda$ connects $s\Gamma_1$ to $s\Gamma_2$. It is instructive to check that $\mathbf{T}$ is indeed a tree. (Check that it is connected and has no nontrivial closed paths.) The group Γ acts on $\mathbf{T}$ by left multiplication. The vertex stabilizers are isomorphic to either Γ_1 or Γ_2, and the edge stabilizers are isomorphic to Λ.

Example E.13. Let $\Gamma^* = \langle \Gamma, z \mid z^{-1}az = \theta(a) \text{ for } a \in \Lambda\rangle$ be the HNN-extension associated with $(\Gamma, \Lambda, \theta)$. Then the associated Bass-Serre tree is the graph $\mathbf{T}$ whose vertex set is Γ^*/Γ and whose edge set is Γ^*/Λ, where $s\Lambda$ connects $s\Gamma$ to $sz\Gamma$. It is again instructive to check that $\mathbf{T}$ is a tree, that Γ^* acts on $\mathbf{T}$ by left multiplication and the vertex stabilizers are isomorphic to Γ, while the edge stabilizers are isomorphic to Λ.

These are the two archetypes of actions on trees. Indeed, Bass-Serre theory provides a recipe for reconstructing a group acting on a tree from its vertex and edge stabilizers by successive amalgamated free products and HNN-extensions.

References. For more on amalgamated free products and Bass-Serre theory see [**13, 121, 172**] . Expanders receive more attention in [**122**].

Appendix F

Bimodules over von Neumann Algebras

In this appendix we study bimodules over von Neumann algebras, restricting our attention to the finite case. Hence, every von Neumann algebra M appearing in this appendix comes together with a distinguished faithful normal tracial state τ (or τ_M, if the context isn't clear). For an inclusion of von Neumann algebras $N \subset M$, we assume $\tau_N = \tau_M|_N$.

Examples.

Definition F.1. Let M and N be von Neumann algebras. A *(left) M-module* is a Hilbert space $\mathcal{H}$ together with a normal unital $*$-homomorphism $\pi\colon M \to \mathbb{B}(\mathcal{H})$. (NB: Injectivity of π is not assumed.) A *right M-module* is a Hilbert space $\mathcal{H}$ together with a normal unital $*$-homomorphism $\rho\colon M^{\mathrm{op}} \to \mathbb{B}(\mathcal{H})$. To reduce clutter, we rarely write π and ρ (but keep them in mind, of course).

An *M-N-bimodule*[1] is a Hilbert space $\mathcal{H}$ together with normal unital $*$-homomorphisms $\pi\colon M \to \mathbb{B}(\mathcal{H})$ and $\rho\colon N^{\mathrm{op}} \to \mathbb{B}(\mathcal{H})$ whose ranges commute. We refer to π as the left action and to ρ as the right action. We use the intuitive notation

$$a\xi x := \pi(a)\rho(x^{\mathrm{op}})\xi$$

for $a \in M$, $x \in N$ and $\xi \in \mathcal{H}$.

By the representation theory of von Neumann algebras, every M-module has a rather simple structure – it's isomorphic to a submodule of $\bigoplus L^2(M)$ (with the diagonal M-action).

[1] Also called a *correspondence* from N to M.

Example F.2. We recall that the formula $\|x\|_2 = \tau(x^*x)^{1/2}$ defines a Hilbertian norm on M and the GNS Hilbert space $L^2(M)$ is the completion of M with respect to this norm. The vector in $L^2(M)$ associated with $x \in M$ is denoted by $\widehat{x}$. The *identity bimodule* (or the trivial bimodule) over M is $L^2(M)$ with the action given by $a\widehat{x}b = \widehat{axb}$ for $a, b, x \in M$.

Example F.3. Let $\mathcal{H}$ be an M-module. A Hilbert subspace $\mathcal{K} \subset \mathcal{H}$ is called an *M-submodule* if $M\mathcal{K} \subset \mathcal{K}$. Let $e \in \mathbb{B}(\mathcal{H})$ be a projection. Observe that $e\mathcal{H}$ is an M-submodule if and only if $e \in M'$. In passing, we note that if $e \in M$, then $e\mathcal{H}$ is an eMe-module.

Example F.4. The *coarse* M-N-bimodule is $L^2(M)\otimes L^2(N)$ with the action given by $a(\eta\otimes\zeta)b = (a\eta)\otimes(\zeta b)$ for $a \in M$, $b \in N$, $\eta \in L^2(M)$ and $\zeta \in L^2(N)$.

Example F.5. Let Γ be a group and $M = L(\Gamma)$. For every unitary representation $(\pi, \mathcal{H})$ of Γ, we associate the M-M-bimodule $\tilde{\mathcal{H}} = \mathcal{H} \otimes \ell^2(\Gamma)$ with left and right actions given by

$$\Gamma \ni s \mapsto (1 \otimes \lambda)(s) \in \mathbb{B}(\tilde{\mathcal{H}}) \text{ and } \Gamma \ni t \mapsto (\pi \otimes \rho)(t^{-1}) \in \mathbb{B}(\tilde{\mathcal{H}}).$$

Notice that $\pi \otimes \rho$ is unitarily equivalent to $1 \otimes \rho$ (cf. Fell's absorption principle) and thus we have a canonical isomorphism $(\pi \otimes \rho(\Gamma))'' \cong M^{\mathrm{op}}$.

Example F.6. For a normal u.c.p. map $\varphi\colon M \to N$, we define an M-N-bimodule $\mathcal{H}_\varphi$ via the minimal Stinespring dilation of φ: Equip $M \odot L^2(N)$ with the semi-inner product

$$\langle \sum_i a_i \otimes \eta_i, \sum_j b_j \otimes \zeta_j \rangle = \sum_{i,j} \langle \varphi(b_j^* a_i)\eta_i, \zeta_j \rangle$$

and promote it to a Hilbert space $\mathcal{H}_\varphi$ by separation and completion. Abusing notation, denote by $b \otimes \zeta$ the vector in $\mathcal{H}_\varphi$ which it represents. The action is given by

$$a(b \otimes \zeta)x = (ab) \otimes (\zeta x).$$

For the unit vector $\xi_\varphi = 1 \otimes \widehat{1} \in \mathcal{H}_\varphi$ we have

$$\langle a\xi_\varphi x, \xi_\varphi y \rangle_{\mathcal{H}_\varphi} = \langle \varphi(a)\widehat{x}, \widehat{y} \rangle_{L^2(N)}$$

for every $a \in M$ and $x, y \in N$. In particular, $\langle \xi_\varphi x, \xi_\varphi \rangle = \tau_N(x)$ for every $x \in N$. For the converse direction, see Exercise F.1.

Comparison of projections and finite bimodules. Comparison theory for projections in von Neumann algebras culminates in the existence of the center-valued trace on a finite von Neumann algebra. The center of M is denoted by $\mathcal{Z}(M)$. As usual, for projections $e, f \in M$, we write $e \precsim f$ if there exists a partial isometry $v \in M$ such that $v^*v = e$ and $vv^* \leq f$. We refer to Theorem V.2.6 and Corollary 2.8 in [**183**] for the following result.

Theorem F.7. *Let M be a finite von Neumann algebra. Then, there exists a unique conditional expectation* ctr *(called the* center-valued trace*) from M onto $\mathcal{Z}(M)$ such that* $\mathrm{ctr}(xy) = \mathrm{ctr}(yx)$ *for every $x, y \in M$. The center-valued trace* ctr *is normal and faithful. Any (normal) tracial state τ on M is of the form $\mu \circ \mathrm{ctr}$, where μ is a (normal) state on $\mathcal{Z}(M)$. For projections $e, f \in M$, one has $e \precsim f$ if and only if* $\mathrm{ctr}(e) \leq \mathrm{ctr}(f)$.

Corollary F.8. *Let A be a maximal abelian $*$-subalgebra of M. Then, every projection in M is equivalent to a projection in A.*

For the proof, we need a lemma.

Lemma F.9. *Let A be a maximal abelian $*$-subalgebra of M. Then, for any nonzero projection $q_0 \in A$ with $Aq_0 \neq q_0Mq_0$, there exists a nonzero projection $q_1 \in A$ such that $q_1 \leq q_0$ and* $\mathrm{ctr}(q_1) \leq \mathrm{ctr}(q_0)/2$.

Proof. Replacing $A \subset M$ with $Aq_0 \subset q_0Mq_0$, we may assume $q_0 = 1$. Note that $\mathcal{Z}(M)$ is a proper subalgebra of the maximal abelian subalgebra A. Choose a nonzero projection $p \in A$ which does not belong to $\mathcal{Z}(M)$. We observe that the center trace $\mathrm{ctr}(p)$ is not a projection (otherwise, we would have $p = p\,\mathrm{ctr}(p) = \mathrm{ctr}(p)$). Hence, replacing p with $1 - p$ if necessary, we may assume that the spectral projection $z = \chi_{(0,1/2]}(\mathrm{ctr}(p)) \in \mathcal{Z}(M)$ is nonzero. Letting $q_1 = pz$, we are done. □

Proof of Corollary F.8. We first show that for any nonzero projection $p \in M$, there is a nonzero projection $q \in A$ such that $q \precsim p$. Truncating by the nonzero central projection $\chi_{(\varepsilon,1]}(\mathrm{ctr}(p))$ if necessary, we may assume that $\mathrm{ctr}(p) \geq \varepsilon$ for some $\varepsilon > 0$ (e.g., $\varepsilon = \|\mathrm{ctr}(p)\|/2$ does the job). If there is a nonzero projection $q \in A$ such that $Aq = qMq$, then q is an abelian projection and $q \precsim p$. (See, e.g., Lemma V.1.25 in [**183**].) Otherwise, we can use Lemma F.9 several times and find projections $1 = q_0 \geq q_1 \geq \cdots \geq q_n = q$ in A such that $\mathrm{ctr}(q) \leq 2^{-n} \leq \varepsilon$.

Now, let a projection $e \in M$ be given and take a maximal projection $f \in A$ such that $f \precsim e$. After conjugating by a unitary element, we may assume that $f \leq e$. Suppose by contradiction that $p = e - f \neq 0$. Applying the above result to $p \in (1-f)M(1-f)$, we can find a nonzero projection $q \in A(1-f)$ such that $q \precsim p$. It follows that $f + q \precsim f + p = e$, which is impossible, so we have $f = e$. □

Let $\mathcal{H}$ be a right M-module and denote by $N \subset \mathbb{B}(\mathcal{H})$ the commutant of the right M-action. By the representation theory of von Neumann algebras, $\mathcal{H}$ is isomorphic to a right M-submodule of $\ell^2 \otimes L^2(M)$ – i.e., there exists an isometry $V\colon \mathcal{H} \to \ell^2 \otimes L^2(M)$ such that $V(\xi x) = (V\xi)x$ for every $\xi \in \mathcal{H}$ and $x \in M$. Since the projection $P = VV^* \in \mathbb{B}(\ell^2 \otimes L^2(M))$ commutes with the right M-action, we have $P \in \mathbb{B}(\ell^2) \bar{\otimes} M$, where M acts on $L^2(M)$ from the left. The isometry V determines a $*$-isomorphism $N \cong P(\mathbb{B}(\ell^2) \bar{\otimes} M)P$. The von Neumann algebra $\mathbb{B}(\ell^2) \bar{\otimes} M$ is of type II_∞ and has a canonical faithful normal semifinite trace $\tilde{\tau}$, defined by

$$\tilde{\tau}([x_{i,j}]_{i,j}) = \sum_i \tau(x_{i,i}) \in [0, \infty],$$

where positive elements $x \in \mathbb{B}(\ell^2) \bar{\otimes} M$ are viewed as infinite matrices $[x_{i,j}]_{i,j}$ with entries in M. We define $\dim_M \mathcal{H} = \tilde{\tau}(P) \in [0, \infty]$ and call it the *dimension* of the right M-module $\mathcal{H}$. It depends on the choice of tracial state τ but not on the choice of V. Indeed, if $W \colon \mathcal{H} \to \ell^2 \otimes L^2(M)$ is another right M-module isometry, then we have $WV^* \in \mathbb{B}(\ell^2) \bar{\otimes} M$ and

$$\tilde{\tau}(VV^*) = \tilde{\tau}(VW^*WV^*) = \tilde{\tau}(WV^*VW^*) = \tilde{\tau}(WW^*).$$

If $\dim_M \mathcal{H} < \infty$, then there exists an increasing sequence of projections $z_n \in \mathcal{Z}(M)$ converging strongly to 1 such that $P(1 \otimes z_n) \precsim E_n \otimes 1_M$, where E_n is a rank-n projection in $\mathbb{B}(\ell^2)$. To see this, take $z_n \in \mathcal{Z}(M)$ to be the maximal projection such that $\sum_i \mathrm{ctr}(P_{i,i} z_n) \leq n$. (It might be easier to consider $\mathcal{Z}(M)$ as a function space on the spectrum of $\mathcal{Z}(M)$.) We note that $\mathcal{H} z_n$ is isomorphic to a right M-submodule of $\ell^2_n \otimes L^2(M)$.

Proposition F.10. *Let $\mathcal{H}$ be an N-M-bimodule with $\dim_M \mathcal{H} < \infty$. Then, there exist a nonzero projection $f \in N$ and a nonzero fNf-M-sub-bimodule $\mathcal{K}$ of $f\mathcal{H}$ such that $\mathcal{K}$ is isomorphic, as a right M-module, to a right M-submodule of $L^2(M)$.*

Proof. Let $\mathcal{H}$ be an N-M-bimodule with $\dim_M \mathcal{H} < \infty$. Truncating by a central projection if necessary, we may identify $\mathcal{H}$ with $P(\ell^2_n \otimes L^2(M))$, as a right M-module, for some n and some projection $P \in \mathbb{M}_n(M)$. It follows that $N \subset P\mathbb{M}_n(M)P$. Let $z \in \mathcal{Z}(N)$ be the projection such that Nz is of type II and $N(1-z)$ is of type I.

Suppose first that $z \neq 0$ and let $q_1, \ldots, q_n \in N$ be mutually equivalent projections with sum z. Then, $f = q_1 \in N$ and $\mathcal{K} = f\mathcal{H}$ satisfies the requirement. Suppose next that $z = 0$ and N is of type I. Then, there exists a nonzero projection $f \in N$ such that fNf is abelian. Choose j such that $(e_{1,j} \otimes 1_M)f \neq 0$ and let p be the right support of $(e_{1,j} \otimes 1_M)f$. It follows that $p \in f\mathbb{M}_n(M)f$ and $p \precsim e_{1,1} \otimes 1_M$ in $\mathbb{M}_n(M)$. By Corollary F.8, the projection p is equivalent to a projection $q \in (fNf)' \cap f\mathbb{M}_n(M)f$. (Consider a maximal abelian subalgebra of $f\mathbb{M}_n(M)f$ which contains fNf.) Now, $f \in N$ and $\mathcal{K} = q\mathcal{H}$ satisfies the requirement. In any case, we are done. □

$L^2(M)$ as unbounded operators and Popa's Theorem. Let M be a von Neumann algebra with a distinguished faithful normal tracial state τ. An important fact about finite von Neumann algebra is that one can interpret the identity bimodule $L^2(M)$ as a space of square integrable closed operators affiliated with M.

A densely defined operator T on $L^2(M)$ is said to be *affiliated with* M if $u'T = Tu'$ (which includes the requirement that $\mathrm{dom}(T) = u'\,\mathrm{dom}(T)$) for every unitary element u' in M'. We observe that if T is a closed operator affiliated with M and $T = u|T|$ is the polar decomposition of T, then $u \in M$ and $|T|$ is a positive self-adjoint operator affiliated with M. A closed operator T which is affiliated with M is said to be *square integrable* if $\widehat{1} \in \mathrm{dom}(T)$. Although we won't need it, note that T is square integrable if and only if for the spectral decomposition $|T| = \int_0^\infty t\,dE(t)$ of $|T|$, one has

$$\tau(T^*T) := \int_0^\infty t^2\,d(\tau \circ E)(t) < \infty.$$

Moreover, for square integrable T, one has $\tau(T^*T) = \|T\widehat{1}\|_2^2$. For every $\xi \in L^2(M)$, we define a densely defined linear operator L_ξ^0 affiliated with M by

$$L_\xi^0 \colon L^2(M) \supset \widehat{M} \ni \widehat{x} \mapsto \xi x \in L^2(M).$$

Proposition F.11. *For every $\xi \in L^2(M)$, the operator L_ξ^0 is closable and its closure L_ξ is a square integrable operator affiliated with M. Moreover, the map $\xi \mapsto L_\xi$ gives a bijective correspondence between $L^2(M)$ and the space of closed square integrable operators affiliated with M.*

Proof. Let $J \colon L^2(M) \to L^2(M)$ be the conjugate linear isometry defined by $J\widehat{x} = \widehat{x^*}$ (cf. Section 6.1). For $\xi \in L^2(M)$, we have

$$\langle L_\xi^0 \widehat{x}, \widehat{y}\rangle = \langle J\widehat{y}, J(\xi x)\rangle = \langle \widehat{1}y^*, x^* J\xi\rangle = \langle x\widehat{1}, (J\xi)y\rangle = \langle \widehat{x}, L_{J\xi}^0 \widehat{y}\rangle.$$

This implies that $(L_\xi^0)^*$ is densely defined and $L_{J\xi}^0 \subset (L_\xi^0)^*$. It follows that L_ξ^0 and $L_{J\xi}^0$ are closable and their closures L_ξ and $L_{J\xi}$ satisfy $L_{J\xi} \subset L_\xi^*$. Since L_ξ^0 is affiliated with M, so is the closure L_ξ. Let $L_\xi = u|L_\xi|$ be the polar decomposition. (We may assume that $u \in M$ is unitary.) Then, we have $L_\xi^* = |L_\xi|u^* = u^* L_\xi u^* = L_{u^*\xi u^*}$. Since $L_{J\xi} \subset L_{u^*\xi u^*}$, we have $J\xi = u^*\xi u^*$ and $L_\xi^* = L_{J\xi}$.

Let T be a closed square integrable operator affiliated with M. We first show that T^* is also square integrable. Let $T = u|T|$ be the polar decomposition. We have that $\widehat{u^*} \in \mathrm{dom}(|T|)$ since $|T|$ is a square integrable operator affiliated with M. It follows that for every $\xi \in \mathrm{dom}(T)$,

$$\langle T\xi, \widehat{1}\rangle = \langle |T|\xi, \widehat{u^*}\rangle = \langle \xi, |T|\widehat{u^*}\rangle.$$

This implies that $\widehat{1} \in \mathrm{dom}(T^*)$. Let $\xi = T\widehat{1}$ and $\eta = T^*\widehat{1}$. Since T and T^* are affiliated with M, it is easy to see that $L_\xi^0 \subset T$ and $L_\eta^0 \subset T^*$. This implies that $L_\xi \subset T \subset L_{J\eta}$. It is clear that $\xi = J\eta$, so $T = L_\xi$. □

We briefly review Jones's *basic construction.* Let $A \subset M$ be a von Neumann subalgebra and denote by E_A the trace-preserving conditional expectation from M onto A and by $e_A \in \mathbb{B}(L^2(M))$ the orthogonal projection onto $L^2(A)$. We note that $e_A x e_A = E_A(x)e_A$ for every $x \in M$. It follows that

$$\mathcal{A} = \{\sum_{k=1}^{n} x_k e_A y_k^* : n \in \mathbb{N},\ x_k, y_k \in M\}$$

is a $*$-subalgebra of $\mathbb{B}(L^2(M))$. It is not hard to show that $\mathcal{A}$ acts nondegenerately on $L^2(M)$ and hence

$$\langle M, A\rangle := \mathcal{A}'' = (M \cup \{e_A\})''.$$

The von Neumann algebra $\langle M, A\rangle$ is called the basic construction of $A \subset M$. We have

$$\langle M, A\rangle' = M' \cap \{e_A\}' = \rho(A),$$

where $\rho(A)$ is the right action of A on $L^2(M)$. The von Neumann algebra $\langle M, A\rangle$ is semifinite with a canonical faithful normal semifinite trace Tr such that $\mathrm{Tr}(x e_A y^*) = \tau(xy^*)$ for $x, y \in M$. (See Exercise F.6.) We note that if $\mathcal{H} \subset L^2(M)$ is a right A-submodule, then the orthogonal projection $P_{\mathcal{H}}$ onto $\mathcal{H}$ belongs to $\langle M, A\rangle$ and $\dim_A \mathcal{H} = \mathrm{Tr}(P_{\mathcal{H}})$. See **[158]** for more on the basic construction.

The following very useful theorem is due to Popa.

Theorem F.12. *Let $A \subset M$ be finite von Neumann algebras with separable predual and let $p \in M$ be a nonzero projection. Then, for a von Neumann subalgebra $B \subset pMp$, the following are equivalent:*

(1) *there is no sequence (w_n) of unitary elements*[2] *in B such that $\|E_A(b^* w_n a)\|_2 \to 0$ for every $a, b \in M$;*

(2) *there exists a positive element $d \in \langle M, A\rangle$ with $\mathrm{Tr}(d) < \infty$ such that the ultraweakly closed convex hull of $\{w^* d w : w \in B \text{ unitary}\}$ does not contain 0;*

(3) *there exists a B-A-submodule $\mathcal{H}$ of $pL^2(M)$ with $\dim_A \mathcal{H} < \infty$;*

(4) *there exist nonzero projections $e \in A$ and $f \in B$, a unital normal $*$-homomorphism $\theta\colon fBf \to eAe$ and a nonzero partial isometry $v \in M$ such that*

$$\forall x \in fBf, \quad xv = v\theta(x)$$

and such that $v^*v \in \theta(fBf)' \cap eMe$ and $vv^* \in (fBf)' \cap fMf$.

[2] A unitary element w in B is a partial isometry in M such that $w^*w = p = ww^*$.

Proof. (1) $\Rightarrow$ (2): By condition (1), there exist a finite subset $\mathfrak{F} \subset M$ and $\varepsilon > 0$ such that $\max_{a,b\in\mathfrak{F}} \|E_A(b^*wa)\|_2 \geq \varepsilon$ for all unitary elements $w \in B$. Let $e_A \in \mathbb{B}(L^2(M))$ be the orthogonal projection onto $L^2(A)$ and let $d = \sum_{b\in\mathfrak{F}} be_Ab^* \in \langle M, A\rangle$. One checks that $\mathrm{Tr}(d) = \sum_{b\in\mathfrak{F}} \tau(bb^*) < \infty$ and

$$\sum_{a\in\mathfrak{F}} \langle w^*dw\widehat{a}, \widehat{a}\rangle = \sum_{a,b\in\mathfrak{F}} \langle e_A\widehat{b^*wa}, \widehat{b^*wa}\rangle = \sum_{a,b} \|E_A(b^*wa)\|_2^2 \geq \varepsilon^2$$

for all unitary elements $w \in B$. This yields condition (2).

(2) $\Rightarrow$ (3): Let d be as in condition (2) and let $\mathcal{C}$ be the ultraweakly closed convex hull of $\{w^*dw : w \in B \text{ unitary}\}$. Since $\mathcal{C}$ can be regarded as a closed convex subset in $L^2(\langle M, A\rangle, \mathrm{Tr})$, we can consider the circumcenter d_0 of $\mathcal{C}$ (cf. Exercises D.1 and F.3). By uniqueness of the circumcenter, we have $d_0 \in p\langle M, A\rangle p \cap B'$. It can be shown that $\mathrm{Tr}(d_0) \leq \mathrm{Tr}(d) < \infty$. (NB: This fact is not entirely trivial since Tr is not bounded.) Therefore, there exists a nonzero spectral projection q of d_0 such that $\mathrm{Tr}(q) < \infty$. It follows that $\mathcal{H} = qL^2(M)$ is a nonzero B-A-submodule such that $\dim_A \mathcal{H} = \mathrm{Tr}(q) < \infty$.

(3) $\Rightarrow$ (4): By Proposition F.10, there exists a nonzero projection $f \in B$, a nonzero fBf-A-sub-bimodule $\mathcal{K}$ of $\mathcal{H}$ and a right A-module isometry $V\colon \mathcal{K} \to L^2(A)$. Let $x \in fBf$ be given. Since VxV^* commutes with the right A-action, we have $VxV^* \in eAe$, where $e = VV^* \in A$. Hence $\theta(x) = VxV^*$ defines a unital normal $*$-homomorphism from fBf into eAe. Let $\xi = V^*\widehat{1}_A \in \mathcal{K}$ and observe that $\xi \neq 0$, since $V\xi = \widehat{e}$. Then we have

$$x\xi = V^*\theta(x)\widehat{1}_A = V^*\widehat{\theta(x)} = V^*\widehat{1}_A\theta(x) = \xi\theta(x)$$

for every $x \in fBf$. Since $\mathcal{K} \subset L^2(M)$, we may view ξ as a square integrable operator L_ξ affiliated with M and we have $xL_\xi = L_\xi\theta(x)$ for every unitary element $x \in fBf$ (see Exercise F.4). It follows that

$$|L_\xi|^2 = (xL_\xi)^*(xL_\xi) = (L_\xi\theta(x))^*(L_\xi\theta(x)) = \theta(x)^*|L_\xi|^2\theta(x)$$

for every unitary element $x \in fBf$; hence $|L_\xi|$ commutes with $\theta(fBf)$. Let $L_\xi = v|L_\xi|$ be the polar decomposition of L_ξ. Then we have

$$xv|L_\xi| = xL_\xi = L_\xi\theta(x) = v|L_\xi|\theta(x) = v\theta(x)|L_\xi|$$

and hence $xv = v\theta(x)$ for every (unitary) element $x \in fBf$. The claims that $v^*v \in \theta(fBf)' \cap eMe$ and $vv^* \in (fBf)' \cap fMf$ are automatic.

(4) $\Rightarrow$ (1): Let e, f and v be as in condition (4). Let E_θ be the trace-preserving conditional expectation from eMe onto $\theta(fBf)$. We note that $E_\theta(v^*v)$ is a nonzero positive element in the center of $\theta(fBf)$ and that $vE_\theta(v^*v)^2v^* \in (fBf)' \cap fMf$. Let $\{f_i\}$ be a maximal family of mutually orthogonal projections in B such that $f_i \precsim f$ in B. Then, $\sum f_i$ coincides with the central support of f in B (cf. [**183**, Lemma V.1.7]). Let $u_i \in B$ be a partial isometry such that $u_iu_i^* = f_i$ and $u_i^*u_i \leq f$, and set $v_i = u_iv$. We

may assume that $u_0 = f_0 = f$. Then, for every unitary element $w \in B$, we have

$$\begin{aligned}\sum_i \|E_A(v_i^* w v_0)\|_2^2 &\geq \sum_i \|v^* v E_\theta(v_i^* w v_0)\|_2^2 \\ &= \sum_i \|v^* v E_\theta(v^* v)\theta(u_i^* w u_0)\|_2^2 \\ &= \sum_i \|E_\theta(v^* v) v^* u_i^* w u_0 v\|_2^2 \\ &= \sum_i \tau(v^* u_0^* w^* u_i v E_\theta(v^* v)^2 v^* u_i^* w u_0 v) \\ &= \sum_i \tau(u_0^* w^* u_i u_i^* w u_0 v E_\theta(v^* v)^2 v^*) \\ &= \sum_i \tau(v E_\theta(v^* v)^2 v^*) \\ &= \tau(E_\theta(v^* v)^3).\end{aligned}$$

Since $\sum_i \|v_i^*\|_2^2 \leq 1$ and $\|E_A(v_i^* w v_0)\|_2 \leq \|v_i^*\|_2$, we can choose a finite subset $\mathfrak{F}$ of $\{v_i\}$ such that

$$\sum_{v_i \in \mathfrak{F}} \|E_A(v_i^* w v_0)\|_2^2 \geq \tau(E_\theta(v^* v)^3)/2 > 0$$

for all unitary elements w in B. This completes the proof. □

Definition F.13. Let $A \subset M$ and $B \subset pMp$ be finite von Neumann algebras. We say B *embeds in A inside M* if one of (and hence all of) the conditions in Theorem F.12 holds.

Note that if there is a nonzero projection $p_0 \in B$ such that $p_0 B p_0$ embeds in A inside M, then B embeds in A inside M (as condition (4) in Theorem F.12 evidently implies). Recall that a (nonzero) projection $f \in B$ is minimal if and only if $fBf = \mathbb{C}f$, and a von Neumann algebra B is *diffuse* if it has no minimal projections.

Corollary F.14. *Let M be a finite von Neumann algebra with separable predual and (A_n) be a sequence of von Neumann subalgebras. Let $N \subset pMp$ be a von Neumann subalgebra such that N does not embed in A_n inside M for any n. Then, there exists a diffuse abelian von Neumann subalgebra $B \subset N$ such that B does not embed in A_n inside M for any n.*

Proof. We may assume that each A_k appears infinitely often in the sequence (A_n). Let $\{x_n\}$ be a $\|\ \|_2$-norm dense sequence in the closed unit ball of M. We will construct an increasing sequence $B_1 \subset B_2 \subset \cdots$ of finite-dimensional abelian von Neumann subalgebras of N with unitary elements

$w_n \in B_n$ such that $\|E_{A_n}(x_j^* w_n x_i)\|_2 \le n^{-1}$ for every $i, j \le n$. Let $B_1 = \mathbb{C}1_N$ and suppose we have constructed $B_1, \cdots, B_n$. Let $\{p_k\}$ be the set of minimal projections in B_n. Since $p_k N p_k$ does not satisfy condition (1) in Theorem F.12, there exists a unitary element $v_k \in p_k N p_k$ such that $\|E_{A_{n+1}}(x_j^* v_k x_i)\|_2 \le ((n+1)\dim B_n)^{-1}$ for every $i, j \le n+1$. We may assume that every v_k has finite spectrum. It is easy to see that $w_{n+1} = \sum_k v_k$ and the finite-dimensional abelian subalgebra B_{n+1} generated by B_n and w_{n+1} satisfy the required property. Since $\lim_n \|E_{A_n}(x_j^* w_n x_i)\|_2 = 0$ for every i, j, the abelian von Neumann subalgebra $B = \bigvee B_n \subset N$ does not satisfy condition (1) for any A_n. Finally, it is clear that B is diffuse. $\square$

There are several instances in which local embeddability implies global embeddability. The first example involves the notion of a Cartan subalgebra.

Definition F.15. Let M be a II_1-factor. A *Cartan subalgebra* is a maximal abelian von Neumann subalgebra $A \subset M$ such that the normalizer

$$\mathcal{N}(A) = \{u \in M : \text{ a unitary element such that } uAu^* = A\}$$

generates M as a von Neumann algebra.

Lemma F.16. *Let $A \subset M$ be a Cartan subalgebra of a II_1-factor. If projections $e, f \in A$ have the same trace, then there is $u \in \mathcal{N}(A)$ such that $ueu^* = f$.*

Proof. Let $(\{e_i\}, \{u_i\})$ be a maximal pair of sequences of nonzero projections $e_i \in A$ and unitary elements $u_i \in \mathcal{N}(A)$ such that $\sum e_i \le e$ and $\sum u_i e_i u_i^* \le f$. We claim that $e' := e - \sum e_i = 0$. Indeed, if e' is nonzero, then $f' = f - \sum u_i e_i u_i^*$ is also nonzero since it has the same trace as e'. Since the projection $\bigvee_{u \in \mathcal{N}(A)} u^* f' u \in M$ commutes with $\mathcal{N}(A)$, it has to be 1 by factoriality of M. Therefore, there is $u_0 \in \mathcal{N}(A)$ such that $e_0 := u_0^* f' u_0 e' \ne 0$. The pair (e_0, u_0) contradicts the maximality of $(\{e_i\}, \{u_i\})$; hence $e = \sum e_i$, as claimed.

Thus $v = \sum u_i e_i \in M$ is a partial isometry such that $v^*v = e$, $vv^* = f$ and $vAv^* = Af$. Applying the same argument to $e^\perp$ and $f^\perp$, we obtain a partial isometry $w \in M$ such that $w^*w = e^\perp$, $ww^* = f^\perp$ and $wAw^* = Af^\perp$. Setting $u = v + w$, we are done. $\square$

Lemma F.17. *Let A and B be Cartan subalgebras of a II_1-factor M. If B embeds in A inside M, then A and B are unitarily conjugate, i.e., there exists a unitary element $u \in M$ such that $uAu^* = B$.*

Proof. Let $e \in A$, $f \in B$, $\theta\colon fBf \to eAe$ and $v \in M$ be as in condition (4) of Theorem F.12:

$$\forall x \in fBf, \quad xv = v\theta(x) \text{ and } v^*v \in \theta(fBf)' \cap eMe, \ vv^* \in (B' \cap M)f.$$

Since B is maximal abelian, we have $f_1 := vv^* \in B$. Let $N = \theta(fBf)' \cap eMe$. Since $Ae \subset N$ is maximal abelian, Corollary F.8 implies that there exists a partial isometry $w_1 \in N$ such that $w_1 w_1^* = v^*v$ and $e_1 := w_1^* w_1 \in Ae$. Let $v_1 = vw_1$. For every $x \in B$ and $a \in A$, we have $w_1 a w_1^* \in N$ and

$$xv_1 a v_1^* = v\theta(fxf)w_1 a w_1^* v^* = vw_1 a w_1^* \theta(fxf)v^* = v_1 a v_1^* x.$$

It follows that $v_1 A v_1^* \subset B' \cap f_1 M f_1 = Bf_1$. Hence,

$$Bf_1 = Bvv^* = v\theta(fBf)v^* = v_1\theta(fBf)v_1^* \subset v_1 A v_1^* \subset Bf_1.$$

Consequently $e_1 = v_1^* v_1 \in A$, $f_1 = v_1 v_1^* \in B$ and $v_1 A v_1^* = Bf_1$. Trimming e_1 if necessary, we may assume that $\tau(e_1) = 1/k$ for some $k \in \mathbb{N}$. Take equivalent projections $e_2, \ldots, e_k \in A$ (resp. $f_2, \ldots, f_k \in B$) such that $\sum_{j=1}^k e_j = 1$ (resp. $\sum_{j=1}^k f_j = 1$). By Lemma F.16, we can find partial isometries $v_2, \ldots, v_k \in M$ such that $e_j = v_j^* v_j$, $f_j = v_j v_j^*$ and $v_j A v_j^* = Bf_j$. Setting $u = \sum_{j=1}^k v_j$, we are done. □

Another case where local embedding implies global embedding is the following.

Lemma F.18. *Let $A, B \subset M$ be diffuse finite von Neumann algebras such that A and $B' \cap M$ are factors. (This implies that M and B are also factors.) Assume that $A_0' \cap M \subset A$ for any diffuse von Neumann subalgebra $A_0 \subset A$. If B embeds in A inside M, then there exists a unitary element $u \in M$ such that $uBu^* \subset A$.*

Proof. Let $e \in A$, $f \in B$, θ, $v \in M$ be given as in condition (4) of Theorem F.12. Since $vv^* \in (fBf)' \cap fMf = f(B' \cap M)f$ (cf. Exercise F.2), $vv^* = ff'$ for some projection $f' \in B' \cap M$. We take projections $f_0 \in fBf$ and $f_0' \in f'(B' \cap M)f'$ such that $\tau(f_0) = 1/n$ and $\tau(f_0') = 1/n'$ for some $n, n' \in \mathbb{N}$. Let $v_0 = f_0' f_0 v$ and $e_0 = \theta(f_0) \in A$. Then, for any $x \in f_0 B f_0$, we have

$$xv_0 = f_0' f_0 x v = f_0' f_0 v \theta(x) = v_0 \theta(x).$$

This implies that $v_0^* v_0 \in A_0' \cap M \subset A$, where $A_0 = \theta(f_0 B f_0) \oplus e_0^\perp A e_0^\perp$ is a diffuse subalgebra of A. Let $u_1, \ldots, u_n \in B$ be partial isometries such that $u_k u_k^* = f_0$ and $\sum u_k^* u_k = 1$, likewise for $f_0' \in B' \cap M$ and $u_1', \ldots, u_{n'}' \in B' \cap M$. We note that $\tau(v_0 v_0^*) = \tau(f_0 f_0') = (nn')^{-1}$. Let $w_{k,k'} \in A$ be partial isometries such that $w_{k,k'}^* w_{k,k'} = v_0^* v_0$ and $\sum w_{k,k'} w_{k,k'}^* = 1$. Then, $u = \sum w_{k,k'} v_0^* u_k u_{k'}'$ is the desired unitary. Indeed, one has

$$\begin{aligned} uxu^* &= \sum_{j,j',k,k'} w_{j,j'} v_0^* u_j u_{j'}' x (u_{k'}')^* u_k^* v_0 w_{k,k'}^* \\ &= \sum_{j,k,k'} w_{j,k'} v_0^* u_j x u_k^* v_0 w_{k,k'}^* \end{aligned}$$

$$= \sum_{j,k,k'} w_{j,k'}\theta(u_j x u_k^*) w_{k,k'}^* \in A$$

for every $x \in B$ (and, in particular, $uu^* = \sum_{k,k'} w_{k,k'} e_0 w_{k,k'}^* = 1$). □

Lemma F.19. *Let M be a finite von Neumann algebra. For every $x \in M$ and projections $e_k \in M$ with $\sum e_k = 1$, one has*

$$|\tau(x)|^2 \le \sum_k \tau(e_k x^* e_k x).$$

Proof. We may assume that $\tau(y) = \langle y\xi, \xi\rangle$ for $y \in M$. Then, we have

$$|\tau(x)|^2 = |\sum_k \tau(e_k x e_k)|^2 = |\sum_k \langle e_k x e_k \xi, e_k \xi\rangle|^2$$
$$\le \sum_k \|e_k x e_k \xi\|^2 \sum_k \|e_k \xi\|^2 = \sum_k \tau(e_k x^* e_k x)$$

by the Cauchy-Schwarz inequality. □

A subgroup $\Lambda \subset \Gamma$ is called *malnormal* if for every $s \in \Gamma \setminus \Lambda$ one has $s\Lambda s^{-1} \cap \Lambda = \{e\}$.

Theorem F.20. *Let $\Lambda \subset \Gamma$ be a malnormal subgroup and $A_0 \subset L(\Lambda)$ be a diffuse von Neumann subalgebra. Then, $A_0' \cap L(\Gamma) \subset L(\Lambda)$. More generally, if $u \in L(\Gamma)$ is a unitary element such that $uA_0u^* \subset L(\Lambda)$, then $u \in L(\Lambda)$.*

Proof. Let $u \in L(\Gamma)$ be a unitary element such that $uA_0u^* \subset L(\Lambda)$. It suffices to show $\tau(\lambda(s)^*u) = 0$ for every $s \in \Gamma \setminus \Lambda$. Let $s \in \Gamma \setminus \Lambda$ be given and take $\varepsilon > 0$ arbitrary. Let $\Lambda' = s\Lambda s^{-1}$. We observe that $\Lambda \cap \Lambda' = \{e\}$ implies that $\tau(ax) = \tau(a)\tau(x)$ for every $a \in L(\Lambda)$ and $x \in L(\Lambda')$. Let's apply this observation to $uA_0u^* \subset L(\Lambda)$ and $\lambda(s)A_0\lambda(s)^* \subset L(\Lambda')$. Since A_0 is diffuse, we may find projections $e_k \in A_0$ such that $\sum e_k = 1$ and $\tau(e_k) < \varepsilon$ for every k. Then, we have

$$\tau(e_k u^* \lambda(s) e_k \lambda(s)^* u) = \tau(u e_k u^* \lambda(s) e_k \lambda(s)^*)$$
$$= \tau(u e_k u^*)\tau(\lambda(s) e_k \lambda(s)^*) = \tau(e_k)^2.$$

Therefore, by Lemma F.19, we have

$$|\tau(\lambda(s)^*u)|^2 \le \sum_k \tau(e_k u^* \lambda(s) e_k \lambda(s)^* u) = \sum_k \tau(e_k)^2 < \varepsilon.$$

Since $\varepsilon > 0$ was arbitrary, we conclude that $\tau(\lambda(s)^*u) = 0$. □

Remark F.21. It can be shown that the subgroup $\Lambda \subset \Gamma$ is malnormal in the following cases: $\Lambda = \Gamma_i \subset \Gamma_1 * \Gamma_2 = \Gamma$, and $\Lambda \subset \Upsilon \wr \Lambda = \Gamma$ (wreath product).

Exercises

Exercise F.1. Let $\mathcal{H}$ be an M-N-bimodule and $\xi \in \mathcal{H}$ be a unit vector such that $\langle \xi x, \xi \rangle = \tau_N(x)$ for every $x \in N$. Prove that the formula

$$\langle \varphi(a)\widehat{x}, \widehat{y} \rangle_{L^2(N)} = \langle a\xi x, \xi y \rangle_{\mathcal{H}}$$

defines a normal u.c.p. map $\varphi\colon M \to N$.

Exercise F.2. Let $A \subset \mathbb{B}(\mathcal{H})$ be a von Neumann algebra and $e \in A$ be a projection. Prove that $(eAe)' = A'e$ in $\mathbb{B}(e\mathcal{H})$.

Exercise F.3. Let N be a semifinite von Neumann algebra with a faithful normal semifinite trace Tr and let

$$\Omega = \{x \in N : \|x\|_2 = \mathrm{Tr}(x^*x)^2 \leq 1\} \subset N.$$

Prove that the formal inclusion $\Omega \hookrightarrow L^2(N, \mathrm{Tr})$ is ultraweak-weak continuous. As a corollary, deduce that if $\mathcal{C}$ is an ultraweakly closed subset of N which is bounded in both the operator norm and the L^2-norm, then $\mathcal{C}$ is closed in $L^2(N, \mathrm{Tr})$.

Exercise F.4. Let $\xi \in L^2(M)$ and $u, v \in M$ be a partial isometry such that $u^*u\xi vv^* = \xi$. Prove that $L_{u\xi v} = uL_\xi v$.

Exercise F.5. Let $A \subset M$ be finite von Neumann algebras and $\langle M, A \rangle$ be the basic construction. Prove that the conditional expectation $E_A\colon M \to A$ extends to $\langle M, A \rangle$ by the relation $e_A x e_A = E_A(x)e_A$.

Exercise F.6. Let $A \subset M$ be finite von Neumann algebras and take a family $\{v_k\}_k$ of partial isometries in $\langle M, A \rangle$ such that $v_k^*v_k \leq e_A$ and $\sum_k v_k v_k^* = 1$. (Such a family exists since the central support of e_A is 1.) Define a normal weight Tr on $\langle M, A \rangle$ by

$$\mathrm{Tr}(z) = \sum_k \langle z v_k \widehat{1}, v_k \widehat{1} \rangle.$$

Prove that Tr is a faithful semifinite trace such that $\mathrm{Tr}(xe_Ay^*) = \tau(xy^*)$ for every $x, y \in M$.

Exercise F.7. Let $A \subset M$ be finite von Neumann algebras and Tr be the canonical faithful normal semifinite trace on $\langle M, A \rangle$. Prove that $\mathrm{Tr}(P) = \dim_A PL^2(M)$ for every projection $P \in \langle M, A \rangle$.

Exercise F.8. Let $\Lambda_1, \Lambda_2 \subset \Gamma$ be groups and suppose that for every $s \in \Gamma$ one has $s\Lambda_1 s^{-1} \cap \Lambda_2 = \{e\}$ (e.g., $\Gamma = \Lambda_1 * \Lambda_2$). Let $A_0 \subset L(\Lambda_1)$ be a diffuse von Neumann subalgebra. Prove that there is no unitary element $u \in L(\Gamma)$ such that $uA_0u^* \subset L(\Lambda_2)$.

References. The study of bimodules was emphasized by A. Connes [**44**] and developed by S. Popa [**156**]. Lemma F.9 is taken from [**94**]. Proposition F.11 is classical. The rest of the results are due to S. Popa. Theorem F.12 and its corollaries are obtained in a series of papers starting with [**160**]. Corollary F.14 was communicated to the authors by S. Vaes. Theorem F.20 is taken from [**154**].

Bibliography

1. C.A. Akemann, J. Anderson and G.K. Pedersen, *Excising states of C^*-algebras*, Canad. J. Math. **38** (1986), 1239–1260.
2. C.A. Akemann and P.A. Ostrand, *On a tensor product C^*-algebra associated with the free group on two generators*, J. Math. Soc. Japan **27** (1975), 589–599.
3. C. Anantharaman-Delaroche, *Systèmes dynamiques non commutatifs et moyennabilité*, Math. Ann. **279** (1987), 297–315.
4. C. Anantharaman-Delaroche, *Amenable correspondences and approximation properties for von Neumann algebras*. Pacific J. Math. **171** (1995), 309–341.
5. C. Anantharaman-Delaroche, *Amenability and exactness for dynamical systems and their C^*-algebras*, Trans. Amer. Math. Soc. **354** (2002), 4153–4178.
6. C. Anantharaman-Delaroche and J. Renault, *Amenable groupoids*, Monographies de L'Enseignement Mathématique 36. L'Enseignement Mathématique, Geneva, 2000.
7. J. Anderson, *A C^*-algebra A for which Ext(A) is not a group*, Ann. of Math. **107** (1978), 455–458.
8. R.J. Archbold and C.J.K. Batty, *C^*-tensor norms and slice maps*, J. London Math. Soc. **22** (1980), 127–138.
9. W.B. Arveson, *Subalgebras of C^*-algebras*, Acta Math. **123** (1969), 141–224.
10. W.B. Arveson, *An invitation to C^*-algebras*. Graduate Texts in Mathematics, No. 39. Springer-Verlag, New York-Heidelberg, 1976.
11. W.B. Arveson, *Notes on extensions of C^*-algebras*, Duke Math. J. **44** (1977), 329–355.
12. G.N. Arzhantseva, J. Burillo, M. Lustig, L. Reeves, H. Short and E. Ventura, *Uniform non-amenability*. Adv. Math. **197** (2005), 499–522.
13. G. Baumslag, *Topics in combinatorial group theory*. Lectures in Mathematics ETH Zürich. Birkhäuser Verlag, Basel, 1993.
14. M.B. Bekka, *On the full C^*-algebras of arithmetic groups and the congruence subgroup problem*, Forum Math. **11** (1999), 705–715.
15. B. Bekka, P. de la Harpe and A. Valette, *Kazhdan's property (T)*. New Mathematical Monographs 11. Cambridge University Press 2008.

16. G.C. Bell, *Property A for groups acting on metric spaces*, Topology Appl. **130** (2003), 239–251.

17. B. Blackadar, *Nonnuclear subalgebras of C^*-algebras*, J. Operator Theory **14** (1985), 347–350.

18. B. Blackadar and E. Kirchberg, *Generalized inductive limits of finite-dimensional C^*-algebras*, Math. Ann. **307** (1997), 343–380.

19. B. Blackadar and E. Kirchberg, *Inner quasidiagonality and strong NF algebras*, Pacific J. Math. **198** (2001), 307–329.

20. E.F. Blanchard and K.J. Dykema, *Embeddings of reduced free products of operator algebras*, Pacific J. Math. **199** (2001), 1–19.

21. D.P. Blecher, *The standard dual of an operator space*, Pacific J. Math. **153** (1992), 15–30.

22. D.P. Blecher and V.I. Paulsen, *Tensor products of operator spaces*. J. Funct. Anal. **99** (1991), 262–292.

23. F.P. Boca, *A note on full free product C^*-algebras, lifting and quasidiagonality*, Operator theory, operator algebras and related topics (Timisoara, 1996), 51–63, Theta Foundation, Bucharest, 1997.

24. M. Bożejko and G. Fendler, *Herz-Schur multipliers and completely bounded multipliers of the Fourier algebra of a locally compact group*. Boll. Un. Mat. Ital. A (6) **3** (1984), 297–302.

25. M. Bożejko and M.A. Picardello, *Weakly amenable groups and amalgamated products*. Proc. Amer. Math. Soc. **117** (1993), 1039–1046.

26. N.P. Brown, *AF embeddability of crossed products of AF algebras by the integers*, J. Funct. Anal. **160** (1998), 150–175.

27. N.P. Brown, *Herrero's approximation problem for quasidiagonal operators*, J. Funct. Anal. **186** (2001), 360–365.

28. N.P. Brown, *On quasidiagonal C^*-algebras*, Operator algebras and applications, 19–64, Adv. Stud. Pure Math., 38, Math. Soc. Japan, Tokyo, 2004.

29. N.P. Brown, *Excision and a theorem of Popa*, J. Operator Theory **54** (2005), 3–8.

30. N.P. Brown, *Invariant means and finite representation theory of C^*-algebras*, Mem. Amer. Math. Soc. **184** (2006), no. 865, viii+105 pp.

31. N.P. Brown and M. Dadarlat, *Extensions of quasidiagonal C^*-algebras and K-theory*, Operator algebras and applications, 65–84, Adv. Stud. Pure Math., 38, Math. Soc. Japan, Tokyo, 2004.

32. M. Burger, *Kazhdan constants for* $\mathrm{SL}(3,\mathbb{Z})$. J. Reine Angew. Math. **413** (1991), 36–67.

33. P.-A. Cherix, M. Cowling, P. Jolissaint, P. Julg and A. Valette, *Groups with the Haagerup property. Gromov's a-T-menability*. Progress in Mathematics, **197**. Birkhauser Verlag, Basel, 2001.

34. M. Choda, *Group factors of the Haagerup type*. Proc. Japan Acad. Ser. A Math. Sci. **59** (1983), 174–177.

35. M. Choda, *Reduced free products of completely positive maps and entropy for free product of automorphisms*, Publ. Res. Inst. Math. Sci. **32** (1996), 371–382.

36. M.D. Choi, *A simple C^*-algebra generated by two finite-order unitaries*, Canad. J. Math. **31** (1979), 867–880.

37. M.D. Choi, *The full C^*-algebra of the free group on two generators*, Pacific J. Math. **87** (1980), 41–48.

38. M.D. Choi and E.G. Effros, *The completely positive lifting problem for C^*-algebras*, Ann. of Math. **104** (1976), 585–609.
39. M.D. Choi and E.G. Effros, *Nuclear C^*-algebras and injectivity: the general case*, Indiana Univ. Math. J. **26** (1977), 443–446.
40. M.D. Choi and E.G. Effros, *Nuclear C^*-algebras and the approximation property*, Amer. J. Math. **100** (1978), 61–79.
41. A. Connes, *Classification of injective factors: cases II_1, II_∞, III_λ, $\lambda \neq 1$*, Ann. Math. **104** (1976), 73–115.
42. A. Connes, *On the cohomology of operator algebras*, J. Functional Analysis **28** (1978), 248–253.
43. A. Connes, *A factor of type* II_1 *with countable fundamental group*. J. Operator Theory **4** (1980), 151–153.
44. A. Connes, *Noncommutative geometry*, Academic Press, Inc., San Diego, CA, 1994. xiv+661 pp.
45. A. Connes and V. Jones, *Property T for von Neumann algebras*. Bull. London Math. Soc. **17** (1985), 57–62.
46. M. Cowling, *Harmonic analysis on some nilpotent Lie groups (with application to the representation theory of some semisimple Lie groups)*. Topics in modern harmonic analysis, Vol. I, II (Turin/Milan, 1982), 81–123, Ist. Naz. Alta Mat. Francesco Severi, Rome, 1983.
47. M. Cowling and U. Haagerup, *Completely bounded multipliers of the Fourier algebra of a simple Lie group of real rank one*. Invent. Math. **96** (1989), 507–549.
48. J. Cuntz, *Simple C^*-algebras generated by isometries*, Comm. Math. Phys. **57** (1977), 173–185.
49. M. Dadarlat, *On the approximation of quasidiagonal C^*-algebras*, J. Funct. Anal. **167** (1999), 69–78.
50. M. Dadarlat, *Nonnuclear subalgebras of AF algebras*, Amer. J. Math. **122** (2000), 581–597.
51. M. Dadarlat, *Residually finite dimensional C^*-algebras and subquotients of the CAR algebra*, Math. Res. Lett. **8** (2001), 545–555.
52. M. Dadarlat, *On the topology of the Kasparov groups and its applications*, J. Funct. Anal. **228** (2005), 394–418.
53. K.R. Davidson, *C^*-algebras by example*, Fields Institute Monographs 6, American Mathematical Society, Providence, RI, 1996.
54. J. De Cannière and U. Haagerup, *Multipliers of the Fourier algebras of some simple Lie groups and their discrete subgroups*. Amer. J. Math. **107** (1985), 455–500.
55. Y. De Cornulier, Y. Stalder and A. Valette, *Proper actions of lamplighter groups associated with free groups*. Preprint, 2007 (arXiv:0707.2039).
56. B. Dorofaeff, *Weak amenability and semidirect products in simple Lie groups*. Math. Ann. **306** (1996), 737–742.
57. K.J. Dykema, *Exactness of reduced amalgamated free product C^*-algebras*, Forum Math. **16** (2004), 161–180.
58. K.J. Dykema and D. Shlyakhtenko, *Exactness of Cuntz-Pimsner C^*-algebras*, Proc. Edinb. Math. Soc. **44** (2001), 425–444.
59. E.G. Effros and U. Haagerup, *Lifting problems and local reflexivity for C^*-algebras*, Duke Math. J. **52** (1985), 103–128.

60. E.G. Effros and E.C. Lance, *Tensor products of operator algebras*, Adv. Math. **25** (1977), no. 1, 1–34.

61. E.G. Effros, N. Ozawa and Z.-J. Ruan, *On injectivity and nuclearity for operator spaces.* Duke Math. J. **110** (2001), 489–521.

62. E.G. Effros and Z.-J. Ruan, *A new approach to operator spaces*, Canad. Math. Bull. **34** (1991), 329–337.

63. E.G. Effros and Z.-J. Ruan, *Operator spaces*, London Mathematical Society Monographs. New Series, 23. The Clarendon Press, Oxford University Press, New York, 2000.

64. G.A. Elliott, *Automorphisms determined by multipliers on ideals of a C^*-algebra*, J. Functional Analysis **23** (1976), 1–10.

65. G.A. Elliott and G. Gong, *On the classification of C^*-algebras of real rank zero II*, Ann. of Math. **144** (1996), 497–610.

66. N.J. Fowler, P.S. Muhly and I. Raeburn, *Representations of Cuntz-Pimsner algebras*, Indiana Univ. Math. J. **52** (2003), 569–605.

67. A. Furman, *Orbit equivalence rigidity.* Ann. of Math. (2) **150** (1999), 1083–1108.

68. L. Ge, *Applications of free entropy to finite von Neumann algebras. II.* Ann. of Math. **147** (1998), 143–157.

69. E. Ghys and P. de la Harpe, *Sur les groupes hyperboliques d'aprés Mikhael Gromov.* Progress in Math., 83, Birkaüser, 1990.

70. M. Gromov, *Hyperbolic groups*, Essays in group theory, 75–263, Math. Sci. Res. Inst. Publ. 8, Springer, New York, 1987.

71. M. Gromov, *Random walk in random groups*, Geom. Funct. Anal. **13** (2003), 73–146.

72. E. Guentner, *Exactness of the one relator groups*, Proc. Amer. Math. Soc. **130** (2002), 1087–1093.

73. E. Guentner, N. Higson and S. Weinberger, *The Novikov conjecture for linear groups*, Publ. Math. Inst. Hautes Etudes Sci. No. 101 (2005), 243–268.

74. E. Guentner and J. Kaminker, *Exactness and the Novikov conjecture*, Topology **41** (2002), 411–418.

75. U. Haagerup, *An example of a nonnuclear C^*-algebra, which has the metric approximation property*, Invent. Math. **50** (1978/79), 279–293.

76. U. Haagerup, *All nuclear C^*-algebras are amenable*, Invent. Math. **74** (1983), 305–319.

77. U. Haagerup, *A new proof of the equivalence of injectivity and hyperfiniteness for factors on a separable Hilbert space*, J. Funct. Anal. **62** (1985), 160 – 201.

78. U. Haagerup, *Group C^*-algebras without the completely bounded approximation property.* Preprint, 1986.

79. U. Haagerup, *Connes' bicentralizer problem and uniqueness of the injective factor of type* III_1, Acta. Math. **158** (1987), 95–148.

80. U. Haagerup and J. Kraus, *Approximation properties for group C^*-algebras and group von Neumann algebras.* Trans. Amer. Math. Soc. **344** (1994), 667–699.

81. U. Haagerup and S. Thorbjørnsen, *A new application of random matrices:* $\mathrm{Ext}(C_r^*(\mathbb{F}_2))$ *is not a group*, Ann. of Math. **162** (2005), 711–775.

82. P.R. Halmos, *Ten problems in Hilbert space*, Bull. Amer. Math. Soc. **76** (1970), 887–933.

83. P. de la Harpe, A.G. Robertson and A. Valette, *On the spectrum of the sum of generators for a finitely generated group.* Israel J. Math. **81** (1993), 65–96.

84. P. de la Harpe and A. Valette, *La propriété (T) de Kazhdan pour les groupes localement compacts*, Astérisque **175** (1989).

85. N. Higson and E. Guentner, *Group C^*-algebras and K-theory*, Noncommutative geometry, 137–251, Lecture Notes in Math., **1831**, Springer, Berlin, 2004.

86. N. Higson and J. Roe, *Analytic K-homology*, Oxford Mathematical Monographs. Oxford Science Publication. Oxford University Press, Oxford, 2000.

87. T.W. Hungerford, *Algebra*, Reprint of the 1974 original. Graduate Texts in Mathematics, 73. Springer-Verlag, New York-Berlin, 1980.

88. T. Huruya, *On compact completely bounded maps of C^*-algebras*, Michigan Math. J. **30** (1983), 213–220.

89. A. Ioana, J. Peterson and S. Popa, *Amalgamated Free Products of w-Rigid Factors and Calculation of their Symmetry Groups.* Acta. Math., to appear.

90. P. Jolissaint, *Haagerup approximation property for finite von Neumann algebras.* J. Operator Theory **48** (2002), 549–571.

91. P. Jolissaint, *On property (T) for pairs of topological groups.* Enseign. Math. (2) **51** (2005), 31–45.

92. V. Jones and V.S. Sunder, *Introduction to subfactors*, London Mathematical Society Lecture Note Series, 234. Cambridge University Press, Cambridge, 1997. xii+162 pp.

93. M. Junge and G. Pisier, *Bilinear forms on exact operator spaces and $B(H) \otimes B(H)$*, Geom. Funct. Anal. **5** (1995), 329 – 363.

94. R.V. Kadison, *Diagonalizing matrices.* Amer. J. Math. **106** (1984), 1451–1468.

95. R.V. Kadison and J.R. Ringrose, *Fundamentals of the theory of operator algebras*, Vol. II, Graduate Studies in Mathematics, 16. American Mathematical Society, Providence, RI, 1997.

96. V. Kaimanovich, *Boundary amenability of hyperbolic spaces.* Discrete geometric analysis, 83–111, Contemp. Math., **347**, Amer. Math. Soc., Providence, RI, 2004.

97. G.G. Kasparov, *Hilbert C^*-modules: theorems of Stinespring and Voiculescu*, J. Operator Theory **4** (1980), 133–150.

98. T. Katsura, *AF-embeddability of crossed products of Cuntz algebras*, J. Funct. Anal. **196** (2002), 427–442.

99. T. Katsura, *On C^*-algebras associated with C^*-correspondences*, J. Funct. Anal. **217** (2004), 366–401.

100. E. Kirchberg, *C^*-nuclearity implies CPAP*, Math. Nachr. **76** (1977), 203–212.

101. E. Kirchberg, *On the matricial approximation property.* Preprint, 1991.

102. E. Kirchberg, *On nonsemisplit extensions, tensor products and exactness of group C^*-algebras*, Invent. Math. **112** (1993), 449–489.

103. E. Kirchberg, *Commutants of unitaries in UHF algebras and functorial properties of exactness*, J. Reine Angew. Math. **452** (1994), 39–77.

104. E. Kirchberg, *Exact C^*-algebras, tensor products and the classification of purely infinite algebras*, Proceedings of the International Congress of Mathematicians, Vol. 1, 2 (Zurich, 1994), 943–954.

105. E. Kirchberg, *Discrete groups with Kazhdan's property T and factorization property are residually finite*, Math. Ann. **299** (1994), 551–563.

106. E. Kirchberg, *On subalgebras of the CAR-algebra*, J. Funct. Anal. **129** (1995), 35–63.

107. E. Kirchberg and N.C. Phillips, *Embedding of exact C^*-algebras in the Cuntz algebra $\mathcal{O}_2$*, J. Reine Angew. Math. **525** (2000), 17–53.

108. E. Kirchberg and S. Wassermann, *Exact groups and continuous bundles of C^*-algebras*, Math. Ann. **315** (1999), 169–203.

109. E. Kirchberg and S. Wassermann, *Permanence properties of C^*-exact groups*, Doc. Math. **4** (1999), 513–558.

110. E. Kirchberg and W. Winter, *Covering dimension and quasidiagonality*, Internat. J. Math. **15** (2004), 63–85.

111. J. Kraus, *The slice map problem and approximation properties.* J. Funct. Anal. **102** (1991), 116–155.

112. A. Kumjian, D. Pask and I. Raeburn, *Cuntz-Krieger algebras of directed graphs*, Pacific J. Math. **184** (1998), 161–174.

113. E.C. Lance, *On nuclear C^*-algebras*, J. Funct. Anal. **12** (1973), 157–176.

114. E.C. Lance, *Hilbert C^*-modules. A toolkit for operator algebraists*, London Mathematical Society Lecture Note Series, 210. Cambridge University Press, Cambridge, 1995.

115. C. Le Merdy, *On the duality of operator spaces.* Canad. Math. Bull. **38** (1995), 334–346.

116. H. Lin, *Tracially AF C^*-algebras*, Trans. Amer. Math. Soc. **353** (2001), 693–722.

117. H. Lin, *Classification of simple C^*-algebras of tracial topological rank zero*, Duke Math. J. **125** (2004), 91–119.

118. H. Lin, *AF-embedding of crossed products of AH algebras by $\mathbb{Z}$ and asymptotic AF embedding.* Preprint, 2006 (math/0612529).

119. H. Lin, *AF-embedding of the crossed products of AH-algebras by a finitely generated abelian groups.* Preprint, 2007 (arXiv:0706.2229).

120. J. Lindenstrauss and L. Tzafriri, *Classical Banach spaces. I. Sequence spaces*, Ergebnisse der Mathematik und ihrer Grenzgebiete, Vol. 92. Springer-Verlag, Berlin-New York, 1977.

121. R.C. Lyndon and P.E. Schupp, *Combinatorial group theory.* Reprint of the 1977 edition. Classics in Mathematics. Springer-Verlag, Berlin, 2001.

122. A. Lubotzky, *Discrete groups, expanding graphs and invariant measures.* With an appendix by Jonathan D. Rogawski. Progress in Mathematics, 125. Birkhäuser Verlag, Basel, 1994.

123. A. Lubotzky and Y. Shalom, *Finite representations in the unitary dual and Ramanujan groups*, Discrete geometric analysis, 173–189, Contemp. Math., 347, Amer. Math. Soc., Providence, RI, 2004.

124. H. Matui, *AF embeddability of crossed products of AT algebras by the integers and its application*, J. Funct. Anal. **192** (2002), 562–580.

125. I. Moerdijk and J. Mrčun, *Introduction to foliations and Lie groupoids.* Cambridge Studies in Advanced Mathematics, 91. Cambridge University Press, Cambridge, 2003.

126. P.S. Muhly, J.N. Renault, D.P. Williams, *Equivalence and isomorphism for groupoid C^*-algebras*, J. Operator Theory **17** (1987), 3–22.

127. G.J. Murphy, *C^*-algebras and operator theory*, Academic Press, Inc., Boston, MA, 1990.

128. F.J. Murray and J. von Neumann, *On rings of operators. IV*, Ann. of Math. **44** (1943), 716–808.

129. A. Olshanskii, *SQ-universality of hyperbolic groups*, Sb. Math. **186** (1995), 1199 – 1211.

130. N. Ozawa, *Amenable actions and exactness for discrete groups*, C.R. Acad. Sci. Paris Ser. I Math. **330** (2000), 691–695.

131. N. Ozawa, *Homotopy invariance of AF-embeddability*, Geom. Funct. Anal. **13** (2003), 216–222.

132. N. Ozawa, *Weakly exact von Neumann algebras*. J. Math. Soc. Japan, to appear.

133. N. Ozawa, *Solid von Neumann algebras*, Acta Math. **192** (2004), 111–117.

134. N. Ozawa, *About the QWEP conjecture*, Internat. J. Math. **15** (2004), 501–530.

135. N. Ozawa, *A Kurosh type theorem for type* II_1 *factors*, Int. Math. Res. Not. (2006), Art. ID 97560, 21 pp.

136. N. Ozawa, *Boundary amenability of relatively hyperbolic groups*, Topology Appl. **153** (2006), 2624–2630.

137. N. Ozawa and S. Popa, *Some prime factorization results for type* II_1 *factors*, Invent. Math. **156** (2004), 223–234.

138. N. Ozawa and S. Popa, *On a class of* II_1 *factors with at most one Cartan subalgebra*. Preprint, 2007 (arXiv:0706.3623).

139. A. Paterson, *Amenability*, Mathematical Surveys and Monographs, 29. American Mathematical Society, Providence, RI, 1988.

140. A. Paterson, *Groupoids, inverse semigroups, and their operator algebras*, Progress in Mathematics, 170. Birkhauser Boston, Inc., Boston, MA, 1999.

141. V. Paulsen, *Completely bounded maps and operator algebras*, Cambridge Studies in Advanced Mathematics, 78. Cambridge University Press, Cambridge, 2002.

142. G.K. Pedersen, *C^*-algebras and their automorphism groups*, London Mathematical Society Monographs, 14. Academic Press, Inc., London-New York, 1979.

143. J. Peterson, *L^2-rigidity in von Neumann algebras*. Preprint, 2006 (math.OA/0605033).

144. J. Peterson and S. Popa, *On the notion of relative property (T) for inclusions of von Neumann algebras*. J. Funct. Anal. **219** (2005), 469–483.

145. M.V. Pimsner, *Embedding some transformation group C^*-algebras into AF-algebras*, Ergodic Theory Dynam. Systems **3** (1983), 613–626.

146. M.V. Pimsner, *A class of C^*-algebras generalizing both Cuntz-Krieger algebras and crossed products by Z*, Free probability theory (Waterloo, ON, 1995), 189–212, Fields Inst. Commun., 12, Amer. Math. Soc., Providence, RI, 1997.

147. M.V. Pimsner, *Embedding covariance algebras of flows into AF-algebras*, Ergodic Theory Dynam. Systems **19** (1999), 723–740.

148. M. Pimsner and D. Voiculescu, *Imbedding the irrational rotation C^*-algebra into an AF-algebra*, J. Operator Theory **4** (1980), 201–210.

149. G. Pisier, *Exact operator spaces*, Recent advances in operator algebras (Orléans, 1992). Astérisque No. 232 (1995), 159–186.

150. G. Pisier, *A simple proof of a theorem of Kirchberg and related results on C^*-norms*. J. Operator Theory **35** (1996), 317–335.

151. G. Pisier, *Similarity problems and completely bounded maps*. Second, expanded edition. Includes the solution to "The Halmos problem". Lecture Notes in Mathematics, 1618. Springer-Verlag, Berlin, 2001.

152. G. Pisier, *Introduction to operator space theory*, London Mathematical Society Lecture Note Series, 294. Cambridge University Press, Cambridge, 2003.

153. G. Pisier, *Remarks on $B(H) \otimes B(H)$*. Preprint, 2005 (math.OA/0509297).

154. S. Popa, *Orthogonal pairs of $*$-subalgebras in finite von Neumann algebras*. J. Operator Theory **9** (1983), 253–268.

155. S. Popa, *A short proof of "injectivity implies hyperfiniteness" for finite von Neumann algebras*, J. Operator Theory **16** (1986), 261–272.

156. S. Popa, *Correspondences*. INCREST Preprint, 56/1986.

157. S. Popa, *On amenability in type II_1 factors*, Operator algebras and applications, Vol. 2, 173–183, London Math. Soc. Lecture Notes Ser., 136, Cambridge Univ. Press, Cambridge, 1988.

158. S. Popa, *Classification of subfactors and their endomorphisms*. CBMS Regional Conference Series in Mathematics, 86. American Mathematical Society, Providence, RI, 1995.

159. S. Popa, *On local finite-dimensional approximation of C^*-algebras*, Pacific J. Math. **181** (1997), 141–158.

160. S. Popa, *On the fundamental group of type* II_1 *factors*. Proc. Natl. Acad. Sci. USA **101** (2004), 723–726.

161. S. Popa, *On the Superrigidity of Malleable Actions with Spectral Gap*. J. Amer. Math. Soc., to appear.

162. T. Pytlik and R. Szwarc, *An analytic family of uniformly bounded representations of free groups*. Acta Math. **157** (1986), 287–309.

163. I. Raeburn, *Graph algebras*, CBMS Regional Conference Series in Mathematics, 103. American Mathematical Society, Providence, RI, 2005.

164. J. Renault, *A groupoid approach to C^*-algebras*, Lecture Notes in Mathematics, 793. Springer, Berlin, 1980.

165. E. Ricard and Q. Xu, *Khintchine type inequalities for free product and applications*. J. Reine Angew. Math. **599** (2006), 27–59.

166. A.G. Robertson, *Property (T) for II_1 factors and unitary representations of Kazhdan groups*, Math. Ann. **296** (1993), 547–555.

167. J. Roe, *Lectures on coarse geometry*, University Lecture Series, 31. American Mathematical Society, Providence, RI, 2003.

168. M. Rørdam, *Classification of nuclear, simple C^*-algebras* Classification of nuclear C^*-algebras. Entropy in operator algebras, 1–145, Encyclopaedia Math. Sci., 126, Springer, Berlin, 2002.

169. M. Rørdam, *A purely infinite AH-algebra and an application to AF-embeddability*, Israel J. Math. **141** (2004), 61–82.

170. J. Rosenberg and C. Schochet, *The Kunneth theorem and the universal coefficient theorem for Kasparov's generalized K-functor*, Duke Math. J. **55** (1987), 431–474.

171. N. Salinas, *Homotopy invariance of Ext(A)*, Duke Math. J. **44** (1977), 777–794.

172. J.-P. Serre, *Trees*. Translated from the French original by John Stillwell. Corrected 2nd printing of the 1980 English translation. Springer Monographs in Mathematics. Springer-Verlag, Berlin, 2003.

173. Y. Shalom, *Bounded generation and Kazhdan's property (T)*. Inst. Hautes Etudes Sci. Publ. Math. No. **90** (1999), 145–168.

174. Y. Shalom, *Elementary linear groups and Kazhdan's property (T)*, in preparation.

175. A.M. Sinclair and R.R. Smith, *The Haagerup invariant for tensor products of operator spaces.* Math. Proc. Cambridge Philos. Soc. **120** (1996), 147–153.

176. G. Skandalis. *Une notion de nucléarité en K-théorie (d'après J. Cuntz).* K-Theory **1** (1988), 549–573.

177. G. Skandalis, J.L. Tu and G. Yu, *The coarse Baum-Connes conjecture and groupoids*, Topology **41** (2002), 807–834.

178. R.R. Smith, *Completely bounded module maps and the Haagerup tensor product*, J. Funct. Anal. **102** (1991), 156–175.

179. J.S. Spielberg, *Embedding C^*-algebra extensions into AF algebras*, J. Funct. Anal. **81** (1988), 325–344.

180. A. Szankowski, *$B(\mathcal{H})$ does not have the approximation property*, Acta Math. **147** (1981), 89–108.

181. S. Szarek, *An exotic quasidiagonal operator*, J. Funct. Anal. **89** (1990), 274–290.

182. M. Takesaki, *On the cross-norm of the direct product of C^*-algebras*, Tohoku Math. J. **16** (1964), 111–122.

183. M. Takesaki, *Theory of operator algebras I.* Encyclopedia of Mathematical Sciences, 124. Operator Algebras and Non-commutative Geometry, 5. Springer-Verlag, Berlin, 2002.

184. M. Takesaki, *Theory of operator algebras II.* Encyclopedia of Mathematical Sciences, 125. Operator Algebras and Non-commutative Geometry, 5. Springer-Verlag, Berlin, 2002.

185. M. Takesaki, *Theory of operator algebras III.* Encyclopedia of Mathematical Sciences, 127. Operator Algebras and Non-commutative Geometry, 5. Springer-Verlag, Berlin, 2003.

186. J. Tomiyama, *Applications of Fubini type theorem to the tensor products of C^*-algebras.* Tôhoku Math. J. (2) **19** (1967), 213–226.

187. J.L. Tu, *Remarks on Yu's "property A" for discrete metric spaces and groups*, Bull. Soc. Math. France **129** (2001), 115–139.

188. D.V. Voiculescu, *Almost inductive limit automorphisms and embeddings into AF-algebras*, Ergodic Theory Dynam. Systems **6** (1986), 475–484.

189. D.V. Voiculescu, *A note on quasidiagonal operators*, Operator Theory: Advances and Applications, Vol. 32, Birkhauser Verlag, Basel, 1988, 265–274.

190. D.V. Voiculescu, *On the existence of quasicentral approximate units relative to normed ideal. Part I*, J. Funct. Anal. **91** (1990), 1–36.

191. D.V. Voiculescu, *A note on quasi-diagonal C^*-algebras and homotopy*, Duke Math. J. **62** (1991), 267–271.

192. D.V. Voiculescu, *Around quasidiagonal operators*, Integr. Equ. and Op. Thy. **17** (1993), 137–149.

193. D.V. Voiculescu, K.J. Dykema and A. Nica, *Free random variables*, A noncommutative probability approach to free products with applications to random matrices, operator algebras and harmonic analysis on free groups. CRM Monograph Series, 1. American Mathematical Society, Providence, RI, 1992.

194. S. Wassermann, *Injective W^*-algebras*, Math. Proc. Cambridge Philos. Soc. **82** (1977), 39–47.

195. S. Wassermann, *C^*-algebras associated with groups with Kazhdan's property T*, Ann. of Math. **134** (1991), 423–431.

196. S. Wassermann, *A separable quasidiagonal C^*-algebra with a nonquasidiagonal quotient by the compact operators*, Math. Proc. Cambridge Philos. Soc. **110** (1991), 143–145.

197. S. Wassermann, *Exact C^*-algebras and related topics*, Lecture Notes Series, no.19, GARC, Seoul National University, 1994.

198. N.E. Wegge-Olsen, *K-theory and C^*-algebras. A friendly approach.* Oxford Science Publications. The Clarendon Press, Oxford University Press, New York, 1993.

199. J. Zacharias, *On the invariant translation approximation property for discrete groups.* Proc. Amer. Math. Soc. **134** (2006), 1909–1916.

200. R.J. Zimmer, *Ergodic theory and semisimple groups*, Monographs in Mathematics, 81. Birkhauser Verlag, Basel, 1984.

201. A. Żuk, *Property (T) and Kazhdan constants for discrete groups*, Geom. Funct. Anal. **13** (2003), 643–670.

Notation Index

Subject Index

Titles in This Series

88 **Nathanial P. Brown and Narutaka Ozawa,** C*-algebras and finite-dimensional approximations, 2008

87 **Srikanth B. Iyengar, Graham J. Leuschke, Anton Leykin, Claudia Miller, Ezra Miller, Anurag K. Singh, and Uli Walther,** Twenty-four hours of local cohomology, 2007

86 **Yulij Ilyashenko and Sergei Yakovenko,** Lectures on analytic differential equations, 2007

85 **John M. Alongi and Gail S. Nelson,** Recurrence and topology, 2007

84 **Charalambos D. Aliprantis and Rabee Tourky,** Cones and duality, 2007

83 **Wolfgang Ebeling,** Functions of several complex variables and their singularities (translated by Philip G. Spain), 2007

82 **Serge Alinhac and Patrick Gérard,** Pseudo-differential operators and the Nash–Moser theorem (translated by Stephen S. Wilson), 2007

81 **V. V. Prasolov,** Elements of homology theory, 2007

80 **Davar Khoshnevisan,** Probability, 2007

79 **William Stein,** Modular forms, a computational approach (with an appendix by Paul E. Gunnells), 2007

78 **Harry Dym,** Linear algebra in action, 2007

77 **Bennett Chow, Peng Lu, and Lei Ni,** Hamilton's Ricci flow, 2006

76 **Michael E. Taylor,** Measure theory and integration, 2006

75 **Peter D. Miller,** Applied asymptotic analysis, 2006

74 **V. V. Prasolov,** Elements of combinatorial and differential topology, 2006

73 **Louis Halle Rowen,** Graduate algebra: Commutative view, 2006

72 **R. J. Williams,** Introduction the the mathematics of finance, 2006

71 **S. P. Novikov and I. A. Taimanov,** Modern geometric structures and fields, 2006

70 **Seán Dineen,** Probability theory in finance, 2005

69 **Sebastián Montiel and Antonio Ros,** Curves and surfaces, 2005

68 **Luis Caffarelli and Sandro Salsa,** A geometric approach to free boundary problems, 2005

67 **T.Y. Lam,** Introduction to quadratic forms over fields, 2004

66 **Yuli Eidelman, Vitali Milman, and Antonis Tsolomitis,** Functional analysis, An introduction, 2004

65 **S. Ramanan,** Global calculus, 2004

64 **A. A. Kirillov,** Lectures on the orbit method, 2004

63 **Steven Dale Cutkosky,** Resolution of singularities, 2004

62 **T. W. Körner,** A companion to analysis: A second first and first second course in analysis, 2004

61 **Thomas A. Ivey and J. M. Landsberg,** Cartan for beginners: Differential geometry via moving frames and exterior differential systems, 2003

60 **Alberto Candel and Lawrence Conlon,** Foliations II, 2003

59 **Steven H. Weintraub,** Representation theory of finite groups: algebra and arithmetic, 2003

58 **Cédric Villani,** Topics in optimal transportation, 2003

57 **Robert Plato,** Concise numerical mathematics, 2003

56 **E. B. Vinberg,** A course in algebra, 2003

55 **C. Herbert Clemens,** A scrapbook of complex curve theory, second edition, 2003

54 **Alexander Barvinok,** A course in convexity, 2002

53 **Henryk Iwaniec,** Spectral methods of automorphic forms, 2002

TITLES IN THIS SERIES